K kinetic energy; K_s, spring scale; K_p, equilibrium constant.

k the ratio c_p/c_v.

k   Newton's proportionality constant.

L distance; length; stroke of piston; represents unit of length; L'', stroke in inches.

M   Mach number; magnetic moment.

M molecular mass; unless otherwise specified, lb/mole = pmole; M_a, molecular mass of gas A, etc. Moment.

$\mathcal{M}$ momentum.

m mass; percentage steam bled in regenerative cycles; exponent in $pV^m = C$ for an irreversible process.

N number of anything; number of power cycles per minute completed by an engine.

N_A Avogadro's number.

n revolutions per minute (rpm); polytropic exponent; number of moles.

P gravitational potential energy; represents the unit of a pound.

p unit pressure; p_m, mean effective pressure (mep); p_{mi}, indicated mep; p_{mB}, brake mep; p_0, stagnation or impact pressure; $p°$, pressure in reference state; p_r, relative pressure; p_R, reduced pressure.

Q heat; Q_A, heat added; Q_R, heat rejected; Q_r, radiated heat; etc.

q heating value; q_l, lower heating value; q_h, higher heating value; q_v, heating value at constant volume; $\bar{q}$, heating value per mole of fuel.

$\mathcal{Q}$ electrical charge, coulombs of electricity.

R specific gas constant, $R = \bar{R}/M = 1545/M$ ($=pv/T$ for ideal gas).

$\bar{R}$ universal gas constant.

R   Reynolds number.

r radius; ratios; reheat factor; r_k, compression ratio; r_c, cut-off ratio; r_p, pressure ratio; r_e, expansion ratio; $r_{f/a}$, fuel/air ratio; $r_{a/f}$, air/fuel ratio; r_h, hydraulic radius.

$\mathcal{R}$ electrical resistance.

S total entropy.

s specific entropy; $s°$, absolute entropy in a standard state; $\bar{s}$, entropy per mole.

s_1^a absolute entropy in state 1; etc.

$\mathcal{S}$ surface tension.

T absolute temperature, usually degrees Rankine; T_R, reduced temperature; T_0, stagnation temperature; T_0, sink temperature; $T°$, standard or reference temperature.

t temperature, usually in degrees Fahrenheit; time.

U total internal energy; U_p, internal energy of products, etc.; $U°$, internal energy at standard state.

u specific internal energy; $\bar{u}$, internal energy of one mole of substance; u_{rp}, internal energy of reaction; $u°$, in standard state.

V total volume; V_D, displacement volume.

v specific volume; v_r, relative volume; v_R, reduced volume; v_{Ri}, ideal reduced volume; $\bar{v}$, volume of one mole.

$\mathcal{V}$ speed.

MARK SHEARER 260-2406

THERMODYNAMICS

THERMODYNAMICS

SIXTH EDITION

VIRGIL MORING FAIRES
Late Professor of Mechanical Engineering
U.S. Naval Postgraduate School

CLIFFORD MAX SIMMANG
Professor and Head
Department of Mechanical Engineering
Texas A & M University

MACMILLAN PUBLISHING CO., INC.
New York
COLLIER MACMILLAN PUBLISHERS
London

Macmillan Publishing Co., Inc.
866 Third Avenue, New York, New York 10022

Collier Macmillan Canada, Ltd.

Library of Congress Cataloging in Publication Data

Faires, Virgil Moring, (date)
  Thermodynamics.

  Includes index.
  1. Thermodynamics. I. Simmang, Clifford M., joint
author. II. Title.
TJ265.F3 1978   536'.7   77-3662
ISBN 0-02-335530-1 (Hardbound)
ISBN 0-02-978910-9 (International Edition)

Printing: 6 7 8   Year: 8   4

ISBN   0-02-335530-1

PREFACE

The previous five editions of this text, authored by the late Professor Virgil Moring Faires, have been invaluable in the preparation of this one. I believe that the extensive rearrangement of material and the presentation of new ideas and methods reflected in this edition exclude no salient features of these earlier editions.

The foreword has been designed to give the student a "feel" for thermodynamics through historical facts and pictures of some equipment related to energy systems. A major departure is made from the previous editions by introducing the pure substance early in the text, thereby permitting the study of processes to be independent of the nature of the fluid; that is, both ideal gases and vapors are treated concurrently in the chapter on processes.

Definitions and usage of terms have been tightened. In particular, the section on mass and weight has received much attention in order to clarify the relationship between these quantities.

The chapter on the second law is introduced earlier in the text in order that greater use may be made of the concept of entropy. A full chapter has been devoted to the gas compressor because of its important role in industry.

In view of the trend away from separate courses in heat transfer by departments other than that of mechanical engineering, a chapter on heat transfer was added. This single chapter is not, of course, intended to supplant completely the normal three-hour course on heat transfer; however, the basic fundamentals of this mode of energy transfer are given adequate coverage to assure a grasp of their significance.

Special emphasis has been given to the increasing role of the SI (Scientific International) metric system of units. Problems at the end of each chapter have been designed to give the student adequate familiarity with its uniqueness and terminology. For ease and convenience in selecting problems with particular units in mind, the problems have been separated and grouped under one of three general headings: SI Units, English Units, and Mixed Units. Approximately half (47%) of the problems involve SI units.

The principal sources of problems is still *Problems on Thermodynamics* (Faires, Simmang, and Brewer), and it reflects the exact arrangement of this sixth edition.

Throughout the text, the engineering flavor is retained. The student is soon made aware that he is studying material relating to the profession for which he is preparing. He is reminded frequently and in various ways of the shortcomings of the ideal models, including evaluations of answers obtained from theoretical equations. He should understand at the end of the course that, no matter how beautiful, every theory needs to be verified in practice. The control volume is likened to the free body in mechanics, and emphasis on it is continued.

College Station, Texas

C. M. Simmang

ACKNOWLEDGMENTS

It is with deep appreciation that I express thanks to the many friends and colleagues of the late Professor Virgil M. Faires who so ably assisted him with the preparation of the first five editions. The fifth edition was an excellent base from which to start.

To my numerous friends and colleagues who encouraged and supported me in this revision, I say thanks. In particular I wish to name Dr. Alan B. Alter and Professor Edwin S. Holdredge, for their sympathetic and numerous suggestions, and Professor Louis C. Burmeister for his excellent review of the manuscript. Thanks are also given to those readers, known only to the publisher, who gave many helpful suggestions and noted necessary corrections that were incorporated in this book.

I would be remiss without acknowledging the dedicated support and personal sacrifice of my wife Elnora; without her constant understanding this work would not have been possible.

CONTENTS

GAS CYCLES 206 8

POWER FROM TWO-PHASE SYSTEMS 228 9

IMPERFECT GASES 255 10

RELATIONS OF THERMODYNAMIC PROPERTIES 275 11

12 MIXTURES OF GASES AND VAPORS 316

13 REACTIVE SYSTEMS 341

18 NOZZLES, DIFFUSERS, AND FLOWMETERS 490

19 HEAT TRANSFER 516

FOREWORD TO THE STUDENT

INTRODUCTION   F.1

The purpose of this foreword is to present background material for the interest and edification of the reader. It is hoped that occasional brief reviews of this section will assist the student immeasurably as he progresses through the selected chapters that follow.

REMARKS ON THERMODYNAMICS   F.2

Thermodynamics is that branch of the physical sciences that treats of various phenomena of energy and the related properties of matter, especially of the laws of transformation of heat into other forms of energy and vice versa. Examples of such everyday transformations are the process of converting heat into electrical work (electrical power generation), of converting electrical work into cooling (air conditioning), of converting work into kinetic energy (automotive transportation), and so on. Thermodynamics relates to so much that no single volume covers all current knowledge of the subject.

Thermodynamics is everybody's problem. The recent overall assessment of the energy supply of this nation and of the world coupled with resulting tremendous escalation in energy costs has concerned everyone. The problem is being solved through the combination of physical laws and man-made legislative laws. Everyone will have a hand in the solution of this worldwide thermodynamics problem.

HIGHLIGHTS IN THERMODYNAMICS   F.3

History reveals that thermodynamics is not a new subject relative to our calendar. Its processes are reflected in the use of gunpowder by the early Chinese, in the building of pyramids by the early Egyptians, in the development of the longbow by the Celtics, and in many other historical events. The following tabulation cites a few dates and events that are highlights in the evolution of this interesting subject.

HIGHLIGHTS IN THERMODYNAMICS

c. 400 B.C.	Democritus wrote that all matter consists of tiny material bits called atoms.
c. 200 B.C.	Archimedes discovered laws for the behavior of liquids and levers.
c. 1500	Leonardo da Vinci stated that air contains two gases.
1638	Galileo approached the concept of temperature by use of his thermoscope.
1640	Grand Duke Ferdinand II of Tuscany invented the sealed-stem alcohol thermometer.
c. 1730	D. Bernoulli developed the kinetic theory of gases.
1770	J. Black introduced the caloric theory which was later (1779) buttressed by the postulates of W. Cleghorn.
1776	A. L. Lavoisier renames the air gas phlogiston as oxygen.
1798	Count Rumford disproved the caloric theory.
1824	S. Carnot introduced new concepts to include the cycle.
1844	R. Mayer deduced the relation between heat and work.
1848	Lord Kelvin defined an absolute temperature scale based upon the Carnot cycle.
c. 1850	J. P. Joule discovered that heat and work are interchangeable at a fixed rate which was reconciled by Clausius.
1850	R. Clausius reformulates the thermal quantity of Carnot into the concept of entropy.
1865	Clausius introduced the concept of U, now called internal energy, and stated the first and second laws.
1897	M. Planck demonstrated the relation between the second law and the concept of irreversibility.
1908	H. Poincaré extended the work of Planck and prescribed a complete structure of classical thermodynamics based upon consistent definitions of measurable quantities.
1909	C. Caratheodory presented a structure different from that of Poincaré but used basic concepts of work and an adiabatic wall.

F.4 MACROSCOPIC VERSUS MICROSCOPIC VIEWPOINTS

Classical thermodynamics takes the macroscopic or large scale view, rather than the microscopic. The development of the subject from the microscopic view, a somewhat recent achievement, is called statistical thermodynamics or statistical mechanics. This approach looks at the individual molecules and their internal structure, a prime concern of physicists, but something that in general we shall do qualitatively, only as is appropriate. Historically the science is experimentally based, and it was developed, with the aid of mathematics, without regard to the structure of matter.

We shall be concentrating on the classical approach—simple, intuitively acceptable, and easy to grasp. Our development will involve, to a degree, the historical evolution on the presumption that it will be of interest and just as binding. The number of variables needed are few and the required mathematics simple. There are advantages and disadvantages in each approach. The microscopic viewpoint requires the acceptance of the atomic model of matter and, of necessity, calls for a certain knowledge of sophisticated mathematics. While it permits better understanding of certain material phenomena, it is somewhat limited in application.

F.5 SOME EQUIPMENT RELATED TO THERMODYNAMICS

All thermodynamic processes normally require some sort of equipment or containment for the transformation of heat into work or vice versa. Although one is concerned only with the nature of the process and the fluid or material that is undergoing thermodynamic changes in pressure, temperature, volume, and so on, nevertheless, a knowledge of the equipment in which the process takes place is of utmost importance. The student will have a better feel for his problem if he can visualize what is taking place mechanically. For example, we cannot see the fuel-air mixture that enters an automobile engine or the products that leave its exhaust manifold; although we can always analyze the processes involved, the problem becomes less complex if we can mentally visualize the moving valves, pistons, gearing, and other components, that are necessary for this transformation of heat into work.

The pictures presented here reveal the operation of some pieces of equipment peculiar to thermodynamics. With knowledge of this kind, the reader can apply thermodynamic principles with a comprehension that would not otherwise be possible. It is anticipated that the teacher will discuss this equipment in more detail; thus, brief descriptions only are given for each piece shown.

SUGGESTIONS TO THE STUDENT F.6

This text is written for you, the typical junior (or thereabouts) engineering student. Getting an education is a process, something in being, that ideally continues for a lifetime. No decent amount of education can be acquired passively. In the words of Sophocles, "One must learn by doing the thing." "Doing the thing" for the prospective engineer consists of applying the science—in this case, thermodynamics. To make intelligent applications, you need much more than the feeling that you understand what is said; you need to make many applications yourself. The greater the variety of problems worked, the greater the comprehension. The teacher and the text can help you proceed with your education, principally by directing your efforts into the most productive channels, but the basic action in getting an education is the student learning. Since it is your job, you know why I have tried to develop the subject with you in mind.

Try to solve the examples after what you consider to be adequate study. Since the solution is there in detail, you may check every step. This is a form of programmed learning. Since the subject builds on itself (most of the building in this text occurs in the first nine chapters), frequent quick reviews to keep things in perspective will be profitable. When the problem concerns a machine or other device, be sure to learn something of its operations if you wish to make a knowledgeable solution. Do not ignore the captions to the illustrations; they often contain useful and sometimes essential ideas.

Most students who desire to become engineers find a real interest in thermodynamics. Perhaps you will too.

CLOSURE F.7

Thermodynamics is an anomaly. It involves things and processes that are intimately familiar to us—things that we see and experience daily, such as automobiles, jet planes, boiling water, refrigeration, and so on. Yet, therein lies the difficulty; we are misled into believing that this familiarity leads to an easy solution of the thermodynamic processes that are happening in each thing. Frustration sets in when one knows all about the problem except how to solve it. Hopefully, this text will prepare you for that task.

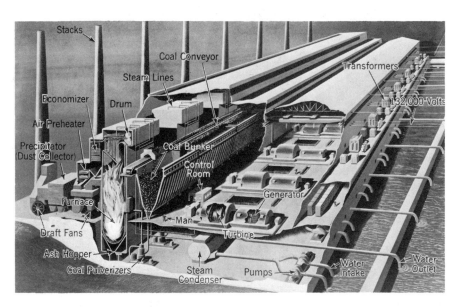

Fig. f/1 *Central Station Power Plant.* An artist's conception of how a certain power plant looks at a capacity of 1,000,000 kW. (*Courtesy the Cleveland Electric Illuminating Co., Cleveland, Ohio.*)

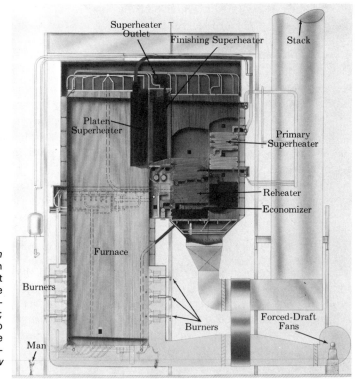

Fig. f/2 *Open-Pass Steam Generator.* In an open-pass unit, the hot gases flow in two or more directions within bounding heating-surface walls; for pressures up to 2650 psi and temperature to 1100°F. (*Courtesy Babcock and Wilcox Co., New York, N.Y.*)

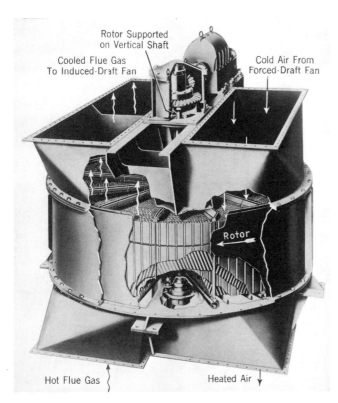

Rotor Supported
on Vertical Shaft

Cooled Flue Gas
To Induced-Draft Fan

Cold Air From
Forced-Draft Fan

Rotor

Hot Flue Gas

Heated Air

Ljungstrom Heater. Typical **Fig. f/3**
operating data for the heater
shown are: boiler rate, 550,000
lb/hr of steam; air temperature
in, 80°F, *out*, 602°F; gas temperature *in*, 720°F, *out*, 346°F.
(*Courtesy Air Preheater Corp.,
New York, N.Y.*)

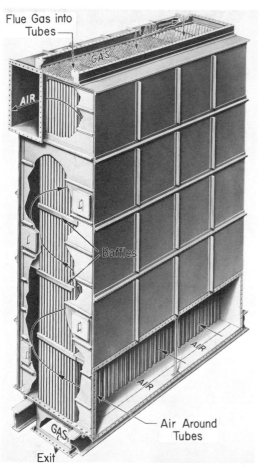

Flue Gas into
Tubes

AIR

GAS

AIR

Baffles

AIR

AIR

GAS

Exit

Air Around
Tubes

Tubular Air Preheater. Typical data for size **Fig. f/4**
shown are: $2\frac{1}{2}$-in. OD tubes; boiler
capacity, 40,000 lb/hr of steam; air temperature *in*, 100°F, *out*, 515°F; flue gas
temperature *in*, 650°F, *out*, 350°F. Steam
generator efficiency is increased about 2%
for each 100°F rise of temperature of the
combustion air.[9.17] (*Courtesy Combustion
Engineering Co., New York, N.Y.*)

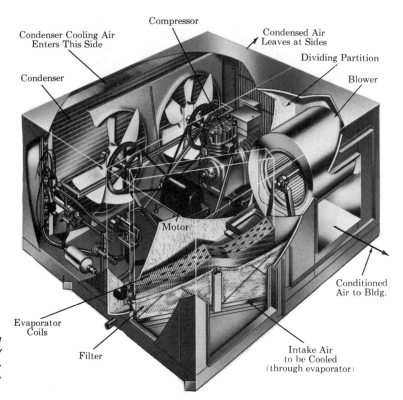

Fig. f/5 *Air Conditioning Unit. (Courtesy Lennox Industries, Inc., Marshalltown, Iowa.)*

Compressor

Condenser Cooling Air Enters This Side

Condensed Air Leaves at Sides

Dividing Partition

Condenser

Blower

Motor

Conditioned Air to Bldg.

Evaporator Coils

Filter

Intake Air to be Cooled (through evaporator)

$t_a = 80°F$
$p_a = 14.7$

Intake

5960 rpm

159 psia
705°F

151 psia
1540°F

34.2 psia
985°F

15 psia
740°F

Exhaust

W_{net}

21,500 hp

8465 rpm

Compressor

Combusters

Turbine

Free Turbine

Gas Generator Section

Fig. f/6   *Gas Turbine to Deliver Shaft Work. (Courtesy Pratt & Whitney Aircraft, East Hartford, Conn.)*

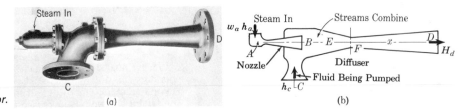

Steam In

D

C

(a)

Steam In

Streams Combine

w_a h_a

B — E

F

x

D

H_d

A

Nozzle

Diffuser

h_c C

Fluid Being Pumped

(b)

Fig. f/7   *Ejector.*

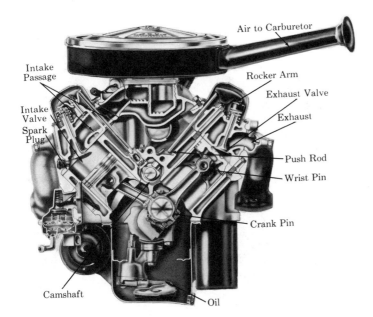

Intake
Passage

Intake
Valve

Spark
Plug

Camshaft

Oil

Air to Carburetor

Rocker Arm

Exhaust Valve

Exhaust

Push Rod

Wrist Pin

Crank Pin

Four-Stroke-Cycle Automotive Engine. (Courtesy General Motors Corp., Warren, Mich.) **Fig. f/8**

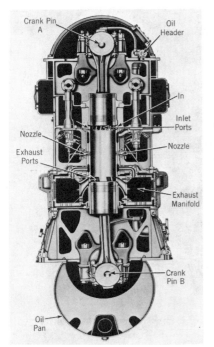

Crank Pin
A

Oil
Header

In

Inlet
Ports

Nozzle

Nozzle

Exhaust
Ports

Exhaust
Manifold

Crank
Pin B

Oil
Pan

Opposed Piston Engine. As the pistons move from the central part of cylinder away from each other, exhaust ports in the cylinder wall (shown exposed at the bottom) begin to open, and exhaust starts. Next, inlet ports (at top) are uncovered and fresh air is blown into the cylinder, scavenging it. (The opening of these ports occurs simultaneously, but the arrangement is such that the exhaust ports begin to open first. Notice that the lower crank *B* leads the upper crank *A* by a small amount.) As the pistons start on compression stroke, ports are covered by the pistons and compression begins. Near the end of the compression stroke, fuel is injected through the injection nozzle, combustion begins, and the two-stroke cycle starts over.

Fig. f/9

The two crankshafts are connected through bevel gears by a vertical shaft, parallel to the axes of the cylinders. The vertical shaft, not visible in this illustration, transmits the power of the upper crankshaft to the lower crankshaft, which delivers the power of the engine. (*Courtesy Fairbanks, Morse and Co., Chicago, Ill.*)

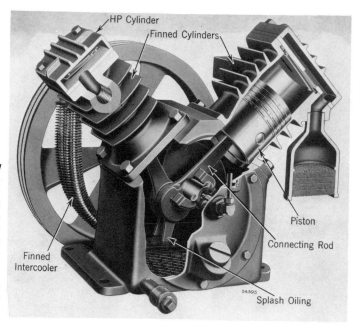

Fig. f/10 *Two-Stage, Air Cooled Compressor.* Observe the finned intercooler, partly visible on the left. Two-stage compression is sometimes recommended for a discharge pressure as low as 80 psia. Several more stages of compression are needed for high discharge pressures, as 100–200 atm. (*Courtesy Ingersoll-Rand Co., New York, N.Y.*)

HP Cylinder

Finned Cylinders

Piston

Connecting Rod

Finned Intercooler

Splash Oiling

54395

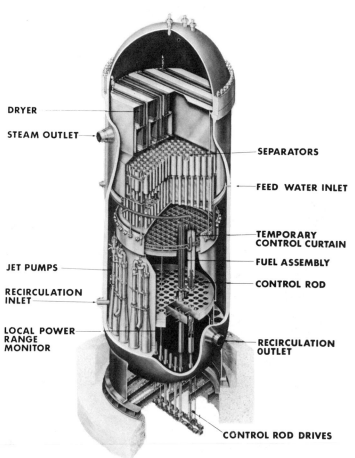

DRYER

STEAM OUTLET

SEPARATORS

FEED WATER INLET

TEMPORARY CONTROL CURTAIN

JET PUMPS

FUEL ASSEMBLY

RECIRCULATION INLET

CONTROL ROD

LOCAL POWER RANGE MONITOR

RECIRCULATION OUTLET

CONTROL ROD DRIVES

Fig. f/11 *Boiling Water Reactor.* The fuel is encased in "rods," labeled *fuel assembly*, arranged in bundles so that water, both the coolant and the working substance for the turbine, flows about them. The jet pumps aid the circulation. The steam passes through separators, and a dryer to remove most of the water drops, and then to the turbine.

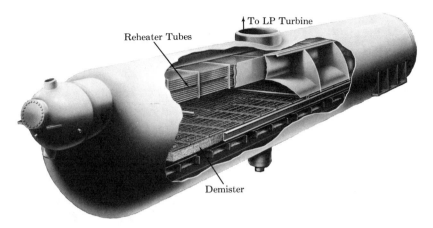

Combination Separator and Reheater. Steam exhausted from the HP turbine **Fig. f/12** with moisture of 9 to 12% passes upward through the wire-mesh demisters, which remove practically all the liquid (mist). The reheater finned tubes are supplied with throttle steam which superheats the steam from the demisters some 100 to 135°F; this steam then enters the LP turbine. (*Courtesy Westinghouse Corp., Philadelphia, Pa.*)

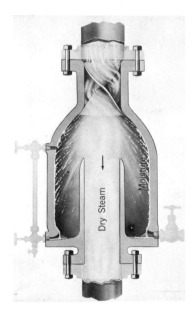

Centrifugal Separator. Steam or other vapor entering at **Fig. f/13** the top is given a swirling motion by the helicoidal surface. Heavier particles, liquid or solid, fly to the sides, while steam of the order of 99% quality departs at the center. (*Courtesy The Swartwout Co., Cleveland, Ohio.*)

1

BASIC PRINCIPLES, CONCEPTS AND DEFINITIONS

The study of thermodynamics, like a journey, must have a beginning. It is assumed that the reader has had first college courses in physics, calculus, chemistry, and mechanics. This chapter is intended to lay the foundation upon which subsequent chapters will build. The student should have crystal clear understandings of the basic principles, concepts, and definitions presented in this chapter before he leaves it.

After covering the basic laws and language and the methods of handling different substances thermodynamically, we will move into application chapters. These not only provide a greater insight into the subject but also give an introduction to specialized applications. Most students will find a personal interest in certain applications. Except for some local needs, the symbols and abbreviations are defined in the inside cover.

Engines (as in automobiles) produce work and refrigerators (as in homes) produce cooling because in each there are actions occurring on a working substance, usually a fluid, in which energy can be stored or from which it can be removed. A *fluid* is a substance that exists, or is regarded as existing, as a continuum characterized by low resistance to flow and the tendency to assume the shape of its container. Examples of fluid working substances are: *steam* in a steam turbine, *air* in an air compressor, *air and fuel mixture* in an internal combustion engine, and *water* in a hydraulic turbine. As we use the word *substance* hereafter, we shall mean something that usually is made up of molecules; sometimes atoms may be involved (in reactive systems). Hence, it will not mean radiation or electrons or other subatomic particles unless they are specifically included.

We shall also speak of it as being *pure* and/or being *simple*. A *pure substance* is one that is homogeneous in composition and homogeneous and invariable in chemical aggregation. For example, if pure water exists as a solid, a liquid, a vapor, or any combination of these, it is a pure substance. On the other hand if air exists as a combination of liquid and vapor, it is not a pure substance since the liquid will be richer in nitrogen than the vapor.

A *simple substance* is one whose state is defined by two independently variable intensive thermodynamic properties; see § 1.5 for discussion of properties and state. The state postulate (or principle), § 3.2, will reveal that a simple substance will have only one relevant, reversible work mode.

1.3   THE SYSTEM

A *system is that portion of the universe, an atom, a galaxy, a certain quantity of matter, or a certain volume in space, that one wishes to study.* It is a region enclosed by specified **boundaries**, which may be imaginary, either fixed or moving.* We partition off a system or region on paper because we wish to study transformations of energy occurring within the boundaries, and with the passage, if any, of energy or matter or both across the system's boundaries, to or from the surroundings. *The region all about the system is called the **surroundings** or **environment**.* The surroundings contain systems, some of which may affect the particular system under study, such as a source of heat. The *free body* in analytic mechanics is a system in which the mode of analysis is based on Newton's laws of motion. In thermodynamics, the principal mode of analysis is based on the mass-energy balance of the system under study.

Systems may be defined in several ways; for present purposes, we wish to distinguish between three kinds. A *closed system* is one in which there is no exchange of matter with the surroundings—mass does not cross its boundaries. An *open system* is one across whose boundaries there is a flow of mass. Each may have energy crossing its boundary. An *isolated system* is one that is completely impervious to its surroundings—neither mass nor energy cross its boundaries.

In Fig. 1/1 let the gas be the system and the cylinder and movable piston be its boundary. If heat be applied externally to the cylinder, the gas will undergo an increase in temperature, raising the piston. With the rise of the piston, the boundary moves. Energy (heat and work) crosses the boundary during this process while the mass of gaseous matter remains constant within the system.

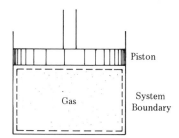

Piston

Gas

System Boundary

Fig. 1/1   *Example of a Closed System.*

---

* It is only natural that as a science develops, there should be semantic troubles involving new ideas—and sometimes it happens with old ideas. On the whole, the classical science of thermodynamics well developed so that most of the technical words have universally recognized meanings. We could say, as Humpty Dumpty said to Alice in Wonderland, "When *I* use a word, it means just what I choose it to mean—neither more nor less." Actually, we do not anticipate deliberately redefining any technical word for our own purposes, but if it happens, it will be with some mention of other existing definitions. We shall in general choose among the current definitions those that most nearly suit our purposes and give you other technical words for the same concepts and properties as we can recall them—that is, where different words meaning the same thing are in current use.

CONTROL SURFACE AND VOLUME   1.4

Often the system under analysis is of the open type, as an automobile engine shown in Fig. 1/2. With open systems we normally specify the boundary as a control surface and the volume encompassed by that surface as the control volume. Thus the control volume may be defined as a volume in space in which one has interest for a particular study or analysis. The mass of working substance within the volume may be constant (though not the same molecular matter at any given instant) as in the case of the automobile engine or that of a simple water nozzle, or it may be varying as in the case of a tire undergoing inflation.

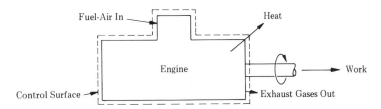

Example of an Open System.   **Fig. 1/2**

PROPERTIES AND STATE   1.5

To compute changes of energy that have occurred in a system or working substance, we must be able to express the behavior of the system in terms of descriptive characteristics that are called ***properties***. Macroscopic properties that are familiar to the reader from prior study include pressure p, temperature T, density ρ, and specific volume v, each of which is to be discussed shortly.

Properties may be classified as intensive or extensive. ***Intensive properties*** *are independent of the mass*; for example, temperature, pressure, density, and voltage. ***Extensive properties*** *are dependent upon the mass of the system and are total values* such as total volume and total internal energy. ***Specific properties*** are those for a unit mass, and are intensive by definition such as specific volume. Thus, thinking generally, we note, as examples, that total volume is an extensive property and that temperature and pressure are inherently intensive.

When we speak of the ***state*** of a pure substance, or system, we are referring to its condition as identified through the properties of the substance; this state is defined generally by particular values of any two independent properties. All other thermodynamic properties of the substance have certain particular values whenever a certain mass of the substance is in this particular macroscopic state. Examples of thermodynamic properties, besides p, v, and T, are: internal energy, enthalpy, and entropy (all to be studied later). Other properties of systems in general include: velocity, acceleration, moment of inertia, electric charge, conductivity (thermal and electrical), electromotive force (emf), stress, viscosity, reflectivity, number of protons, and so on. No matter what happens to a particular mass of a pure substance, be it compressed, heated, expanded, or cooled, if it is returned to the stipulated defining properties, the other *thermodynamic* properties also return to values identical, respectively, with their original values. See Fig. 1/3.

Consider for a moment the qualification, *independent* properties. As we know, the density is the reciprocal of the specific volume; hence these properties are not independent of each other. During the boiling or freezing of a liquid, the pressure and temperature of the two-phase mixture are not independent; the boiling temperature is a certain value for a particular substance, depending upon the value of the pressure.

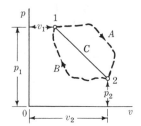

Fig. 1/3   *The Thermodynamic pv Plane.* A substance whose state is represented by point 1 has a temperature T_1. If the pressure and volume are varied as defined by the path 1-*A*-2-*B*-1, returning to their original values, the temperature also returns to value T_1.

From mathematics* one has learned that two coordinates (the values of x and y) locate (or define) a point that is known to be in a given plane (the xy plane). Three coordinates x, y, z locate a point in space. We look upon properties as being coordinates locating a point in space (*defining a state*) and we may picture this point, or any number of state points, projected onto various planes; for example, on the pressure-volume plane of Fig. 1/3, the temperature-entropy plane, and so on. *Any* three properties can be used to define a point in some thermodynamic space. If enough data are at hand, one could construct a thermodynamic *surface* of equilibrium states for any pure substance, using, for example, p, v, T or u, T, p. Then with the thermodynamic surface given, perhaps the p, v, T surface, any two of the three properties may be used to locate the state point. The third property may now be read from the scale along its axis (see Figs. 3/4 and 3/5). Because of the characteristics described, properties are known as ***point functions***. Since space figures are not so easily sketched, it is a blessing that two coordinates generally define the state of a pure substance, making it simple for us to picture such states on a convenient thermodynamic plane.

1.6   SYSTEMS OF UNITS

Isaac Newton† made the momentous statement that the acceleration of a particular body is directly proportional to the resultant force acting on it and inversely proportional to its mass, $a = \mathbf{k}F/m$ where $\mathbf{k}$ is a proportionality constant. Rearranging the equation form of the statement

(1-1A) $F = ma/\mathbf{k}$,

or dimensionally $F \rightarrow ML/\tau^2$, enables us to define a force unit in terms of the mass, length, and time in any system of units.

* See § 11.2 for a more detailed discussion of this topic and its significance.
† Sir Isaac Newton (1642–1727) is often credited with being the greatest scientist of all times. Born of farmer parents, he was soon exercising his mechanical aptitude, devising a water clock and a sun dial during his years in grammar school. Two years after graduating from Cambridge, he had discovered the binomial theorem, started inventing calculus, experimented with color, and speculated on gravity. A few of his achievements: the reflecting telescope, the composite nature of sunlight, a science of optics, the invention of a thermometer (this is long before the discovery of the first law of thermodynamics), and the most monumental—his well-known law of gravitation. He himself credited his scientific successes to hard work, patient thought, and his predecessors—"I have stood on the shoulders of giants," he said.

Those consistent systems of units most commonly used for which **k** is unity, but not without dimensions, include these definitions of force:

cgs system: 1 dyne force accelerates 1 gm mass at 1 cm/sec^2
mks system: 1 newton force accelerates 1 kg mass at 1 m/sec^2
fps system: 1 lb force accelerates 1 slug mass at 1 ft/sec^2

Unfortunately, where the same word is used for both mass and force in a given system, we find the value of **k** is neither unity nor dimensionless. Several definitions of force include:

1 lb force accelerates a 1 lb mass at 32.174 ft/sec^2
1 gm force accelerates a 1 gm mass at 980.66 cm/sec^2
1 kg force accelerates a 1 kg mass at 9.8066 m/sec^2

Rearranging equation (1-1A) produces **k** = ma/F. Applying the preceding definitions of force shows

$$\mathbf{k} = (1\ \mathrm{lb}_m)(32.174\ \mathrm{fps}^2/\mathrm{lb}_f) \rightarrow 32.174\ \mathrm{lb}_m\text{-}\mathrm{ft}/\mathrm{lb}_f\text{-}\mathrm{sec}^2$$

$$\mathbf{k} = (1\ \mathrm{gm}_m)(980.66\ \mathrm{cm}/\mathrm{sec}^2)/\mathrm{gm}_f \rightarrow 980.66\ \mathrm{gm}_m\text{-}\mathrm{cm}/\mathrm{gm}_f\text{-}\mathrm{sec}^2$$

$$\mathbf{k} = (1\ \mathrm{kg}_m)(9.8066\ \mathrm{m}/\mathrm{sec}^2)/\mathrm{kg}_f \rightarrow 9.8066\ \mathrm{kg}_m\text{-}\mathrm{m}/\mathrm{kg}_f\text{-}\mathrm{sec}^2$$

At this point the student must realize that the value of **k** may be other than unity and must have units consistent with the system of units being used.

SI UNITS 1.7

In view of the relative newness, uniqueness, and universal acceptance of this particular set of metric units, it is believed at this point that a short discussion of the SI units is most appropriate. Definitions of the seven base units are given in order to impress upon the reader their physical concepts, realizing that the inadequacies of some definitions may produce a lack of full comprehension at this time.

In 1872 an international meeting was held in France and attended by representatives of 26 countries including the United States. Subsequently, in 1875, 17 countries (including the United States) signed an international treaty, the Metric Convention, to formulate an international metric standard. In 1960 the standard was modernized and called the International System of Units (SI Units).

SI units are divided into three classes: base units, derived units, and supplementary units. See Tables 1.1, 1.2, 1.3.

1.7a Definitions of base units

1. The meter (m) is the unit of length and is equal to 1,650,763.73 wavelengths in vacuum of the radiation corresponding to the transition between the levels $2p_{10}$ and $5d_5$ of the krypton 86 atom.
2. The kilogram (kg) is the unit of mass and is equal to the mass of the international prototype of the kilogram; it is the only base unit with a prefix (kilo).
3. The second (s) is the unit of time and is the duration of 9,192,631,770 periods of the radiation corresponding to the transition between the two hyperfine levels of the ground state of the cesium 133 atom.
4. The ampere (amp) is the unit of electric current and is that constant current which, if

maintained in two straight parallel conductors of infinite length, of negligible circular cross section, and placed 1 m apart in vacuum, would produce between these conductors a force equal to 2×10^{-7} newton per meter (N/m) of length.

5. The Kelvin is the unit of thermodynamic temperature and is the fraction 1/273.16 of the triple point of water.

6. The mole is the unit of substance and is the amount of substance of a system which contains as many elementary entities as there are atoms in 0.012 kg of carbon-12.

7. The candela (cd) is the unit of luminous intensity and is the luminous intensity, in the perpendicular direction, of a surface of $1/600,000$ m^2 of a black body at the temperature of freezing platinum under a pressure 101,325 N/m^2.

TABLE 1.1   SI Base Units

Quantity	Unit Name	Unit Symbol
Length	meter	m
Mass	kilogram	kg
Time	second	s
Electric current	ampere	A
Thermodynamic temperature	Kelvin	K
Amount of matter	mole	mol
Luminous intensity	candela	cd

TABLE 1.2a   Examples of SI Derived Units Expressed in Terms of Base Units

Quantity	SI Unit Name	SI Unit Symbol
Area	square meter	m^2
Volume	cubic meter	m^3
Speed, velocity	meter per second	m/s
Acceleration	meter per second squared	m/s^2
Density	kilogram per cubic meter	kg/m^3
Specific volume	cubic meter per kilogram	m^3/kg
Current density	ampere per square meter	A/m^2

TABLE 1.2b   Examples of SI Derived Units with Special Names

Quantity	SI Unit Name	SI Unit Symbol	Expression in Terms of Other Units	Expression in Terms of SI Base Units
Force	newton	N		$m \cdot kg/s^2$
Pressure	pascal	Pa	N/m^2	$kg/(m \cdot s^2)$
Frequency	hertz	Hz		$1/s$
Energy, work, heat	joule	J	$N \cdot m$	$m^2 \cdot kg/s^2$
Power	watt	W	J/s	$m^2 \cdot kg/s^3$
Quantity of electricity	coulomb	C	$A \cdot s$	$s \cdot A$
Electric potential	volt	V	W/A	$m^2 \cdot kg/(s^3 \cdot A)$
Capacitance	farad	F	C/V	$s^4 \cdot A^2/(m^2 \cdot kg)$
Electric resistance	ohm	Ω	V/A	$m^2 \cdot kg/(s^3 \cdot A^2)$
Conductance	siemens	S	A/V	$s^3 \cdot A^2/(m^2 \cdot kg)$
Magnetic flux	weber	Wb	V/s	$m^2 \cdot kg/(s^2 \cdot A)$
Magnetic flux density	tesla	T	Wb/m^2	$kg/(s^2 \cdot A)$
Inductance	henry	H	Wb/A	$m^2 \cdot kg/(s^2 \cdot A^2)$
Luminous flux	lumen	lm		$cd \cdot sr$
Illuminance	lux	lx		$cd \cdot sr/m^2$

1.7b SI derived units

Derived units are expressed algebraically in terms of base units. A number of derived units have been given special names; subsequently, numerous derived units are expressed by means of these special names. See Tables 1.2a, 1.2b, 1.2c.

Examples of SI Derived Units Expressed by Means of Special Names   TABLE 1.2c

| | SI Unit | | |
Quantity	Name	Symbol	Expression in Terms of SI Base Units
Heat capacity, entropy	joule per kelvin	J/K	$m^2 \cdot kg/(s^2 \cdot K)$
Specific heat capacity	joule per kilogram kelvin	J/(kg · K)	$m^2/(s^2 \cdot K)$
Thermal conductivity	watt per meter kelvin	W/(m · K)	$m \cdot kg/(s^3 \cdot K)$
Dynamic viscosity	pascal second	Pa · s	$kg/(m \cdot s)$
Moment of force	meter newton	N · m	$m^2 \cdot kg/s^2$
Surface tension	newton per meter	N/m	kg/s^2
Molar energy	joule per mole	J/mol	$m^2 \cdot kg/(s^2 \cdot mol)$

1.7c SI supplementary units

There are some units that fall under neither base units nor derived units. Currently two geometrical units have been identified under this supplementary classification.

SI Supplementary Units   TABLE 1.3

| | SI Unit | |
Quantity	Name	Symbol
Plane angle	radian	rad
Solid angle	steradian	sr

The radian is the plane angle between two radii of a circle that cut off on the circumference an arc equal in length to the radius.

The steradian is the solid angle that, having its vertex in the center of a sphere, cuts off an area of the surface of the sphere equal to that of a square with sides of length equal to the radius of the sphere.

ACCELERATION   1.8

Acceleration has the dimensions of length per unit of time squared, L/τ^2. Recall that a dimension is an attribute of something in general terms; thus, length L is also an attribute of volume L^3. Units are dimensional characteristics expressed in terms of defined quantities. The length of a foot is accurately defined (in terms of the international meter = 1,650,763.73 wavelengths from electrically excited krypton-86). Defined units of time are seconds, minutes, hours, days, and so on. In English units, a unit quantity of acceleration is most often taken as 1 foot per second per second (ft/sec^2, or fps^2).

Now we may say, from equation (1-1A), that a unit force is one that produces unit acceleration in a body of unit mass. On the basis of this definition, one can decide upon the units for terms in groups of symbols where it is said that units must be consistent. If one decides to measure mass in pounds and acceleration in fps^2, then *force* in a consistent system, defined in terms of *mass* by (1-1A), must be in poundals ($= lb_m$-ft/sec^2). However, the engineer is accustomed to using the pound as a unit of force. In this case, to be consistent with acceleration in fps^2, *mass* defined in terms of *force* must be in slugs ($= lb$-sec^2/ft). With force in pounds and length in feet, the *consistent unit of energy* is the foot-pound (L-F), from the concept of work in mechanics as being the product of a force times the distance it moves along its direction of action.

Of course, if each term in an equation contains the same unit, it does not matter what unit is used. Since for many purposes the foot-pound is a relatively small quantity of energy, it is traditional in English units to use British thermal units (Btu's), about 778 ft-lb/Btu. Thus, in the application of thermodynamics, you will be continually concerned with conversion constants (Item B 38).* Nevertheless, after the manner of using the conversion constants is explained, we shall tend strongly to write the basic equations without these constants, a policy that requires the reader to be constantly alert. *Specify units for every answer.*

1.9   MASS

The **mass** of a body is the absolute quantity of matter in it, an unchanging quantity for a particular mass when the speed of the mass is small compared to the speed of light (no relativistic effects). Newton's universal gravitation law relates the force of attraction between two masses and, in equation form, is

$$(1\text{-}2) \qquad\qquad F_g = G\frac{m_1 m_2}{r^2} \qquad\qquad [\text{CONSISTENT UNITS}]$$

where F_g is the force of attraction (for the earth's attraction on an earthly system, this is *the* force of gravity) between the masses m_1 and m_2 that are r distance apart, and G is the gravitational constant. For F lb, r ft, and m slugs, a consistent system, $G = 3.44 \times 10^{-8} \, lb_f$-ft^2/slug^2 (ft^4/lb$_m$-sec^4). For F newtons, r meters, m kg, $G = 6.670 \times 10^{-11}$ N-m^2/kg^2. On earth the change in the force of gravity is seldom enough to affect significantly the usual engineering problem. But, for example, at a distance of 1600 mi from the earth, the force of gravity is only half as great as at the earth's surface.

As implied by the foregoing, the attractive force between masses may be used to establish units for measuring mass. A handy and preponderant reference body is the earth; hence a unit of mass is defined by the force interaction between a mass and the earth at a point on the earth's surface where gravity is defined as standard ($g_0 = 32.174$ fps^2)—45° north latitude, close to sea level. A certain platinum mass in

---

* Appendix B contains a number of tables and charts that are needed in the solution of problems. These are lumped together as items, Item B 1, Item B 2, and so on, arranged and numbered in the order in which they are generally referred to in the text.

France is accepted as an international standard kilogram, and the pound is defined in terms of this kilogram. Having such a standard, we may now compare masses on a balance scale (gravitational forces the same on each side and theoretically no friction or air buoyancy). Of course, other means are used to determine the masses of molecules, atoms, and planets.

Since a pound of *mass* located at a point of standard gravity g_0 is subjected to a *force* of gravity of 1 lb, it *weighs* 1 lb there. Then, for a *mass* of $m\,lb$, we write $m/\mathbf{k} = F_g/g = F/a$, or the force in pounds is

(1-3)
$$F = \frac{m}{\mathbf{k}}a$$

where, of course, $m/\mathbf{k}$ is the mass in slugs (a fps^2).

In this text we shall use m for the mass in pounds, kilograms, and so on. In many instances, for example, the many energy balances that we shall make, the use of a consistent system of units is unnecessary; mass units and energy units cancel. Hence, the mass may be in any unit (lb, gm, kg, etc.) and the energy may be in any unit (Btu, cal, J, ft-lb, etc.), but each term must involve the same unit of energy and the same unit of mass.

Where there can be little doubt as to whether the pound abbreviation is for force lb_f or for mass lb_m, we shall not use the subscripts. Remembering that the consistent energy unit comes from the work concept in mechanics (involving force) and keeping the concepts of force and mass distant, one should have no difficulty. Of course, the use of subscripts is optional with the reader.

Example

A car whose mass is 2 metric tons is accelerated uniformly from standstill to 100 kmph in 5 sec. Find the mass in pounds, the acceleration, the driving force in newtons, and the distance travelled.

Solution.

$$m = (2\text{ m tons})(1000\text{ kg/m ton})(2.205\text{ lb/kg}) = 4410\text{ lb}_m$$

$$a = (v_2 - v_1)/t$$

$$= (100 - 0\text{ km/hr})(1000\text{ m/km})/(5\text{ sec})(3600\text{ sec/hr})$$

$$= 5.56\ m/\text{sec}^2$$

$$F = ma/\mathbf{k} = (2000\text{ kg})(5.56\text{ m/sec}^2)/(1\text{ kg-m/N-sec}^2) = 11{,}120\text{ N}$$

$$d = v_{\text{avg}} \cdot t = (1/2)(100\text{ km/hr})(1000\text{ m/km})(5\text{ sec})/(3600\text{ sec/hr})$$

$$= 69.4\text{ m} \rightarrow 227.6\text{ ft}$$

WEIGHT   1.10

The **weight** of a body means the *force of gravity* F_g on the body; it may be determined by a so-called spring scale. Gravity produces a **force field** and a body in this field is subjected to a **body force**. Since the force field on the moon is much less than that on the earth [see equation (1-2)], the weight of a body is less there. In accordance with Newton's law (1-1A) that the acceleration of a particular body is proportional to the resultant force on it, we write $F_g/g = F/a$, where g is the acceleration produced by F_g alone (in vacuum), and a is the acceleration produced by another force F. If the analogous symbols for the moon's gravitational field are F_m and g_m, then $F_f/g = F_m/g_m$.

If we let the acceleration produced by a gravitational field, wherever it is, be represented by g, then Newton's equation (1-1A) says that the force of gravity is

(1-1B) $$F_g = mg/\mathbf{k}$$

in which the units must be consistent. Thus, for the force to be in pounds, the mass must be in slugs when g is in feet per second per second.

Example

Two masses, one of 10 kg and the other unknown, are placed on a scale in a region where $g = 9.67 \text{ m/sec}^2$. The combined weight of these two masses is 174.06 N. Find the unknown mass in kg and lb_m.

Solution.   Using equation (1-1B)

$$F_g = mg/\mathbf{k}$$

$$m = F_g\mathbf{k}/g$$

$$= \frac{(174.06 \text{ N})(1 \text{ kg m/sec}^2\text{-N})}{9.67 \text{ m/sec}^2} = 18 \text{ kg (total mass)}$$

$$\text{unknown mass} = 18 - 10 = 8 \text{ kg} \to (8 \text{ kg})(2.205 \text{ lb}_m/\text{kg})$$

$$= 17.64 \text{ lb}_m.$$

1.11   SPECIFIC VOLUME AND DENSITY

The *density ρ of any substance is its mass* (not weight) *per unit volume.*

(1-4) $$\text{Average density} = \frac{\text{mass}}{\text{volume}}, \quad \rho = \frac{m}{V} \quad \text{or} \quad \rho \equiv \lim_{\Delta V \to 0} \frac{\Delta m}{\Delta V}$$

where, for the density at a point, the volume must contain enough molecules to classify as a **continuum**. If the mass m is measured in pounds and the volume V in cubic feet, then the average density of a particular volume V is $\rho = m/V \text{ lb/ft}^3$; other units include slugs/ft^3, lb/in.^3, gm/cm^3, kg/m^3. The *specific volume v is the volume of a unit mass*, say, cubic feet per pound, and is the reciprocal of the density, $v = V/m = 1/\rho.$

For homogeneous substances a continuum is a quantity of matter involving a very large number of molecules. The density of a cubic centimeter with 2 or 3 molecules in it is not practically useful (except in the sense of "population density"—1 person per square mile). Moreover, if the substance is not homogeneous, the density can only be an average value; in this case, either the whole volume is used for the computation or the sample is large enough to be representative of the whole.

Perhaps the densities of different parts might be definable; for example, in a two-phase system, such as water and steam, we may often be interested in the average density of the water or of the steam, but only occasionally in the overall average density of the mixture. On the other hand, the average specific volumes of such phase mixtures are often used. In a system in a force field, such as the earth's atmosphere to great height, local densities and specific volumes may be useful, but the pattern of variation of density must be established and allowed for if the atmosphere is either the system or the surroundings of a system, as when bodies reenter the atmosphere from space.

SPECIFIC WEIGHT 1.12

*The **specific weight** γ of any substance is the force of gravity on unit volume.*

$$(1\text{-}5) \qquad \text{Average specific weight} = \frac{\text{force of gravity}}{\text{volume}} \qquad \gamma \equiv \lim_{\Delta V \to 0} \frac{\Delta F_g}{\Delta V}$$

usually lb_f/ft^3 or $\text{lb}_f/\text{in.}^3$ in English units. Since the specific weight is to the local acceleration of gravity as the density is to the standard acceleration, $\gamma/g = \rho/\mathbf{k}$, conversion is easily made;

$$(1\text{-}6) \qquad \rho = \frac{\mathbf{k}}{g}\gamma \quad \text{or} \quad \gamma = \frac{g}{\mathbf{k}}\rho$$

If the mass is on or near the surface of the earth, then numerically $g \approx \mathbf{k}$, and the two quantities are nearly the same.

PRESSURE—KINETIC THEORY 1.13

The pressure of a gas, if gravitation and other body forces are negligible (as they generally are for a gas), is caused by the pounding of a large number of gas molecules on the surface. The elementary kinetic theory presumes that the volume of the molecule itself is negligible, that the molecules are so far apart they exert negligible forces on one another, and that the molecules are rigid spheres that do have elastic collisions, with walls and with each other. Elastic collision means, for example, that when molecule *A*, Fig. 1/4, strikes the plane surface *MN* at an angle of incidence of α with the normal *PN*, it rebounds symmetrically on the other side of *PN* with angle α, and with no loss of kinetic energy or momentum; $|v_{A1}| = |v_{A2}|$. The pressure is a consequence of the rate of change of momentum of the molecules striking the surface.

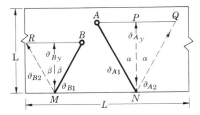

Consider this to be a cubical container, *L* on a side. This   **Fig. 1/4**
assumption simplifies the physical concepts, but the result
is just as general.

Since the momentum is a vector quantity and since *pressure* is defined as *the normal force per unit area*, we note that the change of momentum of molecule A in the y direction is $mv_2 - mv_1 = m_A[v_{A1y} - (-v_{A2y})]$; or for each molecule, it is $2mv_y$. Assume that this molecule moves to the opposite wall and bounces back to strike the wall MN again; the time taken for the round trip is $\tau = 2L/v_y$. Dividing the change of momentum by the time, we get the force produced by one molecule; $2mv_y/\tau = mv_y^2/L$. If all the molecules are alike (same mass m), the total force is

(a)
$$F = \frac{m}{L} \sum_{i=1}^{N} v_{yi}^2 = \frac{Nm}{L} \frac{\Sigma v_{yi}^2}{N}$$

where we have multiplied and divided by N, the number of molecules in the container. The sum of the squares of the y-component velocities Σv_{yi}^2 of all the molecules divided by the total number of molecules N gives the average of the square of v_y, symbolized by a bar thus, $\overline{v_y^2}$. Then, since the average pressure is the total force divided by the area, $p = F/L^2$, we get

(b)
$$p = \frac{F}{A} = \frac{Nm}{L^3} \overline{v_y^2} = \frac{Nm}{V} \overline{v_y^2}$$

where $V = L^3$ = the volume of the container. Inasmuch as the molecules have random motion, as many will strike each of the other surfaces of the container as struck the one studied—evidently true because there must be some pressure on each wall for static equilibrium. This logic is verified by experimental observation that the pressure of the gas "at a point" is the same in all directions. Thus, the component velocities in the x, y, and z directions are such that $\overline{v_x^2} = \overline{v_y^2} = \overline{v_z^2}$. Since these components are orthogonal components, the velocity vector v that has these components is such that

(c)
$$\overline{v^2} = \overline{v_x^2} + \overline{v_y^2} + \overline{v_z^2} = 3\overline{v_y^2}$$

and equation **(b)** becomes $(\overline{v_y^2} = \overline{v^2}/3)$

(1-7)
$$p = \frac{Nm\overline{v^2}}{3V}$$

where $\overline{v^2}$ is the mean-square molecular speed; the square root of this mean square is called the *root mean square*; say $[\overline{v^2}]^{1/2} = v_{rms}$. If a particular molecule strikes another molecule en route to a wall and thus fails to make the round trip, another identical molecule takes its place, producing the same effect, since all collisions are elastic and the average molecular velocity and kinetic energy (relative to the container) do not change.

We speak of the pressure at a *point*, but actual pressure-measuring equipment (Fig. 1/5 shows one type) typically registers not dozens of molecular strikes, but ordinarily millions, in a small fraction of a second. Exceptions to this generalization include extreme vacuums and the outskirts of the earth's atmosphere. At an altitude of 30 mi, the mean free path (MFP) of a molecule is about 1 in., relatively quite far; at 400 mi, the MFP is about 40 mi. This decreasing density means fewer strikes, and if the pressure probe is struck by a molecule only now and then, there is no meaning to the "pressure at a point." A cubic inch of atmosphere (a handful) contains some 4×10^{20} molecules. In dealing with so many particles making up a system, it may occur (microscopically), and probably does, that the pressure on an extremely small area is momentarily quite high (or low) because by chance a number of high velocity (or low velocity) molecules have just happened to strike that area. Such an event involves such a small area that no pressure-measuring instrument could identify it—not to mention the virtually infinitesimal duration of the event. In short, nearly all our gas systems will involve enough molecules that the systems easily qualify as continuums, and the pressure (and other) instruments record a statistical average number that applies to the system (at rest), the macroscopic pressure.

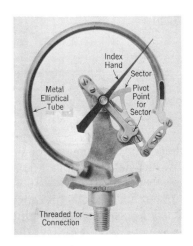

Movement for a Bourdon Pressure Gage. This picture shows the movement in one type of pressure gage. known as the single tube gage. The fluid enters the tube through the threaded connection. As the pressure increases, the tube with an elliptical section tends to straighten, the end that is nearest the linkage moving toward the right. The linkage causes the sector to rotate. The sector engages a small pinion gear. The index hand moves with the pinion gear. The whole mechanism is, of course, enclosed in a case, and a graduated dial, from which the pressure is read, is placed under the index hand. *(Courtesy Crosby Steam Gage and Valve Co., Boston).* **Fig. 1/5**

Barometers are used to measure atmospheric pressure.* It is convenient to have a standard reference atmospheric pressure, which is 760 mm Hg or 29.92 in. Hg at 32°F, or 14.696 psia (14.7 for slide rule), or 1 atm.

Pressure is one of the most useful thermodynamic properties because it is easily measured directly. (Precision measurements of anything are difficult.) Pressure-measuring instruments read a difference of pressures, called *gage pressure*; in pounds per square inch, we shall use the abbreviation psig, the g standing for gage. The absolute pressure, say psia, can be determined from the gage pressure as follows:

(1-8)   Absolute pressure = atmospheric pressure ± gage pressure

where the positive sign applies when absolute pressure is greater than atmospheric, and the negative sign for absolute pressure less than atmospheric. The negative sign

---

* After Evangelista Torricelli (1608–1647) discovered the pressure of the atmosphere, Otto von Guericke (1602–1686) set about producing a vacuum, the first effort by pumping water from a beer keg; but he discovered that tightness with beer is more easily attained than with air. He finally obtained a significant vacuum after having made two hemispheres, known as the Magdeburg hemispheres, which were capable of withstanding atmospheric pressure. Before a large audience of notables, von Guericke placed his hemispheres together and soon had most of the air pumped from the inside. A horse was hitched to each hemisphere, and try as they might, the two of them could not pull the hemispheres apart. The people, who knew nothing of the pressure of the atmosphere, were astounded when von Guericke broke the vacuum and the hemispheres *fell* apart. Had not von Guericke been a public official, renowned for his wisdom and kindness, his magic might have resulted in no good for him. Other scientists of the time were persecuted, even killed, for less.

is for a gage reading called *vacuum pressure* or *pressure*. Each term in (1-8) should, of course, be in the same pressure unit. Equation (1-8), written to apply when the gage proper is located in the atmosphere, may be generalized by this statement: the gage pressure is the difference in pressures of the region to which it is attached (via the "threaded connection," Fig. 1/5) and the region in which the gage is located.

Example

A pressure gage registers 50 psig in a region where the barometer is 14.25 psia. Find absolute pressure in psia, Pa.

Solution. Using equation (1-8)

$$p = p_{atm} + p_g = 14.25 + 50 = 64.25 \text{ psia}$$

also

$$p = (64.25 \text{ psia})(6894.8 \text{ Pa/psi}) = 442,991 \text{ Pa} = 443 \text{ kPa}$$

1.14 FLUID PRESSURE

The foregoing discussion applies particularly to homogeneous systems in equilibrium, imperceptibly affected by body forces (gravity, magnetism). In a liquid system, while the movement of the molecules is considerably more restricted than in a gas, the molecular bombardment does produce pressure but not the total pressure, since the body force of the gravitational field is more likely to produce a significant effect. In a boiler the pressure of the vapor above the water and the pressure at the bottom of the boiler drum (Foreword) are nearly enough the same that the difference is usually ignored. The decision as to whether or not to account for such differences is an engineering decision, to be made in the context of the actual situation, depending on its relative magnitude and on precision requirements.

The following discussion applies to any fluid at rest, but is particularly intended to apply to manometers, a common pressure measuring instrument. There is no pressure gradient in any horizontal direction, but there is in a vertical direction because of gravity; and the pressure p on a cross-sectional area A is uniform. Thus, we may use the differential volume $dV = A\,dz$, shown in Fig. 1/6, as the free body. The force of gravity on this free body is $dF_g = -\gamma A\,dz$ acting through the c.g., where the negative sign is used because z is positive as measured upward, while the force vector dF_g points down. The resultant force of the pressure on the upper side of the element dV is pA; on the lower side, $(p + dp)A$. Summing forces on the element, we get

$$(1\text{-}9) \qquad (p + dp)A - pA - dF_g = A\,dp + \gamma A\,dz = 0 \qquad dp = -\gamma\,dz$$

the basic relation; the units must be consistent. In *short* columns of liquids or gases, the specific weight is virtually constant. If γ varies and its manner of variation as a function of z is known, this equation can be integrated as shown. Integrating (1-9) with constant γ from a liquid surface where the uniform pressure is any value p_a gives

$$(1\text{-}10) \qquad p - p_a = \gamma(z_0 - z) \qquad \text{or} \qquad p = p_a + \gamma(z_0 - z) = p_a + \gamma d$$

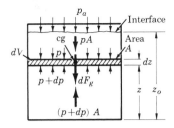

Fluid Pressure. All forces on the element dV act through the center of gravity of dV.   **Fig. 1/6**

where p is the pressure at any level marked by the coordinate z, and p_a is the pressure at z_0 (perhaps the interface of a liquid and a gas or of the liquid and its own vapor); the final form $p_a + \gamma d$ is handy where d represents the vertical dimension (depth) of the fluid, usually a liquid in this application. It often happens that p_a is the ambient atmospheric pressure. The units in (1-9) or (1-10) must be consistent: for example, γ lb/ft^3, z ft, p lb/ft^2; or γ lb/in.^{3}, z in., p lb/in.^{2}.

An infinitesimal area would result in the same basic equation (1-9). Therefore, we can say that the fluid pressure at a point is the same in all directions.

MANOMETERS   1.15

Manometers give a reading as the length of some liquid column: mercury, water, alcohol, etc. If the length of column of liquid, of cross-sectional area A, is d, then the volume is $V = Ad$ and the force of gravity on the column is $F_g = \gamma Ad$, where γ is the specific weight of the fluid; $\gamma = (g/\mathbf{k})\rho$, equation (1-6). The corresponding pressure is $p = F_g/A = \gamma d$. That part of the fluid in the loop HJ, Fig. 1/7, is evidently, from symmetry, in equilibrium and can be ignored. The pressure at B is equal to the pressure p_a at G plus $\gamma_M d = \gamma_M(GH) - \gamma_E(KJ)$, where γ_E is the specific weight of the fluid in the part KJ. If this fluid in KJ is gaseous of ordinary densities (say the fluid in the vessel is a gas), the value of $\gamma_E(KJ)$ may be small enough to be neglected. If this is true, the pressure in the vessel is taken as

(1-11)
$$p = p_a + \gamma d = p_a + \frac{g\rho d}{\mathbf{k}} = p_a + \frac{gd}{\mathbf{k}v}$$

where p_a is the pressure of the surroundings, v is the specific volume, and the units are consistent. See Item B 38, appendix for conversion constants; as (in. Hg) (0.49 psi/in. Hg) $\rightarrow$ lb/in.^{2} The pressure p as it appears in most equations must be absolute pressure. Even when the conversion unit cancels, it is advisable to write it down (and then cancel it) because the *habit* of conversion is important. Also remember to convert gage pressure to absolute pressure.

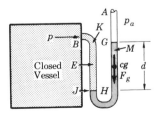

Manometer. If the manometer is open to the atmosphere at A, the gage pressure reading is γd. If instead, another vessel is connected at A, the difference in pressures between the two vessels is γd (fluids of negligible weight except *JHG*).   **Fig. 1/7**

Example

The vessel in Fig. 1/7 contains a gas; the manometer contains a liquid whose density is 300 lb/ft^3 with $d = 50$ in. Find the gas pressure in psi and Pa; local $g = 32.11$ fps^2 and barometer = 14.55 psia.

Solution.   Using equation (1-11)

$$p = p_a + g\rho d/\mathbf{k}$$

$$= 14.55 \text{ psi} + \frac{(32.11 \text{ ft/sec}^2)(300 \text{ lb}_m/\text{ft}^3)(50 \text{ in.})(\text{ft}/12 \text{ in.})}{(32.174 \text{ lb}_m\text{-ft/lb}_f\text{-sec}^2)(144 \text{ in.}^2/\text{ft}^2)}$$

$$= 14.55 + 8.66 = 23.21 \text{ psia}$$

also

$$p = (23.21 \text{ psia})(6894.8 \text{ Pa/psi}) = 160,028 \text{ Paa}$$

1.16   ARCHIMEDES PRINCIPLE

A body immersed in a fluid is subjected to a buoyant (upward) force equal to the weight of the fluid displaced. Since there is no horizontal pressure gradient, this effect is due to the vertical fluid force on surfaces that face downward being greater than the vertical force on the surfaces that face upward. The net force on the body is the force of gravity on the displaced fluid minus the force of gravity on the body. For a floating body at rest, the net force is zero. Observe that, strictly, when a body is weighed in the atmosphere, a correction should be made for the weight of the displaced air, a negligible correction when the density of the body is much greater than the density of the air.

1.17   TEMPERATURE—WITH A MICROSCOPIC VIEW

As Maxwell said,* *the temperature of a body is its thermal state considered with reference to its ability to communicate heat to other bodies.*[1.21] (See Zeroth Law, § 1.20.) It is an intensive property that, as we shall learn, is a measure of the intensity of the stored molecular energy in a system.

Consider for a moment the microscopic view, which necessitates recalling certain things. Let M equal the **molecular mass** (also called *molecular weight, formula mass*); let n equal the number of moles, where, for example, a so-called pound-mole is M lb of a substance. The quantity that is a mole is different in different systems of units:

$$1 \text{ pmole}\dagger \text{ (pound-mole)} = M \text{ lb} \qquad 1 \text{ gmole (gram-mole)} = M \text{ gm}$$

$$1 \text{ kgmole (kilogram-mole)} = M \text{ kg}$$

For instance, for O_2, $M = 32$; therefore 1 pmole of O_2 is 32 lb, 1 gmole is 32 gm, and so on.

---

* Biographical footnote in § 11.3.
† These abbreviations, it is hoped, will reduce the number of times that the beginner cancels the lb as a unit in abbreviations such as lb-mole.

Avogadro's number $N_A = N/n = 6.02252 \times 10^{23}$ molecules/gmole, a fundamental constant of nature, is the number of molecules in a gram-mole; N is the total number of molecules. A **mole volume** $\bar{v}$ is the volume of 1 mole; therefore total volume $V = n\bar{v}$, for n moles. Finally, we ask that you accept, without discussion now, the so-called ideal gas equation, $p\bar{v} = \bar{R}t$, where $\bar{R}$ is the universal gas constant, so that we can perform the following manipulations. Substitute the above value of V into equation (1-7), which is $p = N m \overline{v^2}/(3V)$:

$$p = \frac{N m \overline{v^2}}{3V} = \frac{N m \overline{v^2}}{3n\bar{v}} = \frac{N_A m \overline{v^2}}{3\bar{v}}$$

(1-12A)
$$p\bar{v} = \bar{R}T = \frac{N_A m \overline{v^2}}{3} = \frac{2N_A}{3}\left(\frac{m\overline{v^2}}{2}\right)$$

(1-12B)
$$T = \frac{2}{3}\frac{N_A}{\bar{R}}(\varepsilon) = \frac{2\varepsilon}{3\kappa}$$

where $\varepsilon = m\overline{v^2}/2$ is the average kinetic energy of the ideal gas molecule of mass m and $\kappa = \bar{R}/N_A$, the ratio of the gas constant to Avogadro's number, is another fundamental *constant*, the gas constant per molecule, called **Boltzmann's constant**; $\kappa = 1.38054 \times 10^{-16}$ ergs/°K.* The molecular velocity corresponding to ε in (1-12B) is called the root-mean-square (rms) velocity v_{rms}, § 1.13. It is interesting to note that at a particular temperature (particular v_{rms}) the heavier molecule has more energy. There is much more about the ideal gas later, so all we wish to note now is that the temperature is directly proportional to the average translational kinetic energy of the molecules. Temperature is registered on an instrument by virtue of the exchange of molecular energy until an equilibrium is reached (§ 5.25). Observe that a particular molecule only has energy and does not have a temperature except as it might be a computed number by an equation such as (1-12); it is the gas (the thermometer receives myriad molecular contacts) that has temperature, a macroscopic property.

SCALES OF TEMPERATURE 1.18

A scale of temperature is an arbitrary thing. The Fahrenheit and Celsius (centigrade) scales are based on the **ice point** (the temperature of a mixture of ice and air-saturated water at 1 atm) and the **steam (boiling) point** for water at 1 atm. At the ice point and steam point, the Fahrenheit temperatures are 32°F and 212°F, and the Celsius temperatures are 0°C and 100°C. Thus, between the normal freezing and boiling points of water, there are 100 degrees on the Celsius scale and 180 degrees on the Fahrenheit scale (180/100 = 9/5 = 1.8), giving the relations

(1-13A)
$$t_c = 5/9(t_f - 32)$$

(1-13B)
$$t_f = 9/5 t_c + 32$$

* The symbols adopted for molar properties follow the plan of Keenan and Kaye *Gas Tables*,[0.6] as $\bar{v}$, $\bar{u}$. Most generally, the bar designates an average, as $\overline{v^2}$ is the average of the speeds squared. Considering our particular uses, we believe you will not be confused by this inconsistency.

where t_c and t_f are the temperatures on the Celsius and Fahrenheit scales, respectively. Because there is more than one in use, the scale of a temperature is always stated: as 212°F or 100°C, where F means Fahrenheit and C means Celsius.*

Thermodynamics requires the use of absolute temperature (or thermodynamic temperature), which is measured from a point of absolute zero. While we shall frequently have occasion to say more about temperature, we wish now only to accept the concept of absolute temperature and its relation to the Fahrenheit and Celsius scales. Absolute zero on the Fahrenheit scale is at −459.67°F. Absolute temperatures T on the Fahrenheit scale are called **degrees Rankine** (°R),†

$$(1\text{-}14) \qquad T°R = t°F + 459.67 \approx t°F + 460$$

Absolute temperatures on the Celsius scale are called **degrees Kelvin**, in honor of Lord Kelvin (see footnote in § 6.5), written °K or K (for SI units), and absolute zero is −273.15°C. Thus,

$$(1\text{-}15) \qquad T°K = t°C + 273.15 \approx t°C + 273$$

Temperatures are always measured via the change in some other property (§ 1.19); it took a long time to develop accurate measurements, which are still nonexistent for temperatures outside of the "usual" range. (See the references on this subject at the back of the book for detail much too extensive to include here.) For temperatures measured anywhere in the world to agree, reasonably accurate guide points are necessary for the calibration of the instruments. Since the temperature of the **triple point** of H_2O (§§ 3.4 and 3.7) can be measured with excellent accuracy, it has been agreed upon internationally as the basic point on the absolute scale, to wit, 273.16 K (which is 0.01 deg higher than the ice point). By way of illustration, other temperatures agreed upon by a more or less international consensus, in addition to the ice, steam, and triple points of water, are as follows for two-phase mixtures at 1 atm:

Oxygen (O_2):	−182.970°C, liquid-vapor equilibrium
Mercury (Hg):	−38.87°C, solid-liquid equilibrium
Tin (Sn):	231.9°C, solid-liquid equilibrium
Zinc (Zn):	419.505°C, solid-liquid equilibrium
Sulfur (S):	444.60°C, liquid-vapor equilibrium
Antimony (Sb):	630.5°C, solid-liquid equilibrium
Silver (Ag):	960.8°C, solid-liquid equilibrium
Gold (Au):	1063.0°C, solid-liquid equilibrium
Platinum (Pt):	1774°C, solid-liquid equilibrium
Tungsten (W):	3370°C, solid-liquid equilibrium

The National Bureau of Standards (NBS) uses, among others, the liquid-vapor equilibrium of hydrogen at −253°C and of nitrogen at −196°C. Also, the NBS now

---

* Galileo invented a thermometer in 1592, but it did not have a well-founded scale. Gabriel Fahrenheit of Amsterdam, Holland, was the first (in 1720) to devise an instrument that indicated temperature in degrees, choosing the ice and boiling points of pure water as 32° and 212°, respectively. The centigrade scale was introduced in 1742 by Anders Celsius (1701–1744), a Swedish astronomer and professor at Uppsala.

† See footnote in § 9.3.

calibrates for temperatures between 4 and 14 K by use of acoustical thermometers (the speed of sound in a particular ideal gas is a function of temperature); between 2 and 5 K, a helium-4 vapor pressure scale is used.

Gas thermometers, complex as usual when accuracy is paramount, provide a standard for comparison. Helium has been successfully used to measure temperatures to about 2°R. Platinum resistance thermometers are also used for low temperatures, but at temperatures approaching absolute zero, the resistivity of conductors approaches zero (conductivity approaches infinity) and they therefore cannot be used. Semiconductors have been adapted to this area. If the temperature is above the gold point, definition is in terms of Planck's radiation law, but accuracy of measurements in this range is subject to significant improvement. For that matter, the definition of the entire scale will become more precise as the accuracy of the basic data improves.

Example

A system has a temperature of 250°F. Convert this value to °R, °C, °K.

Solution. Using the various equations.

$$T°R = t°F + 460 = 250 + 460 = 710°R$$

$$t_c = (5/9)(t_f - 32) = (5/9)(250 - 32) = 121°C$$

$$T K = t°C + 273 = 121 + 273 = 394 K$$

MEASURING TEMPERATURE   1.19

The details of the various ways in which temperatures are measured are too extensive for coverage here, but a brief mention of the most common approaches may be informative.

1. *Change in volume.* Nearly everyone is familiar, in a qualitative way at least, with the phenomenon of substances (say, mercury or a gas) that expand with increase in temperature. If the amount of expansion in a particular case is correlated with the freezing and boiling points of water, and the change in volume divided into 100 or 180 parts, the instrument could be used to "read" temperatures. Liquids used include[1.22]: mercury (−38°F to about 600°F—or 900°F with nitrogen above the mercury), alcohol (−100°F to about 300°F), and pentane (−300°F to about 70°F). Glass begins to soften at about 900°F. Upper working limit for gas thermometers is of the order of 2700°F. Since volume changes are unlikely to be exactly proportional to temperature changes, thermometers are calibrated for accuracy.

2. *Change in pressure.* If a gas is confined to a constant volume, its pressure will go up as its temperature increases, and the change in pressure can be correlated with temperature change.

3. *Change in electrical resistivity.* The electrical resistivity of some metals increases almost in direct proportion to temperature increase. Thus, the measured change in resistance of a particular piece of wire can be converted to temperature change. Metals used include nickel, copper, and platinum (for high precision). Also, semiconductors, which have a high sensitivity and a rapid thermal response, are used, especially at low (1 to 60°R) temperatures. Measurements of temperatures by changes of resistivity can be made to be the most precise of all methods between about −180 to 360°F.[1.30]

4. *Change in electrical potential.* The device that measures temperature by the electromotive force is called a ***thermocouple.*** It operates by virtue of the phenomenon occurring when two wires of different materials are joined together at their ends, with different temperatures at the two junctions. The emf is a function of the temperature difference between the junctions, a phenomenon called the *Seebeck effect.* The potentiometer, which measures the emf, can have a scale that reads temperature directly. One junction of the thermocouple is kept at a reference temperature, usually a mixture of ice and water at 32°F. This is one of the favorite means of measuring temperature. Combinations of metals used include: copper and constantan (−300 to 650°F), iron and constantan (−300 to 1500°F), chromel and alumel (−300 to 2200°F).

5. *Optical changes.* A body radiates heat proportional to the fourth power of its absolute temperature ($Q = \varepsilon \sigma T^4$, the Stefan-Boltzmann law). A number of schemes are used to convert this radiation to temperature. By the most common, an optical pyrometer is sighted at the hot body whose *brightness* is compared with the brightness of an adjustable and calibrated source of light within the instrument. When the brightness of the instrument light has been adjusted to be the same as the body whose temperature is desired, the instrument "reads" the body temperature.

1.20   ZEROTH LAW

A (natural) law is a generalization for which man has found no exceptions and is derived from observations of physical phenomena. If a "hot" body interacts with a "cold" body and the two are isolated from their surroundings, the properties of the bodies (say, the temperature, volume, conductivity, and so on) will change. However, after a time, the various properties cease to change. When property changes cease, the bodies are said to be in ***thermal equilibrium.*** See also § 5.25. The Zeroth law states that *when two bodies, isolated from other environment, are in thermal equilibrium with a third body, the two are in thermal equilibrium with each other.* (For proof of this law, see reference [1.4].)

To illustrate, suppose a thermometer *A* that reads 40°F is inserted into a "hotter" system *B* and the two, isolated from other interactions, arrive at a thermal equilibrium. It is a law of nature (second law) that net heat flows from the hotter body to the colder body when a thermal interaction occurs. Hence, heat flows from system *B* to the thermometer *A* (another system). Observe in passing that, for example, with energy leaving system *B*, its temperature will be lowered; hence, the total amount of energy stored in the system *B* whose temperature is measured must be, for accuracy, quite large compared to the amount of energy exchanged with the thermometer *A*. At equilibrium, let the thermometer reading be $t = 100°F$; we say this is the temperature of system *B*. Now if the thermometer is placed in a system *C* and, at equilibrium, the reading is again 100°F, systems *B* and *C* are in thermal equilibrium with respect to one another (the Zeroth law); that is, if *B* and *C* are brought into contact and isolated, there will be no changes of any of their properties (if no chemical reaction occurs). There may, however, be other types of nonequilibrium; for example, a piece of steel may be in thermal equilibrium with its surroundings but may be rusting (a chemical reaction).

---

* Thomas J. Seebeck (1770–1831), who discovered the thermocouple (1821), was born in Estonia. Although he was educated as a Doctor of Medicine (Göttingen College, Germany), he chose to lecture and experiment in the physical sciences. For his invention of the thermocouple, he was given a citation by the Paris Academy of Sciences.

PROCESSES AND CYCLES   **1.21**

If any one or more properties of a system change, the system is said to have undergone a process; there has been a change of state. Actual processes involve changes in all or nearly all properties, but we study thermodynamics via ideal models in which one of the properties often remains constant. For example, if a change of state occurs during which the pressure does not change, 1-2, Fig. 1/8, the working substance is said to undergo a *constant pressure (isobaric) process*; if the volume of a particular mass remains constant, *ab*, Fig. 1/8, but other properties change, the process is called a *constant volume (isometric) process*. Chemical reactions such as the combining of carbon and oxygen to form CO_2 are processes, as are also the action of a battery delivering electricity, of sugar passing into solution in coffee, and so on.

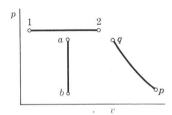

Processes.   **Fig. 1/8**

When a certain mass of fluid in a particular state passes through a series of processes and returns to its initial state, it undergoes a cycle. The processes may be ones that have names, such as constant volume and constant pressure, or they may be a series of state changes without names, as in Fig. 1/3. The closed path of the processes may be any random path 1-*A*-2-*B*-1 or 1-*A*-2-*C*-1, Fig. 1/9; it is only necessary that the substance return to state 1 in order to have completed a cycle. Cycles are studied after a detailed study of processes.

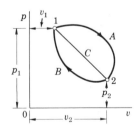

Cycles. By our convention of signs, cycles that trace a clockwise path are delivering work; cycles tracing a counterclockwise path, as 1-*C*-2-*A*-1 or 1-*B*-2-*A*-1, are receiving work.   **Fig. 1/9**

CONSERVATION OF MASS   **1.22**

The law of conservation of mass states that *mass is indestructible*. In applying this law, we must except nuclear processes, during which mass is converted into energy, or vice versa, and recognize that objects moving with a speed close to that of light undergo a significant increase in mass (§ 2.2). However, processes that involve such considerations are mostly in specialized areas that we shall touch only lightly. But we shall often need to make mass balances for "ordinary" processes, and we shall be much concerned with flow processes. Therefore, we shall review these ideas briefly.

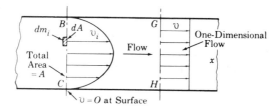

Fig. 1/10   *Laminar flow.*

Consider an elementary case of streamline (laminar) flow, Fig. 1/10, in which the streams are moving in parallel paths in the direction of flow. Within the stream, the molecules have their random motions, previously discussed, and, for example, the reading of a thermometer traveling with an ideal gas stream is an index of the average kinetic energy of the molecules. However, in dealing with the macroscopic systems, we shall mean the velocity (and kinetic energy) of the fluid stream, not of individual molecules. The properties (p, T, ρ, etc.) vary somewhat from point to point in section BC, a type of flow called *two-dimensional flow*. A thin layer of molecules adjacent to the walls of the passageway has zero stream velocity; those molecules moving along the center line have the maximum stream velocity normal to section BC; the theoretical pattern of variation is parabolic, as at BC, Fig. 1/10. Let the ith element with mass dm_i have a velocity v_i that is normal to the cross section at BC; during time $d\tau$, the volume crossing an area dA is $v_i\, d\tau\, dA$. Since volume times density is mass, the differential mass crossing dA is $dm_i = \rho_i v_i\, d\tau\, dA$. Divide both sides by $d\tau$ and get the rate of flow across the entire section BC as the sum of these ith quantities, or

$$(1\text{-}16) \qquad\qquad \dot{m} = \int_A \rho_i v_i \, dA$$

where $\dot{m} = dm/d\tau =$ the mass rate (say, lb/sec) of flow, and ρ_i and v_i represent local densities and velocities, respectively. In many systems involving thermodynamic analysis, it is quite accurate enough to assume that at each point of any cross section the properties are the same [$\rho = \rho(x)$, $p = p(x)$, and so on] and use an average velocity v normal to the section and assumed to be the same at each point in the cross section, as at GH, Fig. 1/10, a pattern called *one dimensional flow*. (You will, of course, meet the most general case in your study of fluid mechanics.) Thus, if the density is the same at all points of the cross section of area A, the mass rate of flow is*

$$(1\text{-}17) \qquad\qquad \dot{m} = \rho v A$$

Even when the flow is turbulent. with the stream having component velocities in x, y, and z directions, the average velocity that gives the correct flow in (1-17) also gives a reasonably accurate value of the kinetic energy.

Accepting the statement of the law of conservation of mass, it follows with respect to any system, that the verbal form of the law is

$$(1\text{-}18\text{A}) \qquad \begin{bmatrix} \text{mass} \\ \text{entering} \end{bmatrix} - \begin{bmatrix} \text{mass} \\ \text{leaving} \end{bmatrix} = \begin{bmatrix} \text{change of mass} \\ \text{stored in system} \end{bmatrix}$$

---

* In handwritten work, the reader may care to adopt the author's practice of using a lettered v *for specific volume and a script v*, somewhat exaggerated perhaps, for speed or velocity.

or we could equally well say: initial stored mass plus mass entering is equal to final stored mass plus mass leaving. Let dm_i equal mass "in," $d(\Delta m)$ equal differential change of stored mass, and dm_e equal mass "exit"; then (1-18A) in symbolic form is

(1-18B) $$dm_i - dm_e = d(\Delta m)$$

(In very short shorthand, we may think of any conservation law as IN − OUT = Δ(stored).)

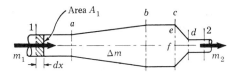

Properties may be substantially the same at all **Fig. 1/11** points in sections such as 1, 2, a, and b; abrupt changes as at c and d increase turbulence, making it probable that states such as e and f are not the same.

In Fig. 1/11, a particular volume, that which is in the passageway between sections 1 and 2, called the control volume (which is some chosen volume in space), has been taken as the open system; m_1 is the mass entering the system at section 1, m_2 is the departing mass at section 2, and Δm is the change in the quantity of mass within the system, *all quantities being measured in the same interval of time*. The symbol Δ shall always indicate a finite change in the sense of the final value minus the initial value. Thus Δm is positive if the mass of the system increases, and negative if it decreases. Equation (1-18) in symbolic form for Fig. 1/11 becomes

(1-18C) $$m_1 - m_2 = \Delta m$$

for a specified time interval.

A **steady-flow system** (more completely defined in §4.8) is an open system in which there is no change of stored mass; $\Delta m = 0$. Since, therefore, $\dot{m}_1 = \dot{m}_2 = \dot{m}$,

(1-19) $$\dot{m} = \rho_1 A_1 v_1 = \rho_2 A_2 v_2 = \frac{A_1 v_1}{v_1} = \frac{A_2 v_2}{v_2}$$

an equation called the **continuity equation of steady flow**. Usual English units are: v fps, ρ lb/ft^3, v ft^3/lb, A ft^2, $\dot{m}$ lb/sec.*

In general, thermodynamics is indifferent to time. A defined process may occur in 1 sec or 1 yr. However, certain events (processes) necessarily involve rates; as the rate of heat flow increases with an increase of temperature difference, mass flow increases with pressure difference, electricity flow increases with voltage difference, chemical reaction rates increase with chemical potential, and so on. These rates must be large enough that particular processes are commercially or humanly feasible. Consequently, we must from time to time include time τ in our calculations.

HEAT RESERVOIR 1.23

A heat reservoir is a thermodynamic system that generally serves as a heat source or heat sink for another system. It is considered to be stable, homogeneous in

---

* For the simple system of Fig. 1/11, equation (1-19) is easily set up. Mass across section 1, $dm_1 = \rho \, dV = \rho_1 A_1 \, dx$. Dividing by $d\tau$, get $\dot{m}_1 = \rho_1 A_1 \, dx/d\tau = \rho_1 A_1 v_1$ as in equation (1-19).

temperature and composition, and uniform in pressure at any level in the gravitation field. It is infinitely large, compared to the system it serves, with a temperature that remains unchanged when it is subjected to a heat interaction. Large rivers and lakes, the oceans, and the atmosphere are excellent natural heat reservoirs and serve a large number of our engineering systems such as power plants, automobiles, ovens, and so on, even though none truly fit the above restrictive definition.

1.24   HEAT ENGINES

A heat engine is a thermodynamic system that operates continuously with only energy (heat and work) crossing its boundaries; its boundaries are impervious to the flow of mass. It may be used to deliver work to external things, as does the Stirling engine, or it may receive work from something external and cause cooling to occur as in the case of a household refrigerator.

1.25   ADIABATIC SURFACE, PROCESS

An adiabatic surface is one that is impervious to heat. It implies perfect insulation. A process that occurs within a system enveloped by an adiabatic surface is called an adiabatic process—there can be no heat flow.

1.26   CLOSURE

Thermodynamics theory is not dependent upon any one system of units. However, numerical answers do depend upon the units specified.

The three properties of pressure, temperature, and volume were discussed in much detail for a good reason. These are properties that may be measured directly; all others are usually implied by them.

Most of the problems for this text are in a separate book, *Problems on Thermodynamics*, by Faires, Simmang, and Brewer. For your convenience, the problems book contains the tables and charts in Appendix B.

> **More problems are available in Faires, Simmang, and Brewer, *Problems on Thermodynamics*, Sixth Edition, published by the Macmillan Publishing Co., which also contains for easy reference all the tables and charts in Appendix B in *Thermodynamics, Sixth Edition*.**

PROBLEMS

(*Note:* Unless otherwise stated, atmospheric pressure should be taken as 14.696 psia or 101.325 kPaa and local gravity acceleration as 32.174 fps² or 9.806 mps²).

SI UNITS

1.1 A 0.1246-m³ system contains 4.535 kg of water vapor at 9.653 MPaa, 671 K. List the values of three intensive properties, one extensive property, and one specific property.

1.2 Express your height, mass, and weight ($g = 9.70$ m/sec²) in terms of SI units (m, kg, N).

1.3 A mass weighing 25 N is suspended from a cord that can be moved vertically, up or down. What are your conclusions regarding the direction and magnitude of acceleration and velocity of the mass when the cord tension is **(a)** 25 N, **(b)** 15 N, **(c)** 35 N?

1.4 For a ballistics study, a 1.9 gm bullet is fired into soft wood. The bullet strikes the wood surface with a velocity of 380 m/s and penetrates 0.15 m. Find **(a)** the constant retarding force, N, **(b)** the time required to stop the bullet, **(c)** the deceleration, m/s².

1.5 Assume 50 kg of mass are placed on the pan of a spring balance located on a freight elevator; local gravity acceleration is 9.70 m/s². **(a)** When the elevator is moving with an upward acceleration of 2.5 m/s², what will the balance read? **(b)** If the elevator is stopped, what will the balance read? **(c)** If the supporting cable breaks (elevator falls freely), what will the balance read? **(d)** If the balance reads 350 N, what are the circumstances?

1.6 A 30-m vertical column of fluid (density 1878 kg/m³) is located where $g = 9.65$ mps². Find the pressure at the base of the column. *Ans.* 543.7 kPa.

1.7 Two liquids of different densities ($\rho_1 = 1500$ kg/m³, $\rho_2 = 500$ kg/m³) are poured together into a 100-ℓ tank, filling it. If the resulting density of the mixture is 800 kg/m³, find the respective amounts of liquids used. Also, find the weight of the mixture; local $g = 9.675$ m/s². *Ans.* $m_1 = 45$ kg.

1.8 Compute the gravitational force between a proton ($m = 1.66 \times 10^{-27}$ kg) and an electron ($m = 9.11 \times 10^{-31}$ kg) in an atom whose radius of electron orbit is 5.29×10^{-11} m. Report answers in units of N and dynes.

1.9 If two thermometers, one reading °C and the other K, are inserted in the same system, under what circumstance will they both have the same numerical reading? What will be the system's temperature when the absolute thermometer reads twice the numerical reading of the Celsius thermometer?

1.10 Calculate the magnitude of the gravity acceleration on the surface of the moon and again at a point 1000 km above the surface of the moon; ignore the gravity effects of the earth. The moon has a mean radius of 1740 km and a mass of 7.4×10^{22} kg. *Ans.* (surface) $g = 1.63$ m/s².

1.11 For a particular thermocouple, if one junction is maintained at 0°C (cold junction) and the other junction is used as a probe to measure the desired Celsius temperature t, the voltage ε generated in the circuit is related to the temperature t as

$$\varepsilon = t(a + bt).$$

Further, for this thermocouple, when ε is in millivolts, the two constants are $a = 0.25$, $b = -5.5 \times 10^{-4}$. **(a)** What are the units of a, b? **(b)** Determine the value of ε for each of the measured temperatures -100°C, 100°C, 200°C, 300°C, 400°C, and plot an εt-curve. *Ans.* **(b)** $\varepsilon = 19.5$ mV for 100°C.

1.12 For the thermocouple in problem 1.11, find the rate of change of ε per °C at each of the temperatures shown -100 to 400°C. *Ans.* 0.14 mV/°C for 100°C.

1.13 If a pump discharges 284 ℓpm of water whose density is 985 kg/m³, find **(a)** the mass flow rate, kg/min, and **(b)** the total time required to fill a vertical cylindrical tank 3.05 m in diameter and 3.05 m high.

ENGLISH UNITS

1.14 A system has a mass of 30 lb. What total force is necessary to accelerate it 15 fps² if **(a)** it is moving on a horizontal frictionless plane or **(b)** it is moving vertically upward at a point where $g = 31.50$ fps²?

1.15 How far from the earth must a body be along a line toward the sun so that the gravitational pull of the sun balances that of the earth? Earth to sun distance is 9.3×10^7 mi; mass of sun is $3.24 \times 10^5 \times$ mass of earth. *Ans.* 1.63×10^5 mi from earth.

1.16 A cylindrical drum (2-ft diameter, 3-ft height) is filled with a fluid whose density is 40 lb_m/ft³. Determine **(a)** the total volume of fluid, **(b)** the total mass of fluid in lb_m, slugs, kg, **(c)** the specific volume of the fluid, **(d)** its specific weight where $g = 31.90$ fps². **(e)** Specify which of the foregoing properties are extensive and which intensive.

1.17 Given the barometric pressure of 14.7 psia (29.92 in. Hg abs), make these conversions: **(a)** 80 psig to psia and to atm, **(b)** 20 in. Hg vac to in. Hg abs and to psia, **(c)** 10 psia to psi vac and to Paa, **(d)** 15 in. Hg gage to psia, to torrs and to Paa.

1.18 **(a)** Define a new temperature scale, say °N, in which the boiling and freezing points of water are 1000°N and 100°N, respectively, and correlate this scale with the Fahrenheit and Celsius scales. **(b)** The °N reading on this scale is a certain number of degrees on a corresponding absolute temperature scale. What is this absolute temperature at 0°N?

Ans. **(a)** $t_N = 9t_c + 100$, **(b)** 2360°N abs, approx.

1.19 A weatherman carried an aneroid barometer from the ground floor to his office atop the Sears Tower in Chicago. On the ground level, the barometer read 30.150 in. Hg abs; topside it read 28.607 in. Hg abs. Assume that the average atmospheric air density was 0.075 lb/ft³ and estimate the height of the building.

Ans. 1451 ft (approx.).

1.20 As illustrated, a mercury manometer is attached to the side of a nearly full water tank. The reading of the mercury column is 15.5 in. Hg gage. Although air on the water maintains the pressure, the other 10-in. leg of the manometer is full of water; for H_2O, $\rho = 62.3$ lb/ft³; for Hg, $\rho = 846$ lb/ft³. If the location is at standard gravity and the temperature of both the water and mercury is 60°F, what is the pressure (psia) in the tank at the level where the manometer is attached? If the tank extends 10 ft below this level, what is the pressure at this depth?   *Ans.* 21.94, 26.27 psia.

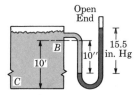

Problem 1.20

1.21 A 51-ft³ tank contains 3 lb of a gas at 80°F and a vacuum pressure of 24 in. Hg. **(a)** What is the absolute pressure in psia and psfa? **(b)** What is the gas' specific volume and density? **(c)** What is its temperature in °C, °R, and °K?

Ans. **(a)** 418 psfa, **(b)** 0.0588 lb/ft³, **(c)** 299.7 K.

1.22 A simple mercury manometer connected into a flow line gives readings as shown in the figure. Local gravity is standard and the mercury density is 0.488 lb/in.³ Find the pressure at points X and Y when the flow line and left leg contain **(a)** air whose density is 0.072 lb/ft³, **(b)** water whose den-

sity is 62.1 lb/ft³. **(c)** Answer **(a)** and **(b)** if the local gravity is $g = 30$ fps².

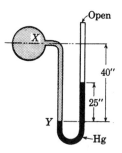

Problem 1.22

Ans. **(a)** 26.90, 26.90, **(b)** 25.46, 26.90, **(c)** 26.10 psia.

1.23 Convert **(a)** 122°F to °C and to K, **(b)** −40°C to °F and to °R, **(c)** 942°R to °C and to K, **(d)** 373 K to °F and to °R.

1.24 A Fahrenheit and a Celsius thermometer are both immersed in a fluid and indicate identical numerical readings. What is the temperature of the fluid expressed as °R and as K?

1.25 The same as problem 1.24 except that the Fahrenheit reading is numerically twice that of the Celsius reading.

Ans. 780°R, 433 K.

1.26 Steam at a pressure of 150 psia and a temperature of 400°F occupies a volume of 3.223 ft³/lb. **(a)** What is its density in lb/ft³ and in slugs/ft³? **(b)** Convert the pressure to in. Hg and ft water (60°F). **(c)** Convert the temperature to °R, K, and °C. Is temperature an intensive or an extensive property?

Ans. **(a)** 0.31 lb/ft³, 0.00965 slugs/ft³; **(b)** 306 in. Hg. 346 ft w.; **(c)** 860°R, 478 K, 204.4°C.

1.27 A fluid moves in a steady flow manner between two sections in a flow line. At section 1: $A_1 = 1$ sq ft, $v_1 = 1000$ fpm, $v_1 = 4$ ft³/lb. At section 2: $A_2 = 2$ ft², $\rho_2 = 0.20$ lb/ft³. Calculate **(a)** the flow (lb/hr) and **(b)** the velocity (fps) at section 2.

Ans. **(a)** 15,000 lb/hr, **(b)** 10.42 fps.

1.28 A 10-ft diameter by 15-ft height vertical tank is receiving water ($\rho = 62.1$ lb/ft³) at the rate of 300 gpm and is discharging through a 6-in. ID line with a constant velocity of 5 fps. At a given instant, the tank is half full. Find the water level and the mass change in the tank 15 min later.

Ans. 3.90 ft, 17,550 lb.

1.29 Write this computer program. It is desired to plot the curve weight (gravitational) versus height (miles above sea level). At equatorial sea level, $g = 32.088$ fps^2; its variation is -0.003 fps^2/1000 ft ascent. Select a given mass, say 100 lb$_m$, and determine its weight variation as it rises from sea level to a height of 1000 miles.

1.30 The conversion of temperature readings from °F to the several scales °C, °R, and K is sought. Write a computer program that will permit this operation.

MIXED UNITS

1.31 What is the mass in kilograms and weight in newtons ($g = 9.65$ m/s^2) of **(a)** a 4000-lb$_m$ automobile, **(b)** a 235-lb$_m$ fullback? Find the mass in grams and the weight in dynes of **(c)** 77 grains of moisture, **(d)** 12 ounces of salt.

1.32 For a given system, two of its independent intensive properties are x and y. State (show proof) which of the following expressions have property characteristics, that is, are point functions:

(a) $6xy$, **(b)** $y\,dx + x\,dy$, **(c)** $y\,dx$, **(d)** $2xy\,dx + x^2y\,dy$, **(e)** $15x^3 + 4xy^2$.

1.33 The mass of a given airplane at sea level ($g = 32.10$ fps^2) is 10 tons. Find its mass in lb$_m$, slugs, and kg and its (gravitational) weight in lb$_f$ and N when it is travelling at a 50,000-ft elevation. The acceleration of gravity g decreases by 3.33×10^{-6} fps^2 for each foot of elevation.

Ans. 9070.3 kg, 19,850 lb$_f$.

1.34 Five masses are as follows: m_1 is 500 gm of mass; m_2 weighs 800 gm; m_3 weighs 4 poundals; m_4 weighs 32.2 lb; and m_5 is 2 slugs of mass. The weights are at standard gravity. What is the total mass expressed **(a)** in pounds, **(b)** in slugs, and **(c)** in grams?

Ans. **(a)** 99.59 lb, **(b)** 3.1 slugs, **(c)** 45,200 gm.

1.35 It is estimated that the mass of the earth is 5.98×10^{24} kg; its mean radius is 6.38×10^6 m. Find its density gm/cm^3 and lb$_m$/ft^3. Compare this value to the density of water (62.4 lb$_m$/ft^3).

Ans. $5.50 \times$ density of water.

2

ENERGY CONCEPTS

2.1 INTRODUCTION

Energy is inherent in all matter. Energy is something that appears in many different forms which are related to each other by the fact that conversion can be made from one form of energy to another. Although no simple definition can be given to the general term energy, E, except that it is the capacity to produce an effect, the various forms in which it appears can be defined with precision. This chapter considers forms of energy with which you are presumably already familiar, and in this sense is something of a review; but a reacquaintance and a fresh point of view should now serve you well.

2.2 RELATION OF MASS AND ENERGY

One of the consequences of Einstein's theory of relativity is that mass may be converted into energy and energy into mass, the relation being given by the famous equation

(2-1)
$$E = mc^2 \qquad \Delta E = c^2\,\Delta m \qquad \text{[CONSISTENT UNITS]}$$

where $c = 2.9979 \times 10^{10}$ cm/sec = speed of light. Since c^2 is a very large number, the energy equivalent to a particular mass is very large; for example, 1 lb of matter is equivalent to nearly 39×10^{12} Btu—but this is not to say that it will ever be possible to convert a particular mass of any substance entirely into energy. Thus, the second form of equation (2-1) is the more informative. See Chapter 5. At this time, we can convert only a very small part of certain matter into energy.

Also, according to Einstein, mass increases with speed. This is the relativistic effect on mass. Let m_0 represent the so-called *rest mass* (with respect to the observer); then the mass m at any speed v is

(2-2)
$$m = \frac{m_0}{[1 - (v/c)^2]^{1/2}}$$

from which it is deduced that the variation of mass of finite bodies with any man made speeds is negligible (not so, however, for particles that travel at speeds approaching that of light).

Example

If a rest mass of 1 kg is accelerated to half the speed of light, what will be its new mass?
Solution. Use equation (2-2) and find

$$m = \frac{m_0}{\left[1 - \left(\dfrac{v}{c}\right)^2\right]^{1/2}} = \frac{1}{\left[1 - \left(\dfrac{0.5c}{c}\right)^2\right]^{1/2}} = 1.155\,\text{kg}$$

MEASURING ENERGY 2.3

The total amount of energy that a system contains cannot be determined. We are accustomed therefore to measuring energy above some arbitrary datum. This practice is satisfactory since, in engineering, we can get results knowing only the *changes* of energy. Thus, we consider the potential energy of 1 lb of water to be the change in potential energy that this pound of water would undergo in falling from the reservoir to the power plant. The level of the power plant is the datum level. In like manner, other forms of energy are measured with respect to some so-called **datum** or **reference state**.

Energy is a *scalar quantity*, not a *vector* quantity. Velocity, a vector, has direction as well as magnitude. Energy has only magnitude. The energy of a system of bodies is simply the sum of the energies of the individual bodies. The total energy of a single system is the sum of the magnitudes of the various forms of energy (such as mechanical kinetic energy, molecular energy, chemical energy) that the system possesses.

Because it is customary, we shall use the British thermal unit (Btu) most frequently as the measure of energy. There are actually several different Btu's (varying by small amounts); our first choice is the International (IT) Btu defined (by the International Steam Table Conference, and then the General Conference on Weights and Measures) basically in terms of joules, an energy unit in mechanics (§ 1.7 and Item B 38). Thus, by definition, Joule's* constant J is

(2-3) $J = 778.16\,\text{ft-lb/Btu}$

use 778 in normal calculations. Larger units of energy that are commonly used include horsepower-hour and kilowatt-hour; on the other hand, for computations of energy of the basic particles of matter, there are the electron volt (eV) and the million electron volts (MeV); 1.6×10^{-12} ergs/eV.

---

* James Prescott Joule (1818–1889), an English scientist, was educated privately by a tutor, being at one time a student assistant to Dalton (§ 6.10). His researches were significant in the budding science of electricity as well as in the science of thermodynamics, in which he established two important fundamental principles: the equivalence of heat and work, and the dependence of the internal energy change of a perfect gas upon temperature change (§ 6.7). As a result of this work, the modern kinetic theory of heat superseded the caloric theory of heat. Joule once remarked, "I believe I have done two or three little things, but nothing to make a fuss about."

2.4   GRAVITATIONAL POTENTIAL ENERGY

If a body in the earth's gravitational field moves away from or toward the center of the earth, work is done by or against the force of gravity. By the convention of signs that we shall use, the change of potential energy is the negative of the work of the force field (a generalization for all conservative force fields).* Although the change of gravitational potential energy is evaluated as a work quantity, we shall find it advantageous to consider such potential energy P separately as energy *stored* in a system (not true of work in general, § 2.9); that is, by virtue of the system's elevation z above a chosen *datum*, it possesses a certain amount of energy that is potentially available for conversion into work under idealized conditions. (Some of the energy change during actual events is necessarily dissipated in frictional effects. There is also the effect of buoyancy of the surrounding medium.) Equating the change of potential energy to the *work* of gravity, we have (with **k** shown as a reminder in one equation)

$$(2\text{-}4) \qquad\qquad dP = F_g\,dz = mg\,dz = \frac{mg}{\mathbf{k}}\,dz$$

where dz is the displacement of the center of gravity in the direction of the force of gravity. If the force of gravity F_g is virtually constant, the integration of (2-4) becomes

$$(2\text{-}5) \qquad\qquad P_2 - P_1 = \Delta P = mg(z_2 - z_1) = \frac{mg}{\mathbf{k}}(z_2 - z_1)$$

Also for constant F_g, the potential energy of a system is

$$(2\text{-}6) \qquad\qquad P = mgz = \frac{mgz}{\mathbf{k}} \quad \text{and} \quad P = \frac{gz}{\mathbf{k}}$$

[UNIT MASS]

Thus, symbol P will represent the potential energy for either unit mass or for m lb. The second form of (2-6) is for $m = 1$ lb. The consistent English units for these quantities are usually: local and standard accelerations of gravity fps^2, mass m slugs, mass m lb, elevation z_2, z_1, z ft measured vertically above any chosen datum level, and P and ΔP in ft-lb or ft-lb$_f$/lb$_m$ for F_g lb. If Btu units are desired, $P = mgz/J$, and so on. For rationalizing the units of equations with m slugs involved, remember slug $\rightarrow$ lb$_f$-sec^2/ft.

Although we shall have little occasion to be involved with body forces other than gravity, observe that equation (2-4) may be generalized for some other body force F_b by $\mathbf{F}_b \cdot d\mathbf{z}$ for dz displacement of the mass center. If the change of elevation of a body in a gravitational field is large, as for satellites and long range missiles, the variation of F_g and g must be accounted for; see equation (1-2).

Although we speak of the potential energy of the system with respect to earth, it is also true that the earth similarly has a potential energy with respect to the system. Thus, when two bodies of different masses have a significant attraction for each

---

* Force fields of electrical origin, which have been important since the birth of commercial electricity, are nearly ubiquitous. Unusually large ones have recently been applied.[4.8]

other and relative motion occurs between them, one must decide which of the two is being studied, that is, which is the system—or perhaps consider both (or all, if there are several bodies) of them as the system.

Example

A mass of 5 kg is 100 m above a given datum where local $g = 9.75 \text{ m/sec}^2$. Find the gravitational force in newtons and the potential energy of the mass with respect to the datum.
   Solution.   Use equation (2-6) and find

$$P = mg \frac{z}{\mathbf{k}}$$

$$= \frac{(5 \text{ kg})(9.75 \text{ m/sec}^2)(100 \text{ m})}{(1 \text{ kg-m/N-sec}^2)} = 4875 \, j$$

$$F_g = mg/\mathbf{k} = (5)(9.75)/1 = 48.75 \text{ N}$$

KINETIC ENERGY   2.5

In most of our energy considerations, we shall be involved with rectilinear translational motion and we may therefore write and manipulate Newton's second law in scalar form $F = ma$. By definition

(a)
$$a \equiv \frac{dv}{d\tau} = \frac{dv}{dx}\frac{dx}{d\tau} = \frac{v\,dv}{dx}$$

where x is the displacement in the direction of motion, the direction of vectors F and a, and a is the acceleration of the mass center with respect to coordinates of constant velocity. Substituting this value of a into $F = ma$ and integrating for constant force F and mass m from state 1 to state 2,

(2-7)
$$Fx = \frac{m}{2}(v_2^2 - v_1^2) \qquad \text{[CONSISTENT UNITS]}$$

where $mv^2/2$ is a form of energy called kinetic energy, applicable to a particle with any kind of motion. This expression $mv^2/2$ applies to a finite mass when all particles of mass have the same velocity v (translation); for v fps, $m \rightarrow$ slugs $= \text{lb}_f\text{-sec}^2/\text{ft}$, and $mv^2/2 \rightarrow \text{ft-lb}_f$. Recall from mechanics that equation (2-7) says that the work done on the body is equal to the change of kinetic energy—a generalization that is true if no other kinds of energy are involved. Thus, we define the kinetic energy of a mass m moving with a speed v as

(2-8)
$$K = \frac{mv^2}{2} = \frac{mv^2}{2\mathbf{k}} \qquad \text{and} \qquad K = \frac{v^2}{2} = \frac{v^2}{2\mathbf{k}}$$

$$\text{[UNIT MASS]}$$

where $mv^2/2$ and $v^2/2$ are good for any consistent system of units and $mv^2/(2\mathbf{k})$ and $v^2/(2\mathbf{k})$ are forms to use for mass pounds, force pounds, feet and seconds. Energy being a scalar quantity, the change in kinetic energy is

(2-9A)
$$\Delta K = \frac{m}{2\mathbf{k}}(v_2^2 - v_1^2) = \frac{m}{2}(v_2^2 - v_1^2)$$

(2-9B)
$$dK = \frac{m}{\mathbf{k}}v\,dv = mv\,dv$$

When the mass is in slugs, divide by Joule's constant J to get Btu; that is, $m(v_2^2 - v_1^2)/(2\mathbf{k}J)$ Btu. And a convenient round number is

$$2\,\mathbf{k}J = (2)(32.17)(778) \approx 50{,}000$$

Kinetic energy is *stored* in a system, one of whose mechanical properties is its speed v. However, since velocity is always relative, it is necessarily referred to particular reference axes. Although we shall have occasion to use axes moving with respect to the earth, consider v as "absolute" (with respect to axes attached to the earth) unless otherwise specified. The kinetic energy of a rotating body is $I\omega^2/2$, where I is the moment of inertia of the body about its axis of rotation. The reader should refer to a mechanics text for further detail.

The simple $mv^2/2$ from Newtonian mechanics falls down when the speed is close to that of light, c. In this event (for example, in studies of submolecular particles), relativistic effects must be accounted for; see a physics text.[1.27]

Example

The combined mass of car and passengers travelling at 72 km/hr is 1500 kg. Find the kinetic energy of this combined mass.

Solution. Using equation (2-8) we find

$$K = mv^2/2\mathbf{k}$$

$$= \frac{(1500\,\text{kg})(72\,\text{km/hr})^2(1000\,\text{m/km})^2}{(2)(1\,\text{kg-m/N-sec}^2)(3600\,\text{sec/hr})^2}$$

$$= 300{,}000\,\text{J} = 300\,\text{kJ}$$

2.6   INTERNAL ENERGY

*The sum of the energies of all the molecules in a system, energies that appear in several complex forms, is the internal energy.** Being an energy content, it is an important and continually useful property. While we shall take brief looks into the molecule later, at this time we can profitably think only of those forms of molecular energy that produce the most pronounced macroscopic effects. The dominant form for gases is the kinetic energy of translation because molecules of mass m move with a velocity v. The total amount of this energy is the total number of molecules in the system times the average kinetic energy of a molecule. Within the most usual temperature range, a change of internal energy of a **monatomic** gas (for example, He, A) is almost entirely a change of *translational kinetic energy.*

---

* Some writers use the term *internal energy* or *energy* to mean the total energy stored within a system, that which is called simply *stored energy* in this book.

In other than monatomic molecules, the mass, which is mostly the mass of the protons, is not so nearly concentrated at a "point"; hence, polyatomic molecules (for example, H_2O, NH_3, C_8H_{18}) may also have significant *energy of rotation*; in general, with rotation components with respect to 3 reference axes. For its simplicity, consider a diatomic (two-atom) molecule (for example, H_2, O_2) and think of it in the form of a dumbbell ●—●. If it rotates about axes that are perpendicular to the axis joining the atoms, the rotational energy is significant; if it rotates about the axis joining the atoms, the moment of inertia of the mass with respect to this axis is so small that this rotational energy is negligible. Incidentally, for now, since it has rotational energy about two axes, a diatomic molecule is said to have two degrees of rotational freedom.

As the temperature of a particular gas increases, the number of molecules whose atoms have noticeable *vibrational energy* increases. This kind of energy is most easily visualized with the dumbbell diatomic molecule (see Fig. 2/2). Imagine the axis connecting the atoms to be an ideal spring and the atoms vibrating to and fro; the atoms thus form what is called a *harmonic oscillator*. This phenomenon is a conservative system with the sum of the potential energies of the atoms (with respect to one another) and their kinetic energies remaining constant. Evidently, the atomic vibrations in polyatomic molecules become much more complex, but the idea is the same.

Then there is a potential energy owing to the force of attraction between molecules, which becomes relatively large for a gas. As an illustration, imagine a pound of water evaporated into a pound of steam, at atmospheric pressure. The measured volume increases by some 1600 times. It requires a large amount of energy to move these molecules apart against their attractive forces; the energy is retained in the steam as part of its internal and stored energy. Finally, on a submolecular basis, there are various forms of energy, such as the orbiting energy of the electrons, their spin energy, and energies related to electrical forces.

The sum of the various forms of energy that a molecule has is the *molecular internal energy U, u*, or **internal energy**. The absolute amount of internal energy that a body has is never known, but fortunately we manage very well because we can compute changes of this energy or measure it from any convenient datum; details of the computations come later. Let

$$u = \text{specific internal energy (1 lb)} \qquad \Delta u = u_2 - u_1$$

$$U = mu = \text{total internal energy for } m \text{ lb} \qquad \Delta U = U_2 - U_1$$

where the energy is usually in Btu's in English units and kJ's in SI units.

In some special cases, a system may not contain molecules; it may be made up of electrons or photons (radiant energy), in which case it does not have internal energy as defined, but it does have *stored energy* of the kinds associated with electrons and photons.

WORK 2.7

The vector expression for the mechanical work of a force as you learned it in mechanics is a dot product $dW = \mathbf{F} \cdot d\mathbf{s}$, good along any path. Again, we can profitably pass to the x component of a force acting through a distance dx (with

$\Delta y = 0$, $\Delta z = 0$) and write

(2-10) $dW = F_x\, dx$

which says that the work done by a force is the product of the displacement of the point of application of the force *times* the component of the force in the direction of the displacement. If the reference axes move with the point of application, no work is done. Example: If you push on a fixed wall, the force of your push does no work from the point of view of an observer who is stationary with respect to the wall. From our previous discussions, you recognize that equation (2-10) defines the consistent unit of energy, say ft-lb, but this unit will usually be converted to Btu's. From your mechanics, recall that the work of a couple M (a torque) acting through an angular displacement $d\theta$ is $dW = M\, d\theta$, a form that we shall need only occasionally.

Work is energy *in transition*; that is, it exists only when a force is "moving through a distance." Contrast the concept of work with that of internal energy. The internal energy is *stored* energy; the system contains it. On the other hand, *a system never contains work*, although it may have the capacity to do work, or work may be done on it (we study what happens in detail later).

The symbol W stands for work done by or on a system of 1 lb or any number of pounds; the distinction is defined by the context. Also, there may be a time unit involved with the rate of flow. Typical units for W include ft-lb, N-m, joules, Btu, and hp-hr. Work per unit of time is *power* $\dot{W}$; typical units are Btu/min, ft-lb/sec, hp, and kw.

Let us close this discussion paraphrasing an often-quoted definition[1.2]: *Work is that transitional energy* (not stored in a moving substance) *crossing the boundaries of a system that could conceivably produce the one and only effect of raising a weight.* We recognize the effect of raising a weight as work against the force of gravity. Also observe that it "could conceivably..."; it might in fact all be immediately dissipated by friction as soon as it leaves the system. If this energy will raise a weight, it will also turn a shaft against a resistance; hence a common name to distinguish this kind of work is *shaft work*. Finally, notice that work is a form of energy that crosses a boundary exclusive of any energy carried by a substance flowing across a boundary.

2.8   WORKING ON A MOVING BOUNDARY OF A SYSTEM

A substance that expands against a resistance (or is compressed) does work (or has work done on it). Let the system be a quantity of an expansible fluid, such as a gas or vapor, enclosed within a cylinder and piston, Fig. 2/1. This is a *closed system* in which *nonflow processes* may occur. The volume of the fluid is V_1 and its pressure is p_1. If we consider the state of the fluid on the pV plane (meaning that the coordinates are p and V), the particular coordinates p_1 and V_1 locate point 1, Fig. 2/1. If the working substance expands and moves the piston against a variable resistance, work will be done by the fluid. In a typical expansion of this sort, the pressure drops and the state of the substance changes as suggested by the curve 1-*ef*-2, the *path of the state point*, a *process*. Consider a change of state from *e* to *f*, Fig. 2/1, so small that the pressure is essentially constant during the change. The force acting on the piston will be the uniform pressure times the area of the piston, $F_x = pA$. The distance that the piston moves is dL, and the work for this infinitesimal motion is

(a) $$dW = (pA)\,dL = p(A\,dL) = p\,dV$$

where $A\,dL = dV$. The total work at the moving boundary is

(2-11) $$W = \int_1^2 p\,dV \quad \text{and} \quad W = \int_1^2 p\,dv \qquad \text{[REVERSIBLE]}$$

$$\text{[UNIT MASS]}$$

in which p lb/ft^2 and V ft^3 give W ft-lb. Also, p N/m^2 and V m^3 give W J. If a system has more than one moving boundary, the work at each is given by (2-11); if the system has an irregular boundary that changes shape and volume, work is given by (2-11), all subjected to conditions highlighted below. We may distinguish $p\,dV$ work by calling it *boundary work*; but since the kind of work is ordinarily clear from the context, the adjective will not be used often.

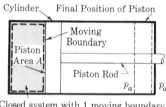

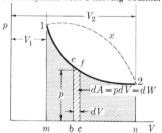

Closed System Work of Expansion. A boundary (piston) of the   **Fig. 2/1**
system with a force on it moves. The differential area
befc $= dA = p\,dV$; and the sum of these areas between 1 and
2 is the total area under the curve 1-*ef*-2. Therefore, for an
internally reversible process: the area "under" the curve on
the pV plane represents the $\int p\,dV$ to some scale.

We should look carefully into the conditions that should be met in order for this equation to be valid. In the first place, if the pressure p on the moving boundary (piston) is uniform at all instants and if the exact manner of this pressure variation with volume $p = p(V)$ is known, the integral of (2-11) gives the work done on (or by) that moving boundary (and by or on the system). Since there are no ready means of measuring pressures at all points of such a surface, we must assume an ideal model that is close enough to reality to give useful results. In this model the pressure is not only uniformly distributed over the piston at any instant, but is also uniform throughout the system, so that the "pressure at a point," any point, is the system's pressure. Moreover, the instantaneous temperature is the same at all points. This uniformity means that at any instant during the process, each property of every small mass of the system is the same no matter where measured. In other words, the process must *proceed through a series of internal equilibrium states* (no pressure or temperature gradients at any instant), a process that we shall say is ***internally*** *reversible* (reversibility is defined in more detail in Chapter 5); it is also called a *quasi-static process.*[1.1]

The implications of the reversibility requirement are best realized by considering some actual related events. The expansion described above would have to take place very slowly, in the limit at an infinitesimal rate; the external resistance to the motion of the boundary must vary (because the internal pressure is varying), yet the force system is to remain in equilibrium. Imagine an extreme case; let the piston in Fig. 2/1 be held in place by, say, pins, with high pressure gas inside the cylinder. Let the pins be removed; the piston accelerates rapidly (as a bullet after the powder has been ignited); the pressure adjacent to the piston drops; then there is a surge of gas in the system toward this lower-pressure region resulting in considerable turbulence (this means internal fluid *friction*, temperature gradients, and pressure gradients). Moreover, if the piston moves so that the gas is expanding, the relative velocity between the molecules and piston is less and therefore the molecular pressure on the piston is less, according to equation (1-7), than the pressure at an interior point; if compression is occurring, the average relative velocity increases and the pressure on the piston would be greater than the pressure "at a point" in the system.

Let the system be the same but let the "piston" be extremely heavy. Because of the large mass and the same force, the acceleration will be much less ($\mathbf{F} = m\mathbf{a}$). If this idea is carried far enough, the expansion rate would be so small that the assumption of internal reversibility would be approached closely. While pistons in actual reciprocating machines inevitably undergo accelerations on every stroke and at times have high velocities, and thus spoil the equilibrium requirements, we can make excellent estimates of the work on moving boundaries in actual engines by starting with the ideal case and then applying correction factors (various kinds of efficiencies) obtained from experience.

Sign Convention.   In general in this book, if a system's work (and $\int p\,dV$) is positive, work is done *by* the system; work done *on* the system is negative (and so is $\int p\,dV$ for reversible processes).*

Finally, to keep the record straight, we should note in Fig. 2/1 that there is some environmental pressure p_a on the opposite side of the moving boundary. If this pressure is constant, as it usually is in engineering situations, the work done against the force $p_a A$ is $-p_a(V_2 - V_1)$. Therefore, thinking of the single stroke 1-2, the amount of *work that can be delivered* by the system in a frictionless movement is $\int p\,dV - p_a(V_2 - V_1)$ because the work of the gas system must be reduced by the amount of work that must be done in pushing back the surroundings. If there is friction, the gas system's delivered work is further reduced by this loss. In problems involving reciprocating machines, analyses are usually made on a cyclic basis, which involves at least two strokes (1 revolution); in this case, the *ideal* $p_a \Delta V$ work in one direction is canceled by the ideal $p_a \Delta V$ work during the other stroke.

2.9   WORK DEPENDS UPON THE PATH

Since the boundary work of a fluid during a quasistatic nonflow process is given by $\int p\,dV$, it depends on the relation between p and V while the process is occurring,

---

* This convention involves an inconsistency in that, usually, energy entering a system is taken as positive. However, this sign convention for work is most common in texts for engineering thermodynamics and moreover it causes no inconvenience. Notice the way the energy balance equations are set up later in this chapter.

that is, on $p = p(V)$. The equation relating p and V can be plotted on the pV plane as a curve. suppose 1-x-2 , Fig. 2/1, represents one function and 1-ef-2 represents another. Since the $\int p\,dV$ would be different for these two paths, even though states 1 and 2 are identical, the work delivered to a shaft is different and therefore depends upon the particular function $p(V)$ connecting states 1 and 2. An indefinite number of functions may be devised relating any two states 1 and 2. No matter what the function, the change of any property x is $x_2 - x_1$; for example, $\int dp = p_2 - p_1$, $\int du = u_2 - u_1$, and so on; dp, du, and so on, are called **exact differentials**. On the other hand, $\int dW \neq W_2 - W_1$; instead we write $\int dW = W$, or $\int dW = W_{1\text{-}2}$ if it is desirable to identify work for a particular state change; dW is said to be an **inexact differential**. Note, however, in $dW = p\,dV$ that dW/p is an exact differential $(= dV)$—that inexact differentials can be made exact by a divisor.

Example—Work of Nonflow Process 2.10

Consider a process 1-2, Fig. 2/1, that passes through a series of equilibrium states in accordance with $pV = C$. Find the expression for the work done by the system during the process.

Solution. From $pV = C$, substitute $p = C/V$ into equation (2-11) and integrate.

(a)
$$\int_1^2 dW = W = \int_1^2 p\,dV = C \int_{V_1}^{V_2} \frac{dV}{V} = C \ln \frac{V_2}{V_1} = p_1 V_1 \ln \frac{V_2}{V_1}$$

where p lb/ft^2, V ft^3 give W ft-lb and p N, V m^2 give W J.

ELASTIC WORK 2.11

There is a class of work called **strain work** that involves a force deforming a solid body. The work that this force does is evaluated by $\int F\,dy$, as previously defined. If the deformations are within the proportional limit, the strain work, also called **elastic energy** or **elastic work**, can be evaluated—and moreover this deformation process is nearly reversible.

A good and useful example is a spring, one of whose properties is its **scale** (also called many other names including *spring constant, spring rate*). If the scale K_s is constant—and many springs are deliberately made with variable rates—the well known Hooke's law applies. In equation form, Hooke's law is $F = K_s y$, where F is the force and y is the deflection. If a spring is deformed, work is done on it in the amount

(2-12)
$$\int F\,dy = \int_1^2 K_s y\,dy = \frac{K_s}{2}(y_2^2 - y_1^2) \quad \overset{\text{DIST FROM LENGTH}}{\underset{\text{FREE}}{\longleftarrow}}$$

where the integral has been made for constant K_s and the units match; for example, K_s lb/in. and y in., or K_s lb/ft and y ft. If the spring is initially at its free length, $y_1 = 0$ and the work done on it is $K_s y^2/2$. The sign of this energy as a work quantity may be made to fit the convention already defined for work; if it is done *by* the system under study, let it be positive; if done *on* the system, negative.

In the deformed state, the work that was done on it is now elastic energy stored in it, provided that there is no friction of any kind. Since the frictional losses are generally quite minor, we consider the energy stored in a spring as being $E_{sp} = K_s y^2/2$, where y is the total deformation, either compression or extension (within the elastic limit). Another point of view is that this energy is potential energy of the spring, because the energy is potentially available to do some work as the spring returns to its free length, or to impart kinetic energy to a mass acted on by it.

Fig. 2/2   *Harmonic Oscillator.* Since the sum of the energies $E_s + K$ is constant, either one defines the other. Therefore, linear vibration would be said to be with one degree of freedom, but the total energy of the system is the sum of the averages, $E_{sav} + K_{av}$.

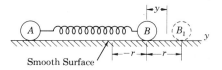

Smooth Surface

Let A and B, Fig. 2/2, be two spheres resting in equilibrium on a frictionless surface by a massless ideal spring with a constant spring scale. For the simplest thought processes, hold A stationary, apply a force to B, displace B a distance r to position B_1, and release it. It will, as you know, oscillate. The force exerted by the spring is always opposite to the displacement from the equilibrium position; hence, $F = -K_s y$. Using Newton's second law, $F = ma$ in algebraic form since we are dealing with rectilinear motion, we make the following transformations:

(a)
$$F = -K_s y = m_B a = m_B \frac{v \, dv}{dy}$$

(b)
$$-K_s \int_{y_1}^{y_2} y \, dy = m_B \int_{v_1}^{v_2} v \, dv$$

(c)
$$-K_s \frac{y_2^2 - y_1^2}{2} = m_B \frac{v_2^2 - v_1^2}{2}$$

(2-13) $-\Delta E_{sp} = \Delta K_B$ or $\Delta E_{sp} + \Delta K_B = 0$ or $E_{sp} + K_B = C$

a simple description of the dynamics of a conservative force system, where the motion is simple harmonic (SHM). It says that the change in the potential (stored) energy of the spring plus the change of kinetic energy of the body B is zero; or that the total energy of the spring (E_{sp}) and body (K_B) as the system is $E_{sp} + K_B =$ constant. If the two bodies A and B had been released simultaneously, the foregoing detail would be a little different, but the final conclusion would be the same—that the total of potential and kinetic energies of the whole system would be constant. This system is an analogue of the vibrational energy of a diatomic (dumbbell) molecule, § 2.6, a ***harmonic oscillator***. [Another example of a conservative system, equation (2-13), is a body falling freely in a vacuum.]

From equation **(a)** we have (subscripts dropped)

(d)
$$a = -\frac{K_s}{m} y \quad \text{or} \quad \frac{d^2 y}{d\tau^2} = -\omega^2 y$$

the differential equation defining harmonic motion. We wish to recall now that the frequency ν of the vibration is given by

(e)
$$\nu = \frac{\omega}{2\pi} = \frac{(K_s/m)^{1/2}}{2\pi} \text{ cps}$$

or $\sec^{-1}$, the units applying for m slugs and K_s lb/ft.

A wire or rod stretched in tension (or a rod in simple compression), within the proportional limit, store the work done on them in a manner analogous to that of the spring. For example, from *strength of materials*, a member subjected to a normal stress σ only, $F = \sigma A = EyA/L$, where $\sigma = E\varepsilon = Ey/L$, E is Young's isothermal modulus of elasticity (elsewhere, E is energy), y is the total strain, $\varepsilon = y/L$ is the unit strain, A is the area of cross section, L is the length of the member, and σ is the unit stress, which must be less than the proportional limit for the assumptions made. Then, for E, A, L constant,

$$(2\text{-}14) \qquad W = \int F\,dy = \frac{EA}{L} \int_{y_1}^{y_2} y\,dy = \frac{EA}{2L}(y_2^2 - y_1^2)$$

where by comparison with equation (2-12) we see that the equivalent spring scale of the wire or rod is $K_s = EA/L$. The stored energy at any strain y is $(y = \varepsilon L)$

$$(f) \qquad \frac{EA}{2L}y^2 = \frac{EAL^2\varepsilon^2}{2L} = V\frac{E\varepsilon^2}{2} = V\frac{\sigma\varepsilon}{2}$$

from which we see that the energy per unit volume is $\sigma\varepsilon/2$. In this discussion, it is assumed that the strain is solely due to force (stress); no strain is caused by temperature change. See also § 11.25.

Similarly, with a knowledge of mechanics of materials, the work (energy) involved in the deformation of beam systems, torsion members, and so on, can be found. We shall explore this area no further at this time.

Example

There are required 203.4 N-m of work to stretch a spring 7.62 cm from its free length. Find the spring constant K_s.

Solution. Using equation (2-12)

$$W = \int F\,dy = \int K_s y\,dy = K_s(y_2^2 - y_1^2)/2$$

$$= K_s y^2/2$$

$$K_s = 2\,W/y^2 = 2(203.4\ \text{N-m})/(0.0762\ \text{m})^2$$

$$= 70{,}060\ \text{N/m}$$

where $y = 7.62\ \text{cm} = 0.0762\ \text{m}$.

SURFACE TENSION 2.12

The molecules on the surface of a liquid (at the interface of a liquid and its surroundings) are not subjected to the same molecular attractions as a bulk molecule. Therefore the properties of the liquid surface (a few molecules thick) are not the same as the bulk properties. This difference results in what is called surface tension $\mathscr{S}$, a force per unit length; $\mathscr{S} = F/L$. A characteristic of surface tension is that it decreases with an increase of temperature, but at a particular temperature it is constant and independent of the area of the surface. If a liquid film is stretched (as a soap bubble), more surface is created and the work done on the film serves to bring molecules from the bulk to the surface. Let the film be stretched a distance dx perpendicular to the length L; then

$$(2\text{-}15) \qquad dW = F\,dx = \mathscr{S}L\,dx = \mathscr{S}\,dA$$

If $\mathscr{S}$ is the same in all directions at any particular instant, the $\int \mathscr{S}\,dA$ gives the surface-tension work for the change of area $\int dA$. The sign of the numerical value may be assigned in accordance with our convention.

Example

A spherical soap bubble of radius r is formed by blowing through a smally soapy blow pipe. If $r = 6$ in. and the surface tension is $\mathscr{S} = 15$ dyne/cm, find the work input in overcoming the surface tension in the bubble.

Solution. Using equation (2-15)

$$W = \int \mathscr{S} \cdot dA = \mathscr{S} \cdot A = \mathscr{S}(4\pi r^2)$$

$$= (15 \text{ dyne/cm})(4\pi)(6 \text{ in.})^2(2.54 \text{ cm/in.})^2$$

$$= 43{,}850 \text{ dyne-cm}$$

$$= 3.235 \times 10^{-3} \text{ ft-lb}$$

2.13   ELECTRICAL WORK

A conservation law associated with electricity is that the total net *charge of an isolated system is constant*. The basic unit of charge is, in magnitude, that of an electron, itself negative, called the **atomic charge** e ($e = 1.60210 \times 10^{-19}$ coulomb, the smallest known charge), but for power circuits, a coulomb is a more practical size; 1 coulomb $= 6.242 \times 10^{18}e$. Electric current is the rate of flow of charge; specifically, 1 amp (ampere) $= 1$ coulomb/sec, or

$$(2\text{-}16\text{A}) \qquad \mathscr{I} = \frac{d\mathscr{Q}}{d\tau} \qquad \text{amp} \to \frac{\text{coulomb}}{\text{sec}}$$

where $d\mathscr{Q}$ is a small quantity of charge crossing any boundary (the system may be a conducting wire) during time $d\tau$. The potential difference $\Delta\mathscr{E}$ volts between two boundaries is work per unit charge, say $dW/d\mathscr{Q}$. For a particular $\Delta\mathscr{E}$, we have

$$(2\text{-}17) \qquad \Delta\mathscr{E} = \frac{dW}{d\mathscr{Q}} \qquad \text{or} \qquad dW = (\Delta\mathscr{E})\,d\mathscr{Q}$$

with W in, say, volt-coulombs = joules = newton-meter = watt-second = volt-ampere-second. [The work crossing a section of a conductor is $\mathscr{E}\mathscr{Q}$, where $\mathscr{E}$ is the potential measured above a suitable datum; thus $(\Delta\mathscr{E})\mathscr{Q}$ is the electrical energy delivered or received when the potential change is $\Delta\mathscr{E}$.] Equation (2-17) is analogous to $p\,dV$ and $F\,dy$, each of which is an intensive property, as $\Delta\mathscr{E}$, times the change of an extensive property, as $d\mathscr{Q}$.

Using the value of $d\mathcal{Q}$ from equation (2-17) in (2-16A), we get

(2-16B) $$\mathcal{I} = \frac{dW}{\Delta\mathcal{E}\,d\tau} \quad \text{or} \quad \frac{dW}{d\tau} = \dot{W} = \mathcal{I}\,\Delta\mathcal{E}$$

say in volt-ampere = watt = joule/second = volt-coulomb/second. Electric circuits have resistances that inevitably result in losses of work. The ohm, the unit of resistance $\mathcal{R}$, is that resistance that permits a current of 1 amp to flow through a potential drop of 1 volt; or $\mathcal{R} = \Delta\mathcal{E}/\mathcal{I}$. In a battery, the difference of potentials at terminals is the no load emf generated by the battery minus the internal loss of potential $\mathcal{I}\mathcal{R}$. In our occasional application, we shall assume that this internal loss is negligible. For a motor, $\Delta\mathcal{E}$ is the back emf, say $\mathcal{E}_b$, and equation (2-16B) becomes $\dot{W} = \mathcal{I}\mathcal{E}_b$.

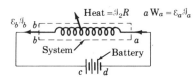

Joule Heat. The resistance wire itself is the system, with   **Fig. 2/3**
energy flows as shown.

Consider the system of electrical resistance in Fig. 2/3 and let the electrical energies entering and departing be $\mathcal{E}_a\mathcal{I}_a$ and $\mathcal{E}_b\mathcal{I}_b$, or a difference of $\Delta\mathcal{E}\mathcal{I}$. This loss of prime energy goes to raising the temperature of the wire in the coil; hence heat begins to pass to the surroundings. Use $\Delta\mathcal{E} = \mathcal{I}\mathcal{R}$ from the definition of resistance, and find the difference of electrical energies at the boundaries to be

(2-18) $$\Delta\mathcal{E}\mathcal{I} = \mathcal{I}^2\mathcal{R} = \text{joule heat}$$

say, in watts = amp²-ohm, a loss that appears as heat leaving the resistance coil. Joule first presented this idea. As shown in the energy diagram of Fig. 2/3, $\mathcal{I}^2\mathcal{R}$ is a positive number and it is shown in the true direction of change (although $\Delta\mathcal{E}$ is algebraically negative).

                                                                          Example   2.14
A battery generates an emf of 18 volts, the $\Delta\mathcal{E}$ at its terminals, Fig. 2/3, while delivering 15 amp. What work is delivered in horse power? What is the joule heat and the resistance a to b if other resistances b to c and d to a are negligible?
   Solution.   Using equation (2-16B), with conversion constants from Item B 38, we have

$$\dot{W} = \mathcal{I}\,\Delta\mathcal{E} = (15)(18) = 270\,\text{W}$$

$$\dot{W} = \frac{270\,\text{wt}}{746\,\text{W/hp}} = 0.362\,\text{hp}$$

which is the power output of an ideal motor (100% efficient) connected to the battery instead of the resistance. If there are no other potential losses, the loss in the resistance ab is balanced by the emf of the battery; therefore,

$$\text{joule heat} = (0.362\,\text{hp})(42.4\,\text{Btu/hp-min}) = 15.3\,\text{Btu/min}$$

$$\mathcal{R} = \frac{\Delta\mathcal{E}}{\mathcal{I}} = \frac{18}{15} = 1.2\,\text{ohms}$$

2.15 GENERALIZED WORK EQUATION

As noted above, expressions for work can be put into analogous forms, further illustrated by the following examples.

Electric charges may be stored in capacitors. Suppose a dielectric is inserted between the parallel plates (of indefinite length) of a capacitor; the work done in changing the charge per unit volume is $dW/V = \mathbf{E} \cdot d\mathbf{p}$, where $\mathbf{E}$ is the intensity of field strength ($\Delta\mathscr{E}$ per meter) at a point in the dielectric and $\mathbf{p}$ is the electric displacement (charge $\mathscr{Q}$ per unit area).

The work of magnetizing or demagnetizing a solid per unit volume is $dW/V = \mu_o \mathbf{H} \cdot d\mathbf{M}$, where $\mathbf{H}$ is magnetic intensity in the material, $\mathbf{M}$ is the magnetic moment, and μ_o is the permeability of free space, a constant.

A review of the equations for work shows that, dimensionally, they are all of the form (intensive property × extensive property). It follows that we could use single symbols for all the pertinent intensive and extensive properties and write

(2-19) $$dW = \mathbf{F}_G \cdot d\mathbf{X}_G$$

where $\mathbf{F}_G$ is a generalized force and $\mathbf{X}_G$ is a generalized displacement. Feel free to consider equation (2-19) as a general definition of W; that is, wherever W appears in a generally applicable equation, it could be any one or any combination of the various forms in which work can be appraised, unless by the context it is evidently a particular one. Most often in this book it will be the work of a fluid system.

2.16 FLOW ENERGY

Flow energy is a special form of work that is significant for a moving stream; it is not dependent on a function $p = p(V)$. **Flow energy** (often called **flow work**) is work done in pushing a fluid across a boundary, usually into or out of a system. It could be included in the sum of all work events for a system in obtaining the net work, but it is more convenient as a separately classified energy term. In Fig. 2/4, let some small quantity V of this substance be on the point of crossing boundary 1 and entering the system. For it to get into the system, work must be done on it in an amount sufficient to move it against the resistance (at pressure $p = p_1$, uniform across section B) offered by the system. The constant resisting force F is pA, and the work done against this resistance in pushing a quantity of fluid of length L across the boundary is $FL = pAL = pV$, where $V = AL$ is the volume of fluid pushed across the boundary. An energy quantity equal to p_1V_1 thus crosses the boundary, Fig. 2/4, and enters the system. Similarly, there is an outgoing energy quantity p_2V_2 if a substance leaves the system, say, at boundary 2. Let the symbol for this energy

Fig. 2/4 *Flow Energy.* One can imagine the flow work being done by a piston at *B* and at *C*. The pressure across a section of the pipe is presumed uniform.

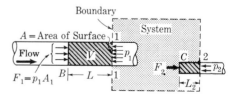

quantity be $E_f = pV$. To get other forms, use the specific volume v and find

(2-20A) $\qquad\qquad E_f = pV = mpv \qquad dE_f = pv\,dm = d(pV)$

(2-20B) $\qquad\qquad E_f = pv \qquad\qquad\quad dE_f = d(pv)$ $\qquad\qquad$ [UNIT MASS]

With time involved, the flow rate is $\dot m$ (say, in pounds per second) and $\dot E_f = pv\dot m = p\dot V$, an instantaneous value if the rate of flow $\dot m$ is varying. Since p, v, and V are properties, pv and pV are also; hence the net flow work of an open system, with one entry and one exit as in Fig. 2/4, is

(a) $\qquad \Delta E_f = E_{f2} - E_{f1} = p_2 V_2 - p_1 V_1 \qquad$ or $\qquad \Delta E_f = p_2 v_2 - p_1 v_1$

$$[\text{UNIT MASS}]$$

where p is the absolute pressure, say, in *pounds per square foot*, $V = mv$ ft^3 is the volume of the mass of substance crossing a boundary and v is the specific volume, each measured at the boundary involved; pv and pV represent energy *only* for the circumstances defined above. If fluid crosses the boundary at several individual places, i, the net flow energy is the algebraic sum: outgoing minus incoming; or

(b) $\qquad\qquad\qquad\qquad E_{f\,\text{net}} = \sum_i p_i V_i$

Since the energy to cause flow in a pipe does not create itself, we may think of the flow work E_f as originating, for example, in a pump located upstream or in the potential energy of a reservoir. However, in its basic concept, it is *not* pump work; it exists because energy is being expended somewhere, somehow, to cause a movement of a fluid across a boundary, a boundary marking a system, drawn anywhere as desired across the stream.*

Example

The flow energy of 5 ft^3 of a fluid passing a boundary to a system is 80,000 ft-lb. Determine the pressure at this point.

Solution.   Using equation (2-20A),

$$E_f = pV$$

$$p = E_f/V = (80,000\,\text{ft-lb}_f)/(5\,\text{ft}^3)$$
$$= 16,000\,\text{psf} = 111\,\text{psi}$$

HEAT—A MICROSCOPIC LOOK   2.17

We have already used the word *heat* several times, perhaps initially in relation to temperature. Like work, heat causes changes in macroscopic properties and for us is

---

* We could have classified flow work and the work of gravity or other body force as part of the work associated with a system. We choose, however, to let the term *work* and the symbol *W* represent work that a system can or may deliver to a shaft.

a technical word. *Heat Q is energy in transit (on the move) from one body or system to another solely because of a temperature difference between the systems.** The interaction occurs by *radiation* or by *conduction*, phenomena whose mechanisms should be understood in an elementary way.

Radiant heat is an electromagnetic emanation; all bodies radiate heat. If two bodies are interacting with radiant heat only, the hotter body radiates more heat than it receives and the colder body radiates less heat than it receives. The accepted concept of radiant energy comes from Planck's (1858–1947) quantum theory, quite well verified, that assumes the radiation to be in discrete amounts, called **photons** or **quanta**; the energy of a photon is $\varepsilon_o = h\nu$, where $h = 6.6256 \times 10^{-34}$ joule-sec is Planck's constant and ν sec^{-1} is the frequency in cycles per second (you recall from physics, a particle can be conceived as having wave properties). The total radiation during any time is a multiple of $h\nu$, but note that $h\nu$ is not a constant, but depends on ν. Since a photon moves with the speed of light $c \approx 3 \times 10^8$ m/sec, its wavelength λ corresponding to a particular frequency is $\lambda = c/\nu$; from which

(a)
$$\varepsilon_o = h\nu = \frac{hc}{\lambda}$$

If a system radiates a quantum of heat without receiving any energy, the allowed state of some molecule has changed so that the energy in the molecule is 1 quantum less than before, which may mean that its vibrational energy, for example, is less by the amount $h\nu$.

In this context we should mention the whole spectrum of radiation. In accordance with Bohr's model of an atom (consisting of negative electrons revolving about a positively charged nucleus of protons and neutrons in planetary fashion), the electron may move only in certain orbits (a simplified way of saying it), which are **allowed states** of the electron. When an electron moves from one orbit to another, the change of orbit (plus any other change in the energy of the atom) must be such that the energy of the atom is changed by the amount of one or more photons. When the electron is in its smallest orbit, it is in the stable or normal state. If, perhaps by collision with another particle or atom, the electron moves to a larger orbit, its energy increases by multiples of $h\nu$, the total amount for a particular frequency depending on whether it moved to the next larger allowed orbit (1 $h\nu$) or skipped to an even larger orbit. The atom with its electrons above its stable state is said to be in an *excited state*.

Systems with excited atoms are capable of emitting radiation of many different wavelengths (or frequencies), where the details do not concern us here; and heat, the radiation with which we shall be most concerned, is only a small part of the spectrum. This is seen from the following typical wavelengths[1.27] in parentheses in meters (the dividing line between named rays is not sharp): cosmic rays (10^{-12} and shorter), gamma rays (10^{-12}–10^{-11}), X rays (10^{-12}–10^{-8}), ultraviolet (10^{-8}–4×10^{-7}), visible light (3.8×10^{-7}–7.8×10^{-7}), infrared or **heat** (7×10^{-7}–10^{-3}), microwaves, radar (10^{-2}–10^{-1}), television, FM radio (1–10), shortwave radio (10–10^2), AM radio (10^2–10^3), and maritime communications (10^3–10^4).

---

* In layman's parlance, *heat* is used to mean molecular energy, that which we here call *internal energy*. If you are accustomed to this notion, make an effort to replace it with the given definition. Also, some writers choose not to classify work and heat as energy. We choose to call it energy because we feel the need of one word that applies to all terms in an energy-balance equation.

A radiant energy of rapidly growing importance from the standpoint of "heating" is the microwave, used also for killing germs, and with possibilities of power transmission. When these waves are focused on salt solutions, water, and some other molecules, the molecules are polarized and become aligned with the electric field. But since the microwave fields are rapidly reversing, the polarized molecules immediately oscillate (atoms vibrate) continuously and rapidly, meaning that their energy content is suddenly increased. Since these waves, contrary to the ordinary infrared, penetrate some solids in depth with high strength, the entire body affected has a sudden temperature rise. This phenomenon, as you know, is being used in "instant" cooking ovens, acting principally on the H_2O molecule, cooking a potato in 5 min.

The principal source of energy for the earth is the radiant energy from the sun, mostly in the infrared spectrum, the earth's part being obviously only a minute part of the total radiation of the sun. Reviewing the basic idea, we see that, microscopically, this form of transition energy (radiation) is a consequence of bits (quanta) of stored energy leaving in the system, the individual bits being so small and the total number of molecules so large that, on the macroscopic scale, changes in the amount of radiation and in the system's properties appear continuous. Compared to metals, gases are poor radiators.

The phenomenon of heat by **conduction** is quite different. In a gas, the molecules in the hotter part are moving faster than in the colder part; conduction in this event is the process of the faster moving ("hotter") molecules colliding with slower ones and communicating some of their energy. (Collision is actually more complex than just elastic response. As the molecules closely approach one another, the attractive force changes to repulsion.) Some of this sort of action also occurs in a liquid where the movement of the molecules is much more restricted. Additionally, some vibrational energy of the molecule (§ 2.6) may also be communicated during a molecular interaction, but this type of energy transfer is most significant and effective for solids, where the molecules do not move around but do vibrate. In most solids the molecular vibrations in the hotter part, by being communicated to adjacent molecules, account for the principal amount of heat transferred by conduction. However, *in metals*, most of the energy moves away from the hotter part *with the movement of free electrons* toward the colder part. It is for this reason that good thermal conductivity goes along with good electrical conductivity.

We also speak of **convected heat**, but this is simply a transport of more energetic molecules from one place to another. If heat enters a gas at the bottom, the hotter gas expands, gets lighter per unit volume, and begins to be replaced by gravitational movements, called *free convection*, by the denser colder parts of the gas. Simultaneously, of course, the more energetic molecules are colliding with less energetic ones, as described above. A similar argument applies to liquids. In a familiar case the air surrounding the furnace of a hot-air heating system receives heat by radiation and conduction. This heated air, being lighter, rises and circulates through the house (or is forced through by a fan, called *forced convection*), giving up energy by radiation and conduction to keep the house and contents warm. When this series of events happens, we say that heat has been convected, although the energy is not heat while it is being transported—only while it is being received or discharged. In our study the convected energy will be cared for by energies associated with the stream.

By its nature, dQ is not a property, not an exact differential; therefore, $\int dQ \neq Q_2 - Q_1$, but $\int dQ = Q$—perhaps $Q_{1\text{-}2}$, meaning the heat for the system while the system underwent a process between states 1 and 2. An enormous difference between work and heat is that work can be converted entirely into heat (or, ideally, entirely into other forms of energy), but heat cannot be converted entirely into work in the most perfect heat engine that the mind can conceive. This idea will be repeated many times in the next few chapters.

2.18   CONSTANT VOLUME SPECIFIC HEAT

We shall first define two specific heats* that are of great utility—those for constant volume c_v, C_v and constant pressure c_p, C_p. The constant volume specific heat of a pure substance is the change of molecular internal energy u for a unit mass (or 1 mole) per degree change of temperature when the end states are equilibrium states of the same volume:

$$(2\text{-}21) \qquad c_v \equiv \left(\frac{\partial u}{\partial T}\right)_v \quad \text{and} \quad C_v \equiv \left(\frac{\partial \bar{u}}{\partial T}\right)_v$$

where c_v is for unit mass and C_v is called the molal specific heat, u is the specific internal energy, $\bar{u}$ is internal energy for 1 mole; $C_v = Mc_v$. In general, a property symbol, as v in (2-21), used as a subscript means that the property is held constant. Because of this, the mathematical notation is then a partial ∂ to indicate that the change (in this case) of u is subject to the restraint of v being constant. In the English system, specific heats are usually expressed as c_v Btu/lb-°R, C_v Btu/pmole-°R.

Fig. 2/5   *Constant Volume System.* Equation (2-22) applies, $dQ = du + dW$ with $dW = p\,dV = 0$; hence $dQ = du$ and $Q = \Delta u$—for this process only. Average $c_v = (\Delta u / \Delta T)_v$, suitable for $\Delta T = T_2 - T_1$.

Consider a closed, nonflow, constant volume system of a fluid, say, a gaseous fluid, Fig. 2/5, for which $dQ = du + dW$. Since $dW = p\,dv$ for an internally reversible process and since v is constant, we have $dQ]_v = du$ for unit mass and $dQ]_v = m\,du = dU$ for mass m. Thus, for any internally reversible $V = C$, $v = C$ process, the heat is given by

$$(2\text{-}22) \quad Q_v = \int dU = m \int_1^2 c_v\,dT \bigg]_v \quad \text{and} \quad Q_v = \Delta u = \bar{c}_v (T_2 - T_1)]_v$$

[INTERNALLY REVERSIBLE]   [UNIT MASS, c_v CONSTANT]

where $c_v = c_v(T)$ and $\bar{c}_v$ is some average value to fit the temperature range T_1 to T_2.† Let us say that any energy equation set up for unit mass of a pure substance is just as well interpreted for 1 mole; $d\bar{u} = C_v\,dT]_v$, $\Delta \bar{u} = \int C_v\,dT]_v$.

---

* Also called *specific heat capacity* and *heat capacity*. The word *capacity* is a carryover from the days of the caloric theory when heat was theorized as being an imponderable substance stored in a body; *heat* is a carryover from the traditional definitions.
† Since there is often a choice of variables in a particular integration, we shall in general use the simple notation 1 and 2, and so on, to designate limits. In equation (2-22), they normally are T_1 and T_2. Elsewhere, limits may be either p_1 and p_2 or V_1 and V_2, and so on.

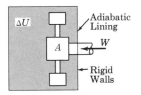

Irreversible, Constant Volume Process. The paddles are "coated with an adiabatic surface"—that is, it is assumed that the device *A* absorbs no heat.   **Fig. 2/6**

Next, consider the irreversible $V = C$ system of Fig. 2/6, where the input energy is work W turning a fan A (no heat). The fan blades striking the molecules transfer energy to them, causing their average molecular speed (and gas temperature) to increase; to be significant, the fanning must be vigorous. Evidently it is possible to bring about the same change of state with a work input as was observed with the heat input of Fig. 2/5. Since there is no heat in Fig. 2/6, $Q = 0$ and thus $Q \neq m \int c_v \, dT$, because m, c_v, and ΔT are all finite quantities; but $\int dU = U_2 - U_1 = m \int c_v \, dT]_v$ for any event between any two equilibrium states 1 and 2—$V_1 = V_2$ and $m = C$. Moreover, $W \neq \int p \, dV$ because no boundary moves, resulting in $\int p \, dV = 0$, but the work W is not zero. This fanning process is not reversible because after the state of the system has been changed and the fan is stopped, the fan is not going to start turning and deliver work from the system. It happens that we can devise on paper a way to get *some* of the input work returned to the surroundings as work but there is no possible way to get all of it back. The system of Fig. 2/5 is internally reversible in the sense that when only heat changes the state, we need only to have the system in an environment where the same amount of heat is rejected and the system returns to its original state, heat being the cause of the change in both directions.

CONSTANT PRESSURE SPECIFIC HEAT   2.19

The constant pressure specific heat of a pure substance is the change of enthalpy (see §§ 3.1 and 4.5 for definition) for a unit mass (or 1 mole) between two equilibrium states at the same pressure per degree change of temperature;

$$(2\text{-}23) \qquad\qquad c_p \equiv \left(\frac{\partial h}{\partial T} \right)_p \qquad \text{and} \qquad C_p \equiv \left(\frac{\partial \bar{h}}{\partial T} \right)_p$$

where $h = (u + pv)$ is the specific enthalpy, and $\bar{h}$ is the enthalpy for 1 mole; $C_p = M c_p$.

For the closed system of Fig. 2/7, the energy equation (2-34A), $dQ = du + dW = du + p \, dv$, for $p = C$ becomes

$$(2\text{-}24) \qquad\qquad dQ_p = du + p \, dv = du + d(pv) = dh$$

for an internally reversible process of unit mass. The nonflow work on the piston, $\int p \, dv = p(v_2 - v_1)$, is the only work of the system. For the restraints specified above,

$$(2\text{-}25) \qquad Q_p = \Delta H = m \int_1^2 c_p \, dT]_p \qquad Q_p = \Delta h = \bar{c}_p (T_2 - T_1)]_p$$

[INTERNALLY REVERSIBLE]   [UNIT MASS, c_p CONSTANT]

where $c_p = c_p(T)$ and $\bar{c}_p$ is some mean value appropriate to the pressure and temperature range T_1 to T_2. To reduce computations, we frequently use mean values of c_v and c_p. When the specific heats c_v, c_p have been significantly affected by the existing pressure, a more elaborate approach is necessary (Chapter 11). Observe that $dh = c_p\,dT]_p$ in any case, but $Q_p = \Delta h$ only for the internally reversible, nonflow system and some steady-flow systems as defined later.

Fig. 2/7   *Constant Pressure System.* The pressure is balanced by a constant load F, plus ambient pressure p_0, on a frictionless, moving piston. The average c_p for a certain temperature range ΔT is $c_p = (\Delta h/\Delta T)_p$.

2.20   RATIO OF SPECIFIC HEATS

Since the **specific heat ratio** k appears often in thermodynamic equations, a symbol for it is convenient;

$$(2\text{-}26)\qquad k \equiv \frac{c_p}{c_v} = \frac{C_p}{C_v}\quad \text{always} > 1 \qquad c_p = kc_v \qquad C_p = kC_v$$

2.21   SPECIFIC HEATS FOR AN IDEAL GAS

For an *ideal gas* $(pv = RT)$, it will be shown (Chapter 6) that the constant volume and constant pressure restraints in, respectively, equations (2-21) and (2-23) are not necessary—also true for many real situations. For such a substance,

$$(2\text{-}27)\qquad du = c_v\,dT \qquad \text{and} \qquad dh = c_p\,dT \qquad [\text{IDEAL GAS}]$$

no matter what happens to the volume or pressure. We conclude that $dh/du = k$, or $\Delta h/\Delta u = \bar{k}$, an average value for an ideal gas.

2.22   MICROSCOPIC ASPECTS OF SPECIFIC HEAT

A look at some of the molecular mechanisms will give insight into why specific heats vary. The ideal gas was defined in words in §1.13 as we derived the elementary kinetic theory equation (1-7) for pressure, $p = Nmv^2/(3V)$, where N is the total number of molecules in a volume V, m is the mass of each molecule, and $\overline{v^2}$ is the mean square of the molecular speed. In §1.17, we stated the ideal gas equation, $p\bar{v} = \bar{R}T$, which can also be written $pv = RT$, where the specific volume $v = \bar{v}/M =$ the mole volume $\bar{v}$ ft^3/pmole divided by the molecular mass M lb/pmole, and $R = \bar{R}/M$, where $\bar{R}$ is the universal gas constant and R is called the **specific gas constant**. At the same time, we found equation (1-12A),

$$(1\text{-}12\text{A})\qquad p\bar{v} = \bar{R}T = \frac{2N_A}{3}\left(\frac{\overline{mv^2}}{2}\right)$$

where N_A is Avogadro's number, the number of molecules in a mole. Since $mv^2/2$ is the average translational kinetic energy per molecule, $N_A mv^2/2$ is the total *translational* kinetic energies of all the molecules in a mole; call this $\bar{u}_t$ for 1 mole. Then from (1-12A)

$$(2\text{-}28) \qquad \bar{u}_t = N_A \frac{\overline{mv^2}}{2} = \frac{3\bar{R}T}{2}$$

The velocity vector v has, of course, three components with respect to the chosen reference axes, and as we have decided (§ 1.13) where the number of molecules is large (the usual system), the average component velocity in the x direction is the same as in the y, or any other, direction. This is in accordance with Boltzmann's *principle of equipartition of energy*, which says in effect that the total translational kinetic energy can be considered in three equal parts ($v^2 = v_x^2 + v_y^2 + v_z^2$). The terminology is that the translating (only) body has three degrees of freedom f, corresponding to the x-, y-, z-locating coordinates. We may say then that the *total* internal molecular energy of the ideal *monatomic* molecule is $\bar{u} = f\bar{R}T/2$ for 1 mole, where $f = 3$, equation (2-28). Furthermore, it is known that the equipartition principle is applicable to other forms of molecular energy (but not electron energy). Generalizing, we have

$$(2\text{-}29) \qquad \bar{u} = C_v T = \frac{f\bar{R}T}{2} \qquad \text{or} \qquad C_v = \frac{f\bar{R}}{2}$$

dividing through by M ($C_v/M = c_v$, $\bar{R}/M = R$), we have

$$(2\text{-}30) \qquad u = c_v T = \frac{fRT}{2} \qquad \text{or} \qquad c_v = \frac{fR}{2}$$

For this *ideal* gas, we can transform the definition of enthalpy $h = u + pv$ to $h = u + RT$, since $pv = RT$. Then with $h = c_p T$ and $u = c_v T$, we have

$$(2\text{-}31\text{A}) \qquad c_p T = c_v T + RT \qquad \text{or} \qquad c_p = c_v + R$$

$$(2\text{-}31\text{B}) \qquad C_p T = C_v T + \bar{R}T \qquad \text{or} \qquad C_p = C_v + \bar{R}$$

where each term of an equation is in the same units. We conclude that

$$(2\text{-}32) \qquad c_p = \left(1 + \frac{f}{2}\right)R \qquad C_p = \left(1 + \frac{f}{2}\right)\bar{R}$$

$$(2\text{-}33) \qquad k = \frac{c_p}{c_v} = \frac{C_p}{C_v} = \frac{1 + f/2}{f/2} = \frac{2}{f} + 1$$

Reconsidering different molecules: the monatomic gas has negligible rotational and vibrational energy and negligible change of electronic energy, except for high temperatures; therefore, its degrees of freedom remain constant over a wide temperature change at $f = 3$ and its specific heats are given by (2-30) and (2-32).

Actual measurements for monatomic gases show that c_v, c_p remain virtually constant at "ordinary" pressures and remarkably in agreement with this theory.

Consider what is known about hydrogen. The variation of its C_p is shown, rather sketchily, in Fig. 2/8. Observe that at low temperature, about 100°R and somewhat below, its $C_p = 4.96$, the value from (2-32) for $f = 3$ degrees of freedom. The reasoning is that at this temperature level, the rotational and other energy forms are essentially zero and the energy of the molecule is mostly translational. If a few quanta of energy ($h\nu$, § 2.17) now enter the gas, some of the molecules assume quanta of rotational energy, but its temperature is determined only by the translational kinetic energy. Since it takes more heat (or other energy) to activate both translational and rotational energies, the specific heat starts to increase. The energy of an individual molecule changes by the discrete amount of a quantum, as previously mentioned, but when so many millions of molecules are involved, the change appears macroscopically as a smooth curve, say, AB, Fig. 2/8. Since the diatomic molecule can have significant rotational kinetic energy with respect to only two axes, its number of degrees of freedom $f = 5$; 3 translational plus 2 rotational. According to (2-32), $C_p = 3.5\bar{R} = 6.95$ Btu/pmole-°R, the theoretical value of the specific heat if all or nearly all of the molecules are in their allowed rotational state (no vibrational energy). Reassuringly, this is seen to be the approximate value in the vicinity BC, Fig. 2/8. Observe that the temperatures here are "ordinary" and that there is a significant temperature range during which C_p varies little, a fact that can be checked in the literature.[0.6] Not only that, but the actual C_p is quite close to this theoretical value, Item B 1. The principal physical phenomenon as energy (quanta) enters the gas is that between A and B; more and more molecules take on their rotational energies; at the level part between B and C, all the allowed rotational states are occupied.

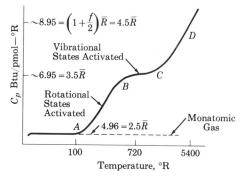

Fig. 2/8   *Constant Pressure Specific Heat for* H₂. *The abscissa is a log scale. All specific heats approach zero as* $T \to 0$. *Other gases might have a similar curve from B to A if they do not condense first (after King*[2.7]*).*

Next comes the vibrational mode of energy (§§ 2.6, 2.11).[2.9] Inasmuch as the molecules act as oscillators when this mode is activated, the atoms have a certain *average* amount of kinetic energy, which is another degree of freedom; but with this vibration, there is also an equal *average* amount of potential energy (with respect to each other, harmonic motion), which means that f in equations (2-29), (2-30), (2-32), and (2-33) is the number of degrees of freedom plus 1, in order to count the potential form. Therefore, for a diatomic gas, we have $f = 3$ trans + 2 rot + 2 vib = 7. Thus, the average internal energy of a mole is $\bar{u} = f\bar{R}T/2 = 7\bar{R}T/2$, equation (2-33), and $C_p = 4.5\bar{R} = 8.95$ Btu/pmole-°R, equation (2-32). As the temperature continues to increase, many complex things begin to happen, and in general the larger the molecule, the lower the temperature at which the complexities become macroscopically apparent—complexities such as torsional vibration, electronic energy changes (the atoms become excited, their electrons begin moving into larger orbits), spin energy of the electron, ionization, and dissociation of the molecule into its atoms. For more detail, see books on statistical thermodynamics and plasmas.

A polyatomic molecule may have rotational kinetic energy with respect to 3 axes; hence, its degrees of freedom for translation and rotation are only $3 + 3 = 6$. Its vibrational mode of energy also begins to be apparent at relatively low temperatures, but the more the number of atoms in the molecule, the more complex this mode becomes. Note how abruptly the curve for CO_2 rises in Fig. 2/9. The simple limited

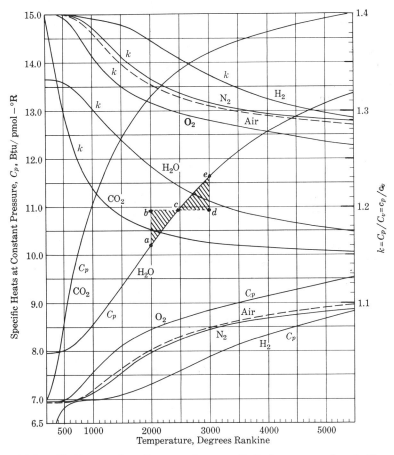

Molal Specific Heats at Low Constant Pressure. Dotted curves are for air. The molal specific heat at constant volume of $R = 1.986$ Btu/pmole-°R is **Fig. 2/9**

[Equation (2-31)] $C_v = C_p - 1.986$ or $C_v = C_p/k$ Btu/pmole-°R,

where the values of k are on the right hand ordinate.

$$c_p = C_p/M \quad \text{and} \quad c_v = c_p - R = C_v/M,$$

where R must be in the same units as c_p. Since $\int C_p \, dT = \Delta\bar{h}$, the area under *ace* is Δh between $T_1 = 2000°R$ and $T_2 = 3000°R$ for the $H_2O(g)$. It follows that the proper mean C_p for evaluating $\Delta\bar{h}$ is one chosen so that area under *abc* is the same as area under *cde*; if the curve between said states is nearly straight, the arithmetic average for T_1 and T_2 is nearly the mean C_p (or k). *(Plotted from data in Keenan and Kaye, Gas Tables, John Wiley & Sons, Inc.)*

agreement between actual and kinetic theory specific heats at least becomes less tangible. You might note in Item B 1 how widely different those values of C_p are for the polyatomic variety.

In the *absence of internal vibrations* in gaseous molecules, the value of k from the degrees of freedom (kinetic theory) are: monatomic, $k = 1.667$; diatomic, $k = 1.4$; polyatomic, $k = 1.333$. Compare with values in Item B 1. See § 11.23 for more on the specific heats of solids.

2.23   ACCOUNTING FOR THE VARIATION OF SPECIFIC HEATS

The best experimental values of specific heats have been found as a consequence of the quantum theory and are called spectrographic specific heats if it is desired to distinguish them from those determined by calorimeter tests. The functions $c_p(T)$ in Table I are designed to fit such data with good accuracy *for the temperature limits specified* and apply for the substance at low pressure. (Serious inaccuracies may occur in extrapolations.) Instantaneous and average values of specific heats can be found from Fig. 2/9 as a function of temperature; see its caption for average values.

2.24   MEAN SPECIFIC HEATS

The use of an equation for specific heat as a regular practice becomes rather tedious unless a computer is handy for the desired calculations. To ease the burden, tables and charts of properties, such as enthalpy and entropy, are available for many common substances (see references in the rear with numbers 0.XX); but if repeated calculations are to be made in a certain temperature range for gaseous substances for which there are no tabulated properties, mean specific heats may be a boon. The proper mean values are a little different, depending on whether specific heat is used for computing Δh or Δu, or used for Δs:

(a)
$$\bar{c} = \frac{\int c\,dT}{T_2 - T_1} \quad \text{or} \quad \bar{C} = \frac{\int C\,dT}{T_2 - T_1} \qquad \text{[FOR HEAT]}$$

(b)
$$\bar{c} = \frac{\int c\,dT/T}{\int dT/T} \quad \text{or} \quad \bar{C} = \frac{\int C\,dT/T}{\int dT/T} \qquad \text{[FOR ENTROPY]}$$

where $\int dT/T = \ln(T_2/T_1)$ between states 1 and 2. See §§ 3.1, 5.5 for definition of entropy(s).

2.25   OTHER FORMS OF ENERGY

Manifestations of energy occur in several forms with which we shall have little concern in this book. However, the laws of thermodynamics apply to all forms of energy and are often useful in specialized fields of study. Recapitulating, we have defined the energy forms:

P, potential energy; *stored*; change is $P_2 - P_1$;
K, kinetic energy; *stored*; change is $K_2 - K_1$;

Variable Specific Heats at Low Pressure   TABLE I

All equations derived from spectrographic data: $c_v = c_p - R$; $C_v = C_p - 1.986$. See Item B 1 for values of R.

(a) This value is derived from Spencer and Justice[2.2]; (b) from Spencer and Flannagan[2.3]; (c) from Chipman and Fontana[2.6]; (d) from Sweigert and Beardsley[2.1]; (e) from Spencer.[2.7]

Substance (Temp. range)	M (Mol. mass)	Btu/lb-°R	Btu/pmole-°R
(a) Air (500–2700°R)	28.97	$c_p = 0.219 + 0.342T/10^4$ $- 0.293T^2/10^8$	$C_p = 6.36 + 9.92T/10^4$ $- 8.25T^2/10^8$
(b) SO_2, sul. diox. (540–3400°R)	64.07	$c_p = 0.1875$ $+ 0.0944T/10^4$ $- 1.336 \times 10^4/T^2$	$C_p = 11.89 + 6.05T/10^4$ $- 85.6 \times 10^4/T^2$
(b) NH_3, ammonia (540–1800°R)	17.03	$c_p = 0.363 + 2.57T/10^4$ $- 1.319T^2/10^8$	$C_p = 6.19 + 43.8T/10^4$ $- 22.47T^2/10^8$
(c) H_2, hydrogen (540–4000°R)	2.016	$c_p = 2.857 + 2.867T/10^4$ $+ 9.92/T^{1/2}$	$C_p = 5.76 + 5.78T/10^4$ $+ 20/T^{1/2}$
(d) O_2, oxygen (540–5000°R)	32	$c_p = 0.36 - 5.375/T^{1/2}$ $+ 47.8/T$	$C_p = 11.515 - 172/T^{1/2}$ $+ 1530/T$
(d) N_2, nitrogen (540–9000°R)	28.016	$c_p = 0.338 - 123.8/T$ $+ 4.14 \times 10^4/T^2$	$C_p = 9.47 - 3470/T$ $+ 116 \times 10^4/T^2$
(d) CO, carb. mon. (540–9000°R)	28.01	$c_p = 0.338 - 117.5/T$ $+ 3.82 \times 10^4/T^2$	$C_p = 9.46 - 3290/T$ $+ 107 \times 10^4/T^2$
(d) H_2O, steam (540–5400°R)	18.016	$c_p = 1.102 - 33.1/T^{1/2}$ $+ 416/T$	$C_p = 19.86 - 597/T^{1/2}$ $+ 7500/T$
(d) CO_2, carb. diox. (540–6300°R)	44.01	$c_p = 0.368 - 148.4/T$ $+ 3.2 \times 10^4/T^2$	$C_p = 16.2 - 6530/T$ $+ 141 \times 10^4/T^2$
(e) CH_4, methane (540–2700°R) (d) (540–1500°R)	16.04	$c_p = 0.211 + 6.25T/10^4$ $- 8.28T^2/10^8$ $c_p = 0.282 + 4.598T/10^4$	$C_p = 3.38 + 100.2T/10^4$ $- 132.7T^2/10^8$ $C_p = 4.52 + 0.00737T$
(b) C_2H_4, ethylene (540–2700°R) (d) (350–1100°R)	28.04	$c_p = 0.0965 + 5.78T/10^4$ $- 9.97T^2/10^8$ $c_p = 0.151 + 4.2T/10^4$	$C_p = 2.706 + 162T/10^4$ $- 279.6T^2/10^8$ $C_p = 4.23 + 0.01177T$
(e) C_2H_6, ethane (540–2700°R) (d) (400–1100°R)	30.07	$c_p = 0.0731 + 7.08T/10^4$ $- 11.3T^2/10^8$ $c_p = 0.1334 + 5.44T/10^4$	$C_p = 2.195 + 212.7T/10^4$ $- 340T^2/10^8$ $C_p = 4.01 + 0.01636T$
(e) C_4H_{10}, n-butane (540–2700°R)	58.12	$c_p = 0.075 + 6.94T/10^4$ $- 11.77T^2/10^8$	$C_p = 4.36 + 403T/10^4$ $- 683T^2/10^8$
(e) C_3H_8, propane (540–2700°R)	44.09	$c_p = 0.0512 + 7.27T/10^4$ $- 12.32T^2/10^8$	$C_p = 2.258 + 320T/10^4$ $- 543T^2/10^8$
(b) C_2H_2, acetylene (500–2300°R)	26.04	$c_p = 0.459 + 0.937T/10^4$ $- 2.89 \times 10^4/T^2$	$C_p = 11.94 + 24.37T/10^4$ $-75.2 \times 10^4/T^2$
(d) C_8H_{18}, octane (400–1100°R)	114.22	$c_p = 0.0694 + 5.27T/10^4$	$C_p = 7.92 + 0.0601T$

U, u, internal energy; stored; change is $U_2 - U_1$, or $u_2 - u_1$;

E_f, flow energy; in *transition*;

W, work; in *transition*; depends on $p(V)$ for a fluid;

Q, heat; in *transition*; a function of the manner in which properties change.

Then there is chemical energy E_{ch} (resulting from a change in molecular structure—as in combustion of fuels), work done in shearing a fluid, electromagnetic emanations other than heat (light, radio waves, and so on), acoustic energy (sound waves), nuclear energy (resulting from a change in the structure of the atom's nucleus and a conversion of mass into energy), energy stored by surface tension or a magnetic field, and others. In subsequent discussions, all energy forms mentioned in this paragraph *are assumed to be absent unless specifically included*. All the energy forms represented by the group of symbols above are rarely involved in a particular event.

2.26   CONSERVATION OF ENERGY

The ***law of conservation of energy*** states that *energy can be neither created nor destroyed*.* It is a law based on physical observations and is not subject to mathematical proof. In its application to energy transformations on this earth, there is no known exception, except as mass is converted into energy and vice versa. In accordance with this exception, we can say that mass is a form of energy and the law still holds. Or, more pragmatically, the law can be amended to say that mass plus energy is conserved. If a nuclear process is not involved, the amount of mass converted to energy (for example, in a combustion process) is so miniscule that it cannot be measured. Hence, we may ignore the exception unless thermodynamics is being applied to nuclear processes.

For any kind of system, the following statements follow logically from the statement that energy is neither created nor destroyed [compare with equation (1-18)]:

$$(2\text{-}34\text{A}) \qquad \begin{bmatrix} \text{energy} \\ \text{entering} \end{bmatrix} - \begin{bmatrix} \text{energy} \\ \text{leaving} \end{bmatrix} = \begin{bmatrix} \text{change of energy} \\ \text{stored within system} \end{bmatrix}$$

$$(2\text{-}34\text{A}) \qquad\qquad\qquad E_{in} - E_{out} = \Delta E_S$$

* This law is not an idea that burst suddenly upon the scientific world. After scientists began working with energy, it was years before the law was comprehended. Benjamin Thompson (Count Rumford), 1753–1814, who has been called an arrogant and insufferable genius, really discovered the equivalence of work and heat in the course of manufacturing cannon (1797) by boring solid metal submerged in water. He was intrigued by the water boiling because of the mechanical work of boring, yet no heat had been added to the water. He convinced himself, but not the world, that the then-accepted caloric theory of heat (a theory that supposed heat to be a substance without mass) did not explain all known phenomena of heat, and that work and heat were in some manner related phenomena. In his words, "Is it possible that such a quantity of heat as would have caused five pounds of ice cold water to boil could have been furnished by so inconsiderable a quantity of metallic dust merely in consequence of a change in its capacity for heat?" Other experimenters later discovered more evidence, until some fifty years after Rumford's cannon experiments, Joule, with assistance from Lord Kelvin, showed conclusively that mechanical work and heat are equivalent. (We are now just a few years from Joule. See the footnote on p. 29. Considering the age of the earth in billions of years and the age of man as over 1,000,000 years, thermodynamics is a new science—as is all science.) Rumford was teaching school in Rumford, Mass. (now Concord, N.H.), when he met, wooed, and won a wealthy widow. He was sympathetic with the "other" side at the time of the Revolutionary War and decided that it would be smart to leave Boston with the British, which he did, deserting his wife and daughter. He gained fame and honors in Europe and England. Now that the American Revolutionary War is long gone, we in the United States like to claim him as our own.

$$(2\text{-}34\text{B}) \qquad \begin{bmatrix} \text{initial} \\ \text{stored} \\ \text{energy} \end{bmatrix} + \begin{bmatrix} \text{energy} \\ \text{entering} \\ \text{system} \end{bmatrix} - \begin{bmatrix} \text{energy} \\ \text{leaving} \\ \text{system} \end{bmatrix} = \begin{bmatrix} \text{final} \\ \text{stored} \\ \text{energy} \end{bmatrix}$$

$$(2\text{-}34\text{B}) \qquad E_{s1} + E_{in} - E_{out} = E_{s2}$$

where E_s represents any and all appropriate kinds of *stored* energies as usually evaluated for a particular mass, for a particular change of mass, or for changes during a particular time interval.

PERPETUAL MOTION OF THE FIRST KIND   2.27

If a device should continuously and indefinitely discharge more energy than it receives, it would violate the law of conservation of energy because it would be creating energy. Such a device is called a perpetual motion machine of the first kind, and in the light of all experience, it is absurd and is impossible to achieve.

CLOSURE   2.28

In its most general aspects, the law of conservation of energy is one of the most useful discoveries ever made. In approaching a real engineering problem, one must be acquainted with all the various forms in which energy appears since none can be ignored. Further, each form is an entity within itself and should be accounted for but once in the solution of a problem. You are encouraged to review this chapter often until its contents literally become a part of you.

PROBLEMS

SI UNITS

2.1 Use dimensional analysis and show that the expression $e = mc^2$ has units of energy.

2.2 Scientists have recently developed a powerful pulse laser for research in materials. Find its energy output in watts for each of the following pulse conditions: **(a)** 20 J in 10 psec, **(b)** 200 J in 35 nsec, **(c)** 800 J in 1 msec. **(d)** At its peak pulse, it will produce 10 terawatts (TW) for a period of 10 psec. Find its discharge in joules.

Ans. **(a)** 20 TW, **(b)** 7.72 GW, **(c)** 800 kW, **(d)** 100 J.

2.3 An electron has a rest mass of 9.11×10^{-28} gm. What is its mass when moving with a speed of $0.95\,c$?

2.4 It is estimated that the United States consumes annually about 1.75×10^{15} W-hr of electrical energy. In accordance with Einstein's theory, how many kilograms of matter would have to be destroyed to yield this energy?

2.5 Much like Newton's universal law of gravity (see §1.9) Coulomb's law states that the force between two electrical charges varies as the product of the charges and inversely as the square of the distance between them: $F = kq_1q_2/r^2$ where F newtons (N), q coulombs (C), r meters, and $k = 9.0 \times 10^9$ N-m^2/C^2. Given two small electrical charges q_1 and q_2 positioned on the x-axis as follows: $q_1 = -4\,\mu C$ at $x_1 = -3$ m; $q_2 = +1\,\mu C$ at $x_2 = +2$ m. Find the position on the x-axis of a third electrical charge q_3 that experiences no net force from these two.

2.6 A girl weighing 470 N hangs suspended on the end of a rope 8 m long. What will be her gain in potential energy when a friend swings her to one side so that the rope makes an angle of 35° with the vertical? If local $g = 9.70$ m/sec^2, what is her mass in kg? in lb$_m$? *Ans.* $\Delta P = 679.5$ N-m.

2.7 The 600-kg hammer of a pile driver is lifted 2 m above a piling head. What is the change of potential energy? If the hammer is released, what will be its velocity at the instant it strikes the piling. Local g = 9.65 m/s².   *Ans.* 11.58 km, 6.21 m/s.

2.8 There are 400 kg/min of water being handled by a pump. The lift is from a 20-m deep well and the delivery velocity is 15 m/s. Find **(a)** the change in potential energy, **(b)** the kinetic energy, **(c)** the required power of the pumping unit; g = 9.75 m/s².

2.9 When an automobile is travelling at 60 km/hr, its engine is developing 25 hp. **(a)** Find the total resisting force in newtons. **(b)** Assuming that the resisting force is directly proportional to the speed, what horsepower must the engine develop to drive the automobile at 100 km/hr?   *Ans.* **(a)** 1118.3 N, **(b)** 69 hp.

2.10 A force F measured in the x-direction is given as $F = a/x^2$ where the constant $a = 9 \text{ N} \cdot \text{m}^2$. Find the work in joules as F moves from $x_1 = 1$ m to $x_2 = 3$ m.

2.11 The force in newtons required to stretch a spring beyond its free length is given by $F = 200x$ where x is in meters. Find the force and work required to stretch the spring 0.1 m; 0.5 m; 1 m.

2.12 If $6\,\ell$ of a gas at a pressure of 100 kPaa are compressed reversibly according to $pV^2 = C$ until the volume becomes $2\,\ell$, find the final pressure and the work.   *Ans.* 900 kPaa, 1200 J.

2.13 An areal soap film with surface tension σ is formed by wetting a wire frame (initially closed) and then moving the slide wire S away from leg b by means of a constant force F. See sketch. **(a)** Show that the work done against the resisting surface tension is $W = \sigma \cdot l \cdot b = \sigma \cdot A$. **(b)** Find the work done when $b = 10$ cm, $l = 6$ cm, and $\sigma = 25$ dyne/cm.

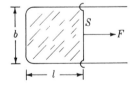

Problem 2.13

Ans. **(b)** 1500 dyne-cm or 1.107×10^{-4} ft-lb.

2.14 Let an areal surface film be formed on a circular wire frame of radius r_1. **(a)** Show that the work done to form this area against the resisting surface tension σ is $W = \pi r_1^2 \sigma = \sigma A$. **(b)** If $\sigma = 50$ dyne/cm, and the work input of 3300 dyne-cm is required to increase the radius (hence the area) from $r_1 = 2$ cm to r_2, find r_2.

2.15 A spherical soap bubble of radius r is formed by means of blowing through a small soapy blow pipe. If $r = 6$ in. and $\sigma = 15$ dyne/cm, find the work input in overcoming the surface tension in the bubble.   *Ans.* $W = 3.235 \times 10^{-3}$ ft-lb$_f$.

2.16 An electric current of 15 amp flows continuously into a resistor of 20 ohms. Compute the power input in kilowatts and horsepower.   *Ans.* 4.5 kW or 6.03 hp.

2.17 A 12-V automobile battery is receiving a constant charge from a generator. The voltage across the terminals is 12.5 V; the current is 8 amp. Determine the power input in watts and horsepower.

2.18 A constant force moves an 18-in. electrical conductor with a velocity of 25 fps orthogonally across a magnetic field whose flux density is 2 weber/m²; note that 1 weber is equivalent to 1 N-sec-m/C. The conductor carries a current of 20 amp. Find the force and the rate of work produced.   *Ans.* 18.30 N, 139.4 W.

2.19 Heat is transferred to an elastic sphere containing a gas at 105 kPaa; the diameter of the sphere is 2 m. Because of heating the sphere diameter increases to 2.2 m and the gas pressure increases in direct proportion to the sphere diameter. Find the work of the gas during this heating process.

2.20 A 12-V battery receives a quick 20-min charge during which time it receives a steady current supply of 50 amp. In this period it experiences a heat loss of 127 kJ. Find the change of internal energy in the battery during this period.

2.21 The flow energy of 124 ℓ/min of a fluid passing a boundary to a system is 108.5 kJ/min. Determine the pressure at this point.   *Ans.* 764.1 kPaa.

ENGLISH UNITS

2.22 A 100-lb mass has a potential energy of −4 Btu with respect to a given datum within the earth's standard gravitational field. **(a)** Find its height relative to the datum. **(b)** If the gravitational field is suddenly disturbed such that the local gravity becomes 25 fps^2, what will be the effect on the potential energy of the mass?

Ans. **(a)** −31.12 ft, **(b)** −2420 ft-lb.

2.23 A 6-slug pile driver is released 20 ft above the head of a piling. For the pile driver at the instant of impact, find **(a)** the change of potential energy, **(b)** the kinetic energy. Frictional effects are negligible and the local gravity is $g = 32.2$ fps^2.

Ans. 1.383 fps, 178.7 ft-lb.

2.24 A system composed of a 10,000-lb elevator moving downward with $v = 5$ fps, a 6000-lb counterweight moving upward with $v = 5$ fps, and a braking pulley with connecting cables. Assume the kinetic energy of the cable and rotating parts to be negligible and determine the frictional energy absorbed by the brake when the elevator is uniformly stopped in 4 ft. *Ans.* 22,220 ft-lb.

2.25 **(a)** A 64,400-lb$_m$ airplane is traveling at 1000 fps (682 mph). How much is its kinetic energy in hp-hr? **(b)** If it suddenly noses vertically upward at this speed, with power off and in the absence of atmospheric resistance, through what vertical distance will it move? Let the average gravity acceleration be $g = 32$ fps^2.

2.26 An experimental nose cone whose mass is 100 lb is projected 200 miles above the earth's surface. What gravitational work was required assuming that the gravity acceleration varies in accordance with $g = A − Bh$, where $A = 32.174$ fps^2 and $B = 3.31 \times 10^{-6}$ for the height h in feet.

Ans. 99.85×10^6 ft-lb.

2.27 Every 6 hr a small satellite orbits the earth; the apogee is triple the perigee. Assume plane motion and no effect of other heavenly bodies. The radius of the earth is approximately 20.91×10^6 ft; let $g = 32.17$ fps^2 and remain constant. For the satellite find **(a)** its minimum altitude and **(b)** its minimum velocity. *Hint:* Review Kepler's three laws of planetary motion in your mechanics text.

Ans. **(a)** 1249 mi, **(b)** 6300 mph.

2.28 Work is done by a substance in reversible nonflow manner in accordance with $V = 100/p$ ft^3, where p is in psia. Evaluate the work done on or by the substance as the pressure increases from 10 psia to 100 psia. *Ans.* −33,150 ft-lb.

2.29 Evaluate the nonflow work in terms of p_1, V_1, p_2, V_2 of a fluid undergoing a reversible state change in accordance with each of the following defining relations: **(a)** $p = C$, **(b)** $V = C$, **(c)** $pV = C$, **(d)** $pV^3 = C$, **(e)** $pV(\ln V) = C$, **(f)** $p = 200/V^2 + 2$psia.

2.30 Determine the atmospheric work done as a 2-in. cube of ice melts in a region of 1 atm. At 32°F these densities obtain for water: liquid, 62.42 lb/ft^3; solid, 57.15 lb/ft^3.

2.31 During the execution of a reversible nonflow process the work is −148.1 Btu. If $V_1 = 30$ ft^3 and the pressure varies as $p = −3V + 100$ psia, where V is in ft^3, find V_2.

2.32 The scale K of a tension spring is variable and is related to its length y such that $K = cy^n$ where c is a constant and n an exponent. Find the work required to stretch the spring from y_1 to y_2.

2.33 There are required 124 ft-lb of work to compress a spring from its free length y_1 to that of $y_2 = 2.5$ in.; the constant scale is $K = 100$ lb$_f$/in. Find the free length.

Ans. $y_1 = 7.96$ in.

2.34 Demonstrate that the work required to stretch a wire within the elastic region is $W = −0.5AEl(\varepsilon)^2$ where, for the wire, l is the initial length, A its cross-sectional area, E is Young's modulus of the material, and ε is the unit strain.

2.35 In problem 2.34 let the wire be steel ($E = 30 \times 10^6$ psi) with $A = 0.01$ in.^{2}, $l = 10$ ft, and a force be gradually applied until its pulling effect on the wire is 1200 lb$_f$. Find the work using the results in problem 2.34. Check your solution by solving for work as simply the product of average force and distance.

Ans. $W = −288$ in.-lb$_f$.

2.36 A centrifugal air compressor compresses 200 cfm from 12 psia to 90 psia. The initial specific volume is 12.6 ft^3/lb and the final specific volume is 3.25 ft^3/lb. If the inlet suction line is 4-in. ID and the discharge line is 2.5-in. ID, determine **(a)** the change in flow work between the boundaries, ft-lb/min, **(b)** the mass rate of flow, lb/min, and **(c)** the change in velocity.

Ans. **(a)** 323,000, **(b)** 15.88, **(c)** −12.9 fps.

2.37 A closed system executes a series of processes for which two of the three quantities W, Q, and ΔU are given for each process. Find the value of the unknown quantity in each case.

(a) $W = +10$ hp, $Q = +500$ Btu/min, $\Delta U = ?$

(b) $W = +65$ Btu, $Q = ?$, $\Delta U = -25$ Btu.

(c) $W = ?$, $Q = +25$ kW, $\Delta U = 0$.

(d) $W = -389$ ft lb, $Q = -1.5$ Btu, $\Delta U = ?$

(e) $W = +2 \times 10^8$ gm-cm, $Q = +5000$ gm-cal, $\Delta U = ?$ *Ans.* **(a)** $\Delta U = +76$ Btu/min.

2.38   A closed gaseous system undergoes a reversible process during which 25 Btu are rejected, the volume changing from 5 ft^3 to 2 ft^3, and the pressure remains constant at 50 psia. Find the change of internal energy.
Ans. 2.8 Btu.

2.39   Assume 8 lb of a substance receive 240 Btu of heat at constant volume and undergo a temperature change of 150°F. Determine the average specific heat of the substance during the process.
Ans. 0.20 Btu/lb-°F.

2.40   For a constant pressure system whose mass is 80 lb, 1 hp-min is required to raise its temperature 1°F. Determine the specific heat for the system, Btu/lb-°F.

2.41   The ratio of specific heats is $k = c_p/c_v$ and, for an ideal gas, their difference is $c_p - c_v = R$, a constant. Combine these two expressions and show that $c_v = R/(k - 1)$ and $c_p = kR/(k - 1)$.

2.42   The following expressions relate to a particular gaseous mass: $pv = 95T$, $h = 120 + 0.6T$ where these units obtain: p in psf, v in ft^3/lb, T in °R and h in Btu/lb. If the specific heats are temperature dependent only, find c_p and c_v.
Ans. 0.6, 0.478 Btu/lb$_m$-°R.

2.43   Compare values of the specific heat c_p for air at 3000°R as obtained from each of three sources: Item B1, Table I, and Fig. 2/9. Do the variations justify the use of the last two sources at this elevated temperature?

2.44   **(a)** Steam flows through a turbine at the rate of 100 lb/min with $\Delta K = 0$ and $Q = 0$. At entry, its pressure is 175 psia, its volume is 3.16 ft^3/lb, and its internal energy is 1166.7 Btu/lb. At exit, its pressure is 0.813 psia, its volume is 328 ft^3/lb, and its internal energy is 854.6 Btu/lb. What horsepower is developed? **(b)** The same as **(a)** except that the heat loss from the turbine is 10 Btu/lb of steam.
Ans. **(a)** 861 hp, **(b)** 839 hp.

2.45   A 1-lb fluid system initially at $p_1 = 100$ psia, $v_1 = 1$ ft^3, executes a reversible expansion in a frictionless piston-cylinder arrangement in accordance with $pV^n = c$. Let the final pressure p_2 range from 100 to 10 psia while the final volume is $v_2 = 5$ ft^3 in all cases. Under these constraints write a computer program that will select values of the final pressure, then solve for the respective values of the exponent n and the corresponding work W, and finally will plot a curve of W versus n.

3

THE PURE SUBSTANCE

A pure substance has been defined as one that is homogeneous and invariable in chemical composition. When it exists in a multiphase mixture, the chemical composition is the same in all phases. For example, ice, a mixture of ice and liquid water, and steam are all pure substances. On the other hand, consider an initially pure substance that is a uniform mixture of gaseous oxygen and nitrogen undergoing a cooling process. If some of the gaseous mixture should liquefy, the liquid portion would have a composition different from that of the remaining gaseous mixture and the whole would no longer be a pure substance.

In the previous chapters we have considered the four familiar properties pressure, temperature, volume, and internal energy. Now studying the pure substance, we must include two additional properties, enthalpy and entropy. Each of these will be discussed in subsequent chapters. All are needed for a better understanding of the pure substance. For the present, let it be sufficient to define enthalpy (h) and entropy (s) as simply:

(3-1)
$$h = u + pv$$

(3-2)
$$s = \int (dQ/T)_{\text{rev}} + s_0$$

Consider the various ways in which energy in the form of work may be transferred to a substance under study. If the substance is compressible, its energy may be increased by a boundary change ($p\,dV$ work). If it is elastic, its energy may be increased through solid extension ($V\sigma\,d\varepsilon$ work). No matter what mode (manner) of work is used to transfer energy to a substance, there will be at least one independently variable property for each mode (V in the first case, ε in the second, and so on).

In addition, these properties can be held fixed and energy transferred to the substance by heat, which will vary the temperature; thus, another independent variable is noted. In summary, for each of the independent ways of transferring energy to a substance, there is one independently variable thermodynamic property.

The preceding ideas may now be formalized in the state postulate (principle): *The number of thermodynamic properties that may be varied independently for a given system is equal to the number of possible reversible work modes plus one.*

Note that no reference is made to irreversible work modes since we can always accomplish the resulting irreversible effects through the combination of reversible work and heat. Thus, if there are N possible reversible work modes for a given system, there are only $N + 1$ thermodynamic properties that may be varied independently. For a simple system, only two thermodynamic properties are required to determine its state (§ 1.2).

3.3   PHASES

In general, a pure substance may exist in any of three phases or in a mixture of *phases*: the solid phase, the liquid phase, and the vapor or gaseous phase. *Melting*, or *fusion*, is the change of phase from solid to liquid. The change in the opposite direction is *freezing*, or *solidifying*. The change of phase from the liquid to the gaseous phase is called *vaporization*, and the liquid is said to *vaporize* (or *boil*). The change from vapor (gaseous phase) to liquid is *condensation*, and during the process the vapor is said to be *condensing*.

Not all substances pass through these three phases; some normally pass directly from the solid to the gaseous phase (or vice versa), a change of phase called *sublimation*. Moreover, many substances that ordinarily pass through the three phases during heating of the solid phase may sublimate under certain conditions. For example, a piece of ice exposed to the atmosphere at temperatures below 32°F will sublimate and, if given time, will pass entirely into the atmosphere as water vapor (steam). In its solid state at 1 atm pressure, solid carbon dioxide (dry ice) sublimates while it receives heat and does refrigeration.

3.4   CHANGES OF PHASE AT CONSTANT PRESSURE

Think of a pure solid substance (say, ammonia) in a state where the "high" pressure is p_1 and temperature T_1, Fig. 3/1. Let the pressure remain constant and add heat; 1-a represents heating the solid. At a the solid is at its melting point for

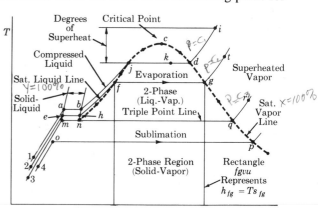

Fig. 3/1   *Ts   Diagram   for   a   Typical Hypothetical Substance.* For one a little more like H_2O, see Fig. 3/7. See Fig. 3/2 for some saturation curves closely to scale.

the pressure p_1, and while it melts at this pressure, the two-phase mixture of solid and liquid remains at a constant temperature. As usual, we are assuming that the heating proceeds slowly enough that the matter goes through consecutive equilibrium states. The amount of heat to change the solid to a liquid, or vice versa, to freeze the liquid, is called the ***heat of fusion***, by definition at constant pressure. It is also the *enthalpy of fusion* h_{if}, where subscript i suggests ice (not as on Fig. 3/1) and subscript f suggests fluid (really liquid). Notice that the melting ab (or freezing ba) occurs while both temperature and pressure remain constant; for the states along ab, they are *not* independent variables; when one is set, so is the other. Between a and b, part of the substance is solid, part liquid, the amount of solid decreasing in the direction a to b. Next, as heat flows in, the liquid is heated at $p = C$ along bj. At j, the liquid has arrived at its boiling point for the pressure p_1. The temperature at j (along jd) is called the ***saturation temperature*** for the specified pressure; the pressure at this boiling temperature is called the ***saturation pressure***. For other than amorphous materials, there is always a certain temperature at which boiling occurs for a given pressure, and like the fusion (melting) point, it is different for different pressures. For example, water boils at 212°F (100°C) at 1 atm pressure and at 327.82°F at 100 psia pressure. For ammonia at 15 psia, the corresponding saturation temperature is $-27.29°F$; at 56.05°F, the corresponding saturation pressure is 100 psia. The process of boiling jd is ***evaporation***; in the opposite direction dj, it is ***condensation***. The substance in state j is entirely a ***saturated liquid*** with an enthalpy h_f that can often be found in tables; after all of it is evaporated and still at the same temperature at d, it is ***saturated vapor***, with an enthalpy h_g. Between j and d, part is liquid and part vapor, the amount of liquid decreasing in the direction j to d. Recognize at the outset that the liquid in the equilibrium mixture anywhere along jd, as at k, is saturated liquid, and the vapor mixed with it is saturated vapor (it is a *heterogeneous* mixture, rather than *homogeneous*). But when it is said without qualification that *the liquid is saturated liquid* or that *the vapor is saturated vapor*, it is generally meant that 100% of the fluid is composed of saturated liquid or vapor, as the case may be. If the context does not make it clear that one means *only* saturated vapor, we may say *dry-and-saturated vapor*, meaning no liquid present. When the system is not entirely in one phase or the other, its ***quality*** x, *which is the fraction or percentage by mass that is vapor*, or ***percentage moisture*** y, *which is the fraction or percentage by mass that is liquid*, is specified. A 1-lb mixture with a quality of 75% contains 0.25 lb of liquid and 0.75 lb of vapor. The heat for evaporation of a saturated liquid into a saturated vapor, or vice versa, is called the ***latent heat***, or the ***enthalpy of evaporation*** h_{fg} (subscripts fg are standard—incidental relation to f and g on Fig. 3/1), by definition at constant pressure. It is *latent* rather than *sensible* because a thermometer senses no change. After all of the substance is in a saturated vapor state as at d, further heat results in ***superheated vapor***. If the temperature rise is carried far enough, the substance may become quite a different substance because of dissociation of the molecule. Suppose the state of the substance is represented by i, Fig. 3/1; the difference between the temperature t_i and the saturation temperature corresponding to $p_i = p_1$ is called the ***degrees of superheat*** ($= t_i - t_d$). Observe the shape of this constant pressure curve 1-*abjdi*.

If the same experiment is run at another lower pressure p_2, we find a constant pressure curve 2-*ehfgt*, fusion occurring along *eh*, evaporation along *fg*, and superheating along *gt*. State f is a saturated liquid, and state g is a saturated vapor. Similar experiments could be run for an indefinite number of pressures with observations of saturation pressures and temperatures, and then corresponding entropies could be computed so that many saturation states as *n*, *g*, *j*, and *d*, *h*, *q* are plotted, with smooth curves drawn. The curve *nfjc* is called the **saturated liquid curve**, the locus of all states of saturated liquid, and the curve *cdgqp* is called the **saturated vapor curve**.

As the experiment is run for lower and lower pressures, a pressure such as p_3, where, as heat is added, the substance melts as usual *mn*; but after it becomes all liquid and further heat is added, the temperature does not change. Instead of the liquid getting hotter, as it did *bj*, it begins to evaporate, and does so without temperature change until the saturated vapor state at *q* is reached. As a matter of fact, if an equilibrium system is achieved at this pressure and temperature with a mixture of some solid, some liquid, and some vapor, these three phases steadily coexist and can continue to do so as long as the p_3 and t_{mnq} remain the same. For this reason, this line *mnq* is called the **triple point** (for the reason it is called a *point*, see Fig. 3/6). Examples of triple-point states are: H_2O, $p = 0.08865$ psia, $t = 32.018°F$ (by international agreement, the basic point of the temperature scale, § 1.18); ammonia NH_3, $p = 0.88$ psia, $t = -107.86°F$; nitrogen N_2, 1.086 psia, $114.1°R$; carbon dioxide CO_2, 5.1 atm, $-56.6°C$; which is to say, for example, that liquid NH_3 will not exist in a stable equilibrium state at $p < 0.88$ psia.

If the foregoing experiment is continued with the pressure reduced to something less than that at the triple point, as p_4, Fig. 3/1, the solid **sublimates**, the process being called **sublimation**, which is the phenomenon of a change of phase from a solid to a vapor. The heat required to change the state from saturated solid at *o* to saturated vapor at *p* is the *heat of sublimation*, or the enthalpy h_{ig} of sublimation. Observe, at the end of the previous paragraph, that since the triple-point pressure of CO_2 is 5.1 atm, it will sublimate at 1 atm (so-called dry ice).

Assuming that Fig. 3/1 is qualitatively reasonable, we observe that the latent heat $(Q = \Delta h)$ of evaporation h_{fg} decreases as the pressure increases (area under *jd* less than area under *gh*). Therefore, it might be expected that in carrying the experiment to higher and higher pressures, one will be found where the enthalpy of evaporation is zero. At this point *c*, the liquid and vapor curves meet and a vapor is indistinguishable from a liquid. This point is called the *critical point* and the properties there are critical properties: **critical temperature** T_c, **critical pressure** p_c, **critical volume** v_c, an important thermodynamic state. See Item B 16 for some critical-point properties. One may correctly conclude that at any state where the temperature is greater than the critical temperature, there is no phase of the substance that is liquid. It is interesting to note that a gas may be more dense than its liquid. For example, gaseous nitrogen at 15,000 atm has $\rho = 0.1301$ gm/cm^3; the density of its normal liquid is 0.071 gm/cm^3; of its solid, 0.081 gm/cm^3.[1.16]

3.5   COMPARISON OF LIQUID AND VAPOR CURVES

The saturated liquid and vapor lines, as plotted for 1 lb, for several substances are shown in Fig. 3/2. Notice the variability of the latent heat of evaporation, proportional to the distance between the liquid and vapor lines at a particular temperature. Sulfur dioxide, carbon dioxide, ammonia, and Freon 12 are refrigerants (H_2O is also used as a refrigerant)—Items B 31 and B 33 through B 36. Mercury (Hg) may be used to generate power in turbines (§ 11.13).

Not all saturated-vapor curves slope downward toward the right. Some show a double curvature, for example, benzene; and several such curves slope downward toward the left as, for example, the saturated-vapor curve of acetic acid, Fig. 3/3. If a substance with this characteristic undergoes an isentropic expansion, it becomes dryer or more highly superheated, whereas other substances ·discussed here become wetter or lose superheat.

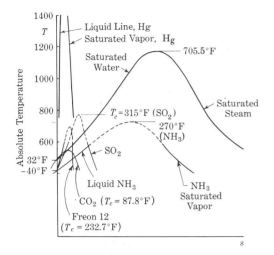

Various Fluids on Ts Plane. All curves are plotted to the same scales.   **Fig. 3/2**

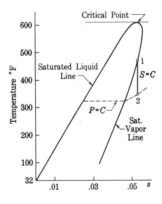

Saturation Curve for Acetic Acid.   **Fig. 3/3**

THERMODYNAMIC SURFACES **3.6**

As you recall, a surface is defined by a function such as $z = z(x, y)$ relating the co-ordinates. Since properties are the mathematical equivalent of coordinates, any three properties may be used if a suitable function can be found. The equation of the ideal-gas surface is $pv = RT$. Moreover, if equilibrium properties of any substance are in abundance, any three of them, as p, v, and T or u, s, and p, may be plotted in space to produce a thermodynamic surface, as in Figs. 3/4 and 3/5. Point c is the critical point; *mck* in Fig. 3/5 is the critical isobar; *qcr* in Fig. 3/4 is the critical isotherm. In Fig. 3/4, lines of $T = C$, $p = C$ in the two-phase region are straight lines parallel to the $v = C$ axis; *abndef* is an isobar; *ghij* is an isotherm. Imagine the surfaces of Fig. 3/4 projected onto the pv plane and compare with Fig. 3/7(a). While studying § 3/7, compare the projections of the surfaces in Fig. 3/4 on the pT plane with Fig. 3/6.

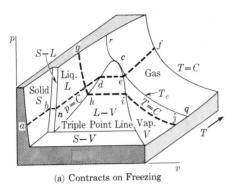

(a) Contracts on Freezing

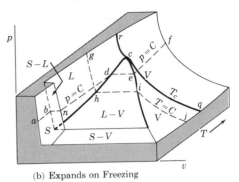

(b) Expands on Freezing

Fig. 3/4   *Thermodynamic Surfaces in pvT Coordinates.* Simplified. Such surfaces can be plotted to the extent of our knowledge of the equilibrium properties of a particular substance, and a point on the surface represents an internally stable equilibrium state. Substances other than water that expand on freezing are bismuth, antimony, and gallium, figure (b). Code: *S* = solid, *L* = liquid, *V* = vapor.

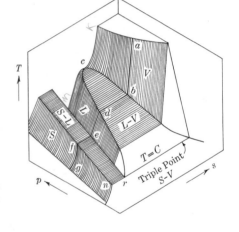

Fig. 3/5   *Thermodynamic Surface in Tps Coordinates.*[83] Substance contracts on freezing. Imagine this surface projected on the *Ts* plane and compare with previously sketched *Ts* planes.

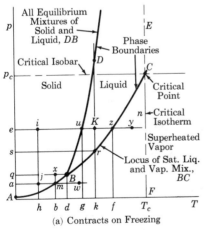

(a) Contracts on Freezing

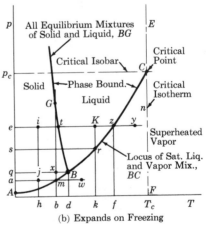

(b) Expands on Freezing

Fig. 3/6   *Phase Diagrams for One-Component System.* Such curves are defined by experimental values for a particular substance, the curves shown indicating typical trends. For example, p_s is the saturation pressure corresponding to the saturation temperature T_k. AB is locus of equilibrium mixtures of solid and vapor.

Because *hi*, *de*, and so on, in Fig. 3/4 are parallel to the *v* axis, a projection on the *pT* plane of the saturated liquid and vapor curves *hdcei* shows as only one line, *BrzC* in Fig. 3/6, a line called a *phase boundary*. All phase boundaries are characterized by latent heat. Notice that the triple-point line projects to a point *B*, Fig. 3/6. At temperatures and pressures below those at *B*, the solid sublimates and the phase boundary is curve *AmB*. Most substances contract on freezing, for which the phase boundary between liquid and solid is something like *BuD* in Fig. 3/6(a). Fortunately, water expands on freezing with a phase boundary as *BtG*, Fig. 3/6(b). An isobar is *iKz*; an isotherm is *krK*, *hji*, and so on.

Given a solid in state *i*, heat it at constant pressure; its temperature rises as heat flows in until the phase boundary *S-L* is reached at *u* (or *t*); the temperature remains constant during melting, but when it becomes all liquid, more heat raises the temperature from *u* to *z* (or from *t* to *z*), where evaporation starts. Again, the point *z* on this plane does not change as long as there is an equilibrium two-phase mixture. When all the liquid is evaporated, additional heat then superheats the vapor, say to state *y*.

At the triple point *B*, having started from a solid at *x* with $p = C$, there may be three phases simultaneously existing; superheating *BH* does not occur in an equilibrium mixture until only saturated vapor remains at *B*. A solid in state *j* at a still lower pressure never passes through a liquid phase; at *m*, sublimation starts; after all the solid has evaporated (equilibrium), then superheating, say *mw*, may occur. Curve *AB* is the locus of all equilibrium states of solid-vapor mixtures. If we have a mixture of phases at $p_q = p_B$ and $T_d = T_B$, and the pressure is lowered to $p_a = p_m$, the liquid phase will have disappeared after the mixture arrives again at an equilibrium condition. Similarly, with all three phases coexisting at *B*, if the pressure is increased to $p_s = p_r$, the solid phase will have disappeared at internal equilibrium; the total stored energy is assumed to remain constant.

All points between the phase boundaries *BC* and *BD* (or *BG*), as *K* will be liquid, compressed or subcooled liquid, § 3.12. In state *K*, Fig. 3/6, it is a **compressed liquid** because $p_e = p_K$ is greater than the saturation pressure $p_r = p_s$ at temperature $T_K = T_k$; it is a **subcooled liquid** because $T_k < T_z$ is the saturation temperature for pressure p_e. A liquid can be made to evaporate (boil) by lowering its pressure. Suppose the substance is entirely saturated liquid in state *z* (a mixture of liquid and vapor will do) and the pressure is lowered to $p_s = p_r$; then in order for the substance to reach internal equilibrium with no change in stored energy, some of the liquid must evaporate until the temperature of the mixture is $T_k = T_r$. If at *z* the substance is 100% saturated vapor, a lowering of the pressure with constant stored energy results in a superheated vapor state.

PHASE RULE 3.8

Considering Willard Gibbs' famous phase rule with respect to a pure substance as discussed above, called a one-component system, we observed that at the triple point no intensive property can be changed without the disappearance of at least one phase; there are no independent properties. For this reason, a system in a triple-point state is said to have no degrees of freedom and is **invariant**.

At other points on the phase boundaries of a two-phase one-component system, the pressure (or temperature) may be changed independently and there still may be two phases. Such a system has one degree of freedom and is said to be **univariant**. For a gas, or superheated vapor, two intensive properties, both p and T, may be varied without a change of phase. This type of system has two degrees of freedom and is **divariant**.

If the substance consists of more than one component, as air, then there are more degrees of freedom because equilibrium conditions will be also related to composition. But we shall not proceed any further on this now. Gibbs' phase rule is

(a) $$\phi + F = C + 2$$

where ϕ is the number of phases that can coexist in equilibrium, F is the number of degrees of freedom, and C is the number of components involved. Equation **(a)** has its greatest usefulness in multicomponent systems. Applied to the triple point discussed above, we have $C = 1$ component, $\phi = 3$ phases; then $F = C + 2 - \phi = 1 + 2 - 3 = 0$, no degrees of freedom as previously stated.

Significantly, Gibbs' rule says that no more than three phases of a single component can coexist, even though a substance may exist in more than three phases. For example, eight phases of solid H_2O have been discovered by Bridgman,[1.1] not all stable. At about 282,000 psia, H_2O solidifies at 126°F, a different phase from ice as we know it. Helium between the temperatures of 3.9 and 9.5°R has been found to have two different phases of liquid.[1.3]

Speaking of Gibbs and phases, let us look into an equilibrium mixture of a liquid and its vapor. Let an amount dm of liquid evaporate; the Gibbs function of the liquid decreases by an amount $G_f\,dm$ and the Gibbs' function of the vapor increases by the amount $G_g\,dm$; the change dG is

(b) $$dG = dm(G_g - G_f) = dm(h_g - Ts_g - h_f + Ts_f) = dm(h_{fg} - Ts_{fg})$$

where $G = h - Ts$, another property, and called the Gibbs function.

Since Ts_{fg} is equal to h_{fg}, the data in parentheses are equal to zero, and the change $\Delta G = 0$. Therefore, the mixture of saturated liquid and vapor is in a stable equilibrium state (internally), §§ 5.25, 5.26. It follows that the specific Gibbs functions for saturated liquid and for saturated vapor are equal. By similar logic, we find that at the triple point $G_{\text{solid}} = G_{\text{liq}} = G_{\text{vap}}$. Of course, we could have said that the equilibrium condition of §§ 5.25, 5.26 required $dG = 0$; with the same conclusions. See also § 13.44.

3.9   EQUATIONS OF STATE

One important underlying characteristic of the pure compressible substance (in the absence of electricity, magnetism, gravity, and capillarity) is that its state may be defined by two independent properties. This is excellent since thermodynamics concentrates heavily upon such a substance. It follows from this state postulate that, given any three properties (1, 2, 3),

property 1 = f (property 2, property 3).

In some special cases this set of equations can be expressed in explicit algebraic form; however, in general it is easier to represent it graphically or by tables. These equations, however they may appear, algebraically, graphically, or in tabular form, which relate the intensive thermodynamic properties of any pure substances are called the equations of state of that substance. For a more complete discussion of several selected equations of state, please glance at Chapter 10.

It is noteworthy that for gases (vapors) at low density ($p \doteq 0$), experimental data substantiates the p-v-T behavior to be closely

$$pv = RT$$

where R is the universal gas constant and depends upon the units selected for p, v, and T. Some frequently used units of R are:

$$\bar{R} = 1545 \text{ ft-lb}_f/\text{pmole-°R}$$

$$\bar{R} = 1.9859 \text{ Btu/pmole-°R}$$

$$\bar{R} = 8.3143 \text{ kJ/kgmole-K}$$

We also use units for R on the fps-system; see Item B1. A more detailed discussion of this particular equation of state is given in Chapter 6.

GAS TABLES 3.10

At this time we will but briefly introduce the gas tables (Items B 2–B 10). A detailed discussion of the tables will follow in Chapter 6 wherein the ideal gas is fully presented. These tables are for gases at relatively low pressure, that is, they are temperature dependent only. One enters the appropriate table knowing the temperature and, for the time being, reads values of enthalpy (h) and internal energy (u). The significance of the remaining three values p_r, v_r, and ϕ will be brought out in Chapter 6.

Example

Two moles of oxygen at 500°R undergo a process until the temperature becomes 900°R. Find the change of internal energy and enthalpy.

Solution. From Item B7 at 500°R we read $\bar{h}_1 = 3466.2$ Btu/mol, $\bar{u}_1 = 2473.2$ Btu/mol; at 900°R we read $\bar{h}_2 = 6337.9$, $\bar{u}_2 = 4550.6$

$$\Delta H = n(\bar{h}_2 - \bar{h}_1) = (2)(6337.9 - 3466.2) = 5743.4 \text{ Btu}$$

$$\Delta U = n(\bar{u}_2 - \bar{u}_1) = (2)(4550.6 - 2473.2) = 4154.8 \text{ Btu}$$

LIQUID–VAPOR TABLES 3.11

Tables and charts of thermodynamic properties of a number of pure substances in or about a two-phase region are available (see Items B 13 through B 16, B 26 through B 31, and B 33 through B 36, inclusive, Appendix; also Table II, § 3.12, and references [0.7], [0.9], [0.10], [0.32], and others). In most, the symbols for the properties are standardized: subscript g, vapor; subscript f, liquid; subscript fg, change from liquid to vapor; subscript i, frequently solid. In most tables the datum for the extensive properties is specified arbitrarily; for properties at states below the datum, they are simply negative and the sign should be held and handled algebraically, a frequent happening in work with refrigerants.

We shall use H_2O for our detailed discussion, inasmuch as it is the most common of the working substances; vapor tables bear a strong family resemblance. The ASME *Steam Tables*, abbreviated hereafter as ASME S.T., take the entropy and internal energy* as zero at the triple point, 32.018°F (formerly in United States tables, enthalpy was taken as zero for saturated liquid at 32°F). See Items B 13 through B 15. In refrigerant tables, the datum is most often saturated liquid at −40°F. A plot to scale (almost) of the saturated liquid and vapor lines for H_2O is shown in Fig. 3/7 on the *pv* and *Ts* planes. The volume scale for the liquid has been distorted (shown too large) because of the vast difference between liquid and vapor volumes at low pressure (see the tables). In Fig. 3/7(a), observe how the saturated vapor line flattens out toward the right, the volume increasing at an increasing rate. The dotted constant pressure line *zr* in Fig. 3/7(b) is merely representative, the quantitative difference between it and the saturation curve being exaggerated. Being perhaps the most atypical of substances, water has some unique attributes[3.5]; one is that the $p = C$ line crosses the saturated liquid line at the point of maximum density, point *n* at 39°F. [See Fig. 3/4(b).] Contrary to the more typical substance suggested by Fig. 3/1, the fusion temperature of H_2O decreases with increase of pressure (ordinary states—see § 3.8).

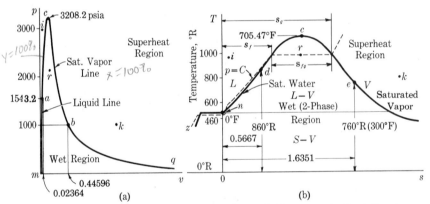

Fig. 3/7   *Liquid and Vapor Lines on pv and Ts Planes for* H_2O. The numerically defined points *d* and *e* were chosen arbitrarily.

Evidently from the definition of the symbols,

$$v_g = v_f + v_{fg}, \quad h_g = h_f + h_{fg} \quad s_g = s_f + s_{fg} \quad \text{and} \quad u_g = u_f + u_{fg}$$

If a state point is located in the two-phase region, say *r*, Fig. 3/7, the quality *x* or fraction of liquid *y* must be known to find its properties; $x + y = 1$. If in 1 lb there is *x* lb of saturated vapor and *y* lb of saturated liquid, the entropy of the mixture, which the reader can check as abscissa on Fig. 3/7(b), is

(a) $$s = xs_g + ys_f = xs_g + (1 - x)s_f = s_f + x(s_g - s_f) = s_f + xs_{fg}$$

---

* The international agreement is to take entropy and the Helmholtz function ($A = u - Ts$) as zero at the triple point.

Eliminating x in favor of y, we get

(b) $s = ys_f + xs_g = ys_f + (1 - y)s_g = s_g - y(s_g - s_f) = s_g - ys_{fg}$

For slide rule work, the expression **(b)** with y gives more accurate answers when the quality is high (above about 75%); the expression **(a)** with x is better for low quality (below 25%). There is little difference in the intermediate range. All specific properties of a two-phase system can be obtained by analogy with equations (a) and **(b)**. For 1 lb,

(c) $h = h_g - yh_{fg}$ $h = h_f + xh_{fg}$

(d) $v = v_g - yv_{fg}$ $v = v_f + xv_{fg}$

[HIGH QUALITY] [LOW QUALITY]

For a solid-vapor mixture, $s = s_i + xs_{ig}$, $h = h_i + xh_{ig}$, $v = v_i + xv_{ig}$, where, for example, h_{ig} is the enthalpy change from solid to saturated vapor (sublimation), $h_g - h_i$.

Usually, tables of the kind we are now interested in do not give values of the internal energy; hence, when needed, it is computed from $u = h - pv$, where the units in United States tables are generally Btu per lb. In the case of a two-phase mixture, the properties h and v must be in accord, say, $h = h_g - yh_{fg}$ and $v = v_g - yv_{fg}$.

At low pressures, $u_f \approx h_f$. For example, at the triple point, 32.018°F, $u_f = 0$; therefore

(e) $h_f = u_f + p_{sat}v_f = p_{sat}v_f = \dfrac{(144)(0.08865)(0.016022)}{778} = 0.00026 \text{ Btu/lb}$

where p_{sat} is the saturation pressure at 32.018°F and numerical values are taken from Item B 13.

Example—Internal Energy of Superheated Steam

Find u for steam at 100 psia and 600°F.
Solution.   From Item B 15, we find $h = 1329.6$ and $v = 6.216$; hence

(a) $u = h - \dfrac{pv}{J} = 1329.6 - \dfrac{(100)(144)(6.216)}{778} = 1214 \text{ Btu/lb}$

COMPRESSED LIQUID   3.12

A *compressed liquid* is one whose pressure is higher than the saturation pressure corresponding to its temperature. A *subcooled liquid* is one whose temperature is below the saturation temperature corresponding to its pressure. These two definitions define identical states, mean the same thing, and the names are customarily used interchangeably. In Fig. 3/8, imagine a saturated liquid in state d, and let it cool at constant pressure to either state B, c or b. It has become subcooled. On the other hand, imagine a saturated liquid at a, Fig. 3/8; let it be pumped to a higher pressure *bcBd*. If it is pumped isothermally, the end state is *b*; if isentropically, the end state is *c*; if isometrically ($v = C$), the end state is *B*. Each of these states represents compressed liquid.

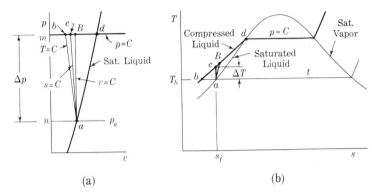

Fig. 3/8   *Compressed   Liquid. Differences   in   state are   magnified   for clarity.*
                                                           (a)   (b)

Typically, vapor tables do not give properties of compressed liquid (an exception, the ASME S.T.[0.32]), which means that a good estimation is convenient. The easiest assumption is that the liquid is incompressible, $v = C$, suitable for pressures that are not too high and for liquids not too compressible, in which case, points *b*, *c*, and *B* are quite close together. First from equation (2-34, 4-12) for a reversible pumping where $\Delta K = 0$, $\Delta P = 0$, we have

(a) $dW = -dh$ or $W = -\Delta h_s$ [ac]

TABLE II   Compressed Water

Extracted with permission from the ASME *Steam Tables*, published by the American Society of Mechanical Engineers. The ASME S.T. contain many more states of compressed liquid. See also Item B 15.

Absolute Pressure (sat. °F)	Temperature:						
	32°F	100°F	200°F	300°F	400°F	500°F	600°F
200 (381.8)							
v =	0.01601	0.01612	0.01663	0.01744			
h =	0.59	68.52	168.51	269.96			
s =	−0.0000	0.1294	0.2938	0.4369			
600 (486.20)							
v =	0.01599	0.01610	0.01660	0.01741	0.0186		
h =	1.8	69.58	169.42	270.70	375.49		
s =	0.0000	0.1292	0.2933	0.4362	0.5657		
1000 (544.58)							
v =	0.01597	0.01608	0.01658	0.01738	0.01855	0.02036	
h =	3.00	70.63	170.33	271.44	375.96	487.79	
s =	0.0001	0.1289	0.2928	0.4355	0.5647	0.6876	
2000 (635.80)							
v =	0.01591	0.01603	0.01653	0.01731	0.01844	0.02014	0.02332
h =	5.99	73.26	172.60	273.32	377.19	487.53	614.48
s =	0.0002	0.1283	0.2916	0.4337	0.5621	0.6834	0.8091
3000 (695.33)							
v =	0.01586	0.01599	0.01648	0.01724	0.01833	0.01995	0.02276
h =	8.95	75.88	174.88	275.22	378.47	487.52	610.08
s =	0.0002	0.1277	0.2904	0.4320	0.5597	0.6796	0.8009

where the subscript s indicates constant entropy, process ac, Fig. 3/8. Then from equation (4-15) with $T\,ds = 0$, we have

(b) $$dh = v\,dp \quad \text{or} \quad \Delta h \approx v \int dp = v\,\Delta p$$

Observe that $v\,\Delta p$ is the rectangular area $naBm$, Fig. 3/8(a); $v = v_a = v_f$ is the saturated liquid volume available in vapor tables. Let $p_a = p_{sat}$, meaning the saturation pressure from which the liquid was imagined to have been pumped to the actual pressure p_{act}. Equation **(b)** can then be written

$$(3\text{-}3) \qquad h_c \approx h_B = h_f + \frac{v_f(p_{act} - p_{sat})}{J} \quad \text{or} \quad h_B - h_f \approx \frac{v_f(p_{act} - p_{sat})}{J}$$

an approximation of the state after isentropic pumping if the liquid is nearly incompressible. (See § 11.17 for more detail.)

In dealing with a compressed liquid, we first decide whether or not any adjustment of saturated-state properties is needed. If the answer to this question is yes, we then decide whether the approximation involved in equation (3-3) is appropriate (which could also be almost the same as h_b). These are engineering decisions easily made with a background of experience. In the meantime, for pedagogical purposes for H_2O, let us say: when $p \lessgtr 400$ psia, use saturated liquid properties, *at the specified temperature*, for compressed liquid properties; when $p > 400$ psia, make an accurate or an approximate correction, depending on the accuracy needed and the facilities at hand. See Table II.

Example—Comparison of Enthalpy Changes of Water During Compression   3.13

Given saturated water at 100°F, determine its enthalpy if it is compressed to 3000 psia **(a)** isothermally, **(b)** isentropically, and **(c)** with constant volume.

Solution. **(a)** If it is compressed isothermally, its properties after compression are 100°F and 3000 psia. From Table II, we find $h = 75.88$ Btu/lb for this state, represented by b, Fig. 3/8.

(b) From the steam tables, we find the original entropy of $s_f = 0.1295$ Btu/lb-°R. Interpolating at 3000 psia in the full steam tables for this entropy, the temperature is about 101°F, and the enthalpy is 76.9 Btu/lb to one decimal place. This state is represented by c in Fig. 3/8. Since $h_f = 67.999$ Btu/lb, the steady-flow isentropic work of this compression is $76.9 - 68 = 8.9$ Btu/lb as a positive number ($\Delta K = 0$, $\Delta P = 0$).

(c) The constant volume compression is imaginary, since the water is not truly incompressible. By equation (3-3), we have

$$h_B = 67.999 + \frac{(0.01613)(3000 - 0.949)(144)}{778} = 67.999 + 8.96 = 76.96 \text{ Btu/lb}$$

from which we note that the work of the compression is 8.96 Btu/lb, not too different from that found for an isentropic compression. However, in states where water is more compressible and for other more compressible liquids, the agreement deteriorates.

CHARTS FOR PROPERTIES   3.14

Using any two point functions of substances,, such as temperature and entropy, we may construct to scale a diagram containing a series of lines, each representing a constant pressure; another series where each line represents a particular volume; and another series for constant quality; and so on. On the Ts plane, Fig. 3/9, several constant pressure, constant volume, and constant quality (also superheat) lines have been plotted. Thus, a point may be located by any two of these coordinates, after which all other properties pictured may be read from the diagram. The degree of accuracy of the reading will depend upon the spacing of the lines and the size of the diagram. A large Ts chart for H_2O is included in the ASME S.T.[0.32]

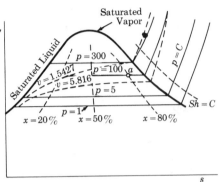

Fig. 3/9   *Temperature-Entropy Diagram for Steam.* The point *a*, for example, represents steam at a pressure of 100 psia and a quality of 80%. These two coordinates could be used to locate *a*, and then the volume could be determined from the constant volume lines (dotted); in this case, between $v = 1.54274$ and $v = 5.816$. (There are not enough constant volume lines on this figure for accurate interpolation.)

3.15   THE MOLLIER DIAGRAM

The *Mollier diagram* is a chart on which enthalpy is the ordinate and entropy the abscissa. On this chart, Fig. 3/10, a series of constant pressure lines, a series of constant quality and superheat lines, and a series of constant temperature lines are plotted. The constant temperature lines, of course, coincide with the constant pressure lines in the wet region (below the saturated-vapor line), but bend toward the right away from the constant pressure lines in the superheat region. A large Mollier chart that covers a section similar to that marked off by the dotted lines of Fig. 3/10 is included in the ASME S.T.,[0.32] a smaller one in *Problems*.

The Mollier chart is most useful in connection with steady-flow processes. The constant quality lines for steam are labeled according to their percentage of moisture, $y = 1 - x$. The constant pressure lines in Fig. 3/10 are straight in the wet region, the break shown being the result of a change of scale for entropy. See the examples in the caption to Fig. 3/10 for the method of using the chart.

Example

Use of Item B 16, Mollier Chart (SI).
1. One kg of steam is at 100 bar, 600°C. Find these properties, h, v, s. Using Item B 16 (SI):

$$h = 3622 \text{ kJ/kg} \qquad v = 0.0384 \text{ m}^3/\text{kg} \qquad s = 6.9 \text{ kJ/kg K}$$

2. Three kg of steam change state from 1 bar, $x_1 = 90\%$ (pt 1) to 300°C (pt 2), and $s = c$. Find s_1, p_2, ΔH_{12}, and ΔV_{12}. Using Item B 16 (SI):

$$s_1 = ms_1 = (3)(6.755) = 20.265 \text{ kJ/K}$$

$$p_2 = 20.5 \text{ bar}$$

$$\Delta H_{12} = m(\Delta h) = (3)(3025 - 2450) = 1725 \text{ kJ}$$

$$\Delta V_{12} = m(\Delta v) = (3)(0.125 - 1.5) = -4.125 \text{ m}^3$$

3. Two kg of steam occupy 4 m^3 at $250°C$. What is its state? Using Item B 16 (SI):

$$v_1 = V_1/m = 4/2 = 2 \text{ m}^3/\text{kg}$$

Locating the intersection of $v_1 = 2 \text{ m}^3/\text{kg}$ and $t_1 = 250°C$ shows the state to be superheated with the degree of superheat being $(t_1 = t_{\text{sat}}) = (250 - 100) = 150°C$.

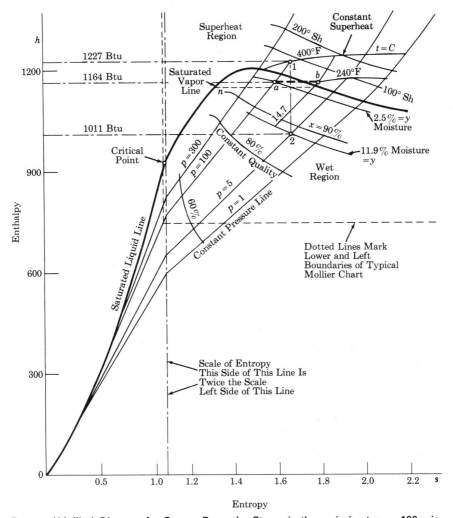

Entropy

Enthalpy-Entropy (Mollier) Diagram for Steam. Example: Steam in the main is at $p_a = 100$ psia. The sample in calorimeter is at 14.7 psia and 240°F. What is the moisture content y_a in the main?   **Fig. 3/10**
   Solution. By state in calorimeter, locate *b* at intersection of curves $p_b = 14.7$ psia and $t_b = 240°F$. Since throttling into calorimeter is according to $h_a = h_b$, move along horizontal line until the pressure line $p_1 = 100$ is reached, and locate *a*. Read answer $y_a = 2.5\%$. If initial pressure is too high, $p_1 > p_n \approx 1800$ psia, expansion does not reach superheat at 14.7 psia and a throttling calorimeter could not be used.
   Example. Steam enters turbine at 100 psia and 400°F, point 1, expands isentropically to 5 psia. What work is done if $\Delta K = 0$.
   Solution. Using properties of entering steam, locate 1 at intersection of curves $p_1 = 100$ psia and $t_1 = 400°F$. Follow constant entropy (vertical line) to 5-psia line, and locate 2. Move to left ordinate from 1, read $h_1 = 1227$ Btu/lb; move to left ordinate from 2, read $h_2 = 1011$ Btu/lb. Equation (**d**), §7.16, gives $W = 1227 - 1011 = 216$ Btu/lb; $y_2 = 11.9\%$.

3.16   THE *ph*-CHART

The *ph*-chart is another convenient method for displaying a mass of thermodynamic data on a single page. There are several *ph*-charts in the Appendix, each displaying selected thermodynamic properties for the fluid under study. The advantage of a chart, compared to a table, is that visual interpolation may be accomplished directly on the chart for odd-valued entry properties whereas tables require straight-line type interpolation that is often time consuming and sometimes inaccurate.

Figure 3/11 is a *ph*-chart for Freon 12 (see Item B 35). Note how the isothermals drop off sharply in the superheat region. Under the two-phase dome the isothermals are horizontal lines and follow the pressure lines. See examples in the caption to Fig. 3/11 for the method of using the chart.

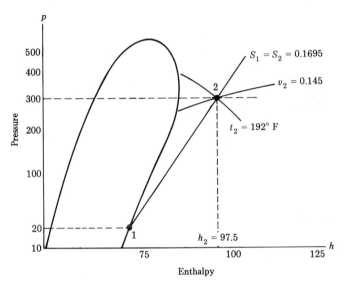

Fig. 3/11   *The ph-Chart for Freon 12. Example*: Saturated freon vapor at 20 psia is compressed along a constant entropy line until the pressure becomes 300 psia. Find these final properties t_2, h_2, s_2, v_2.
Solution: Locate initial point 1 as intersection of the 20 psia-line and the saturated vapor line. Now move diagonally upward along the $s_1 = s_2$ line until it intersects with the 300 psia-line. Read respective values as $t_2 = 192°F$, $h_2 = 97.5$ Btu/lb, $s_2 = s_1 = 0.1695$ Btu/lb-°R, $v_2 = 0.145$ ft³/lb.

3.17   CLOSURE

By now the reader should realize that substances, not unlike people, are all different. A table, chart, or equation of state that specifically suits the working substance under analysis must be used. This chapter should have given a reasonable facility in handling liquid-vapor tables, in using charts, and in quickly locating either in Appendix B. Familiarity and use of this knowledge will be required immediately in succeeding chapters.

PROBLEMS

SI UNITS

3.1 If 1 kg of water is at 10^6 Paa (10 bar), 300°C, use Item B 16 (SI) and find its volume, enthalpy, and entropy; determine its internal energy. Show *hs*-sketch.

Ans. U = 2796 kJ/kg.

3.2 Assume 2 kg of air at 3.03975 MPaa (30 atm), 143 K undergo a constant pressure process until the temperature becomes 293 K. From Item B 26, determine the total change of enthalpy and entropy. Show *Ts*-sketch of process.

Ans. 351.7 kJ, 1.758 kJ/K.

3.3 There are 500 ℓ helium in a container at 1.01325 MPaa (10 atm), 180 K. For the helium (use Item B 30) find the total enthalpy and entropy; show *Ts*-sketch.

3.4 Liquid air undergoes a phase change at constant pressure of 20 atm until it is saturated vapor. Find the change in enthalpy and entropy for a mass of 10 kg of air.

3.5 Assume 2 kg of hydrogen (Item B 28) are initially at 40 atm, 100 K. **(a)** If the hydrogen changes pressure to 1 atm without a change of temperature, find the changes of enthalpy and entropy. **(b)** If the hydrogen changes temperature to 35 K with no change of entropy, find the change of pressure and enthalpy. *Ans.* **(a)** 66.989 kJ, 23.027 kJ/K.

3.6 Say that 1 kg-mole of water is heated at constant pressure ($p = 5$ bar) from a wet state ($x_1 = 85\%$) to 400°C. Use Item B 16 (SI) and find the changes of volume, enthalpy, and entropy. Show *hs*-sketch.

3.7 Cool 1500 ℓ of water at constant volume from 40 bar, 612°C to 400°C. Use Item B 16 (SI), show sketch of process, and find **(a)** mass of water processed, **(b)** change of enthalpy, **(c)** change of entropy, **(d)** change of internal energy.

3.8 A 500-ℓ drum contains a gaseous substance at 10 atm, 140 K. Find the mass if the substance is **(a)** oxygen (Item B 27), **(b)** nitrogen (Item B 29), or **(c)** helium (Item B 30).

3.9 If 1 mole (kg) of water is at 20 bar, 400°C (see Item B 16-SI), calculate the Gibbs property $G = H - TS$ and the internal energy U.

Ans. G = −27,936 kJ, *U* = 53,028 kJ.

3.10 Calculate the value of c_p for water vapor in the region of 30 bar, 450°C; use Item B 16 (SI) and finite differences. Suggest use these two points: 30 bar, 440°C and 30 bar, 460°C for the finite differences of T and h.

3.11 A cooling process takes place at constant $p = 4$ atm for 0.32 kg oxygen initially at $T_1 = 190$ K; the final specific entropy is $s_2 = 135$ J/gmol-K. Use Item B 27, show sketch thereof, and find **(a)** final temperature T_2, °K, **(b)** total ΔS_{12}, J/K, **(c)** total ΔV_{12}, ℓ, **(d)** total ΔH_{12}, J; **(e)** calculate the total ΔU_{12}, J.

Ans. **(b)** −450 J/K, **(d)** −54,450 J.

3.12 Assume 5 kg of nitrogen are heated at constant volume from 95 K to 250 K; the specific volume is $v = 10$ cm^3/gm. Use Item B 29 (show sketch thereof) and determine **(a)** density ρ, gm/cm^3, **(b)** total ΔS_{12}, J/K, **(c)** total ΔH_{12}, J, **(d)** p_1, and p_2, atm; **(e)** total ΔU_{12}, J. *Ans.* **(b)** 8.5 kJ/K, **(c)** 1.450 MJ.

3.13 According to Gibbs' phase rule, see equation **(a)** § 3.7, how many degrees of freedom does a system consisting of a pure substance have if that substance is a superheated vapor or gas? If it is a mixture of vapor and liquid? If it is at the triple point (§ 3.3)?

3.14 Steam undergoes an isenthalpic process ($h = c$) from 15 bar, 350°C to 100 bar. Use Item B 16 (SI) and find final temperature t_2, Δv_{12}, Δs_{12}.

3.15 A 100-ℓ rigid tank with adiabatic walls is divided into equal parts *A* and *B* by a partition. On one side is steam at 1 bar, 200°C; on the other side is steam at 20 bar, 400°C. The partition is removed and thorough mixing occurs. Determine the equilibrium state (p, t) and ΔS.

3.16 Saturated steam vapor at 250°C moves along its isotherm until the pressure becomes 1 bar. Locate the end state points on the Mollier chart (Item B 16-SI) and for each pound of steam processed find **(a)** p_1, **(b)** Δh_{12}, **(c)** Δs_{12}, **(d)** Δv_{12}.

ENGLISH UNITS

3.17 Use several data points selected from Item B 14 and plot the saturated liquid-vapor curve for water on the *pv*-plane. The results will afford the viewer a perspective of the true shape of this curve.

3.18 The same as 3.17 except that the data points are to be selected from Item B 13 and the saturated liquid-vapor curve is to be plotted for water on the TS-plane.

3.19 If 1 lb of saturated water vapor is at 100 psia, find its temperature, volume, enthalpy, entropy, and internal energy. Also, find the change in these properties from saturated liquid to saturated vapor at 100 psia. Sketch the pv and TS diagrams.

3.20 Freon 12 undergoes a change of state from saturated liquid at 200 psia to superheated vapor at 200 psia, 240°F; see Item B 35. For each pound of Freon 12, find (**a**) h_{fg}, (**b**) Δh_{12}, (**c**) final entropy s_2 and volume v_2. Show ph sketch.

3.21 Air at 100°F undergoes a temperature change to 1500°F while its pressure remains relatively low. Use Item B 2 and for $m = 10$ lb air find ΔH_{12}, ΔU_{12}, and $\Delta\phi_{12}$. What are the ratios p_{r_2}/p_{r_1} and v_{r_1}/v_{r_2}?
Ans. $\Delta H_{12} = 3598$ Btu, $p_{r_2}/p_{r_1} = 102$.

3.22 A gaseous fluid with properties similar to those given in Item B 8 is at 260°F. It is compressed until $p_{r_2}/p_{r_1} = 10$. For 2 pmoles find (**a**) final temperature T_2, (**b**) ΔH_{12}, (**c**) ΔU_{12}, (**d**) $\Delta\phi_{12}$, (**e**) ratio v_{r_1}/v_{r_2}.

3.23 For each of the gases listed in Item B 2–B 10, compare the respective values of enthalpy $\bar{h}$ at 1000°R and based upon 1 pmole of the gas. Which has the larger $\bar{h}$? Which the smaller? Which two are closest in $\bar{h}$-values?

3.24 Air at 2350°R and relatively low pressure has its v_{r_1} increased tenfold; find T_2. Also find ΔH_{12}, ΔU_{12}, $\Delta\phi_{12}$ for 4 lb air. Use Item B 2.

3.25 If 1 lb of saturated water vapor is at 100 psia, find its temperature, volume, enthalpy, entropy and internal energy; also, find the change in these properties from saturated liquid to saturated vapor at 100 psia. Show pV and TS diagrams. See Item B 14.

3.26 A 10-ft³ drum contains saturated vapor at 100°F. What are the pressure and mass of vapor in the drum if the substance is (**a**) H_2O, (**b**) NH_3, (**c**) Freon 12?
Ans. (**a**) 0.94924 psia, 0.02854 lb_m.

3.27 Complete the following table for water; see Items B 13, B 14, and B 15.

State	(a)	(b)	(c)	(d)	(e)
Pressure, psia	120		50		1200
Temperature, °F		500		900	
Volume, ft³/lb	4.361				
Volume, ft³					
Mass, lb	2	1	3	1.5	4
Quality, %			70		
Moisture, %					20
Enthalpy, Btu		1245.1			
Entropy, Btu/°R				2.1450	

3.28 (**a**) What volume is occupied by 5 lb of steam at 2000 psia and 60% quality? (**b**) What volume is occupied by 5 lb of steam at 20 psia and 60% quality? (**c**) Is it permissible to omit the volume of the liquid in either of the foregoing cases?

3.29 (**a**) Sulfur dioxide is at 180°F, 0.40 Btu/°R-lb. Locate this state point on Item B 36 and find the pressure and enthalpy. (**b**) Two lb of mercury are at 200 psia and have a total enthalpy of 280 Btu. From Item B 34, describe its state—include the temperature, entropy, and quality or degrees of superheat.

3.30 (**a**) If 10 lb of carbon dioxide occupy 4 ft³ at 150 psia, find the enthalpy, temperature, quality or degrees of superheat. Use Item B 31. (**b**) Five lb of Freon 12 are at 100 psia, 250°F. Locate this state on Item B 35 and find the volume, enthalpy, and entropy.

3.31 On each of the four charts, Item B 31 (CO_2), Item B 33 (NH_3), Item B 35 (F12), Item B 36 (SO_2), locate the saturated vapor state at 50°F. Now follow the constant entropy line from this saturated vapor state to the full extent of the right side of the chart. For 1 lb of each of the four substances and between the two state point locations, give the changes in pressure, temperature, enthalpy, and volume as read from the charts.

3.32 (**a**) Compute the specific values for the Helmholtz and Gibbs functions for saturated water vapor at 100 psia. (**b**) Now let the water be saturated liquid at 100 psia and compute these functions. Compare.

3.33 Evaluate the constants A and B in the relation $h = A + Bpv$ for steam in the vicinity of 100 psia and 500°F; say, 100 psia, 450°F and 100 psia, 550°F. Then check the validity of the resulting relation for 200 psia and 500°F. Compute the percentage deviation from steam table value of h.

3.34 Dühring's rule states that if the saturation temperature of one fluid is plotted (rectangular coordinates) versus that of another fluid for the same pressures, a straight line (approximately) will result. Using water and ammonia, check this statement for the pressure range 10 psia to 100 psia, and write an equation for the resulting curve.

3.35 Water leaves a pump at 3000 psia, 300°F. For this compressed state, use Table II and find its volume v, enthalpy h, and entropy s. From Item B 13 find these values for saturated liquid water at 300°F and compare.

3.36 The water you drink from a fountain at 50°F is in a compressed (subcooled) state. At 50°F its saturation pressure is 0.17796 psia, yet it leaves the fountain at 14.7 psia. Does this compressed condition have much effect on its volume or enthalpy?

3.37 Assume 10 lb/sec of steam undergo a constant entropy process' from 250 psia, 700°F to atmospheric pressure. Sketch this process on the Mollier chart (Item B 16) and find t_2, s_2, percentage moisture, ΔH_{12}.

Ans. 212°F, 17.00 Btu/°R-sec, 4%, −2600 Btu/sec.

3.38 The two specific heats c_v, c_p (see §§ 2.18, 2.19) are desired for steam in the general region of 90 psia, 800°F. To calculate c_p, use the two state points 90 psia, 700°F and 90 psia, 900°F. To calculate c_v, use points 85 psia, 700°F and 100 psia, 900°F; note the volume constancy for these points. Compare these values with those found in Item B 1 given for low pressure water vapor.

3.39 The variation of c_v, c_p, and $k = c_p/c_v$ with pressure and temperature are being studied for superheated steam. Let the pressure range be 15–100 psia and the temperature range be 250–900°F. See §§ 2.18, 2.19 for discussion of the specific heats. Write a computer program that will produce the respective values throughout the p and t ranges described. Assume that the steam table data are stored in the memory of the computer.

4

THE FIRST LAW, ENERGY

4.1 INTRODUCTION

With the contents of the foregoing chapters fixed in mind, we will now proceed to apply this knowledge. Quite naturally a discussion of the first law of thermodynamics is apropos at this point. In addition to applying this law under varying circumstances, we shall also develop the general energy equation for the open system. You will find that these are simple yet powerful tools for solving sophisticated thermal problems.

4.2 FIRST LAW OF THERMODYNAMICS

Historically,* many experiments were run, accompanied by measurements of work and heat. Whenever the measurements were made on a cyclic basis (at first, principally by Joule), it was found that the net heat for the system was equal to the net work. In the form of equation (2-34A), we may generalize

(a) $\quad Q_{in} + W_{in} = Q_{out} + W_{out} \quad$ or $\quad Q_{in} - Q_{out} = \sum Q = W_{out} - W_{in} = \sum W$

the basis of the first law. Paraphrasing early statements of this law, we may say that *when a system undergoes a cyclic change, the net heat to or from the system is equal to the net work from or to the system.* In symbolic form, let

(b) $\qquad\qquad \sum Q = \oint dQ \quad$ and $\quad \sum W = \oint dW$

where the circle on the integral sign means that the integration (sum) is made entirely around the cycle (along a closed path), a symbol that will often be convenient. Thus,

$$(4\text{-}1) \quad \oint dQ = \oint dW \quad \text{or} \quad \oint dQ - \oint dW = 0, \quad \text{or} \quad \oint (dQ - dW) = 0$$

---

*Most physical laws stem from observations of things that happened repetitively. Unlike theorems, corollaries, and postulates, physical laws can be demonstrated but never proved in a mathematical sense. The first law falls into this category.

a mathematical expression of the early concept of the first law, where, of course, both heat and work must be expressed in the same energy units and signs are in accord with the previously defined convention. Since there is an advantage in broadening the statement to include all forms of energy, we may consider the law of conservation of energy as covering the traditional first law and restate it: *one form of energy may be converted into another.*

INTERNAL ENERGY—A CONSEQUENCE OF THE FIRST LAW 4.3

We have already discussed internal energy as being a reflection of the molecular state of a given fluid system; see § 2.6. Now, with the acceptance of the first law, we shall prove that as a consequence of it, internal energy exists and is a property.

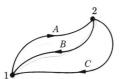

System Executing a Cyclic Process　**Fig. 4/1**

Let a closed system change by some process A from state 1 to state 2 and by some process B from state 2 to state 1; see Fig. 4/1. Then, according to equation (4-1) we may write

$$(4\text{-}2A) \qquad \oint_{1A2B1} (dQ - dW) = \int_{1A2} (dQ - dW) + \int_{2B1} (dQ - dW) = 0$$

Now let process C be a different means by which the system may change from state 2 to state 1. For the cyclic change $1A2C1$ we may also write

$$(4\text{-}2B) \qquad \oint_{1A2C1} (dQ - dW) = \int_{1A2} (dQ - dW) + \int_{2C1} (dQ - dW) = 0$$

Combining these two equations and simplifying results in

$$(4\text{-}2C) \qquad \int_{2B} (dQ - dW) = \int_{2C} (dQ - dW)$$

We conclude that when a system changes from one state to another, the value of the integral $\int (dQ - dW)$ is independent of the process involved and dependent only upon the end states; hence, it follows that this integral has the characteristics of a property. Since the two quantities involved (dQ and dW) are both energy values, we shall call this integral internal energy and designate it by the symbol E. It follows that

$$(4\text{-}2) \qquad E_2 - E_1 = \int_{1A2} (dQ - dW) = Q - W$$

where Q and W are the net heat and work experienced by the system as it executes a change of state between 1 and 2.

Written in the differential form, we have

$$dE = dQ - dW$$

or

(4-3) $$dQ = dW + dE$$

Example

Let a closed system execute a state change for which the heat is $Q = 100\,J$ and the work is $W = -25\,J$. Find ΔE.

Solution.

$$\Delta E = Q - W$$

$$= 100 - (-25)$$

$$= 125\,J$$

4.4   RELATION BETWEEN *E, U*

Consider a closed system consisting of a pure substance in the absence of electricity, magnetism, capillarity, gravity, and motion. On the basis of its mass we would assign a value to its internal energy and call it U, the thermal internal energy, fixed by any two independent properties.

If suddenly the substance were given motion within the closed boundaries and simultaneously was subjected to a gravitational field, the internal energy would now include these energy effects and be assigned the symbol E where

(4-4) $$E = U + P + K$$

On the basis of a 1-lb_m mass system, this becomes

$$e = u + P + K$$

In the absence of gravity and motion

$$E = U$$

4.5   ENTHALPY

Any combination of properties can be used to define another property, but only a few combinations would be useful. Perhaps the most useful is the property **en-thalpy**,* pronounced en-thal'-py, defined by (enthalpy h not related to Planck's constant, § 2.17)

(a) $$h \equiv u + pv \quad \text{and} \quad H = mh \equiv U + pV$$

---

* Other names in use for this property are *total heat* and *heat content*, both of which should be avoided.

Since all terms must have the same units, the usual conversion applies to pv, pV; for p lb/ft^2, V ft^3, use $J = 778$ ft-lb/Btu as in pv/J and pV/J to get Btu. Enthalpy has the units of energy, but *it is not a form of energy*. In various forms, we have the change of enthalpy given by

(b) $$dh = du + d(pv)$$

(c) $$\Delta h = \Delta u + \Delta(pv) \quad \text{or} \quad h_2 - h_1 = u_2 - u_1 + p_2 v_2 - p_1 v_1$$

for unit mass and good for any substance. Like internal energy u, enthalpy is measured from a convenient datum. Most often we shall be computing changes of enthalpy of a pure substance.

CLOSED SYSTEMS   4.6

The closed system, defined in § 1.3 as one that does not exchange matter with the surroundings, may be concerned with many types of energies; examples: a radio tube, a transistor, a thermocouple, a gas thermometer, a structural beam, and innumerable others. The type of closed system that we shall study most frequently is one consisting of a fluid undergoing one or more processes. The processes may be a nonflow variety, as the fluid enclosed by the piston and cylinder of Fig. 1/1, called a *closed nonflow system*, or just *nonflow* with the *closed* understood; or the fluid may be flowing, as in a steam power plant, Fig. 4/2. In any system, any form of energy may be stored except work and heat (which should be generalized to include other radiant energy, if appropriate). However, the forms of stored energy E_s that we

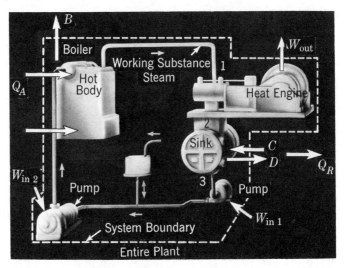

Closed Cyclic System—Steam Power Plant. Let the system be the H$_2$O in pipes, engine, and other equipment. Heat is added to the H$_2$O in the boiler. It comes from the chemical reaction of the fuel and air stream (another system) entering at *A*. Water is evaporated in the boiler; the steam flows to the prime mover (engine), where some of its energy is converted into work W_{out}; exhaust steam enters the condenser where it rejects heat to the cooling water (that enters at *C*, leaves at *D*—another system); the condensed steam (water) is pumped back into the boiler via two pumps, with works $W_{in} = W_{in1} + W_{in2}$. The net cycle work is $W = W_{out} - W_{in}$.

Fig. 4/2

meet most often here are kinetic K, potential (gravitational) P, and molecular U:

(4-5) $E_s = U + K + P$ or $E_s = u + K + P$

[UNIT MASS]

but other forms must not be overlooked, principally chemical energy E_{ch}, introduced later in this book. By definition of a property, $\Delta E_s = 0$ for one complete cycle. If the substance is flowing, as in Fig. 4/2, or is being stirred, the value of stored U, K, and P may be different for each element of mass m_1, which means that the total energy is the sum of the energies of each of the i elements comprising the system:

(a) $$E_s = \sum_i (m_i u_i + m_i K_i + m_i P_i)$$

or some average estimate may be satisfactory for a purpose.

   Either a nonflow or closed system may or may not be cyclic. A closed system may be traveling in space, possessing various kinds of stored energy, including kinetic energy of the whole if the reference axes are not attached to the system; it may also be accelerated by gravitational attraction, decelerated by atmospheric resistance; but if a jet is used to change its attitude, the system is not closed during this event.

   An *isolated system* (previously defined, § 1.3) is one with rigid boundaries exchanging neither energy nor mass with its surroundings ($\Delta m = 0$, $W = 0$, $Q = 0$), a special form of closed system, a useful thermodynamic concept.

4.7   ENERGY EQUATION FOR CLOSED SYSTEMS

   Energy equations are symbolized statements of the law of conservation of energy, with defined sign conventions. If heat Q is placed on the *in* side, equation (2-34), it is positive for heat added. If work W is placed on the *out* side, it is positive when work is done *by* the system. Therefore, according to equation (2-34), the energy balance for a nonflow process is

(4-6) $dQ = dE_s + dW$ or $Q = \Delta E_s + W$

[NONFLOW ENERGY EQUATION]

where Q is the net heat to (+) or from (−) any kind of closed nonflow system, W is the net work, and ΔE_s is its change of stored energy. The stored energies may be any storable varieties, as chemical energy.

   Consider a particular nonflow closed system, the gaseous substance in the cylinder of Fig. 4/3, in which the only stored energy is internal molecular energy U, u. Let heat Q flow into this system, which causes the fluid to expand pushing the movable piston against any kind of resistance, say, the pressure p_0 of the surroundings, a weight, and a spring that exerts a variable force F, depending on the type and amount of compression of the spring. Thus, work $W = \int p\, dV$, § 2.8, is done by the fluid in overcoming these resistances. The nonflow energy equation becomes

(4-7A) $dQ = dU + dW$ or $Q = \Delta U + W$

$$(4\text{-}7\text{B}) \qquad dQ = du + dW \qquad \text{or} \qquad Q = \Delta u + W \qquad [\text{UNIT MASS}]$$

where each energy quantity must be in the same unit; $\Delta u = u_2 - u_1$ is positive for an increase; the lowercase u says that all terms are for unit mass, say, 1 lb (or 1 mole, if this is convenient); $U = mu$. For finite changes, states 1 and 2 *must be internal equilibrium states* in order to evaluate $u_2 - u_1$.

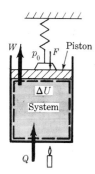

Energy Diagram of a Nonflow System. If heat both enters and departs, $\int dQ = Q_{\text{in}} - Q_{\text{out}} = Q$, where Q represents the net heat of the process. Similarly, $\int dW = W_{\text{out}} - W_{\text{in}} = W$, where W is the net work. NOTE: The work of the system $W = \Delta P + \Delta E_{\text{sp}} + p_o \Delta V$, where ΔP is the change of potential energy of the weight of the piston and attached mass, § 2.4, ΔE_{sp} is the change of stored energy in the spring, § 2.11, and $p_o \Delta V$ is the work done by the system in pushing back the surrounding atmosphere. **Fig. 4/3**

OPEN SYSTEMS AND STEADY FLOW 4.8

Many of the systems related to the manufacture of power are open systems (mass crosses its boundaries), and there are innumerable others. It is common in studying closed systems, such as the diagrammatic power plant of Fig. 4/2, to draw boundaries making a particular piece of equipment the system. For example, suppose the engine is a steam turbine. To study the turbine, draw boundaries at the entrance and exhaust sections, Fig. 4/4, and consider the steam as the system, deciding in detail the kind (and amount, if possible) of energy crossing each section. Then write an energy balance that accords with the conservation law. The boundary surface of an open system (turbine casing, and so on) is called the ***control surface***, and the volume within this surface is the ***control volume***. Ordinarily, we use an idealized model that is not too difficult to analyze. In particular we often assume that it is a ***steady-flow system***, defined as follows: (1) The mass rate of flow into the system is equal to that from the system; there is neither accumulation nor diminution of mass within the system. (2) There is neither accumulation of diminution of energy within the system; it follows that the rate of flow of heat $\dot{Q}$ and work $\dot{W}$ are constant; for example, the

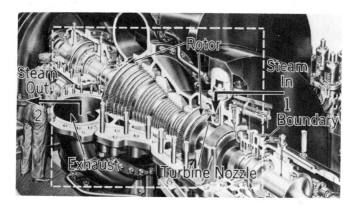

Turbine as a System. **Fig. 4/4**

turbine of Fig. 4/4 delivers work and radiates heat at a constant rate while operating under steady-flow conditions. (3) The state of the working substance at any point in the system remains constant. (4) For the determination of properties, one-dimensional flow (§ 1.22) at inlet and exit boundaries of the system is assumed; properties are then handled as though uniform across these sections.

For some purposes, a reciprocating engine may be analyzed as a steady-flow system when: (1) The intake is from a large plenum (theoretically, infinitely large) where the properties of the substance entering the engine are uniform and can be measured at a plenum boundary where they are constant. (2) The discharge is into a large plenum where presumably uniform properties at its exit state can be measured and found steady. (3) The temperature of each point on the engine varies periodically and is the same every time the engine completes its series of events. (4) The heat and work are the same for each complete series of events.

Unsteady-state and therefore transient systems are of several forms, each one characterized by accumulation or diminution of either mass or energy, or both, within the system. For example: (1) Fluid may flow outward only from a system (as a system of air in a tank from which air flows to "pump up" a tire). (2) Fluid may flow inward only (as in refilling the air tank with the discharge valve closed). (3) Fluid may be flowing both inward and outward but at different rates so that the amount of fluid within the system changes. Or (4) the engine is warming up (from a cold start or increasing power) or cooling off (decreasing power). Energy balances for such systems will be made and applied later (Chapter 7); for the present, we shall concentrate on the simpler steady-flow control volume.

4.9 GENERAL ENERGY EQUATION FOR THE OPEN SYSTEM

Let us consider the open system A in Fig. 4/5 contained by the control surface S with a tiny mass dm thereon.

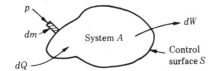

Fig. 4/5   *Open System*

The total initial and final internal energies of the system A are denoted by E_1 and E_2, respectively, and the specific internal energy of the tiny mass dm is e. Let the tiny mass dm be pushed across the surface S to join system A. During this joining process, a heat and work interaction occur resulting in dQ and dW.

While dm is crossing control surface S, it will be subjected to a pressure p which when exerted on a small part of dm gives rise to the force acting on dm. If we think of dm as a cube of side l, then the work done on it in terms of the pressure p acting on surface l^2 and moving it the distance l across the boundary is

$$pl^2 \cdot l = pV = pv\,dm$$

where v is the specific volume of the mass dm as it crosses S. We are now prepared to audit the various energies for the system A.

For the multisystem A and dm we can apply equation (4-2) and find

(4-8A) $$E_2 - E_1 = dQ + e\,dm + pv\,dm - dW$$

or $$dQ = E_2 - E_1 - (e + pv)\,dm + dW$$

(4-8B) $$dQ = E_2 - E_1 - (u + K + P + pv)\,dm + dW$$

(4-8C) $$dQ = E_2 - E_1 - (h + K + P)\,dm + dW$$

For a steady stream of masses dm across surface S, equation (4-8C) becomes

(4-8) $$Q = E_2 - E_1 - \sum (h + K + P)\,dm + W$$

where h denotes the specific enthalpy of the mass as it crosses the control surface. If a mass element were to pass outward from A across the control surface, then dm is negative. You are cautioned to survey these signs carefully.

Example

Let an insulated empty pressure vessel, Fig. 4/6, be filled with steam from a flowing steam main. Find the relative temperature of the steam in the vessel.

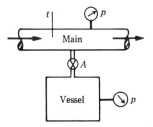

Filling Process **Fig. 4/6**

Solution. Valve A is permitted open until the steam pressure in the vessel reaches that of the steam main. For the flow process of steam from main to vessel, $E_1 = 0$, $Q = 0$, $W = 0$, $\Delta K = 0$, $\Delta P = 0$.

Applying equation (4-8) there results

$$Q = E_2 - E_1 - \sum (h + K + P)\,dm + W$$

$$hm = E_2 \quad \text{or} \quad h_{\text{line}} = u_{\text{vessel}}$$

Since h is greater than u at any given state and since the pressure in the vessel is the same as the main pressure, then the vessel temperature must exceed that of the main.

STEADY-STATE, STEADY-FLOW OPEN SYSTEMS 4.10

We now set up an energy balance equation involving those energy forms with which we shall be most concerned, an equation that can be applied to many flow systems. Since there is no change of stored mass or stored energy, the conservation law, equation (2-34), and equation (4-8) reduce to

(a) Energy entering system = energy leaving system

Thus the problem is simply one of accounting for the energy forms crossing the boundary: potential energy P, kinetic energy K, internal energy U, all energy forms *stored in the stream* (not system) as it crosses a boundary, flow energy E_f, heat Q, and shaft work W, Fig. 4/7. Flow energy $p_1V_1 = E_{f1}$ is entering the system because of the work done at boundary 1 against a pressure p_1 to force the fluid into the system.

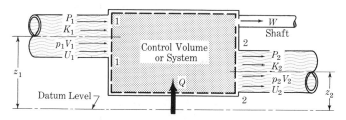

Fig. 4/7 *Energy Diagram of a Steady-Flow System.* At section 1, the pressure, specific volume, and velocity of the fluid are p_1, V_1, and v_1, respectively; at section 2, they are p_2, V_2, and v_2. An *energy diagram* is a representation of a system with an indication thereon of all changes of energy. In a particular application, one or several of the energy terms may be zero or negligible; or some other form of energy may be significant. Observe that W = the *net* work and Q = the *net* heat. The energies that are carried by the moviing fluid are *convected energies*.

Similarly, $E_{f2} = p_2V_2$ is the work to force the fluid out against the outside pressure p_2. Since Q is shown as heat *in* and W as work *out*, net Q will be positive when heat is added and negative when heat is rejected; net W is positive when the system does the work (energy out) and negative when work is done on the system. The energy equation is then written, Fig. 4/7, according to equation **(a)**,

(4-9) $P_1 + K_1 + E_{f1} + U_1 + Q = P_2 + K_2 + E_{f2} + U_2 + W$

which is illustrative of the kind of thinking one would do in making an energy balance for any kind of system. Even though mass and energy are moving across boundaries at steady rates, we nevertheless find it convenient to make computations for 1 lb, imagining the events happening to this unit mass as it passes through; if the rate of flow is $\dot{m}$, then for example, $\dot{W} = \dot{m}W$, where W is the work of unit mass. Also, we must assume that the properties at a boundary section are uniform. Now solve for Q from equations (4-9) for a unit mass and manipulate it.

(4-9A) $Q = \Delta u + \Delta(pv) + \Delta K + \Delta P + W$

(4-9B) $dQ = du + d(pv) + dK + dP + dW$

(4-9C) $dQ = dh + dK + dP + dW$

(4-9D) $dQ = du + p\,dv + v\,dp + dK + dP + dW$

where all terms are in the same units, all Δ values are the final property value minus the initial, and the flow rate is constant, and where $dh = d(u + pv) = du + d(pv)$ from § 4.5 is used in (4-9C); and in the last form, the flow energy term is replaced by a mathematical equivalent. It is good to realize now that a change in the value of pv may, and usually does, occur in a nonflow system, but it has significance as a separate energy quantity only in flow across a boundary. Since it turns out to be just as easy to compute Δh as Δu, there is an advantage of the form (4-9C) for steady-flow situations.

Although the energy equations (4-7) and (4-9) are appropriate for a large number of typical problems, the beginner would serve his own interest to form a habit of sketching an energy diagram, even for these simple systems. The correlative thinking processes are needed for the numerous situations that do not fit these equations. Let the first decision for a problem be the location of the system's boundaries; then indicate on a sketch the significant energy "flows" and changes. Place complete confidence in the conservation-of-energy law.

APPLICATIONS OF THE STEADY-FLOW EQUATION 4.11

It will often be true, unless the working substance is a liquid (as in a hydroelectric power plant) or unless there are great changes of altitude, that the change of gravitational potential energy is negligible. If so, the steady-flow energy equation reduces to

(4-10) $$dQ = dh + dK + dW \qquad [\Delta P = 0]$$

In any case, to make energy balances, one must know what the system does, more or less how it does it, and what energy forms are significant.

Consider a turbine, gas or steam. It receives a stream of fluid, more or less steadily when the load is constant, at high pressure; the fluid then expands to a low pressure doing work (the objective of the system is to manufacture work). The energy that the entering stream contains is u_1 and K_1; and as a unit mass is pushed across boundary 1, the energy of flow work $p_1 v_1$ enters the control volume. See the energy diagram, Fig. 4/8. Similarly, the energy departing with the stream is $u_2 + p_2 v_2 + K_2$, where $K = v^2/(2\mathbf{k})$ Btu for a mass of $m = 1$ lb. Evidently, we may let $h = u + pv$ at each boundary. Since the fluid passing through the turbine is certain to be at a temperature different from the surroundings, heat Q is unavoidable. The delivered work W is the work of the fluid in its passage through the control volume. Since ΔP is quite insignificant, this energy balance results in equation (4-10).

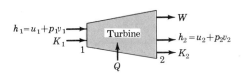

Fig. 4/8

Control Volume for Turbine. For steam turbines and the usual gas turbines, the working substance is hotter than the surroundings, and therefore heat Q leaves the system. If Q should be the unknown in equation (4-9) for such systems, it will be a negative number.

Now it happens that the overall change of kinetic energy ΔK for the typical turbine is so small that, except for the most exacting tests, it is usually ignored; $\Delta K \approx 0$. Equation (4.10) thus becomes

(4-11) $$dQ = dh + dW \qquad [\Delta K = 0, \Delta P = 0]$$

Central station steam turbines, for example, are quite well insulated. Moreover, the time it takes for a particular unit of mass to pass through is short, so that there is not much time for heat to leave. It follows that in some circumstances, especially in a first approximation, the heat may be neglected, in which case, we have

(4-12) $dW = -dh$ $[Q = 0, \Delta K = 0, \Delta P = 0]$

If these various decisions had all been made at the outset, the K_1, K_2, and Q of Fig. 4/8 would not show on the applicable energy diagram. Observe that the foregoing energy equations are obtained without bothering with any of the complex details of the events within the turbine, but that we do need to know "something" about the system.

For another illustration, consider a nozzle, a necessary device in a turbine (in the "black box" of Fig. 4/8), in a jet engine, and in many other places. It receives a fluid and guides its expansion in an orderly manner to a lower pressure, with the objective of converting some of the entering energy into kinetic energy at exit. No shafts are turned; $W = 0$. The time of passage of a particular particle is only a fraction of a second, and very closely, $Q = 0$. Since u and E_f are involved at both sections, use $h = u + pv$ and get the energy diagram of Fig. 4/9. Equating "energy in" to "energy out" and solving for ΔK, we get

(4-13) $\Delta K = -\Delta h = -(h_2 - h_1) = h_1 - h_2$ $[Q = 0, W = 0, \Delta P = 0]$

from which we note that here the enthalpy change is the negative of the kinetic energy change. In many cases the entering kinetic energy is negligible as compared to the final value, $K_1 \approx 0$ (especially if we back up a bit from the entrance section to where the properties are p_0, T_0, etc., Fig. 4/9), in which case

(a) $K_2 = \dfrac{v_2^2}{2\mathbf{k}} \approx h_1 - h_2$ $[K_1 = 0, Q = 0, W = 0]$

Btu for a mass of 1 lb.

Finally, for a system where $W = 0$ but the heat is not negligible, the conventional energy balance becomes

(b) $dQ = dh + dK$ $[W = 0, \Delta P = 0]$

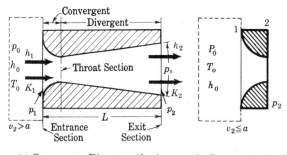

(a) Convergent—Divergent Nozzle   (b) Convergent Nozzle

Fig. 4/9   *Energy Diagram for Nozzle.* The control surface is the internal surface of the nozzle and the imaginary planes at sections 1 and 2. If a is the local acoustic speed, the speed in the throat of figure (a) is a; the speed at the exit section of (a) is $v_2 > a$. In the convergent nozzle, the exit speed is never greater than a; $v_2 \leqq a$ in figure (b). See Chapter 18.

Do not forget: as a fluid flows across a boundary, enthalpy represents energy, but *two different kinds of energy*, internal energy and flow energy. It is recommended for purposes of clear thinking that, in other than routine steady-flow systems, the energy quantities u and pv be considered individually at the outset and then be combined into h when it is proper.

Example

Steam flow in a nozzle.

Steam at 50 bar, 500°C enters a nozzle, similar to that in Fig. 4/9, and expands steady-state, steady-flow to 20 bar with $s = c$; $K_1 \approx 0$, $\Delta P = 0$, $Q = 0$, $W = 0$. Find the exit kinetic energy K_2 and the velocity v_2.

Solution. Refer to equation 4-13(b),

$$dW = dh + dK$$

and $dK = -dh$

$$K_2 = h_1 - h_2$$

From Item B 16(SI), we get

$$h_1 = 3550 \qquad h_2 = 3243 \, \text{kJ/kg}$$

For 1 kg/s flowing,

$$K_2 = (3550 - 3243) = 307 \, \text{kJ/s}$$

Now $mK_2 = mv^2/2$

or

$$v = (2K_2)^{0.5}$$

$$= \left[(2)\left(\frac{307 \, \text{kJ}}{\text{kg}}\right)\left(\frac{1000 \, \text{J}}{\text{kJ}}\right)\left(\frac{1 \, \text{Nm}}{\text{J}}\right)\left(\frac{1 \, \text{kg m}}{\text{s}^2\text{N}}\right) \right]^{0.5}$$

$$= 784 \, \text{m/s}$$

PROPERTY RELATIONS FROM ENERGY EQUATIONS 4.12

Consider systems made up of fluids. Now that we know $dQ = T\,ds$ for any reversible process and $dW = p\,dv$ for the boundary work of a reversible nonflow system, we may make some mathematical changes. Using these expressions in the nonflow energy equation (4-7), we have

(4-14)
$$T\,ds = du + p\,dv \qquad \text{or} \qquad ds = \frac{du}{T} + \frac{p}{T}\,dv$$

and, significantly, this is an equation containing only point functions. Therefore it is valid between any two equilibrium states for any kind of process, nonflow or steady flow, reversible or irreversible, but if the process joining the states is *not* internally reversible, then $T\,ds$ and $p\,dv$ do *not* represent heat and work, respectively. In other words, the integrations of (4-14) can be made, as we shall learn, via some chosen reversible path joining the points, but the energy interactions with the surroundings along those reversible paths will not be the same as the actual energy exchange.

Next, differentiate $h = u + pv$, the definition of enthalpy, and use $du + p\,dv = T\,ds$; $dh = du + p\,dv + v\,dp$, which gives

$$(4\text{-}15) \qquad dh = T\,ds + v\,dp \qquad \text{or} \qquad T\,ds = dh - v\,dp$$

where $dQ = T\,ds$ for a reversible process. Equations (4-14) and (4-15) are repeatedly useful, applying as they do to any pure substance and to any equilibrium states.

Letting $dQ = T\,ds$ in the steady-flow equation (4-9), we get

(a) $\qquad\qquad\qquad T\,ds = du + p\,dv + v\,dp + dK + dP + dW$

or, since $du + p\,dv$ cancels $T\,ds$,

$$(4\text{-}16) \qquad\qquad\qquad -v\,dp = dK + dP + dW$$

which is to say that in a reversible steady-flow process, $-\int v\,dp$ equals the shaft work W plus the changes of kinetic and potential energies, equations (2-9) and (2-5). If ΔP is negligible,

$$(4\text{-}17) \qquad\qquad\qquad -\int_1^2 v\,dp = W + \Delta K \qquad\qquad [\text{UNIT MASS}]$$

If both ΔP and ΔK are negligible, the shaft work of a steady-flow process is

(b) $\qquad\qquad\qquad W = -\int_1^2 v\,dp \qquad\qquad [\text{UNIT MASS}]$

Similarly, if $W = 0$ and $\Delta P = 0$, $\Delta K = -\int v\,dp$. If the process is irreversible, $-\int v\,dp$ does not have the meaning of (4-16).

The integral $-\int v\,dp$ (or $-\int V\,dp$) represents an area on the pv (or pV) plane, as indicated in Fig. 4/10, spoken of as the area "behind" the curve. From equation (4-15), we see that it represents $\int(T\,ds - dh)$; if the process is reversible, $\int T\,ds = Q$; if $Q = 0$ for the reversible process, the area $(-\int v\,dp)$ represents the change of enthalpy Δh. From equation (4-16), we find that it also represents $\Delta K + \Delta P + W$, where any one or more of these terms may be zero.

For one more useful mathematical consequence, differentiate pv to get $d(pv) = p\,dv + v\,dp$, and observe that

$$(4\text{-}18) \qquad \int p\,dv = -\int v\,dp + \Delta(pv) \qquad -\int v\,dp = \int p\,dv - \Delta(pv)$$

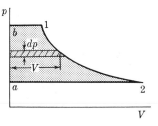

Fig. 4/10 *Area for $\int V\,dp$ (or $\int v\,dp$). Area 1-2-ab represents the work of a reversible steady-flow process when $\Delta K = 0$ and $\Delta P = 0$, and $-\Delta H$ when $Q = 0$. Observe that $-\int V\,dp$ is a positive number when integrated from 1 to 2 (that is, for an expansion). To make the integration, one must have a relation between p and V (or v) as $pv^n = C$.*

where $\Delta(pv) = p_2v_2 - p_1v_1$. The relation in (4-18) is handy whenever one of the integrals is easier to make than the other.

UNSTEADY STATE, TRANSIENT FLOW, OPEN SYSTEMS   4.13

We may define a **transient process** as an unbalanced one in search of an equilibrium state or a steady state. In a transient process in nature, time becomes an important coordinate, but many transient systems can be defined as an event in a limited time so that certain thermodynamic computations may be made—albeit for idealized models, as usual.

Examples of some transient situations will help to clarify what is meant. Suppose a steam power plant is operating in a steady state (constant output) and the demand for power decreases. Among the many complex things which begin to happen, the steam flow to the turbine and from the steam generator must be reduced; consequently, the amount of fuel and air fed to the furnace must be decreased. It is easy to visualize that there will be changes of pressures and temperatures. Not only will the furnace temperatures change, but there will be changes throughout the electric generator. If the new load on the plant remains constant, the plant will eventually arrive again at a new steady state; in the meantime the process of change is a transient one. It is unlikely that an automobile can be operated on the road in a steady state, but on a test block the engine may approach such conditions. A particular point on a cylinder wall may be repeatedly varying through the same temperature range during every engine cycle. If the load is changed, every temperature cycle changes, including that of the cooling water and oil. These ideas may be extrapolated to all power producing or power consuming systems.

Most of the problems on nonsteady state are beyond the scope of this text, but a few may be made simple enough that, via an understanding of these easier cases, one can achieve a deeper insight into thermodynamics.

To set up a general equation that does not look too cluttered with symbols, let E_s represent *all* energy quantities stored in the control volume per pound of substance, including chemical energy, and so on, except that our discussion at this point will be limited. Moreover, we shall find it convenient to assume one dimensional flow wherever the fluid crosses a boundary (§ 1.22), the fluid having uniform properties on a cross section; and, in computing stored energy, to assume that, for the states concerned, the properties are uniform throughout the system. In the beginning, Fig. 4/11, the total stored energy is $m_{s1}E_{s1}$, where m_{s1} is the mass within the control volume at that instant. After some time interval, the stored energy will be $m_{s2}E_{s2}$. Let the time interval be infinitesimal $d\tau$, the corresponding change of stored energy

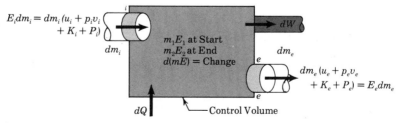

Energy Diagram—Variable Flow. All events are for the same time interval. If some part of the control surface moves (as a piston), it will be moving against a resistance, which involves work and is considered to be included in dW.   **Fig. 4/11**

is $d(m_sE_s)$, or as a rate, $d(m_sE_s)/d\tau$. The energy quantities entering and departing the control volume with the mass flows are the stored varieties, for example, $u + K + P$, as well as the flow energy $E_f = pv$ as the masses cross the boundaries. Thus, let $E_i = u_i + p_iv_i + K_i + P_i = h_i + K_i + P_i$, the specific energy *in* and $E_e + h_e + K_e + P_e$, the exit energy. In the time $d\tau$, the corresponding masses are dm_i and dm_e; the rates are $\dot{m}_i = dm_i/d\tau$ and $\dot{m}_e = dm_e/d\tau$. The conservation law, equation (2-21), applied to the energy diagram of Fig. 4/11 then gives

(4-19A) $$E_i dm_i + dQ = d(m_sE_s) + E_e\, dm_e + dW$$

(4-19B) $$E_i\dot{m}_i + \dot{Q} = \frac{d(m_sE_s)}{d\tau} + E_e\dot{m}_e + \dot{W}$$

where (4-19B) is energy conservation on a rate basis, dQ, $\dot{Q}$ and dW, $\dot{W}$ are summed over the entire control surface and are the *net* amounts. If m_{s1}, m_{s2}, E_{s1}, and E_{s2} can be determined at the beginning and end of the time involved, the change in stored energy is $\Delta m_sE_s = m_{s2}E_{s2} - m_{s1}E_{s1}$. If mass crosses the boundary at a number of points, say j points, the net energy crossing the boundary with the streams is the sum of the energies for j crossings, $\sum E_j\, dm_j$; if entering streams carry positive energy, the more general form of equation (4-19) becomes

(4-19C) $$dQ + \sum_j E_j\, dm_j = d(m_sE_s) + dW$$

where $E_j = (u + pv + K + P)_j$ plus any other energy transported as energy stored in the fluid stream. Let $m_{s2} - m_{s1} = \Delta m_s$ and write the conservation of mass as

(4-20) $$dm_i = d(\Delta m_s) + dm_e \quad \text{or} \quad \dot{m}_i = \frac{d(\Delta m_s)}{d\tau} + \dot{m}_e$$

This balance can often be made in finite quantities for a particular though unknown time interval. If the stored energy is molecular only, and the masses crossing the boundaries involve only internal energy and flow work, equation (4-19) becomes

(4-21) $$dQ = d(m_su_s) + dW + h_e\, dm_e - h_i\, dm_i$$

The enthalpies and internal energies should in general be measured above a zero datum, for example $u = c_vT$, $h = c_pT$ for ideal gases, or values of u and h taken from gas tables.

4.14   MATTER CROSSING MORE THAN TWO BOUNDARY POINTS

By implication, the open systems discussed have had one entry and one exit boundary. If mass crosses boundaries at more than two places, the problem is no more complicated for steady-flow situations, only longer. Let the energy crossing a boundary be represented by E, as E_A for boundary A. Since the energy stored in the stream is $u + K + P$ and since flow energy pv crosses each boundary, $E = u + pv + K + P = h + K + P$. Suppose there are two entries and two exits, with anything happening in the control volume, Fig. 4/12. If the potential energy changes (not shown in Fig. 4/12) are significant, z_A, z_B, and so on, are simply measured from any convenient datum to the boundaries at A, B, and so on, equation (2-5).

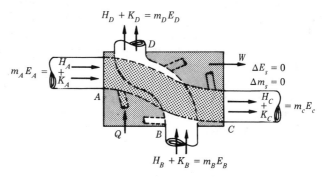

Energy Diagram of Control Volume with Four Streams. **Fig. 4/12**

Entering: $\dot{m}_A E_a = \dot{m}_A(h_A + K_A + P_A)$ $\dot{m}_B E_B = \dot{m}_B(h_B + K_B + P_B)$

Leaving: $\dot{m}_C E_C = \dot{m}_C(h_C + K_C + P_C)$ $\dot{m}_D E_D = \dot{m}_D(h_D + K_D + P_D)$

where it is essential to include the mass of each stream, all for the same unit of time. By conservation laws,

(4-22) $$\dot{m}_A + \dot{m}_B = \dot{m}_C + \dot{m}_D \qquad \text{[MASS]}$$

(4-23A) $$\dot{Q} + \dot{m}_A E_A + \dot{m}_B E_B = \dot{m}_C E_C + \dot{m}_D E_D + \dot{W}$$

Generalized, with $\sum \dot{m}_i E_i$, for i streams, representing all outgoing energy minus all incoming energy, except heat and work, equation (4-20) becomes

(4-23B) $$\dot{Q} = \sum_i \dot{m}_i E'_i + \dot{W}$$

where $\dot{Q}$ and $\dot{W}$ are net values, the convention of signs is as previously defined, and the properties for evaluating the E's are measured at the boundaries concerned; the mass flows may be computed from $\dot{m} = \rho A v$, § 1.22.

BOUNDARIES OF THE SYSTEM 4.15

It is fitting to consider a few different locations of boundaries, and, incidentally, to observe the conversion of energy in the form called work, the finest kind. Because of its cost and value, man does what is feasible to keep from wasting work.

Consider Fig 4/13. System (1) is an expansible fluid. With thermodynamic idealizations, we can determine ***ideal work*** for this fluid as it changes state (and perhaps moves into and out of the cylinder). Because of inevitable friction and other irreversibilities (§ 5.9), the actual delivered work is something less (if "done by") or more (if "done on") than it would be for a frictionless, perfect machine; moreover, the delivered work depends on where it is measured. The actual net work of system (1) for complete cycles of events is the ***indicated work*** W_I, so called because the measuring instrument used is called an *indicator*. This work W_I crosses the boundary (the piston face) into system (2), which consists entirely of mechanical elements, bounded by the heavy dashed line. At some boundary C on the connecting rod, the work flux is something less than W_I because of friction at the piston rings and cylinder walls, at the packing glands and pin at B, but the next work that we are interested in is the **brake work** W_B, so called because it was first measured by a brake that absorbed the shaft output of the engine. It was also called **shaft work**; $W_B < W_I$ for an engine manufacturing power, as a Diesel engine, because of losses already mentioned plus more frictional losses at the crank pin, shaft bearings, and with the fanning of air, say, by a flywheel, and so on. The **mechanical efficiency** of a reciprocating, power-delivering engine is taken as $\eta_m = W_B/W_I$ (in the sense of output/input).

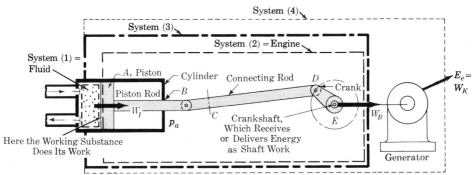

Fig. 4/13   *System Boundaries*. The shaft work (or any kind of work) is energy *in transition*; it exists by virtue of the movement of an element (force) against a resistance. Shaft work may be transformed into kinetic energy *of* and *stored in* the attached rotating parts to a limited extent. If this were a compressor, the net work flows would be reversed.

The shaft work, of course, is used to do innumerable things. If it drives a generator, schematically indicated in Fig. 4/13, and is directly connected (no intermediate gearing or belting), virtually the entire brake work is delivered to the generator. The generator delivers electrical energy E_e, usually called **combined work** W_K or **overall work** in this context. Because of electrical and mechanical losses, $W_K < W_B$ and the ratio $\eta_g = W_K/W_B$ is the **generator efficiency** where W_B is the *input* shaft work to the generator.

System (3), within the heavy dot-dash line, includes energy changes of the fluid and the shaft output. System (4), the heavy dotted boundary, takes in the generator. Other parts can be bounded as systems, as appropriate for a particular study. Notice that Fig. 4/13 is not an energy diagram, because it does not indicate all energy changes.

4.16   Example—Air Compressor

An air compressor takes in air at 15 psia and discharges it at 100 psia; $v_1 = 2\,\text{ft}^3/\text{lb}$ and $v_2 = 0.5\,\text{ft}^3/\text{lb}$. The increase of internal energy is 40 Btu/lb and the work is 70 Btu/lb; $\Delta P = 0$ and $\Delta K = 0$. If steady flow exists, how much heat is transferred?

Solution by Energy Equation.   The forms of energy to be considered are flow energy, internal energy, shaft work, and, of course, heat. For 1 lb, Δpv is

(a) $$\Delta E_f = E_{f2} - E_{f1} = \frac{(100)(144(0.5))}{778} - \frac{(15)(144)(2)}{778} = +3.7\,\text{Btu/lb}$$

Since work is done *on* the air to compress it, the work term W in equation (4-9) is negative; $W = -70$ Btu/lb. Since the internal energy increases, $\Delta u = u_2 - u_1 = 40$ Btu/lb. Using these various values in equation (4-9), solve for Q and find

(b) $Q = \Delta u + \Delta E_f + W = (+40) + (+3.7) + (-70) = -26.3$ Btu/lb

where the negative sign indicates that heat is rejected by the system. Some air compressors have water jackets for the purpose of cooling the air during compression; hence the negative sign is not unexpected.

Solution by Energy Diagram.   To promote thinking with respect to an energy diagram, we shall repeat the solution from this viewpoint. The procedure is to decide upon what the system is, say, the control volume that includes the compressor. Figure 4/14 sets the boundaries and shows what is known. Since the work is actually an input, it is so shown. The internal energy may be measured above the datum of the entering air, which gives $u_1 = 0$ and $u_2 = 40$ because $u_2 - u_1 = +40$. At this stage, it may not be known whether heat flows in or out. If we choose to show the heat Q passing in, as in Fig. 4/14, a plus sign for the answer says that the *direction assumed is correct*; and a negative sign says that the assumed *direction is wrong*. Hence, it does not matter if the decision is correct at this point. As in mechanics, if an unknown force is assumed in the wrong sense, its value, as solved for, is negative.) Now setting up the energy balance in accordance with Fig. 4/14, we get

$$\text{Energy in} = \text{energy out}$$

(c)

$$5.55 + 70 + 0 + Q = 9.25 + 40$$

from which $Q = -26.3$ Btu/lb, heat abstracted as before.

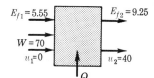

In this approach, if it is known, for example, that heat passes out of the system, the direction for Q could be reversed, in which case, the answer for Q would be positive. Rationalize this point of view with the one in the text.   **Fig. 4/14**

FRICTIONAL ENERGY   4.17

One frequently hears the term *frictional energy* or its equivalent, meaning work transformed into heat by frictional effects, but it is not a form of energy different from those already defined. To allow for energy losses due to friction, engineers generally use efficiency values accumulated through experience, a practice to be explained later as occasions arise.

If the frictional loss is caused by the rubbing together of two solid parts, with or without lubricant, what happens is that work is expended to excite the molecules in the vicinity of the surfaces being rubbed, increasing their energy, and raising the temperatures. This is to say that work is converted into internal energy of the rubbing bodies (of a bearing, say). As soon as the surfaces get hotter than the surroundings, heat is conducted through the parts, and energy is radiated and conducted away as heat. That portion of this energy carried off by oil in a bearing will also be given off as heat when the oil is cooled. Thus, the frictional energy usually appears as heat—heat dissipated into the atmosphere, which is such an immense reservoir that it does not increase in temperature, except locally. (The earth exists in virtually "steady state" with its surroundings.)

If the frictional loss is that caused by fluid friction, the substance has more internal energy at the end of the process that it would have had in the absence of fluid friction. If the substance also gets hotter than its environment, at least a part of the "frictional energy" eventually leaves as heat to the surroundings. As "frictional energy" becomes stored energy, it is *always at the expense of work.*

Suppose we let E_F represent the frictional losses at bearings, and so on; in the system of Fig. 4/15, W_s is the shaft work. If $Q = 0$, except as E_F is heat, the energy balance is   actual

(a) $$h_1 + K_1 = h_2 + K_2 + W_s + E_F$$

where the fluid work is $W = W_s + E_F$.

↑
Ideal

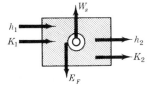

Fig. 4/15   *Energy Diagram with Friction.* If there had been no friction ($E_F = 0$), the fluid work W would have been all shaft work.

4.18   ENERGY EQUATION FOR FLOW OF INCOMPRESSIBLE FLUIDS

The reader may have derived the Bernoulli equation in a study of fluid mechanics, using a force analysis and the principles of mechanics. Heat and internal energy were not involved for a good historical reason. Daniel Bernoulli (1700–1782), a young contemporary of Newton (1642–1727), derived the equation around 1735, whereas the law of conservation of energy was not accepted until about 1850—some 100 years later. However, the energy law throws additional light on the Bernoulli equation. Consider the steady-flow energy equation in the following form [from equation (4-9B)]:

(a) $$Q = \Delta u + \Delta(pv) + \Delta K + \Delta P + W$$

as applied to Fig. 4/16, an open system of an incompressible fluid flowing steadily in a pipe. With no shafts, $W = 0$; if the flow is frictionless (viscosity is zero) and $Q = 0$, the change of molecular internal energy $\Delta u = 0$. Considering the flow to be effectively one dimensional, § 1.22, let $v = 1/\rho$ and find the energy relations between boundaries 1 and 2 as

(b) $$\Delta\left(\frac{p}{\rho}\right) + \Delta K + \Delta P = 0$$

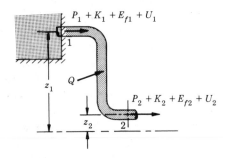

Fig. 4/16   *Energy Diagram of Pipe Line with Incompressible Fluid Crossing Boundaries.* If the fluid entering the control volume at boundary 1 is at the environmental temperature and if the flow is not frictionless, the natural flow of heat Q would be from the system (§ 4.17); normally for this situation, heat is quite negligible.

(c) $$\frac{p_2}{\rho_2} - \frac{p_1}{\rho_1} + \frac{v_2^2}{2\mathbf{k}} - \frac{v_1^2}{2\mathbf{k}} + \frac{g}{\mathbf{k}}z_2 - \frac{g}{\mathbf{k}}z_1 = 0 \text{ ft-lb/lb}$$

where, for length in feet, force in pounds, and mass in pounds, each term has the unit ft-lb/lb. Substituting for the density $\rho = \mathbf{k}\gamma/g$, equation (1-6), multiplying each term by $\mathbf{k}/g$, and letting $\gamma_1 = \gamma_2$, equation **(c)** becomes one form of Bernoulli's equation that applies to incompressible flow,

(4-24) $$\frac{p_2}{\gamma_2} - \frac{p_1}{\gamma_1} + \frac{v_2^2}{2g} - \frac{v_1^2}{2g} + z_2 - z_1 = 0 \text{ ft}$$

By inspection, this equation can be written in differential form, sometimes called the Euler equation,

(d) $$\frac{dp}{\gamma} + \frac{v\,dv}{g} + dz = 0$$

which, of course, can be integrated for variable specific weight γ, and which is obtained from the momentum principle in mechanics (Chapter 18). In English units, both the force-pound and the mass-pound are in each term (just as in the metric system there may be a force-kilogram and a mass-kilogram). If the pounds are canceled (not really possible since $\text{lb}_f \neq \text{lb}_m$), each term can be *and is* interpreted as a *head* in feet, a common name in hydraulics. The $v^2/(2g)$ term is called the **velocity head**; p/γ is called the **pressure head**. If the system of Fig. 4/16 should lead to a hydraulic turbine at 2, the events would be that some of the potential energy of the reservoir, approximately Δz, would be converted into kinetic energy and then some of the kinetic energy would be converted into work in the turbine. (This is not to say that the pressure head at level z_2 may not be much larger than at z_1.)

Reverting to the energy form **(c)**, using all the energy terms shown in Fig. 4/16, we have

(e) $$\frac{p_2}{\rho_2} - \frac{p_1}{\rho_1} + \frac{v_2^2}{2\mathbf{k}} - \frac{v_1^2}{2\mathbf{k}} + \frac{g}{\mathbf{k}}z_2 - \frac{g}{\mathbf{k}}z_1 = Q - (u_2 - u_1) \text{ ft-lb/lb}$$

The Bernoulli equation with friction added has a **friction head** or **head loss** term in it, representing the work of overcoming fluid friction including that on the inside surfaces, in place of the $Q - \Delta u$ term of equation **(e)**; that is, if E_F represents the friction head, $-E_F = Q - \Delta u$. The force analysis shows how to compute E_F (if enough data are at hand); thermodynamics tells what becomes of it.

Comparing equations **(d)** and **(e)**, we may point out that equation **(c)** does not preclude a flux of heat, but for incompressible frictionless flow, $Q = \Delta u$ and the right hand side of **(e)** becomes zero. An incompressible fluid undergoing a small change of temperature will have virtually constant density, $\rho_1 \approx \rho_2$.

INCOMPRESSIBLE FLOW THROUGH FAN OR PUMP 4.19

Liquids act as incompressible fluids in most engineering situations, but sometimes even gas processes are such that the gas density changes little. For example, a *fan*,

Figs. 4/17 and 4/18, is a device for moving a gas, with a small pressure change of the order of 0.25 in. H_2O to a few inches of water (written, for example, 5 in. wg, where wg stands for water gage). Here, too, the practice of the industry is to consider the various energy quantities as *heads*. The energy quantities of equation **(a)**, §4.18, in the ordinary application are: negligible heat, $Q = 0$; the change of internal energy, except as included in the friction head, is small compared to other energy changes, $\Delta u \approx 0$; the pressures are read at points between which $\Delta z \approx 0$ and $\Delta P \approx 0$; but, of course, shaft work enters the control volume to turn the fan (control surfaces on each end of the fan). The total work to drive the fan is called the *total head*, represented locally by H_t, and is usually a positive number. Then from equation **(a)**,

(4-25)
$$H_t = -W = \frac{p_2 - p_1}{\rho} + \frac{v_2^2 - v_1^2}{2\mathbf{k}} \text{ ft-lb/lb}$$

where usually $\rho_1 \approx \rho_2$. Since the fan is designed or selected to overcome the various resistances to flow in the duct, the work input must be great enough to include these losses, caled the **friction head** E_F (ft-lb/lb).

(4-26)
$$H_t = -W = \frac{p_2 - p_1}{\gamma} + \frac{v_2^2 - v_1^2}{2g} + E_F.$$

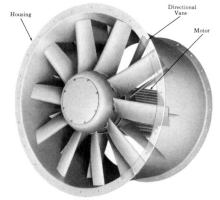

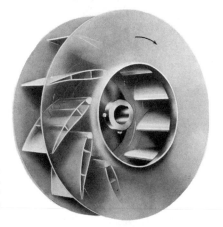

Fig. 4/17   *Axial-Flow Fan.* The motor in this particular unit is inside the housing, directly connected to the fan shaft. Behind the fan blades are seen some stationary vanes, which serve the purpose of "removing" the whirling motion given to the gas stream by the fan blades. (*Courtesy Buffalo Forge Co., Buffalo, N.Y.*).

Fig. 4/18   *Rotor for Centrifugal Fan.* A centrifugal fan operates on the same principle as centrifugal blowers. The impellers of this rotor are designed on airfoil principles (for greater efficiency and quietness). (*Courtesy The Green Fuel Economizer Co., Beacon, N.Y.*).

Reflection will suggest that (4-26) applies even better to a liquid passing through a pump when the control volume boundaries are as specified, on each side of the pump ($\Delta z \approx 0$). However, it should be instructive to return again to fundamentals, as for instance, equation (4-16), in which $-\int v\, dp$ gives $\Delta K + \Delta P + W$ for a reversible process. With $v = 1/\rho = $ constant, the integral is $-\Delta p/\rho$, and we have

$$(4\text{-}27) \qquad -W_p = \left(\frac{p_2}{\rho} - \frac{p_1}{\rho}\right) + \left(\frac{v_2^2}{2\mathbf{k}} - \frac{v_1^2}{2\mathbf{k}}\right) + \frac{g}{\mathbf{k}}(z_2 - z_1) \qquad \text{ft-lb/lb,}$$

[PRESSURE HEAD] [VELOCITY HEAD] [GRAVITATIONAL HEAD]

which is the same as would be obtained from an energy diagram as in Fig. 4/19. The velocity head ΔK may be negligible, so that when the pressures are measured at the intake and discharge boundaries, $-W_p \approx (p_2 - p_1)/\rho = v(p_2 - p_1)$. If the *pump efficiency* is η_p, the actual pump work would be $W_p' = W_p/\eta_p$. To get the power requirements from the foregoing equations, multiply by $\dot{m}$, the mass rate of flow.

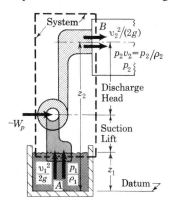

Energy Diagram for Adiabatic Pumping of Liquid. Since the pump work W_p is shown in its correct sense, it will be a positive number for the energy balance from this diagram. If the pressure gage on the discharge side is close to the pump, this would permit omitting the discharge gravity head z_2; at this location the gage measures the total static head, which is the pressure head plus the gravitational head, and includes the frictional head. **Fig. 4/19**

CLOSURE 4.20

It would be superb if everything about the thermodynamics could be said at once, with the reader comprehending fully. Not only is this impossible, but it is also often difficult to write about one idea without involving another. Thus, we have referred to several ideas that are to be explained in more detail later. A review of these first four chapters now and then, while studying the next few, may reveal some meanings missed during the first study.

The primary aim is to become familiar with energy and energy balances, and acquire a handy knowledge of certain concepts (definitions). Except for energy balances, the two most useful equations in this chapter are probably (4-14) and (4-15), often called $T\, ds$ equations:

$$T\, ds = du + p\, dv \qquad \text{and} \qquad T\, ds = dh - v\, dp$$

PROBLEMS

SI UNITS

4.1 Show that as a consequence of the first law, internal energy E exists and is a property.

4.2 A closed system executes a series of processes for which two of the three quantities, W, Q, and ΔE are given for each process. Find the value of the unknown quantity in each case.

(a) $W = -35$ kJ, $Q = ?$, $\Delta E = -35$ kJ.
(b) $W = +1.2$ MJ, $Q = +645$ kJ, $\Delta E = ?$
(c) $W = ?$, $Q = 5$ kJ, $\Delta E = 4.22$ kJ.

4.3 There are 2 kg of fluid mass in a closed container at rest on a given datum; local gravity is $g = 9.65$ m/sec^2. The container is now raised vertically 1000 m and the fluid mass given a swirling velocity of 50 m/sec. Initially the internal energy of the fluid was $E_1 = U_1 = 20$ kJ. Find E_2.

4.4 The internal energy of a certain closed system is given by $U = A + BpV$. Show that if it undergoes a reversible nonflow process with $Q = 0$, the relation between p and V is $pV^k = C$, where C is some constant and $k = (B + 1)/B$.

4.5 The internal energy E of a given system is a function of temperature $(t, °C)$ only and is given as $E = 30 + 0.3\,t$ IT cal. During the execution of a given process, the work done per degree temperature change is $dW/dt = 0.11$ kgm/°C. Find the heat Q as the temperature changes from 200°C to 400°C. Also, find E_1 and E_2.
Ans. $Q = 467$ J, $E_1 = 377$ J, $E_2 = 628$ J.

4.6 The work and the heat per degree change of temperature for a system executing a nonflow process are given by $dW/dt = 80$ W-sec/°C   and $dQ/dt = 15$ IT cal/°C, respectively. Determine the change of internal energy for the system as its temperature increases from 150°C to 250°C.
Ans. -172 J.

4.7 During a reversible process executed by a nonflow system, the pressure increases from 344.74 kPaa to 1378.96 kPaa in accordance with $pV = C$, and the internal energy increases 22,577 J; the initial volume is $V_1 = 85\ell$. Find the heat. *Ans.* $-18,045$ J.

4.8 Begin with the definition of enthalpy, $h = u + pv$, and show that for a reversible process, $-v\,dp = T\,ds - dh$.

4.9 Develop the expression $-v\,dp = dK + dP + dW$ stating all restrictions.

4.10 Show that for a steady flow process, $\int p\,dv = \Delta(pv) + \Delta K + \Delta P + W_{sf}$.

4.11 Apply each of the integrals, $\int p\,dv$ and $-\int v\,dp$, to a process wherein the pressure and the volume vary between two state points in accordance with $pv^n = c$ and find that one integral is n times the other. Here n is an exponent and c a constant.

4.12 An analysis of the flow of a compressible fluid through a frictionless nozzle

will result in the expression $dK = -v\,dp$. Often this analysis is based upon the nozzle as being adiabatic. Show that this expression obtains for a frictionless nozzle, independent of heat consideration. State all restrictions involved.

4.13 An internally reversible process occurs in a system during which $Q = -12$ kJ, $\Delta U = -79$ kJ, and $\Delta H = -111$ kJ. **(a)** Find the work if the system is nonflow. **(b)** Determine the shaft work and the change of flow energy if the system is steady-state, steady-flow with $\Delta K = 4$ kJ. **(c)** Using the conditions stated in **(b)**, evaluate $\int p\,dV$ and $-\int V\,dp$ in kJ.
Ans. **(a)** 67 kJ, **(b)** 95 kJ, -32 kJ, **(c)** 67 kJ, 99 kJ.

4.14 A 1460-kg automobile is brought to rest in a distance of 122 m from a speed of 113 km/hr. The kinetic energy of rotation of the wheel is negligible. **(a)** How much frictional energy is absorbed by the brakes? **(b)** If we imagine the stopping being brought about by a constant collinear force resisting the motion, how much is this force? Use only energy principles.
Ans. **(a)** 719 kJ, **(b)** 5893 N.

4.15 A closed system executes a reversible process wherein the pressure and volume vary in accordance with $pV^n = C$; $Q = 16.247$ kJ, $\Delta U = 47.475$ kJ. If $p_1 = 138$ kPaa, $V_1 = 141.6\,\ell$,   and $p_2 = 827.4$ kPaa, find n and V_2.

4.16 Steam at 15 bar, 300°C is contained in a 1-m^3 rigid sphere. A valve is opened to allow steam to escape slowly while heat is added to the steam in the sphere at a rate to maintain its temperature constant. What heat has been supplied when the pressure in the sphere reaches 1 bar? Hint: Plot a curve of enthalpy versus mass flow out and approximate the integral $\int h\,dm$.

ENGLISH UNITS

4.17 The volume of a compressible fluid system changes from $V_1 = 1$ ft^3 to $V_2 = 5$ ft^3 during an internally reversible process in which the pressure varies as $p = (100/V + 50)$ psia when V is in ft^3. **(a)** For the process find $-\int V\,dp$ and $\int p\,dV$. **(b)** If the process is steady flow with $\Delta K = 5$ Btu, $\Delta P = -2$ Btu and $\Delta H = 120$ Btu, find the work and heat. **(c)** If the process is nonflow, find W, Q and ΔU.

4.18 A gaseous substance whose properties are unknown, except as specified below, undergoes an internally reversible process during which

$$V = (-0.1p + 300) \text{ ft}^3 \text{ when } p \text{ is in psfa.}$$

(a) For this process, find $-\int V\,dp$ and $\int p\,dV$, both in Btu, if the pressure changes from 1000 psfa to 100 psfa. **(b)** Sketch the process (realistically) on the pV plane and compute the area "back" of the curve (no integration). **(c)** If the process is steady-state, steady flow with the kinetic energy increasing 25 Btu, $\Delta P = 0$, and the enthalpy decreasing 300 Btu, determine the work and the heat. **(d)** If the process is nonflow, what is the work and the change of internal energy?

Ans. **(a)** 283, 63.7; **(c)** 258, −17; **(d)** 63.7, −80.7 Btu.

4.19 Assume 5 lb/sec of fluid enter a steady-state, steady flow system with $p_1 =$ 100 psia, $\rho_1 = 0.2$ lb/ft^3, $v_1 = 100$ fps, $u_1 =$ 800 Btu/lb, and leave with $p_2 = 20$ psia, $\rho_2 = 0.05$ lb/ft^3, $v_2 = 500$ fps, and $u_2 =$ 780 Btu/lb. During passage through the open system, each pound rejects 10 Btu of heat. Find the work in horsepower. *Ans.* 168 hp.

4.20 **(a)** In a steam nozzle (Fig. 4/9, § 4.11 *Text*), no work is done and the heat is zero. Apply the steady-state, steady flow energy equation, and find the expression for the final velocity: (1) if the initial velocity is not negligible and (2) if the initial velocity is negligible. Show the energy diagram for 1 lb of steam. **(b)** There are supplied to a nozzle 2400 lb/hr of steam at an absolute pressure of 200 psia. At the entrance, $v_1 = 6000$ fpm, $v_1 = 2.29$ ft^3/lb, and $u_1 = 1113.7$ Btu/lb. At exit, $p_2 = 14.7$ psia, $v_2 = 26.77$ ft^3/lb, and $u_2 = 1077.5$ Btu/lb. Calculate the exit velocity.

4.21 A compressor draws in 500 cfm of air whose density is 0.079 lb/ft^3 and discharges it with a density of 0.304 lb/ft^3. At the suction, $p_1 = 15$ psia; at discharge, $p_2 = 80$ psia. The increase in the specific internal energy is 33.8 Btu/lb, and the heat from the air by cooling is 13 Btu/lb. Neglecting changes of potential and kinetic energy, determine the work done on the air in Btu/min and in hp. *Ans.* −2385 Btu/min.

4.22 A steady-state, steady flow thermodynamic system receives 100 lb/min of a fluid at 30 psia and 200°F and discharges it from a point 80 ft above the entrance section at 150 psia and 600°F. The fluid enters with a velocity of 7200 fpm and leaves with a velocity of 2400 fpm. During this process, there are supplied 25,000 Btu/hr of heat from an external source, and the increase in enthalpy is 2.0 Btu/lb. Determine the work done in horsepower. *Ans.* 5.48 hp.

4.23 An adiabatic vessel contains steam at 1025 psia, 750°F. A valve is opened slowly permitting an outflow of steam; it is closed quickly when the steam remaining in the vessel becomes saturated vapor. What fractional part of the original steam remains? Assume that this part underwent a reversible adiabatic expansion. *Ans.* 30.3%.

4.24 A 10-in. diameter, adiabatic cylinder C (see figure) is fitted with a frictionless, leakproof piston P that has an attached unstressed spring S whose scale is $k = 1000$ lb/in. The cylinder is connected by a closed valve A to line L in which steam flows at 500 psia, 700°F; the spring side of the cylinder is evacuated. Valve A is now opened slowly permitting steam to move the piston through distance x, thus compressing the spring. At this point valve A is closed, trapping steam in the cylinder at 150 psia. Find the temperature and the mass of steam in the cylinder.

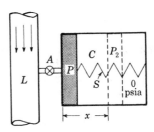

Problem 4.24

4.25 An evacuated adiabatic cylinder has a volume of 0.535 ft^3 and is connected by a closed valve to a line in which steam flows at 500 psia, 700°F. The valve is slowly opened, permitting an inflow of steam, and quickly closed when the cylinder pressure becomes 150 psia. Determine the temperature and mass of steam in the cylinder.

4.26 Steam at 1000 psia, 1200°F is in a rigid container which is fitted with a closed valve; the container is immersed in a liquid bath which is maintained at a constant temperature of 1200°F by an electrical input of I^2R. The valve on the container is slowly opened permitting steam to escape to the atmosphere until the container pressure becomes 100 psia. Find the electrical input per cubic foot of container volume. The only heat interaction is between the bath and the container of steam.

4.27 A 200-ft³ adiabatic and frictionless piston-cylinder arrangement C contains air at 200 psia, 440°F; valve B is closed. See figure. The piston P is capable of subjecting the air to a constant pressure of 200 psia and is resting on stops. Valve B is slowly opened permitting air to flow through the isentropic turbine T and in turn be exhausted to the atmosphere; piston P remains on the stops. Find the turbine work W after flow ceases.

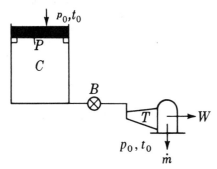

Problems 4.27, 4.28

4.28 The same as 4.27 except that the stops are withdrawn at the instant valve B is opened thus permitting the piston to maintain the air pressure in the cylinder C at 200 psia at all times. The piston comes to rest at the bottom of the cylinder when flow ceases.

4.29 The feasibility of fusion welding is being studied. This action utilizes the energy destroyed by friction. For this study, these data are available. Kinetic energy is stored in a 322-lb flywheel, where moment of inertia is $I = 5$ slug-ft², rotating at $\omega = 30$ rad/sec. Two 0.5-in. diameter steel rods are to be butt-welded (end-to-end) with one turning with the flywheel, the other stationary. The rods are pressed together with appropriate force and friction between the ends brings

the flywheel to rest. Will the molecular energy of the rod ends be increased sufficiently to cause the steel to fuse? Steel has a density of 0.286 lb/in.³, a melting point of 2600°F, and a specific heat of 0.12 Btu/lb-°R. The active length of each rod affected by friction is $\frac{1}{16}$ in. and no significant energy leaves the rod as heat. Find the temperature rise. Do you judge this idea feasible?

4.30 A portion of an inclined 8-in. (ID) pipe containing flowing water ($\rho = 1.90$ slugs/ft³) has two pressure gages attached 2000 ft apart. The upstream gage reads 21.1 psig and is 50 ft above the datum; the downstream gage is 38 ft above the datum. The flow rate is 1.55 cfs, and it is estimated that the friction head for the pipe is 1.1 ft of H_2O per 100 ft of pipe length; $Q = 0$. Using the energy equation, estimate the reading on the down-stream gage.

Ans. 16.86 psig.

4.31 Apply the steady flow equation (energy balance) to liquid water flowing without friction in a pipe and obtain the classical Bernoulli equation which states that the sum of these three head-differences is zero: pressure head, velocity head, and gravitational head.

4.32 A small blower handles 1530 cfm of air whose density is $\rho = 0.073$ lb/ft³. The static and velocity heads are 6.45 and 0.48 in. wg (at 60°F), respectively. Local gravity acceleration is $g = 31.95$ fps.² **(a)** Find the power input to the air from the blower. **(b)** If the initial velocity is negligible, find the final velocity.   *Ans.* **(a)** 1.66 hp, **(b)** 2800 fpm.

4.33 Air is removed from a large space and given a velocity of 63 fps by a fan. The air density is $\rho = 0.075$ lb/ft³ and the work done on the air is 0.0155 hp-min/lb air. Find the static head on the fan, in. wg (at 100°F).

4.34 A large forced-draft fan is handling air at 14.7 psia, 110°F under a total head of 10.5 in. wg (at 110°F). The power input to the fan is 300 hp and the fan is 75% efficient. Compute the volume of air handled each minute. Local gravity acceleration is $g = 31.85$ fps.²   *Ans.* 138,000 cfm.

4.35 Water flows steadily through the pipe system shown in the figure. These data apply: $D_1 = 1$ ft, $D_2 = D_3 = 0.5$ ft; $p_1 = 12$ psig, $v_1 = 10$ fps; propeller input energy

$W_p = 3.91$ hp. Assume that the internal energy u and the density $\rho = 62.4$ lb/ft^3 remain constant. Solve for **(a)** v_2, **(b)** p_2, and **(c)** p_3.

   Ans. **(a)** 40 fps, **(b)** 1.9 psig. **(c)** 3.8 psig.

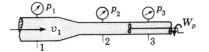

Problem 4.35

5

THE SECOND LAW, ENTROPY

5.1 INTRODUCTION

"The Moving Finger writes; and, having writ,/ Moves on: nor all your Piety nor Wit/ Shall lure it back to cancel half a Line/ Nor all your Tears wash out a Word of it" (Omar Khayyám, *Rubáiyát*).

Where the first law provided the basis for internal energy, the second law provides the basis for the concept of entropy. Continuing, the first law allows the possibility that heat and work are 100% interchangeable without constraints; the second law harnesses the first law by restricting the amount of heat that can be converted into work.

In this chapter we will examine numerous consequences of the second law. Actually its ramifications are so extensive that we shall only touch on a few aspects of immediate importance to an engineer. As the poet says, it is so ubiquitous in nature that mankind cannot avoid it. So we proceed from the sublime to the pragmatic.

We now need to present briefly a unique cycle, the Carnot cycle; see § 8.3 for a complete discussion. On the Ts-plane the cycle appears as a rectangle, four processes, two isotherms and two isentropics; see Fig. 5/1.

The Carnot cycle is unique in that it is reversible externally as well as being reversible internally since heat is accepted and rejected across vanishingly small temperature differences; see § 5.9.

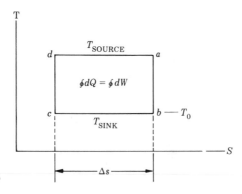

Fig. 5/1   *Carnot Cycle*

Because we are working with rectangular areas on the Ts-plane, we may deduce that

(a)
$$Q_a/T_{\text{sou}} = Q_r/T_{\text{sink}} = Q_r/T_0 = \Delta S$$

and

(b)
$$dQ_a/T_a = dQ_r/T_0 = ds$$

where $T_{\text{sink}} = T_0$.

Other observations that may be made at this time concerning the Carnot engine (which operates on this cycle) are that it is reversible, is most efficient, and is often used as the standard for all engines (and cycles). Further, we also note that the area $abcd\,(\oint dQ)$ represents the maximum work $(\oint dW)$ produced under these constraints—we call this maximum work availability (§ 5.7).

STATEMENTS OF THE SECOND LAW OF THERMODYNAMICS 5.2

Over the years, many statements of the second law have been made, and actually the beginner gets considerably more information from a number of statements than from any one. The second law, like the first, is an outcome of experience, and its logical discovery and refinement began with the work of Carnot. Some significant statements are:

A. Clausius (§ 5.9): *It is impossible for a self-acting machine unaided by an external agency to move heat from one body to another at a higher temperature.*

B. Kelvin-Planck[1.26]: *It is impossible to construct a heat engine which, while operating in a cycle produces no effects except to do work and exchange heat with a single reservoir.*

C. *All spontaneous processes result in a more probable state* (§ 5.27). The natural trend of events is from a less probable to a more probable state.

D. *The entropy of an isolated system never decreases.* If all processes that occur within the system are reversible, its entropy remains the same; otherwise, its entropy increases.

E. *No actual or ideal heat engine operating in cycles can convert into work all the heat supplied to the working substance; it must reject some heat.* This statement highlights the important concept of the ***degradation of energy*** as it passes from a higher temperature state to one of lower temperature. The idea of degradation in a process is that all actual processes result in the degradation of energy that might have been converted into work into energy that cannot be so converted.

F. Caratheodory: *In the vicinity of any particular state 2 of a system, there exist neighboring states 1 that are inaccessible via an adiabatic change from state 2.* For example, in the Joule-Thomson porous plug experiment, Fig. 6/5, § 6.11, the state may change *adiabatically* from 1 to 2, but not from 2 to 1, no matter how small the pressure drop becomes. The significance of this statement is made evident by a mathematical theorem, invented and proved by Caratheodory (1909). He set out to define the second law and entropy without reference to Carnot or any heat engine—and succeeded.[1.5]

From certain of these statements, the truth of others may be proved with mathematics and logic[1.2]; notice how easy it is to see the equivalence of statements A and B in Fig. 8/16(b) and (c).

5.3   CLAUSIUS INEQUALITY

This concept, basically another statement of the second law, may be evolved in a number of ways. Consider any kind of system B, Fig. 5/2, that completes one or more whole cycles. The heat dQ_B and work dW_B may be either positive or negative; they are shown in the conventional positive sense for system B. For a cycle, the first law applied to system B gives

(a)
$$\oint dQ_B = \oint dW_B$$

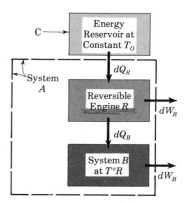

Fig. 5/2   *Clausius Inequality.* For heat to flow between reservoir C and system B, a temperature difference is necessary. An imaginary system, such as B, that receives heat from an external source, such as C, while B's temperature varies or is different from T_0, can be made externally reversible with the imaginary use of a reversible engine R, which receives the heat reversibly at temperature T_0 and delivers it at temperature T of system B.

To provide a reversible exchange of heat with the reservoir C, insert reversible engine R and let it make an integral number of cycles to one of B; R receives heat dQ_R and does work dW_R. Equation (a, § 5.1) applied to the engine R, with dQ_B positive as shown, gives

(b)
$$\frac{dQ_R}{T_0} = \frac{dQ_B}{T} \quad \text{or} \quad dQ_R = T_0 \frac{dQ_B}{T}$$

The first law applied to R (any term either positive or negative) is

(c)
$$dQ_R = dW_R + dQ_B \quad \text{or} \quad dW_R = dQ_R - dQ_B$$

Substituting dQ_R from equation **(b)** into equation **(c)**, we get

(d)
$$dW_R = T_0 \frac{dQ_B}{T} - dQ_B$$

Let systems R and B together be system A. The total work, system A, with $\oint dQ_B = \oint dW_B$ and all cyclic sums for the same duration, is then

(e)
$$\oint dW_A = \oint (dW_B + dW_R) = \oint \left(dQ_B + T_0 \frac{dQ_B}{T} - dQ_B \right) = T_0 \oint \frac{dQ_B}{T}$$

Since this system A exchanges heat with a single reservoir, its work must be zero or less, § 10.13 and the second law statement **B**. Therefore,

(f)
$$\oint dW_A = T_0 \oint \frac{dQ_B}{T} \gtreqless 0$$

Since T_0 is always positive, it follows that $\oint dQ_B/T \gtreqless 0$. Inasmuch as B is any kind of a system, we may drop the subscript B, let dQ be the heat for any system, and get the famous **Clausius Inequality**,

(5-1)
$$\oint \frac{dQ}{T} \gtreqless 0$$

ENTROPY FROM THE SECOND LAW 5.4

If the processes of system B are all reversible, $\oint dW_A = 0$ in equation **(f)** above and

(a)
$$\oint \frac{dQ}{T}\bigg]_{rev} = 0$$

which conforms to a basic requirement for a property. Clausius therefore concluded that dQ/T was a change of property, one that we have already defined as entropy, $ds = dQ/T]_{rev}$, equation (3-2), often called the **second law equation**.

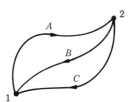

Closed System Executing a Cyclic Process **Fig. 5/3**

Now let us approach the concept of entropy from a different viewpoint using equation (5-1). Let a closed system change by some reversible process A from state 1 to state 2 and by some some reversible process B from state 2 to state 1; see Fig. 5/3. Then according to the Clausius inequality,

(a)
$$\oint_{1A2B1} dQ/T = \int_{1A2} dQ/T + \int_{2B1} dQ/T = 0$$

Now let the reversible process C be a different means by which the system may change from state 2 to state 1. For the cyclic change $1A2C1$ we may also write

(b)
$$\oint_{1A2C1} dQ/T = \int_{1A2} dQ/T + \int_{2C1} dQ/T = 0$$

Subtracting equation **(b)** from equation **(a)** results in

$$\int_{2B1} dQ/T - \int_{2C1} dQ/T = 0$$

or

(c)
$$\int_{2B1} dQ/T = \int_{2C1} dQ/T = \int_{21(\text{rev})} dQ/T$$

We conclude that the integral $\int_{\text{rev}} dQ/T$ has the characteristics of a property, (that is, it is independent of path and dependent only upon state point locations), and call it entropy. Thus

(d)
$$ds = (dQ/T)_{\text{rev}}$$

or

$$\Delta S = \int_{\text{rev}} dQ/T$$

5.5   ENTROPY PRODUCTION

If a cycle has some internal irreversibility in it, $\oint dS = 0$ and equation (5-1) becomes $\oint dQ/T < 0$, which can then be made an equality by the addition of a term,

(5-2)
$$\oint dS = \oint \frac{dQ}{T} + \Delta S_p = 0$$
[WITH INTERNAL IRREVERSIBILITY]

where ΔS_p, always positive, is the increase in entropy assignable to irreversibilities *internal* to the cyclic system, which we may call the *internal entropy production* or *entropy growth*. Equation (5-2), the cyclic change of entropy, is equal to zero because entropy is a property.

Imagine any three-process cycle for explanation purposes, processes *ab*, *bc*, and *ca*. Suppose just one of these processes *ca* is irreversible. The cyclic integral of dQ/T is simply the sum of the integrals of all of the processes of the cycle; for the cycle *abc*, equation (5-2) becomes

(a)
$$\oint \frac{dQ}{T} + \Delta S_p = \int_a^b \frac{dQ}{T} + \int_b^c \frac{dQ}{T} + \int_c^a \frac{dQ}{T} + \Delta S_p = 0$$

Entropy being a point function,

(b)
$$\Delta S_{ab} + \Delta S_{bc} + \Delta S_{ca} = 0$$

Since, for the internally reversible processes,

(c)
$$\Delta S_{ab} = \int_a^b \frac{dQ}{T} \quad \text{and} \quad \Delta S_{bc} = \int_b^c \frac{dQ}{T}$$

it follows that for the single irreversible process *c* to *a*,

(5-3A)
$$\Delta S' = \int \frac{dQ}{T} + \Delta S_p \quad \text{or} \quad dS' > \frac{dQ}{T}$$

where the particular limits c to a have been dropped because the result applies to any process and dS' is the actual entropy change for an irreversible process. The conclusion is that *the change of entropy during an irreversible process is greater than* $\int dQ/T$ and the amount by which it is greater is the internal growth ΔS_p. The idea of the increase of entropy with irreversibility fits with statement **D** of the second law. Summarizing, we have

$$(5\text{-}3\text{B}) \qquad\qquad dS \geqq \frac{dQ}{T} \quad \text{and} \quad \Delta S \geqq \int \frac{dQ}{T}$$

For an *isolated* system, § 1.3, for which $dQ = 0$, we have, from equation (5-3), $ds \geqq 0$. For this generalization to apply to processes as we have studied them, the isolated system may be called the **universe**, which includes not only *the* system (sys) but also all portions of the affected surroundings (sur), say, a source (sou) and sink. See statement D, § 5.2. Then in accordance with the second law, the change of entropy of the universe (isolated system) is

$$(\text{d}) \qquad\qquad \Delta S = \Delta S_{\text{sys}} + \Delta S_{\text{sur}} = \Delta S_{\text{sys}} + \Delta S_{\text{sou}} + \Delta S_{\text{sink}} \geqq 0$$

where the equality applies only when the whole operation is reversible, externally and internally with respect to the system. With irreversibility, $\Delta S > 0$ and its value is the increase in entropy of this universe;

$$(5\text{-}4) \qquad\qquad \Delta S_p = \Delta s_{\text{sys}} + \Delta S_{\text{sou}} + \Delta S_{\text{sink}}$$

the *entropy production*, often expressed as a rate (as Btu/°R-min, kw/°R), and each term must have its proper algebraic sign. Equation (5-4) applies either to processes or to cycles. When the term *entropy production* is used unmodified by an adjective, the entropy production of the universe is intended.

Example—Entropy Production Within a System   **5.6**

Let 1 lb of air undergo a cycle $abc'd$, Fig. 5/4, where it is heated isobarically ab at $p_{ab} = 160$ psia from 140 to 540°F, expands adiabatically bc' to 40°F and 10 psia, rejects 73.15 Btu/lb of heat isothermally $c'd$ to the sink, and completes the cycle isentropically da. Compute (a) $\oint dQ/T$ and the *internal* entropy production, (b) the entropy production of process bc', and (c) the thermal efficiency. (d) For the cycle $abcd$ which is internally reversible, calculate the thermal efficiency. (e) What would be the thermal efficiency of a Carnot engine for the same temperature limits?

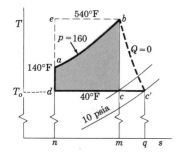

The shaded area *abcd* represents the work, $\oint dQ$, for all   **Fig. 5/4**
internally reversible processes. When one or more processes
of a cycle are irreversible, as with bc', the enclosed area does
not represent $\oint dQ$ but $\oint dQ = \oint dW$. External irreversibilities
are not considered in this example.

Solution.   (a) For process ab, with Item B 2 data, we have

(a) $\Delta s_{ab} = \int \left[\dfrac{dQ}{T}\right]_{rev} = \int_{a}^{b} \dfrac{c_p \, dT}{T} = \Delta\phi = 0.75042 - 0.62607 = 0.12435 \text{ Btu/lb-}^\circ\text{R}$

For bc', $Q = 0$. For $c'd$,

(b) $\Delta s_{c'd} = \int \left[\dfrac{dQ}{T}\right]_{rev} = \dfrac{Q_{c'd}}{T_o} = \dfrac{-73.15}{500} = -0.1463 \text{ Btu/lb-}^\circ\text{R}$

For da, $Q = 0$. Therefore

(c) $\displaystyle\oint \dfrac{dQ}{T} = 0.12435 - 0.1463 = -0.02195 \text{ Btu/lb-}^\circ\text{R}$

By equation (5-2), $\Delta S_p = +0.02195 \text{ Btu/lb-}^\circ\text{R}$, which, in this example, is not the entropy production of the *universe* because the entropy change of the surroundings is not included.
   (b) For process bc', $R \ln (160/10) = 0.19006$, and

$$\Delta s_{bc'} = \Delta\phi_{bc'} - R \ln \dfrac{p_{c'}}{p_b}$$

(d)
$$= 0.58233 - 0.75042 + R \ln \dfrac{160}{10} = 0.02197 \text{ Btu/lb-}^\circ\text{R}$$

Since bc' is adiabatic, $\int dQ/T = 0$ and the change of entropy is the entropy production, by equation (5-3). This answer is in agreement with that of equation **(c)**, as it should be.
   (c) The work of the cycle is $\oint dQ = \oint dW$.

(e) $Q_A = \int c_p \, dT = \Delta h = 240.98 - 143.47 = 97.51 \text{ Btu/lb}$

The only other heat is the given $Q_R = -73.15$. Therefore

(f) $e = \dfrac{97.51 - 73.15}{97.51} = \dfrac{24.36}{97.51} = 25\%$

   (d) The corresponding reversible process from state b is an isentropic bc. In this case $\Delta s_{cd} = -s_{ab} = -0.12435$ from equation **(a)**. Then the heat rejected at $T_0 = 500^\circ$R is

(g) $Q_R = T_0 \, \Delta s = (500)(-0.12435) = -62.175 \text{ Btu/lb}$

(h) $e = \dfrac{97.51 - 62.175}{97.51} = \dfrac{35.335}{97.51} = 36.2\%$

   (e) The Carnot cycle operating through the same temperature range is $ebcd$; its thermal efficiency is

(i) $e = \dfrac{T_1 - T_2}{T_1} = \dfrac{1000 - 500}{1000} = 50\%$

5.7   AVAILABILITY, CLOSED SYSTEM[1,2]

   The availability of a closed system in a particular state is the maximum work that the system could conceivably deliver to something other than the surroundings as its state changes to the dead state (a stable thermodynamic equilibrium with the environment), *exchanging heat only with the environment.* At this point it is easy to accept the idea that the maximum possible work would be obtained if the system should proceed reversibly, externally and internally, from its given state to the dead state, because if some other process resulted in more work, the second law would be violated.

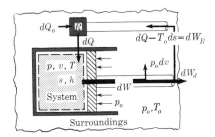

Fig. 5/5

Availability—Closed System. The energy flows are shown in the conventional positive sense, *dQ* added, *dW* delivered, but these are algebraic quantities and may actually be in either sense. Similarly, the work of the surroundings may be in either sense.

Consider the closed system of Fig. 5/5 in the state defined by p, v, T, or other properties, surrounded by an environment at p_0, T_0. Since the temperature of the system will be different from that of the surroundings or will probably change during a process, there will occur a temperature difference with a resulting heat interaction. If heat flows simply from the higher to the lower temperature, there is irreversibility. This heat interaction is made reversible via a reversible engine R. The reversible engine R of Fig. 5/5 completes a very large number of cycles for a movement of the boundary, one for each dT change in the system's temperature, so that each small quantity of heat dQ to the system is transferred reversibly. The work to drive the reversible engine, by the first law, is

(a)
$$dW_R = dQ - dQ_0 = dQ - \frac{T_0}{T} dQ = dQ - T_0 \, ds_{sys}$$

where the ratio $dQ_0/T_0 = dQ/T$, equation (a, § 5.1), and the definition $ds = dQ/T$ have been used (T is the temperature of the system while it receives heat dQ).

Of the total work dW delivered by the system, the amount dW_R is used to drive R and the amount $p_0 \, dv$ is used to displace the surroundings. Hence, the *maximum work* that can be delivered during the *process* to something other than the surroundings is

$$W_d = \int [dW - (dQ - T_0 \, ds) - p_0 \, dv]$$

(b)
$$= -\int (dQ - dW - T_0 \, ds + p_0 \, dv)$$

For a closed system, the energy equation (4-7) is

(4-7) $dQ = dE_s + dW$ or $dQ - dW = dE_s$

Use this value of dE_s in **(b)** and integrate from the given state to the dead state (§ 5.25) to obtain the maximum work $W_{d\max}$;

$$W_{d\max} = -\int^0 (dE_s + p_0 \, dv - T_0 \, ds) = -[E_{s0} - E_s + p_0(v_0 - v) - T_0(s_0 - s)]$$

where the state of the surroundings (p_0, T_0) is taken as constant; p_0, T_0, and v_0 are properties of the system after the system has reached the dead state. Let

$$(5\text{-}5) \qquad \mathscr{A}_n = E_s + p_0 v - T_0 s \qquad \text{or} \qquad \mathscr{A}_n = u + p_0 v - T_0 s$$

$$[\text{MOLECULAR STORED}$$
$$\text{ENERGY ONLY}]$$

which is seen to be defined by properties only and is therefore a property: call this property the nonflow **availability function**, after Keenan.[1.2] Then the maximum possible delivered work of a closed-system *process* is

$$(5\text{-}6\text{A}) \qquad\qquad\qquad W_{d\max} = -\Delta\mathscr{A}_{n/0} \qquad\qquad\qquad [\text{AVAILABILITY}]$$

where $-\Delta\mathscr{A}_{n/0}$, called the nonflow **availability**, represents the negative of the change of $\mathscr{A}_n$ from the given state to the dead state. If a pure substance with stored energy in the molecular form is involved,

$$(5\text{-}6\text{B}) \qquad W_{d\max} = -\Delta\mathscr{A}_{n/0} = (u + p_0 v - T_0 s) - (u_0 + p_0 v_0 - T_0 s_0)$$

By similar logic, if a closed system undergoes a change from any state 1 to any state 2, the maximum work that can be delivered with only reversible heat interaction with the surroundings is

$$(5\text{-}7) \qquad\qquad W_{d\max} = -\Delta\mathscr{A}_n = -(\mathscr{A}_{n2} - \mathscr{A}_{n1}) = \mathscr{A}_{n1} - \mathscr{A}_{n2}$$

where the availability function $\mathscr{A}_n$ is as defined in equation (5-5). If $-\Delta\mathscr{A}_n$ comes out negative, this is the *minimum* conceivable input work to bring about a state change from 1 to 2.

Example

A closed system consists of a kilogram of water vapor at 30 bar, 450°C in an environment where $p_0 = 1$ bar, $t_0 = 35°C$. What is its availability?

Solution.   From Item B 16 (SI Chart), the system properties are $h = 3342$ kJ/kg, $v = 0.108$ m^3/kg, $s = 7.08$ kJ/kg K. At the dead state, $h_0 = 147$ kJ/kg, $v_0 = 0.00101$ m^3/kg, $s_0 = 0.505$ kJ/kg K.

Using equation (5-6B),

$$\mathscr{A}_{n/0} = (u + p_0 v - T_0 s) - (u_0 + p_0 v_0 - T_0 s_0)$$

$$= [h + v(p_0 - p) - T_0 s] - [h_0 - T_0 s_0]$$

$$= (h - h_0) + v(p_0 - p) - T_0(s - s_0)$$

$$= (3342 - 147) + (0.108)(1 - 30)(10^5) - (308)(7.18 - 0.505)$$

$$= 3195 - 313.2 - 2025.1$$

$$= 857 \text{ kJ/kg}$$

a measure of the maximum useful work that can be performed by this system and environment combination.

AVAILABILITY, STEADY-FLOW SYSTEM 5.8

*The **availability** of a steady-flow system is the maximum work that can conceivably be delivered by, say, a unit mass of the system to something other than the surroundings as this unit mass changes from its state at the entrance to the control volume to the dead state at the exit boundary, meanwhile exchanging heat only with the surroundings.* The concepts are much the same as before; a reversible engine R is used to effect a reversible heat interaction with the surroundings. The work of this engine R, Fig. 5/6, is given by exactly the same expression as before, for the same reasons; it is $dW_R = dQ - T_0\,ds$, where dQ and ds apply to the system. Whatever work is involved on or by the surroundings is cared for, as we have learned, in the energy balance, equation (4-9);

$$(4\text{-}9) \qquad\qquad dQ - dW = dh + dK + dP$$

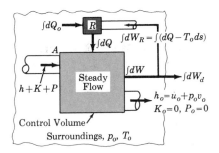

Availability—Steady Flow. Energies entering and departing are shown for unit mass; $\int dW$ and $\int dQ$ are work and heat for the time interval it takes for unit mass to pass through the control volume.   **Fig. 5/6**

Write the equation for the delivered work, use the value of $dQ - dW$ in (4-9), and get

$$dW_d = dW - dW_R = dW - (dQ - T_0\,ds)$$

(a)

$$= -(dh + dK + dP - T_0\,ds)$$

Integration from the given (entrance) state to the dead state gives the maximum possible work, called the steady-flow **availability**, $-\Delta\mathscr{A}_{f/0}$;

$$W_{d\,\text{max}} = -\Delta\mathscr{A}_{f/0} = -\int^{0} (dh + dK + dP - T_0\,ds)$$

$$(5\text{-}8)$$

$$= -[h_0 - T_0 s_0 - (h + K + P - T_0 s)]$$

where $K_0 = 0$ and $P_0 = 0$ for the dead state. Let $\mathscr{A}_f = h + K + P - T_0 s$, another property in a particular environment. Then the maximum possible delivered steady-flow work during a change between random states 1 and 2 is

$$(5\text{-}9) \qquad\qquad W_{d\,\text{max}} = -\Delta\mathscr{A}_f = -(\mathscr{A}_{f2} - \mathscr{A}_{f1}) = \mathscr{A}_{f1} - \mathscr{A}_{f2}$$

Notice that the Carnot cycle provides a measure of the maximum possible work to be obtained from a heat engine cycle operating between particular temperatures. The availability functions permit a computation of the maximum possible work for a *process* between any two states with the exchange of heat only with the surroundings. Let the symbol $\mathscr{A}$ without subscript represent the availability function for either the closed or steady-flow system.

If kinetic and potential energies are negligible. $\mathscr{A}_f$ reduces to $\mathscr{A}_f = b = h - T_0 s$, also called the *availability function* or *Darrieus function*. The degree of perfection of a particular steady-flow process between any two states of a pure substance may reasonably be indicated by the ***effectiveness***,[1.2] defined as follows with $\Delta K = 0$ and $\Delta P = 0$:

$$(5\text{-}10) \qquad \varepsilon = \frac{W'}{b_1 - b_{2'}} \quad \text{and} \quad \varepsilon = \frac{b_1 - b_{2'}}{W'} \qquad [\Delta K = 0]$$

$$[\text{WORK OUT}] \qquad\qquad [\text{WORK IN}]$$

where W' is the actual fluid work and $b_1 - b_{2'}$ is the change of $h - T_0 s$ between the actual end states of the process.

Example

Consider the same water vapor system and environment combination of the previous example (§ 5.7) now moving in an open steady-flow system with a velocity of 150 m/s. What is its availability?

Solution.   Applying the data from the previous example to equation (5-8) we get

$$\mathscr{A}_{f/0} = (h + K + P - T_0 s) - (h_0 - T_0 s_0)$$

$$= (h - h_0) - T_0(s - s_0) + K \qquad [P = 0]$$

$$= (3342 - 147) - 308(7.08 - 0.505) + \frac{150^2}{2(1000)}$$

$$= 3195 - 2025.1 + 11.3$$

$$= 1181 \text{ kJ/kg}$$

a measure of the maximum useful work that can be performed by this system and environment combination.

5.9   REVERSIBILITY

One of the many facets provided by the second law is the basis for the concept of reversibility. Although our knowledge* of thermodynamics may appear meager at this point, we can get some help on reversibility from a few examples and a definition. The importance of the idea of reversibility is that a process or cycle that is reversible in every respect is one that is as perfect as the mind can conceive. *Such idealizations provide both a standard of perfection and a means of mathematical simplifications.* Hence, whenever we can define the reversible process, we have a chance to find the irreversibilities and perhaps correct them to some extent. All actual processes are irreversible, as mankind has long realized.†

---

* Knowledge is helpful to straight thinking. However, it has been said by Ambrose Bierce, "Learning: The kind of ignorance distinguishing the studious."
† No use crying over spilled milk. All the king's horses and all the king's men could not put Humpty Dumpty together again.

If a glass marble is dropped in a vacuum (no air friction) onto a firm glass surface, it will bounce almost back to its original height; the process is nearly reversible; all the loss of potential energy has been converted into kinetic energy at the instant before impact; the coefficient of restitution being nearly unity, the speed after contact is only a little smaller than the speed before contact; at the instant that the marble is at rest, nearly all the energy from the work of gravity has been stored in the marble and surface by elastic deformation of the parts. With no irreversibility (loss), bouncing would go on forever and the sum of the kinetic and gravitational potential energies of the marble would be constant, the familiar *conservative system*. If the marble is lead, the process is relatively irreversible; the bounce is not nearly so high as the starting point; much of the work of gravity lost during the contact period has gone to increase the molecular activity (temperature) of the impacting bodies and will be eventually transferred to the surroundings as heat. The bouncing will soon cease and *all* the marble's original potential energy will have been dissipated to the surroundings; the marble will have arrived at the *dead state*.

We can say loosely that any process that will not occur in the opposite sense is irreversible, and you readily recognize many such irreversibilities. If a gas is heated by electricity, the heat comes from a hot wire ($\mathscr{I}^2\mathscr{R}$), and we feel confident from elementary experience that the hot gas will not return the electricity to the electrical system (§ 7.4) upon cooling. If the brakes on an automobile bring it to rest we feel confident that the energy absorbed by the brakes will not propel the car when the brakes are released. When two automobiles collide, we feel confident that they will not of their own accord return to their pristine shape; nature may repair the injured people, but not by the reverse of the process by which the injury occurred. We may generalize these examples by saying that irreversibility is present whenever: (1) heat flows because of a temperature difference, (2) an inelastic impact occurs (all are), (3) frictional work (work done by a force of friction) is involved. Case (2) is a subhead of (3), but the the idea is important enough to justify separate listing.

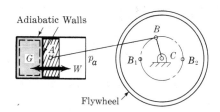

Reversible System. No friction of any kind. By defini-   **Fig. 5/7**
tion, adiabatic means that the heat is zero ($Q = 0$),
and adiabatic walls are ones that are *perfectly* insulated (there is no such thing).

Consider one more example for its informational value: A nonflow system of an expansible fluid G in a cylinder, Fig. 5/7, is surrounded by ***adiabatic walls***. Imagine that the piston is at an infinitesimal distance past headend, dead-center position (crank at B_1) and that the fluid G is at a pressure above atmospheric. If the engine is free to move, G expands and does work at the expense of its internal energy [$Q = \Delta u + W = 0$, equation (4-7)]. If the start is from rest, this work can be used to speed up the flywheel, storing kinetic energy in it. At the end of the stroke of the piston (crank at B_2), the kinetic energy in the flywheel is just enough to compress G back to its original state (p_1, v_1, T_1), where the engine again comes to rest. As already mentioned, the gas must expand slowly so that there is internal equilibrium at all times. Actually, of course, we cannot wait indefinitely for the piston to complete a round trip; and we cannot eliminate friction.

Finally, the reversible process is a *controlled* process, as contrasted to an uncontrolled expansion, as in opening a valve and letting the fluid flow through without doing work. On the other hand, a *guided expansion* in a nozzle may closely approach a reversible process in which the kinetic energy of a jet is reconverted in an ideal diffuser (no fluid friction, turbulence) to bring the substance back to its original state.

(a) *External irreversibility* is some irreversibility external to the system. When the system is the *working substance G*, Fig. 5/7, for example, friction at the bearings A, B, C, and so on, are external irreversibilities; all absorb some of the W output of the gas system. Instinctively you know that you cannot use the "frictional energy" to recompress the fluid to its original state (really an aspect of the second law of thermodynamics).

Another external irreversibility is due to the flow of heat through the containing walls (system = gas), inasmuch as an adiabatic wall is a fiction of the mind. Since the temperature of the substance probably changes during the expansion, there will be a temperature difference ΔT with the surroundings. If there is a temperature difference, heat is necessarily transferred. As Clausius* said in 1850, "It is impossible for a self acting machine unaided by external agency to move heat from one body to another at a higher temperature" (another aspect of the second law). Actually, then, whenever net heat flows, there is irreversibility, but the transfer approaches reversibility as ΔT approaches zero.

(b) *Internal irreversibility* is any irreversibility within the system. These may include the friction losses at A, B, C, etc., of Fig. 5/7 if the boundaries are drawn to include these parts. If the gas G is the system, internal irreversibilities are associated with fluid friction (turbulence, friction of fluid on the walls of a pipe or other container, rubbing of one layer of fluid on another), pressure, and temperature gradients within the fluid. The diffusion (mixing) of two gases or liquids is an irreversible process, as is combustion. After a gas expands irreversibly, some of the energy that might have been converted into work remains stored in the substance; in practice, it is eventually discharged to the sink.

If the systems are gas systems, *the external irreversibility of interest will be a transfer of heat through a temperature drop, and internal irreversibility will be principally fluid friction.* To summarize: *If, after a process is completed, it can be made to retrace in the reverse order the various states of the original process, if all energy quantities to and from the surroundings can be returned to their original states (work returned as work, heat returned as heat, etc.), then the **process is externally and internally reversible**.* If a reversible process occurs and is then reversed to the original state, there is no evidence that it ever occurred.

5.10   IRREVERSIBILITY

Irreversibility, in addition to its ordinary meaning that we have been using, has more than one defined technical meaning in the literature. We shall use the one

---

* Rudolf Julius Emmanuel Clausius (1822–1888), born in northern Germany, a professor of physics, was a genius in mathematical investigations of natural phenomena. He elaborated and restated the work of Carnot, deducing the principle of the second law of thermodynamics. His mathematical work in optics, electricity, and electrolysis was significant. James Clerk Maxwell credits him with being the founder of the kinetic theory of gases, on the basis of which Clausius made many important calculations, one of which was the mean free path of a molecule. He is the author of an exhaustive treatise on the steam engine, wherein he emphasized the new concept *entropy*. As put by M. Flanders, *At the Drop of Another Hat*, "Heat can't flow from a cooler to a hotter,/You can try it if you like but you'd far better notter." (Here taken from Montgomery, reference 5.1.)

after Keenan.[5.3] *The irreversibility I is the maximum possible work $W_{d\max} = -\Delta\mathscr{A}$ that a system can do in passing from one state to another* minus *the delivered work W_d that can be used to do something other than displace the environment.* This delivered work may be actual fluid work, for example, or some other energy that is 100% available. Also inasmuch as the definition of I is algebraic, the reversible work may be the numerical *minimum* conceivable *input work.*

IRREVERSIBILITY OF CLOSED SYSTEM 5.11

Consider the general case of heat interactions between a source, the system, and the sink, Fig. 5/8. At the outset, observe that total quantities, not quantities per unit mass, are involved. Let the system be closed and let E equal the stored energy. By § 5.7 and equation (5-7)

(a) $$\Delta\mathscr{A}_{\text{sys}} = \Delta E_{\text{sys}} + p_0\,\Delta V_{\text{sys}} - T_0\,\Delta S_{\text{sys}}$$

By the first law, equation (4-7), the *net heat for the system is*

(b) $$Q_{\text{sys}} = \Delta E_{\text{sys}} + p_0\,\Delta V_{\text{sys}} + W_{d\,\text{sys}}$$

where the total system work is $p_0\,\Delta v_{\text{sys}}$ against the surroundings plus delivered $W_{d\,\text{sys}}$. In an energy balance for the system, equation **(b)**, the heat Q_{sou} from the source minus the heat $|Q_{\text{sink}}|$ to the sink is the net heat Q_{sys} for the system; but in the final equation, Q_{sou} is to be with respect to the source, negative in this case, and Q_{sink} is to be with respect to the sink, which receives it, and therefore positive. On this basis, the net heat to the system is $Q_{\text{sys}} = -Q_{\text{sou}} - Q_{\text{sink}}$. Using this value of Q_{sys} in equation **(b)**, we find

(c) $$\Delta E_{\text{sys}} + P_0\,\Delta V_{\text{sys}} = -Q_{\text{sou}} - Q_{\text{sink}} - W_{d\,\text{sys}}$$

This result substituted into **(a)** gives

(d) $$\Delta\mathscr{A}_{\text{sys}} = -Q_{\text{sou}} - Q_{\text{sink}} - W_{d\,\text{sys}} - T_0\,\Delta S_{\text{sys}}$$

For the source, we write the change of availability as

(e) $$\Delta\mathscr{A}_{\text{sou}} = \Delta E_{\text{sou}} + p_0\,\Delta V_{\text{sou}} - T_0\,\Delta S_{\text{sou}}$$

The first law applied to the source (no delivered work) gives

(f) $$Q_{\text{sou}} = \Delta E_{\text{sou}} + p_0\,\Delta V_{\text{sou}}$$

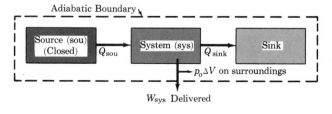

Adiabatic Boundary

Source (sou) (Closed) — Q_{sou} — System (sys) — Q_{sink} — Sink

$p_0\Delta V$ on surroundings

W_{sys} Delivered

Irreversibility. Consider that all events happen simultaneously the same duration of time; or convert all finite quantities to infinitesimals, with the same development in the text.

Fig. 5/8

which used in **(e)** gives

(g)
$$\Delta \mathcal{A}_{\text{sou}} = Q_{\text{sou}} - T_0\,\Delta S_{\text{sou}}$$

Now apply the definition of reversibility, using equations **(d)** and **(g)**;

$$I = -(\Delta \mathcal{A}_{\text{sys}} + \Delta \mathcal{A}_{\text{sou}}) - W_{d\,\text{sys}}$$

(h)
$$= -(-Q_{\text{sou}} - Q_{\text{sink}} - W_{d\,\text{sys}} - T_0\,\Delta S_{\text{sys}} + Q_{\text{sou}} - T_0\,\Delta S_{\text{sou}}) - W_{d\,\text{sys}}$$

Since $Q_{\text{sink}} = T_0\,\Delta S_{\text{sink}}$, substitute this equivalent into **(h)**, and simplify;

(5-11A)
$$I = T_0(\Delta S_{\text{sys}} + \Delta S_{\text{sou}} + \Delta S_{\text{sink}}) = T_0\,\Delta S_p$$

which, in words, says that the irreversibility of a particular process is equal to the absolute sink temperature *times* the entropy production (the change of entropy of the associated universe). This entropy change of the universe is the entropy production, equation **(d)**, § 5.5.

Example

A closed system accepts reversibly 1000 kJ of heat while its temperature remains at 400 K; the source temperature is 800 K; $T_0 = 300$ K. Find the entropy production and irreversibility.
   Solution.   From equation (5-4)

$$\Delta S_p = \Delta S_{\text{sys}} + \Delta S_{\text{sou}} + \Delta S_{\text{sink}}$$

$$= \frac{1000}{400} + \frac{-1000}{800} = 2.5 - 1.25$$

$$= 1.25 \text{ kJ/K}$$

From equation (5-11A)

$$I = T_0\,\Delta S_p$$

$$= (300)(1.25) = 375 \text{ kJ}$$

5.12   IRREVERSIBILITY OF A STEADY-FLOW SYSTEM

The change of availability for a steady-flow system, equation (5-9), is

(a)
$$\Delta \mathcal{A}_{\text{sys}} = \Delta H_{\text{sys}} + \Delta K_{\text{sys}} + \Delta P_{\text{sys}} - T_0\,\Delta S_{\text{sys}}$$

From the steady-flow energy equation (4-9), we have

(b)
$$Q_{\text{sys}} - W_{d\,\text{sys}} = \Delta H_{\text{sys}} + \Delta K_{\text{sys}} + \Delta P_{\text{sys}}$$

which, used in **(a)** gives

(c)
$$\Delta \mathcal{A}_{\text{sys}} = Q_{\text{sys}} - W_{d\,\text{sys}} - T_0\,\Delta S_{\text{sys}}$$

$$= -Q_{\text{sou}} - Q_{\text{sink}} - W_{d\,\text{sys}} - T_0\,\Delta S_{\text{sys}},$$

where $Q_{\text{sys}} = -Q_{\text{sou}} - Q_{\text{sink}}$ as explained in the previous article.

For the steady-flow source, assumed to deliver no work, the energy balance, by equation (4-9), is

(d) $$Q_{sou} = \Delta H_{sou} + \Delta K_{sou} + \Delta P_{sou}$$

and its change of availability is

(e) $$\Delta \mathscr{A}_{sou} = \Delta H_{sou} + \Delta K_{sou} + \Delta P_{sou} - T_0 \Delta S_{sou} = Q_{sou} - T_0 \Delta S_{sou}$$

where the result in **(d)** has been utilized. Observe that if the source is a closed system, the same result is obtained [equation **(g)**, § 5.11] and therefore the result is general. Recalling that $W_{d\max} = -\Delta \mathscr{A}$, apply the definition of irreversibility, use equations **(c)** and **(e)**, and get

(f) $$I = -(\Delta \mathscr{A}_{sys} + \Delta \mathscr{A}_{sou}) - W_{d\,sys}$$

$$= -(-Q_{sou} - Q_{sink} - W_{d\,sys} - T_0 \Delta S_{sys} + Q_{sou} - T_0 \Delta S_{sou}) - W_{d\,sys}$$

Use $Q_{sink} = T_0 \Delta S_{sink}$ as before, simplify, and find,

(5-11B) $$I = T_0(\Delta S_{sys} + \Delta S_{sou} + \Delta S_{sink}) = T_0(\Delta S_{sys} + \Delta S_{sur}) = T_0 \Delta S_p$$

the same as found for the closed system; see equation (5-4). Consider equation (5-11) to be generally applicable. If the system is cyclic and computations are for an integral number of cycles, $\Delta S_{sys} = 0$. If the system undergoes an adiabatic process only, $\Delta S_{sou} = 0$ and $\Delta S_{sink} = 0$.

In the foregoing discussion, the entropy increase of the sink has been Q_{sink}/T_0; but one should be aware that all energy "dumped" into the sink increases the sink's entropy. We should generalize then that

(g) $$\Delta S_{sink} = \frac{E_{sink}}{T_0}$$

where E_{sink} may include energy, say kinetic energy from a jet engine, other than heat.

AVAILABLE PORTION OF HEAT   5.13

As has already been said in one way or another, *available energy* is work itself, or a form that is 100% convertible into work in ideal machines, as kinetic energy, electricity, and so on. We also have learned from Carnot and elsewhere, that *not all* of the heat passing into (or from) a system is available for conversion into work on a cyclic basis. The question is, in general, how much can conceivably be so converted.

Let *ab*, Fig. 5/9, be any internally reversible process. Consider the differential strip 1-*mn*-4, where the temperature change as heat enters is infinitesimal; $T_1 = T_4 = T$. After equation [(a), §5-1], we write

(a) $$\frac{dQ}{T} = \frac{dQ_R}{T_0} \quad \text{or} \quad dQ_R = T_0 \frac{dQ}{T}$$

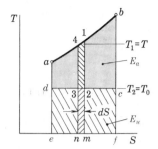

Fig. 5/9 *Available Portion of Heat.* The amount is determined by imagining a cycle being completed *bcd* to *a* with reversible processes; heat rejection is reversible, $T_2 = T_0$. The work of this cycle is E_a.

because 1-2-3-4, Fig. 5/9, is an infinitesimal Carnot cycle: dQ is the heat added; dQ_R is the heat rejected, area 2-3-*mn*. If $T_2 = T_0$, the sink temperature, the dQ_R is the minimum amount of heat that must be rejected, inasmuch as the processes 1-2 and 3-4 are isentropic. Therefore this particular dQ_R has a distinctive meaning; it is the *unavailable part of the rejected heat*; accordingly it is given a special symbol E_u (= minimum Q_R as defined above). Therefore, equation **(a)** for the internally reversible heat transfer becomes

(b)
$$E_u = T_0 \int \frac{dQ}{T}\bigg]_{\text{rev}} = T_0\,\Delta S$$

where $\Delta S = \int dQ/T$ for the system. It follows that the available part of the heat Q, which is the maximum conceivable work that can be obtained from the heat is

(5-12)
$$E_a = Q - E_u = Q - T_0\,\Delta S$$

If heat is negative, E_a is that portion of the heat that is available as it is departing the system. In Fig. 5/9, the unavailable part of the heat is represented by area *cdef*; the available part by area *abcd*. If the energy supplied for the same state change is paddle work W_p, instead of heat, an analogous portion of the input is available after it has entered the system; $E_a = W_p - T_0\,\Delta S$.

5.14 Example—Irreversibility of an Adiabatic Process

 A steady-flow compressor for a gas turbine engine receives air at 14 psia and 520°R, adiabatically compresses it to 98 psia, with a compressor efficiency of 83%; no significant change ΔK. The available sink is at 510°R. For the system of 1 lb of air, determine (a) the change of availability and work for isentropic compression. For the actual compression, find (b) the change of availability and work, (c) the change of availability of the immediate surroundings, and (d) the irreversibility. Use the table of Item B 2. See Fig. 5/10.

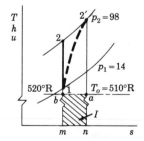

Fig. 5/10 *Air as an Ideal Gas.*

Solution. From Item B 2 at 520°R, $p_{r1} = 1.2147$ and $h_1 = 124.27$ Btu/lb. Then

(a)
$$p_{r2} = p_{r1}\frac{p_2}{p_1} = (1.2147)\left(\frac{98}{14}\right) = 8.503$$

Corresponding to this p_{r2} in Item B 2, we find the temperature as 903°R, to the nearest whole degree. At this temperature, $h_2 = 217.00$ Btu/lb.

(a) For the isentropic, $\Delta s = 0$. By equation (5-9),

(b)
$$\Delta\mathscr{A}_f = \Delta h - T_0\,\Delta S = 217.00 - 124.27 - 0 = 92.73 \text{ Btu/lb}$$

The work $W_c = -\Delta h = -92.73$. The increase of availability of this reversible process is numerically equal to the work input.

(b) The actual work is $W_c' = W_c/\eta_c = h_1 - h_{2'}$;

(c)
$$W_c' = -\frac{92.73}{0.83} = -111.72 = 124.27 - h_{2'}$$

from which $h_{2'} = 236$ Btu/lb. Find this value in Item B 2 and read the temperature at 2' to be 979°R; also $\phi_{2'} = 0.74514$. Then for $r_p = 98/14 = 7$ and $\phi_1 = 0.59173$, we get

(d)
$$\Delta s_{1\text{-}2'} = \Delta\phi - R \ln r_p = 0.74514 - 0.59173 - R \ln 7 = 0.02002 \text{ Btu/lb-°R}$$

say, 0.02; $\phi_{2'} - \phi_2$ is *slightly* different because the temperatures were rounded off to whole numbers, $\bar{R} = 1.986$ Btu/lb-°R.

(e)
$$\Delta\mathscr{A}_f = \Delta h - T_0\,\Delta s = 235.77 - 124.27 - (510)(0\,02) = 101.3 \text{ Btu/lb}$$

a *greater increase* than in the reversible compression.

(c) The change of availability of the immediate surroundings with only work leaving the surroundings is a decrease equal in magnitude to the work into the system, -111.7 Btu/lb—no entropy associated with work itself.

(d) Since the process is adiabatic, $\Delta S_{\text{sou}} = 0$ and $\Delta S_{\text{sink}} = 0$; therefore

(f)
$$I = T_0\,\Delta s_{\text{sys}} = (510)(0.02) = 10.2 \text{ Btu/lb}$$

Looking at this part from the standpoint of the definition of I, and letting $\Delta K = 0$, $\Delta P = 0$, we have (all terms applicable to the system)

(g)
$$\Delta\mathscr{A}_f = \Delta H + \Delta K + \Delta P - T_0\,\Delta S \qquad \text{[EQUATION (5-9)]}$$

(h)
$$Q = \Delta H + \Delta K + \Delta K + W_d = 0 \qquad \text{[EQUATION (4-9)]}$$

Using the value of $\Delta H + \Delta K + \Delta P$ from **(h)** in **(g)**, we get

(i)
$$\Delta\mathscr{A}_f = -W_d - T_0\,\Delta S$$

(j)
$$I = -\Delta\mathscr{A}_f - W_d = W_d + T_0\Delta S - W_d = T_0\Delta S$$

or, as expected, $T_0\,\Delta s$ for 1 lb as in this example.

Example—Irreversibility of a Heat Exchange with Sink 5.15

An ideal gas is compressed isothermally 1-2, Fig. 5/11, at 1600°R, with heat being transferred at the rate of 3200 Btu/min to a sink *ab* at 600°R; let the process be steady flow with no ΔK or ΔP. Compute (a) the change of availability of the system, (b) the work of the system, and (c) the irreversibility of the event. See Fig. 5/11.

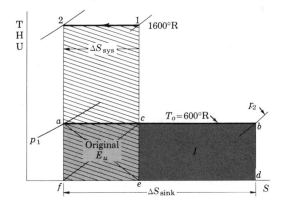

Fig. 5/11   The area *acef* under T_o for ΔS_{ab} represents original unavailable part of heat from 1–2 as $T = C$. The area *cbde* represents the increase of unavailable energy with heat transfer, and, in this example, the irreversibility *I*.

Solution.   (a) Since the temperature is constant for this unknown ideal gas, $\Delta H = 0$, $\Delta U = 0$, and

(a)
$$\Delta S_{1\text{-}2} = \int_1^2 \frac{dQ}{T} = \frac{Q}{T} = \frac{-3200}{1600} = -2 \text{ Btu/°R-min}$$

(b)
$$\Delta \mathscr{A}_{1\text{-}2} = \Delta H - T_0 \Delta S = 0 - (600)(-2) = 1200 \text{ Btu/min}$$

(b) For steady flow, equation (4-9), $Q = \Delta H + W$, or $Q = W_d$, since $\Delta H = 0$; thus, $W_d = Q = -3200$ Btu/min.

(c) The entropy change of the sink is Q/T_0, where Q is positive for the sink,

(c)
$$\Delta S_{ab} = \frac{Q}{T_0} = \frac{3200}{600} = 5.333 \text{ Btu/min}$$

(d)
$$\Delta S_p = \Delta S_{\text{sys}} + \Delta S_{\text{sink}} = -2 + 5.333 = 3.333 \text{ Btu/min-°R}$$

and $I = T_0 \Delta S_p = (600)(3.333) = 2000$ Btu/min.

Again, considering the definition of irreversibility, we write

(e)
$$Q_{\text{sys}} = \Delta H_{\text{sys}} + \Delta K_{\text{sys}} + \Delta P_{\text{sys}} + W_{d\,\text{sys}}$$

(f)
$$\Delta \mathscr{A}_{\text{sys}} = \Delta H_{\text{sys}} + \Delta K_{\text{sys}} + \Delta P_{\text{sys}} - T_0 \Delta S_{\text{sys}}$$

$$= Q_{\text{sys}} - W_{d\,\text{sys}} - T_0 \Delta S_{\text{sys}}$$

by using equation **(e)**. Note that $Q_{\text{sys}} = -Q_{\text{sink}} = -T_0 \Delta S_{\text{sink}}$, and apply the definition of I:

$$I = -\Delta \mathscr{A}_{\text{sys}} - W_{d\,\text{sys}} = -(-T_0 \Delta S_{\text{sink}} - W_{d\,\text{sys}} - T_0 \Delta S_{\text{sys}}) - W_{d\,\text{sys}}$$

(g)
$$= T_0(\Delta S_{\text{sink}} + \Delta S_{\text{sys}})$$

Also note that $I = W_{d\,\text{max}} - W_d = -\Delta \mathscr{A}_{\text{sys}} - W_d = -1200 - (-3200) = 2000$ Btu/min as found above.

Example—Irreversibility Because of a Heat Exchange   **5.16**

In the heat exchanger of Fig. 5/12 (see Fig. 7/9), 100 lb/min of water (specific heat $c = 1$ Btu/lb-°R) are to be heated from 140°F (600°R) to 240°F (700°R) by hot gases ($c_p = 0.25$ Btu/lb-°R) that enter the exchanger at 440°F (900°R) with a flow rate of 200 lb/min. Let the heat interaction between water and gas be the only heat, with no significant changes ΔK and ΔP. The available sink is at 500°R. (a) Compute the change of availability of the gas and of the water, and the net change. (b) What portion of heat is available as it leaves the gas? What portion is available with respect to the water? (c) What is the irreversibility? (d) Calculate the steady-flow availability of the hot gas in state a.

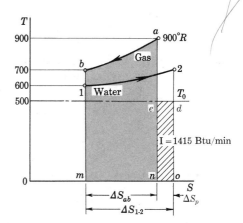

(a) Energy Diagram for Counterflow
    Heat Exchanger.

(b) Irreversibility on TS plane.

Entropy Growth via Heat Exchange. In figure (b), it is quite unlikely that the   **Fig. 5/12**
absolute entropies S_b and S_1 are the same; the curves are drawn that way
for convenience.

Solution.   First make preliminary calculations. No work is done. Therefore the energies entering and leaving the system are the internal energies plus the flow works (enthalpies) of the flowing fluids, as shown in Fig. 5/12(a). We may think of two subsystems, the gas and the water. In accordance with the first law, the heat from the gas is numerically equal to the heat into the water, and the algebraic sum of these heats is equal to zero. For the water, $\Delta H_w = mc\,\Delta T$, very nearly (more accurate means of getting Δh for water are given in Chapter 3). For the gas, $\Delta H_g = mc_p\,\Delta T$. Therefore, we have (Fig. 5/12)

(a)
$$Q_{\text{water}} + Q_{\text{gas}} = 0 \quad \text{or} \quad Q_{\text{gas}} = -Q_{\text{water}}$$

(b)
$$\Delta H_w = Q_{\text{water}} = mc(T_2 - T_1) = (100)(240 - 140) = 10{,}000 \text{ Btu/min}$$

(c)
$$\Delta H_g = Q_{\text{gas}} = mc_p(T_b - T_a) = (200)(0.25)(T_b - 900) = -10{,}000$$

from which $T_b = 700°R$, the final temperature of the gas that is necessary to satisfy the first law. The entropy changes are

(d)
$$\Delta S_g = mc_p \ln \frac{T_b}{T_a} = -(200)(0.25) \ln \frac{900}{700} = -12.57 \text{ Btu/°R-min}$$

the negative sign indicating a decrease of entropy, and

(e)
$$\Delta S_w = mc \ln \frac{T_2}{T_1} = 100 \ln \frac{700}{600} = +15.4 \text{ Btu/°R-min}$$

the positive sign indicating an increase of entropy.

(a) The availability changes are

(f)
$$\Delta \mathcal{A}_g = \Delta H_g - T_0 \Delta S_g = -10,000 - (500)(-12.57) = -3715 \text{ Btu/min}$$

(g)
$$\Delta \mathcal{A}_w = \Delta H_w - T_0 \Delta S_w = +10,000 - (500)(15.4) = +2300 \text{ Btu/min}$$

The net change is $\sum \Delta \mathcal{A} = -1415$ Btu/min, the negative sign indicating a decrease.

(b) Of the 10,000 Btu of heat leaving the gas ab, Fig. 5/12, we find

(h)
$$E_{ug} = T_0 \Delta S = (500)(-12.57) = -6285 \text{ Btu/min, a decrease}$$

(i)
$$E_{ag} = -10,000 + 6285 = -3715 \text{ Btu/min, a decrease}$$

Of the 10,000 Btu received by the water 1-2,

(j)
$$E_{uw} = (500)(15.4) = 7700 \text{ Btu/min, an increase}$$

(k)
$$E_{aw} = 10,000 - 7700 = 2300 \text{ Btu/min, an increase}$$

The available portions are then 3715/10,000 or 37.15% for the gas; 2300/10,000 or 23% for the water. The available energy lost in the heat transfer is $-3715 + 2300 = -1415$ Btu/min, which is 1415/3715 = 38.1% of the original available energy lost without using a single Btu of heat.

(c) The entropy production (no energy to the sink) is

(l)
$$\Delta S_p = \Delta S_{1\text{-}2} + \Delta S_{ab} = +15.4 + (-12.57) = +2.83 \text{ Btu/°R-min}$$

(m)
$$I = T_0 \Delta S_p = (500)(2.83) = 1415 \text{ Btu/min}$$

which is observed to be the same as the increase of unavailable energy and, numerically, as the decrease of the availability. These are typical results for heat exchangers, but further generalization is not advised.

(d) To get the availability at state a, compute ΔS from a to the dead state. All pressures are essentially ambient.

(n)
$$-\Delta S_{a\text{-}0} = mc_p \ln \frac{T_a}{T_0} = (200)(0.25) \ln \frac{900}{500} = 29.39 \text{ Btu/°R-min}$$

$$-\Delta \mathcal{A}_{fa/0} = H_a - H_0 - T_0(S_a - S_0) = mc_p(T_a - T_0) - T_0(S_a - S_0)$$

(o)
$$= (200)(0.25)(900 - 500) - (500)(29.39) = 5305 \text{ Btu/min}$$

In short, if the gas at *a* could undergo a reversible process to the dead state, exchanging heat only with the environment, 53% of it could be converted into work.

GENERAL REMARKS ABOUT ENTROPY, AVAILABILITY AND   5.17
IRREVERSIBILITY

We may now say that entropy is that property whose change is a measure of irreversibility of an event or a measure of the amount of energy that becomes unavailable, a measure that sometimes can be evaluated numerically. Since electricity is available energy, the use of electricity for heating water, for heating metal for heat treatment, and so on, is using the highest grade of energy (100% available) for a job that can be done by relatively low grade energy. Nevertheless, such use of high grade energy is often *economical* in industry or is so convenient that the purchaser is willing to pay the extra cost. As implied by the second law, energy may be graded. The energy from a source at high temperature is of higher grade (a larger percentage is available if it is used directly in a heat engine) than energy from a source at a low temperature. This situation highlights the desirability of transferring heat with as small a temperature difference as practicable. Since the *rate* of heat flow is proportional to the temperature difference, other things being equal, the difference must be large enough that the heat flows at a rate to satisfy human needs and desires, real or imagined.

Ultimately the available energy that is evolved becomes unavailable energy. All actual work produced is eventually dissipated by friction, or the equivalent, and becomes stored energy in the sink. For example, all the millions of horsepower produced in automotive engines arrive at the sink level via friction between the vehicle and the air, friction associated with the engine, and friction at numerous other points; the remainder of the energy released by the fuel reaches the sink via the exhaust, the air across the radiator, and so on. High-grade energy is continuously being degraded. See statement E, *degradation of energy*, § 5.2.

From time to time, we have mentioned the availability of kinetic energy and potential energy. Since speed is always relative, kinetic energy is also relative. When we think of kinetic energy in its "absolute" sense, we simply mean that the v in $mv^2/2$ is with respect to the earth. There are situations in thermodynamics where it is desired to reckon the kinetic energy from another datum—with respect to a moving body (meaning one moving with respect to the earth). In this case, the speed v is with respect to the moving body that contains the reference axes, and the corresponding energy is available relative to this moving body (relative to the frame of reference for speed). Similarly, the total gravitational potential energy with respect to the earth should be reckoned by using the distance to the center of the earth, but the effectively available part of this energy is with respect to the surface of the earth. However, changes in kinetic and potential (all kinds) energies can all conceivably be converted into work and are therefore available.

The sun emits energy that reaches the earth, and a part of this energy is available; that is, some of it can be converted into work in a typical heat engine. (Other kinds of engines may be devised to convert some of the sun's rays directly into electricity.) Various fuels are consumed in such ways that we can utilize some of the available parts of the energies released. In any case, it takes organization, the expenditure of mental and physical effort, in order for us to be able to utilize as work any portion of available energy that evolves.

5.18   HEAT vs. ENTROPY FOR IRREVERSIBLE PROCESS

We have seen how heat and entropy are related for a reversible process. To determine if there is any relation between these two for an irreversible process, let a closed system change reversibly along path R from state 1 to state 2 and then return to state 1 via an irreversible process I; see Fig. 5/13.

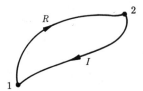

Fig. 5/13   Closed System Executing a Cyclic Process.

By equation [(a), § 5.4] there results

(a)
$$\oint_{1R2I1} ds = \int_{1R2} ds + \int_{2I1} ds = 0$$

By equation (5-1) there results

(b)
$$\oint_{1R2I1} dQ/T = \int_{1R2} dQ/T + \int_{2I1} dQ/T \leqq 0$$

But

$$\int_{1R2} dQ/T = \int_{1R2} ds$$

Therefore

(c)
$$\oint_{1R2I1} dQ/T = \int_{1R2} ds + \int_{2I1} dQ/T < 0$$

Now combining equations **(a)** and **(c)** we have

(d)
$$\int_{2I1} dQ/T - \int_{2I1} ds < 0$$

or

(e)
$$\int_{irr} ds > \int_{irr} dQ/T$$

a relation that will bear keeping in mind.

5.19   ENTROPY CHANGE, OPEN SYSTEM

As will be observed in § 7.27 for definable equilibrium states, the change of entropy within a control volume, the system, during time $\Delta\tau$ is $\Delta S_{sys} = m_2 s_2^a - m_1 s_1^a$, where s_2^a and s_1^a are absolute entropies of equilibrium states 1 and 2 at the beginning

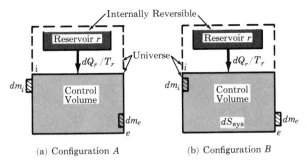

(a) Configuration A   (b) Configuration B

The development of equation **Fig. 5/14**
(5-13) is such that the sign of dQ_r
is determined with respect to the
system, positive as shown.

and end of the time interval. To apply the principle of increase of entropy, $dS \gtreqqless 0$, to a control volume, all reservoirs with heat interactions must be included to constitute a universe. In Fig. 5/14(a), the mass dm_i is about to enter the system, and, simultaneously, mass dm_e is about to leave at boundary e; both events occur in time $d\tau$, after which dm_i is inside and dm_e is outside the control volume, Fig. 5/14(b). Concurrently, there is heat dQ_r from a reservoir at temperature T_r. For configuration A, Fig. 5/14, the entropy of dm_i and the material in the system is $S_A = S_{1sys} + S_i^a \, dm_i$. In configuration B, the entropy within the control volume plus the entropy of mass dm_e is $S_B = S_{2sys} + S_e^a \, dm_e$. While this entropy change occurs in time $d\tau$, the reservoir, assumed internally reversible, undergoes an entropy change of $dS_r = -dQ_r/T_r$, negative because heat is shown leaving r and dQ_r is to be positive when it enters the control volume. Let $(S_2 - S_1)_{sys}$ be dS_{sys} for time $d\tau$, and write the change of entropy of the isolated system (universe) as

(5-13A) $$dS_{sys} + s_e^a \, dm_e - s_i^a \, dm_i + dS_r \gtreqqless 0$$

in accordance with equation (5-3), and the amount by which it exceeds zero is the entropy growth. If elements of mass cross at several points, say at j points, and if there are heat interactions with r reservoirs, we can rewrite equation (5-13) as

(5-13B) $$d(\Delta S)_p = dS_{sys} - \sum_j s_j^a \, dm_j - \sum_r \frac{dQ_r}{T_r}$$

where $d(\Delta S_p)$ is the infinitesimal entropy production, dS_{sys} is the change of entropy "content", both during $d\tau$, and dm_j is positive as it enters and negative as it departs the control volume. On a rate basis, equation (5-13) takes the form

(5-13C) $$\frac{d(\Delta S_p)}{d\tau} = \frac{dS_{sys}}{d\tau} - \sum_j s_j^a \dot{m}_j - \sum_j \frac{\dot{Q}_r}{T_r}$$

If the different streams are different substances and if the mass varies, all entropies should be absolute, as set up; but if the system is a pure substance and the mass is constant, any datum is all right.

Example—Mechanical Losses   5.20

A gear box receives 100 kW and delivers 95 kW while operating in steady state at 140°F. The surroundings are at 500°R. What is the entropy production of the universe and the irreversibility?

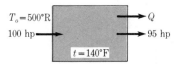

Fig. 5/15   *Gear Box.*

Solution.   See the energy diagram of Fig. 5/15. The rate at which work is lost is

(a) $E_w = 100 - 95 = 5\,\text{kW}$

Since the operation is steady state, heat $\dot{Q} = -5\,\text{kW}_t$ leaves the box at 600°R. Considering equation (5-13B) in connection with this system, we decide: there is no change in the entropy content of the system (steady state); since no masses enter or depart, $\int s_{aj}\,dm_j = 0$ for this unit time; the heat enters the sink (reservoir), which is at 500°R. Therefore

$$\Delta \dot{S}_p = -\frac{\dot{Q}_r}{T_r} = -\frac{-5}{500} = 0.01\,\text{kW/°R}$$

Considering the system (gear box), the lost work occurs at 600°R, assuming all parts of the box are at the same temperature (unlikely); hence it adds to the system's entropy at the rate of $5/600 = 0.0083\,\text{kW/°R}$. But the heat leaves at 600°R, which is to say that the decrease of entropy with heat cancels that produced because of friction. The irreversibility is $I = T_0\,\Delta \dot{S}_p = (500)(0.01) = 5\,\text{kW}$.

5.21   Example—Irreversibility of Flow in Pipe

An uninsulated pipeline is used to transport nitrogen in steady flow at the rate of 5 lb/sec with $\Delta K \approx 0$ and $\Delta P = 0$. At section 1, Fig. 5/16, $p_1 = 200\,\text{psia}$, $T_1 = 800°\text{R}$; at section 2, $p_2 = 160\,\text{psia}$, $T_2 = 700°\text{R}$. The local sink is at 520°R. Determine the irreversibility.

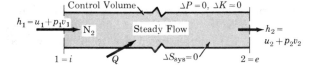

Fig. 5/16   *Open Steady-Flow System*, $\Delta K = 0$, $\Delta P = 0$, $W = 0$.

Solution.   In Item B 6, we find the following properties: $\bar{h}_1 = 5564.4$, $\bar{\phi}_1 = 48.522$, $\bar{h}_2 = 4864.9$, $\bar{\phi}_2 = 47.588$. By the energy diagram of Fig. 5/16, we get

(a) $Q = \bar{h}_2 - \bar{h}_1 = 4864.9 - 5564.4 = -699.5\,\text{Btu/pmole}$

Between sections 1 and 2, by equation (6-13),

(b) $\bar{s}_2 - \bar{s}_1 = \Delta \bar{s} = \Delta \bar{\phi} - \bar{R}\ln\dfrac{p_2}{p_1} = 47.588 - 48.522 + \bar{R}\ln\dfrac{200}{160} = -0.4909\,\text{Btu/°R-pmole}$

which turns out to be a decrease of entropy despite the frictional losses because of the relatively large heat loss. Equation **(b)** gives the net entropy crossing the boundaries i and e (1 and 2) for 1 pmole of flow. In equation (5-13), $\Delta S_{\text{sys}} = 0$ because there is steady-state flow. Since $\dot{m}_i = \dot{m}_e = \dot{m} = 5\,\text{lb/sec}$, $-(S_{\text{in}} - S_{\text{out}})$ is

(c) $-\displaystyle\sum_j s_j^a \dot{m}_j = \dot{m}(s_2 - s_1) = \frac{5}{28.016}(-0.4909) = -0.0876\,\text{Btu/°R-sec}$

where for N_2, $M = 28.016$ lb/pmole. Because the heat is leaving the system, $Q = -699.5$ Btu/pmole enters the sink at 520°R, or

(d)
$$-\frac{\dot{Q}_r}{T_r} = -\frac{(-699.5)(5)}{(520)(28.016)} = +0.24 \text{ Btu/°R-sec}$$

(e)
$$\Delta\dot{S}_p = -0.0876 + 0.24 = 0.1524 \text{ Btu/°R-sec}$$

(f)
$$\dot{I} = T_0\Delta\dot{S}_p = (520)(0.1524) = 79.25 \text{ Btu/sec}$$

For frictionless flow with no heat loss, there would be zero irreversibility.

Example—Entropy Production for Transient Change **5.22**

A drum D, Fig. 5/17, containing 0.00266 moles of hydrogen at 20 psia and 80°F, is recharged with 0.00686 moles of H_2 from a steady-state supply at 600 psia and 140°F. The state in the drum after adiabatic charging ceases is $p_2 = 100$ psia and $T_2 = 756$°R. Determine the entropy growth of this process.

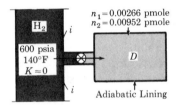

The boundary ii is curved into the supply line just to emphasize that the state of the incoming H_2 is that in this line, assumed uniform across any section of the pipe. **Fig. 5/17**

Solution. Using the values of $\bar{\phi}$ in Item B 5 as absolute entropies at the particular temperature and $p_0 = 1$ atm, we find

600 psia, 600°R:
$$\bar{s}_i = \bar{\phi} - \bar{R}\ln\frac{p}{p_0} = 31.959 - 1.986\ln\frac{600}{14.7}$$
$$= 24.595 \text{ Btu/pmole-°R}$$

20 psia, 540°R:
$$\bar{s}_1 = 31.232 - 1.986\ln\frac{20}{14.7} = 30.6214 \text{ Btu/pmole-°R}$$

100 psia, 756°R:
$$\bar{s}_2 = 33.565 - 1.986\ln\frac{100}{14.7} = 29.7553 \text{ Btu/pmole-°R}$$

In equation (5-13), $\sum Q_r/T_r = 0$; the operation is adiabatic. Since no streams are leaving the drum, no entropy is being convected out; hence, with $\bar{s}_i$ constant and $n_2 = n_1 + n_i = 0.00266 + 0.00686 = 0.00952$ moles,

(a)
$$S_{in} = \int \bar{s}_i\,dn_i = \bar{s}_i n_i = (0.00686)(24.595) = 0.1687 \text{ Btu/°R}$$

(b)
$$\Delta S_{sys} = n_2\bar{s}_2^a - n_1\bar{s}_1^a = (0.00952)(29.7553) - (0.00266)(30.6214) = 0.2018 \text{ Btu/°R}$$

(c)
$$\Delta S_p = \Delta S_{sys} - S_{in} = 0.2018 - 0.1687 = 0.0331 \text{ Btu/°R}$$

5.23   SECOND LAW ANALYSIS

A first law analysis, which is basically an energy accounting, is an action that is not only helpful but essential; yet it does not provide the engineer with a complete answer to what may be a primary endeavor—to wit, obtaining as much work as possible from a work-manufacturing machine and, sometimes even more importantly using the minimum work to accomplish a particular purpose. For this objective, a second law analysis should designate areas which are subject to significant improvement, because its aims are, first, to discover where major destructions of available energy occur and, second, to help decide if there are practical actions that can be taken to reduce these losses. While technically it may be quite feasible to save available energy, it may not be economically feasible. (We might keep in mind that the economics of things change.)

A second law analysis may be made in several different ways. For example, one could determine the entropy growth of each process, considering the steps that could be taken for improvement. Or one could compute the changes of availability, comparing these changes with actual works.

5.24   HELMHOLTZ AND GIBBS FUNCTIONS

Some chemical processes can be continuous, or effectively so for a period of time. Therefore, it is appropriate to inquire briefly into work possibilities for such cases. For a closed, nonflow system, the first law equation (4-6) gives $dW = dQ - dE_s$, wherein the work dW may be in any form, §2.15. For a reversible process $dQ = T\,dS$, or $dW_{rev} = T\,dS - dE_s$, which of course is the maximum work for the process. The actual work will be less than this; or

(a) $$dW \leqq T\,dS - dE_s$$

If this equation is integrated for constant T, we get

(b) $$W \leqq T(S_2 - S_1) - (E_{s2} - E_{s1}) = -[(E_{s2} - TS_2) - (E_{s1} - TS_1)]$$

which is the gross work output of the closed system, or the net delivered work of a constant volume process proceeding at constant temperature. The stored energy E_s includes chemical energy, as well as molecular energy, if a chemical reaction occurs (plus other relevant forms of stored energy), although equation **(b)** may be applied to a pure substance with $E_s = U$ (or u) if nonflow. The combination of properties $E_s - TS = A$, a property called the **Helmholtz function**. Thus,

(5-14) $$W_{T\max} = -\Delta A = A_1 - A_2 = -\Delta(E_s - TS) \qquad [T = C]$$

which is the *maximum* work that can be delivered, or the numerical minimum work input. For a pure substance with only molecular stored energy, the specific Helmholtz function is

(5-15) $$A = u - Ts$$

If a boundary, perhaps an imagined one, moves against the surroundings at a constant pressure, the work done in displacing the surroundings by the constant boundary pressure is $p(V_2 - V_1)$; it follows that the maximum work $(p = C,$ $T = C)$ that can be delivered to something other than the surroundings is

(c) $$W_{T,p} \leqq [E_{s2} - TS_2 - (E_{s1} - TS_1)] - (pV_2 - pV_1)$$

(5-16A) $$W_{T,p\max} = -[(E_{s2} + pV_2 - TS_2) - (E_{s1} + pV_1 - TS_1)]$$

$$= -(G_2 - G_1) = -\Delta G$$

where $G = E_s + pV - TS$, a property called the **Gibbs* function**, whose unit is a unit of energy. Let $E_s + pV = H$, the total (§ 4.4) enthalpy. Then total $G = H - TS$, and equation (5-16) becomes

(5-16B) $$W_{T,p\max} = -[(H_2 - TS_2) - (H_1 - TS_1)]$$

Imagine a chemical process taking place in a reaction chamber; let the reactants flow in, each in the standard state of 77°F and 1 atm; let the products depart (with help from permeable membranes, § 13.24), each in the standard state. Then equation (5-16) may be written

(5-16C) $$W_{T,p\max} = -\Delta G^\circ = -[H_p^\circ - H_R^\circ - T^\circ(S_p^\circ - S_r^\circ)]$$

where H_p° is the enthalpy of all the products, H_R° is the total enthalpy of the reactants all at 77°F, and

(d) $$S_p^\circ = \sum_p n_i \bar{s}_i^\circ \quad \text{and} \quad S_R^\circ = \sum_r n_i \bar{s}_i^\circ$$

the sum in each case being for n_i moles of each component of products and reactants, respectively, and $\bar{s}_i^\circ$ represents the absolute entropy of each at 1 atm and 77°F.

NOTE: Most of the chemical processes of this chapter are restricted only to standard states because we wish to put the work of such processes in context with cycles and the second law, and to say that heat engines are not the only means of manufacturing power. The subject of combustion is presented in more detail in Chapter 13. It is presumed that any chemical process referred to here is one that goes to completion; the reactants combine into stable forms of molecules.

For a pure substance with only molecular stored energy, the specific Gibbs function is $G = h - Ts$ and

(5-17A) $$W_{T,p\max} = -\Delta G = -[(h_2 - Ts_2) - (h_1 - Ts_1)]$$

---

* J. Willard Gibbs (1839–1903), born in Connecticut, a professor of mathematics at Yale, undoubtedly contributed more to the science of thermodynamics than any other American, and perhaps he was the world's major contributor to the mathematics of the science. As a mathematical physicist, he developed the thermodynamic principles of chemical equilibrium, much of the mathematics of the science, and a general approach to statistical mechanics. He also was an originator of vector analysis and made contributions to the electromagnetic theory of light. See *The Collected Works of J. Willard Gibbs*, Yale University Press, New Haven, Connecticut.

The steady-flow equation (4-10) with $\Delta K = 0$ and $\Delta P = 0$ only for convenience becomes $dQ = dH + dW$. As above, apply this to a reversible process, $dQ = T\,dS$, to obtain the maximum work at constant temperature; or, as before,

$$(5\text{-}17\text{B}) \qquad W_{max} = T(S_2 - S_1) - (H_2 - H_1)$$

$$= -[(H_2 - TS_2) - (H_1 - TS_1)] = -\Delta G$$

The changes of both the Helmholtz and Gibbs functions are called *free energy*, that is, energy free to be converted into work. The $T\,\Delta S$ term represents the heat exchanged with the surroundings, when the maximum (minimum) work is being produced (consumed), at $T = C$, a temperature which is not necessarily the same as the sink temperature. An example of the reversed process is the electrolysis of water, where the minimum of work input is desired. The property relations, $A = u - Ts$, $A = E_s - TS$, $G = H - TS$, may be used mathematically in any correct manner, but the negative changes $-\Delta A$, $-\Delta G$ represent ideal work only for the conditions specified. Total quantities for each function were used at the outset because these properties are quite useful for chemical reactions where various masses of different substances are involved. The useful work of a battery, say an ordinary lead-acid battery, is the electricity it delivers; it gives off gas, with a work output that does nonuseful work in pushing back the surrounding atmosphere.

5.25   EQUILIBRIUM

Since the word *equilibrium* must convey different ideas (in detail) in different contexts and situations, we shall necessarily be furthering its definition from time to time.

Mechanical equilibrium, the condition for *statics*, means that the system is not accelerating, $\Sigma\mathbf{F} = 0$ and $\Sigma\mathbf{M} = 0$. There are degrees or classes of equilibrium, as suggested by Fig. 5/18. The ball in the spherical bowl (a) is in *stable equilibrium*; if the ball is displaced, it returns of its own accord to its original position (in the presence of the same force field, gravity). The short stubby column is stable in the sense that it would take a relatively large disturbance to overturn it. At the other extreme (c), the ball balanced on top of a spherical surface or the column balanced on a pointed end is in *unstable equilibrium* because it takes only a minute disturbance to bring about a profound change, and there is no tendency for a return to these precarious positions. In the *metastable equilibrium*, Fig. 5/18(b), it takes somewhat more than a very small disturbance to cause the system to proceed further toward stable equilibrium, but the configuration is easily destroyed beyond the point of return. Then there is *neutral equilibrium* (d) as represented by a ball on a horizontal surface.

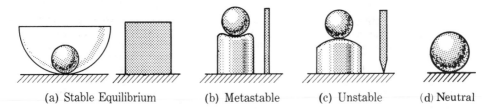

(a) Stable Equilibrium   (b) Metastable   (c) Unstable   (d) Neutral

Fig. 5/18   *Kinds of Equilibrium.*

A system is in internal ***thermal equilibrium*** (§ 1.20) when all its parts are at the same temperature. It is in thermal equilibrium with another system when both systems have the same uniform temperature. As you know, steam condenses during heat rejection when a certain temperature has been reached (the temperature depending on the pressure). However, under laboratory conditions and in some actual processes, steam (and similarly other vapors) can be cooled below the temperature at which condensation normally begins and can be held there (in a static situation). Such a state is called *supersaturated state*. Condensation occurs easily when there are nuclei, for example, ions or dust, to attract and hold the gaseous molecules; this attraction becomes quite positive once the accumulation of molecules begins and the liquid drop quickly reaches a typical size. The supersaturated condition is an example of a *thermally metastable equilibrium*. A small disturbance (as a small impulse) may start condensation and a rapid approach toward stable equilibrium.

A system is in ***chemical equilibrium*** if it has no tendency to undergo further chemical reaction. If a bit of sodium is dropped into water, a chemical reaction starts; but when the reaction is completed, chemical equilibrium is established. An isolated volume of the atmosphere is in *stable* chemical equilibrium, because there is virtually no tendency for a reaction of the air's constituents. A certain mixture of gasoline and air is in *metastable* equilibrium, since a spark (a small disturbance) may cause it to explode. Other mixtures of substances may be caused to react by the addition of a catalyst.

There are a few other "forces" that are involved in equilibrium, such as electrical and magnetic. When these, as well as the mechanical, thermal, and chemical forces, are in equilibrium in an isolated system, the system is said to be in ***thermodynamic equilibrium***; *there is no tendency towards spontaneous change*—meaning a change without outside influence.

When a system is in thermodynamic equilibrium with the natural environment (sink), it is in the ***dead state***. The ***sink*** is basically the local environment. In space it is eventually space itself. On the earth it is the atmosphere, a river, lake or ocean, and so on (we may even use ocean water at great depths for sink purposes). Thus, it is evident that, depending on location, the temperature of the available sink may be any within a wide range of temperatures, but in thermodynamic studies, the sink is assumed constant—as it virtually is for a particular system in a particular geographical locality.

QUANTITATIVE CONSIDERATION OF EQUILIBRIUM 5.26

There are several ways in which computations may be made to decide whether or not a process may occur, that is, whether an isolated system is in stable equilibrium. For example, if, for any process imagined to be toward a more stable state, the change of availability $-\Delta \mathscr{A} \gtreqless 0$, the system is stable. In short, there will *not* be a spontaneous change to a state of greater availability, in violation of the second law.

Entropy is also used. For an isolated system (constant total stored energy E and constant mass m) in a stable state, it is true that

(5-1B) $$dS]_{E,m} \leqq 0$$

for any possible variation of state (statement D, § 5.2). In words, a closed system without heat or work interaction with the surroundings is in stable equilibrium if any allowed change of state indicates a decrease in (or no change of) entropy. As suggested by the previous example, part (b), the isolated system includes all affected parts of the universe. If $\Delta S]_{E,m} = 0$ for any imagined change, the change may be reversible or the equilibrium may be neutral with respect to the two states. If $\Delta S]_{E,m} > 0$, a spontaneous change to a more probable (more stable) state is possible.

The criterion most used by chemical engineers for constant mass systems is with respect to the Gibbs function. If a constant temperature/constant pressure process naturally results in the possibility of a work output, $(-\Delta G)_{T,p} > 0$, the process will occur or tend to occur spontaneously. If an input of work, or its equivalent, is necessary, $(-\Delta G)_{T,p} < 0$, the process will not take place of its own accord. The usual situation is one for which $\Delta K = 0$ and $\Delta P = 0$. Thinking of work as being $-(\Delta G)_{T,p}$, we may say:

If $dG_{T,p} = 0$, there is chemical equilibrium; processes that occur do so reversibly; not in thermodynamic equilibrium unless in mechanical and thermal equilibrium with the local sink; see Fig. 13/10;

If $-\Delta G > 0$, a spontaneous process (reaction) occurs or is tending to occur;

If $-\Delta G < 0$, no reaction will occur in an isolated system; the system is in stable equilibrium.

From a practical standpoint, a reaction when $-\Delta G < 0$ may be brought about by an *input.* Dodge[1.16] says that at 1-atm pressure: if $-\Delta G > 0$, the reaction is promising; if $-10^4 < -\Delta G < 0$, the reaction is of doubtful promise; if $-\Delta G < -10^4$ cal/gmole, the reaction will be feasible only under unusual circumstances. In the same manner, the Helmholtz function may be used as a criterion and is applicable to systems with v, T constant.

5.27   SECOND LAW AND PROBABILITY

Of his famous equation of entropy in terms of thermodynamic probability, Max Planck said[5.1] "I simply postulated that $S = \kappa \ln \Omega$, where κ is a universal constant ...," but of course, out of context, this is a vast oversimplification. As for all great scientific generalities, there was a buildup for years prior to this, followed by years of testing afterward. Moreover, the equation is called Boltzmann's* relation; Ω is the thermodynamic probability, and we should have at least a qualitative idea of its meaning and significance.

Take a system more or less familiar to all, two dice (unloaded) A and B. If A is rolled at random, one face is just as likely to turn up as another; all have the same probability; this is also true of die B. Let the number up on A be designated as $A1$, $A3$, and so on; the number up on B as $B1$, $B3$, and so on. Now if the *pair* is rolled, the combination $A6$ and $B6$ (a 12) is just as likely as the combination $A3$ and $B4$ (a 7). Call a particular combination a "microstate." Since each die can be in 6 different states, the number of combinations is $(6)(6) = 36$; that is, the dice may be in any one of 36 different microstates, where, for example, the state $A3$, $B4$ is different from the state $A4$, $B3$ (though both add to 7). *One microstate is as probable as another.*

In rolling the dice, one is much more likely to roll a 7 than a 12, because, while there is only one microstate for a 12 ($A6$, $B6$), there are 6 microstates that produce

---

* Ludwig Boltzmann (1844–1906), an Austrian physicist, made important contributions to the kinetic theory of gases and enunciated the Stefan–Boltzmann law of black-body radiation.

a 7 ($A1, B6; A2, B5; A3, B4; A4, B3; A5, B2; A6, B1$). It is in this sense that the macrostate that we observe is the most probable state.

In the system of 2 dice, we have 2 particles, each of which has 6 microstates. A system of atoms (particles) may exist in immeasurably more microstates. Consider an atom as a *harmonic oscillator* (§ 2.11) with respect to its mean position in an ideal crystal at absolute zero temperature, where the atom is in its ground level (lowest) energy state. All the atoms that make up the solid (system) are called collectively an *ensemble*. If each atom in the ensemble is in its ground level energy state, there is only 1 microstate inasmuch as each one has its unique position in space, the position being the only feature that distinguishes atom *A* from atom *B*. We now define Ω as the number of microstates a system may have without appearing macroscopically different. For example, in the case at hand, since the various atoms are identical in all respects, it does not make any difference where atom *A* is, in the upper left-hand corner of a crystal or anywhere else. With $\Omega = 1$, $\ln \Omega = 0$, and $S = 0$ in $S = \kappa \ln \Omega$, a consideration that led to the third law, § 6.12. Now suppose one of Planck's quanta of energy, $\varepsilon_0 = h\nu$, § 2.17 (where again in this article, *h* is Planck's constant, not enthalpy, and ν is frequency), enters this system; what happens is that as one of the particles receives it, its energy of oscillation increases by the amount $h\nu$. But if there are *N* particles (oscillators), there are now *N* different microscopic arrangements, depending on which particle receives the energy, that "look" the same macroscopically, and one microstate is just as probable as another. If the system receives *n* quanta of energy, these quanta may be distributed among the particles in many different ways; if $n < N$, then *n* particles may receive 1 quantum each (say, the first excited state); but some particles may require 2 (in the second excited state) or more quanta, so that the possible number of microstates greatly increases with *n*. The **allowed states** are those where the energy of one or more particles changes by the amount $h\nu$ or multiples thereof; no intermediate changes are possible. Also, the lower excited states are always more heavily populated than the higher excited states.

Consider an assembly consisting of *N* particles; of these *N* particles, say N_1 is in a certain quantum state (the energy ε_1 of each is a certain number of quanta), N_2 with energy ε_2 in another quantum state, and so on. The total energy E_t and the total number of particles *N* are each constant; summed for *i* particles, we have

(a) $$N = \sum_i N_i \quad \text{and} \quad E_t = \sum_i N_i \varepsilon_i$$

Then the total number of microstates or the thermodynamic probability is

(b) $$\Omega = \frac{N!}{N_1! N_2! \cdots N_i!} = \frac{N!}{\prod_i N_i!} \qquad [0! = 1]$$

where $N_1!$ is the number of arrangement of the particles N_1, $N_i!$ is the number of arrangements of the N_i particles with energy ε_i, and so on; the denominator is the product of these factorials. Since *the state of maximum probability is the observable macrostate*, the values of $N_1! \cdots N_i!$ that result in the maximum Ω are the ones used for computing entropy. We have

(5-19) $$S = \kappa \ln \Omega$$

where Boltzmann's constant , $\kappa = \bar{R}/N_A = 1.38 \times 10^{-16}$ ergs/K-molecule, the gas constant per molecule, equation (6-6). The units for S are the same as for κ, and the equation may be easily converted to mole quantities, as seen. With considerably more development, equations **(b)** and (5-19) can be put into more usable forms, but since the classical manner of computing entropy is adequate for by far most engineering problems, we shall leave further detail to another course.

Next, consider two ensembles of gases A and B with thermodynamic probabilities Ω_A and Ω_B; the total number of microstates Ω_t available if the ensembles combined is $\Omega_A\Omega_B$ (analogous to the dice with 6 states each, the total for the two is 36). Thus,

(c) $$S = \kappa \ln \Omega_t = \kappa \ln \Omega_A\Omega_B = \kappa \ln \Omega_A + \kappa \ln \Omega_B$$

that is, although the probabilities are multiplicative, the $\ln \Omega$ is additive, as is S.

Before leaving this phase of the discussion, we should include the concepts of order and disorder. As already implied, the order in a crystalline solid is good. The atoms and molecules have no mobility; only their energy levels are different. In a liquid, molecules move about, with their movements more restricted than in a gas, but with more disorder (more allowed microstates) than in the solid. In the gaseous phase, the motion is quite random. As we learned before, the state of a monatomic molecule is defined by its location (x, y, z) or velocity (v_x, v_y, v_z), the change of any one meaning a change of the molecule's state. A polyatomic molecule in addition may have states in which there are rotational and vibrational energies, which increase the possible number of microstates, and, of course, the randomness or disorder becomes more pronounced. Let a gas be brought to a high pressure and temperature. Since this is a state imposed on the substance, it represents a certain orderliness. It is in a condition which if left alone long enough will proceed to a more probable state, one of greater disorder, until eventually it arrives at the dead state, the most probable state of all (see statement C, § 5.2), the one of maximum entropy.

Summarizing, the observed macrostate corresponds to the microstate that is so probable that it is the only one we "see"; the variations on each side of the average are imperceptible. In the case of the two dice first mentioned, the most likely number to show is 7, but of course its probability is not so high as to preclude the observance of some others, on occasion with chagrin. If one tossed a billion dice at a time, the average total of the dots showing would be so nearly the same number that it is safe to say that that observed number is the macroscopic state. Yet a billion molecules is not very many (10^{20} or more are in a good handful).

On a microscopic basis, a violation of the second law is conceivable but, so far, undiscovered. It was Maxwell who invented a "demon" to circumvent the second law. Imagine a box containing a gas separated into two parts by a partition in which is a small hole and a cover over it. The gas is macroscopically at the same pressure and temperature everywhere. Now the "demon" is stationed on one side of the hole with instructions to open it when fast-moving molecules come along, but to close it to slow ones. The result is that the side of the box receiving the speedy molecules begins to experience a rise in temperature, while the other side experiences a drop (average molecular velocity decreasing). In their most probable state, the molecules have random disordered motion. Hence, if the hole is left unattended, as many molecules will go through in one direction as the other, and the energies of the molecules passing through will average the same in each direction. An inequality of molecular movement from one part to the other *may* occur, and the probability of this event happening increases as the duration of observation decreases; that is, for a very short instant, more molecules may pass to the right than to the left, but for any finite time this difference is not observable macroscopically. The significance of Maxwell's demon is that there might be some organism in the universe that has an equivalent selective power, resulting in a violation of the second law.

Man's efforts are generally directed toward producing ordered arrangements, toward minimizing the natural degradation of the second law whether the system be himself or a carrot or a space vehicle. Once an ordered configuration has been obtained, whether it is an educated mind or a steam power plant, then the system must be controlled and guided; effort must be expended to have its change to the dead state occur as efficiently as possible.

PERPETUAL MOTION OF THE SECOND KIND 5.28

A proposed machine that violates the second law of thermodynamics is called a *perpetual motion machine of the second kind*. For example, if it can be shown that such a machine or cycle produces more work from a given supply of energy than the available part, the machine will not operate as proposed.

THE SINK 5.29

In practice, over a short period of time, the sink temperature T_0 is constant; that is, the temperature of such a vast reservoir as the atmosphere, the ultimate repository, is not noticeably affected by all the heat discharged from all systems. The most economical T_0 is the one naturally at hand. If the discharge temperature T_0 were lowered by refrigeration, there would be a net loss because the work to run the refrigeration cycle would be more than the work gained from the power cycle—as the second law says. Many of our big central station power plants operate more efficiently during the winter than during the summer because of the lower sink temperature naturally available during the winter.

CLOSURE 5.30

On a microscopic scale, the system, say an ensemble of molecules, is indifferent as to whether a change of state occurs because of a work or a heat interaction. For example, the temperature of a certain closed, constant volume system may be raised a particular amount ΔT either by a paddle-work or by a heat input, causing the average energy of the molecules to increase the same amount in either case, and the final microstates are identical. But the external phenomena are significantly different. If the ΔT is brought about by paddle work, the work input to the system is accompanied by no change in the entropy of the surroundings (ignoring the entropy changes that occurred as the work was manufactured—in another system). If the ΔT is brought about by heat, some reservoir (source) undergoes a concurrent *decrease* in entropy, which means an overall smaller irreversibility, a profound dissimilarity from the thermodynamic point of view.

PROBLEMS

SI UNITS

5.1 There are supplied 3.60 MJ of heat to a Carnot power cycle operating between 900 K and 300 K. Sketch the *TS* diagram and find **(a)** total change of entropy during the heating or cooling process, **(b)** heat rejected from the cycle, **(c)** evaluation of $\oint dQ$, **(d)** total change of entropy during any one cycle.

Ans. **(a)** 4 kJ/K, **(b)** 1.20 MJ, **(c)** 2.40 MJ.

5.2 An inventor proposes a heat engine that receives heat (160 kJ/min) at 800 K. For a sink temperature of 500 K, he claims that 65 kJ/min of work are delivered. Do you think that this claim is justifiable? Is any thermodynamic principle violated? Explain.

5.3 Your automobile battery exchanges heat with only one reservoir, its surroundings. Is the work obtained from it in violation of the second law? Explain.

5.4 The ordinary household refrigerator is a system that receives work (electrical energy) and exchanges heat with a single reservoir (the kitchen, say). Why does this not refute the Kelvin-Planck statement of the second law as given in § 5.2(B) Explain.

5.5 Demonstrate that if the Clausius statement of the second law, § 5.2, were suddenly discovered invalid, a consequence would be the violation of the Kelvin-Planck statement.

5.6 Demonstrate that the expression $\Delta S \geqq 0$ results from the application of Clausius' inequality to an isolated system.

5.7 A Carnot power cycle 1–2–3–4 has superimposed an actual cycle 1–2'–3–4' (actual adiabatic expansion 1–2' and actual adiabatic compression 3–4'). See figure. Demonstrate the validity of the Clausius

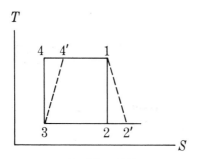

Problem 5.7

inequality $\oint dQ/T \leqq 0$ by applying it first to the Carnot cycle and noting that $\oint dQ/T = 0$ and then to the actual cycle and noting that $\oint dQ/T < 0$.

5.8 An inventor claims that not only has he developed a truly reversible adiabatic refrigerant compressor, but also has invented a throttling valve that will cause the state of the fluid to move along a path of diminishing entropy. As a consequence, the capacity (refrigeration) of a refrigerating system is increased. Refute the claims of this inventor, especially the one relevant to the throttling action, by applying the Clausius inequality to a reversed cycle employing these claims.

5.9 An industrial space heater (air blown over an electrical grid) uses 30 kW hr of electrical energy to heat the incoming atmospheric air from 10°C to 48.9°C. Calculate the entropy production of the air system and of the universe.

5.10 Steam with $h = 2442.6$ kJ/kg flows at the rate of 22.7 kg/s into a water-steam condenser at the constant temperature of 48.9°C. The condensed steam leaves with $h = 181.4$ kJ/kg. The cooling water ($c_p = 4.187$ kJ/kg-K) enters at 24.4°C and leaves at 37.8°C. Consider the condenser as an adiabatic steady flow system with $\Delta P = 0$ and $\Delta K = 0$. Compute the entropy production for the system; $t_0 = 21.1°C$.

Ans. $\Delta S_p = 9.30$ kJ/K-s.

5.11 There are transferred 500 kJ of heat from reservoir *A* (at 1000 K) to reservoir *B* (at 500 K); in each case the reservoirs remain at constant temperature (see § 1.23). According to the second law, what is the net change of entropy of the two reservoirs?

5.12 A liquid system whose mass is *m* kg at temperature *T* was formed by bringing together adiabatically two equal masses of the liquid; initially, one mass was at T_1 and the other at T_2. Show that the entropy production for the system (and of the universe) is $\Delta S_p = mc_p(T_1 + T_2)/2(T_1T_2)^{1/2}$ where the specific heat of the liquid is c_p. Also, prove that ΔS_p must be positive.

5.13 An electric current of 15 amp flows continuously through a 30-ohm resistor being maintained at a constant temperature of 28°C by moving cool air (initially at 15°C) across it.

For each minute find the entropy production (a) in the resistor, (b) in the air if its temperature rise is $\Delta t = 10°C$, and (c) of the universe.

5.14 Find the availability of a 1-kg mass of water in a closed container at a pressure of 13.79 bar and in the following state: (a) its temperature is 427°C; (b) it is saturated vapor; (c) it is saturated liquid. In all cases, the surrounding atmosphere is at $p_0 = 1$ atm, $t_0 = 26.7°C$ ($h_0 = 111.75$ kJ/kg, $v_0 = 0.231$ m^3/kg, $s_0 = 0.3903$ kJ/kg-K).

Ans. (a) 818 kJ.

5.15 Find the ratio of the availabilities of hydrogen to air, each pound of gas in a closed system at $p_1 = p_0$, T_1; the environment is at p_0, T_0. *Ans.* 14.24 to 1.0.

5.16 During a constant pressure nonflow process there are abstracted 787 kJ of heat from 2.27 kg of nitrogen initially at 1378.96 kPaa, 427°C; the environmental conditions are $p_0 = 101.325$ kPa, $t_0 = 15.6°C$. Find (a) the change in availability, (b) the maximum work the system could do either initially or finally, (c) the entropy production for the universe.

5.17 It is believed that a closed system executes a process during which its entropy change is +100 J/K while it receives 55 kJ of heat from a reservoir at 500 K. There are no other effects. State if this process is reversible or irreversible. Or possibly is it impossible?

5.18 During a reversible process, a closed system does 50 kJ of work while it receives 50 kJ of heat. State if the entropy change is positive, negative, or zero. Show tangible proof.

5.19 Find the irreversibility in each of the following situations: (a) 200 kJ of heat are transferred to the atmosphere directly from a constant-temperature reservoir that is at $p_A = p_0$ and 400 K; (b) 200 kJ of heat are transferred directly from the atmosphere to a constant temperature reservoir that is at $p_B = p_0$ and 200 K; (c) 200 kJ of heat are transferred directly from reservoir A to reservoir B. In all cases, the atmosphere is at $p_0 = 1$ atm, 300 K.

5.20 Differentiate the Gibbs function $G = H - Ts$ and note that if the temperature remains unchanged, there results $(\partial p/\partial G)_T = 1/v = \rho$, density. This states that for a given fluid, the slope of its isotherm on the pG plane is solely a function of its density. Apply this idea to a liquid-vapor water system in equilibrium at 1 atm, 100°C, and discuss the merits of the criterion $(\Delta G)_{p,T} \geqq 0$. Note that at 1 atm, 100°C, the density of the liquid water is 1600 times that of the water vapor.

5.21 Five cards each have a single number written thereon: 1, 2, 3, 4, and 5. Let two cards be withdrawn in succession. What is the probability that the sum of the two card numbers will be odd? How many microstates are represented by the problem statements?

Ans. 3 to 5, 12 microstates.

5.22 The same as **5.21** except let the probability relate to the sum being even.

5.23 A system is comprised of two dice, A and B. The pair of dice is now rolled. What is the probability of rolling an 8 (combined sum of the numbers on the two upper faces)? How many microstates are represented by this statement?

Ans. 5 to 36, 5 microstates.

5.24 The same as **5.23** except let the probability relate to the number (sum) 5.

5.25 All reversible heat engines have identical efficiencies when operating within the same temperature limits. Demonstrate that if this were not so, a perpetual motion machine of the second kind would result. Accomplish this by selecting two reversible engines, assign different efficiencies to them, and let the engine with the higher efficiency drive the one with the lower as a heat pump.

5.26 The entropy production of a gaseous system interacting with a constant temperature (T_s) heat source is being studied. Three gases (H_2, N_2, and CO_2) are used; see Table I. Let the source temperature be 2778 K and let each gaseous system interact independently and be heated at constant pressure from 300 K to 2222 K. First account for the variation of specific heats and find the entropy production of the gas/source system. Second, use the constant specific heats found in Item B 1 and find the entropy production. Compare results. Write a program for the foregoing problem.

ENGLISH UNITS

5.27 A 50-gal insulated water heater undergoes a recovery cycle from 90°F to

170°F in 30 min; the mass of water involved is 415 lb. Find the entropy production of the universe **(a)** if the heating is done by electricity, **(b)** if the heating is done by a constant temperature source at 840°F with all heat being received by the water ($c_p = 1.0$ Btu/lb-°F).

Ans. **(a)** 1.873 Btu/°R-min, **(b)** 1.027 Btu/°R-min.

5.28 If an ideal gas, whose equation of state is $pv = RT$ (see § 2.21), undergoes a reversible process, show that its change of specific entropy between the two state points is given by

$$\Delta s = c_v \ln T_2/T_1 + R \ln v_2/v_1$$

Begin with the expression for entropy change as given by the second law, that is, $\Delta S = \int_{rev} dQ/T$.

5.29 From a constant temperature reservoir (in the surroundings) at 3000°F, there are transferred 2000 Btu of heat to a Carnot engine. The engine receives the heat at 440°F and discharges at the sink temperature of 80°F. **(a)** Are the heat transfers reversible? Compute the change of entropy of the engine system accompanying the heat added process and the heat rejection process. **(b)** What is the net change of entropy of the universe in one cycle? of the engine in one cycle? **(c)** When all the Carnot work produced has been used, what is ΔS_p for the universe? **(d)** If the input to the engine is 2000 Btu of paddle work (other events remaining as first described), what is ΔS_p for the universe?

Ans. **(a)** 2.222, −2.222 Btu/°R, **(b)** 1.644 Btu/°R, 0, **(c)** 3.124, **(d)** 2.222 Btu/°R.

5.30 Assume 2 mol of oxygen are confined in a container at 100 psia, 500°F; the surroundings are at $p_0 = 14.7$, $t_0 = 80°F$. Find the availability of the oxygen. Use Item B 7.   *Ans.* 2466 Btu.

5.31 Say 1 lb of air at 120°F receives energy at constant volume, nonflow, with a pressure increase from $p_1 = 20$ psia to $p_2 = 75$ psia. If the environmental conditions are $p_0 = 15$ psia and $t_0 = 60°F$, determine the resulting change of availability of the air if the energy is **(a)** heat, **(b)** paddle-wheel work, and **(c)** electrical energy; heat source is at minimum temperature.   *Ans.* **(a)** 160 Btu.

5.32 Find the availability of a 1-lb mass of water that is flowing at 10 fps, is subjected

to a pressure of 200 psia, and is in the following state: **(a)** its temperature is 800°F; **(b)** it is saturated vapor; **(c)** it is saturated liquid. In all cases, the surrounding atmosphere is at $p_0 = 14.7$ psia, $t_0 = 80°F$. Compare answers with **5.17**.   *Ans.* **(a)** 474.0 Btu/lb.

5.33 If 1 lb of air is heated in an isobaric steady-state, steady-flow process at $p = 73.5$ psia from $t_a = 140°F$ to $t_b = 540°F$; $p_0 = 14.7$ psia and $t_0 = 40°F$, compute **(a)** its change of availability during the process and **(b)** the maximum work that can be delivered as the system changes from state *a* to the dead state.

Ans. [Item B 2.] **(a)** 35.335, **(b)** 92.62 Btu/lb.

5.34 Air leaves a nozzle at 14.7 psia, 740°F, 2000 fps in an environment where $p_0 = 14.7$ psia, $t_0 = 80°F$. For each 1 lb/sec, what is the maximum work that this jet is capable of producing; $\Delta P = 0$. Use Item B 2.

Ans. 137 Btu/sec.

5.35 A system consisting of 2 lb of oxygen receives heat at constant pressure of 20 psia, the state changing from 20 psia and 150°F to 640°F in an environment where $p_0 = 15$ psia and $t_0 = 40°F$. **(a)** For both nonflow and steady-state, steady flow processes ($\Delta P = 0$, $\Delta K = 0$) to the dead state, determine the availability of the oxygen before and after heating. **(b)** What is the change of availability? Solve by gas tables.

Ans. (Steady flow) **(a)** 22.5, 113 Btu, **(b)** 90.5 Btu.

5.36 Saturated steam vapor at 90°F enters the shell side of a surface type condenser and leaves as saturated liquid at 90°F. Cooling water flows through the condenser tubes entering at 60°F and in the mass ratio of 57.9 lb water/1 lb steam. Find **(a)** exit temperature of the water, **(b)** change in availability of the steam, **(c)** change in availability of the water, **(d)** entropy production of the universe per pound of steam. For the surroundings, $p_0 = 14.7$ psia, $t_0 = 50°F$.

5.37 A frictionless piston cylinder arrangement contains 1 lb of water initially at 100 psia, 400°F; the surrounding atmosphere is $p_0 = 14.7$ psia, $t_0 = 80°F$. Find the change in availability, the entropy production of the universe, and the irreversibility in each of the following cases: **(a)** the system receives heat isothermally until its volume doubles; the heat source is at the minimum possible temperature; **(b)** the system receives heat at constant pressure until its volume doubles; the heat source is at the minimum possible temperature; **(c)** the same as **(b)** except that the process occurs because of internal stirring rather than a heat interaction.

5.38 For each of the following described processes, a steady flow of water vapor at 120 psia, 500°F enters the device at low velocity and leaves at 35 psia; $p_0 = 14.7$ psia, $t_0 = 40°F$. For each case find the change of availability, the entropy production of the universe, and the irreversibility; **(a)** it expands reversibly and adiabatically in a turbine and leaves with a low velocity; **(b)** it expands reversibly and adiabatically in a nozzle; **(c)** it expands adiabatically through a porous membrane and leaves with a low velocity.

5.39 Steam at 456°F with a specific heat of $c_p = 0.6$ Btu/lb-°R is to be heated to 708°F in a heat exchanger by heat from a gas whose $c_p = 0.24$ and whose initial temperature is 1500°F. The rate of flow of the gas is 52.5 lb/sec, and of the steam 25 lb/sec, all in steady flow. Compute **(a)** the final temperature of the gas; **(b)** the change in entropy of the steam, of the gas, and the net change of entropy; and **(c)** the change of availability and the irreversibility of the exchanger for a sink at 530°R.

Ans. **(a)** 1200°F, **(b)** $\Delta S_p = 1.5523$ Btu/°R-sec, **(c)** $I = 823$ Btu/sec.

5.40 During the execution of a cycle, 2 lb of air receive heat at constant pressure while the temperature increases from 900°F to 1200°F in an internally reversible manner. The natural sink temperature is 95°F. Determine the amount of heat that can be converted into work.

5.41 The temperature of 5 lb of air is raised from 200°F to 500°F at constant pressure with an electrical coil . **(a)** How much energy is required? Is this input reversible? **(b)** If the environmental temperature is $t_0 = 40°F$, what part of the above energy was available before it entered the gaseous system? After it entered? **(c)** Find ΔS of the gaseous system; of the surroundings. **(d)** What is the entropy production?

Ans. **(a)** 360, **(b)** 360, 135 Btu, **(c)** $\Delta S_g = 0.45$, **(d)** 0.45 Btu/°R.

5.42 Assume 1 lb of wet steam initially at 200 psia, quality $x_1 = 50\%$, is contained in a frictionless piston-cylinder arrangement; $p_0 = 14.7$ psia, $t_0 = 60°F$. Heat is received $(p = c)$ until the steam becomes dry vapor at 600°F; the heat reservoir is at the lowest possible constant temperature. **(a)** For the process find the change in availability of the steam, the irreversibility, and the available portion of the heat transferred. **(b)** If the energy input had been by means of internal paddle work instead of heat, find the change of availability and the irreversibility.

5.43 Develop the relation between heat and entropy that occur during an irreversible process, namely $\int_{irr} ds > \int_{irr} dQ/T$.

5.44 There are added 50 Btu of heat to 2 lb of oxygen at constant temperature. Determine the change in **(a)** the Helmholtz function, **(b)** the Gibbs function. **(c)** Does either of the above answers signify anything? **(d)** If the constant temperature is 600°R and the sink temperature is 500°R, what portion of the heat is available? Is unavailable?

5.45 Expand 2 lb/sec of carbon monoxide at 120 psia, 100°F to a state of 15 psia, 100°F; $t_0 = 100°F$, $\Delta P = 0$, $\Delta K = 0$. **(a)** Find the maximum work possible for any process between these state points. **(b)** What is the heat if the process is a reversible isothermal? *Ans.* **(a)** 165, **(b)** 165 Btu/sec.

5.46 Assume 3 lb of air are at 100 psia, 1000°R. Using Item B 2, determine **(a)** the Helmholtz function, **(b)** the Gibbs function. **(c)** If the air expands to 20 psia and 1000°R, what maximum work can be expected and what will be the total entropy of the air?

Ans. **(a)** -4340, **(b)** -4134, **(c)** 331 Btu, 5.188 Btu/°R.

5.47 A system consisting of a stoichiometric mixture of 1 mol of CH_4 and 2 mol of O_2 at 1 atm and 77°F reacts to produce a products-mixture of 1 mol of CO_2 and 2 mol of H_2O vapor also at 1 atm and 77°F. Using Item B 11, determine ΔS and the maximum amount of work per mol of CH_4 that can be developed from this reaction.

Ans. 344,504 Btu/mol CH_4, -1.235 Btu/mol CH_4-°R.

5.48 In § 5.26, we have, as one criterion of equilibrium for an isolated system of constant mass, $\Delta S)_{E,m} \leq 0$. Under these same conditions, another criterion must also obtain, $\Delta E)_{S,m} \geqq 0$. This simply states that if a system of constant mass is in stable equilibrium, the energy of the system must increase, or remain constant, for all possible adiabatic spontaneous variations. Demonstrate that these two criteria are equivalent.

5.49 Apply the criterion $\Delta E)_{S,m} \geqq 0$ to a system composed of a marble resting on the bottom of a bowl and demonstrate that the system is in a state of static equilibrium.

6

THE IDEAL GAS

From previous descriptions and from the defining equation $pv = RT$ that we have previously needed in our discussions, the reader already has a good definition of an *ideal gas* or *perfect gas*. The consequences include certain laws, called ideal gas laws, that we could now evolve quite compactly, but since these laws are so serviceable in dealing with ideal gases, the emphasis of looking at them individually is worth the time.

In general, the equilibrium states of substances are defined by an equation of state relating three properties, usually p, v, and T because they are directly measurable. The ideal gas is ideal in that its equation of state $pv = RT$ is so simple that the resultant mathematics is simple, which makes it a fitting substance to learn about while the ideas of the previous chapters are being assimilated.

Not only is idealization a practical thing to do in engineering and science, but it also happens that many actual gases behave very nearly as an ideal gas. *All gases approach the ideal gas behavior as pressure decreases*, because the molecules are then getting farther apart so that forces of attraction between molecules are approaching zero and the molecules themselves are occupying a negligible part of the volume. Hence, we think of a gas at "low" pressure as acting as an ideal gas, but "low" must be interpreted in terms of the substance. The gases with the smaller molecules, monatomic and diatomic gases, approach ideal gas behavior most closely; so "low" pressure for He, H_2, air, and so on, may be as much as several hundred psi, but for steam at atmospheric temperatures, "low" pressure is of the order of 1 psia or less.

Since there is no distinct line of demarcation for an actual gas between states where it acts "ideally" and where it does not, the engineer must often make a decision based on his experience and know-how. *If the ideal-gas laws yield sufficiently accurate results for the purpose, the substance is considered as an ideal or perfect gas; otherwise it is a nonideal or imperfect gas.* At this state the reader accepts the judgment of his test; in practice, an enormous number of engineering problems can be satisfactorily solved with the ideal-gas constraint if realistic values of specific heats are used.

6.2   BOYLE'S LAW

Robert Boyle (1627–1691), during the course of experiments with air, observed the following relation between the pressure and volume. *If the temperature of a given quantity of gas is held constant, the volume of gas varies inversely with the absolute pressure during a quasistatic change of state (**Boyle's law**).* In mathematical form, if a gas is in a condition represented by state 1, Fig. 6/1, and undergoes a change of state at constant temperature (***isothermal process***) to state 2, then

$$(6\text{-}1) \qquad \frac{p_1}{p_2} = \frac{V_2}{V_1} \qquad \text{or} \qquad p_1 V_1 = p_2 V_2 \qquad \text{or} \qquad pV = C \qquad \text{or} \qquad pv = C$$

$$[\text{BOYLE'S LAW}, \ T = C]$$

applicable to a particular mass in any two equilibrium states 1 and 2, where C is a constant in a generic sense (we say, when $T = C$, then $pv = C$; but $T \neq pv$), and equation $pv = C$, for example, represents a family of curves, a different curve for each different constant.

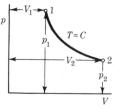

Fig. 6/1   *Boyle's Law.* The equation on the curve on the pV plane is $pV = C$ or $pv = C$, an equilateral hyperbola. A point such as 1 is a *state point.* The curve joining 1 and 2 is the *path of the state point* as the state of the substance changes, a *process.*

6.3   CHARLES' LAW

About a hundred years after the discovery of Boyle's law, two Frenchmen, Jacques A. Charles (1746–1823) and Joseph L. Gay-Lussac (1778–1850), each without knowledge of the other's work, discovered the law that we usually call ***Charles' law***. This law is in two parts, one for constant pressure and one for constant volume. For any two equilibrium states 1 and 2 of a particular mass for which the pressure is the same,

$$(6\text{-}2) \qquad \frac{V_1}{V_2} = \frac{T_1}{T_2} \qquad \text{or} \qquad \frac{T_1}{V_1} = \frac{T_2}{V_2} \qquad \text{or} \qquad \frac{T}{V} = C \qquad \text{or} \qquad \frac{T}{v} = C$$

$$[\text{CHARLES' LAW}, \ p = C]$$

For any two equilibrium states 1 and 2 of a particular mass for which the volume is the same,

$$(6\text{-}3) \qquad \frac{p_1}{p_2} = \frac{T_1}{T_2} \qquad \text{or} \qquad \frac{T_1}{p_1} = \frac{T_2}{p_2} \qquad \text{or} \qquad \frac{T}{p} = C$$

$$[\text{CHARLES' LAW}, \ V = C]$$

* Edme Mariotte, a Frenchman, independently discovered this same principle at approximately the same time that Boyle did. Although Mariotte, therefore, is due fully as much credit for the discovery, the law is more commonly called Boyle's in this country.

where the temperatures T are necessarily absolute temperatures. These equations are seen to be straight lines ($y = Cx$), Fig. 6/2, that pass through the origin with slope C. This would be the origin for all *ideal* gases, but curves for actual gases only approach the origin as the pressure is reduced, and moreover, actual gases undergo changes of phase at low temperatures. Experiments on nearly perfect gases furnish a clue to absolute, ideal gas zero temperature.

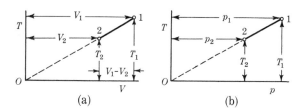

Charles' Law. In (a), the line 1-2 represents a constant pressure process (isobaric) for an *ideal gas* as it appears on the *TV* plane. In (b), the line 1-2 represents a constant volume (isometric) process as it appears on the *Tp* plane. **Fig. 6/2**

For example, the slope of the line in Fig. 6/2(a) is

(a)
$$\frac{T_1 - T_2}{V_1 - V_2} = \frac{\Delta T}{\Delta V} = \frac{\Delta t}{\Delta V} = C$$

Experimental data for air at $p = 14.696$ psia are: at $t_2 = 32°F$, the specific volume is $v_2 = 12.39 \text{ ft}^3$; at $t_1 = 100°F$, $v_1 = 14.1 \text{ ft}^3$. Then

(b)
$$C = \frac{t_1 - t_2}{v_1 - v_2} = \frac{100 - 32}{14.1 - 12.39} = 39.76$$

(c)
$$T = Cv = 39.76v$$

This is the equation of the line on Fig. 6/2(a) for air and the data given. Thus, the absolute temperature at 32°F (where $v = 12.39$), as computed for this state, is $T = (39.76)(12.39) = 492.6°$ abs., a rough estimate of the absolute temperature on the Fahrenheit scale corresponding to the freezing point of water, 32°F. With more accurately measured volumes, we might have obtained a better answer. If the data for some other gas were used as above, we should find a slightly different absolute zero point. Thus, we may have as many absolute gas scales as gases. However, extrapolating to zero pressure and by using a more refined technique[1.71] with a more nearly perfect gas (helium at low pressure), we find that 32°F is 491.69°R, the thermodynamic temperature. We have already learned that the *ideal* gas temperature and the thermodynamic temperature are the same (Chapter 5). Hence, from now on, we shall call it simply the **absolute temperature.**

AVOGADRO'S LAW 6.4

An Italian physicist, Amadeo Avogadro (1776–1856), stated that: *Equal volumes of all ideal gases at a particular pressure and temperature contain the same number of molecules.* This statement is strictly true only for an ideal gas.

The number of molecules in a mole of any substance is a constant N_A, called **Avogadro's number**, representing a logical conclusion; the molecular mass M is proportional to the mass m of a molecule, or $M_X/M_Y = m_X/m_Y$, say for any substances X and Y, or $M_X/m_X = M_Y/m_Y = M/m = N_A = 6.0225 \times 10^{23}$ gmole^{-1}.

Returning to Avogadro's law, we may write for any two gases X and Y, when each is occupying the same volume at the same p and T.

(a) $$\frac{M_X}{M_Y} = \frac{\rho_X}{\rho_Y} = \frac{v_Y}{v_X} \quad \text{or} \quad M_X v_X = M_Y v_Y = Mv$$

Since the ideal gases X and Y are randomly chosen, Mv is the same for all gases and is called the *mole volume* $\bar{v} = Mv$. The following numbers suggest how the mole volumes vary for actual gases, all for $p = 1$ atm and $t = 32°F$: H_2, 359 ft^3; He, 358.8 ft^3; CO_2, 356.4 ft^3; ammonia NH_3, 353.5 ft^3.

6.5   THE EQUATION OF STATE

We may use any two of the three Boyle and Charles laws to derive the ideal gas equation, more or less the historical evolution. For *any* two states 1 and 2 (see Fig. 6/3), we imagine them connected by isobaric 1-*a* and isometric *a*-2 lines, and apply Charles' law as follows ($v_a = v_2$, $p_a = p_1$):

(a) $$\frac{v_1}{v_a} = \frac{T_1}{T_a} \quad \text{or} \quad T_a = T_1\left(\frac{v_a}{v_1}\right) = T_1\left(\frac{v_2}{v_1}\right) \qquad [p = C]$$

(b) $$\frac{p_a}{p_2} = \frac{T_a}{T_2} \quad \text{or} \quad T_a = T_2\left(\frac{p_a}{p_2}\right) = T_2\left(\frac{p_1}{p_2}\right) \qquad [v = C]$$

Equating the two values of T_a and collecting like subscripts, we get

(c) $$\frac{p_1 v_1}{T_1} = \frac{p_2 v_2}{T_2} = R \qquad \text{a constant}$$

Since a like relation could be derived for any other random pair of points (as 1 and 3, $p_1 v_1/T = p_3 v_3/T_3$), we may generalize and say that pv/T must be a constant for a particular ideal gas. In English units, for p lb/ft^2, v ft^3/lb, $T°$ R, the constant R is called the *Specific gas constant* or just the *gas constant*.

(6-4A) $$pv = RT \qquad p = \rho RT \qquad \text{and} \qquad pV = mRT$$

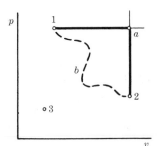

Fig. 6/3   Equilibrium states 1 and 2 are randomly located (also 3); path 1-*b*-2 is any kind of path; for 1-*a*, $p = C$; for *a*-2, $v = C$.

where $\rho = 1/v$ = density and $V = mv$ for m lb. If each side of $pv = RT$ is multiplied by M, we get

(d)
$$\bar{R} = MR = \frac{p(Mv)}{T} = \frac{p\bar{v}}{T}$$

Since $Mv = \bar{v}$ varies inversely with pressure (Boyle) and directly with T (Charles) and since Mv is the same for all ideal gases, it follows that MR is the same for all ideal gases. This constant $\bar{R}$ is called the **universal gas constant**, and its accepted values are

(e) $$\bar{R} = 1545.32 \frac{\text{ft-lb}}{\text{pmole-°R}} \quad \text{and} \quad \bar{R} = \frac{1545.32}{778.16} = 1.9859 \frac{\text{Btu}}{\text{pmole-°R}}$$

also 1.9859 cal/gmole-K; use 1545 and 1.986 for ordinary work. See Item B 38 for values of $\bar{R}$ in various units. The equation of state for moles is

(6-4B)
$$p\bar{v} = \bar{R}T \quad \text{or} \quad pV = n\bar{R}T$$

where n is the number of moles in total volume, $V = n\bar{v}$.

An ideal gas is defined as one that accords with equation (6-4) *and* has constant specific heats, certainly the most easily handled expansible substance. It happens that more and more tabulations of gas properties that account for the variation of specific heat are becoming available. If such properties are at hand, § 6.15, the principal reason for using constant specific heats is pedagogic simplicity—unless they really are nearly constant or unless some principles of thermodynamics are to be presented in their simplest form. A gas that follows $pv = RT$ but whose specific heats vary could be considered a slightly tainted ideal gas. We shall let the context say whether the specific heats vary or not, or simply specify it.

THE GAS CONSTANT   6.6

The value of R for any gas may be determined from accurate experimental observations of simultaneous values of p, v, and T. See Item B 1 for others. Considering the units of R, we may write

(a)
$$R \rightarrow \frac{(\text{pressure unit})(\text{volume unit/unit mass})}{\text{absolute temperature}}$$

Thus, R *could* be computed to be in any combination of units matching these dimensions. From another point of view, since the unit of pv is an energy unit $[(\text{lb/ft}^2)(\text{ft}^3/\text{lb}) = \text{ft-lb/lb}]$, we see that

(b)
$$R \rightarrow \frac{\text{energy unit}}{(\text{mass})(\text{absolute temperature})}$$

which could be Btu/lb-°R (Item B 1). Note that R ft-lb/lb-°R has a pound of force in the numerator and a pound of mass in the denominator. In all cases, the units must be rationalized.

Since R is considered an ideal gas constant and since the accepted value of $\bar{R} = 1545.32$, the gas constants in Item B 1 are computed from $1545.32/M$, where M is the molecular mass.

On the level of a molecule, the gas constant $\bar{R}$ for 1 mole divided by the number of molecules N_A in a mole is

$$(6\text{-}5) \qquad \kappa = \frac{\bar{R}}{N_A} = \frac{8.3143 \times 10^7}{6.02252 \times 10^{23}} = 1.38054 \times 10^{-16} \frac{\text{ergs}}{\text{K-molecule}}$$

the gas constant per molecule, called **Boltzmann's constant**, a fundamental constant of science that we previously used (§ 1.17); $\bar{R} = 8.3143$ J/gmole-K, or 8.3143×10^7 ergs/gmole-K. Also $\kappa = 1.38054 \times 10^{-23}$ J/K-molecule.

Two additional values of $\bar{R}$ that will be used extensively are $\bar{R} = 1.9859$ Btu/pmole-°R and $\bar{R} = 8.3143$ kJ/kgmole-K (for SI use.)

6.7   JOULE'S LAW

Joule, using an experiment first performed by Gay-Lussac, submerged two copper vessels A and B in water and performed the experiment suggested by Fig. 6/4. From instruments then available, Joule observed that the temperature of the water surrounding the containers was the same after as before the test, a simple observation that leads to an important deduction.

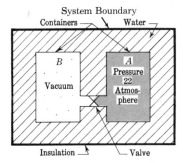

Fig. 6/4   *Joule's Experiment.* Vessel *A* held air at 22 atm; vessel *B* of the same volume was evacuated. The contents of the insulated system were in thermal equilibrium. After the valve was opened, and the gas was at rest, the pressure was measured to be 11 atm.

There are several facts to be noted. First, there was a momentary flow of air, giving rise to energy terms characteristic of flow, but the gas was at rest in the initial and final states. Second, inasmuch as there was no connection by which energy could leave the system as work, $W = 0$. Third, since the temperature of the water was the same, the conclusion is that no flow of heat from the air to the water, or from the water to the air, occurred; $Q = 0$. Fourth, the temperature of the air in container A must have dropped, since the air in it at any instant during flow was doing work in pushing gas into B (§ 7.25). The temperature of the gas in B underwent a rise because work was being done on it. However, when these short-lived effects again equalized thermally, the temperature of the air in the containers must have been closely the same as initially, since otherwise, because of the temperature difference, there would have been a net transfer of heat.

From the energy equation for A and B, $\Delta U = Q - W = 0$. The pressure changed, the volume changed, but the internal energy did *not* change; that is, $(\partial u/\partial v)_T = 0$ and $(\partial u/\partial p)_T = 0$. Consequently, we conclude with Joule that the internal energy of this gas was not a function of the pressure or volume. However since the temperature did not change, Joule deduced that *the change of internal energy of an ideal gas is a function of only the temperature change, u = u(T)—**Joule's law**.* Joule was not unaware of the vagaries of the experiment, including not only the flow phenomena, but probably of the fact that since the mass of the container and water was very much larger than the mass of air, a small change in air temperature might go undetected in water temperature. Nevertheless, the law can now be proved mathematically for the ideal gas [that is, $(\partial u/\partial v)_T = 0$); see equation (11-24), § 11.16]; and all gases approach conformity to Joule's law as their pressure is decreased.

SPECIFIC HEATS OF AN IDEAL GAS   6.8

In accordance with Joule's law, that Δu is a function of temperature only, we now drop the constant volume restriction of § 2.19 and write

$$(6\text{-}6) \qquad c_v \equiv \frac{du}{dT} \qquad du = c_v\,dT \qquad u = \int_0^T c_v\,dT = \int_0^T c_v(T)\,dT$$

[IDEAL GAS ONLY, ANY PROCESS, EQUILIBRIUM STATES]

If it is permissible to use a constant c_v, $\Delta u = c_v(T_2 - T_1)$. If a suitable value of c_v is used, we also have $u = c_v T$ from the integral above, which we would think of as being the internal energy measured from a datum of $0°R$. Sometimes we use $u = c_v t$, which is from $0°F$ datum.

Similarly, from equation (2-23),

$$(6\text{-}7) \qquad c_p = \frac{dh}{dT} \qquad dh = c_p\,dT \qquad h = \int_0^T c_p\,dT = \int_0^T c_p(T)\,dT$$

[IDEAL GAS, ANY PROCESS, EQUILIBRIUM STATES]

When c_p is constant, $\Delta h = c_p(T_2 - T_1)$. For a proper value of c_p, $h = c_p T$ (or $h = c_p t$ from $0°F$ datum). If instantaneous values of c_p, c_v are used in this manner, the corresponding values of h and u are somewhat high; but the result is often a convenient first approximation.

From the definition of $h \equiv u + pv$ and $pv = RT$, we have $h = u + RT$, ideal gas; and $dh = du + R\,dT$. Using equations (6-6) and (6-7) in this equation, we have

$$(6\text{-}8) \qquad c_p\,dT = c_v\,dT + R\,dT \qquad \text{or} \qquad c_p = c_v + R$$

repeated here from § 2.22 for convenient reference. Since $k = c_p/c_v$, use $c_p = kc_v$ in equation (2-31) and find

$$(6\text{-}9) \qquad c_v = \frac{R}{k-1} \qquad \text{and} \qquad c_p = \frac{kR}{k-1}$$

but since k varies when the specific heats vary, these values of c_v and c_p apply in general at particular states, or are suitable average values. Since c_p, c_v, and R must have the same units, use R/J Btu/lb-°R values from Item B 1 for typical problem work in this course. Any of the foregoing equations are appropriate for a unit mole;

(a) $d\bar{u} = C_v\,dT \qquad \bar{u} = C_v T \qquad d\bar{h} = C_p\,dT \qquad \bar{h} = C_p T$

(b) $C_p - C_v = \bar{R} \qquad C_v = \dfrac{\bar{R}}{k-1} \qquad C_p = \dfrac{k\bar{R}}{k-1}$

where our usual $\bar{R} = 1.986$ Btu/pmole-°R.

6.9   Example

A gas, initially at $p_1 = 517.2$ kPaa and $V_1 = 142\,\ell$, undergoes a process to $p_2 = 172.4$ kPaa and $V_2 = 274\,\ell$ during which the enthalpy decreases 65.4 kJ. The specific heats are constant; $c_v = 3.157$ kJ/kg K. Determine **(a)** ΔU, **(b)** c_p, and **(c)** R.

Solution.   Since we do not know the nature of the process, we must consider various basic relations. In solving for ΔU, normally we would use $\Delta U = mc_v\,\Delta T$. We could eliminate T by using $T = pV/mR$ but we know neither m or R. Thus we resort to another relation, $\Delta H = \Delta U + \Delta(pV)$.

(a) Being a decrease, ΔH is negative:

$$\Delta H = -65.4 = \Delta U + [(142.4)(274) - (517.2)(142)]$$

$$= \Delta U - 26.2$$

$$\Delta U = -65.4 + 26.2 = -39.2 \text{ kJ}$$

(b) From equations (6-6) and (6-7) we note

$$\frac{\Delta h}{\Delta u} = \frac{c_p}{c_v} = k = \frac{-65.4}{-39.2} = 1.668$$

(c) From equation (2-31), we have

$$R = c_p - c_v = 5.26 - 3.157 = 2.103 \text{ kJ/kg K}$$

where $c_p = kc_v = (1.668)(3.157) = 5.26$ kJ/kg K.

6.10   DALTON'S LAW OF PARTIAL PRESSURE

It was John Dalton (1766–1844) who first stated that *the total pressure p_m of a mixture of gases is the sum of the pressures that each gas would exert were it to occupy the vessel alone at the volume V_m and temperature T_m of the mixture*—but this law too turns out to be strictly true only for the ideal gas. If p_a, p_b, and p_c represent, respectively, the individual pressures of the mixed gases A, B, C, for i constituents of a mixture, Dalton's law states

(6-10) $p_m = p_a + p_b + p_c + \cdots = \sum_i p_i$

$$[T_m = T_a = T_b = T_c \qquad V_m = V_a = V_b = V_c]$$

Apply $pV = n\bar{R}T$ to an individual gas and to the mixture; for example,

(a) $$p_a V_m = n_a \bar{R} T_m \quad \text{and} \quad p_m V_m = n_m \bar{R} T_m \qquad \left[n_m = \sum_i n_i \right]$$

where n is the number of moles. By division of these equations, we get

(b) $$\frac{p_a V_m}{p_m V_m} = \frac{n_a \bar{R} T_m}{n_m \bar{R} T_m} \quad \text{or} \quad \frac{p_a}{p_m} = \frac{n_a}{n_m} = X_a$$

where X_a is called the volumetric or mole fraction (or percentage) of gas A (it is the fraction of the total volume V_m that gas A would occupy if it were separated from the other gases at p_m and T_m). We may generalize that for ideal gases, the partial pressure of the i component of the mixture is its volumetric fraction times the total pressure p_m;

(6-11) $$p_i = X_i p_m \quad \text{where} \quad \sum_i X_i = 1$$

THE JOULE-THOMSON EXPERIMENT   6.11

To obtain more accurate data than provided by Joule's experiment of Fig. 6/4, Joule and William Thomson (Lord Kelvin, § 8.5) devised in 1853 the famous porous-plug experiment diagrammed in Fig. 6/5. It consisted of a well-insulated pipe with a porous plug that offered considerable resistance to flow, so that there could be a relatively large drop in pressure from p_1 to p_2. In checking off the various energies, we conclude that $Q = 0$, $W = 0$, $\Delta P = 0$, and the rate of flow is low so that K is small and, virtually, $\Delta K = 0$. As for the remaining energies, we find $u_1 + p_1 v_1$ crossing boundary 1 and $u_2 + p_2 v_2$ crossing 2; in short, from the conservation of energy, $h_1 = h_2$, but the process itself is irreversible with its details undefinable. This kind of process, as the opening of a water faucet and letting the water flow, is called a **throttling process** when it is *steady flow*. Whenever a fluid flows in an uncontrolled manner from a higher pressure region to a lower pressure region with no work being done, we say that it is *throttling*, but it is not a throttling process as defined (steady flow).

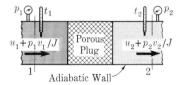

Joule–Thomson Flow. Steady flow; upstream and downstream properties, p, t, etc., are maintained constant at boundaries 1 and 2 where turbulence of any sort approaches zero.   **Fig. 6/5**

Since $h = h(T)$ for an ideal gas, equation (6-7), we would expect no difference between temperatures T_1 and T_2 if an ideal gas were flowing. However, even for the more nearly ideal gases, Joule and Thomson found a difference. As a matter of fact, it turned out that a gas has an inversion point I, Fig. 6/6; they found for air that $t_2 < t_1$, as, say $t_d < t_c$, Fig. 6/6. But they found that the hydrogen's temperature increased, $t_2 > t_1$, as, say, $t_b > t_a$. Of course, the curve $abcd$ is not good for both H_2

and air; we mean that at the usual atmospheric temperatures or thereabouts, the state of H_2 is outside of its inversion point, while that of air is inside. The *isenthalpic* (constant enthalpy) curve is simply the locus of all states with the same enthalpy; it is not a process.

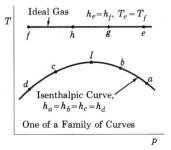

Fig. 6/6   *Isenthalpic Curves.* For an ideal gas *ef*, $T_e = T_f$ as well as $h_e = h_f$. For an actual gas *abcd*, *I* is the inversion point. Throttling *a* to *b*, temperature increases; throttling *c* to *d*, temperature decreases.

The Joule-Thomson discovery is that for actual gases, h and u *are* functions of pressure; and the Joule-Thomson coefficient $\mu = (\partial T / \partial p)_h$, which is the change of temperature with respect to pressure while the enthalpy is constant ($dh = 0$), has become a measure of the degree of departure of the gas from ideal. As simple as this experiment is, together with the mathematical relations of thermodynamics (Chapter 11), it can result in an amazing amount of information (§ 11.19).

6.12   THIRD LAW OF THERMODYNAMICS

The quantum theory and statistical thermodynamics suggest that the entropy of a pure substance approaches zero as the absolute zero temperature is approached; $s \to 0$ as $T \to 0$ for any process. This point has been so intensely investigated during this century that a forthright statement of the *third law* of thermodynamics is: *at absolute zero, the entropy of a pure substance (in equilibrium at 0°R) in some "perfect" crystalline form becomes zero.* This law is the outcome of many contributions, especially of Einstein, Nernst (whose *heat theorem* was probably the first formal statement), Boltzmann, and Planck. It follows that all substances as they actually exist have finite positive entropies.

Numerical values of absolute entropies are now available for many substances; often, values at a standard state only are given, as $\bar{s}°$ in Table III and Item B 11, Appendix. The most common standard state for this purpose is $p° = 1$ atm and $t° = 77°F$ (25°C).

TABLE III   Absolute Entropy of Some Gases

Values of $\bar{s}°$ at 1 atm and 77°F in Btu/pmole-°R = cal/gmole-K = kcal/kgmole-K. See Item B 1 for molecular masses; Item B 11 for more absolute entropies.

Argon, 36.983	Fluorine, 48.6	Hydrogen, 31.208
Chlorine, 53.298	Helium, 30.126	Neon, 34.948

* Many scholars object to this statement. Perhaps the proper inference from the evidence is *the change of entropy during an isothermal process approaches zero as ·the temperature approaches zero*; $\lim_{T \to 0} (\partial s / \partial v)_T = 0$ and $\lim_{T \to 0} (\partial s / \partial p)_T = 0$.

If a change of state of a system takes place irreversibly, as along the dotted path 1-*b*-2, Fig. 6/7, the change of entropy can be found from the definition of entropy in terms of a reversible process ($\Delta s = \int dQ/T$) by devising a reversible path between the states. This path may consist of any series of internally reversible processes (Chapter 7), such as 1-*d*-2, Fig. 6/7. However, for the ideal gas, we can use either equation (4-14) or (4-15) and $pv = RT$ and obtain equations that automatically set up the path. In equation (4-14), substitute $R/v = p/T$,

(a)
$$ds = \frac{du}{T} + \frac{p\,dv}{T} = c_v\frac{dT}{T} + \frac{R\,dv}{v}$$

[ANY SUBSTANCE]   [IDEAL GAS]

The ideal-gas form can be integrated with a $f(T)$ for c_v from Table I, or with a suitable *constant* value;

(6-12A)
$$\Delta s = \int_1^2 \frac{c_v(T)\,dT}{T} + R\ln\frac{v_2}{v_1} \qquad \Delta s = c_v\ln\frac{T_2}{T_1} + R\ln\frac{v_2}{v_1}$$

$[c_v = f(T)]$ [CONSTANT c_v]

usually in Btu/lb-°R, but observe that *it is only necessary for s, c_v, and R to be in the same units*. Now that we have the entropy change in (6-12A) specifically in terms of point functions, this demonstrates at least for an ideal gas that entropy is a point function, as previously asserted. From (6-12A), we see that if $v = C$ with c_v constant, then $\Delta s = c_v\ln(T_2/T_1)$; if $T = C$, $\Delta s = R\ln(v_2/v_1)$. A clear physical meaning of equation (6-12A) is given in Fig. 6/7.

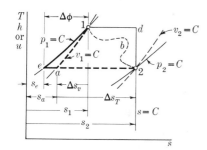

Entropy Change. The scale of the ordinate can also be for *h* and *u* as well as *T* for *ideal gases only*. Notice that the $c_v\ln(T_2/T_1)$ term of (6-12) evaluates the change along an internally reversible $v = C$ path 1-*a*, and that the $R\ln(v_2/v_1)$ term evaluates it along an internally reversible $T = C$ path *a*-2, the algebraic sum is $\Delta s = s_2 - s_1$. In equation (6-13), the $c_v\ln(T_2/T_1)$, or $\Delta\phi$, term evaluates the change along the quasistatic $p = C$ process 1-*e*; $R\ln(p_2/p_1)$ evaluates it *e*-2, the net change being $s_2 - s_1$ again.

Fig. 6/7

Similarly, from equation (4-15), and $v/T = R/p$, we have

(b)
$$ds = \frac{dh}{T} - \frac{v\,dp}{T} = c_p\frac{dT}{T} - \frac{R\,dp}{p}$$

[ANY SUBSTANCE]   [IDEAL GAS]

(6-13A)
$$\Delta s = \int_1^2 \frac{c_p(T)\,dT}{T} - R\ln\frac{p_2}{p_1} \qquad \Delta s = c_p\ln\frac{T_2}{T_1} - R\ln\frac{p_2}{p_1}$$

$[c_p = f(T)]$ [CONSTANT c_p]

whose physical significance is explained in Fig. 6/7. This equation is generally more convenient to use than (6-12) because pressures are easier to measure than are volumes. Moreover, we find in the literature tabulated values for some gases of the integral

$$(6\text{-}14) \qquad \phi = \int_0^T \frac{c_p(T)\, dT}{T} \qquad \text{from which} \quad \Delta\phi = \int_1^2 \frac{c_p(T)\, dT}{T}$$

integrals that have taken into consideration the variation of specific heat. Using the symbol ϕ as in (6-14), we write

$$(6\text{-}13\text{B}) \qquad \Delta s = \Delta\phi - R \ln\frac{p_2}{p_1} = \phi_2 - \phi_1 - R \ln\frac{p_2}{p_1}$$

Recall that $- \log A/B = + \log B/A$, and $\ln N = \log_e N = 2.3 \log_{10} N$. As in previous instances, the foregoing equations may as well be interpreted for 1 mole of a pure substance;

$$(6\text{-}12\text{B}) \qquad \Delta\bar{s} = \int_1^2 \frac{C_v(T)\, dT}{T} + \bar{R} \ln\frac{v_2}{v_1}$$

$$(6\text{-}13\text{C}) \qquad \Delta\bar{s} = \Delta\bar{\phi} - \bar{R} \ln\frac{p_2}{p_1}$$

where in (6-12) and (6-13) the volumes must be in the same units and the pressure must also be in the same units.

The **absolute entropy** s_1^a for any state 1 of an ideal gas is some known absolute entropy, as $s°$ in the standard state, plus the change of entropy Δs from the known state to state 1. Using equation (6-13) without its subscripts, we have

$$(6\text{-}15\text{A}) \qquad s_1^a = s° + \Delta s = s° + \int_{T°}^{T_1} \frac{c_p\, dT}{T} - R \ln\frac{p_1}{p°}$$

where the superscript $°$ designates a property in a standard state. Since the standard state pressure is nearly always a standard atmosphere, $p° = 1$ atm, we may express the particular pressure p_1 in atmospheres, use $\Delta\phi$, and get

$$(6\text{-}15\text{B}) \qquad s_1^a = s° + \phi_1 - \phi° - R \ln p_{1\text{atm}}$$

If $\phi° = s°$, as is true for the Keenan and Kaye *Gas Tables*,* (except the ones for mixtures, § 6.15) this equation becomes (see § 6.15 concerning the use of *Gas Tables*)

$$(6\text{-}16) \qquad\qquad s_1^a = \phi_1 - R \ln p_{1\text{atm}} \qquad\qquad [\text{ONLY WHEN } \phi° = s°]$$

Either equation (6-15) or (6-16) may be for 1 lb or 1 mole. It is only essential that s,

---

* Where $\bar{\phi}°$ (77°F) in the *Gas Tables* and $\bar{s}°$ in Item B 11 do not agree exactly, attribute the difference to different fundamental data.

ϕ, and R be in the same units. To find the change of entropy during a chemical process, one must use a datum common to all the substances involved, and the only one is the third law datum (§ 6.12). Also, when the masses of systems vary, it may be necessary to work from a zero datum.

Example—Irreversible Mixing of Gases   6.14

Given: 2 moles of hydrogen and 2 moles of oxygen, each at $p = 4$ atm and 100°F, separated by a membrane, as seen in Fig. 6/8. If the membrane is broken and the gases diffuse, what is the change of entropy of the system if the gases act ideally? The system is isolated.

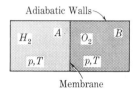

Irreversible Diffusion. Equations (6-12) and (6-13) are equally con-   **Fig. 6/8**
venient in this application.

Solution.   With like molecules together, there is a certain order in the arrangement. As the O_2 and H_2 diffuse, it is most unlikely that these molecules will again segregate themselves into the space they previously occupied; hence the process is irreversible. However, it is only necessary to have the initial and final equilibrium states defined. Since these gases are at the same temperature at the start and since no energy enters or leaves, the internal energy and temperature remain constant. With 2 moles of each gas at the same p and T, $V_A = V_B$ (Avogadro's law). Therefore, when H_2 occupies the whole volume, its pressure is $p_2 = p_1 V_1/V_2 = p_1 V_A/(V_A + V_B) = p_1/2 = 2$ atm (Boyle's law). Mixed with the O_2, the pressure of the H_2 is still 2 atm (Dalton's law). By the same argument, the pressure and volume changes of the O_2 are the same. Then from equation (6-13) we get

$$\Delta \bar{s} = C_p \ln \frac{T_2}{T_1} - \bar{R} \ln \frac{p_2}{p_1}$$

$$= 0 - 1.986 \ln 2/4 = 1.986 \ln 2 = +1.376 \text{ Btu/mole-°R}$$

for each gas. The total increase is $(4)(1.376) = 5.504$ Btu/°R for 4 moles of mixture. The total mass of mixture is $(2)(32) + (2)(2.016) \approx 68$ lb; $\Delta s = 5.504/68 = 0.81$ Btu/°R-lb of mixture.

GAS TABLES   6.15

Tabulated properties of gases (and other substances) are great time savers. Such properties for a few gases at low pressure are given in Items B 2 through B 10, extracted from Keenan and Kaye, *Gas Tables.* See also references numbered 0.1–0.32. Since in these tabulations the variations of specific heats is cared for, much better answers are obtained without tedious details of integrating with specific heat equations (Table I). In using any set of tables or any chart, one must note the units. For air, Item B 2, specific properties are given (for 1 lb); in the other gas tables, properties are for 1 pmole (symbols with bar: $\bar{h}, \bar{u}, \bar{\phi}$). With known molecular masses (Item B 1) and other tables for convenience), one can convert from pounds to moles, or vice versa. (ADVICE: If a slide rule is being used, do the converting *after* a problem has been solved, if possible and if conversion is needed, *so that the number of significant figures given in the tables is not lost in a subtraction.*) The enthalpies and internal energies are pseudoabsolute values, in effect computed from

(a)
$$h = \int_0^T c_p\, dT \quad \text{and} \quad u = h - \frac{pv}{J} = h - RT$$

where the integration is from absolute zero temperature to T. Thus the $\int c_p\, dT$ between limits of 1 and 2 is simply the difference of the values read from the table, $h_2 - h_1$. All the other symbols in the tables represent point functions too.

The symbol ϕ, $\bar{\phi}$, called the **entropy function**, is defined by equation (6-13), $\phi(T)$. The authors say that ϕ is computed from an arbitrary zero at temperature T_0, accounting for the fact that $c_p \to 0$ as $T \to 0$. If the values of ϕ in the *Gas Tables* are compared to published values of absolute entropies in the standard state, they are found to be closely the same. Therefore, one is probably safe in using ϕ as the absolute entropy at 1 atm and the temperature for which ϕ is selected—*except from the air table*. In the air table, $\phi + 1$ is close to s_{abs}, but it does *not* account for the change of entropy with change of pressure (the constituents of air are each at a partial pressure less than 1 atm).

There are two other properties in the tables that are quite helpful, the relative pressure p_r and the relative volume v_r. Let $\Delta s = 0$; then equation (6-12) gives

(b)
$$\ln \frac{p_2}{p_1} = \frac{\phi_2 - \phi_1}{R}$$

If $\ln p_r = \phi/R$, we see from **(b)** that $\ln(p_{r2}/p_{r1}) = (\phi_2 - \phi_1)/R$; therefore, since **(b)** is for two states at the same entropy, we have

(6-17)
$$\left(\frac{p_2}{p_1}\right)_s = \frac{p_{r2}}{p_{r1}} \qquad \text{[SAME ENTROPY]}$$

The actual numbers recorded for p_r are proportional to ϕ/R. Since the relative volumes v_r were computed from $v_r = RT/p_r$, the ratio

(6-18)
$$\left(\frac{v_2}{v_1}\right)_s = \frac{v_{r2}}{v_{r1}} \qquad \text{[SAME ENTROPY]}$$

also holds true only for states with the same entropy.

The full set of tables not only has properties at much closer intervals, giving more accurate results with straight-line interpolations, but also has a number of other convenient tables, including numerical values of the term $R \ln N$, equation (6-13).* To convert the properties of air to a mole basis, use the equivalent molecular mass of air as $M = 28.970$ (29 on the slide rule).

6.16   Example—Gas Table Properties

Hydrogen, initially at 15 psia and 80°F, is compressed at the rate of 20 cfm in a hot environment to a pressure of 75.3 psig. The details of the process are unknown but it is estimated that heat enters at the rate of 20 Btu/min. For a change of entropy of

---

* In the solutions of examples in this text, use will be made of the full *Gas Tables*, which may sometimes be slightly different from an interpolation in the Appendix tables.

0.0268 Btu/min-°R, determine (a) the final temperature, (b) the work if the system is nonflow, and (c) the work for a steady-flow system if $\Delta P = 0$ and $\Delta K = 0$. Use Item B 5.

Solution. (a) Since the table, Item B 5, gives properties per mole, we find from equation (6-4)

$$\dot{n} = \frac{p_1 V_1}{\bar{R} T_1} = \frac{(15)(144)(20)}{(1545)(460 + 80)} = 0.0518 \text{ moles/min}$$

$$\Delta \bar{s} = \frac{0.0268}{0.0518} = 0.518 \text{ Btu/pmole-°R}$$

At $T = 540°R$, from the table, $\bar{h}_1 = 3660.9$, $\bar{u}_1 = 2588.5$, and $\bar{\phi}_1 = 31.232$. Since Δs is given and with the pressures known, we think of equation (6-13). For $p_2 = 75.3 + 14.7 = 90$ psia, $\bar{R} \ln p_2/p_1 = \bar{R} \ln 90/15 = 3.5582$ Btu/pmole-°R. Then substituting known values in (6-13), we have

$$\Delta \bar{s} = \bar{\phi}_2 - \bar{\phi}_1 - \bar{R} \ln r_p$$

$$0.518 = \bar{\phi}_2 - 31.232 - 3.5582$$

from which $\bar{\phi}_2 = 35.308$ Btu/pmole-°R; $r_p = p_2/p_1$, the pressure ratio. Return to Item B 5 and find $T_2 = 970°R$ (to the nearest whole degree—usually quite accurate enough for our purposes) corresponding to $\bar{\phi}_2 = 35.308$. Now at state 2 from the table of Item B 5,

$$\bar{h}_2 = 6654.6 \quad \text{and} \quad \bar{u}_2 = 4728.3 \text{ Btu/pmole}$$

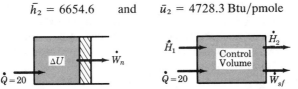

(a) Nonflow   (b) Steady Flow   *Energy Diagrams.* **Fig. 6/9**

(b) From the energy balance, Fig. 6/9(a), or from equation (4-7), the nonflow work is ($Q = 20/0.0518 = 386$ Btu/pmole)

$$W_n = -(\bar{u}_2 - \bar{u}_1) + Q = -4728.3 + 2588.5 + 386 = -1753.8 \text{ Btu/pmole}$$

or $\dot{W}_n = (-1753.8)(0.0518) = -90.6$ Btu/min power input; equivalent to $90.6/42.4 = 2.14$ hp.

(c) The energy balance, Fig. 6/9(b), or the equation (4-11) gives the steady-flow work as[*]

$$W_{sf} = -(\bar{h}_2 - \bar{h}_1) + Q = -6654.6 + 3660.9 + 386 = -2607.7 \text{ Btu/pmole}$$

corresponding to $(-2607.7)(0.0518) = -135$ Btu/min, or $135/42.4 = 3.19$ hp.

CLOSURE 6.17

This chapter establishes the attributes of an ideal gas, information that will be continually useful. At this time, one should have a fairly good idea as to when ideal gas assumptions are valid from an engineering viewpoint.

---

[*] It is not always handy or convenient to add dots and lines over the symbols. If the units are given for a numerical value, there is no need. But they may be very useful when the symbols are alone.

As already established, dense gases may be quite nonideal. Also, at high temperatures, the molecules undergo a significant thermal dissociation, breaking up into atoms ($0.5N_2 \rightarrow N$). While the resulting mixture may behave closely as an ideal gas, it has become a different substance. In certain circumstances, mostly at very high temperatures, the gas becomes highly ionized, forming a plasma[6.1] consisting of charged particles (the major portion of matter in the universe). Finally, gases (and matter) exist in various isotopes; unless otherwise specified, it is tacitly presumed that only the predominant and most stable isotope is involved.

PROBLEMS

SI UNITS

6.1   For an ideal gas expanding according to Boyle's law, integrate the expressions **(a)** $\int p\,dV$ and **(b)** $-\int V\,dp$. **(c)** Sketch the process on the pV plane and shade the areas which represent these integrals. Discuss their significance.

6.2   The temperature of an ideal gas remains constant while the absolute pressure changes from 103.4 kPaa to 827.2 kPaa. **(a)** If the initial volume is 80 ℓ, what is the final volume? **(b)** For 160 gm of the gas, determine the change of density expressed as a percentage of the initial density.
          Ans. **(a)** 10 ℓ, **(b)** 700% increase.

6.3   The atmospheric pressure at the base of a mountain is 730 mm Hg and at its top is 365 mm Hg. The atmospheric temperature is 15.55°C, constant from base to top. Local gravity is g = 9.75 m/s². Find the height of the mountain.

6.4   The pressure on 142 ℓ/min of air is boosted reversibly from 2,068.44 kPaa to 6,205.32 kPaa while the temperature remains constant at 24°C; $\Delta U = 0$. **(a)** Find $\int p\,dV$ and $-\int V\,dp$. **(b)** For a nonflow process, find W, Q, ΔH, and ΔS. **(c)** For a steady flow process during which $\Delta K = 5$ kJ/min and $\Delta P = 0$, find $\int p\,dV$, $-\int V\,dp$, W, ΔE_f, ΔH, Q, and ΔS.
     Ans. **(b)** $W = -322.6$ kJ/min, $\Delta S = -1.086$ kJ/K-min.

6.5   Sulfur dioxide at a temperature of 204°C occupies a volume of 0.3 m³. **(a)** If the volume is increased to 0.9 m³ while the pressure is maintained constant (ideal gas behavior), what is the final temperature in K? In °C? **(b)** If the initial volume is maintained constant and the pressure tripled, what is the final temperature in K? In °C?

6.6   Using Avogadro's principle, compare the relative densities of gases, each at the same pressure and temperature: oxygen to methane; helium to hydrogen; xenon to propane.

6.7   At 2,068.44 kPaa, 37.8°C, 0.142 m³ of methane have a total mass of 1.82 kg. Using Avogadro's principle find the mass of carbon dioxide contained in a 0.85-m³ tank at 2,068.44 kPaa, 37.8°C.   *Ans.* 30 kg.

6.8   Which is the heavier—dry air (per properties in Item B 1) or wet air (mixture of dry air and water vapor—as atmospheric air)? *Suggestion*: Compare the density of dry air with that of water vapor (considered to act as an ideal gas) at the same pressure and temperature.

6.9   Express the ideal gas equation $pv = RT$ in each of the following differential forms:

$$dp/p + dv/v = dT/T$$

$$dp/p - d\rho/\rho = dT/T$$

6.10   Prove that the trace of an isothermal process on the pV-plane for an ideal gas must be negative in slope with a curvature that is concave upwards.

6.11   A steel company plans on using 17 m³ of O_2 in processing one metric ton of steel. If this volume is measured at 101.325 kPaa and 21°C, what mass of O_2 is needed for a 20,000 metric ton/month furnace?

6.12   It is planned to lift and move logs from almost inaccessible forest areas by means of balloons. Helium at atmospheric pressure (101.325 kPaa) and temperature (21.1°C) is to be used in the balloons. What minimum balloon diameter (assume spherical shape) will be required for a gross lifting force of 20 metric tons?

6.13 A 283-ℓ tank $\mathcal{A}$ contains air at 689.48 kPaa and 37.8°C. Another 283-ℓ tank $\mathcal{B}$ contains air at 6894.8 kPaa and 37.8°C. Which system is capable of producing more work? Does the product pV represent energy stored in the tank? Explain.

6.14 For an ideal gas, show the difference in the specific heats c_p and c_v is simply the gas constant R, that is, $c_p - c_v = R$.

6.15 The internal energy of a certain ideal gas is related to its temperature as $u = RT/2$. Determine the specific heat ratio $k = c_p/c_v$ for this gas.

6.16 For a certain ideal gas, $R = 0.277$ kJ/kg-K and $k = 1.384$. **(a)** What are the values of c_p and c_v? **(b)** What mass of this gas would occupy a volume of 0.425 m^3 at 517.11 kPaa and 26.7°C? **(c)** If 31.65 kJ are transferred to this gas at constant volume in **(b)**, what are the resulting temperature and pressure?

Ans. (a) $c_p = 0.9984$ kJ/kg-K, **(b)** 2.647 kg, **(c)** 43.27°C, 545.7 kPaa.

6.17 A mixture is formed at 689.48 kPaa, 37.8°C by bringing together these gases—each volume before mixing measured at 689.48 kPaa, 37.8°C: 3 mol CO_2, 2 mol N_2, 4.5 mol O_2. Find the partial pressures of the components after mixing.

Ans. $p_{CO_2} = 217.7$ kPaa.

6.18 In a multicomponent gaseous mixture, two of the constituents (neon and xenon) are included volumetrically as 12% and 25%, respectively. If the neon exerts a partial pressure of 331 kPaa, find the mixture pressure and the partial pressure of the xenon.

6.19 An air mass of 0.454 kg and an unknown mass of CO_2 occupy an 85-ℓ tank at 2068.44 kPaa. If the partial pressure of the CO_2 is 344.74 kPaa (ideal gas), determine its mass. *Ans.* 0.138 kg.

6.20 Show that the Joule-Thomson coefficient, $\mu = (dT/dp)_h$, must be zero when applied to an ideal gas.

6.21 Under given conditions, a certain gas with a Joule-Thomson coefficient of $\mu = -0.03$ undergoes a pressure decrease of 3.45 M Pa when processed slowly through an adiabatic porous plug. What is its temperature change?

6.22 **(a)** Derive the following equation for a change of entropy between any two states of an ideal gas with constant specific heats:

$$s = c_p \ln \frac{v_2}{v_1} + c_v \ln \frac{p_2}{p_1}$$

(b) Helium changes its state from 586.06 kPaa and 194 ℓ to 9,459.67 kPaa and 28.3 ℓ. What is ΔS for 0.2 kg mol? *Ans.* -4.623 kJ/K.

6.23 After a series of state changes, the pressure and volume of 2.268 kg of nitrogen are each doubled. What is ΔS? *Ans.* $+2.805$ kJ/K.

6.24 Air (an ideal gas) undergoes an adiabatic throttling process from 538.99 kPaa to 103.42 kPaa. What is the change of specific entropy? *Ans.* $\Delta s = 0.497$ kJ/kg-K.

6.25 Two vessels $\mathcal{A}$ and $\mathcal{B}$ of different sizes are connected by a pipe with a valve (similar to Joule's set-up). Vessel $\mathcal{A}$ contains 142 ℓ of air at 2,767.92 kPaa, 93.33°C. Vessel $\mathcal{B}$, of unknown volume, contains air at 68.95 kPaa, 4.44°C. The valve is opened, and when the properties have been determined, it is found that $p_m = 1,378.96$ kPaa, $t_m = 43.33$°C. **(a)** What is the volume of vessel $\mathcal{B}$? **(b)** Compute the heat and entropy change. **(c)** What work is delivered?

6.26 For two or more separated ideal gases at the same temperature and pressure diffusing to a uniform mixture, show that the net entropy change is $\Delta S = -\bar{R} \sum_i \eta_i \ln x_i$, where x_i is the mole fraction of the ith component in the mixture; η_i, moles of i.

6.27 It is desired to compare the lifting characteristics of the several gases given in Item B 1 when each has been placed in a balloon that is subject to a normal atmosphere. It is known that all data given in this table have been stored in the memory of the computer. Write the program.

ENGLISH UNITS

6.28 The temperature of 4.82 lb of oxygen occupying 8 ft^3 is changed from 110°F to 200°F while pressure remains constant at 115 psia. Determine **(a)** the final volume and **(b)** the change in density expressed as a percentage of the initial density. **(c)** Now with the pressure varying, but with the volume constant, determine the final pressure if the absolute temperature is quadrupled.

Ans. **(a)** 9.26 ft³, **(b)** −13.61%, **(c)** 460 psia.

6.29 An automobile tire contains a certain volume of air at 30 psig and 70°F. The barometric pressure is 29.50 in. Hg. If, due to running conditions the temperature of the air in the tire rises to 160°F, what will be the gage pressure? Assume that the air is an ideal gas and that the tire does not stretch.

Ans. 37.5 psig.

6.30 Let the tire in **6.29** initially contain 0.25 lb of air and have an automatic valve that releases air whenever the pressure exceeds 34 psig. Under certain running conditions, the steady-state temperature becomes 180°F. **(a)** What mass of air escapes? **(b)** What is the tire pressure when the remaining air has returned to 70°F?

6.31 At atmospheric pressure and 62°F, the approximate density of hydrogen is determined as $\rho = 0.00529$ lb/ft³; at $t = 32$°F, $\rho = 0.00562$ lb/ft³. From these data compute the approximate absolute zero point on the Fahrenheit scale.

6.32 Assume 10 ft³/min ($m = 0.988$ lb/min) of hydrogen initially at $p_1 = 400$ psia, $t_1 = 300$°F are cooled at constant volume to 120°F in an internally reversible manner. **(a)** Evaluate the integrals $\int p\,dV$ and $-\int V\,dp$. **(b)** For a nonflow process, find p_2, ΔU, ΔH, Q, and ΔS. **(c)** Solve part **(b)** if the process is steady flow with $\Delta P = 0$, $\Delta K = 0$.

Ans. **(a)** 0, +176 Btu/min, **(b)** $p_2 = 305$ psia, $\Delta H = -610$ Btu/min, $Q = -434$ Btu/min, $\Delta S = -0.649$ Btu/min-°R, **(c)** $W = 176$ Btu/min.

6.33 The same as **6.32** except that the process occurs at constant pressure to 120°F.

6.34 **(a)** Assuming that all gases at 1 atm, 32°F occupy the standard mol volume of 359 ft³, find the densities of these gases at 1 atm, 32°F; xenon; argon; mercury; ozone; air. **(b)** For this state, compare each of the densities of the first four gases to that of the last one, air.

6.35 **(a)** What is the specific volume of a gas at 180 psia and 90°F when its density is 0.0446 lb/ft³ at 14.7 psia, and 32°F? **(b)** Calculate its gas constant and approximate molecular weight.

Ans. **(a)** 2.05 ft³/lb, **(b)** 95.5, 16.

6.36 A balloon, considered spherical, is 30 ft in diameter. The surrounding air is at 60°F and the barometer reads 29.60 in. Hg. What gross load may the balloon lift if it is filled with **(a)** hydrogen at 70°F and atmospheric pressure, **(b)** helium at 70°F and atmospheric pressure? **(c)** Helium is nearly twice as heavy as hydrogen. Does it have half the lifting force of hydrogen?

Ans. **(a)** 997.1 lb, **(b)** 925.2 lb.

6.37 There are withdrawn 200 ft³ of air measured at 15 psia and 90°F from a 50-ft³ tank containing air initially at 100 psia and 140°F. What is the pressure of the air remaining in the tank if its temperature is 130°F?

6.38 Two spheres, each 6 ft in diameter, are connected by a pipe in which there is a valve. Each sphere contains helium at a temperature of 80°F. With the valve closed, one sphere contains 3.75 lb and the other 1.25 lb of helium. After the valve has been open long enough for equilibrium to obtain, what is the common pressure in the spheres if there is no loss or gain of energy?

Ans. 32 psia.

6.39 The same as **6.38** except that the spheres contain hydrogen.

6.40 A closed vessel A contains 3 ft³ (V_A) of air at $p_A = 500$ psia and a temperature of $t_A = 120$°F. This vessel connects with vessel B, which contains an unknown volume of air V_B at 15 psia and 50°F. After the valve separating the two vessels is opened, the resulting pressure and temperature are $p_m = 200$ psia and $t_m = 70$°F, respectively. What is the volume V_B?

Ans. 4.17 ft³.

6.41 A system consists of two vessels A and B with a connecting valve. The vessel A contains 15 ft³ of nitrogen at 220 psia and 110°F. Vessel B contains 2 lb of nitrogen at 80 psia and 60°F. After the valve separating the two vessels is opened, the resulting equilibrium temperature becomes 90°F. What is the final pressure p_m? *Ans.* 180.5 psia.

6.42 Calculate the values of c_p, k, and the molar specific heats C_v and C_p of certain gases having values of R ft-lb/lb-°R and c_v Btu/lb-°R, given as follows: **(a)** $R = 60$, $c_v = 0.2$, **(b)** $R = 400$, $c_v = 0.76$, **(c)** $R = 250$, $c_v = 0.42$.

Ans. **(a)** c_p: 0.277 Btu/lb-°R, 7.13 Btu/mol-°R; $k = 1.385$.

6.43 An unknown gas at $p_1 = 95$ psia and $V_1 = 4$ ft^3 undergoes a process to $p_2 = 15$ psia and $V_2 = 16.56$ ft^3, during which the enthalpy decreases 83 Btu; $c_v = 0.1573$ Btu/lb-°R. Determine **(a)** c_p, **(b)** R, and **(c)** ΔU.
Ans. **(a)** 0.223 Btu/lb-°R, **(b)** 51.1, **(c)** −58.6 Btu.

6.44 A gas initially at $p_1 = 120$ psia and $V_1 = 9.87$ ft^3 changes state to a point where $p_2 = 40$ psia and $V_2 = 23.4$ ft^3, during which the internal energy decreases 116.6 Btu; $c_p = 0.24$ Btu/lb-°R. Determine **(a)** c_v, **(b)** R, **(c)** ΔH.

6.45 Assume 4 lb of a certain gas with $R = 80$ ft-lb/lb-°R undergo a process that results in these changes: $\Delta U = 1382$, $\Delta H = 2075$ Btu. Find **(a)** k, c_v, c_p; **(b)** the temperature change. **(c)** If the process had been internally reversible with $p = c$, find Q, W, $\int p\, dV$, $-\int V\, dp$ for both nonflow and steady flow ($\Delta P = 0$, $\Delta K = 0$).

6.46 For a certain ideal gas, $R = 77.8$ ft-lb/lb-°R and $c_p = 0.2 + 0.0002T$ Btu/lb-°R. It is heated from 40°F to 140°F. For 1 lb, find **(a)** the change of internal energy, **(b)** the change of enthalpy, **(c)** the change of entropy if the heating is at constant volume, **(d)** the change of entropy if the heating is at constant pressure, and **(e)** the value of k at 140°F.
Ans. **(a)** 21 Btu/lb, **(d)** 0.0565 Btu/lb-°R, **(e)** 1.455.

6.47 The partial pressures of a four-component gaseous mixture (CO, CH$_4$, N$_2$, A) at 90°F are respectively: 7, 21, 14, 28 psia. For the mixture find its pressure and volumetric analysis.

6.48 When 10 lb of hydrogen initially at 80°F and occupying 6 ft^3 undergo a state change to 150°F, an entropy increase of 8.02 Btu/°R is observed. Find the final total volume.

6.49 Helium changes its state from 85 psia, 272°R and 6.86 ft^3 to 1372 psia, 0.425 ft^3. For 0.2 pmole, find the entropy change and the final absolute entropy. See Table III.   *Ans.* −1.10, 3.552 Btu/°R.

6.50 If 2 lb of a certain gas, for which $c_v = 0.1751$ and $c_p = 0.202$ Btu/lb-°R, are throttled adiabatically from an initial volume of 4 ft^3 until its entropy increase is 0.189 Btu/°R, what is the final volume?
Ans. $V_2 = 134.2$ ft^3.

6.51 The entropy increase of 1 lb of hydrogen is 0.125 Btu/lb-°R when it is throttled adiabatically from 200 psia. What is the final pressure?

6.52 The pressure and absolute temperature of 1 lb argon are each tripled from 1 atm, 77°F. What is the final absolute entropy? Use table III.

6.53 **(a)** If 1 lb of air at 80°F, occupying 6 ft^3, changes state until the temperature is 150°F and the volume is 10 ft^3, determine ΔS. **(b)** Ten ft^3 of air at 60°F and 25 psia change state until $p_2 = 100$ psia and $V_2 = 8$ ft^3. Determine ΔS. **(c)** Ten ft^3 of air at 150 psig and 40°F undergo a series of state changes until $t_2 = 140$°F and $p_2 = 300$ psig. Determine ΔS.
Ans. **(a)** 0.0557, **(b)** 0.266, **(c)** −0.00522 Btu/°R.

6.54 Assume 2 lb of air undergo a state change from 1 atm, 560°R to 960°R in a frictionless manner and without a change of entropy. Use Item B 2 and find **(a)** ΔU, **(b)** ΔH, **(c)** the final pressure, **(d)** the volume ratio V_1/V_2.

6.55 There are expanded 1.5 pmoles of CO from 60 psia, 800°R to 15 psia, 800°R. Use Item B 4 and find the change of entropy. Are the values of p_r and v_r relevant in this problem? Why?   *Ans.* 4.13 Btu/°R.

6.56 There are compressed 60 lb/sec of N$_2$ from 140°F without entropy change; the volume ratio is $V_1/V_2 = 5$. If the process is a reversible steady flow wherein $\Delta K = 0$, $\Delta P = 0$, use Item B 6 and find **(a)** T_2, **(b)** the ratio p_2/p_1, **(c)** ΔpV, **(d)** the work.
Ans. **(a)** 1128°R, **(b)** 9.41, **(c)** 2247 Btu/sec, **(d)** −7992 Btu/sec.

6.57 Say 4 mol of O$_2$ undergo a cooling process from 1 atm, 1940°R to 1 atm, 860°R. **(a)** Find ΔS. **(b)** If the system is of the closed type, find Q and W. Use Item B 7.

6.58 A gas turbine compresses 100 lb/sec of air from 1 atm, 40°F through a pressure range of $p_2/p_1 = 10$; the steady flow process is frictionless and without entropy change. Use Item B 2 and find **(a)** T_2, **(b)** ΔH, **(c)** ΔU, **(d)** the change in flow energy, **(e)** W where $\Delta P = 0$, $\Delta K = 0$.
Ans. **(a)** 960°R, **(b)** 11,158, **(c)** 8,006, **(d)** 3152, **(e)** −11,158 Btu/sec.

7

PROCESSES OF FLUIDS

7.1 INTRODUCTION

Of the infinite variety of processes, we shall present a few named ones, usually defined as internally reversible (quasistatic, if nonflow). Reversible processes are the ideal prototypes that actual processes approach or processes that we wish real processes would closely approach.

Acquire the habit of sketching each simple process on the pV and TS planes; this will provide the skill for more complex situations later. Solve each problem using appropriate charts, tables, and thermodynamic relations; remember that $pv = RT$ and its many spinoffs are accurate only for low density gases and vapors.

One should consider the details in this chapter about processes as examples of the application of the basic ideas presented in previous chapters. Watch the wording and definitions; for example, isometric implies reversibility *only*, whereas constant volume could mean reversibility or irreversibility. Observe how the thought process starts with the basic energy equations followed by an energy diagram and thence to the pV and TS diagrams. Definitions and conservation laws are very important; seek to relate these to the defined process. By all means avoid trying to commit the detailed process equations to memory. Perhaps a helpful plan would be to make up a personal reference sheet as you go through this chapter, noting the laws and equations to which you are referrred and, at the end, discovering the beautiful consistency of it all.

Finally, you will be handling fluids of all sorts with both closed (nonflow) and open systems. Be alert in all ways as you attack each problem.

7.2 ISOMETRIC PROCESSES

Consider an isometric (or isochoric, or constant volume) process that is internally reversible (quasistatic, if nonflow), involving a pure substance, Fig. 7/1(c). For a $v = c$ nonflow closed system, $dQ = du + dW_n$ and $dW_n = p\,dv = 0$; therefore $Q = \Delta U$. The specific heat equation (2-22) also gives $Q = \int c_v\,dT$. If varieties of stored energy other than molecular are involved, we have

(a) $$dQ = dE \quad \text{or} \quad Q = E_2 - E_1$$

where $E = U + P + K$. In an isometric, steady flow, steady state process (say with an incompressible fluid), equation (4-9A) $Q = \Delta U + \Delta E_f + \Delta K + \Delta P + W_{sf}$ reduces to

(b) $$W_{sf} = -(\Delta E_f + \Delta K + \Delta P)$$

because with $Q = \Delta U$, these terms cancel. It is important to note here that heat Q is independent of whether there is flow or nonflow for a given process. Once the process and end state locations have been set, Q is the same for a given fluid irrespective of whether there is flow.

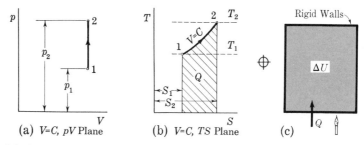

(a) *V=C, pV* Plane (b) *V=C, TS* Plane (c)

Reversible Isometric Process. On the *pV* plane, there is no area under 1-2; **Fig. 7/1** $\int p\,dV = 0$.

For an ideal gas between any two states we observe that $\Delta u = \int c_v\, dT$; but $Q = \Delta u$ only for the quasistatic isometric.

The equation of the process on the pV (or pv) plane is $V = c$ (or $v = c$), Fig. 7/1(a). Integrate $ds = dQ/T = c_v\, dT/T$ with c_v constant and get

(c) $$s = c_v \ln T + C$$

where C is a constant of integration and **(c)** is the equation of an isometric curve on the *Ts* plane when the specific heat is constant.

Where we have a superheated vapor (state point 1) that is cooled at constant volume until part of it condenses, our pV and TS sketches must reflect the saturated liquid-vapor lines; see Fig. 7/2. In the solution of problems, it is often necessary to use the defining relation of the process; in this case,

(d) $$v_1 = v_2 = (v_g - y_2 v_{fg})_2 = (v_f + x_2 v_{fg})_2 \quad \text{ft}^3/\text{lb}$$

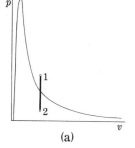

(a)

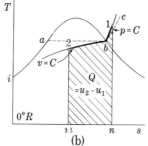

(b)

Reversible Constant Volume **Fig. 7/2** *(Isometric) Process.* The area *m-2-1-n* represents heat and also Δu in this process. The dotted line *abc* represents a constant pressure line.

From this equation, the quality x_2 may be found, with v_1 obtained from a vapor table. Knowing this quality and an additional property, determine other properties of the vapor at 2. Since the $\int p\,dV$ is zero, the heat is

(e)
$$Q = \Delta u = u_2 - u_1 = \left(h_2 - \frac{p_2 v_2}{J}\right) - \left(h_1 - \frac{p_1 v_1}{J}\right)$$

7.3   Example—Isometric (Reversible) Process

A pound of fluid at 15 psia, 300°F is heated $(v = c)$ reversibly until its temperature becomes 600°F; see Fig. 7/1. Find p_2, Q, W, Δh, and Δs if **(a)** the process is nonflow and the fluid is (1) air, as ideal gas, or, (2) water; also solve if the process is **(b)** steady state, steady flow with $\Delta P = 0$, $\Delta K = 0$ and the fluid is (1) air, as ideal gas, or, (2) water.

   Solution.   (a), (1) Air as ideal gas (nonflow): From $pv = RT$ with $v = c$

$$p_2 = p_1(T_2/T_1) = (15)(1060/760) = 20.92 \text{ psia}$$

By Item B 1:

$$Q = \Delta u = c_v(T_2 - T_1) = (0.1714)(1060 - 760) = 51.42 \text{ Btu}$$

By *Gas Tables*:

$$Q = \Delta u = (183.29 - 129.99) = 53.30 \text{ Btu}$$

$$W_n = \int p\,dv = 0 \qquad dv = 0$$

By Item B 1:

$$\Delta h = c_p(T_2 - T_1) = 0.24(1060 - 760) = 72 \text{ Btu}$$

By *Gas Tables*:

$$\Delta h = (h_2 - h_1) = (255.96 - 182.08) = 73.88 \text{ Btu}$$

By Item B 1:

$$\Delta s = \int_{\text{rev}} dQ/T = c_v \int dT/T = c_v \ln T_2/T_1$$

$$= 0.1714 \ln 1060/760 = .0570 \text{ Btu/°R}$$

By Gas Table:

$$\Delta s = \Delta \phi - R \ln p_2/p_1$$

$$= (0.76496 - 0.68312) - (53.34/778) \ln 20.92/15 = 0.059 \text{ Btu/°R}$$

Solution (a) (2) water (nonflow): Using Item B 15, at state 1, we find:

$$v_1 = 29.899 \text{ ft}^3/\text{lb} \qquad h_1 = 1192.5 \text{ Btu/lb} \qquad s_1 = 1.8134 \text{ Btu/lb} - °\text{R}$$

For state 2, $v_2 = v_1 = 29.899$, $t_2 = 600°F$. Interpolating in Item B 15, we find

$$p_2 = 21.24 \qquad h_2 = 1334.8 \qquad s_2 = 1.9335$$

$$\Delta h = h_2 - h_1 = 1334.8 - 1192.5 = 142.3 \text{ Btu/lb}$$

$$\Delta s = 1.9335 - 1.8134 = 0.1201 \text{ Btu/°R}$$

$$\Delta u = \Delta h - \Delta pv = 142.3 - (29.899)(144/778)(21.24 - 15) = 107.8 \text{ Btu/lb}$$

$$Q = \Delta u = 107.8 \text{ Btu/lb} \qquad W_n = 0$$

Solution (b)(1) steady state, steady flow ($\Delta P = 0$, $\Delta K = 0$) air: From (a)(1), $p_2 = 20.92$ psia, $Q_{sf} = Q_n = 51.42(53.30$ by Tables) Btu

$$\Delta s = 0.0570(0.0590 \text{ by Tables}) \text{ Btu/°R}, \Delta h = 72(73.88 \text{ by Tables}) \text{ Btu}$$

$$W_{sf} = -\int v\, dp = -v(p_2 - p_1) = -R(T_2 - T_1)$$

$$= -(53.34/778)(1060 - 760) = -20.56 \text{ Btu}$$

also $\qquad W_{sf} = -(\Delta E_f + \Delta P + \Delta K) = -\Delta E_f = -\Delta pv$

$$= -(\Delta h - \Delta u) = -(72 - 51.42) = -20.58 \text{ Btu}$$

Solution (b)(2) Steady state, steady flow ($\Delta P = 0$, $\Delta K = 0$) water: From (a)(2) $p_2 = 21.24$ psia, $Q_{sf} = Q_n = 107.8$ Btu

$$\Delta s = 0.1201 \text{ Btu/°R} \qquad \Delta h = 142.3 \text{ Btu} \qquad \Delta u = 107.8 \text{ Btu}$$

$$W_{sf} = -(\Delta E_f + \Delta P + \Delta K) = -\Delta E_f = -\Delta pv$$

$$= -(\Delta h - \Delta u) = -(142.3 - 107.8) = -34.5 \text{ Btu}$$

Example—Irreversible Constant Volume 7.4

A 10-ft³ vessel of hydrogen at a pressure of 305 psia is vigorously stirred by paddles, Fig. 7/3, until the pressure becomes 400 psia. Determine (a) the change of internal energy, (b) the work, (c) the heat, and (d) the change of entropy of 1 lb of the system. The system is the hydrogen in the tank, which has an adiabatic lining. Let specific heats be constant.

Solution. (a) This is evidently an irreversible constant volume process, inasmuch as we cannot expect the system to deliver the same quantity of paddle work that it receives. We know that $\Delta U = m \int c_v\, dT$ for any process of an ideal gas; $\Delta U = mc_v \Delta T$ for c_v constant, but we know no temperatures. Therefore, eliminate temperatures with $pV = mRT$ and find

(a) $$\Delta U = \frac{c_v V}{R}(p_2 - p_1) = \frac{(2.434)(10)}{766.54}(400 - 305)(144)$$

or $\Delta U = 434$ Btu, where c_v and R have been taken from Item B 1 and where we remembered (144) that p must be in pounds per square foot.

(b) Although $\int p\, dV = 0$ because $dV = 0$, the work is not zero; $W = \int p\, dV$ only for a reversible nonflow process, a system that has a moving boundary. Since the walls do not permit flow of heat, the nonflow energy equation becomes

(b) $Q = \Delta U + W = 0$ or $W = -\Delta U$

or $W = -434$ Btu, where the minus sign says that the work is done *on* the system.

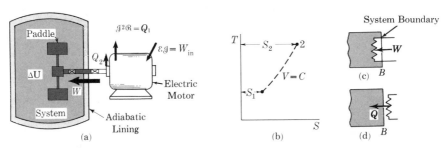

Fig. 7/3   *Irreversible Constant Volume Process.* The area under the curve in (b) does not
represent heat for the arrangement in (a), where there is no heat; besides, the
exact locations of the intermediate points between 1 and 2 are unknown for an
irreversible process. In (c) and (d), the electricity is used in resistance coils,
instead of driving a paddle. If in (c) the system's boundary is taken as *B*, the
energy crossing it is in electrical form, the equivalent of work. On the other
hand, if in (c) the system is taken as the *gas*, with an irregular boundary
excluding the coil, the energy entering the gas is heat from the coil, as it also is
in (d), with a boundary at *B*. If the motor is the system: $\mathcal{E}_b\mathcal{I}$ is the input
electricity, $Q_1 = \mathcal{I}^2\mathcal{R}$ is the electricity lost in the windings, Q_2 is the energy lost
in bearings, and so on.

NOTE: If electricity is to be used in a motor to cause work to flow into the system, Fig.
7/3 shows the hard way—driving a paddle wheel. The easy way would be to install an
electrical resistance coil inside the system and feed electricity directly to it, using Joule heat
$\mathcal{I}^2r$ to change the state. But note that in this case, the electric energy *crossing the boundary*
fits the general definition of work. Evidently, the *specified state change* in this problem can be
brought about entirely by heat, entirely by work, or by part heat and part work in any
proportions.

(c) The heat $Q = 0$ by statement of the problem.
(d) For a reversible process, $\Delta s = \int dQ/T]_{rev}$, and in this process, $Q = 0$. Since entropy
is a point function, we can compute its change along any reversible path, say, an isometric
between 1 and 2. Thus, from $ds = c_v\, dT/T$ or from equation (6-12),

(c) $$\Delta s = \int_1^2 c_v \frac{dT}{T} = c_v \ln \frac{T_2}{T_1} = c_v \ln \frac{p_2}{p_1} = 2.434 \ln \frac{400}{305} = 0.659 \text{ Btu/lb-°R}$$

in which Charles' law, $T_2/T_1 = p_2/p_1$, was used. The sign is positive; the entropy increases.
Not enough data are given to determine the total mass and total entropy change.

7.5 ISOBARIC PROCESS

An *isobaric process* is an internally reversible (quasistatic, if nonflow) process of a
pure substance during which the pressure remains constant. If the system is a closed
system, Fig. 7/4, $dQ = dE_s + dW$. With only molecular stored energy, $dQ = du + dW$.

(a) $$\int p\, dv = p \int dv = p\Delta v = p_2v_2 - p_1v_1 = \Delta(pv) \quad \text{and} \quad -\int v\, dp = 0$$

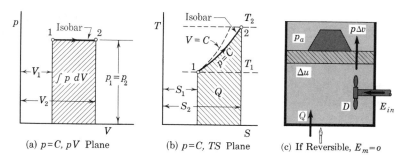

(a) $p=C, pV$ Plane   (b) $p=C, TS$ Plane   (c) If Reversible, $E_m = o$

Isobaric Process. Observe that the shaded area of figure (a) is $W_n = p(V_2 - V_1)$, the work for a reversible nonflow process, which is the same as the result obtained from $\int p\,dV$. The area under a curve on the pV plane of an irreversible process is $\int p\,dV$, but this integral is not then the work.   **Fig. 7/4**

If the process is nonflow and reversible, its work $W_n = \int p\,dv = p\Delta v$. For an internally reversible $p = C$, $Q = \int c_p\,dT$, which is also Δh; $\Delta h_p = \int c_p\,dT$. If the system is an open, steady-flow control volume, equation (4-9C) becomes (because $dQ = dh$)

(b) $$W_{sf} = -(\Delta K + \Delta P)$$

also apparent from equation (4-16).

If the system is an *ideal gas*, Charles' law, $V_2/V_1 = v_2/v_1 = T_2/T_1$ may be handy, as well as $pv = RT$. Between *any* two ideal-gas states, $\Delta h = \int c_p\,dT$, and for constant c_p, $\Delta h = c_p\Delta T$; $Q = \Delta h$ for the nonflow isobaric process.

The $p = C$ curve is called an **isobar**, and its equation on the pv (or pV) plane is $p = C$, Fig. 7/4(a). To get the equation on the Ts plane, integrate $ds = c_p\,dT/T$ and find, for constant c_p,

(c) $$s = c_p \ln T + C$$

where C is a constant of integration. Sketched alone on the Ts plane, a $v = C$ and a $p = C$ curve look alike; but since the area under the isobar $\int c_p\,dT$ must be greater than that under an isometric curve $\int c_v\,dT$ between two particular temperatures, the $v = C$ curve is steeper than the $p = C$ curve *at a particular state.*

If the system is a vapor whose state point crosses the saturated vapor line, we are again reminded of some differences that will be reflected in the traces on the pV and TS planes (compared to those $p = c$ traces for an ideal gas); see Fig. 7/5.

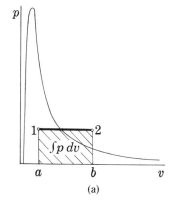
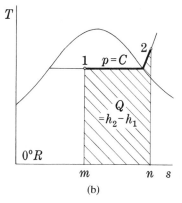

(a)   (b)

Reversible Constant Pressure (Isobaric) Process. The area *a-1-2-b* in (a) represents work. The area *m-1-2-n* in (b) represents the heat.   **Fig. 7-5**

Suppose in this case that point 1 is in the wet region and point 2 is in the superheat region, Fig. 7/5. With the pressure known, a point in the wet region is generally defined by giving its quality x or percentage moisture y. (State 1 could be defined also by giving the entropy or the internal energy.) Similarly, state 2 in the superheat region is generally, but not necessarily, defined by giving its temperature and pressure. The work of a reversible nonflow process at $p = C$ is

(a)
$$\int p\,dv = p(v_2 - v_1) \quad \text{ft-lb/lb vapor}$$

where $v_1 = (v_g - y_1 v_{fg})_1 = (v_f + x_1 v_{fg})_1$; the specific volume v_2 of the superheated vapor is taken directly from the superheat tables. The heat is

(b)
$$Q = h_2 - h_1 \quad \text{Btu/lb}$$

$$[\text{FLOW OR NONFLOW, } p = C \text{ (STAGNATION ENTHALPIES)}]$$

as for any $p = C$ process, where h_2 is taken from the superheat tables for Fig. 7/5, and h_1 is found from $h_1 = (h_f + x_1 h_{fg})_1 = (h_g - y_1 h_{fg})_1$.

7.6   Example—Isobaric Process, Ideal Gas

An ideal gas with $R = 386$ ft-lb/lb-°R and a constant $k = 1.659$, undergoes a reversible $p = c$ process during which 500 Btu are added to 5 lb of the gas; the initial temperature is $t_1 = 100°F$. Find t_2, W_n, W_{sf} ($\Delta P = 0, \Delta K = 0$), ΔS.

Solution. The specific heat c_p is found from equation (6-8)

$$c_p = kR/J(k-1) = \frac{(1.659)(386)}{778(1.659-1)} = 1.25 \text{ Btu/lb-°R}$$

Now $Q = 500 = \Delta H = mc_p(T_2 - T_1)$

$$T_2 = \frac{Q}{mc_p} + T_1 = \frac{500}{(5)(1.25)} + 560 = 640°R$$

$$W_n = \int p\,dV = p(V_2 - V_1) = mR(T_2 - T_1)$$

$$= (5)(386/778)(640 - 560) = 198.5 \text{ Btu}$$

$$W_{sf} = -\int V\,dp = 0 \qquad \Delta P = 0, \Delta K = 0, dp = 0$$

$$\Delta S = \int dQ/T = mc_p \ln T_2/T_1$$

$$= (5)(1.25) \ln 650/560 = 0.8346 \text{ Btu/°R}$$

7.7   Example—Isobaric Process, Vapor

If 2 lb of freon (F12) vapor at 100 psia, 150°F are reversibly cooled until the vapor becomes saturated, find properties on the *ph*-chart (Item B 35) and determine t_2, ΔH, ΔV, W_n, $W_{sf}(\Delta P = 0, \Delta K = 0)$; Q. See Fig. 7/6.

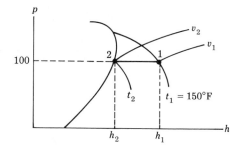

ph-Sketch for Freon F-12. **Fig. 7/6**

Solution. From Item B 35 (sketch) we read these values: $t_2 = 81°F$, $h_1 = 97.2$, $h_2 = 85.4$ Btu/lb, $v_1 = 0.49$, $v_2 = 0.41$ ft³/lb.

$$\Delta H = m\,\Delta h = 2(85.4 - 97.2) = -23.6 \text{ Btu}$$

$$\Delta V = m\,\Delta v = 2(0.41 - 0.49) = -0.16 \text{ ft}^2$$

$$W_n = \int p\,dV = p(\Delta V) = (100)(144/778)(-0.16) = -2.96 \text{ Btu}$$

$$W_{sf} = -\int V\,dp = 0 \qquad (\Delta P = 0, \Delta K = 0, dp = 0)$$

$$Q_n = Q_{sf} = \Delta H = -23.6 \text{ Btu}$$

Example—Irreversible Constant Pressure **7.8**

An insulated system of oxygen is at an initial state of 50 psia, 2 ft³, and 140°F. The pressure of the surroundings is $p_a = 14.7$ psia. Paddle work W_p is done on the system until $t_2 = 260°F$ and $V_2 = 2.4$ ft³; $p_2 = p_1$. Let the specific heats be constant and use Item B 1. Determine (a) the change of entropy, (b) the change of enthalpy, and (c) the following work quantities: the work done by the system, the paddle work, the work done in displacing the surroundings, and the work done against gravity. See Fig. 7/7.

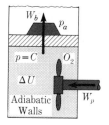

Irreversible Constant Pressure Process. **Fig. 7/7**

Solution. (a) An irreversible system is perhaps more informative of energy relations than a reversible one, the best approach (in the absence of an efficiency value) being a detailed energy analysis. Let the system be the O_2; its mass is

(a)
$$m = \frac{pV}{RT} = \frac{(50)(144)(2)}{(48.29)(600)} = 0.497 \text{ lb}$$

Since entropy is a point function, its change is dependent only on states 1 and 2. In equation

(6-13), $R \ln (p_2/p_1) = 0$ for $p_2 = p_1$; and ΔS for the system is

(b) $$\Delta S = mc_p \int_1^2 \frac{dT}{T} = (0.497)(0.219) \ln \frac{720}{600} = 0.0198 \text{ Btu/}°\text{R}$$

(b) The change of enthalpy is

(c) $$\Delta H = mc_p(T_2 - T_1) = (0.497)(0.219)(720 - 600) = 13.1 \text{ Btu}$$

(c) If the pressure remains $p = 50$ psia on the piston, a reasonable approximation for present purposes despite the turbulence produced by the paddles, the work by the fluid on the moving boundary, which is the piston, Fig. 7/7, is

(d) $$W_b = \frac{p(V_2 - V_1)}{J} = \frac{(50)(144)(2.4 - 2)}{778} = 3.7 \text{ Btu}$$

The paddle work can now be obtained from an energy balance for the system.

(e) $$W_p = \Delta U + W_b = \Delta U + \Delta pV = \Delta H = 13.1 \text{ Btu}$$

The work done in displacing the surroundings is

(f) $$W_a = \frac{p_a \Delta V}{J} = \frac{(14.7)(144)(2.4 - 2)}{778} = 1.09 \text{ Btu}$$

Since the total work W_b *by* the system, if there is no friction, goes to displacing the surroundings W_a and the work W_g of lifting the weight and piston against the force of gravity, we have

(g) $$W_g = \Delta P = W_b - W_a = 3.7 - 1.09 = 2.61 \text{ Btu}$$

This is not the change of potential energy of our defined gas system but of another system that is a moving boundary.

7.9 STEADY FLOW WITH $\Delta K = 0$, $\Delta P = 0$, $W = 0$

These restraints closely define many actual flow processes, the most common being flow through *heat exchangers*, Figs. 7/8 and 7/9. As seen in Fig. 7/8, no work is done, effectively $\Delta P = 0$, and in general the change of kinetic energy is relatively

Fig. 7/8 *Heat Exchanger, Extended Surface.* Used for heating viscous oils for the purpose of reducing the pumping work. Oil enters from the oil tank at the open left end. Steam circulates through the tubes; two-pass flow through tubes, single-pass flow through the shell. (*Courtesy Brown Fintube Co., Elyria, Ohio.*)

small. Equation (4-9) then reduces to

$$dQ = dH \qquad\qquad Q = \Delta H = wc_p(T_2 - T_1)$$

(a) $\begin{bmatrix}\text{ANY FLUID, STEADY FLOW} \\ dK = dP = dW = 0\end{bmatrix}$ $\begin{bmatrix}\text{IDEAL GAS, STEADY FLOW,} \\ \Delta K = \Delta P = W = 0, c_p \text{ CONSTANT}\end{bmatrix}$

equations that hold *even if the pressure changes significantly* because of friction.

(a) Circumferential extended surface.   (b) Longitudinal extended surface.

Extended Surface Heating Elements. One fluid flows inside, which is the   **Fig. 7/9**
control volume in this example, and the other surrounds the element and is
therefore the surroundings. "Extended surface," rather than a smooth external
pipe surface, is especially good on the gas side because of the relatively large
resistance to heat flow offered by the gas film on the surface. (a) *Courtesy
Foster Wheeler Corp, New York.* (b) *Courtesy Brown Fintube Co., Elyria, Ohio.*

Example—Steady Flow, Steady State   **7.10**

   Flowing oil is heated by condensing steam in a heater similar to that in Fig. 7/8. These data
are known for the oil: flow rate is 284 ℓ/min, enters at 32°C, leaves at 77°C, specific
gravity = 0.88, specific heat c_p = 2.136 kJ/kg K. The steam enters as saturated vapor at
10 bar and leaves as saturated liquid (h_f = 763 kJ/kg). Find the steam requirement, mass and
volume.

   Solution.   Oil density = $(0.88)(998) = 878.2$ kg/m^3

$$Q_{\text{steam}} = Q_{\text{oil}}$$

$$\Delta H_{\text{steam}} = \Delta H_{\text{oil}}$$

$$\dot{m}_s(h_g - h_f) = \dot{m}_o c_p(t_2 - t_1)$$

From Item B 16(SI), we read these properties of the steam entering the heater:

$$h = 2777 \text{ kJ/kg} \qquad v = 0.195 \text{ m}^3\text{/kg} \qquad s = 6.59 \text{ kJ/kg K}$$

Then

$$\dot{m}_s(2777 - 763) = \left(\frac{284}{1000}\right)(878.2)(2.136)(77 - 32)$$

$$\dot{m}_s = 11.90 \text{ kg/min}$$

$$\dot{V}_s = \dot{m}_s v = (11.9)(0.195) = 2.321 \text{ m}^3\text{/min}$$

ISOTHERMAL PROCESS   **7.11**

   An *isothermal process* is an internally reversible (quasistatic, if nonflow) constant
temperature process of a pure substance; a curve that is the locus of points

representing a particular temperature is called an *isotherm*, Fig. 7/10. Analogous to other process names, the word *isothermal* is often used to mean only *constant temperature*; but if the process does not fit the definition, the difference will be clear. In the energy equations, we can only say that

(a)
$$\int dQ = Q = \int T\,ds\Big]_{rev} = T\int_1^2 ds = T(s_2 - s_1)$$

good for any substance, nonflow or steady flow. Then the work of a nonflow process is

(b)
$$W = Q - \Delta u = u_1 - u_2 + Q$$

The work of a steady-flow isothermal is

(c) $W = -\Delta h - \Delta K + Q$ or $W = Q - \Delta h = h_1 - h_2 + Q$

$$[\Delta K = 0, \Delta P = 0]$$

where the signs should be carefully reckoned.

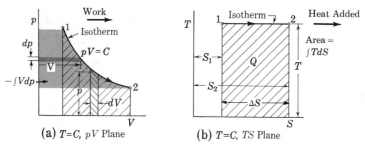

(a) $T=C, pV$ Plane   (b) $T=C, TS$ Plane

Fig. 7/10   *Reversible Isothermal Process.* Observe that the differential area in (a), $p\,dV$, is a differential work quantity dW_n in a nonflow system; area "under" equals area "behind" the curve on the pV plane for this process.

If the system is an *ideal gas* we think of Boyle's law, $pv = C = p_1v_1 = p_2v_2$, and so on, and, of course, $pv = RT$. From equation (4-18), $\int p\,dv = -\int v\,dp + \Delta(pv)$, we find that $\int p\,dv = -\int v\,dp$, since $pv = C$. Then, for $p = C/v$ (or $v = C/p$),

(d)
$$\int p\,dv = -\int v\,dp = C\int_1^2 \frac{dv}{v} = p_1v_1 \ln\frac{v_2}{v_1} = RT\ln\frac{p_1}{p_2}$$

where typical units would be ft-lb/lb, or, for the last form, the unit is the same as the energy unit for R. Reversible nonflow work W_n is given by equation **(b)**. In a steady-flow process with $\Delta P = 0$, equations **(b)** and (4-16) give

(e)
$$-\int_1^2 v\,dp = p_1v_1 \ln\frac{v_2}{v_1} = \Delta K + W_{sf}$$

Since $\Delta T = 0$, $\Delta u = \int c_v\,dT = 0$ and $\Delta h = \int c_p\,dT = 0$ for an ideal gas. Therefore, this process, for an ideal gas only, may be said to be a constant internal energy process.

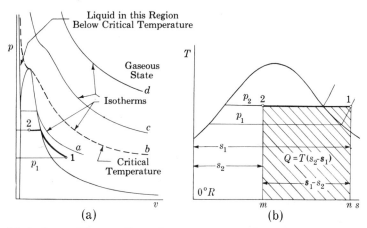

Reversible Isothermal Process. Several constant temperature lines (isotherms), a, b, c, d, are drawn in (a). As the temperature rises above the critical temperature, the curve d approaches the typical curve of the isothermal for a perfect gas.

Fig. 7/11

When the system is a vapor, we should again examine the pV and TS traces as the state point moves across the saturated vapor line; see Fig. 7/11. For the $T = C$ process 1-2, the point moves from the superheat region to the wet region. The area $m - 2 - 1 - n$, $\int T\, ds = T \int ds$, represents heat

(f)
$$Q = T(s_2 - s_1)$$

where $s_2 - s_1$ is negative in this case. In applying equation **(f)**, find the entropy at 1, a point in the superheat region, in the superheat tables, and compute the entropy at 2, a point in the wet region, from $s_2 = (s_g - y_2 s_{fg})_2 = (s_f + x_2 s_{fg})_2$. The change in internal energy is found as usual for vapors. In passing, recall that Joule's law does *not* hold for an imperfect gas; pv is *not* a constant.

Example—Isothermal Process, Air **7.12**

If $2 \, \text{kg/sec}$ of air are compressed isothermally (and reversibly) from $p_1 = 103.42 \, \text{kPaa}$, $t_1 = 30°\text{C}$ to $p_2 = 597.11 \, \text{kPaa}$, find Q, W, ΔS if process is (a) nonflow, (b) steady-state steady flow, $\Delta P = 0$, $\Delta K = 0$.

Solution. From Item B 38, $R = 8.3143 \, \text{kJ/kgmole-K}$.
(a) Nonflow:

$$W_n = \int p\, dV = p_1 v_1 \ln V_2/V_1 = mRT_1 \ln p_1/p_2$$

$$= (2/28.97)(8.3143)(303) \ln (103.42/517.11) = -279.91 \, \text{kJ}$$

$$Q_n = W_n = -279.91 \, \text{kJ}$$

$$\Delta S = \int dQ/T = Q/T = -279.91/303 = -0.9238 \, \text{kJ/K}$$

(b) Steady-state, steady flow:

$$Q_{sf} = Q_n = -279.91 \text{ kJ}$$

$$W_{sf} = -\int V \, dp \qquad (\Delta P = 0, \Delta K = 0)$$

$$= \int p \, dV = -279.91 \text{kJ} \qquad \text{since } pV = C$$

$$\Delta S = -0.9238 \text{ kJ/K}$$

7.13   ISENTROPIC PROCESS

An *isentropic process* is a reversible adiabatic process, detailed in this article for a pure substance (§ 3.1). Since *adiabatic* simply means *no heat*, there are innumerable adiabatics, special varieties you have already met. The definition of entropy being $ds = dQ/T]_{rev}$, a reversible adiabatic is one of constant entropy. One might meet an irreversible process where $s_1 = s_2$ and $Q \neq 0$ with no telling what has happened in between, but we do not call this an isentropic. The energy equations (4-6), (4-7A), and (4-9) for any adiabatic, reversible or irreversible, reduce to

(a) $W = -\Delta E_s \qquad W_n = -\Delta u$

[CLOSED SYSTEM]   [NONFLOW]

(b) $W_{sf} + \Delta P + \Delta K = -\Delta h$ [STEADY FLOW]

For an *ideal gas*, we can derive a pressure-volume relation using equations (4-14) and (4-15);

(c) $T \, ds = du + p \, dv = 0 = c_v \, dT + p \, dv$

(d) $T \, ds = dh - v \, dp = 0 = c_p \, dT - v \, dp$

By division, we get

(e) $\dfrac{c_p}{c_v} = -\dfrac{v \, dp}{p \, dv} = k \qquad \text{or} \qquad -k \dfrac{dv}{v} = \dfrac{dp}{p}$

If k is taken as constant, perhaps some mean value, integration gives

(f) $-k \ln \dfrac{v_2}{v_1} = \ln \left(\dfrac{v_1}{v_2}\right)^k = \ln \dfrac{p_2}{p_1}$

Taking the antilogarithm, we get

(7-1) $\left(\dfrac{v_1}{v_2}\right)^k = \left(\dfrac{V_1}{V_2}\right)^k = \dfrac{p_2}{p_1} \quad \text{or} \quad p_1 v_1^k = p_2 v_2^k \quad \text{or} \quad pv^k = C \quad pV^k = C$

$$p = C\rho^k$$

[IDEAL GAS, $k = C$]

Since states 1 and 2 were chosen at random on any isentropic line, we have the *pv* relation for an ideal-gas isentropic process. Equation (7-1) applies to a particular mass for any two points of the same entropy and it describes the path of an $s = C$ line on the *pv* (*pV*) plane, Fig. 7/12 (see caption).

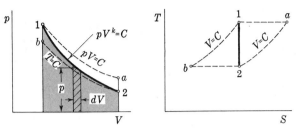

Fig. 7/12

Isentropic Process. For a comparison of the isothermal and the isentropic curves, isothermals 1-*a* and *b*-2 are shown dotted. Point *a* is related to 2 in being on the constant volume line through 2. Points 1 and *b* are also on a constant volume line. Observe that, at a point common to the two curves, the isentropic line on the *pV* plane is steeper than the isothermal; also that there is no area under the curve 1-2 on the *TS* plane, which accords with the definition that $Q = 0$ and $S = C$.

Using the equation of state in the form $p_1 v_1/T_1 = p_2 v_2/T_2$, and the process equation $p_1 v_1^k = p_2 v_2^k$ to eliminate first the *p*'s and then the *v*'s, we find the temperature-volume and the temperature-pressure relations for *constant specific heats*:

$$(7\text{-}2) \qquad \frac{T_2}{T_1} = \left(\frac{V_1}{V_2}\right)^{k-1} = \left(\frac{V_2}{V_1}\right)^{1-k} = \left(\frac{v_2}{v_1}\right)^{1-k} \qquad \text{and} \qquad TV^{k-1} = C$$

$$(7\text{-}3) \qquad \frac{T_2}{T_1} = \left[\left(\frac{p_2}{p_1}\right)^{1/k}\right]^{k-1} = \left(\frac{p_2}{p_1}\right)^{(k-1)/k} \qquad \text{and} \qquad \frac{T}{p^{(k-1)/k}} = C$$

The integrals $\int p\,dV$ and $-\int V\,dp$ can now be completed. Substitute $p = C/V^k$ in $\int p\,dV$, which represents the area under the *pV* curve of Fig. 7/12, to get

$$(g) \qquad \int p\,dV = C\int_{V_1}^{V_2} \frac{dV}{V^k} = \left[\frac{CV^{-k+1}}{-k+1}\right]_{V_1}^{V_2} = \frac{CV_2^{1-k} - CV_1^{1-k}}{1-k}$$

Setting the first $C = p_2 V_2^k$ and the second $C = p_1 V_1^k$, we find

$$(7\text{-}4) \qquad \int_1^2 p\,dV = \frac{p_2 V_2 - p_1 V_1}{1-k} = \frac{mR}{1-k}(T_2 - T_1)$$

$$[\text{WHENEVER } pV^k = C]$$

where $pV = mRT$ is used; this equation gives the work W_n of a nonflow isentropic process, ideal gas—and also $-\Delta u$, by equation (a) in this section. Our normal units for the *pV* part are ft-lb; in the last expression, the unit is the same as the energy unit in R. By way of illustration, for n moles ($V = n\bar{v}$),

(h) $$\int p \, dV = \frac{n\bar{R}}{1 - k}(T_2 - T_1)$$ [n MOLES]

where $\bar{R}$ is the universal gas constant. Similar transitions of other equations are easily made. The $-\int V \, dp$, which represents the area behind the curve, Fig. 7/13, can be obtained from $-\int V \, dp = \int p \, dV - \Delta(pv)$, or by integrating again;

(7-5) $$-\int_1^2 V \, dp = -C \int_1^2 \frac{dp}{p^{1/k}} = \frac{k(p_2 V_2 - p_1 V_1)}{1 - k}$$

$$= \Delta K + \Delta P + W_{sf}$$ [BY EQUATION (4-16)]

$$= Q - \Delta H = -\Delta H$$ [BY EQUATION (4-15)]

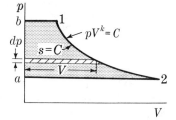

Fig 7/13   *Internally Reversible Adiabatic Process.* In general for a reversible process, with $\Delta P = 0$, $-\int V \, dp = \Delta K + W_{sf} = Q - \Delta H$, § 4.12. For an isentropic process, $Q = 0$.

For a pure vapor passing isentropically from a state of superheat (point 1) across the saturated vapor curve into the wet region (point 2), we must again view this trace on the pV and TS planes; see Fig. 7/14.

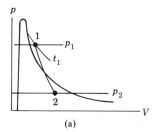   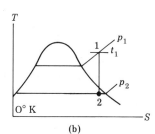

Fig. 7/14   *Isentropic Process for Vapor.*   (a)   (b)

In the solutions of problems, it is often necessary to use the defining relation of the process; in this case

(7-6) $$s_1 = s_2 = (s_g - y_2 s_{fg})_2 = (s_f + x_2 s_{fg})$$

From this equation, the quality x_2 may be found, with s_1 obtained from a vapor table. Knowing this quality and an additional property, other properties at point 2 are determined.

Charts such as the Mollier (hs) and the ph-chart are very handy for solving isentropic process problems.

Example—Isentropic Process 7.14

Three pounds per second of steam expand in a turbine isentropically from $p_1 = 300$ psia and $t_1 = 700°F$ to $t_2 = 200°F$. Find x_2 and the steady-flow work ($\Delta K = 0$). If the process is nonflow, what is W?

Solution. See Fig. 7/14. From Item B 15, we find $s_1 = 1.6758$, $h_1 = 1368.9$, and $v_1 = 2.2263$. From Item B 13, we get $p_2 = 11.526$ psia and

$$s_{f2} = 0.2940 \qquad h_{f2} = 168.09 \qquad v_{f2} = 0.016637$$

$$s_{fg2} = 1.4824 \qquad h_{fg2} = 977.9 \qquad v_{fg2} = 33.622$$

$$s_{g2} = 1.7764 \qquad h_{g2} = 1146.0 \qquad v_{g2} = 33.639$$

(One often saves time by writing down all the table values at once.) Using $s_1 = s_2 = (s_g - y_2 s_{fg})_2$, we have $1.6758 = 1.7764 - 1.4824y_2$; from which

$$y_2 = 6.83\% \qquad \text{and} \qquad x_2 = 93.17\%$$

$$h_2 = (h_g - y_2 h_{fg})_2 = 1146.0 - (0.0683)(977.9) = 1079.4 \text{ Btu/lb}$$

The work of this steady-flow isentropic is

$$W = -\Delta h = h_1 - h_2 = 1368.9 - 1079.2 = 289.7 \text{ Btu/lb}, \qquad \text{or} \qquad 869.1 \text{ Btu/sec}$$

for 3 lb/sec. Next, find the internal energies at states 1 and 2.

$$u_1 = h_1 - \frac{p_1 v_1}{J} = 1368.9 - \frac{(300)(144)(2.226)}{778} = 1245.3 \text{ Btu/lb}$$

$$v_2 = v_{g2} - y_2 v_{fg2} = 33.64 - (0.0679)(33.62) = 31.36 \text{ ft}^3\text{/lb}$$

$$u_2 = 1079.2 - \frac{(11.526)(144)(31.36)}{778} = 1012.3 \text{ Btu/lb}$$

The nonflow isentropic work is

$$W = -\Delta u = u_1 - u_2 = 1245.3 - 1012.3 = 233 \text{ Btu/lb}$$

or 699 Btu/sec for 3 lb/sec.

Example—Isentropic Process with Variable Specific Heats 7.15

The molar specific heat of a mixture of gases in an internal combustion engine has been found to be given by (see § 12.3)

$$C_v = 7.536 + \frac{3.16T}{10^4} - \frac{3.28T^2}{10^8} - \frac{2027.5}{T} + \frac{(80.2)(10^4)}{T^2} - \frac{37.1}{T^{1/2}} \text{ Btu/pmole-°R}$$

The compression ratio $r_k = V_1/V_2 = 6$ for an isentropic compression 1-2 in this engine and the initial temperature is $t_1 = 140°F$. Compute the temperature t_2.

Solution. With C_v given, use equation (6-12) and get

(a) $\quad \Delta \bar{s} = 0 = \int \frac{C_v\, dT}{T} + \bar{R} \ln \frac{V_2}{V_1} \qquad \text{or} \qquad \int \frac{C_v\, dT}{T} = \bar{R} \ln \frac{V_1}{V_2} = 3.558 \text{ Btu/pmole-°R}$

for $V_1/V_2 = 6$. Integrate the left side with the given C_v;

$$\int_{600}^{T_2} C_v \frac{dT}{T} = 7.536 \ln T_2 - 7.536 \ln 600 + \frac{3.16 T_2}{(10)^4} - \frac{(3.16)(600)}{(10)^4} - \frac{3.28 T_2^2}{(10)^8(2)}$$

$$+ \frac{(3.28)(600)^2}{(10)^8(2)} + \frac{2027.5}{T_2} - \frac{2027.5}{600} - \frac{(80.2)(10)^4}{2T_2^2}$$

$$+ \frac{(80.2)(10)^4}{(2)(600)^2} + \frac{(37.1)(2)}{T^{1/2}} - \frac{(37.1)(2)}{(600)^{1/2}}$$

The practical way to find T_2 is to substitute different assumed values of it into the foregoing expression (simplified as much as possible) until one is found that gives 3.558, the required value according to equation **(a)**. Getting help for a start, compute T_2 for the case of constant specific heats;

$$\frac{T_2}{T_1} = \left(\frac{V_1}{V_2}\right)^{k-1} \qquad \text{or} \qquad T_2 = (600)(6^{0.4}) = 1228°\text{R}$$

Since k decreases with temperature, this value is necessarily greater than the answer. One might also compute T_2 for some value of k which is certain to be below its true value. Then a range is defined between which the answer should be found. After several guesses, we find for $T_2 = 1180°\text{R}$ that

$$\int_{600}^{1180} C_v \frac{dT}{T} \approx 3.558$$

which is the desired value; therefore $t_2 = 1180 - 460 = 720°\text{F}$.

7.16   ADIABATIC PROCESSES, REVERSIBLE AND IRREVERSIBLE

Adiabatic processes are frequently assumed to be a reasonable approximation of some actual process. The energy relations of equations **(a)** and **(b)**, § 7.13, apply as well to irreversible adiabatics. One extreme adiabatic is with no work done, called the ***throttling process***, § 6.11. The steady-flow equation then reduces to

(a) $h_1 + K_1 = h_2 + K_2$

where states 1 and 2 are defined in Fig. 7/15. If the system boundaries can be chosen so that $K_1 \approx K_2$, the defining equation becomes

(b) $dh = 0$ $h_1 = h_2$ or $c_p(T_2 - T_1) = 0$

[ANY SUBSTANCE]   [IDEAL GAS]

Concerning the semantics, observe that as far as the end states 1 and 2 are concerned, a throttling process of an *ideal* gas is at constant temperature, but it is not an isothermal process as defined here. The throttling process, or any so-called *free expansion*, is not organized for the production of work or for conserving the kind of energy that might be converted into work (as the kinetic energy generated in a nozzle). Since the end states are at the same temperature, Boyle's law, $p_1v_1 = p_2v_2$, applies; the entropy change may be computed from either equation (6-11) or (6-12), $\Delta s = R \ln v_2/v_1 = R \ln p_1/p_2$, always greater than zero; also $\phi_2 - \phi_B$, Fig. 7/15.

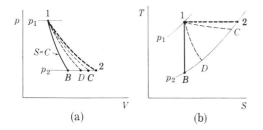

(a)   (b)

Adiabatics. The area under 1-*D* in (a) is $\int p\,dV$,   **Fig. 7/15** if it could be made, but does not represent W_n. The area under 1-*D* in (b) is $\int T\,dS$, if it could be done, but does not represent Q.

The other extreme is the reversible adiabatic, 1-*B*, Fig. 7/15, just discussed in detail. In numerous adiabatic processes, as steady flow through a steam turbine, a gas turbine, and a centrifugal compressor, there are some fluid frictional losses, with the process represented by a dotted line such as 1-*D*, Fig. 7/15. The smaller the losses, the closer *D* approaches *B*. While the examples given are steady-flow machines, the process may be nonflow in a closed system.

In a general way, we shall use prime marks to indicate actual values and actual states; for example, W' is the actual work (the work of the fluid unless otherwise indicated); $\Delta H'$ is the actual change of enthalpy; state 2', Fig. 7/16, is the actual final state of the process (as differentiated from 2, which is the ideal final state). In accordance with the second law (Chapter 5), it can be safely said that *the entropy of an adiabatic process remains constant (ideal) or increases*—never decreases. Thus, in both the expansion and the compression, Fig. 7/16(a) and (b), states 2' are to the right of 2.

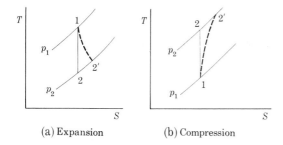

(a) Expansion   (b) Compression

Irreversible Adiabatic Processes. Entropy   **Fig. 7/16** always increases.

It should also be pointed up here that isentropics and irreversible adiabatics can either be nonflow or steady flow.

Directing our attention to a vapor system undergoing an adiabatic process, let the second state along an isentropic process be designated 2 and along an irreversible adiabatic, 2', Fig. 7/17. Then, from the nonflow energy equation (4-7), with $Q = 0$,

(c) $W = -\Delta u$ or $W = u_1 - u_2$ or $W' = u_1 - u_{2'}$

[NONFLOW]   [REVERSIBLE]   [IRREVERSIBLE]

From the steady-flow energy equation (4-10), with $\Delta P = 0$, or by equation (4-12),

(d) $W = -(\Delta h + \Delta K)$ or $W = h_1 - h_2$ or $W' = h_1 = h_{2'}$

    [STEADY FLOW]   [REVERSIBLE, $\Delta K = 0$] [IRREVERSIBLE, $\Delta K = 0$]

An example of steady flow to which this equation applies is a steam turbine. For steady flow with $W = 0$ and $Q = 0$, as in a nozzle or diffuser,

(e) $\Delta K = -\Delta h$ $K_2 = \dfrac{v_2^2}{2kJ} = h_1 - h_2$ $K_1 = \dfrac{v_1^2}{2kJ} = h_2 - h_1$

          $[W = 0]$ $[v_1 \approx 0]$ $[v_2 \approx 0]$

If the process is reversible (isentropic),

(f) $s_1 = s_2 = (s_g - y_2 s_{fg})_2 = (s_f + x_2 s_{fg})_2$

                    [REVERSIBLE NONFLOW OR STEADY FLOW]

in which it is presumed that state 2 is in the wet region, Fig. 7/17.

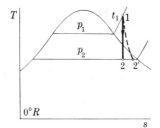

Fig. 7/17   *Isentropic Process.*

7.17   Example—Reversible and Irreversible Adiabatics

There are compressed 3500 cfm of air from $p_1 = 14.5$ psia and $t_1 = 75°F$ to $p_2 = 25$ psia. Determine the work for each of the following processes: (a) nonflow reversible adiabatic (isentropic), (b) nonflow irreversible adiabatic to a temperature of 190°F (650°R), (c) steady-flow reversible adiabatic (isentropic) for which $v_1 = 40$ fps and $v_2 = 120$ fps, (d) steady-flow adiabatic to a temperature of 190°F, velocities as in (c). (e) What is the change of entropy of the process in (b) and in (d)?

Solution.

(a) CONSTANT k.   We shall solve this problem both by constant specific heats, Item B 1, and by gas tables, Item B 2. The temperature after isentropic compression is, Fig. 7/16(b),

(a) $\dfrac{T_2}{T_1} = \left(\dfrac{p_2}{p_1}\right)^{(k-1)/k}$ or $T_2 = 535\left(\dfrac{25}{14.5}\right)^{(1.4-1)/1.4} = 625°R$

The amount of air involved is

(b) $\dot{m} = \dfrac{p_1 V_1}{RT_1} = \dfrac{(14.5)(144)(3500)}{(53.3)(535)} = 256 \text{ lb/min}$

For a nonflow isentropic, the work on the fluid supplied from the outside is

(c) $\dot{W}_n = -\Delta U = \dot{m}c_v(T_1 - T_2) = (256)(0.1714)(535 - 625) = -3950\,\text{Btu/min}$

GAS TABLE.   The properties at state 1 (75°F or 535°R) are

$$h_1 = 127.86 \qquad p_{r1} = 1.3416 \qquad u_1 = 91.19 \qquad \phi_1 = 0.59855$$

Using equation (6-17), applicable to an isentropic process, we get

(d) $$p_{r2} = \frac{p_2}{p_1}p_{r1} = \frac{25}{14.5}(1.3416) = 2.313$$

Entering Item B 2 at $p_{r2} = 2.313$, find $T_2 = 625°R$, $h_2 = 149.49$, and $u_2 = 106.64$. Then

(e) $\dot{W}_n = \dot{m}(u_1 - u_2) = 256(91.19 - 106.64) = -3950\,\text{Btu/min}$

(This and other answers will check with those for constant k because in this problem the temperature range is small enough that the specific heats are virtually constant.)
  (b) CONSTANT k.   For the nonflow irreversible adiabatic, Fig. 7/16(b),

(f) $\dot{W}'_n = \dot{m}c_v(T_1 - T_{2'}) = (256)(0.1714)(535 - 650) = -5046\,\text{Btu/min}$

GAS TABLE.   Enter the table for $T_{2'} = 190 + 460 = 650°R$ and find $u_{2'} = 110.94$, $h_{2'} = 155.50$, and $\phi_{2'} = 0.64533$.

(g) $\dot{W}'_n = \dot{m}(u_1 - u_{2'}) = 256(91.19 - 110.94) = -5060\,\text{Btu/min}$

  (c) CONSTANT k.   For a steady-flow process with $Q = 0$ and $s = C$,

(h) $$\dot{W}_{sf} = -\Delta K - \Delta H = \dot{m}\left(\frac{v_1^2 - v_2^2}{2kJ} + h_1 - h_2\right)$$

$$= 256\left[\frac{40^2 - 120^2}{50,000} + 0.24(535 - 625)\right] = -5595\,\text{Btu/min}$$

GAS TABLE.   For table values found above,

(i) $$\dot{W}_{sf} = \dot{m}\left(\frac{v_1^2 - v_2^2}{2kJ} + h_1 - h_2\right)$$

$$= 256\left(\frac{40^2 - 120^2}{50,000} + 127.86 - 149.49\right) = -5602\,\text{Btu/min}$$

  (d) CONSTANT k.   For an irreversible steady-flow adiabatic, $-\Delta H' = mc_p(T_1 - T_{2'})$

(j) $$\dot{W}'_{sf} = 256\left[\frac{40^2 - 120^2}{50,000} + 0.24(535 - 650)\right] = -7131\,\text{Btu/min}$$

GAS TABLE

(k) $$\dot{W}'_{sf} = 256\left(\frac{40^2 - 120^2}{50,000} + 127.86 - 155.5\right) = -7140\,\text{Btu/min}$$

(e) CONSTANT k.   To find the entropy change of the system, use equation (6-13), since the p's and T's are known.

(l)
$$\Delta \dot{S} = \dot{m}c_p \ln \frac{T_{2'}}{T_1} - \dot{m}R \ln \frac{p_2}{p_1}$$

$$= 256\left(0.24 \ln \frac{650}{535} - 0.0686 \ln \frac{25}{14.5}\right) = +2.4 \text{ Btu/°R-min}$$

GAS TABLE

(m)
$$\Delta \dot{S} = \dot{m}\left(c_p \ln \frac{T_{2'}}{T_1} - R \ln \frac{p_2}{p_1}\right) = \dot{m}\left(\phi_{2'} - \phi_1 - R \ln \frac{p_2}{p_1}\right)$$

$$= 256(0.64533 - 0.59855 - 0.0373) = 2.425 \text{ Btu/°R-min}$$

If the full *Gas Tables* are at hand, one can look up the value of $R \ln (p_2/p_1)$; otherwise, it must be computed. If you observe that 2 and 2' are on the same pressure line [$\ln (p_2/p_{2'}) = 0$] and that $s_2 - s_2 = s_{2'} - s_1$; then $\Delta S = m(\phi_{2'} - \phi_2)$. Try it. The steady-flow computation gives the entropy change of the air as it passes the entrance boundary until it reaches the exit.

7.18   Example—Irreversible Adiabatic

Let the data be the same as in the previous example (7.14) except that an irreversible adiabatic expansion occurs in a turbine to state 2', Fig. 7/18, on the saturated-vapor line. Determine (a) the horsepower output, (b) the irreversibility for a sink temperature of $200°F = t_{01} = t_2$ and the irreversibility for a sink temperature of $t_{02} = 80°F$, (c) the engine efficiency (turbine efficiency) and the effectiveness for sink t_{01}, and (d) the effectiveness for sink t_{02}.

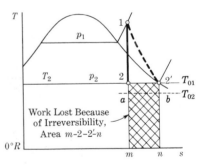

Fig. 7/18   For T_{01}, the irreversibility I is represented by area m-2-2'-n. For T_{02}, it is represented by *mabn*.

Solution.   (a) Look up $h_{2'} = h_g$ and $s_{2'} = s_g$ in Item B 13 and use other properties as already found in § 7.14.

(a)
$$W' = -\Delta h = h_1 - h_{2'} = 1368.9 - 1146.0 = 222.9 \text{ Btu/lb}$$

For a flow of 3 lb/sec and 42.4 Btu/hp-min, we get

(b)
$$\dot{W}' = \frac{(3)(60)(222.9)}{42.4} = 946 \text{ hp}$$

(b) By equation (5-11), the irreversibilities at 660 and 540°R are

(c)
$$I_{660} = T_0(s_{2'} - s_1) = (660)(1.7764 - 1.6758) = 66.4 \text{ Btu/lb}$$

(d) $I_{540} = (540)(1.7764 - 1.6758) = 54.3 \text{ Btu/lb}$

The irreversibility at 540°R is less than at 660°R because at the higher temperature there is available energy that could be delivered as work by a Carnot engine between these temperatures. In equation **(c)**, we could properly think of this as the internal irreversibility of the process itself.

(c) The engine (turbine) efficiency, $\eta_t = W'/W$, is

(e) $$\eta_t = \frac{h_1 - h_{2'}}{h_1 - h_2} = \frac{222.9}{289.7} = 77\%$$

For effectiveness, equation (5-10), § 5.8, we have

(f) $\Delta\mathscr{A}_{1\text{-}2'} = \Delta b = \Delta h - T_0\Delta s = -222.9 - 66.4 = -289.3 \text{ Btu/lb}$

(g) $$\varepsilon = \frac{W'}{-\Delta\mathscr{A}} = \frac{222.9}{289.3} = 77\%$$

the same as η_t for this particular t_{01}. However, 200°F is not a realistic sink temperature.

(d) At $T_{02} = 60 + 460 = 540°\text{R}$,

(h) $\Delta\mathscr{A}_{1\text{-}2'} = \Delta b = \Delta h - T_0\Delta S$

$$= -222.9 - (540)(1.7764 - 1.6758) = -277.2 \text{ Btu/lb}$$

(i) $$\varepsilon = \frac{W'}{-\Delta\mathscr{A}} = \frac{222.9}{277.2} = 80.4\%$$

The effectiveness at the lower sink temperature 80.4% is greater than that at the higher sink temperature (77%) because in effect it assumes that the available part as represented by 2'-2-*ab*, Fig. 7/18, can be converted into work. Because of this, the engine efficiency is more intimately related to the engine's performance unless its processes proceed to the sink temperature.

THROTTLING CALORIMETER 7.19

The *throttling process*, § 7.16, equation **(b)**, § 7.16, $h_1 = h_2$, is commonly used in steam-power practice to determine the quality of steam. Suppose it is desired to know the quality of the steam at state 1, Fig. 7/19, in a steam line. If a sample of this high-pressure steam is throttled to a lower pressure in a throttling calorimeter,

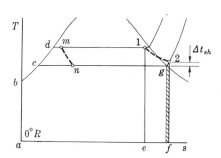

Throttling Processes (Irreversible). (Area *abd*-1-*e*) ≈   **Fig. 7/19**
(area *abc*-2-*f*) because these areas approximately represent, respectively, h_1 and h_2. If the vapor is quite wet, it will not reach a superheated state, as evident for the throttling *mn*.

Fig. 7/20, then, within limitations, it will become superheated, point 2. In the superheated condition, we can measure its pressure and temperature, thus defining state 2. With p_2 and t_2 known, look up the enthalpy h_2 and compute y_1 from

(a)
$$h_2 = (h_g - y_1 h_{fg})_1 \text{ Btu/lb}$$

where h_g and h_{fg} correspond to some known pressure p_1, Fig. 7/19.

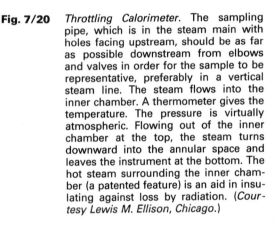 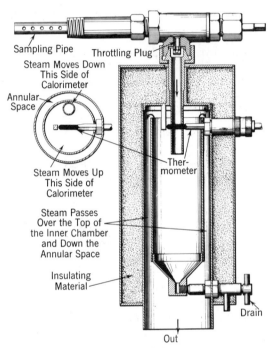

Fig. 7/20   *Throttling Calorimeter.* The sampling pipe, which is in the steam main with holes facing upstream, should be as far as possible downstream from elbows and valves in order for the sample to be representative, preferably in a vertical steam line. The steam flows into the inner chamber. A thermometer gives the temperature. The pressure is virtually atmospheric. Flowing out of the inner chamber at the top, the steam turns downward into the annular space and leaves the instrument at the bottom. The hot steam surrounding the inner chamber (a patented feature) is an aid in insulating against loss by radiation. (*Courtesy Lewis M. Ellison, Chicago.*)

The drawback in using equation **(a)** is that the actual temperature and pressure at 2 is nearly always such that a double interpolation in the superheat tables is necessary. Since the highest degree of accuracy is not usually required, the computation may be simplified with the use of the specific heat of steam at constant pressure. In Fig. 7/19, note that the enthalpy at 2 is equal to the enthalpy of saturated steam h_g plus the enthalpy to superheat the steam from g to 2. The heat to superheat the steam Q_{sh} may be computed from $Q_{sh} = mc_p\Delta t$ because the points g and 2 are on a constant pressure curve. For the conditions encountered in a calorimeter, c_p for steam may be taken as 0.48 Btu/lb-°F. Thus, for 1 lb,

(b)
$$h_2 = h_g + c_p(t_2 - t_g)$$

7.20   POLYTROPIC PROCESSES

A *polytropic process*, Fig. 7/21, is an internally reversible process during which

(7-7) $$pV^n = C \quad \text{and} \quad p_1V_1^n = p_2V_2^n = p_iV_i^n \qquad \text{[ANY FLUID]}$$

where n is any constant. Because $pV^n = C$ and $pV^k = C$ are mathematically the same, the integrals $\int p\,dV$ and $-\int V\,dp$ are alike except that n is in place of k; so from § 7.13,

(7-8) $$\int_1^2 p\,dV = \frac{p_2V_2 - p_1V_1}{1-n} = \frac{mR(T_2-T_1)}{1-n} = \frac{nR(T_2-T_1)}{1-n}$$

(7-9) $$-\int_1^2 V\,dp = \frac{n(p_2V_2 - p_1V_1)}{1-n} = \frac{nmR(T_2-T_1)}{1-n} = \frac{(n)nR(T_2-T_1)}{1-n}$$

Whenever the process equation is of the form $pV^n = C$, $-\int V\,dp = n\int p\,dV$. The foregoing applies to any substance in the defined process, and the integrals have the meanings previously defined: $\int p\,dV = W_n$, $-\int V\,dp = \Delta K + W_{sf} = Q - \Delta H$, for $\Delta P = 0$.

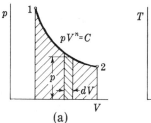

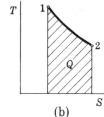

(a)   (b)

Reversible Polytropic Process. These curves show the general appearance of this process when $1 < n < k$. Since, in theory, the polytropic exponent may have any value $-\infty < n < +\infty$, curves sloping in any general direction may be obtained (see Fig. 7/24). In practice, n for the process called polytropic does not vary greatly from k.

Fig. 7/21

If the substance is an *ideal gas*, use $p_1V_1^n = p_2V_2^n$ and $p_1V_1/T_1 = p_2V_2/T_2$ as in § 7.13 for equations (7-2) and (7-3) and find the TV and Tp relations:

(a) $$\frac{T_2}{T_1} = \left(\frac{V_2}{V_1}\right)^{1-n} = \left(\frac{V_1}{V_2}\right)^{n-1} = \left(\frac{v_1}{v_2}\right)^{n-1} = \left(\frac{p_2}{p_1}\right)^{(n-1)/n} \qquad \text{[IDEAL GAS]}$$

If an ideal gas process is known or assumed to be polytropic, and if two state points are completely defined (say, pressures and volumes are known), the value of n can be found by use of logarithms (or the log-log scales on the slide rule) on equation **(a)**.

For a constant mass, internally reversible, nonflow system, the heat can be obtained from equation (4-14), $dQ = du + p\,dv$. From equation (7-8), we deduce that $p\,dv = R\,dT/(1-n)$; also we know that $du = c_v\,dT$ for an ideal gas. Substitution in (4-14) gives

(b) $$dQ = c_v\,dT + \frac{R\,dT}{1-n} = \frac{c_v + R - nc_v}{1-n}\,dT$$

Then, using $c_p = c_v + R = kc_v$, we get from equation **(b)**

(7-10) $$dQ = \left[c_v\frac{k-n}{1-n}\right]dT \qquad \text{or} \qquad Q = \left[c_v\left(\frac{k-n}{1-n}\right)\right](T_2-T_1) = c_n\Delta T$$

$$\underbrace{\phantom{\left[c_v\left(\frac{k-n}{1-n}\right)\right]}}_{C_n}$$

[CONSTANT k]

where the bracketed term is often called the ***polytropic specific heat*** $c_n(C_n)$. Observe also that c_n is negative when $k > n > 1$. A negative specific heat means that heat is rejected by the substance even though the temperature increases, or that heat is added even though the temperature decreases.

In all previous processes involving a vapor system, the integral $\int p\,dV$ was not used except for the isobaric process $p = c$. This was because complicated equations of state (for vapors) make this approach more difficult.

However, we define an internally reversible *polytropic* process for a vapor by the equation $pv^n = C$. Using $pv^n = C$, with n as some constant average value, we get

(c)
$$\int_1^2 p\,dv = \frac{p_2 v_2 - p_1 v_1}{1 - n}$$

which is the work of a nonflow process. The given data, including the definition of the process, must be sufficient to locate precisely the state points marking the extremities of the process, say 1-2, Fig. 7/22. Then, equation (4-14) may be applied with $T\,ds = dQ$; $Q = \Delta u + \int p\,dv$.

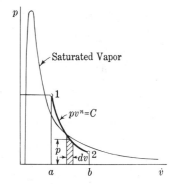

Fig. 7/22   *Polytropic Process.* The location of the starting point 1, the extent of the expansion or compression, and the value of n determine whether or not the curve $pv^n = C$ crosses the saturated vapor line.

Similarly, from equation (4-15), and perhaps more conveniently,

(d)
$$Q = \Delta h - \int_1^2 v\,dp = h_2 - h_1 + \frac{n(p_2 v_2 - p_1 v_1)}{J(1 - n)}$$

where the integral is as given in § 7.13, equation (7-5).

7.21   Example, Polytropic Process—Solutions with Constant k and by Gas Table

Hydrogen expands from a state of 125 psia, 600°R, and 9.6 ft³ to $p_2 = 14.7$ psia with $n = 1.3$ during a polytropic process, Fig. 7/23. (a) Compute the work, heat, and change of entropy of a nonflow process. (b) Compute the same items for a steady-flow process during which $\Delta K = -2$ Btu. Does the area "behind" the curve on the pV plane represent the steady-flow work?

Solution.   (a) By either approach, the mass and any unknown properties at state 2 must be computed by ideal gas equations.

(a)
$$m = \frac{p_1 V_1}{RT_1} = \frac{(125)(144)(9.6)}{(766.54)(600)} = 0.376 \text{ lb}$$

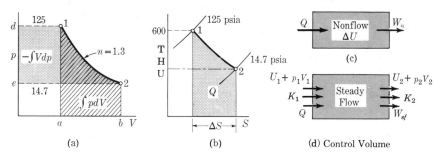

(a)   (b)   (d) Control Volume   **Fig. 7/23**

The work can be obtained by algebraic maneuvers without computing another property at 2, but the temperature will be needed anyway for obtaining table values;

(b)
$$T_2 = T_1\left(\frac{p_2}{p_1}\right)^{(n-1)/n} = 600\left(\frac{14.7}{125}\right)^{(1.3-1)/1.3} = 366°R$$

or $-94°F$. From equation (7-8), by either approach,

(c)
$$W_n = \int p\,dV = \frac{mR(T_2 - T_1)}{J(1 - n)} = \frac{(0.376)(766)(366 - 600)}{(778)(-0.3)} = +288\text{ Btu}$$

represented by the area a-1-2-b, Fig. 7/23. Use the conversion 778 or the R/J from Item B 1.

CONSTANT k

(d)
$$c_n = c_v\frac{k - n}{1 - n} = 2.434\left(\frac{1.405 - 1.3}{1 - 1.3}\right) = -0.852\text{ Btu/lb-°R}$$

(e)
$$Q = mc_n\,\Delta T = (0.376)(-0.852)(-234) = +74.9\text{ Btu}$$

represented by the shaded area on the TS plane.

(f)
$$\Delta S = m\int\left[\frac{dQ}{T}\right]_{\text{rev}} = mc_n\ln\frac{T_2}{T_1} = (0.376)(-0.852)\ln\frac{366}{600} = 0.1585\text{ Btu/°R}$$

GAS TABLE.   In the gas table solution, the heat is found from $Q = \Delta U + \int p\,dV$, or from $Q = \Delta H - \int V\,dp$. We shall use the former. From the table of Item B 5: $\bar{h}_1 = 4075.6$, $\bar{u}_1 = 2884.1$, $\bar{\phi}_1 = 31.959$, $\bar{h}_2 = 2486$, $\bar{u}_2 = 1759.1$, and $\bar{\phi}_2 = 28.608$.

(g)
$$\Delta U = m\left(\frac{\bar{u}_2 - \bar{u}_1}{M}\right) = 0.376\left(\frac{1759.1 - 2884.1}{2.016}\right) = -210\text{ Btu/lb}$$

(h)
$$Q = \Delta U + \int p\,dV = -210 + 288 = +78\text{ Btu}$$

(i)
$$\Delta S = \frac{m}{M}\left(\bar{\phi}_2 - \bar{\phi}_1 - \bar{R}\ln\frac{p_2}{p_1}\right) \qquad\qquad\text{[EQUATION (6-13)]}$$

$$= \frac{0.376}{2.016}\left(28.608 - 31.959 - 1.986\ln\frac{14.7}{125}\right) = +0.168\text{ Btu/°R}$$

The best answers are obtained with gas table properties.

(b) The change of entropy between the control volume boundaries, where the properties are assumed to be uniform, is the same as before, justified on the basis of s being a point function. The heat is also the same as before, as may be decided from $Q = mc_n \Delta T$, though other rationalizations may be used. To get the steady-flow work by the shortest route, use

(j)
$$-\int V\,dp = n\int p\,dV = (1.3)(288) = 374.4 \text{ Btu} = \Delta K + W_{sf}$$

by equation (4-16), from which, for $\Delta K = -2$ Btu, we get $W_{sf} = 376.4$ Btu. By direct attack from the energy diagram of Fig. 7/23(d), we have

(k)
$$W_{sf} = \frac{0.376(4075.6 - 2486)}{2.016} + 2 + 78$$

from which $W_{sf} = 376.5$ Btu. The area behind the path does not represent the process work; rather $-\int V\,dp = \Delta K + W$, and $\Delta K \neq 0$.

Time is necessarily involved in a steady-flow process, but it makes no difference in the answers whether the entering 9.6 ft^3 pass through the process in 1 sec or 1 day. It is only necessary for the defined process to occur.

7.22   CURVES FOR DIFFERENT VALUES OF *n*

The helpfulness of being able to visualize the general way in which process curves go on the pV and TS planes is immeasurable. Consider the equation $pV^n = C$ with various values of n and the curves of Fig. 7/24.

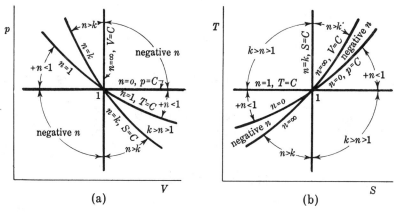

(a)   (b)

Fig. 7/24   *Effect of Varying n.* Expansions or compressions are imagined to take place from some common point 1. All positive values of *n* give curves in the second and fourth quadrants on the *pV* plane (a); positive values of *n* may produce curves in all four quadrants on the *TS* plane (b). Curves with values of *n* between 1 and *k* will fall in the second and fourth quadrants on the *TS* plane and within a narrow region on the *pV* plane.

Let $n = 0$; then $pV^0 = C$ or $p = C$, an isobaric process. This value of $n = 0$ substituted in various polytropic equations will give the correct forms for the constant pressure process. Let $n = \infty$; then, from $pV^n = C$, we have $p^{1/n}V = p^{1/\infty}V = V = C$, an isometric process. This value of $n = \infty$ substituted in various polytropic equations will yield the correct forms for the constant volume process.

Let $n = k$; then $pV^k = C$, which is recognized as the equation for an isentropic process. This value of $n = k$ substituted wherever n appears yields the correct forms for the isentropic process.

Let $n = 1$; then $pV = C$, which is recognized as Boyle's law for an ideal gas and the equation of an isothermal process. This process is the exception, since $n = 1$ yields indeterminate forms in most of the polytropic equations.

STAGNATION PROPERTIES AND MACH NUMBER 7.23

Stagnation properties are those thermodynamic properties that a moving stream of compressible fluid would have if it were brought to rest isentropically (no outside work; the kinetic energy brings about the compression). This compression is the reverse of the process of isentropic expansion through a nozzle from a large reservoir where the speed and kinetic energy are negligible; that is, the properties of the substance in this reservoir are stagnation properties. The reversed process may take place in a diffuser, Fig. 7/25, where the entering kinetic energy K_1 may be large and the exit kinetic energy some lesser value; $p_2 > p_1$. If the final kinetic energy $K_0 = 0$, then the corresponding properties of the gas are stagnation properties when the process is isentropic. Also when a body is in a moving stream of fluid or when it is moving with a velocity relative to the fluid, the fluid is brought substantially to rest relative to the body on the surfaces facing the flow. At the points where the speed of the fluid relative to the body is small, the kinetic energy $K_0 \approx 0$, and in the limit when $K_0 = 0$, we have

$$(7\text{-}11) \qquad h_0 = h + K = h + \frac{v^2}{2kJ}$$

where the conversion constant $J = 778$ for the units of Btu, lb mass, ft, and sec, and where h is the enthalpy of the fluid whose speed relative to the body, or diffuser, is v fps. Stagnation enthalpy is often a convenient concept for compressible-flow problems, but the interested reader must keep in mind that, as an energy quantity, it represents *three* distinct kinds of energy: internal energy, flow energy, and kinetic energy.

If the body is an aircraft (or other device) moving through still air, the speed of the air relative to the body is $v = v_p$, the speed of the plane (or device). If the fluid is a gas with a constant c_p, then $h_0 - h = c_p(T_0 - T)$ and equation (7-11) becomes

$$(7\text{-}12) \qquad T_0 = T + \frac{v^2}{2kJc_p} \qquad \text{and} \qquad \frac{T_0}{T} = \left(\frac{p_0}{p}\right)^{(k-1)/k} = 1 + \frac{v^2}{2kJc_pT}$$

$$[Q = 0, W = 0] \qquad\qquad [s = C]$$

(a) $v_1 > a$; Convergent-Divergent   (b) $v_1 < a$; Divergent

Diffusers. Boundaries of the control volume are at *a* and *b*. When the initial speed v_1 is supersonic, the diffuser is convergent-divergent (if K_o is small), as in (a). If v_1 is subsonic, $v_1 < a$, the diffuser is divergent only, as in (b). Minimum section in (a) is called the throat.   **Fig. 7/25**

where T is the temperature obtained from an instrument at rest with respect to the fluid (moving at the velocity of the moving fluid) and T_0 is the **stagnation** (sometimes called *total*) temperature. However, consider equation (7-11) with respect to the steady-flow energy equation (4-9) and observe that equation (7-11) is just as valid for an irreversible adiabatic as for an isentropic; in fact, it says that, for $Q = 0$, $W = 0$, and $\Delta P = 0$, *the adiabatic flow is one of constant stagnation enthalpy.* Also, *for an ideal gas* with $h = h(T)$, the first equation (7-12) is similarly valid, but $T_0/T = (p_0/p)^{(k-1)/k}$ is correct only when $s = C$, and p_0 is therefore the isentropic stagnation pressure. With irreversibility (friction), the final pressure $p_{0'} < p_0$, Fig. 7/26, and can be estimated with a **pressure coefficient** κ_r, determined from experience, and for the states in Fig. 7/26, defined by

(a)
$$\kappa_r = \frac{p_2 - p_1}{p_0 - p_1} = \frac{p_{0'} - p_1}{p_0 - p_1}$$

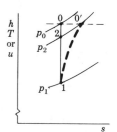

Fig. 7/26   The stagnation $p_0 > p_{0'} = p_2$, but $T_0 = T_{0'}$ for an ideal gas and $h_0 = h_{0'}$ for any gas (from energy relations, $\Delta K = \Delta h = 0$). In actual diffusers, the exit velocity is often small enough that the exit kinetic energy is negligible, so that a stagnation state is virtually achieved.

A form of equation (7-12), handy in some contexts, is obtained by using the Mach* number **M**, which is the ratio of the actual speed *divided by* the local speed of sound a in the fluid;

(7-13)
$$\mathbf{M} \equiv \frac{v}{a} \qquad \mathbf{M} < 1 \qquad \mathbf{M} > 1$$

[SUBSONIC] [SUPERSONIC]

The acoustic speed a in an ideal gas is given by $a = (\mathbf{k}kRT)^{1/2}$, an equation to be derived later (§ 18.3)—the units are, for your convenience, $\mathbf{k}$ ft-lb/lb-sec^2, R ft-lb/lb-°R, T°R, and a fps. Now use $c_p = kR/[J(k-1)]$ in (7-12) and equation (7-13) to find

(7-14)
$$\frac{T_0}{T} = 1 + \frac{k-1}{2}\mathbf{M}^2 \qquad \text{[ADIABATIC]}$$

good for any stagnation process of an ideal gas, but $(p_0/p)^{(k-1)/k} = 1 + (k-1)\mathbf{M}^2/2$ only for the constant entropy case. It is strongly recommended

---

* Ernst Mach (1838–1916), who studied and later taught at Vienna, was a noted physicist with strong interests and competence in psychology and philosophy. He is the author of several books, of which three or more have been translated into English and other languages. His experimental work in ballistics was eventually recognized for its significance in connection with supersonic motion; so the Mach number, an important concept in compressible flow, was named after him.

that, in the solution of the few problems of this course, you start with the basic definition, whenever possible, for example, equation (7-11), rather than using a derived form that might happen to fit the data of a particular problem.

Example—Stagnation Properties 7.24

Suppose a satellite in entering the earth's atmosphere happens to be moving at 20,000 fps when at a point where $p_a = 0.01$ psia and $t_a = 100°F$. Considering the event adiabatic and k constant, estimate (a) the stagnation temperature on the satellite's leading surface, (b) the pressure for a pressure coefficient of 70%, and (c) the satellite's speed as a Mach number.

Solution.   (a) Using equation (7-12), we have a hypothetical temperature of

(a)
$$T_0 = T + \frac{v^2}{2kJc_p} = 560 + \frac{4 \times 10^8}{5 \times 10^4 \times 0.24} = 33{,}893°R$$

(b) The isentropic stagnation pressure, from (7-12), is

(b)
$$p_0 = p\left(\frac{T_0}{T}\right)^{k/(k-1)} = 0.01\left(\frac{33.893}{560}\right)^{1.4/04} = 17{,}300 \text{ psia}$$

For $\eta_r = 0.70$, by equation **(a)**, § 7.23,

(c)
$$p_{0'} = p + 0.7(p_0 - p) = 0.01 + 0.7(17{,}300 - 0.01) = 12{,}110 \text{ psia}$$

(c) For an acoustic speed of

(d)
$$a = (kkRT)^{1/2} = [(32.2)(1.4)(53.3)(560)]^{1/2} = 1160 \text{ fps}$$

we find $\mathbf{M} = 20{,}000/1160 = 17.24$.

Example—Stagnation Properties

Steam vapor, saturated at 10 bar, is flowing in a line at 1000 m/s. What are its stagnation properties?

Solution.   From equation (7-11)

$$h_0 = h + K$$

From Item B 16(SI) we read these properties of the steam, saturated at 10 bar:

$$h = 2778 \text{ kJ/kg} \qquad s = 6.59 \text{ kJ/kg K}$$

$$t = 180°C \qquad v = 0.195 \text{ m}^3/\text{kg}$$

Thus,

$$h_0 = h + K = 2778 + (1/2)(1000)^2 = 3278 \text{ kJ/kg}$$

$$s_0 = s = 6.59 \text{ kJ/kg K}$$

We now find p_0 and t_0 on Item B 16(SI) at h_0 and s_0 intersection:

$$p_0 = 77 \text{ bar} \qquad t_0 = 450°C \qquad v_0 = 0.04 \text{ m}^3/\text{kg}$$

7.25   FLOW FROM A RIGID TANK

Suppose an adiabatic ($Q = 0$) rigid tank system containing a compressible fluid mass m_1 has an outflow only and does no shaft work ($W = 0$); see Fig. 7/27. When the outflow is stopped (or ceases), a certain mass m_2 will remain in the tank. Now think with me for a moment. The mass remaining can be assumed to have undergone an isentropic process from the original partial volume it occupied in the tank at p_1 and t_1 to its final volume of the total tank at p_2, t_2. This assumption includes either a zero heat capacity for the tank or an adiabatic lining in the tank. Otherwise, with the outflow of each particle of gas, the temperature of that remaining falls below the original equilibrium temperature and heat begins to flow from the material of the tank into the fluid.

The defining equation of the fluid remaining in the tank is $s_1 = s_2$. If the fluid is an ideal gas, all p, V, and T isentropic relations apply and we can determine the final temperature of the remaining fluid as being

(a)
$$T_2 = T_1(p_2/p_1)^{\frac{k-1}{k}}$$

For a vapor occupying the tank, we can use $s_1 = s_2$, find s_1 in the appropriate table, and then locate state point 2 from s_2 and the pressure p_2. Problems of this nature are very interesting.

7.26   Example—Gas Flowing Out of Tank

(a) A 50-ft³ tank with an adiabatic lining (without "heat capacity") contains air at 294 psia and 400°F, Fig. 7/27. If the valve is opened a small amount (so that we can ignore the kinetic energy at boundary e and so that the turbulence in the tank is a minimum), allowing air to escape until the inside pressure is equal to $p_0 = 14.7$ psia, what amount of air leaves the tank? How much energy crosses the boundary e? Compute two ways. (b) If it is now desired to reduce the tank pressure to 0.5 atm by reversible pumping, how much work is required?

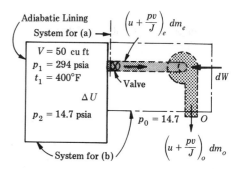

Fig. 7/27   Work is shown as it actually is, an input—not conventionally out. The energy balance should be in accord.

Solution. At this stage, for useful results, we must be able to choose a boundary (or boundaries) where the intensive properties of the mass crossing the boundary are the same at any instant for each element dm crossing the boundary. This requirement is essentially met by the specified conditions at boundary e, where the properties are the same as those in the tank, assumed to be the same throughout at any instant. (a) If heat is negligible, the assumption of isentropic expansion of the gas remaining in the tank at any instant is reasonable. Then by (a), § 7.25, with $T_1 = 400 + 460 = 860°R$,

(a)
$$T_2 = T_1\left(\frac{p_2}{p_1}\right)^{(k-1)/k} = 860\left(\frac{14.7}{294}\right)^{0.4/1.4} = 366°R$$

the final temperature. By Item B 2, $p_{r1} = 7.149$; then $p_{r2} = (14.7/294)(7.149) = 0.3575$. Enter table with $p_{r2} = 0.3575$ and find $T_2 \approx 366°R$, which checks. With initial and final states known, the mass of air at each state can be computed from $pV = mRT$:

(b) $\qquad m_1 = \dfrac{(294)(44)(50)}{(53.3)(860)} = 46.18 \text{ lb} \qquad$ and $\qquad m_2 = \dfrac{(14.7)(44)(50)}{(53.3)(366)} = 5.426 \text{ lb}$

Therefore the mass balance gives the departed mass as $46.18 - 5.426 = 40.75$ lb.

Considering the energy balance of the system (a) directly from Fig. 7/27, or with respect to equation (4-20) where $Q = 0$, $W = 0$, and $m_i = 0$, we have

(c) $\qquad \displaystyle\int d(mu) + \int h_e \, dm_e = 0 \qquad$ or $\qquad \displaystyle\int h_e \, dm_e = m_1 u_1 - m_2 u_2$

which is to say that in this case the energy crossing boundary e is the decrease of internal energy of the system. The initial and final internal energies in the tank are:

$$m_1 u_1 = m_1 c_v T_1 = (46.18)(0.1714)(860) = 6807 \text{ Btu}$$

$$m_2 u_2 = m_2 c_v T_2 = (5.426)(0.1714)(366) = 340 \text{ Btu}$$

$$\Delta(m_s u_s) = \Delta U_s = U_2 - U_1 \qquad\qquad = -6467 \text{ Btu}$$

The energy crossing boundary e is 6467 Btu. Since the specific heats do not change much in this temperature interval, almost the same answer is obtained from the gas table (Item B 2). Try it.

Because the plan of attack may be useful elsewhere, we shall now solve the same problem by $\int h_e \, dm_e$. First at any instant during the process

(d) $$T = T_1 \left(\frac{p}{p_1}\right)^{(k-1)/k} = \frac{T_1}{p_1^K} p^K$$

where $K = (k - 1)/k$ for convenience. Since the mass in the tank at any instant is $m = pV/RT$, where V is constant, the mass having left the tank is $m_e = m_{s1} - pV/RT$. Differentiate this equation, using equation (d).

(e) $$dm_e = -\frac{V}{R} d\left(\frac{P}{T}\right) = -\frac{Vp_1^K}{RT_1} d(p^{-K}p) = -\frac{Vp_1^K}{RT_1}(1 - K)p^{-K} \, dp$$

Using $u + pv = h = c_p T = c_p T_1 p^K / p_1^K$, we get

(f) $$\int h_e \, dm_e = \int \frac{c_p T_1 p^K}{p_1^K} \left(-\frac{Vp_1^K (1 - K) \, dp}{RT_1 p^K}\right)$$

$$= \int_{p_1}^{p_2} -\frac{c_p V(1 - K)}{R} \, dp = -\frac{c_p V}{kR}(p_2 - p_1)$$

where the value of K is substituted and the integration is made from state 1 to state 2. The numerical value, measured from a pseudozero at 0°R, is

(g) $$\int h_e \, dm_e = E_{\text{out}} = -\frac{(0.24)(50)}{(1.4)(53.3)}(14.7 - 294)(144) = 6468 \text{ Btu}$$

a good check with the previous answer.

(b) If $\int h_e \, dm_e$ were evaluated, the system for this part could well be that outlined by the light center line. However, the whole apparatus gives the simpler solution, with the boundary o where the mass moves out; in this case e is an internal boundary and irrelevant. For the system (b), the energy balance from Fig. 7/27 is

(h)
$$W = \int h_o \, dm_o + \Delta U$$

Note that if the system expands isentropically while the compressor reversibly compresses the gas, the drop in temperature in the tank is the same as the rise in temperature through the compressor; therefore the exit temperature at o remains constant at the temperature of the start of pumping, to wit, $T_o = 366°R$. Therefore, $h_o = c_p T_o =$ a constant, and $\int h_o \, dm_o = m_o h_o$.

Let 3 represent the final state in the tank.

(i)
$$T_3 = T_2 \left(\frac{p_3}{p_2}\right)^{(k-1)/k} = 366 \left(\frac{7.35}{14.7}\right)^{0.286} = 300°R$$

(j)
$$m_3 = \frac{p_3 V_3}{RT_3} = \frac{(7.35)(144)(50)}{(53.3)(300)} = 3.31 \text{ lb}$$

(k)
$$U_3 = m_3 c_v T_3 = (3.31)(0.1714)(300) = 170.2 \text{ Btu}$$

(l)
$$\Delta U = U_3 - U_2 = 170.2 - 340 = -169.8 \text{ Btu}$$

The mass departing during this phase of the operation is $m_2 - m_3 = 5.43 - 3.31 = 2.12 \text{ lb}$.

$$\int h_o \, dm_o = m_o h_o = (2.12)(0.24)(366) = 186.2 \text{ Btu}$$

$$W = 186.2 - 169.8 = 16.4 \text{ Btu}$$

Since the answer is positive, the work is an input as shown in Fig. 7/27 (as, of course, we already knew).

7.27   ENTROPY CHANGE IN A CONTROL VOLUME, PURE SUBSTANCE IN TRANSIENT PROCESS

No matter what happens in a constant mass closed system, its change of entropy is its final minus its initial absolute entropy (if the third law, § 6.12, is accepted), or the change computed along any internally reversible path connecting the states [say, equations (6-12) or (6-13) for an ideal gas]. In dealing with a control volume where the system changes from one equilibrium state to another after some net mass has crossed a boundary, or boundaries, the same idea may be used but the different masses before and after must be accounted for. Let s^a equal the absolute entropy; then the change of entropy within the control volume is

(7-15)
$$\Delta S = S_2 - S_1 = m_2 s_2^a - m_1 s_1^a$$

where $s^a = s° + \Delta\phi - R \ln (p/p°)$, equation (6-15); m_1 is the mass in the volume at time τ_1; m_2 is the mass at time τ_2; and, of course states 1 and 2 are equilibrium states. If the amount of substance is measured in moles, we have $\Delta S = n_2 \bar{s}_2^a - n_1 \bar{s}_1^a$, where n is the number of moles in the control volume. If the system is a homogeneous mixture of ideal gases, each component in the mixture is considered individually at its partial pressure; or certain characteristics of the mixture (c_p, c_v) must be known—as is the case the mixture called air.

Example—Entropy, Unsteady Flow 7.28

What is the change of entropy within the control volume of the example of § 7.26(a) (where air at 294 psia and 400°F flows from a 50-ft³ adiabatic tank until the pressure in the tank is 14.7 psia)?

Solution. From § 7.26, we have $m_1 = 46.18$ lb, $m_2 = 5.426$ lb, and $T_2 = 366°R$. Perhaps for present purposes, it will be accurate enough to determine the absolute entropy of air in the standard state, 1 atm and 77°F, by considering it as composed of 21% O_2 and 79% N_2 by volume. (Actually, accounting for 0.99% argon will hardly be detectable in numerical work.) On this basis, the partial pressures of the O_2 and N_2 are by X_{ipm}, § 6.10, 0.21 atm and 0.71 atm, respectively. Taking the values of $\bar{s}°$ for O_2 and N_2 from the table of Item B 11 and noting that $\Delta \phi = 0$ at the constant temperature of 77°F, we get the absolute entropy values at their partial pressures in the atmosphere as

(a) $\qquad O_2$: $\quad \bar{s}^a = \bar{s}° - \bar{R} \ln p = 49.004 - \bar{R} \ln 0.21 = 52.101$ Btu/pmole-°R

(b) $\qquad N_2$: $\quad \bar{s}^a = 45.767 - \bar{R} \ln 0.79 = 46.234$ Btu/pmole-°R

Since volumetric fractions are the same as mole fractions, the absolute entropy of 1 mole of air at 1 atm and 77°F is closely

(c) $\qquad \bar{s}^a$, air $= (0.79)(46.234) + (0.21)(52.101) = 47.47$ Btu/pmole-°R

or $s_0 = 47.47/29 = 1.637$ Btu/lb-°R. The absolute entropies of the air, initially and finally, from equation (6-15), are

(d)
$$S_1^a = m_1(s_0 + \Delta\phi - R \ln p_{1\,\text{atm}})$$
$$= 46.18(1.637 + 0.71323 - 0.59945 - 0.2053) = 43.7 \text{ Btu/°R}$$

(e)
$$S_2^a = m_2(s_0 + \Delta\phi - R \ln 1)$$
$$= 5.426(1.637 + 0.50765 - 0.59945 - 0) = 8.38 \text{ Btu/°R}$$

(f) $\qquad \Delta S = S_2^a - S_1^a = 8.38 - 43.7 = -35.32$ Btu/°R

where $\Delta\phi = \phi - \phi°$ from Item B 2 and $p_1 = 294/14.7 = 20$ atm.

Example—Transient Flow 7.29

A tank contains 1 lb of saturated steam at 250°F, state 1, and is connected to a steam line where the state remains steady at 120 psia and 500°F, Fig. 7/28, with negligible kinetic energy. The valve at boundary ii is opened and the tank is filled. After the valve is closed, the state in the tank is 120 psia and 560°F, state 2. Determine the heat, the change of entropy within the tank, and the entropy production for a sink temperature of 70°F.

Solution. With the tank as the system, the energy balance from Fig. 7/28 is

(a) $\qquad\qquad H_i + Q = \Delta U = U_2 - U_1$

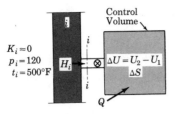

Fig. 7/28

Properties as follows are taken from Items B 13 and B 15 (or full tables):

$$p_1 = 29.825 \qquad v_2 = 4.9538 \qquad v_i = 4.6341$$

$$v_1 = 13.819 \qquad h_2 = 1308.0 \qquad h_i = 1277.4$$

$$h_1 = h_g = 1164.0 \qquad s_2 = 1.7182 \qquad s_i = 1.6872$$

$$s_1 = 1.7000$$

For $m_1 = 1$ lb of saturated steam the volume of the tank is $v_{g1} = 13.819$ ft^3. At the final state, $v_2 = 4.9538$ ft^3/lb; therefore

(b) $\qquad m_2 = \dfrac{V}{v_2} = \dfrac{13.819}{4.9538} = 2.79$ lb $\qquad$ and $\qquad m_{in} = 2.79 - 1 = 1.79$ lb

(c) $\qquad H_i = m_{in}h_i = (1.79)(1277.4) = 2286$ Btu

(d) $\qquad U_2 = m_2 u_2 = (2.79)\left(1308.0 - \dfrac{(120)(144)(4.9538)}{778}\right) = 3342$ Btu

(e) $\qquad U_1 = u_1 = (h - pv)_1 = 1164 - \dfrac{(29.825)(144)(13.819)}{778} = 1087.7$ Btu

Substituting the values found into equation **(a)**, we get

(f) $\qquad\qquad\qquad Q = 3342 - 1087.7 - 2286 = -31.7$ Btu

Using the absolute entropy of water at 32°F as 0.8391 Btu/lb-°R, the absolute entropies for the state in this example in Btu/lb-°R are

$$s_1^a = 1.7000 + 0.8391 = 2.5391 \qquad s_2^a = 1.7182 + 0.8391 = 2.5573$$

$$s_i^a = 1.6872 + 0.8391 = 2.5263$$

The entropy change within the tank is

(g) $\qquad \Delta S_{sys} = m_2 s_2^a - m_1 s_1^a = (2.79)(2.5573) - 2.5391 = 4.5958$ Btu/lb-°R

For the entropy production, consider equation (5-13). The entropy convected into the control volume with the mass flow is

(h) $\qquad\qquad\qquad S_{in} = m_{in}s_i^a = (1.79)(2.5263) = 4.5221$ Btu/°R

Notice that this is a decrease of entropy of the *surroundings*. The entropy change caused by the heat is

(i)
$$S_Q = \int \frac{dQ_r}{T_r} = \frac{-31.7}{530} = -0.0598 \text{ Btu/°R}$$

where $T_r = T_0 = 70 + 460 = 530°R$. By equation (5-13),

(j)
$$\Delta S_p = 4.5958 - 4.5221 - 0.0598 = +0.0139 \text{ Btu/°R}$$

Observe that in equation **(j)**, the conversion to absolute entropy makes no difference in the answer, but that the individual computation of **(g)**, for example, would not be the true entropy change within the control volume if the steam-table-datum values were used.

Example—Transient Flow   7.30

A 10-ft³ vessel contains H_2O at 300°F and 5% quality. A valve in the upper part is kept open until the temperature falls to 250°F. (a) How much H_2O leaves the vessel? (b) Compute the original and final volumes occupied by the vapor. The process is adiabatic and kinetic energies are negligible.

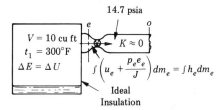

Fig. 7/29

Solution. (a) To solve such a problem, one necessarily makes assumptions of some kind. Perhaps the best plan is to let the liquid be on the bottom and let the rate of flow be small with internal equilibrium existing at all times with only vapor passing out. (Of course, the longer the time of flow, the more heat is transferred, no matter how well the vessel is insulated.) In any case, the decrease of stored energy $-\Delta U$ is equal to the energy departing, $-\int h_e \, dm_e$, as shown by the energy balance,

(a)
$$0 = \Delta U - \int h_e \, dm_e \quad \text{or} \quad U_1 - U_2 = m_1 u_1 - m_2 u_2 = -\int h_e \, dm_e$$

where we have let $u + pv/J = h$ at section e, Fig. 7/29. Since h_e varies with temperature, the $\int h_e \, dm_e$ is not readily evaluated. But if we note that the total variation of the enthalpy of saturated steam between 300 and 250°F is less than 1.5%, we may justifiably for engineering purposes assume that h_e is constant at the average value. Look up the needed specific properties and make the preliminary computations.

$$t_1 = 300°F \qquad p_1 = 67.005 \text{ psia} \qquad t_2 = 250°F \qquad p_2 = 29.825 \text{ psia}$$

$$v_{f1} = 0.01745 \qquad h_{f1} = 269.7 \qquad v_{f2} = 0.017 \qquad h_{f2} = 218.59$$

$$v_{fg1} = 6.4483 \qquad h_{fg1} = 910.0 \qquad v_{fg2} = 13.802 \qquad h_{fg2} = 945.4$$

$$v_{g1} = 6.4658 \qquad h_{g1} = 1179.7 \qquad v_{g2} = 13.819 \qquad h_{g2} = 1164.0$$

$$v_1 = 0.01745 + (0.05)(6.4483) = 0.3399 \text{ ft}^3/\text{lb}$$

$$m_1 = \frac{V}{v} = \frac{10}{0.3399} = 29.4\,\text{lb}$$

$$h_1 = 269.7 + (0.05)(910) = 315.2\,\text{Btu/lb}$$

$$u_1 = 315.2 - \frac{(67.005)(144)(0.3399)}{778} = 311\,\text{Btu/lb}$$

$$U_1 = (29.4)(311) = 9143.4\,\text{Btu}$$

$$h_e \approx \frac{(1179.7 + 1164)}{2} = 1171.9\,\text{Btu/lb}$$

(Incidental intelligence: It happens that a somewhat better average obtained from a plotted curve of h_g is 1172.6.)

(b) $$-\int h_e\,dm_e = -h_e \int_{m_1}^{m_2} dm_e = h_e(m_1 - m_2) = (29.4)(1171.9) - 1171.9m_2$$

$$u_{fg2} = 945.4 - \frac{(29.825)(144)(13.802)}{778} = 869.2\,\text{Btu/lb}$$

$$U_2 = m_2(u_f + x_2u_{fg})_2 = 218.59m_2 + 869.2x_2m_2$$

where we have used $u_{f2} \approx h_{f2} = 218.59$. Note that this approximation is quite justified, especially considering the nature of other assumptions. Substituting appropriate values in the energy balance equation **(a)**, we have

(c) $$9143.4 - 218.59m_2 - 869.2x_2m_2 = (29.4)(1171.9) - 1171.9m_2$$

with two unknowns. Another constraint containing the same unknowns is

(d) $$V = 10 = m_2v_2 = m_2(v_f + x_2v_{fg})_2 = m_2(0.017 + 13.802x_2)$$

Substitute m_2 from this equation **(d)** into **(c)**, solve for x_2, and find $x_2 = 2.67\%$. The mass left in the vessel is

(e) $$m_2 = \frac{10}{v_2} = \frac{10}{0.017 + (0.0267(13.8)} = 25.94\,\text{lb}$$

The mass leaving is $m_1 - m_2 = 29.4 - 25.94 = 3.46\,\text{lb}$. Use these results and check the energy balance of equation **(a)**.

(b) At state 1, the mass of vapor is $x_1m_1 = (0.05)(29.4) = 1.47\,\text{lb}$; since it is all saturated vapor, its volume is $1.47v_{g1} = (1.47)(6.458) = 9.493\,\text{ft}^3$. At 2, $x_2m_2 = (0.0267)(25.94) = 0.693\,\text{lb}$, corresponding to a volume of $(0.693)(13.819) = 9.58\,\text{ft}^3$.

7.31   SOME EFFICIENCIES AND RATIOS

A frequently used index of the efficiency of a turbine, or other engine, is called the *engine efficiency* η, or, when specifically applied, a ***turbine efficiency***, defined as the *actual work delivered divided by the corresponding ideal work*;

(7-16)
$$\eta = \frac{\text{actual work}}{\text{ideal work}} = \frac{W'}{W} \qquad [\text{ENGINE EFFICIENCY}]$$

in which the actual work W' may, of course, be measured at different points, resulting in different engine efficiencies. For example, if it is the brake work W_B, the corresponding engine efficiency is the brake engine efficiency η_b. For the turbine efficiency, W' is usually the *fluid* work. Moreover, quite commonly, a turbine efficiency is estimated on the basis that it is adiabatic; thus, for an actual adiabatic 1-2', Fig. 7/16(a), versus the isentropic 1-2, where boundary 2 is at the exit of the turbine, the **turbine efficiency** would be

(a)
$$\eta_t = \frac{h_1 - h_{2'}}{h_1 - h_2} \qquad [\text{TURBINE EFFICIENCY}]$$

where $\Delta K \approx 0$ or the *enthalpies are stagnation enthalpies*, § 7.23. If ΔK is significant, obtain W, W' from the steady-flow energy equation.

The **nozzle efficiency**, Fig. 7/16(a), where state 2 is at the exit boundary of the nozzle, is

(b)
$$\eta_n = \frac{\text{actual } \Delta K'}{\text{ideal } \Delta K_s} = \frac{h_1 - h_{2'}}{(h_1 - h_2)_s} \qquad [\text{NOZZLE EFFICIENCY}]$$

$\Delta K' = K_{2'} - K_1$, $\Delta K_s = K_2 - K_1$ during an isentropic expansion. Often, the initial kinetic energy is quite negligible. Similarly, the **diffuser efficiency**, *when the final kinetic energy is negligible*, state 2', Fig. 7/16(b), is

(c)
$$\eta_d = \frac{(h_2 - h_1)_s}{K_1} = \frac{(h_2 - h_1)_s}{h_{2'} - h_1} \qquad [K_2 \approx 0]$$

where states 2 and 2' are at the same pressure. See §§ 7.23, 15.14, and 18.15.

The **compression efficiency** is defined as the ideal (isentropic) work of compression divided by the actual work;

$$\eta_c = \frac{\text{ideal work}}{\text{actual work}} = \frac{W}{W'} \qquad [\text{COMPRESSION EFFICIENCY}]$$

The **ratio of expansion** r_e during any particular process is the ratio of the *final volume V_2* divided by the *initial volume V_1*. For instance, the isothermal expansion ratio in Fig. 7/10 is V_2/V_1.

The **ratio of compression** r_k of a process is the ratio of the *initial volume V_1* divided by the *final volume V_2*. For instance, the isentropic compression ratio in Fig. 7/16(b) is $r_k = V_1/V_2$.

The **pressure ratio** r_p is the highest pressure divided by the lowest pressure of a particular process; p_1/p_2 for the isentropic process 1-2, Fig. 7/16(a), and p_2/p_1 for Fig. 7/16(b). Observe that *each of these ratios is greater than* 1.

CLOSURE 7.32

Thermodynamics is exacting—but, to some, fascinating. In the beginning, however, many readers are overcome by the many new technical words and ideas. Frequent reviews are helpful, especially if used where it is possible to codify the knowledge presented and see its relation to the fundamental laws. Although the beginner is prone to depend unduly on memory for details, it is easier and more profitable to recognize the repetitive utility of basic ideas. Concentrate on learning the definitions of words and laws for better understanding and communication. Thus, the summary of Table IV is not intended as a crutch on which to lean, but as an example of review activity.

TABLE IV   Ideal Gas Formulas

For constant mass systems undergoing internally reversible processes.

Process →	Isometric $V = C$	Isobaric $p = C$	Isothermal $T = C$	Isentropic $S = C$	Polytropic $pV^n = C$
p, V, T relations	$\dfrac{T_2}{T_1} = \dfrac{p_2}{p_1}$	$\dfrac{T_2}{T_1} = \dfrac{V_2}{V_1}$	$p_1 V_1 = p_2 V_2$	$p_1 V_1^k = p_2 V_2^k$ $\dfrac{T_2}{T_1} = \left(\dfrac{V_1}{V_2}\right)^{k-1}$ $= \left(\dfrac{p_2}{p_1}\right)^{(k-1)/k}$	$P_1 V_1^n = p_2 V_2^n$ $\dfrac{T_2}{T_1} = \left(\dfrac{V_1}{V_2}\right)^{n-1}$ $= \left(\dfrac{p_2}{p_1}\right)^{(n-1)/n}$
$\displaystyle\int_1^2 p\,dV$	0	$p(V_2 - V_1)$	$p_1 V_1 \ln \dfrac{V_2}{V_1}$	$\dfrac{p_2 V_2 - p_1 V_1}{1 - k}$	$\dfrac{p_2 V_2 - p_1 V_1}{1 - n}$
$-\displaystyle\int_1^2 V\,dp$	$V(p_1 - p_2)$	0	$p_1 V_1 \ln \dfrac{V_2}{V_1}$	$\dfrac{k(p_2 V_2 - p_1 V_1)}{1 - k}$	$\dfrac{n(p_2 V_2 - p_1 V_1)}{1 - n}$
$U_2 - U_1$	$m \displaystyle\int c_v\,dT$ $mc_v(T_2 - T_1)$	$m \displaystyle\int c_v\,dT$ $mc_v(T_2 - T_1)$	0	$m \displaystyle\int c_v\,dT$ $mc_v(T_2 - T_1)$	$m \displaystyle\int c_v\,dT$ $mc_v(T_2 - T_1)$
Q	$m \displaystyle\int c_v\,dT$ $mc_v(T_2 - T_1)$	$m \displaystyle\int c_p\,dT$ $mc_p(T_2 - T_1)$	$m \displaystyle\int T\,ds$ $p_1 V_1 \ln \dfrac{V_2}{V_1}$	0	$m \displaystyle\int c_n\,dT$ $mc_n(T_2 - T_1)$
n	∞	0	1	k	$-\infty$ to $+\infty$
Specific heat c	c_v	c_p	∞	0	$c_n = c_v\left(\dfrac{k - n}{1 - n}\right)$ $[k = C]$
$H_2 - H_1$	$m \displaystyle\int c_p\,dT$ $mc_p(T_2 - T_1)$	$m \displaystyle\int c_p\,dT$ $mc_p(T_2 - T_1)$	0	$m \displaystyle\int c_p\,dT$ $mc_p(T_2 - T_1)$	$m \displaystyle\int c_p\,dT$ $mc_p(T_2 - T_1)$
$S_2 - S_1$	$m \displaystyle\int \dfrac{c_v\,dT}{T}$ $mc_v \ln \dfrac{T_2}{T_1}$	$m \displaystyle\int \dfrac{c_p\,dT}{T}$ $mc_p \ln \dfrac{T_2}{T_1}$ $m \displaystyle\int \dfrac{c_v\,dT}{T} + mR \ln \dfrac{v_2}{v_1}$	$\dfrac{Q}{T}$ $mR \ln \dfrac{V_2}{V_1}$ $m \displaystyle\int \dfrac{c_p\,dT}{T} - mR \ln \dfrac{p_2}{p_1}$	0	$m \displaystyle\int \dfrac{c_n\,dT}{T}$ $mc_n \ln \dfrac{T_2}{T_1}$

PROBLEMS

SI UNITS

7.1 A thermodynamic system consists of a fluid mass within an insulated (adiabatic) rigid container having a set of internal paddles. The mass is stirred by rotating the paddles with external work W_p. For the system, evaluate each of the following expressions stating whether its value is equal to, less than, or greater than zero: **(a)** net work, **(b)** ΔU, **(c)** ΔH, **(d)** ΔS, **(e)** $\int dQ/T$, **(f)** ΔpV, **(g)** $\int p\,dV$, **(h)** $-\int V\,dp$.

7.2 There are 1.36 kg of air at 137.9 kPaa stirred with internal paddles in an insulated rigid container, whose volume is $0.142\,\text{m}^3$, until the pressure becomes 689.5 kPaa. Determine: **(a)** the work input, **(b)** $\Delta(pV)$, **(c)** ΔE_f, and **(d)** Q.

Ans. **(a)** 196.2, **(b)** 78.3 kJ.

7.3 A certain quantity of 30°C saturated water $(v_f = 0.001\,\text{m}^3,\; h_f = 125.7\,\text{kJ/kg},\; p_{sat} = 0.04\,\text{bar})$ is injected into a $0.3\,\text{m}^3$ evacuated drum and then heated until the pressure and temperature become 20 bar and 300°C, respectively. Determine **(a)** the mass of water used and **(b)** the heat required.

7.4 A reversible nonflow constant volume process decreases the internal energy by 316.5 kJ for 2.268 kg of a gas for which $R = 430\,\text{J/kg-K}$ and $k = 1.35$. For the process, determine **(a)** the work, **(b)** the heat, and **(c)** the change of entropy if the initial temperature is 204.4°C.

Ans. **(b)** −316.5 kJ, **(c)** −0.7572 kJ/K.

7.5 The internal energy of a perfectly insulated and rigid 283-ℓ gaseous nitrogen system is increased 2930 kJ by means of internal stirrers (paddles). If the initial pressure is $p_1 = 1379\,\text{kPaa}$ find p_2 and ΔS (per kg mass).

7.6 During a cryogenic experiment, 2 g mole of oxygen are placed in a closed 1-ℓ Dewar flask at 120 K. After some time, the temperature is noted to be 215 K. Use Item B 27 and find **(a)** the increase in pressure, **(b)** Q, **(c)** the irreversibility.

7.7 There are 1.36 kg of a gas, for which $R = 377\,\text{J/kg-K}$ and $k = 1.25$, that undergo a nonflow constant volume process from $p_1 = 551.6\,\text{kPaa}$ and $t_1 = 60°C$ to $p_2 = 1655\,\text{kPaa}$. During the process the gas is internally stirred, and there are also added 105.5 kJ of heat. Determine **(a)** t_2, **(b)** the work input, **(c)** Q, **(d)** ΔU, and **(e)** ΔS.

Ans. **(a)** 999 K, **(b)** −1260.5 kJ, **(e)** 2.253 kJ/K.

7.8 A vertical, frictionless piston and cylinder arrangement contains 1 kg of steam at 25 bar, 260°C. Heating occurs at constant pressure causing the piston to move upward until the initial volume is doubled; local $g = 9.145\,\text{m/s}^2$. Find **(a)** the change in potential energy of the piston, **(b)** the heat. **(c)** If the process had occurred because of an input of paddle work only (no heat), find the net work of the system.

Ans. **(a)** 224, **(b)** 1018, **(c)** −794 kJ.

7.9 There are expanded 0.90 kg/s of steam at constant pressure from 30 bar and 70% quality to a final state. If the process is nonflow for which $W = 121.3\,\text{kJ/s}$, find **(a)** the final temperature, **(b)** Q, **(c)** the available part of Q for a sink temperature of $t_0 = 27°C$.

Ans. **(a)** 282°C, **(b)** 966, **(c)** 414 kJ/s.

7.10 If 2 kg of steam at 18 bar and 288°C undergo a constant pressure process until the quality becomes 50% $(h_2 = 2220,\; v_2 = 0.075,\; S_2 = 5.18)$. Find W, ΔH, Q, ΔU, and ΔS: **(a)** if the process is nonflow, **(b)** if the process is steady flow with $\Delta K = 0$. Use Item B 16(SI).

Ans. **(a)** $W = -220\,\text{kJ}$, $Q = -1566\,\text{kJ}$, $\Delta S = -3.20\,\text{kJ/K}$.

7.11 **(a)** Demonstrate that for a reversible isothermal process (gas or vapor) depicted on the pV-plane, the area beneath the trace is exactly equal to that behind (to the left of) the trace. **(b)** Prove that this trace must be concave upward and have a negative slope on this plane.

7.12 There are 2.27 kg/min of steam undergoing an isothermal process from 27.6 bar, 316°C to 6.9 bar. Sketch the pV and TS diagrams, and determine **(a)** ΔS, **(b)** Q, **(c)** W for nonflow, **(d)** W for steady flow with $\Delta P = 0$, $\Delta K = +42\,\text{kJ/min}$.

Ans. **(a)** 1.6 kJ/K-min. **(b)** 942. **(c)** 846, **(d)** 798 kJ/min.

7.13 Helium at 100 atm, 165 K expands isothermally to 1 atm. For 2 kg, find **(a)** W for a nonflow and for a steady flow process

($\Delta P = 0$, $\Delta K = 0$), **(b)** ΔU, **(c)** Q, and **(d)** ΔS. Use Item B 30.

7.14 During a reversible process there are abstracted 317 kJ from 1.134 kg/s of a certain gas while the temperature remains constant at 26.7°C. For this gas, $c_p = 2.232$ and $c_v = 1.713$ kJ/kg-K. The initial pressure is 586 kPaa. For both nonflow and steady flow ($\Delta P = 0$, $\Delta K \approx 0$) processes, determine **(a)** V_1, V_2 and p_2, **(b)** W, and Q, **(c)** ΔS, and ΔH.

Ans. **(a)** 302, 50 ℓ/s, 3539 kPaa, **(b)** -317 kJ, **(c)** $\Delta S = -1.056$ kJ/K-s.

7.15 A given gaseous system undergoes an isentropic process, state 1 to state 2. **(a)** Combine the two relations $pv = RT$ and $pv^k = c$ and show that $T_2/T_1 = (p_2/p_1)^{(k-1)/k} = (v_1/v_2)^{k-1}$. **(b)** Integrate the two expressions, using $pv^k = c$, and show that $-\int v\, dp$ is k times $\int p\, dv$ by comparison.

7.16 During an isentropic process of 1.36 kg/s of air, the temperature increases from 4.44°C to 115.6°C. For a nonflow process and for a steady flow process ($\Delta K = 0$, $\Delta P = 0$), find **(a)** ΔU, **(b)** W, **(c)** ΔH, **(d)** ΔS, and **(e)** Q.

Ans. **(a)** 108.8, **(b)** -108.8, **(c)** 152.2 kJ/s.

7.17 Steam flows istentropically through a nozzle from 1517 kPaa, 288°C to 965 kPaa. Sketch the *TS* diagram and for $\dot{m} = 454$ g/s, determine **(a)** Δs, **(b)** t_2, **(c)** W, **(d)** ΔK. **(e)** Assume nozzle efficiency $e_n = 94\%$ and solve for the preceding parts.

Ans. **(b)** 232°C, **(d)** 48.3 kJ/s.

7.18 A throttling calorimeter is connected to a steam main in which the pressure is 15 bar. In the calorimeter, $p_2 = 1$ bar and $t_2 = 110$°C. Compute the quality and the percentage moisture of the steam sampled.

Ans. $y = 5.2\%$.

7.19 If 4 kg of steam at 12 bar, 260°C expand into the wet region to 90°C in a polytropic process where $pV^{1.156} = C$, find **(a)** y_2, ΔH, ΔU, ΔS, **(b)** $\int p\, dV$, **(c)** $-\int V\, dp$, **(d)** W for nonflow, and **(e)** W for steady flow if $\Delta K = -3$ kJ/kg. What does $-\int V\, dp$ represent in this process? **(f)** Find Q from steady flow and also from nonflow energy equations. Use Item B 16 (SI).

7.20 The work required to compress a gas reversibly according to $pV^{1.30} = C$ is 67,790 J if there is no flow. Determine ΔU and Q if the gas is **(a)** air, **(b)** methane.

Ans. **(a)** 50.90, -16.99 kJ.

7.21 A stagnation probe was fabricated using ordinary solder (melting point 183°C). Estimate the limiting air velocity to which the probe may be subjected based on normal static conditions of 1 atm, 26.7°C.

7.22 A 142-ℓ drum contains oxygen initially at 552 kPaa and 37.8°C. The drum is fitted with an automatic pressure release valve that maintains the maximum pressure constant at 689.5 kPaa. Determine the heat transferred to the oxygen while the temperature of that which remains in the drum increases to 282.2°C. *Ans.* 173 kJ.

ENGLISH UNITS

7.23 There are 2 lb of a liquid-vapor mixture ($x_1 = 50\%$) of CO_2 contained in a rigid vessel at 90 psia. Heat is absorbed by the CO_2 until its temperature becomes 400°F; the environment is at $p_0 = 14.7$ psia, $t_0 = 80$°F. Use Item B 31 and determine **(a)** the final pressure p_2 and volume, **(b)** Q and that portion of Q which is available ($s_1 = 0.150$ Btu/°R-lb), **(c)** the irreversibility if the heat source is at the minimum constant temperature.

Ans. **(a)** 400 psia, 1 ft³, **(b)** -30.4 Btu, **(c)** 139 Btu.

7.24 A certain ideal gas has a constant $R = 38.9$ ft-lb/lb-°R with $k = 1.4$. **(a)** Find c_p and c_v. **(b)** If 3 lb of this gas undergoes a reversible nonflow constant volume process from $p_1 = 20$ psia, 140°F to 740°F, find p_2, ΔU, ΔH, Q, W. **(c)** If the process in (b) had been steady flow with $\Delta P = 0$, $\Delta K = 0$, find W and ΔS.

Ans. **(a)** 0.125, 0.175 Btu/lb-°R, **(b)** 40 psia, $Q = 226$ Btu, **(c)** -90 Btu, 0.26 Btu/°R.

7.25 A pressure vessel contains 2.5 lb of saturated steam at 70 psia. Determine the amount of heat which must be rejected in order to reduce the quality to 60%. What will be the pressure and temperature of the steam at this new state?

Ans. $Q = -875$ Btu.

7.26 Confine 1 lb of wet steam at 500°F and with an enthalpy of 800 Btu in a rigid container. Heat is applied until the steam becomes saturated. Determine **(a)** the initial pressure p_1, **(b)** quality x_1, **(c)** volume v_1, **(d)** the final pressure p_2; and for the process, **(e)** W, **(f)** Q.

Ans. **(a)** 680.8 psia, **(b)** 43.75%, **(c)** 0.3064 ft^3, **(d)** 1383 psia, **(f)** 336.4 Btu.

7.27 Ammonia vapor at 100 psia and 320°F is cooled at constant pressure to a saturated vapor during steady flow with $\Delta P = 0$, $\Delta K = 0$. For 3 lb, find **(a)** Q and W, **(b)** ΔV, **(c)** ΔU, and **(d)** ΔS. **(e)** The same as before except that the process is nonflow. See Item B33.

Ans. **(a)** -468 Btu, **(b)** -5.55 ft^3, **(c)** -365.2 Btu, **(d)** -0.732 Btu/°R.

7.28 A perfectly insulated system contains 2 ft^3 of hydrogen, at 75°F and 80 psia, which receives an input of paddle work at constant pressure until the temperature is 150°F. See figure. Determine **(a)** the heat, **(b)** the change of internal energy, **(c)** the work input, **(d)** the net work, **(e)** the change of entropy, and **(f)** the change of enthalpy.

Ans. (Item B 1) **(b)** 10.25, **(c)** 14.40, **(d)** -10.25 Btu, **(e)** 0.02525 Btu/°R, **(f)** 14.40 Btu.

7.29 Two lb of hydrogen simultaneously reject heat and receive paddlework input in a nonflow change of state at constant pressure from an initial temperature of 250°F to a final temperature of 90°F. See figure. If the heat rejected is thrice the paddle work, determine **(a)** ΔU, ΔH, and ΔS for the system, **(b)** Q, **(c)** W_{net}.

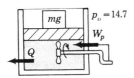

Problems 7.28, 7.29

Ans. (Item B 5) **(a)** -786, -1102 Btu, -1.762 Btu/°R, **(b)** -1653, **(c)** -867 Btu.

7.30 Evaporate 5 lb/sec of Freon 12 at constant pressure, $p = 20$ psia, from saturated liquid to a state where $t_2 = 60$°F; for the initial state, $t_1 = -8.1$°F, $s_{fl} = 0.01551$ Btu/°R-lb, $v_{fl} = 0.01345$ ft^3/lb. For nonflow reversible conditions find **(a)** Δs, **(b)** Q, **(c)** W. **(d)** If the process is steady flow with $\Delta P = 0$, $\Delta K = 0$, find Q and W. Use Item B 35.

7.31 Carbon dioxide is being considered as a possible coolant for an unmanned orbiting space platform. Saturated vapor enters a constant pressure steady flow heat absorber and leaves at 30 psia, 450°F. Find Q and ΔS for each 1-lb/sec. Use Item B 31.

7.32 There are expanded isothermally 3 lb/sec of NH$_3$ from saturated vapor to the state of 10 psia, 100°F. Determine **(a)** Q, **(b)** W (nonflow), **(c)** W (steady flow with $\Delta K = 0$), **(d)** ΔG, **(e)** the unavailable part of Q for $t_0 = 60$°F. See Item B 33.

7.33 There are compressed isothermally 800 cfm of air measured at 80°F and 200 psia; $p_2 = 600$ psia. For both nonflow and steady flow ($v_1 = 75$ fps, $v_2 = 150$ fps, $\Delta P = 0$) processes, compute **(a)** $\int p\,dV/J$ and $-\int V\,dp/J$, **(b)** ΔU, ΔH, and ΔS, **(c)** work and heat.

Ans. **(a)** $-32,500$ Btu/min, **(b)** $\Delta S = -60.24$ Btu/°R, **(c)** $W_n = -32,500$, $W_{sf} = -32,770$, $Q = -32,500$ Btu/min.

7.34 Mercury is being considered as a working fluid for a heat engine aboard an orbiting space platform. Radiant heat is absorbed by the mercury initially at 100 psia, $y_1 = 45\%$; the final state is 0.5 psia, 900°F. For each pound of mercury being circulated, find **(a)** Δs, Δh, **(b)** Q, **(c)** W where $\Delta P = 0$, $\Delta K = 0$. **(d)** If the constant temperature radiant source is at 10,000°R, find the entropy production for this heat interaction. Use Item B 34.

7.35 Saturated Freon 12 liquid is evaporated at constant temperature in a refrigeration system; it leaves as saturated vapor at 15 psia. For steady flow of 1 lb/sec find Q. What is the final entropy? The final volume? Use Item B 35.

7.36 A two-fluid H$_2$O/NH$_3$ system (see figure) is composed of two weighted, frictionless pistons A and B, a conducting rigid partition P, 10 lb of saturated H$_2$O vapor at 300°F in part S, and 5 lb of saturated NH$_3$ liquid at 60°F ($v_f = 0.026$ ft^3/lb) in part N; $p_0 = 14.7$ psia. Heat now flows from the

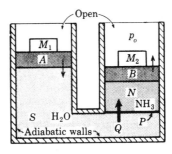

Problem 7.36

condensing steam through partition P to boil the NH_3. At the instant the last drop of liquid NH_3 is evaporated, find for the process **(a)** Q, **(b)** the amount of steam condensed, **(c)** ΔS for the H_2O, **(d)** ΔS for the NH_3, **(e)** ΔS for the total system, **(f)** the net work for S and N, **(g)** the available part of Q as it left the H_2O ($t_0 = 60°F$), and after it entered the NH_3, **(h)** the net change of potential energy of the piston-weight systems.

7.37 Assume 10 lb/sec of steam at 110 psia, 600°F enter a small ideal steady-state, steady-flow, turbine and, after isentropic expansion, are exhausted at 20 psia; $\Delta P = 0$, $\Delta K = 0$. Sketch the process on the pV and TS planes, use Item B 15, and find **(a)** exhaust temperature t_2, **(b)** $\dot{W}$ in kW and in MJ/hr. **(c)** Now use the Mollier chart (Item B 16) and check answers in **(a)** and **(b)**.
   Ans. **(a)** 250°F, **(b)** 1707 kW, 6145 MJ/hr.

7.38 The internal energy of a certain ideal gas is given by the expression $u = 850 + 0.529pv$ Btu/lb where p is in psia. Determine the exponent k in $pv^k = C$ for this gas undergoing an isentropic process.
   Ans. $k = 1.35$.

7.39 During an isentropic process of 4 lb/sec of air, the temperature increases from 80°F to 440°F. Compute **(a)** ΔU, ΔH, Q, and ΔS, **(b)** $\int p\,dV$ and $-\int V\,dp$, **(c)** the nonflow work, **(d)** the steady flow work where $\Delta P = 0$, $\Delta K = -10$ Btu/sec. **(e)** For an irreversible adiabatic process from the same initial state to the same final pressure, the final temperature is 500°F. Find the works, nonflow and steady flow ($\Delta P = 0$, $\Delta K = -10$ Btu/sec). What is ΔS for this process?

7.40 During a refrigerating cycle, 1 lb/sec of saturated Freon 12 vapor is compressed from 19 psia to 200 psia, 170°F. The process is a steady flow irreversible adiabatic ($\Delta P = 0$, $\Delta K = 0$). **(a)** Find W and the compressor efficiency. **(b)** The compressed Freon 12 flows through a long pipe to a condenser, arriving at 180 psia, 160°F; $t_0 = 80°F$. Compute the loss of availability and the irreversibility. Was there any heat interaction during flow in the pipe?

7.41 Assume 5 lb/sec of NH_3 are compressed adiabatically from 100 psia, 170°F to 220 psia in steady flow; $\Delta K = 0$. The entropy increases 0.1950 Btu/sec-°R because of irreversible effects. Determine **(a)** $t_{2'}$, **(b)** W',

(c) the extra work needed because of irreversible effects.
   Ans. **(a)** 340°F, **(b)** 454, **(c)** -151 Btu/sec.

7.42 Say 8 lb/sec of CO_2 gas are compressed polytropically ($pV^{1.2} = C$) from $p_1 = 15$ psia, $t_1 = 140°F$ to $t_2 = 440°F$. Assuming ideal gas action, find p_2, W, Q, ΔS **(a)** as a nonflow process, **(b)** as a steady flow process where $\Delta P = 0$, $\Delta K = 0$.

7.43 As part of a refrigeration cycle, saturated Freon 12 vapor at $-10°F$ is compressed polytropically to 200 psia, 200°F. **(a)** for nonflow conditions find n in $pV^n = C$, Δs, W, Q; the irreversibility if $t_0 = 40°F$ and if all rejected heat is received by the sink. **(b)** For steady flow conditions, $\Delta P = 0$ and $\Delta K = 0$, find W, Q, and the irreversibility. See Item B 35.

7.44 Helium undergoes a nonflow polytropic process from $V_1 = 2$ ft^3 and $p_1 = 14.4$ psia to $p_2 = 100$ psia. Assuming that 9.54 Btu of work are done during the process, determine the value of n in $pV^n = C$, and find Q.   *Ans.* 1.559, -1.517 Btu.

7.45 Hydrogen at 2 lb/min undergoes a steady flow polytropic process from $p_1 = 50$ psia, $v_1 = 90$ ft^3/lb, $v_1 = 300$ fps, $\Delta P = 0$, until the pressure, volume, and velocity are doubled. Determine **(a)** n in $pV^n = C$, **(b)** T_1 and T_2, **(c)** ΔH, ΔU, and ΔS, **(d)** $\int p\,dV/J$ and $-\int V\,dp/J$, **(e)** W and Q.
   Ans. (Item B 5) **(a)** -1, **(b)** 846°R, 3384°R, **(c)** 18,710, 13,710 Btu/min, 8.705 Btu/min-°R, **(d)** 2500, -2500, **(e)** -2511, 16,210 Btu/min.

7.46 A turbojet plane flies at an altitude of 25,000 ft where the atmospheric conditions are 6.20 psia and $-20°F$. Compute stagnation temperature and pressure of the layer of air wetting the plane's outer surface if the plane speed is **(a)** 600 mph, **(b)** 1000 mph, **(d)** 3000 mph. Plot a curve of stagnation temperature t_0 and pressure p_0 as a function of the Mach number. Use Item B 2.   *Ans.* **(a)** 504.6°R, 10.03 psia.

7.47 Flowing steam at 800 psia, 800°F, moving at 24,000 fpm in a line, is suddenly stopped by an instantaneous closing of a valve. Find the pressure and temperature of the steam initially interfaced at the valve surface.

7.48 A portable oxygen tank A for use in a space project has a volume of 0.2 ft^3. It is to be charged from a supply line in which the

O$_2$ is flowing steadily at 100 fps at 180 psia and 60°F. When charging started, the tank contained O$_2$ at 40°F and 1 atm. Charging ceases when the mass m_2 in the tank is 6 times the original amount. If the tank walls are adiabatic, compute the final temperature and pressure.

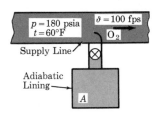

Problem 7.48

Ans. [Item B 1.] 240°F, 8.38 atm.

7.49 A 100 ft^3 rigid tank T containing air at 1 atm and 60°F is evacuated to zero pressure, theoretically, by a vacuum pump P (see figure). Heat Q_T is added to the air in the tank at a rate to maintain a constant tank temperature; heat Q_p is rejected at the pump so that compression is isothermal. Find **(a)** Q_T, **(b)** the pump work W_p, and **(c)** Q_p. Solve first by using Item B 1; then by Item B 2.

Ans. (Item B 1) **(a)** 272, **(b)** −272, **(c)** −272 Btu.

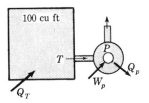

Problem 7.49

7.50 An adiabatic vessel contains steam at 1025 psia, 750°F. A valve is opened slowly permitting an outflow of steam; it is closed quickly when the steam remaining in the vessel becomes saturated vapor. What fractional part of the original steam remains? Assume that this part underwent a reversible adiabatic expansion. *Ans.* 30.3%.

7.51 Steam at 1000 psia, 1200°F is in a rigid container which is fitted with a closed valve; the container is immersed in a liquid bath which is maintained at a contant temperature of 1200°F by an electrical input of I^2R. The valve on the container is slowly opened permitting steam to escape to the atmosphere until the container pressure becomes 100 psia. Find the electrical input per cubic foot of container volume. The only heat interaction is between the bath and the container of steam.

8

GAS CYCLES

8.1 INTRODUCTION

To this point our discussion has been mostly about processes, thermodynamic events of great practical importance. It happens though that most of the power generated in our civilization is the output of cycles or of engines whose actual operation can be paraphrased by an ideal cyclic model; also cyclic devices generate by far most of our refrigeration. Not the least of the benefits from a study of cycles will be a better appreciation of the second law of thermodynamics. In this chapter we pay particular attention to the **heat engine** or **thermal engine**, which may be defined as *a closed system* (no mass crosses its boundaries) *that exchanges only heat and work with its surroundings and that operates in cycles.*

In the analysis of gas cycles, we shall be particularly interested in: (1) the heat supplied to the cycle, (2) the heat rejected, (3) the net work, (4) the efficiency, and (5) the mean effective pressure. Other items command our attention in special cases. For ideal cycles that are composed of *internally* reversible processes, the net work is found from a cyclic integral, $\oint dQ$, $\oint p\,dv$, or $-\oint v\,dp$. The typical units of these integrals being different, one must be sure in connection with ratios, such as thermal efficiency and mep, that *proper units are used.*

8.2 EXPANSIBLE FLUID IN A HEAT ENGINE

The essential elements of a thermodynamic heat engine with a fluid as the working substance are: (a) first, naturally, a **working substance**, matter that receives heat, rejects heat, and does work; (b) a **source of heat** (also called a *hot body*, a *heat reservoir*, or just *source*), from which the working substance receives heat; (c) a **heat sink** (also called a *receiver*, a *cold body*, or just **sink**), to which the working substance can reject heat; and (d) an **engine**, wherein the working substance may do work or have work done on it. Various other devices and accessories are essential for the completion of any particular cycle. Study Fig. 9/1 for the basic elements of a steam power plant.

A cycle is completed when the properties of a system have returned to their original values (§ 1.21). The equations to be obtained for cycles may be thought of as applying to a single cycle, but from a practical angle, cycles are repeated continuously. For the work and heat obtained from computations for one cycle to apply continuously, the system must operate in steady state. In *steady-state operation, the energy and mass content of the closed system remains constant and the properties of the system at any stage of the cycle are always the same as the system passes through this stage.*

CYCLE WORK, THERMAL EFFICIENCY, AND HEAT RATE

We have already observed from the first law, § 4.2 and equation (4-1), that for cycles, reversible or irreversible, for any substance,

$$(4\text{-}1) \qquad \oint dQ = \oint dW$$

in which $\oint dW$ is the net work W_{net} of the cycle, sometimes represented by the symbol W alone; in Fig. 9/1, it is seen to be the work of the engine W_{out} minus the work to the pumps W_{in}. Also, $\oint dQ$ is the *net heat*, which may be written ΣQ for the cycle and which is composed of heat added Q_A and heat rejected Q_R. If any previously given equation for Q is used, the resulting sign tells whether heat is added (+) or rejected (−). However, most of the time, we know beforehand whether heat moves in or out; hence we frequently use the scalar value of $|Q_R|$ without regard to sign, saying, for instance, the heat added minus the heat rejected. Thus, when it is clear that cycles are meant, we may write the cycle net work as

$$(8\text{-}1) \qquad W_{net} = \Sigma Q = \oint dQ \qquad \text{or} \qquad W = Q_A - |Q_R|$$

[ALGEBRAIC SUM]   [ARITHMETIC DIFFERENCE]

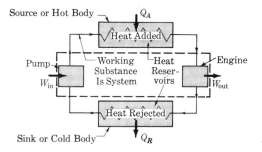

Power Cycle, $Q_A > Q_R$. **Fig. 8/1**

Efficiency may usually be thought of in its simplest form, *output* divided by *input*. The output of a power cycle, which is a thermodynamic cycle for the production of power, is the net work; the input is the heat supplied to the system *from an external source* of heat, Q_A in Fig. 8/1. For a net work $W = \oint dW = W_{out} - |W_{in}|$, the **thermal efficiency** of a **power cycle** is

$$(8\text{-}2) \qquad e = \frac{\oint dW}{Q_A} = \frac{W_{net}}{Q_A} = \frac{\oint dQ}{Q_A} = \frac{Q_A - |Q_R|}{Q_A} \qquad \text{[CYCLE ONLY]}$$

When the heat Q_A supplied to a cycle (or engine) from an external source corresponds to some unit of work, as a hp-hr or kw-hr, it is called the **heat rate**, a term quite commonly used to indicate the efficacy of an engine or cycle delivering power. In terms of heat rate, we have the thermal efficiency as

(a) $\qquad e = \dfrac{2544}{\text{heat rate}} \qquad e = \dfrac{2544\ \text{Btu/hp-hr}}{Q_A\ \text{Btu/hp-hr}}$ or $e = \dfrac{3412\ \text{Btu/kW-hr}}{Q_A\ \text{Btu/kW-hr}}$

In equation (8-2), we find the net work either from $\oint dW$ or from $\oint dQ$, as convenient in a particular situation.

8.4   THE CARNOT CYCLE*

It is fitting that the first cycle analyzed be the most efficient cycle conceivable. There are other ideal cycles as efficient as the Carnot cycle, but none more so. Such a perfect cycle forms a standard of comparison for actual engines and actual cycles and also for other less efficient ideal cycles, permitting us to judge how much room there might be for improvement. Too great a difference between actual and ideal efficiencies suggests a search for means by which the actual efficiency can be raised.

The mechanism for carrying out the Carnot cycle is described in Fig. 8/3. Since the Carnot cycle consists of two isothermals *ab* and *cd*, Fig. 8/2, and two isentropic processes *bc* and *da*, it is a rectangle on the *TS* (or *Ts*) plane for any kind of working substance. All heat exchanges occur at constant temperature. The significant feature of all reversible cycles, namely, that all processes are externally as well as internally reversible, means that $\Delta T = 0$ in Fig. 8/2; or we say that the cycle pictured approaches reversibility as $\Delta T \to 0$.

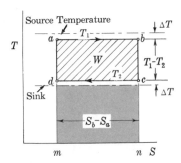

Fig. 8/2   *Carnot Cycle, Any Substance.* Let *a* be the initial point, clockwise cycle, which, according to our convention of signs, is a power cycle. The net work *W* will be positive. For the state point to return to *a*, thus completing the cycle, it *must* move toward the left in some manner after *b* is reached. Since a leftward movement of the state point on the *TS* plane indicates heat rejected, we may say now that *some heat must be rejected*, that all the heat supplied cannot be converted into work.

Let us now say that *T* is the *thermodynamic temperature* (§ 8.5) and we continue to accept entropy as a property (confirmed as §§ 5.3 and 5.4); $Q = \int T\,dS]_{\text{rev}} = T\int dS = T\,\Delta S$ [a reminder that areas on the *TS* (or *Ts*) plane for reversible processes represent heat].

* Nicholas Leonard Sadi Carnot (1796–1832), a quiet, unassuming Frenchman who lived during the turbulent Napoleonic period, devised and analyzed the Carnot cycle in his *Reflections on the Motive Power of Heat* at the early age of only twenty-three or twenty-four. It did not matter that the caloric theory of heat was the accepted theory in his time. Carnot's cycle is independent of the theory of heat as well as of the working substance. His life was unspectacular. He loved mottoes. One of his favorites, "Speak little of what you know, and not at all of what you do not know," reveals something of the man's nature.

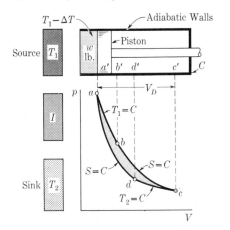

Operation of the Carnot Engine. The *pV* diagram is **Fig. 8/3** for a noncondensing working substance. A cylinder *C* contains *w* lb at a temperature $T_1 - \Delta T$, where ΔT approaches zero. The cylinder head, the only place where heat may enter or leave the system, is placed in contact with the source that has a constant temperature T_1. Heat flows from the source into the substance in the cylinder, which therefore undergoes an isothermal process *ab*, and the piston moves from *a'* to *b'*. If the temperature difference ΔT is infinitesimal, the transfer of heat is reversible. Next, the cylinder is removed from the hot body and the insulator *I* is placed over the head of the cylinder, so that there is *no heat*; any further process is adiabatic. The isentropic change *bc* now occurs, wherein the temperature drops from $T_1 - \Delta T$ to $T_2 + \Delta T$ because of the work being done at the expense of the internal energy, and the piston moves from *b'* to *c'*. When the piston reaches the end of the stroke *c'*, the insulator *I* is removed and the cylinder head is placed in contact with the sink, which remains at a constant temperature T_2. Heat then flows reversibly to the sink, and the isothermal *cd* occurs while the piston moves from *c'* to *d'*. Finally, the insulator *I* is again placed over the head and the isentropic compression *da* returns the substance to its initial condition, the temperature increasing from $T_2 + \Delta T$ to $T_1 - \Delta T$ because the work of compression increases the store of internal energy.

In practice, heat would flow very slowly for a small temperature difference ΔT, and therefore the movement of the piston and the *rate* of doing work would be infinitesimal. A finite temperature difference precludes external reversibility. The mechanical friction of the moving parts of the machine, the internal friction due to turbulence within the substance, and the heat transferred through the cylinder walls (it is impossible to make a nonconducting substance) also preclude reversibility in any real engine.

(a) $$Q_A = T_1(S_b - S_a) = T_1 \Delta S = \text{area } abmn$$

(b) $$Q_R = T_2(S_d - S_c) = -T_2(S_c - S_d) = -T_2 \Delta S = \text{area } cdmn$$

(c) $$W_{\text{net}} = \oint dQ = T_1 \Delta S - T_2 \Delta S = (T_1 - T_2) \Delta S$$

where $\Delta S = S_b - S_a = S_c - S_d$, Fig. 8/2. Observe that since cyclic work is equal to $\oint dQ$, areas enclosed on the *TS* plane by *internally reversible processes* completing a cycle represent work (of course not ordinarily to the same scale as those on the *pV* plane). The thermal efficiency of the Carnot cycle is

(8-3) $$e = \frac{W_{\text{net}}}{Q_A} = \frac{(T_1 - T_2)(S_b - S_a)}{T_1(S_b - S_a)} \quad \text{or} \quad e = \frac{T_1 - T_2}{T_1}$$

THERMAL EFFICIENCY

[CARNOT AND OTHER REVERSIBLE CYCLES ONLY]

We shall demonstrate shortly that this efficiency is the highest conceivable thermal efficiency for the particular temperature limits of operation, T_1 and T_2. Only cycles that are both externally and internally reversible will have an efficiency this good. From

(d) $$e = \frac{Q_A - |Q_R|}{Q_A} = 1 - \frac{|Q_R|}{Q_A} = \frac{T_1 - T_2}{T_1} = 1 - \frac{T_2}{T_1}$$

we have

(8-4)
$$\frac{|Q_R|}{Q_A} = \frac{T_2}{T_1} \quad \text{and} \quad \frac{Q_A}{T_1} = \frac{|Q_R|}{T_2} \qquad \text{[CARNOT CYCLE]}$$

ratios that also will be true for other reversible cycles exchanging heat reversibly with constant temperature reservoirs. Equation (8-4) expresses a characteristic of the Carnot engine for which we shall find use.

8.5 THERMODYNAMIC TEMPERATURE

Lord Kelvin* deduced a temperature scale, called the thermodynamic temperature, that is independent of the thermometric substance. We have previously specified that T was this temperature, but we had to speak of measuring T by changes in other properties (§ 1.20). Keeping in mind that the mathematical definition of entropy $ds = dQ/T]_{rev}$ necessarily involves the thermodynamic temperature, we may imagine a scale of temperatures designated by θ and say it is independent of the thermometric substance; that is, $ds = dQ/\theta]_{rev}$. A constant temperature process is a $\theta = C$ process, the Carnot cycle is a rectangle on the θs plane, and the efficiency of the Carnot cycle is, after § 8.4,

(a)
$$e \equiv \frac{Q_A - |Q_R|}{Q_A} = 1 - \frac{|Q_R|}{Q_A} = \frac{\theta_1 - \theta_2}{\theta_1} = 1 - \frac{\theta_2}{\theta_1} \quad \text{and} \quad \frac{Q_A}{\theta_1} = \frac{|Q_R|}{\theta_2}$$

[ANY CYCLE]   [REVERSIBLE CYCLE]

in accordance with equation (8-4). Thus the temperature scale T used in § 8-4 is the same as the scale θ.

To appreciate the significance of this kind of temperature, consider Fig. 8/4 (and caption). Observe from the Carnot efficiency, equation (8-3), that $W = eQ_A$. Then for engines A and B, Fig. 8/4, we have

(b)
$$W_A = e_A Q_1 = \left(\frac{\theta_1 - \theta_2}{\theta_1}\right) Q_1 = \frac{Q_1}{\theta_1}(\theta_1 - \theta_2)$$

* William Thomson (Lord Kelvin) (1824–1907), who was a professor of physics at Glasgow University, is credited by some as being the greatest English physicist. Certainly he possessed a rare combination of talents. His early education was received from his father, who also was a professor at Glasgow University. As a youth, he was robust, an active participant in athletics and student affairs at Cambridge, yet he was most distinguished in his studies, and before his graduation at twenty-one from Cambridge, he had established an enviable reputation in scientific circles by his original contributions. An excellent mathematician, a genius at inventing and designing laboratory apparatus and models, he claimed that he could not understand his own ideas until he saw them at work in models. He contributed most to the science of thermodynamics, having established a thermometric scale of absolute temperatures which is independent of the properties of any gas, having aided in establishing the first law of thermodynamics on a firm foundation, and having stated signficantly the second law. He was the inventor of some 56 instruments and machines, and in addition to all this, he was interested in the arts and was himself a musician. He was knighted for his indispensable services in laying the first successful transatlantic cable, and later was made a peer, Baron Kelvin of Larg. He vigorously denounced the "absurd, ridiculous, time-wasting, brain-destroying British system of weights and measures," favoring the metric system. He received honorary degrees from nearly every important university in Europe and was elected a member of every foreign academy of science and art.

(c)
$$W_B = e_B Q_2 = \left(\frac{\theta_2 - \theta_3}{\theta_2}\right) Q_2 = \frac{Q_2}{\theta_2}(\theta_2 - \theta_3)$$

Let $W_A = W_B$ in these equations and find

(d)
$$\theta_1 - \theta_2 = \theta_2 - \theta_3$$

because $Q_1/\theta_1 = Q_2/\theta_2$ by equation **(a)** above. By specifying that each of the Carnot engines of Fig. 8/4 do the same work, we find all the temperature intervals $\theta_3 - \theta_4 = \theta_4 - \theta_5$, and so on, equal to each other. Thus, we may let some unit of work in a certain Carnot engine define a particular temperature interval, and the definition can be such as to match degrees on the Kelvin or Rankine or any other absolute scale. No matter what size interval is chosen, $Q_1/\theta_1 = Q_2/\theta_2 = Q_i/\theta_i$.

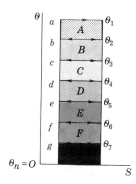

Kelvin Temperature. Carnot engine *A* receives heat Q_1 from a **Fig. 8/4** source at temperature θ_1; *A* discharges heat Q_2 reversibly at temperature θ_2 to Carnot engine *B*; then *B* similarly interacts (Q_3, θ_3) with Carnot engine *C*; and so forth: $\Delta\theta = 0$.

On this energy scale, we visualize an absolute zero temperature that, if violated, also results in a violation of the first law. For example, let $\theta_2 = \theta_n$ in equation **(a)** be a negative absolute temperature, and find the thermal efficiency $e > 100\%$. This would mean that the engine delivers more energy than it receives; it would create energy. While a rigorous study of the logic on this point is more complex than indicated,[1.32] the above suggests that we do not expect to find a sink at a temperature below zero degrees absolute, and that we do have an absolute scale of temperature.

CARNOT CYCLE WITH IDEAL GAS   8.6

Now let T represent the ideal-gas temperature, as in $pv = RT$, and analyze a Carnot cycle for this substance. Although the Carnot cycle is always a rectangle on the Ts plane, it cannot be pictured on the pv plane without some equation of state (or other knowledge of the substance's actual properties). For the ideal gas at $T = C, pv = C$; and when $s = C, pv^k = C$ for constant k (§§ 7.11 and 7.13), from which we obtain the cycle on the pv plane of Fig. 8/5. From § 7.11, the heat of an isothermal process is $Q = RT \ln(v_2/v_1)$. The cycle efficiency for an ideal gas working substance with states as defined in Fig. 8/5 is

(a)
$$e \equiv \frac{Q_A - |Q_R|}{Q_A} = \frac{RT_1 \ln (v_b/v_a) - RT_2 \ln (v_c/v_d)}{RT_1 \ln (v_b/v_a)}$$

[ANY CYCLE]   [CARNOT CYCLE]

By equation (7-2) applied to the isentropic processes bc and da,

(b) $\quad \dfrac{v_c}{v_b} = \left(\dfrac{T_b}{T_c}\right)^{1/(k-1)} = \left(\dfrac{T_1}{T_2}\right)^{1/(k-1)} \quad$ and $\quad \dfrac{v_d}{v_a} = \left(\dfrac{T_a}{T_d}\right)^{1/(k-1)} = \left(\dfrac{T_1}{T_2}\right)^{1/(k-1)}$

from which $v_c/v_d = v_b/v_a$. Substitute this value of v_c/v_d into **(a)**, cancel identical terms, and find

(c) $\qquad\qquad\qquad\qquad\qquad e = \dfrac{T_1 - T_2}{T_1}$

This result, being in the same form as equation **(a)** of § 8.5, says that the ideal gas temperature is the same as the Kelvin temperature. Now that we have a definition of an energy scale of temperature, we are not relieved of the necessity of measuring temperatures via other properties, and equation **(c)** suggests that a nearly ideal gas would be useful for this purpose, as it was historically.

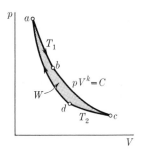

Fig. 8/5   *Carnot Cycle, Ideal Gas.*

8.7   CYCLE WORK BY $\oint p\,dV$ OR $-\oint V\,dp$

The net work of a series of processes is the algebraic sum of the works of the individual processes. If the processes are internally reversible and the series constitutes a cycle, the net work is represented by the enclosed area on the pV plane, say, $abcd$, Fig. 8/5, and may be evaluated by either

(8-5) $\qquad\qquad W_{\text{net}} = \oint p\,dV \qquad$ or $\qquad W = -\oint V\,dp$

For $-\oint V\,dp$, the kinetic energy returns to its original value upon the completion of each cycle—see equation (4-17). Using the integrals of $\int p\,dV$ as found in Chapter 7 and applying them to the Carnot cycle, we find

(a) $\qquad \oint p\,dV = p_a V_a \ln \dfrac{V_b}{V_a} + \dfrac{p_c V_c - p_b V_b}{1-k} + p_c V_c \ln \dfrac{V_d}{V_c} + \dfrac{p_a V_a - p_d V_d}{1-k}$

the net work. We previously found that for any substance the Carnot work is $(T_1 - T_2)(s_b - s_a)$. An exercise for the reader is to show that equation **(a)** reduces to this form.

Since areas on the pV plane represent energy, any work quantity can be pictured as a certain size of rectangle on this plane. If a cycle of internally reversible processes is given on the pV plane, the enclosed area, as *abcd*, Fig. 8/6, can be reproduced as a rectangle whose length is $V_{max} - V_{min}$ for the cycle and whose height is such as to make the areas the same; area *abcd* = area *efgh*, Fig. 8/6. Let $V_{max} - V_{min} = V_D$, a volume called the **displacement volume**. Then the height of the rectangle is the **mean effective pressure** p_m, known in abbreviation as the mep. The concept of mep originated with, and is still applied to, reciprocating engines for which it is that average constant pressure that, acting through *one stroke*, will do on the piston the net work of a single cycle. Because of its general significance, it is extended to cycles as explained, and is defined by

(8-6A) $$p_m \equiv \frac{W}{V_D} = \frac{\oint p\,dV}{V_D} \left[\frac{\text{work/cycle}}{\text{displacement/stroke}} \quad \text{or} \quad \frac{\text{work/cycle}}{V_{max} - V_{min}}\right]$$

where the units should be such as to result in some commonly recognized pressure unit; for W ft-lb and V_D ft³, then p_m is lb/ft², preferably converted to lb/in.² Looking at equation (8-6A) in the form $W = p_m V_D$, we see that the higher mep results in more work per cycle for a particular size of engine, as represented by V_D. Since the most important parameter is power, which depends upon the number of cycles per minute, we have

(8-6B) $$\dot{W} = p_m \dot{V}_D$$

These equations can be applied as well to rotary engines as to reciprocating engines when the definition of mep is extended to cover cycles. The more the volume of the engine, the more costly to build; but some rotary engines operating in cycles with low mep's offset this low mep with much higher speed (larger $\dot{V}_D$, say, in cfm). See Chapter 14 for more on $\dot{V}_D$. The mean effective pressure of the Carnot engine, Fig. 8/6, is

(a) $$p_m = \frac{(T_1 - T_2)mR \ln(V_b/V_a)}{V_c - V_a}$$ [CARNOT]

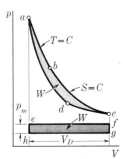

Mean Effective Pressure. **Fig. 8/6**

Example—Carnot Cycle **8.9**

A Carnot cycle operates on air, considered as an ideal gas with constant k. At the beginning of isothermal expansion, $p_1 = 300$ psia, $V_1 = 5$ ft³, and $t_1 = 540°$F. The ratio of isothermal expansion is $r_t = 2$ $(= V_2/V_1 = V_3/V_4)$ and the isentropic compression ratio $r_k = 5$ $(= V_4/V_1 = V_3/V_2)$. Find (a) the sink temperature and the pressure at each corner of the cycle, (b) the change of entropy during an isothermal process, (c) the heat supplied to the cycle, the heat rejected, (d) the horsepower developed if the volume V_1 is 5 cfm, (e) the efficiency, and (f) the mep.

Fig. 8/7 The enclosed area on each plane represents work. The unit of pV areas is ft-lb; of TS areas, Btu. Since the Btu is 778 times larger than the foot-pound, it would be inconvenient to show these areas to relative size. This pV diagram is drawn approximately to scale for the data in § 8.9. Notice that the enclosed area is relatively long and slender.

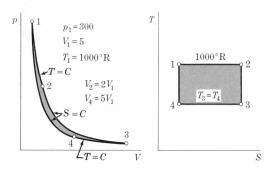

Solution.   (a) See Fig. 8/7. For $T_1 = 540 + 460 = 1000°R$, $k = 1.4$, and $r_k = V_4/V_1 = 5$, we get

$$T_4 = T_1\left(\frac{V_1}{V_4}\right)^{k-1} = \frac{1000}{5^{0.4}} = 525°R \qquad \text{or } t_4 = t_3 = 65°F$$

$$p_4 = p_1\left(\frac{V_1}{V_4}\right)^k = \frac{300}{5^{1.4}} = 31.6\,\text{psia} \qquad p_2 = p_1\left(\frac{V_1}{V_2}\right) = \frac{300}{2} = 150\,\text{psia}$$

$$p_3 = p_2\left(\frac{V_2}{V_3}\right)^k = \frac{150}{5^{1.4}} = 15.8\,\text{psia}$$

(b) The change of entropy, $\Delta S_{1\text{-}2} = -\Delta S_{3\text{-}4}$, is

$$\Delta S = \frac{mR}{J}\ln\frac{V_2}{V_1} = \frac{p_1 V_1}{J T_1}\ln\frac{V_2}{V_1} = \frac{(300)(144)(5)}{(778)(1000)}\ln 2 = 0.1925\,\text{Btu/°R}$$

(c) The heats are

$$Q_{1\text{-}2} = \frac{mRT_1}{J}\ln\frac{V_2}{V_1} = (S_2 - S_1)T_1 = (0.1925)(1000) = 192.5\,\text{Btu}$$

$$Q_{3\text{-}4} = \frac{mRT_3}{J}\ln\frac{V_4}{V_3} = -(S_2 - S_1)T_3 = -(0.1925)(525) = -101\,\text{Btu}$$

(d) The work is $\oint dQ$, or $W = 192.5 - 101 = 91.5$ Btu for $V_1 = 5\,\text{ft}^3$. If we consider, for example, that one cycle is completed each minute with $V_1 = 5\,\text{ft}^3$, the work is 91.5 Btu/min and $hp = 91.5/42.4 = 2.16$ (Item B 38).

(e) The efficiency is W/Q_A or $(T_1 - T_2)/T_1$;

$$e = \frac{91.5}{192.5} = 47.5\% \qquad \text{or} \qquad e = \frac{1000 - 525}{1000} = 47.5\%$$

(f) The mean effective pressure, for $V_D = V_3 - V_1$, $V_2 = 2V_1$, and $V_3 = 5V_2$, is

$$p_m = \frac{W}{V_D} = \frac{(778)(91.5)}{(50-5)(144)} = 11 \text{ psi}$$

Note the conversion of units. Compared to most practical engines this mep is low, especially in relation to the highest pressure of 300 psia for which the engine parts must be designed; but it is characteristic of Carnot engines.

REGENERATION 8.10

Regeneration is a process internal to a thermodynamic system whereby the external irreversibility of the system is either obviated or lessened. External irreversibility occurs in most engines because of two temperature differentials—source temperature above that of the working substance and sink temperature below that of the working substance; both are generally necessary for the operation of an engine.

The addition of a regenerator (and its process) may completely obviate these two differentials by internally cooling the working substance on one hand while it is being heated on the other. This may sound a bit like one becoming richer because he has shifted money from one pocket to another. Thermodynamically this internal interchange of energy does improve the efficiency of the system under study. Its overall operation tends to approach that of the Carnot system. Through regeneration two engines (Stirling and Ericsson) have attained the Carnot efficiency; none have exceeded it. Like the two isentropic processes in the Carnot cycle, see Fig. 8/3, temperature changes have occurred through regeneration because of internal happenings without benefit of an external heat interaction (which we don't want).

STIRLING CYCLE* 8.11

There are several other reversible cycles. Of these we shall describe the Stirling cycle because it incorporates a feature of particular importance in modern power plants, and was probably the first to do so; this feature is *regeneration.*

The ideal cycle is composed of two isothermal and two isometric processes, the regeneration occurring at constant volume, Fig. 8/8. As in the Carnot cycle, heat is received from a source (along *ab*) and rejected to the sink (along *cd*) at constant temperature, with the heat interaction approaching reversibility as the temperature difference during the heat flow approaches zero. After heat has been received *ab*, the working substance (originally air) passes to the regenerator. In the ideal cycle the volume remains constant as the gas passes through the regenerator. This type of regenerator is a chamber in which heat flows from the gas and is stored as molecular energy of the contents, which may be such as a maze of brickwork, some wire mesh, or a quantity of crimped wire. One end of the ideal regenerator is at temperature T_1, the other at T_2, and there is a gradual temperature gradient from one end to the other.

* Robert Stirling (1790–1878) was an esteemed Scottish minister, so well regarded that he was awarded an honorary Doctor of Divinity degree. He spent considerable time designing, improving, and selling his "air engine," with his brother James working with him for some 29 years. The Stirlings did not understand the thermodynamic reasons for regeneration, because thermodynamics itself had not yet evolved very far (Joule was a younger contemporary). Rankine (footnote, § 9.3) explained it in 1854. It was James who had the idea of closing and pressurizing the system, using gas at all times with pressure greater than atmospheric.

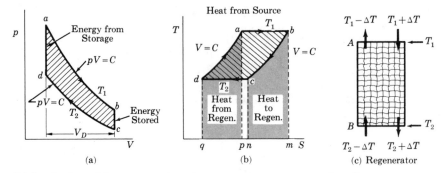

Fig. 8/8   *Stirling Cycle and Regenerator.* Heat is dumped into the sink at T_2 along *cd*.

To store the energy reversibly, we imagine the gas entering the regenerator at end A at a temperature $T_1 + \Delta T$, Fig. 8/8(c), so that heat begins to flow to the regenerator at A. As the gas passes through the regenerator, its temperature drops gradually to $T_2 + \Delta T$ at the exit B; at all points in the regenerator, the gas temperature is higher than that of the regenerator by an amount ΔT.

After this storage operation, the gas rejects heat to an external receiver, the sink, during the isothermal compression *cd*. Then this gas reenters the regenerator at B, Fig. 8/8(c), at a temperature $T_2 - \Delta T$, and being slightly cooler than the contents, it receives heat and continues to receive heat until it leaves at temperature $T_1 - \Delta T$ at A. As in the previous passage in the opposite direction, there is always a temperature difference of ΔT between the gas and the regenerator. In the limit, as $\Delta T \rightarrow 0$, the whole operation becomes reversible. It is essential to observe that the heat to the gas in moving from B to A is the same as the energy stored when the gas passed from A to B.

> *This heat does not involve an external source;* it is an *interchange* of heat *within* the system. Now, if the source and sink are at temperatures T_1 and T_2, respectively, all processes are reversible *externally and internally*.

In the real engine, energy exchanges are at finite rates. Nevertheless, we shall consider m lb, and with ΔP and ΔK changes negligible. For states as defined in Fig. 8/8, we have

(a)
$$Q_A = mRT_1 \ln \frac{V_b}{V_a} \quad \text{along } ab$$

(b)
$$Q_R = mRT_2 \ln \frac{V_d}{V_c} = -mRT_2 \ln \frac{V_c}{V_d} \quad \text{along } cd$$

Since $V_c/V_d = V_b/V_a$, we find the work and thermal efficiency to be

(c)
$$W_{\text{net}} = \oint dQ = (T_1 - T_2)mR \ln \frac{V_b}{V_a}$$

(d)
$$e = \frac{W_{\text{net}}}{Q_A} = \frac{T_1 - T_2}{T_1} \qquad \text{[REVERSIBLE CYCLE]}$$

which is the same as that of the Carnot cycle, the highest possible efficiency for any system operating between temperatures T_1 and T_2. The work may also be found from $\oint p\,dV$,

(e)
$$W_{\text{net}} = \oint p\,dV = p_a V_a \ln \frac{V_b}{V_a} + p_c V_c \ln \frac{V_d}{V_c}$$

The mean effective pressure is

(f)
$$p_m = \frac{W_{\text{net}}}{V_D} = \frac{(T_1 - T_2)mR \ln (V_b/V_a)}{V_b - V_d}\,\text{psf}$$

The interested student might set up the equation for thermal efficiency of a cycle *abcd* as in Fig. 8/8 except that no regenerator is used, showing that the resulting efficiency is *not* as high as the Carnot efficiency.

The actual prototype of the Stirling closed cycle engine runs quietly. For this reason, such engines were manufactured and sold until sometime after World War I. Until electricity became well distributed, many of these "air engines" were used as pumping engines, even in private residences, because of their quietness. Also because of this feature, their use in submarine propulsion systems and elsewhere is being studied.[18.2]

Other uses under investigation include: power for satellite space stations, portable generating sets, marine engines, and no one knows but that someday they might be

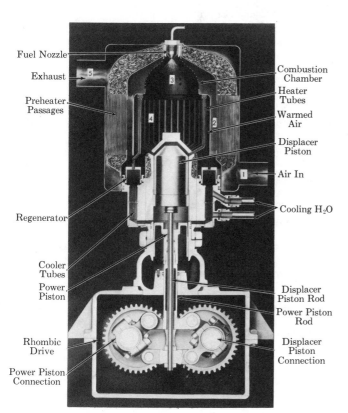

Fuel Nozzle
Exhaust
Preheater Passages
Regenerator
Cooler Tubes
Power Piston
Rhombic Drive
Power Piston Connection

Combustion Chamber
Heater Tubes
Warmed Air
Displacer Piston
Air In
Cooling H_2O
Displacer Piston Rod
Power Piston Rod
Displacer Piston Connection

Artist's Schematic Diagram of Stirling Engine. This is a closed system. The power piston compresses and expands the gas. The displacer piston moves the gas from cold to hot regions through the regenerators on its downward stroke, and oppositely on its up movement. More heat is transferred from the regenerators than during the "constant" temperature process. No valves and no explosions mean low noise level. Unique with this engine, if an input work is supplied to the shaft, it begins to act on a thermodynamic reversed cycle, permitting the "pumping" of heat from a low temperature to a higher one. (*Courtesy General Motors Laboratories, Warren, Mich. and SAE.*)   **Fig. 8/9**

used for automotive power, thereby reducing or eliminating this source of smog. The sources of heat include: combustion of fuels in the usual manner, combustion of hydrocarbons with hydrogen peroxide (used for torpedo propulsion), nuclear reactors and isotopes, sun (perhaps the sun's rays beamed through a quartz window), and any system in which energy with availability is stored. See Figs. 8/9, 8/10, and 8/11. There are versions of Stirling engines other than that of Fig. 8/9, one of which is used as a heat pump, with temperatures as low as 30 K.

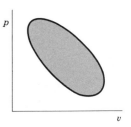

Fig. 8/10   *Actual pV Diagram for Stirling Engine.* It is informative to be acquainted with the departures of actual events as compared to the ideal, the lack of resemblance in this case being perhaps more than usual. Yet the actual engine operates with a good thermal efficiency, the best being better than 30% (after General Motors Corp.[6.2]).

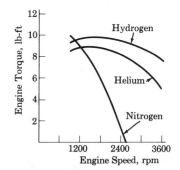

Fig. 8/11   *Output of Stirling Engine, Different Working Substances.* As speeds increase, the heavier molecules (N$_2$ vs. H$_2$) with greater viscosity result in greater friction and a rapid decrease of output. At the lowest speeds the larger molecules resulted in a greater output because of less leakage about the piston, which became the dominant factor. Hydrogen and helium have much larger specific heats than air or N$_2$, and are not so deleterious to the lubricant. All the foregoing suggests that there is more to an engineering thermodynamics problem than thermodynamics (after General Motors Corp.[8.1] with SAE permission).

8.12   ERICSSON CYCLE*

The ideal Ericsson cycle consists of two isothermal and two isobaric processes, with the regeneration occurring during constant pressure. On the *TS* plane, a sketch of it looks like that in Fig. 8/8(b), and the analysis is quite similar.

8.13   REVERSED AND REVERSIBLE CYCLES

At the outset, we should distinguish between a *reversed* and a *reversible* cycle. A *reversed cycle* is a general term including all for which the net work is an energy

---

* John Ericsson (1803–1889), born a Swede, methodical, stubborn, irascible, designer and inventor extraordinary, had a passion for machinery. In England, he built a locomotive to compete with one designed by Stephenson, called the Rocket. Although the Rocket won the prize on tractive effort and dependability, Ericsson's locomotive established the amazing speed record of 30 mph. The *London Times* said, "It seemed to indeed fly, presenting one of the most sublime spectacles of human ingenuity and human daring the world ever beheld." Because he was unable to interest anyone in England in his screw propeller—a substitute, at the time, for paddle wheels—he came to the United States to promote it. He was the designer of the *Monitor*, which battled the Confederate ironclad *Merrimac*. He designed and built a hot-air engine for a boat, launched in 1853, sunk shortly afterward in a squall. The weight of the engine for the power developed was tremendous, and the boat was not capable of carrying it in rough weather. He was a prolific producer that space does not permit even a list of his inventions and designs. During the last few years of his life, he was a recluse in New York City, irate at the elevated, a disbeliever in the telephone.

input; the net heat is rejected; $|Q_R|$ is numerically greater than Q_A. *A reversible cycle is one composed of processes that conform to the definition of § 5.9; they are reversible internally and externally*; there is no friction, and heat is transferred with an infinitesimal temperature drop.

Customarily, in a *power cycle*, the state point is pictured as moving in a clockwise sense on the pV and TS planes. In a reversed cycle, the state point conventionally moves in a counterclockwise sense. We can reverse cycles that receive and reject heat while the temperature of the working substance varies, but since the external irreversibility is inevitable in a natural heat transfer with varying temperature, such a cycle, though *reversed, is not a reversible cycle*. See Fig. 8/12.

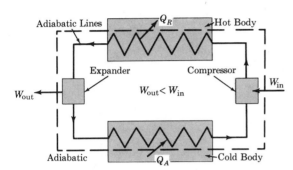

Reversed Cycle—Heat Pump. This **Fig. 8/12** figure represents a reversed cycle; heat is rejected at the higher temperature and added at the lower temperature. The net work is $W = W_{out} - W_{in} = \oint dQ = Q_A - |Q_R|$, a negative number that indicates work is done *on* the system. The system is the circulating working substance.

Reversed cycles are used for two purposes: (1) to provide a cooling effect (a refrigerating machine) and (2) to provide a heating effect (Chapter 17). If the system receives work from the outside, heat can be made to flow into the system from the cold reservoir and flow from the system into the hot reservoir, Fig. 8/12. Thus, a general name for reversed cycles is **heat pump**, but in common usage, *heat pump* is applied to the cycle used to obtain a heating effect.

REVERSED CARNOT CYCLE 8.14

Inasmuch as each process of the Carnot cycle is reversible internally and externally, the cycle itself can be operated as a heat pump. An examination of Fig. 8/13 shows that for a given temperature range and a particular isothermal curve *ad*, the work must necessarily be the same as in the power cycle; the heat rejected by the reversed cycle at the higher temperature must be equal to the heat added in the power cycle, and so on. The parameter used to indicate the efficiency of a reversed

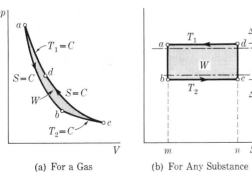

(a) For a Gas   (b) For Any Substance

Reversed Carnot Cycle. An isentropic **Fig. 8/13** expansion *ab* lowers the temperature to where heat may be added to the system reversibly from a cold body at $T_2 + \Delta T$ along an isothermal *bc*. Isentropic compression *cd* brings temperature slightly higher than the hot body (at, say $T_1 - \Delta T$), so that heat may be rejected along *da*. The refrigeration is represented by the area *mbcn*. If the cycle is used for heating instead of refrigeration, the energy represented by the area *madn*, being the heat rejected, is the warming effect.

cycle is called the *coefficient of performance*, abbreviated COP and represented by γ. According to the objective of the reversed cycle, the COP (output/input), always written as a positive number, is

$$(8\text{-}7) \qquad \gamma_c \equiv \frac{\text{refrigeration}}{\text{net work}} = \frac{Q_A}{|W|} \qquad \text{or} \qquad \gamma_h \equiv \frac{\text{heat out}}{\text{net work}} = \frac{|Q_R|}{|W|}$$

$$\text{[USED FOR COOLING]} \qquad\qquad\qquad \text{[USED FOR HEATING]}$$

Suppose that a Carnot engine receives heat from a source at T_H while the engine temperature is T_1, where $T_H > T_1$, Fig. 8/14(a). Similarly, let the system at T_2 discharge heat to a sink at T_0, $T_2 > T_0$. The efficiency of the engine is $(T_1 - T_2)/T_1$, as usual, but as described, the cycle is externally irreversible with finite ΔT's. As a result, it falls short of the maximum possible efficiency of $(T_H - T_0)/T_H$ with the given heat reservoirs. The reversed cycle can be made to operate between any two temperatures to serve any desired purpose, but the normal temperature relations are shown in Figs. 8/14(b) and (c). The refrigeration cycle rejects heat to a naturally available sink T_0 and receives heat from the system to be cooled to T_c. The heating cycle receives heat from a naturally available sink and rejects heat to the room to be maintained at temperature T_r.

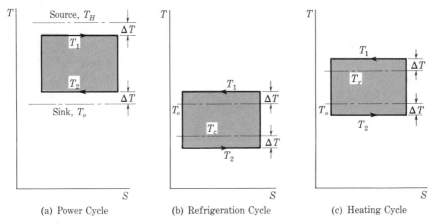

(a) Power Cycle   (b) Refrigeration Cycle   (c) Heating Cycle

Fig. 8/14   *Comparison of Carnot Cycles for Different Purposes.* Each cycle approaches external reversibility as $\Delta T \to 0$.

Example—Reversed Carnot Cycle

A reversed Carnot cycle operating as a refrigerator, as shown in Fig. 8/14(b), has a refrigeration capacity of 600 kJ/min; its temperatures are $T_2 = 260$ K and $T_1 = 340$ K. (a) How much heat is rejected? (b) Find its COP, γ_c. (c) Find the work. (d) If this system is used for heating, find its COP, γ_h. (e) What would its efficiency be if it ran as an engine?
Solution.

(a)
$$Q_R = Q_A \frac{T_1}{T_2} = (600)\left(\frac{340}{260}\right) = 785 \text{ kJ/min}$$

(b)
$$\text{COP:} \quad \gamma_c = \frac{Q_A}{Q_A - Q_R} \doteq \frac{600}{600 - 785} = 3.24$$

(c) $\dot{W} = Q_A - Q_R = 600 - 785 = -185\,\text{kJ/min} = -3.08\,\text{kW}$

(d) $\text{COP:} \quad \gamma_h = \dfrac{Q_R}{Q_A - Q_R} = \dfrac{785}{600 - 785} = 4.24$

(e) As an engine $e = \dfrac{T_1 - T_2}{T_1} = \dfrac{340 - 260}{340} = 23.5\%$

Example—Cycle Analysis **8.15**

   Suppose a cycle is composed of the following processes: from $p_1 = 20\,\text{psia}$ and $t_1 = 200°\text{F}$, 1 lb/sec of air is compressed polytropically with $pV^{1.2} = C$ until $p_2 = 100\,\text{psia}$; it then undergoes an isobaric process, and the cycle is closed by an isothermal process. Determine (a) the heat added, the heat rejected, the work and horsepower, (b) the thermal efficiency, and (c) the mep. Write the equation for the work from the pV plane in terms of the state names in Fig. 8/15.

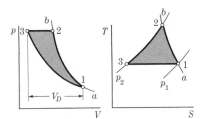

Three-Process Cycle.   **Fig. 8/15**

   Solution.   (Use table Item B 1.) The cycle defined is like none that has ever been used; such cycles are designed to include the most practice with thermodynamics per unit of working time. First, sketch the cycle on the pV and TS planes. The defined polytropic compression slopes upward toward the left on both planes because $1 < n < k$, for $n = 1.2$ for air; see Fig. 7/24. Sketch curves ab on both planes of indefinite extent, Fig. 8/15. Name two states for a compression, say 1 and 2. Since the process 2-3 is isobaric, draw a constant pressure line through 2 of indefinite extent (both sides of 2 because at this point it may not be clear where state 3 is relative to 2). If the location of 3 is certain, all that is necessary is to draw an isothermal curve through 1 and 3. Perhaps the safest step is to return to state 1 and draw an isothermal through it, letting it intersect the $p = C$ line where it will. Depending on what processes are involved, it may be easier to complete the sketch on one plane before trying the other plane. For example, in this case, an isothermal through 1 on the TS plane is obvious; where the horizontal line through 1 on the TS plane intersects the isobaric previously drawn of indefinite extent is state 3. Now observe that on the TS plane the cycle 1-2-3 proceeds counterclockwise. (If there is no error to this point, we know that the net work will be negative.) It will always be true that *the cycle goes CC on both planes or CL on both*. This observation may help in sketching a process about which one is uncertain. It does not matter that the final sketch is far from a scale representation, but it is important that the general direction of the process be right, as per Fig. 7/24.

   Assuming that the states are in correct relative positions, we see from the TS plane that heat is added along 3-1 and rejected along 1-2-3. Likewise, work is delivered (positive) during 3-1 and received during 1-2-3. This cycle is *reversed* but not *reversible*. We could by the use of reversible Carnot engines make this cycle reversible, but, as described, it does not accord with our definition of a reversible cycle.

   (a) First, compute some needed properties (cfs = ft³/sec).

$$T_2 = T_1\left(\frac{p_2}{p_1}\right)^{(n-1)/n} = (660)\left(\frac{100}{20}\right)^{0.2/1.2} = 863°R$$

(a)

$$\dot{V}_1 = \frac{\dot{m}RT_1}{p_1} = \frac{(53.3)(660)}{(20)(144)} = 12.2 \text{ cfs} \qquad \dot{V}_3 = \dot{V}_1\left(\frac{p_1}{p_3}\right) = 12.2\left(\frac{20}{100}\right) = 2.44 \text{ cfs}$$

(b) $$\dot{Q}_{3-1} = \dot{Q}_A = \frac{p_1\dot{V}_1}{J}\ln\frac{V_1}{V_3} = \frac{(20)(144)(12.2)}{778}\ln\frac{12.2}{2.44} = 72.6 \text{ Btu/sec} \qquad \text{or } 76.6 \text{ kW}_t$$

$$\dot{Q}_{1-2} = \dot{m}c_v\left(\frac{k-n}{1-n}\right)(T_2 - T_1)$$

(c)

$$= (0.1714)\left(\frac{1.4 - 1.2}{1 - 1.2}\right)(863 - 660) = -34.8 \text{ Btu/sec}$$

(d) $$\dot{Q}_{2-3} = \dot{m}c_p(T_3 - T_2) = 0.24(660 - 863) = -48.7 \text{ Btu/sec}$$

(e) $$\dot{Q}_R = -34.8 - 48.7 = -83.5 \text{ Btu/sec} \qquad \text{or} \qquad -\frac{83.5}{0.948} = 88.1 \text{ kW}_t$$

where kW_t means *kilowatt thermal* to distinguish it from kW of power output.

$$\dot{W} = \oint d\dot{Q} = 72.6 - 83.5 = -10.9 \text{ Btu/sec}$$

(f) $$\dot{W} = 76.6 - 88.1 = -11.5 \text{ kW} \qquad \text{or} \qquad 11.5\text{-kW input}$$

Notice that the algebraic signs confirm our observations made at the outset.

(b) For the thermal efficiency, imagine the cycle now proceeding clockwise as a power cycle; then the heat added is $Q_{3\text{-}2\text{-}1} = 83.5$, and

(g) $$e = \frac{10.9}{83.5} = 13.05\%$$

This reversed cycle is no good for refrigeration or heating.

(c) The mep is (observe how low it is)

(h) $$p_m = \frac{W}{V_D} = \frac{W}{V_1 - V_3} = \frac{10.9 \times 778}{(12.2 - 2.44)(144)} = 6.03 \text{ psi}$$

(d) The work equation by $\oint p\, dV$, from Fig. 8/15, is

(i) $$W_{net} = \oint p\, dV = \frac{p_2 V_2 - p_1 V_1}{1 - n} + p_2(V_3 - V_2) + p_3 V_3 \ln\frac{V_1}{V_3}$$

an algebraic sum. As for the heat computations, the signs are automatically negative where they should be if correct numbers have been inserted. In important calculations, all available checks are made; for example, compute the work from equation **(i)** and see that it is the same as already found in **(f)**.

REVERSIBLE ENGINE MOST EFFICIENT   8.16

For particular temperature limits, no heat engine can be more efficient than a reversible engine. The truth of this statement, known as ***Carnot's principle***, is demonstrated by a process of reasoning. The reasoning will be clearer if numbers instead of symbols are used. Imagine a reversible engine R, Fig. 8/16(a), taking 100 Btu from the hot reservoir ($T_1 = C$) in any chosen time, converting 40 Btu into work, and rejecting 60 Btu to the cold reservoir ($T_2 = C$). The thermal efficiency of the engine is therefore $40/100 = 40\%$. Now we have just learned from the discussion of the Carnot cycle (§ 8.16) that if this engine is reversed, 40 Btu will be necessary to drive it, 60 Btu will be taken from the cold reservoir, and 100 Btu will be discharged to the hot reservoir. In Fig. 8/16(b), an irreversible engine I is driving the reversible engine. For the moment, assume that the irreversible engine I is more efficient than the reversible engine R, say $e_I = 50\%$. Then, since it takes 40 Btu to drive R, engine I will need to take

(a)
$$Q_A = \frac{W}{e} = \frac{40}{0.5} = 80 \text{ Btu}$$

from the source and discharge 40 Btu to the cold reservoir. Observe that Fig. 8/16(b) is an isolated system, wherein the reversible engine R discharges $100 - 80 = 20$ Btu more to the source than the irreversible engine I takes from the source. Moreover, the reversible engine R takes $60 - 40 = 20$ Btu more from the cold reservoir than the irreversible engine I discharges to the cold reservoir. In other words, for the *assumed condition* that I is more efficient than R, we find in Fig. 8/16(b) that heat is being moved continuously from a cold reservoir to a hot reservoir without external aid. All our experience indicates that heat will not flow in any such manner in an isolated system. We can cause heat to move from a cold reservoir to a hot reservoir by supplying work from a source external to the system (a Diesel engine or electric motor driving the compressor of a refrigerating system) but *net heat will not flow of its own accord from a cold to a hot body* (Clausius, § 5.2), a statement of the second law of thermodynamics.

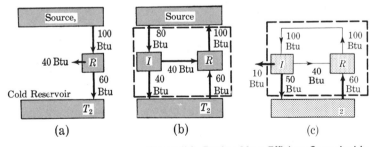

Reversible Engine Most Efficient Conceivable.   **Fig. 8/16**

Instead of simply moving heat, as in Fig. 8/16(b), we could direct the flow of energy from the reversible engine directly into the irreversible engine, as in Fig. 8/16(c), whose 50% efficiency would allow it to drive engine R and at the same time deliver 10 Btu of work to something outside the system. Thus, this system *exchanges heat with a single reservoir* (sink) and delivers work (Kelvin-Planck prohibition, § 5.2). These events, Figs. 8/16(b) and (c), are a violation of the second law, discussed further in the next chapter, and have never been known to happen; they are contrary to all man's experience with energy. Therefore, we say that the assumption that engine I is more efficient than engine R is absurd and impossible.

Engine I could be taken as another reversible engine as easily as not, and the same reasoning would lead to the conclusion that *one reversible heat engine cannot be more efficient than another when both engines operate between the same temperature limits*; *all reversible engines operating between the same temperature limits have the same thermal efficiency*; namely $(T_1 - T_2)/T_1$, which is the highest conceivable thermal efficiency for a heat engine.

8.17   CLOSURE

The Carnot cycle is not seriously considered today as a means of manufacturing power, but it is a monumental concept that, as we have seen, has led to important deductions regarding the limitations of any heat power machine. The Ericsson and Stirling cycles involve an idea, regenerative heating, used in modern steam power and gas turbine plants, as well as in various other heat saving ways. Also, in this chapter, we have learned of the essential elements of any cycle and how to analyze cycles. Notwithstanding the visionary or impractical aspects of these cycles, a study of them yields results of great practical value.

Note that a particular heat interchanged between systems is negative for one and positive for the other, which is to say that one concentrates (via energy diagrams) on the system being studied.

PROBLEMS

SI UNITS

8.1   The execution of a cyclic process by a closed system involves the following four heat interactions: $Q_1 = 50$, $Q_2 = -75$, $Q_3 = 125$, $Q_4 = -40$ kcal. Concurrently, four work interactions occur during the cyclic process: $W_1 = 200$, $W_2 = -124$, $W_3 = 100$, $W_4 = ?$ kJ. Find **(a)** the fourth work W_4 and **(b)** the efficiency e. **(c)** What are ΔU and ΔS for this cyclic process?

Ans. **(a)** 75 kJ, **(b)**, 34.26%, **(c)** 0.

8.2   The thermal efficiency of a particular engine operating on an ideal cycle is 30%. Determine **(a)** the heat supplied per 1200 W-hr of work developed, **(b)** the ratio of heat supplied to heat rejected, and **(c)** the ratio of work developed to heat rejected.

Ans. **(a)** 14,399 kJ, **(b)** 1.43, **(c)** 0.428.

8.3   An engine operating on the Carnot principle receives heat from a large reservoir of water-vapor mixture in equilibrium at 1 atm and discharges 5275 kJ/hr to a large sink reservoir of water-ice mixture in equilibrium at

1 atm. If the power developed is 536 W, determine the number of Kelvin degrees separating the ice point and the absolute zero temperature. *Ans.* 273 K.

8.4   Show that the thermal efficiency of the Carnot cycle in terms of the isentropic compression ratio r_k is given by $e = 1 - 1/r_k^{k-1}$

8.5   Represent the Carnot cycle as a square on the TS plane. Now inscribe within this square a circle of maximum diameter D which represents a cycle. Noting that both cycles operate between the same temperature limits T_1 (source) and T_2 (sink), demonstrate that the Carnot cycle is the more efficient.

8.6   Gaseous nitrogen actuates a Carnot power cycle in which the respective volumes at the four corners of the cycle, starting at the beginning of the isothermal expansion, are $V_1 = 10.10$, $V_2 = 14.53$, $V_3 = 226.54$, and $V_4 = 157.73$ ℓ. What is the thermal efficiency?

Ans. 66.7%.

8.7   Demonstrate that the efficiencies of the three cycles—Carnot, Stirling and Ericsson—are identical under the same temperature restrictions and are equal to $e = 1 - T_{min}/T_{max}$.

8.8 Astronauts are proposing a portable Stirling cycle engine to furnish power in an environment where the heat is rejected only by radiation and subsequently is proportional to the area and the fourth power of the temperature of the radiating surface as $Q_r = KA_r T_r^4$ where K is a constant. The source of heat for the engine will be a small portable reactor. In view of the necessity for minimum mass requirements, prove that for a given power output and a predetermined maximum temperature of the engine, the radiating area A_r will be minimum when the ratio of engine temperatures is $T_{min}/T_{max} = 3/4$.

8.9 The Ericsson cycle is composed of two isothermal processes and two isobaric processes, with regenerative heat exchange during the isobaric processes. Properties at the beginning of isothermal expansion are 689.48 kPaa, 142 ℓ, and 282.2°C. For a ratio of isothermal expansion of 2 and a minimum temperature of 4.4°C, find **(a)** ΔS during the isothermal process; **(b)** Q_A, Q_R, W, and e; **(c)** the volume at the end of isothermal expansion and the overall ratio of expansion; and **(d)** p_m.
Ans. **(a)** 0.122 kJ/K; **(b)** $Q_A = 67.7$ kJ, $e = 50\%$; **(c)** $r_e = 4$; **(d)** 159 kPa.

8.10 Given these processes: isobaric, isometric, isothermal, and isentropic. Invent four different three-process cycles. Show the corresponding pV and TS diagrams for each and write expressions for Q_A, Q_R, and W from the pV-plane.

8.11 (*STp*) A thermodynamic cycle is composed of the following reversible processes: isentropic expansion 1–2; isothermal 2–3; constant pressure 3–1. Sketch the pV and TS diagrams for the cycle and write expressions for **(a)** Q_A and Q_R, **(b)** net work from the pV and TS planes, and **(c)** e and p_m.

8.12 (*STp*) The cycle described in **8.11** operates on 113 g of nitrogen; the expansion ratio 1-2 is $r_e = 5$; $t_1 = 149$°C, $p_1 = 689.5$ kPaa. Determine **(a)** p, V, and T at each point, **(b)** Q_A and Q_R, **(c)** W from the pV plane (check by net Q), **(d)** e and p_m. **(e)** If 100 Hertz are completed, find the power in kW.
Ans. **(b)** 23.53, -16.8 kJ, **(c)** 6.73 kJ, **(d)** 73.3 kPa, **(e)** 673 kW.

8.13 (*S, V, n*) A three-process cycle of an ideal gas (for which $c_p = 1.064$, $c_v = 0.804$ kJ/kg-K) is initiated by an isentropic

compression 1-2 from 103.4 kPaa, 27°C to 608.1 kPaa. A constant volume process 2-3 and a polytropic 3-1 with $n = 1.2$ complete the cycle. Circulation is at a steady rate of 0.907 kg/s. Compute $\oint p\,dV$, $\oint dQ$, the thermal efficiency, and mep.
Ans. 9.5 kJ/s, 19.1%, 18.8 kPa.

8.14 A substance executes a reversed Carnot cycle during which it receives 105.5 kJ/min of heat. Determine the work required, if the adiabatic compression process triples the initial absolute temperature.
Ans. 3.52 kW.

8.15 A refrigeration cycle operates on the Carnot cycle between 244.4 K and 305.6 K with an input of 7.46 kW. Sketch the cycle on the TS plane, and determine **(a)** COP, **(b)** the tons of refrigeration (1 ton = 211 kJ/min.
Ans. **(a)** 4, **(b)** 8.48.

8.16 The COP (γ) of a reversed Carnot cycle is 5.35 when refrigeration is done at 255 K; $\Delta S = 0.38$ kJ/K during the isothermal heat interactions. Find **(a)** the refrigeration, **(b)** the temperature at which heat is rejected, **(c)** the net work. **(d)** If this cycle is used for heating (same temperature limits), what is the COP? **(e)** Prove that $\gamma_H = \gamma_C + 1$ where $\gamma =$ COP.

8.17 A Carnot system is to be used first as a power engine and then reversed and used as a refrigerator. Show that the ratio of its coefficient of performance (COP) as a refrigerator to its efficiency, e, as an engine, in terms of the respective temperatures T_{max} and T_{min}, is COP$/e = T_{max}T_{min}/(T_{max} - T_{min})^2$.

8.18 A reversible engine operating between a high temperature source and a low temperature sink receives 53 kJ and produces 350 W-min of work. Determine **(a)** the thermal efficiency, **(b)** the ratio of Rankine temperatures of source and sink, **(c)** the ratio of Kelvin temperatures of source and sink, and **(d)** the sink temperature if the source temperature is 260°C.

8.19 Using a heat source, a cold body, and the two reversible engines as shown in the figure, let $Q_A = 60$ IT cal, $W_2 = 40$ W-sec, and $T_2 = 333.3$ K. If the engines have

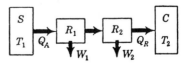

Problems 8.19; 8.34

equal thermal efficiencies, calculate **(a)** the temperature at which R_2 receives heat, **(b)** the thermal efficiencies, **(c)** T_1, **(d)** W_1, and **(e)** Q_R.

Ans. **(a)** 416.1 K, **(b)** 19.8%, **(d)** 49.8 W-sec.

8.20 Temperatures on the moon are known to exist at 394 K on the dayside to 117 K on its nightside. Devise a heat engine to be used on the moon which employs these thermal conditions and discuss its feasibility to include anticipated efficiency.

8.21 A computer programmed study is to be conducted on a Stirling engine operating on nitrogen. At the beginning of compression (point 1), $p_1 = 103$ kPaa, $T_1 = 310$ K, $m = 1$ kg. Use compression ratios of 6, 8, 10, and 12 and find the work for each of the following two cases: first, account for the variation of the specific heats using data from Table I; second, use the constant specific heats as given in Item B 1. Write the program and compare results.

ENGLISH UNITS

8.22 A Carnot engine operating between 900°F and 90°F produces 40,000 ft-lb of work. Determine **(a)** the heat supplied, **(b)** the change of entropy during heat rejection, and **(c)** the thermal efficiency of the engine.

Ans. **(a)** 86.3 Btu, **(b)** −0.0635 Btu/°R, **(c)** 59.55%.

8.23 A Carnot power cycle operates on 2 lb of air between the limits of 70°F and 500°F. The pressure at the beginning of isothermal expansion is 400 psia and at the end of isothermal expansion is 185 psig. Determine **(a)** the volume at the end of the isothermal compression, **(b)** ΔS during an isothermal process, **(c)** Q_A, **(d)** Q_R, **(e)** the work, **(f)** e, **(g)** the mep.

Ans. **(a)** 7.82 ft³, **(b)** 0.0948 Btu/°R, **(c)** 91 Btu, **(f)** 44.7%, **(g)** 15.88 psi.

8.24 **(a)** A cycle with regeneration is the Stirling cycle; it is composed of two isothermal processes and two regenerative constant volume processes. See § 8.11. Sketch this cycle on the pV and TS planes and write expressions for Q_A, Q_R, W, e, and p_m. **(b)** Air is made to pass through a Stirling cycle. At the beginning of the isothermal expansion, $p_1 = 105$ psia, $V_1 = 2$ ft³, and $t_1 = 600$°F. The isothermal expansion ratio is $r_e = 1.5$, and the minimum temperature in the cycle is

$t_3 = 80$°F. Calculate **(a)** ΔS during the isothermal processes, **(b)** Q_A, **(c)** Q_R, **(d)** W, **(e)** e, and **(f)** p_m.

Ans. **(a)** 0.01487 Btu/°R, **(b)** 15.75 Btu, **(e)** 49.1%, **(f)** 41.7 psi.

8.25 Assume 2 lb of nitrogen actuate a Stirling cycle (see **8.24** for a description) between the temperature limits of 240°F and −160°F for cryogenic use. If the maximum pressure in the cycle is 200 psia and the isothermal compression ratio is $r_k = 4$, determine **(a)** Q_A, **(b)** Q_R, **(c)** W, **(d)** e, and **(f)** p_m.

8.26 An Ericsson cycle, § 8.12, operates on 0.75 lb oxygen from 60 psia and 1200°F at the beginning of the isothermal expansion process to the lower temperature limit of 200°F. If the isothermal expansion ratio is 3, determine **(a)** Q_A, **(b)** Q_R, **(c)** W, **(d)** e, and **(e)** p_m. **(f)** What is the efficiency of the same cycle without regeneration? Compare with **(d)**.

8.27 The same as **8.26**, except that the working fluid is 0.5 lb nitrogen.

Ans. **(a)** 64.6, **(b)** −25.7, **(c)** 38.9 Btu, **(d)** 60.1%, **(e)** 15.28 psi, **(f)** 20.60%.

8.28 (*TVn*) A thermodynamic cycle is composed of the following reversible processes: isothermal compression 1–2; constant volume heating 2–3; polytropic process 3–1, with $pV^{1.45} = C$. Sketch the pV and TS diagrams for the cycle operating on a diatomic gas and write expressions for **(a)** Q_A and Q_R, **(b)** W from the pV plane and **(c)** e and p_m.

8.29 (*TVn*) The cycle described in **8.28** operates on 5 lb of oxygen with $p_1 = 16$ psia and $t_1 = 105$°F. During the isothermal process 315 Btu are transferred from the working substance. Determine **(a)** the isothermal compression ratio, V_1/V_2; **(b)** p_2, p_3, and T_3; **(c)** Q_A and Q_R, **(d)** W and e; and **(e)** p_m.

Ans. **(a)** 6.02, **(b)** $p_3 = 217$ psia, 1270°R, **(c)** 555, −384.1, **(d)** 171 Btu, 30.77%, **(e)** 18.71 psi.

8.30 Assume 4 lb of carbon monoxide execute a reversed Carnot cycle between the temperature limits of 40°F and 500°F. The minimum pressure in the cycle is 15 psia and the isothermal compression ratio is $r_k = 3$. Find **(a)** the volumes and pressures at the four corners, **(b)** Q_A, Q_R, W, and **(c)** the entropy change during the isothermal processes.

8.31 Two reversible heat engines are operating between the same temperature limits. One engine develops 50 hp and has an efficiency of 40%; the other engine receives 4240 Btu/min from the hot body. Determine the work of the second engine and the heat which each engine rejects to the cold body.

Ans. 40 hp, 2544, 3180 Btu/min.

8.32 Prove that all reversible engines have identical efficiencies when operating within the same temperature restrictions.

8.33 The designer of a new type of engine claims an output of 225 hp with a fuel that releases 19,250 Btu/lb when burned. If the heat is supplied and rejected at average temperatures of 750°F and 145°F, respectively, estimate the lowest conceivable necessary amount of fuel for the rated output of the engine. *Ans.* 59.5 lb/hr.

8.34 Two reversible engines R_1 and R_2 are connected in series between a heat source S and a cold body C as shown in Fig. 8.19. If $T_1 = 1000°R$, $T_2 = 400°R$, $Q_A = 400$ Btu, and the engines have equal thermal efficiencies, determine **(a)** the temperature at which heat is rejected by R_1 and received by R_2, **(b)** the work W_1 and W_2 of each engine, and **(c)** the heat rejected, Q_R, to the cold body.

Ans. **(a)** 632°R; **(b)** 147.2 Btu, 93 Btu; **(c)** −160 Btu.

9

POWER FROM
TWO-PHASE SYSTEMS

9.1 INTRODUCTION

The Carnot cycle, which consists of two isothermals and two isentropics, can operate with any substance; hence, the limits it sets are as appropriate to a vapor cycle as to any, and all the conclusions based on the Carnot cycle are valid (Chapters 5, 7). Nevertheless, in practice, it is advisable to use, as standards of comparison, cycles that are less perfect, but accord more closely with actual events. This chapter presents the fundamental ideas for power cycles whose working substance undergoes a change of phase during the cyclic processes.

A powerful advantage of two-phase cyclic systems is that the fluid being pumped from the low pressure to the high pressure is liquid; and the cost of pumping liquid is considerably less than that for compressing a gaseous substance. For the foreseeable future, by far the major amount of electrical power will be generated with H_2O (water and steam) as the working substance. The source of heat may be either fossil fuels or a nuclear reactor. For special purposes, other sources of heat and other working substances are used. *Examples*: In the SNAP systems for a power source in space, mercury is a working substance (it has already been used in central station power plants, § 9.13), and the source of energy is radioactive matter. In another study with the objective of using solar energy (though other sources would be easily adapted), the working substance is Dowtherm A, a proprietary fluid. The Dowtherm saturation curves on the Ts plane are similar to those in Fig. 3/3, so that the working substance becomes superheated as it expands through the turbine. As we know, steam expands into the wet region, and turbine operation does not improve with liquid drops present. An interesting proposal is to use propane or ammonia as a working substance for large ocean-floating electric power plants, using the warm surface water for a heat source and deep cold water (from about 2000 ft down) as the sink.

Inasmuch as steam power plants are ubiquitous and highly developed, our attentions will be focused in this area; but the ideas presented apply for any working substance. See Foreword for descriptions of some major pieces of equipment.

9.2 IDEAL CYCLE, IDEAL ENGINE

Since in steam-power cycles the steam passes through various devices (the steam generator, the engine (turbine), the condenser, and pumps), it is often advisable to

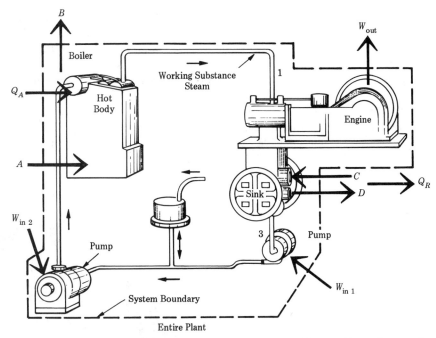

Elements of a Thermodynamic Cycle—Heat Engine. The *working substance* is   **Fig. 9/1**
H_2O; the *source* is the furnace; the *sink* is the cooling water circulated through
the condenser. In this cycle another necessary element is the pump to return
the working substance to its higher pressure. (*Courtesy Power, New York City.*)

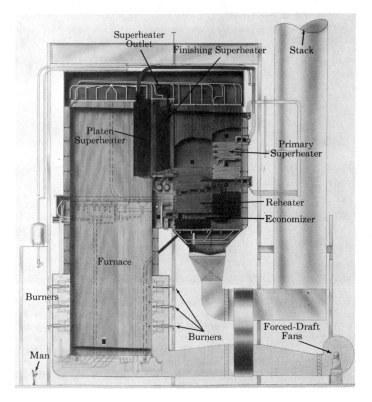

Once-Through Steam   **Fig. 9/2**
Generator. Natural-gas
fired; 3,547,000 lb/hr
delivered at 2640 psig
and 1005°F; reheat to
543 psig and 1005°F;
turbine, 500 Mw at
3600 rpm with 7 closed
regenerative feedwater
heaters, one reheat; shaft
power for pumps,
12,600 kw. (*Courtesy
Foster Wheeler Corp.,
Livingston, N.J.*)

consider separately the cycle and the engine—as well as the other pieces of equipment (in order to study their performance); see Fig. 9/1. There are two inherent characteristics to keep in mind: (1) For the cycle, the state of highest availability is the state of the steam as it *leaves* the steam generator, the significant state for the cycle. But because of inevitable losses in the steam line connecting the steam generator (Figs. 9/2 and 9/3) with the engine, the state of the steam as it *enters* the engine is primary for the engine work and efficiency. (2) The work of the engine is the *gross* work (turbine W_t); the cycle work is the turbine work minus the input work. We shall consider only the H_2O system, for which the input work is the pump work W_p; or $\oint dW = W_t - |W_p|$. Other work is necessary to run various auxiliaries.

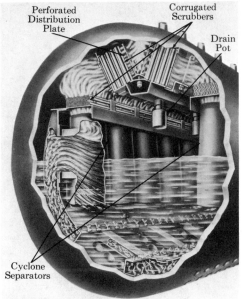

Fig. 9/3   *Inside a Boiler Drum for Large Capacity*. The steam-water mixture, rising from the tubes about the furnace, is guided to the cyclone separators, where high-speed tangential motion causes the water to go to the side (centrifugal force) and then downward to join other water. The lighter steam moves upward in the center, through the corrugated scrubbers that provide a large surface to intercept water particles while the steam flows sinuously between them. (*Courtesy Babcock and Wilcox Co., New York, N.Y.*)

9.3   RANKINE CYCLE AND ENGINE*

The Rankine cycle is shown in Fig. 9/4 on the *pv* and *Ts* planes. The steam leaves the boiler in say, state 1, moves to the engine without loss (ideal cycle), expands isentropically in the ideal engine to state 2, and passes to the condenser (Fig. 9/5). Circulating water condenses the steam to a saturated liquid at 3, from which state it is isentropically pumped into the steam generator in a subcooled liquid state *B*. Normally it enters the economizer (Fig. 9/6) of the steam generator before flowing

---

* William John M. Rankine (1820–1872), a professor at Glasgow University, was contemporary with those giants of thermodynamics—Joule, Maxwell, Kelvin, and Clausius. It was during the lifetime of and largely due to the efforts of these five men that the laws and the science of thermodynamics were formulated and interpreted. Rankine had no small part in these developments. He was a man of versatile genius and a prolific contributor to the engineering and scientific literature of his day. Among his published writings are included an important textbook on mechanics, a civil engineers' manual, and works on water supply, the steam engine, shipbuilding, and many other subjects of a technical nature, not to mention *Songs and Fables*—for he was also a composer of music and a vocalist of good talent. Rankine's modification of Gordon's column formula is in wide use today.

into the boiler drum (Fig. 9/3). Note the irreversible process of mixing cold water at temperature t_B with the hot water in the boiler at temperature $t_4 = t_1$. The compressed liquid at B is heated until it becomes saturated at 4, after which it is evaporated to steam at 1, and the cycle begins again. If the steam is superheated before it leaves the steam generator, the corresponding Rankine cycle would be ef-3-B-4-e. Since the steam generator, turbine, condenser, and pump are each taken as a steady-flow machine, the applicable equation is $Q = \Delta h + \Delta K + W$. For the ideal cycle, ΔK is always taken as zero for each process. The net work $W_{net} = \oint dQ = \oint dW$; choosing $\oint dQ$, we have

(a) $$W_{net} = h_1 - h_B - (h_2 - h_3) = h_1 - h_2 + h_3 - h_B$$

represented by the enclosed area 1-2-3-B-4, Fig. 9/4, where $Q_A = h_1 - h_B$, and

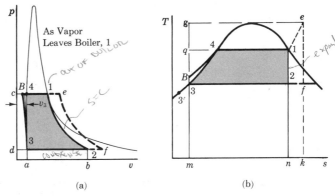

(a)   (b)

Rankine Cycle. The volumes of the liquid in (a) and the temperature rise 3-B in (b) are greatly exaggerated; state 1, or e, is the state of the working substance as it leaves the heat source. The equivalent Carnot cycles for the states 1 and e are 3-q-1-2 and 3-gef, respectively. Typically, the liquid leaving the condenser is subcooled a few degrees to some state 3; the subcooling effect is negligible, but the enthalpy entering the pump corresponds to the actual temperature. **Fig. 9/4**

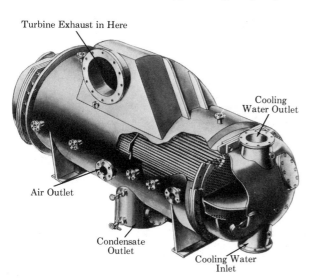

Turbine Exhaust in Here

Cooling Water Outlet

Air Outlet

Condensate Outlet

Cooling Water Inlet

Surface Condenser. A two-pass condenser. The perforated metal sheets, called tube sheets, at the ends of the tubes provide support and keep cooling water and condensing steam separated. (*Courtesy Foster Wheeler Corp., New York, N.Y.*) **Fig. 9/5**

$Q_R = h_3 - h_2$. From the energy diagram of the pump, Fig. 9/7, with $Q_p = 0$, $\Delta K = 0$, and $\Delta P = 0$, the pump work is $W_p = -\Delta h = h_3 - h_B$. Using the value of h_B, which is the enthalpy of the water entering the boiler, from this equation in equation **(a)**, we have

(9-1) $$W_{\text{net}} = h_1 - h_2 + W_p$$ [RANKINE CYCLE]

[h_1 = ENTHALPY AT STEAM GENERATOR]

where W_p is a negative number, obtained in accordance with §§ 3.11 and 4.19, say, $W_p \approx v_{f3}(p_3 - p_B)$. For the ideal engine with an isentropic expansion 1-2, Fig. 9/7, the gross work out is

(9-2) $$W_t = W_{\text{out}} = h_1 - h_2$$ [RANKINE ENGINE]

[h_1 = ENTHALPY AT THROTTLE]

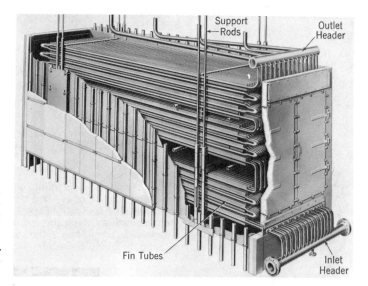

Fig. 9/6   *Fin-Tube Economizer.* Look for the longitudinal fins on the tubes, added to increase the heat transfer surface. Extended surface in some heat exchangers is also provided by circular fins. See Fig. 7/9. (*Courtesy Combustion Engineering Co., New York, N.Y.*)

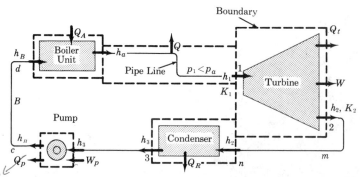

Fig. 9/7   *Energy Diagrams for Steam-Power Plant.* The radiation losses Q, Q_t, and Q_p are small. The loss from the turbine is shown as $-Q_t$ so that this heat is now algebraic in the defined sense. The line *mn* is not really a line at all because the condenser is attached directly to the turbine, Fig. 9/19.

$$W_p = -(h_B - h_3)$$
$$= h_3 - h_B$$

Observe that equation (9-1) for the H_2O system is in the form of $W_{net} = \oint dW = W_{out} - |W_{in}|$, generally the most convenient way of obtaining the net work of steam cycles. The activities of the engine are represented in Fig. 9/8(a), where the areas under b-1 and 2-a represent flow work crossing the boundaries of a steady-flow engine, included, of course, in the enthalpy terms of equation (9-2). On the Ts plane, the only engine activity seen is the expansion 1-2, Fig. 9/8(b).

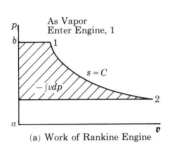

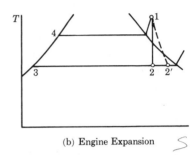

(a) Work of Rankine Engine

(b) Engine Expansion

Rankine Engine. In (a), the area 1-2-ab $= p_1v_1 + (u_1 - u_2) - p_2v_2 = h_1 - h_2$; **Fig. 9/8** state 1 is at the engine throttle. In a steady-flow engine, p_1v_1 and p_2v_2 represent flow work energy crossing boundaries 1 and 2.

When the highest pressure in the system is not very high, say less than 400 psia, depending upon the accuracy desired by the engineer, the amount of pump work W_p may be considered negligible compared to the engine and cycle works.

For the actual turbine, with a compressible fluid as the working substance, it is safe to say that ΔP is always negligible. If ΔK and heat Q_t (turbine casing heat loss) are not negligible, the steady-flow engine work is then given by

(9-3) $$W' = h_1 - h_{2'} + K_1 - K_{2'} + Q_t = h_{01} - h_{02'} + Q_t$$

where h_{01}, $h_{02'}$ are stagnation enthalpies, $K = v^2/(2kJ)$ Btu/lb, and $s_{2'} > s_1$, Fig. 9/8(b). Recall the turbine efficiency $\eta_t = W'/W$, equation (7-31); see §7.31 for other engine efficiencies. See Appendix A for mechanical efficiency, generator efficiency, brake work, and combined work. The pump efficiency, §4.19, is used to obtain the actual pump work on the fluid;

(b) $$W'_p = \frac{W_p}{\eta_p} \quad \text{where} \quad |W_p| \approx v_{f3}(p_B - p_3)$$

as a positive number; the subscripts refer to Fig. 9/4.

RANKINE EFFICIENCIES—WITH STEAM RATES 9.4

The thermal efficiency of any cycle is W_{net}/Q_A. For the Rankine cycle,

(9-4) $$e = \frac{W_{net}}{Q_A} = \frac{h_1 - h_2 - |W_p|}{h_1 - h_3 - |W_p|} \qquad \text{[RANKINE CYCLE]}$$

The thermal efficiency of engines is W/E_c, where W is the engine work, and E_c is the energy chargeable against the engine. The energy chargeable against an engine is largely a matter of what is customary, though there is always supporting logic. For example, the Rankine engine (turbine) is charged with the entering enthalpy, but credited with the enthalpy of the saturated *liquid* at the condenser temperature. Therefore, $E_c = h_1 - h_{f2}$ [$h_{f2} = h_3$, Fig. 9/8(b)]:

(9-5) $$e = \frac{W_{out}}{E_c} = \frac{h_1 - h_2}{h_1 - h_{f2}}$$ [RANKINE ENGINE]

The **steam rate** ω of an engine or power plant is the mass of steam used to perform a unit of work, the work unit usually being either a hp-hr or kw-hr. The **heat rate** of an engine or power plant is the energy chargeable per unit of work. For the Rankine engine the heat rate is $\omega(h_1 - h_{f2})$, where h_1 is the enthalpy at the throttle of the engine. If ω is lb/hp-hr, then the heat rate is in Btu/hp-hr. Observe that for ω lb/hp-hr and 2544 Btu/hp-hr, the work is $W = 2544/\omega$. The steam (or other vapor) rate ω may be ω_b lb/bhp-hr when the work is measured at the shaft, or, for example, ω_k lb/kw-hr when the work is the overall combined output of the turbogenerator combination. For reciprocating engines, we may have ω_i lb/ihp-hr, when the work is the indicated (cylinder) work. Let ω without a subscript represent the steam rate for the ideal case. The following examples give the pattern from which one may write the corresponding expression no matter where the work is measured.

(9-6) $$e = \frac{W}{E_c} = \frac{2544}{\omega(h_1 - h_{f2})}$$ heat rate: $\omega(h_1 - h_{f2}) = \frac{2544}{e}$

(9-7) $$e_k = \frac{W_K}{E_c} = \frac{3412}{\omega_k(h_1 - h_{f2})}$$ $\omega_k(h_1 - h_{f2}) = \frac{3412}{e_k}$

where $\omega_k(h_1 - h_{f2})$ is the *combined heat rate*; ω_k lb/kw-hr. Given an actual steady-flow turbine, the corresponding ideal Rankine engine is one wherein (1) the pressure and quality (or superheat) at the beginning of the engine expansion (state 1) are the same as at the throttle of the actual engine, and (2) the pressure at the end of the isentropic expansion is the same as that of the actual exhaust pressure. Because of this, h_1 and h_{f2} in equations (9-6) and (9-7) are the same in both ideal and actual cycles—but do not draw general conclusions because this is not always true.

9.5   Example—Rankine Engine
   A turbine receives steam at 100 bar, 600°C and exhausts it at 2 bar from which it is used for process heating; ΔP and ΔK are negligible. (a) For the ideal Rankine engine, determine work W, steam rate ω, thermal efficiency e, and mean effective pressure p_m. (b) For the actual engine, the brake efficiency is $\eta_b = 84\%$; the driven generator efficiency is 93% and the rated output of the generator is 30 MW. Estimate the enthalpy $h_{2'}$ and quality (or superheat) of the exhaust. Compute the combined work W_K, combined heat rate ω_K, and the total throttle flow for the rated power. (c) Determine the irreversibility and the change of availability of the actual expansion process for a sink temperature of $t_0 = 27$°C.
   Solution.   From Item B 16 (SI Mollier Chart), we get the following properties:

$$p_1 = 100 \text{ bar} \qquad p_2 = 2 \text{ bar} \qquad h_{f2} = 505 \text{ kJ/kg}$$
$$\qquad\qquad\qquad\qquad\qquad\qquad\qquad\qquad \text{(taken from SI Table)}$$

$$t_1 = 600°C \qquad\qquad t_2 = 120.2°C$$

$$h_1 = 3621 \text{ kJ/kg} \qquad h_2 = 2620 \text{ kJ/kg}$$

$$s_1 = 6.901 \text{ kJ/kg-K} \qquad s_2 = 6.901 \text{ kJ/kg-K}$$

$$v_1 = 0.03837 \text{ m}^3\text{/kg} \qquad v_2 = 0.83 \text{ m}^3\text{/kg}$$

For isentropic expansion, Fig. 9/8, we have,

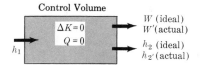

(a) $W = h_1 - h_2 = 3621 - 2620 = 1001 \text{ kJ/kg}$

(b) $\omega = \dfrac{3600}{W} = \dfrac{3600 \text{ kJ/kW-hr}}{1001 \text{ kJ/kg}} = 3.60 \dfrac{\text{kg}}{\text{kW-hr}}$

(c) $e = \dfrac{W}{E_c} = \dfrac{W}{h_1 - h_{f2}} = \dfrac{1001}{3621 - 505} = \dfrac{1001}{3116} = 32.1\%$

where h_{f2} is the enthalpy of the liquid at the exhaust pressure.

(d) $p_m = \dfrac{W}{V_D} = \dfrac{W}{v_2} = \dfrac{1001 \text{ kJ/kg}}{0.83 \text{ m}^3\text{/kg}} = 1206 \text{ kPa}$

Since the mechanical losses in a turbine are relatively small, the fluid work is not greatly different from the brake work. At this stage of our study, assume these works are equal for a turbine unless otherwise specified: $W_B \approx W'$. Using this assumption and the energy diagram of Fig. 9/9, we have

(e) $W' \approx W_B = \eta_b W = (0.84)(1001) = 840.8 \text{ kJ/kg}$

(f) $h_{2'} = h_1 - W' = 3621 - 840.8 = 2780.2 \text{ kJ/kg}$

From Item B 16, read $t_{2'} = 155°C$ (superheated).

Control Volume

| $\Delta K = 0$ |
| $Q = 0$ |

h_1

W (ideal)
W' (actual)
h_2 (ideal)
$h_{2'}$ (actual)

Enthalpy of Exhaust. For steady flow there is no change in the stored energy within the control volume with time. Implicitly, $\Delta P = 0$, $\Delta K = 0$, $Q = 0$.   **Fig. 9/9**

The work output of the generator is

(g) $W_K = \eta_g W_B = (0.93)(840.8) = 782 \text{ kJ/kg}$

The combined heat rate is

(h) $\omega_K (h_1 - h_{f2}) = \dfrac{3600}{782}(3621 - 505) = 14{,}345 \text{ kJ/kW-hr}$

The total steam flow for 30 MW, or 30,000 kW, is this output times the flow per unit power, or

(i) $\dot{m} = (30{,}000)(\omega_K) = (30{,}000)\left(\dfrac{3600}{782}\right) = 138{,}107 \text{ kg/hr}$

(c) By equations (5-11) and (5-9), we find

(j) $I = T_0(S_{2'} - S_1) = (300)(7.310 - 6.901) = 122.7 \text{ kJ/kg}$

(k) $\Delta \mathscr{A}_f = h_{2'} - h_1 - T_0(S_{2'} - S_1) = 2780.2 - 3621 - 122.7 = -963.5 \text{ kJ/kg}$

the decrease of availability of the steam during the expansion of which $W' = 840.8 \text{ kJ/kg}$ was delivered as work.

9.6   IMPROVING THE THERMAL EFFICIENCY OF RANKINE ENGINE

One reason for the importance of the ideal cycles is that more often than not, when certain changes bring about an increase in the efficiency of the ideal cycle, analogous changes in the actual cycle increase the actual thermal efficiency. The following changes tend to increase the Rankine efficiency.

1. *If the condenser temperature is lowered,* the heat rejected will be less, the work will be greater, and therefore the efficiency will be increased. Nature, however, precisely defines the limit of improvement by the value of the temperature of the available sink. Effectively, this temperature is somewhat (say, 20°) above that of condenser water from rivers, lakes, and so on. The best condensers are so well designed that no great thermal improvement is likely here, but efforts at improvement are continually made because in very large plants a small improvement involves a significant dollar saving. Condensing plants commonly operate more efficiently in winter than in summer because of the lower cooling-water temperature in the winter.

2. Up to a boiler pressure of 2500 psia,[9.5] the Rankine efficiency increases, with saturated delivered steam, which accounts in part for the gradually increasing use of higher and higher pressures.

3. The improvement brought about by the use of superheated vapor accentuates the improvement in the *actual* thermal efficiency. Particles of liquid have a lower speed than the vapor, thus getting in the way of steam, creating more chaos and entropy, and reducing the efficiency. Of perhaps greater economic significance is the eroding effect on the turbine blades of these high velocity liquid drops, necessitating periodic replacement of blades when moisture is excessive. Observe in Fig. 9/10 that the higher the pressure for a particular initial temperature and a saturated vapor curve with negative slope, the wetter is the vapor at the exhaust; point *k* versus *n* versus 2. Consequently, the higher the pressure at 1, the more desirable is higher temperature at 1. (In BW nuclear reactors, the turbine typically receives saturated vapor. To keep the liquid content low, mechanical means remove the liquid during the expansion. Also, the highest-pressure steam from the reactor may be used to superheat steam entering the lower-pressure stages. See Fig. f/12, Foreword.)

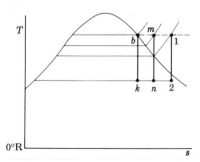

Fig. 9/10   *Effect of Pressure on Exhaust State for Rankine Engine.*

In the Rankine cycle, cold liquid is plunged into the hot fluid in the boiler, there to mix irreversibly. If some way could be devised whereby the liquid could be heated to boiler temperature by a reversible interchange of heat within the system, as in the Stirling cycle (§ 8.11), this irreversible mixing would be prevented, and the resulting ideal cycle could have an efficiency as high as that of the Carnot cycle.

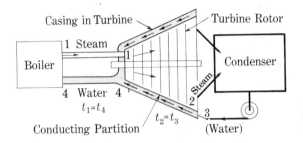

Regenerative Steam Cycle, Diagram- **Fig. 9/11**
matic. The working substance is the system.

A hypothetical reversible cycle is shown diagrammatically in Fig. 9/11, in which the turbine rotor is surrounded by a hollow casing. Let the vapor at the beginning of expansion be saturated (state 1, Fig. 9/12). The feedwater from the condenser is to be pumped through the hollow part of the turbine casing, Fig. 9/11, back to the boiler. The condensate enters the hollow part in condition 3, Fig. 9/12. On the inside of the casing is the steam just about to be exhausted in condition 2. Now, as the condensate or feedwater moves along the casing toward the high pressure end of the turbine, it picks up heat from the steam on the other side through a perfectly conducting partition. Considering the history of a drop of water, we imagine this drop picking up heat at such a rate that it is always at the temperature of the steam opposite it on the other side of the casing, the reversible manner of transferring heat. Thus, 1 lb of liquid will be heated gradually along 3-4 and at the same time 1 lb of the steam in the turbine will lose gradually a like amount of heat during expansion 1-2, a heat exchange within the system, or *regenerative heating.* The liquid now enters the boiler in a saturated state 4. Since the heat rejected during 1-2 (area 1-*ba*-2) is numerically equal to the heat supplied during 3-4 (area *m*-3-4-*n*), they cancel in $\oint dQ$ and the net work is

(a)

$$W_\text{net} = \oint dQ = T_1(s_1 - s_4) - T_2(s_2 - s_3)$$

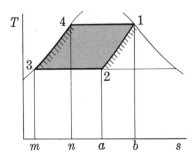

Perfect Regenerative Heating. Pump work effects ignored. **Fig. 9/12**

The heat from an *external source* is $Q_A = T_1(s_1 - s_4)$. Since $s_2 - s_3$ is equal to $s_1 - s_4$ (the curves 1-2 and 4-3 are parallel), the efficiency is the same as for the Carnot cycle,

(b)
$$e = \frac{T_1 - T_2}{T_1}$$

The lesson from the Carnot cycle, Chapter 8, is that to obtain the best efficiency heat must be added at the highest temperature. If all the heat cannot be added at the highest temperature, any step that increases the average temperature at which the system receives heat will result in improvement. That is what actual regenerative feedwater heating does—increases the average temperature at which heat is received from an external source.

9.8   REGENERATIVE CYCLE

Even if the foregoing regenerative cycle could be practically approximated, it is unlikely that it would be used because of the low quality of the steam during the latter stages of the expansion near point 2, Fig. 9/12. A similar effect of regeneration can be obtained by **bleeding** or **extracting** small quantities of steam at various points during the expansion, utilizing as fully as possible the energy of the **bled steam** rather than absorbing the required energy from *all* the steam. In this way, the major portion of the steam, the part that continues to expand and do work, is not subjected to excessive condensation. We shall use a regenerative steam cycle, depicted diagrammatically in Fig. 9/13, with three extraction points and three open heaters (O.H.) as an example; it is represented on the *Ts* plane in Fig. 9/14. The reader's aim should be to acquire from the following examples the understanding needed for independent analyses of different arrangements.

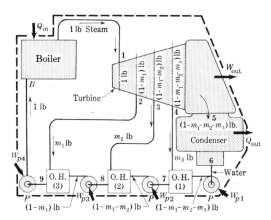

Fig. 9/13   *Diagrammatic Layout for Regenerative Heating. An open heater O.H. is one wherein the vapor and liquid are in direct contact.*

The operations are as follows, on the basis of 1 lb of steam having arrived at the throttle from the steam generator without loss, state 1. After an expansion to state 2, m_1 lb (a fractional part of 1 lb) is bled for feedwater heating. The remainder of the steam $(1 - m_1)$ lb continues to expand in the turbine to state 3, where another quantity of steam m_2 lb is extracted for feedwater heating. The remainder, $(1 - m_1 - m_2)$ lb, continues expansion. At state 4, m_3 lb is extracted for heating, so that there is left in the turbine $(1 - m_1 - m_2 - m_3)$ lb, which expands further and passes to the condenser. This mass of steam is condensed and the condensate is pumped to the heater (1), 6 to B_6, where it is heated from B_6 to 7, Fig. 9/14, by mixing (in an open heater) with the m_3 lb of steam bled for the purpose and condensed from 4 to 7; thus, the amount of water pumped, 7 to B_7, to heater (2) is $(1 - m_1 - m_2)$ lb. In heater (2), this water is heated from condition B_7 to condition 8, Fig. 9/14, by the condensation of m_2 lb from 3 to 8. Leaving heater (2) is $(1 - m_1)$ lb of water, which is pumped 8 to B_8 and heated from B_8 to 9, Fig. 9/14, by the condensation of m_1 lb from 2 to 9 in heater (3). The 1 lb of water leaving heater (3) is pumped 9 to B_9 into the boiler.

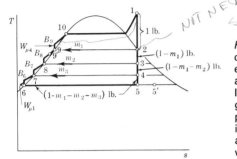

Regenerative Cycle. The saturated liquid and vapor curves are drawn for a unit mass. To account for extracted steam, these curves would be shifted an amount proportional to the mass fraction extracted. Inasmuch as the addition of these details would generally confuse rather than elucidate, the simple plan of indicating on the diagram the mass involved is adopted. It follows that, as sketched, the enclosed area on the Ts plane does *not* represent the cyclic work. Per pound, $s_1 = s_2 = s_3 = s_4 = s_5$.

Fig. 9/14

The first step is to make an energy balance for each heater, ideally with an adiabatic control surface. The energy given up by the steam from the turbine is equal to the energy increase of the feedwater entering the heater, all at constant pressure and/or steady flow. Write the energy loss of the steam as a positive number, and neglect the individual pump works for these energy balances;

$$\text{Heat to water} = \text{heat from steam}$$

(a) Heater (1): $\qquad (1 - m_1 - m_2 - m_3)(h_7 - h_6) = m_3(h_4 - h_7)$

(b) Heater (2): $\qquad (1 - m_1 - m_2)(h_8 - h_7) = m_2(h_3 - h_8)$

(c) Heater (3): $\qquad (1 - m_1)(h_9 - h_8) = m_1(h_2 - h_9)$

where, by way of explanation, $h_2 - h_9$ in equation **(c)** is the energy given up by the condensation of *1 lb* of steam from 2 to 9, and m_1 is the mass of steam condensed; $h_9 - h_8 \approx h_9 - h_{B8} = h_9 - h_8 - W_{p3}$ is the heat required to heat 1 lb of water from 8 to 9, and $(1 - m_1)$ is the mass of water involved. Having found the various specific enthalpies at 2, 3, 4, 5, 6, 7, 8, and 9, we may use equations **(c)** and **(b)** and **(a)** in this order to find m_1, m_2, and m_3.

Next observe that this regenerative cycle may be conceived of as a series of Rankine cycles: 1-2-9-10, 2-3-8-9, 3-4-7-8, and 4-5-6-7. Write the expression for the gross or *engine* work for each of these imaginary cycles, but note that the mass of fluid involved in each is not unit mass. Taking the masses from Fig. 9/14, we find

(d) $\qquad W_{\text{out}} = h_1 - h_2 + (1 - m_1)(h_2 - h_3) + (1 - m_1 - m_2)(h_3 - h_4)$

$$+ (1 - m_1 - m_2 - m_3)(h_4 - h_5) \; \text{Btu/lb throttle steam}$$

for the particular engine of Fig. 9/13. The *net* work of the H_2O system becomes

(e) $$W_{net} = \oint dW = W_{out} - |\Sigma\, W_p| \text{ Btu/lb steam at the throttle}$$

The heat supplied from an *external source* to the ideal cycle is $Q_A = h_1 - h_{B9}$ where $h_{B9} = h_9 + W_{p4}$. Notice that W_{p4} represents the pump work of the *last* feedwater pump. Thus,

(f) $$Q_A = h_1 - h_9 - W_{p4} \text{ Btu/lb}$$

The thermal efficiency of the ideal cycle is $e = W_{net}/Q_A$.

The total pump work $\Sigma\, W_p$ may be found from the following equation **(g)** if properties of a compressed liquid along an isentropic line are available:

(g) $$\Sigma\, W_p = (1 - m_1 - m_2 - m_3)(h_{B6} - h_6) + (1 - m_1 - m_2)(h_{B7} - h_7)$$
$$+ (1 - m_1)(h_{B8} - h_8) + h_{B9} - h_9 \text{ Btu/lb}$$

of steam reaching the throttle. The individual pump works can be approximated by a constant volume compression as before. For example, the pump work from p_9 to p_1 is $|W_{p4}| = h_{B9} - h_9 \approx v_{f9}(p_1 - p_9)/J$ Btu/lb. Moreover, since the total pump work $\Sigma\, W_p$ is small compared with other relevant energies, an approximation that is usually satisfactory is the work to pump 1 lb of saturated liquid at constant volume from p_6 to p_1; thus,

(h) $$\Sigma\, W_p \approx \frac{v_{f6}(p_1 - p_6)}{J} \text{ Btu/lb}$$

Actual regenerative cycles in use have from one to nine or more stages of heating. The number of stages used is purely an economic matter. The use of one stage results in a decided improvement in the thermal efficiency, Fig. 9/15, and would justify a relatively large expenditure for equipment. The incremental improvement due to the addition of a second heater is not so great as that due to the first heater, with further diminishing returns as additional heaters are added. Four states of

Fig. 9/15   *Saving in Heat Rate from Regenerative Heating. These curves are for the operating conditions shown. They indicate that for a particular number of heaters, the saving is a maximum at a certain temperature rise of the feedwater; for example, four heaters give a maximum saving of about 12.3% of the corresponding Rankine cycle when the feedwater temperature increases 340°F.*

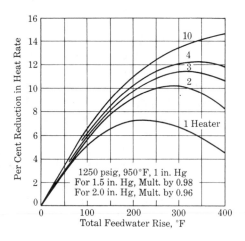

feedwater heating become profitable for a pressure of about 600 psia. The economic number of stages increases as the initial pressure and temperature increase.

Example—Computing Extracted Steam

Let us find the amount of extracted steam m_1 needed for heater (3), Figs. 9/13 and 9/14, by assuming these various states and properties: One kg steam enters the turbine at $p_1 = 50$ bar, $t_1 = 500°C$; expansion ($s = c$) occurs to 20 bar at which point m_1 kg of steam are extracted for the heater. Also entering the heater is the liquid ($1 - m_1$) with an enthalpy of $h_f = 640$ kJ/kg; the 1 kg leaves the heater saturated at 20 bar ($h_f = 909$ kJ/kg).

Solution.   From Item B 16(SI), we read these property values:

$$h_1 = 3433 \text{ kJ/kg} \qquad s_1 = s_2 = 6.97 \text{ kJ/kg K}$$

$$h_2 = 3148 \text{ kJ/kg}$$

From equation **(c)**, § 9.8, we have

$$(1 - m_1)(h_9 - h_8) = m_1(h_2 - h_9)$$

or

$$m_1 = \frac{h_9 - h_8}{h_2 - h_8} = \frac{909 - 640}{3148 - 640} = 0.1073 \text{ kg}$$

The remaining two heaters would be handled similarly if sufficient data were known.

THE REGENERATIVE ENGINE   9.9

The ideal engine is taken to correspond to some actual engine. In the ideal engine, the pressures at the throttle, at the various extraction points, and at the exhaust are the same as those for the actual engine; also, the temperature (or quality) at the throttle of the ideal and actual engines is the same. These conditions, together with the isentropic expansion, define the ideal engine. For three stages of regenerative heating, the engine work is given by equation **(d)**, § 9.8, in which the subscripts are defined in Fig. 9/16. To obtain the thermal efficiency of the engine, charge against it

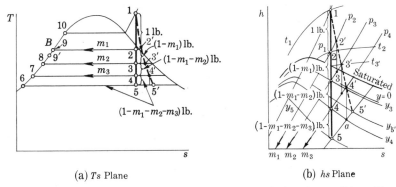

(a) *Ts* Plane   (b) *hs* Plane

Regenerative Engine. State 1 represents the actual throttle condition. The **Fig. 9/16** actual expansion is with increasing entropy, the dotted curve, called the *condition curve* if 2′ and 2 correspond to a stage or a group of stages in a turbine, the stagnation enthalpy difference $h_{2'} > h_2$ is called the *reheat* in this context (no heat involved); $h_{2'} > h_2$ because of friction, not heat if adiabatic conditions prevail, and $h_{2'} > h_2$ (for $\Delta K = 0$) because the actual energy converted into work is less than that along the $s = C$ expansion.

$E_c = h_1 - h_9$, where we observe that *the credit is the enthalpy of the saturated liquid h_9 as it leaves the last feedwater heater*, provided that the bled steam is used only for feedwater heating. The computed percentages of steam bled in the ideal *engine* (m_1, m_2, and m_3) are found from energy balances as for the corresponding ideal *cycle*.

The *actual engine* does not bleed the ideal quantities of steam as computed above. The actual temperature of the water leaving the heaters is below the temperature of the steam used for heating. Thus, the actual temperature of the water leaving the last heater, Fig. 9/16, is some value $t_{9'} < t_9$. Therefore, the energy charged against the actual engine is $E_c' = h_1 - h_{9'}$, $e_k = W_K/E_c' = 3412/[\omega_k(h_1 - h_{9'})]$. The *heat rate* is $\omega_k(h_1 - h_{9'})$ Btu/kW-hr. For a directly connected electric generator, $W_B/W_K = \eta_g$ is the *generator efficiency*, which for large generators may be of the order of 97%. The mechanical losses in the turbine are so small that the fluid work is not very different from the brake work; $W' \approx W_B$, a good first approximation.

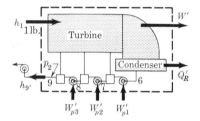

Fig. 9/17   *Energy Diagram for Finding Q_R' and $h_{5'}$, Fig. 9/12.*

Since one may place boundaries anywhere desired, let the boundaries be as shown in Fig. 9/17, with state points defined in Fig. 9/16. If the actual amounts of steam bled are known, the total being $\Sigma m'$, the actual heat rejected via the condenser is

(a)
$$Q_R' = (1 - \Sigma m')(h_{5'} - h_6) \text{ Btu}$$

for 1 lb of throttle steam. This idea is easily extrapolated for any number of extractions. The actual pump work within the boundaries of Fig. 9/17 is $W_{p1}' + W_{p2}' + W_{p3}' = \Sigma W_p'$ (where the sum includes all pumps *except the last one*) and is determined or estimated by methods previously explained. If the objective of this energy balance is to estimate the actual condition of the exhaust steam (say, for designing a condenser), write the energy balance and solve for $h_{5'}$.

9.10   REHEAT

The improvement in the thermal efficiency of the Rankine engine through the use of expanding superheated (dry) vapor is noted in § 9.6 (item 3 in the list). Unfortunately, the complete expansion in a Rankine engine from throttle to exhaust (process *a-b-c*, Fig. 9/18) normally cannot take place and remains always dry;

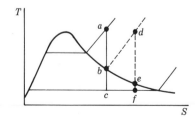

Fig. 9/18   *Effect of Reheat.*

sooner or later the state point must cross into the two-phase (wet) region in order to take advantage of a lower exhaust pressure.

If at point *b*, Fig. 9/18, the saturated vapor be reheated (often called resuper-heated) to its initial throttle temperature, rather than complete its expansion to point *c*, the subsequent expansion *d-e-f* will cross into the wet region at *e*. A smaller expansion period in the two phase region *e-f* will result, compared to *b-c* which is without benefit of reheat.

In normal power plant operations, the expanding steam is withdrawn from the engine when it is within about 25°F of becoming saturated vapor. Next it is piped to a reheater located in the boiler area; see Fig. 9/2. There it is reheated to near initial throttle temperature and finally piped back to the engine for subsequent expansion. A rough rule of thumb for good design says to keep the moisture limit of the engine exhaust to not more than 15% (liquid); see point *f*, Fig. 9/18.

A side advantage of reheat lies in the greater energy derived from each pound of working substance as it passes through the engine. Because of reheat, each pound of working substance produces more work through the addition of the reheat energy.

REHEAT-REGENERATIVE CYCLE AND ENGINE   9.11

As noted in § 9.6, as the pressure increases for a particular initial temperature, the amount of liquid in the exhaust increases. Since the materials used must not be subjected to temperatures that weaken them unduly, a plan used for lower exhaust moisture level in high pressure plants is to reheat the steam to the practical limit temperature after a partial expansion. This idea, combined with regenerative heating for improved thermal efficiency, is common practice in central station power plants. For illustration, we have chosen a three-stage regeneration cycle with reheat occurring at the first bleeding point, Fig. 9/19; the *Ts* diagram is in Fig. 9/20. The various points of the two figures have corresponding numbers. Let *1 lb of throttle steam* be the basic unit. Since m_1 lb of steam go to feedwater heater (3) at the place where reheating starts, the amount of steam to the reheater is $1 - m_1$ lb. By the same approach as before, we have

(a)
$$W_{out} = h_1 - h_2 + (1 - m_1)(h_3 - h_4) + (1 - m_1 - m_2)(h_4 - h_5)$$
$$+ (1 - m_1 - m_2 - m_3)(h_5 - h_6) \text{ Btu/lb throttle steam}$$

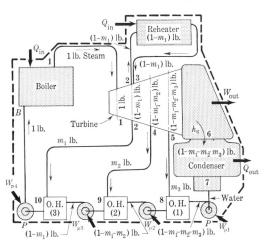

Diagrammatic Layout for Reheat-Regenerative Plant. The working substance is the system.   **Fig. 9/19**

(b)
$$W_{net} = \oint dW = W_{out} - |\Sigma W_p|$$

The total pump work ΣW_p is given by equation **(g)** of §9.8 for the three-stage heating, or an analogous equation for a different number of heaters; but see equation **(h)**, §9.8. The heat from an external source is

(c) $Q_A = h_1 - h_{10} + (1 - m_1)(h_3 - h_2) - |W_{p4}|$ Btu/lb throttle steam

where $h_{10} + W_{p4} = h_{B10}$ and the term $(1 - m_1)(h_3 - h_2)$ represents heat supplied in the reheater. The thermal efficiency is $e = W_{net}/Q_A$.

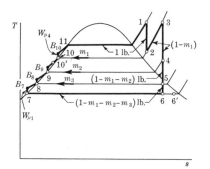

Fig. 9/20   *Reheat-Regenerative Cycle.* In the ideal cycle, the only heat exchanges with the surroundings are at the steam generator, the reheater, and the condenser; $W_{out} - |W_p| = W_{net} = Q_{in} - |Q_{out}| = \oint dQ$.

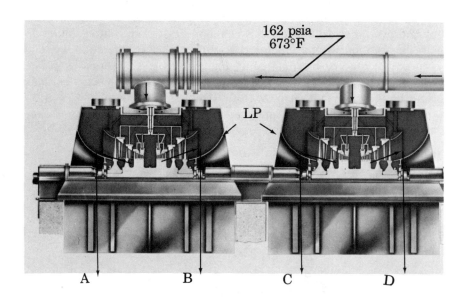

A, B, C, D to Condenser, 2.5 in. Hg
Extr. Nos. 1, 2, 3, 4 Not Shown

Fig. 9/21   *Large Central Station Steam Turbine.* A layout and design for a supercritical turbine in sizes of 500 to 700 MW at 3600 rpm; arrows indicate steam flow. It consists of HP, IP, and two LP cylinders, each double flow, directly coupled on collinear shafts to the generator; one impulse stage for each flow, all other stages reaction; seven extractions for feedwater heating, one reheat. Extr. =

Diagrams for the engine processes matching the cycle previously described are shown in Fig. 9/22. The energy chargeable against the engine is

(d)
$$E_c = h_1 - h_{10} + (1 - m_1)(h_3 - h_2)$$ [IDEAL ENGINE]

/ Re superheat)

(e)
$$E_c' = h_1 - h_{10'} + (1 - m_1')(h_3 - h_{2'})$$ [ACTUAL ENGINE]

each in Btu/lb of steam at the throttle, where, in any case, 10, 10' represent liquid leaving the last feedwater heater and the second term is the heat added during resuperheating. A large reheat-regenerative turbine is shown in Fig. 9/21.

An engineering consideration has to do with the relation of size, initial pressure, and efficiency (or heat rate). In general, it is said that the efficiency of the steam power plant increases with increasing initial pressure (but see § 9.6). However, as the pressure increases, the specific volume decreases, with the result that for a particular size (kW) the volume of steam flow becomes too small for good efficiency (high-pressure turbine blades become so small that the percentage leakage over the tips is excessive). Therefore, if pressures beyond this point are to be used, size must increase, requiring a larger total steam flow. For example, the optimum pressure for a 100-MW (100,000-kW) turbogenerator is about 3500 psia, at which a computed

From Reheater
595 psia 1000°F

From
Reheater

No. 5 Extr.
165 psia
670°F

No. 6 Extr.
320 psia
830°F

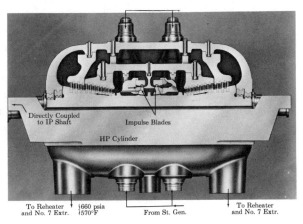

Directly Coupled
to IP Shaft

Impulse Blades

HP Cylinder

To Reheater
and No. 7 Extr.

{660 psia
{570°F

From St. Gen.

To Reheater
and No. 7 Extr.

extraction; St. Gen. = steam generator; the properties shown are rounded but typical for this class of turbine. Extraction points not shown are: 68 psia, 490°F; 38 psia, 390°F; 20 psia, 275°F; 6.9 psia with *h* = 1112 Btu/lb. (*Courtesy Westinghouse Electric, Philadelphia.*)

heat rate is about 7700 Btu/kW-hr. If this same size unit operates on 2000 psia steam, the heat rate approaches 7900 Btu/kW-hr. For a 500-MW unit, the optimum pressure is about 4500 psia, but since the curve is rather flat in the vicinity of the optimum, the heat rate would be little different for a pressure of 4000 psia. Observe that these pressures are above the critical pressure of H_2O, about 3200 psia. The following data are common to all of the foregoing comparative figures: initial temperature, 1100°F; condenser pressure, 1.5 in. Hg abs.; single reheat to 1050°F; 6 stages of regenerative feedwater heating.

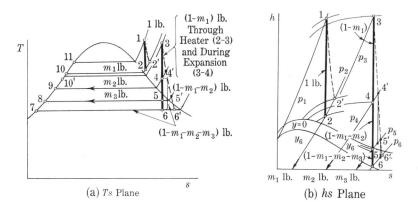

(a) Ts Plane   (b) hs Plane

Fig. 9/22   *Diagrams for Reheat-Regenerative Engine.* Observe the effect of pressure drop in passage through the reheater.

9.12   CLOSED FEEDWATER HEATERS

In the closed type of heater, the hot and cold fluids do not mix as in the open heater, but are kept separated. An advantage of closed heaters is that a single pump may be used for two or more heaters. Entrained gases, especially oxygen and nitrogen, are deleterious to steels. Since hot water can "hold" less gas than cold water, deaerating heaters are used and may be located in the higher-temperature section. If the pressure in the heater is greater than atmospheric, the liberated gas can be vented to the surroundings.

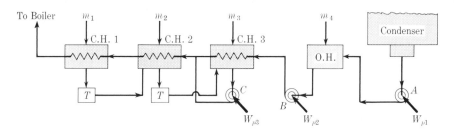

Fig. 9/23   *Closed Heaters.* A trap T is a device that automatically allows accumulated liquid to pass from a higher-pressure to a lower-pressure region.

Four stages of heaters are illustrated in Fig. 9/23. Consider the closed heater 1 (CH 1); m_1 lb of extracted steam at a pressure below that of the water in the tubes arrives from the turbine and is condensed in the process of heating the feedwater. One way of caring for this condensate (called the *drip*) is to use a trap[9.6] and let it pass to a lower-pressure region. A second method of handling the drip is shown in CH 3; a drip pump *C* pumps it into the main boiler feed line. At this point, instead of the pump *C*, another trap could let the condensate flow to OH.

$OMIT$

BINARY VAPOR CYCLES 9.13

The reader has no doubt observed that at typical sink temperatures, say about 85°F, the saturation pressure of the steam is low, so that in the early days of the development of large scale power Professor Josse of Germany suggested the use of a vapor other than steam for the low temperature range. He chose sulfur dioxide because its saturation pressure at typical sink temperatures (71.2 psia at 90°F) presents no problem in maintaining a vacuum. However, progress in condenser design has eliminated the need for this extra complication.

At the top side of the steam cycle, mercury has been used for the high temperature range. This alternative was prompted because at high temperature the saturation steam pressure is quite high and there was great difficulty in keeping leakage in control. Moreover, the higher the pressure, the less is the enthalpy of evaporation, and the smaller the percentage of the heat supplied at constant temperature. Since the saturation pressure of mercury at, for example, 940°F is a moderate 125 psia, Fig. 9/24, it would take relatively inexpensive metal to contain the working substance, and, notably, a large proportion of the heat from an external source would be received at the highest temperature. The thermodynamic advantage of this feature is one of the lessons from the Carnot cycle.

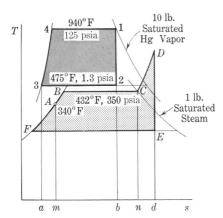

Binary Vapor Cycle. Pump works ignored.   **Fig. 9/24**

Some disadvantages of mercury as a working substance are: it does not wet the metal surface (lower heat transfer rates), its vapor is poisonous, and it is scarce and expensive.

9.14 INCOMPLETE-EXPANSION ENGINE AND CYCLE

James Watt* invented and developed the principal features of the reciprocating steam engine. Since it seldom expands the steam to the exhaust pressure, a fairer standard of comparison would be an ideal incomplete-expansion *engine* without clearance, as in Fig. 9/25. Since the enclosed area 1-2-3-*ab* represents work, we have the engine work as

(a)
$$W = h_1 - h_2 + v_2(p_2 - p_3)$$

where $h_1 - h_2$ is represented by the area *b*-1-2-*d*. For computing the thermal efficiency, use $E_c = h_1 - h_{f3}$, with the credit of the liquid enthalpy at the exhaust h_{f3};

(b)
$$e = \frac{W}{E_c} = \frac{h_1 - h_2 + v_2(p_2 - p_3)}{h_1 - h_{f3}}$$

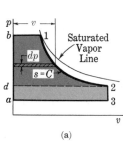

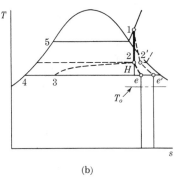

(a)   (b)

Fig. 9/25 *Incomplete-Expansion Engine.* The diagram (a) may be thought of as a conventional indicator card for an engine without clearance. Only 1-2 is a thermodynamic process involving constant mass. Intake *b*-1, where 1 is the *point of cutoff* expansion 1-2; exhaust valve opens at 2, the *point of release,* and pressure drops instantaneously to exhaust pressure (ideally); 3-*a* is exhaust. The actual expansion will be with increasing entropy, say 1–2'-*e'* in (b).

James Watt (1736–1819), born of Scottish middle-class parents, was a delicate child, timid, and had little desire to play as children play. Because of an evident manual dexterity, he was trained to be an instrument maker. As such, he was called upon to repair a model of a Newcomen steam engine. While working on this engine, he decided that it could be improved, with the result that before his life was over he had conceived of most of the basic features of the modern reciprocating steam engine. Watt's education was largely self-directed, but it was a good education. Also he had a tenacity of purpose that is rare in the run of men but so often is found in those who accomplish great things. Fortunately, he had intelligence too. Before the first model of his engine was built, he carried on extended researches on the properties of steam, about which practically nothing was known at the time. After two unsuccessful models, and with his money gone, Watt was about to give up when he persuaded Dr. Roebuck to finance his experiments. Again, two more unsuccessful engines were built, and now Dr. Roebuck was in financial difficulties of his own. After about five years, Roebuck sold his interest to a manufacturer, Mathew Boulton, who was able to furnish further financial aid. Shortly thereafter, with better workmanship, a successful engine was manufactured, and a commercial success was assured. After installing one of his engines, Watt wrote to Boulton as follows: "The velocity, violence, magnitude, and horrible noise of the engine give unusual satisfaction to all beholders.... I have once or twice trimmed the engine to end its strokes gently and make less noise; but Mr. —— cannot sleep unless it seems quite furious ... and by the by, the noise seems to convey great ideas of its power to the ignorant, who seem to be no more taken with modest merit in an engine than in a man."

The *mean effective pressure* of the zero clearance engine of Fig. 9/25 is $p_m = W/V_D$, where $V_D = v_2$ for 1 lb of steam admitted.

The net work of the ideal cycle, Fig. 9/26, is $W_{net} = W - |W_p|$ as before. Incomplete-expansion engines almost always operate with a low initial pressure; if this is so, the pump work is taken as negligible for ordinary calculations.

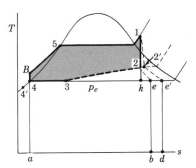

Incomplete-Expansion Cycle. **Fig. 9/26**

If, as we have done before, we imagine that the intake and exhaust are via large, perfectly insulated plenum chambers wherein the state of the working substance is steady and in virtual equilibrium (states 1, e, e'), the average condition of the exhaust can be estimated as for a steady-flow engine, Fig. 9/8; that is, $h_e = h_1 - W$ and $h_{e'} = h_1 - W'$, where the actual fluid work W' is closely obtained as the indicated work W_I; an estimation of state e' is needed for the condenser design. For reciprocating engines, let $W' = W_I$.

CLOSURE   9.15

The ideas presented in this chapter are combined in practice with details varying innumerably, but energy balances are in patterns similar to the foregoing examples. As the reader would expect, no aspect was discussed exhaustively. For fossil-fueled power plants, there is a strong trend toward supercritical plants ($p_1 > 3208$ psia $= p_c$), even though there is no heat of evaporation. With the use of several regenerative heaters, the *average* temperature at which the H_2O receives the heat from an external source is high enough to be satisfactory.

Many plans are always under study; for example, a combination of a gas-turbine and steam-turbine plant. One idea is to seal the boiler furnace, burn the fuel under pressures suitable for a gas turbine, and then lead the partially cooled furnace gases through the gas turbine to generate power. Steam is generated by the partial cooling of gases. Another idea is to lead the exhaust from the gas turbine, which contains considerable uncombined oxygen, to the furnace, burn more fuel, and generate steam. The idea in either case is to get as much power as possible from the products of combustion.

For the purposes of this chapter, the turbine is just a "black box." Knowing certain properties of the working substance as it enters and departs, we can learn or predict much. Since it is fitting for the engineer to have some acquaintance with what happens inside, there is an elementary explanation in Chapter 18, to which the reader may go now if he so desires.

PROBLEMS

SI UNITS

9.1 A thermodynamic cycle with elements as described in § 9.1 operates as follows: engine produces 22.4 kW; hot body supplies 4220 kJ/min; small pump (to circulate working substance) absorbs 2 kW. Sketch the system and calculate **(a)** the net work, **(b)** the heat rejected, and **(c)** the thermal efficiency.

Ans. **(a)** 20.4 kW, **(b)** 2996 kJ/min, **(c)** 28.95%.

9.2 An experimental geothermal energy system in Baja, California, consists of a hot water well, a flasher-separator-collector, and a 10,000-kW Rankine engine. The pressurized ground water at 172.4 bar, 282°C leaves the well to enter the flash chamber maintained at 13.8 bar (h_f = 829, h_{fg} = 1961 kJ/kg). The flashed vapor passes through the separator and collector to enter the turbine as saturated vapor at 13.8 bar; the turbine exhausts at 1 bar. The unflashed water runs to waste. Find the hourly amount of ground water required for continuous operation. Use Table II for the compressed liquid state.

Ans. 379,535 kg/hr.

9.3 The condensing pressure for a Rankine engine is 1 bar (h_f = 417.4 kJ/kg). Calculate the net work for 1 kg/s of steam and the thermal efficiency when the steam at the beginning of expansion is at 50 bar and **(a)** saturated, **(b)** 350°C, **(c)** 640°C. Note the variation of efficiency and of the quality at the end of the expansion.

Ans. **(c)** 1107 kJ/s, 32.9%.

9.4 Steam at 100 bar, 600°C is received by a Rankine engine and exhausts at 2 bar; ΔP and ΔK are negligible. **(a)** For the ideal engine, find W, ω, e, and p_m. **(b)** For the actual engine η_b = 84%, η_{gen} = 93% and the driven generator produces 30 MW. Find W_K, e_K, total throttle flow, and estimate the exhaust enthalpy $h_{2'}$.

Ans. **(a)** 1001 kJ/kg, 3.6 kg/kW-hr, 32.1%, 1206 kPa, **(b)** 782 kJ/kg, 25.1%, 138,107 kg/hr, 2780 kJ/kg.

9.5 Sketch on the Ts plane an ideal regenerative cycle with one stage of extraction feedwater heating, and write the equations

for **(a)** the mass of steam bled, **(b)** the heat supplied, **(c)** the heat rejected, and **(d)** the net work.

9.6 A turbine, with one extraction for regenerative feedwater heating, receives steam with an enthalpy of 3373 kJ/kg and discharges it with an exhaust enthalpy of 2326 kJ/kg. The ideal regenerative feedwater heater receives 11,338 kg/hr of extracted steam at 345 kPaa (whose h = 2745 kJ/kg). The feedwater (condensate from the condenser) enters the heater with an enthalpy of 140 kJ/kg and departs saturated at 345 kPaa (h_f = 582 kJ/kg). Calculate the turbine power.   *Ans.* 18,116 kW.

9.7 Steam at 100 bar, 500°C enters an ideal engine that has one stage of reheat; exhaust is at 0.5 bar and 85% quality. The work produced by the engine is 800 kJ/kg of steam. Determine the thermal efficiency of the engine.   *Ans.* 28%.

9.8 A reheat cycle with two stages of reheating is executed, with steam expanding initially from 200 bar and 540°C. The two reheater pressures are 40 bar and 10 bar, and the steam leaves each reheater at 540°C. Condensation occurs at 60°C where h_f = 251 and h_B = 268 kJ/kg. Sketch the Ts diagram. For the ideal cycle, and 1 kg/s of steam, find **(a)** Q_A and e. **(b)** For the engine, ignore the pressure drop through the reheater, let the engine operate through the same states and compute W and e.

Ans. **(a)** 4166 kJ/s, 43.7%, **(b)** 1839 kJ/s, 44%.

9.9 Assume an ideal reheat-regenerative cycle: after some expansion, steam is extracted for feedwater heating; after further expansion, there is a reheat; then expansion to exhaust. Write the equations for **(a)** the amount of extracted steam, **(b)** the net work, and **(c)** the thermal efficiency. The equations should refer to a Ts diagram with named points.

9.10 A reheat-regenerative engine receives steam at 207 bar and 593°C, expanding it to 38.6 bar, 343°C. At this point, the steam passes through a reheater and reenters the turbine at 34.5 bar, 593°C, whence it expands to 9 bar, 427°C, at which point steam is bled for feedwater heating. Exhaust occurs at 0.7 bar. Beginning at the throttle (point 1), these enthalpies are known (kJ/kg):

$h_1 = 3511.3$ $h_2 = 3010.0$

$h_{2'} = 3082.1$ $h_3 = 3662.5$

$h_4 = 3205.4$ $h_{4'} = 3222.9$

$h_5 = 2308.1$ $h_6 = 162.2$

$h_7 = 742.1$ $h_{7'} = 724.2$

For the ideal engine, sketch the events on the Ts plane, and for 1 kg throttle steam, find **(a)** the mass of bled steam, **(b)** the work, **(c)** efficiency, and **(d)** the steam rate. In the actual case, water enters the boiler at 171°C and the brake engine efficiency is 75%. **(e)** Determine the brake work and the brake thermal efficiency. **(f)** Let the pump efficiency $\eta_p = 65\%$, and estimate the enthalpy of the exhaust steam.

9.11 The effect of throttle temperature on the efficiency of a simple Rankine engine is desired. Select a specific Rankine engine, establish the throttle and condensing pressures, and then permit the throttle temperature to vary from that of saturation to a maximum value of 800°C. Find the respective values of the efficiency under these constraints and plot the curve efficiency versus throttle temperature. Program this problem.

9.12 A study of the economy of a regenerative engine is being conducted. It is desired to learn how the efficiency varies with the number of heaters used. Select a specific engine (this will establish a set of throttle conditions and a condensing pressure) and let the number of heaters vary from one to ten using an appropriate manner for spacing the heaters in the cycle. Find the efficiency in each case and plot the curve efficiency versus number of heaters. Write the program for this problem.

ENGLISH UNITS

9.13 There are received 100,000 lb/hr steam at 310 psia, 900°F by a Rankine engine; exhaust occurs at 15 psia and $\Delta P = 0$,

$\Delta K = 0$. For the engine find **(a)** W, **(b)** ω, **(c)** e. For the cycle find **(d)** W, **(e)** e.

Ans. **(a)** 9,465 kW, **(b)** 10.56 lb/kW-hr, **(c)** 24.95%, **(d)** 9,425 kW, **(e)** 24.86%.

9.14 Water at 1300 psia is delivered to a steam generator. The exit state from the superheater is defined by 1100 psia and 1100°F. The condenser pressure is 2.223 psia. The pump receives the liquid at 110°F. **(a)** For a pump efficiency of 70%, what is the specific pump work? the heat supplied in the steam generator? **(b)** Compute the approximate pump work needed to deliver 3.6×10^6 lb/hr. See caption to Fig. f/3, Foreword.

9.15 The use of mercury as a working substance for the production of power in space vehicles is being studied. Let the Hg undergo a Rankine cycle from saturated vapor at 200 psia at the throttle to a condenser pressure of 3.6319 psia ($t_{sat} = 550°F$). The energy source is an isotope (P_u^{238}). Determine **(a)** the mass rate of flow for a turbine fluid work of 5 kW and turbine efficiency of 55%, **(b)** the energy received from the isotope and the Hg cycle efficiency with pump work neglected, **(c)** the quality of the exhaust. **(d)** If the vapor at state 1 ($t_{sat} = 1017.2°F$) is heated to 1300°F, compute its entropy (datum as in Item B 37). Consider the monatomic vapor an ideal gas with $c_p = 0.02474$; also at 550°F, $h_f = 17$ Btu/lb.

Ans. **(a)** 12.3 lb/min. **(b)** 30 kW, 16.7%, **(c)** 76%, **(d)** 0.1235 Btu/lb-°R.

9.16 For an exhaust pressure of 2.223 psia (130°F) in a Rankine cycle, compare the heats supplied, the heats rejected, and the thermal efficiencies when at the beginning of isentropic expansion the steam is at **(a)** 200 psia and 600°F, **(b)** 400 psia and 600°F, **(c)** 800 psia and 600°F. Note the variation of e.

Ans. **(a)** 1224.1, −881 Btu/lb, 28.0%, **(c)** 1171.3, −767.5 Btu/lb, 34.5%.

9.17 Freon (F-12) has been proposed as the working fluid for an automobile powered by a Rankine engine. Superheated Freon vapor at 500 psia enters the engine and is exhausted to the condenser as saturated vapor at 100°F. Sketch the ph-diagram labeling pertinent points and find **(a)** W, **(b)** e, **(c)** the vapor rate, lb/hp-hr. **(d)** For a net output of 50 hp, compute the rate of flow of

F-12. **(e)** If instead of the F-12, the engine used steam entering at 500 psia but exhausting as saturated vapor at one standard atmosphere, find W, e, and ω and compare with the F-12 values.

Ans. **(a)** 10 Btu/lb, **(b)** 15.15%, **(c)** 254.4 lb/hp-hr, **(d)** 12,720 lb/hr, **(e)** 403 Btu/lb, 29.33%, 6.31 lb/hp-hr, 315.5 lb/hr.

9.18 There are received 150,000 lb/hr of steam by an ideal regenerative engine, having only one heater, of which the heater receives 33,950 lb/hr; the condenser receives the remainder at 1 psia. If the heater pressure is 140 psia, find the state (quality or °Sh) of the steam **(a)** at the heater entrance, **(b)** at the condenser entrance.

Ans. **(a)** °Sh = 6.97, **(b)** 78.5%.

9.19 An ideal supercritical regenerative cycle is executed with 1,000,000 lb/hr of steam initially at 5600 psia and 1500°F. There are five stages of regenerative feedwater heating for which the respective extraction pressures are 1050 psia, 750 psia, 230 psia, 110 psia, and 15 psia. Condensation occurs at 1 psia. Find **(a)** the hourly amounts of steam extracted, **(b)** the net horsepower, and **(c)** the thermal efficiency.

9.20 In an ideal regenerative cycle, the steam expands in the turbine from 1100 psia and 1100°F to 80 psia, where m lb of steam is extracted for feedwater heating (open heater). The remainder continues expansion to a condenser pressure of 0.9492 psia, 100°F. The throttle flow is 100 lb/sec. Compute **(a)** the cycle work and thermal efficiency, **(b)** the percentage improvement of efficiency as compared to the cycle without regenerative heating. **(c)** Consider the properties given as applying to the engine ($\Delta K \approx 0$) whose brake engine efficiency is 80%; generator efficiency is 94%. For the engine with regenerative heating, compute the power output, the combined steam rate, and the combined heat rate. **(d)** Estimate the enthalpy of the exhaust steam for the engine, assuming that the actual bled steam is $m' = 0.15$ lb/lb of throttle steam. The efficiency of the condensate pump is 50%. Show by sketch the energies and boundaries for the system used for this energy balance.

Ans. **(a)** 44.4%, **(b)** 6.2%, **(c)** 45 MW, 7.99 lb/kW-hr, 10,200 Btu/kW-hr, **(d)** 1036 Btu/lb.

9.21 An ideal reheat steam turbine with one stage of reheating develops 280 Btu/lb of work between the throttle valve and the reheater, and 320 Btu/lb between the reheater and the condenser. After absorbing 240 Btu/lb of heat in the reheater, the steam reenters the turbine at 180 psia and 1000°F and expands to the condenser pressure of 1 psia. Find the thermal efficiency of the turbine.   *Ans.* 34.5%.

9.22 Steam is delivered by a steam generating unit at 1100 psia and 900°F. After expansion in the turbine to 300 psia, the steam is withdrawn and reheated to the initial temperature. Expansion now occurs to the condenser pressure of 1 psia. **(a)** for 1 lb/sec of steam in an ideal cycle, find e. **(b)** For an ideal engine operating through the same states, but ignoring the pressure drop in the reheater, find W and e. **(c)** An actual turbine operates between these same state points except that the steam enters the reheater at 300 psia, 600°F and leaves at 290 psia, 900°F. The combined steam rate is 6.3 lb/kW-hr, the generator efficiency is 94%, and the heat loss through the turbine casing is 1% of the throttle enthalpy. Find the turbine efficiency and the approximate quality or °Sh of the exhaust steam.

Ans. **(a)** 41.5%, **(b)** 647.1 Btu/sec, 41.4%, **(c)** 89%, $x_{4'} = 91\%$.

9.23 Sketch the Ts diagram for a reheat-regenerative engine, with one reheat followed by additional expansion and one extraction. Number the points successively, starting with the throttle as point 1. The enthalpies are: $h_1 = 1351.3$, $h_2 = 1201.4$, $h_{2'} = 1214.6$, $h_3 = 1318.6$, $h_4 = 1170$, $h_{4'} = 1185$, $h_5 = 923.7$, $h_{5'} = 972$, $h_6 = 69.7$, $h_7 = 250.1$, $h_{7'} = 244$. The actual engine has a combined steam rate of 8.6 lb per kW-hr; the generator efficiency is 88%. For 1 lb of throttle steam, find **(a)** m_1 and m_1', **(b)** W and the actual fluid work W', **(c)** e and e_k, and **(d)** η_k.

Ans. **(a)** 0.1639, 0.1563, **(b)** 504.5, 450.3, **(c)** 41.4%, 32.75%, **(d)** 78.7%.

9.24 In a supercritical steam power plant, steam leaves the superheater at 4000 psia and 1000°F. Arriving at the turbine throttle without loss, it then expands isentropically to 600 psia, where m_1 lb is extracted for (open) feedwater heating and the remainder is resuperheated to 1000°F. Next, it returns to the turbine and expands isentropically to 20 psia, where m_2 lb of steam is extracted for feedwater heating; the remainder continues the ideal expansion to 94°F. For the ideal cycle, determine **(a)** the mass fractions of bled steam, **(b)** the thermal efficiency, **(c)** the rate of steam flow for 500 MW net output, **(d)** the heat rate. Show a Ts sketch with the supercritical pressure line and pump processes, and with all significant states named. (An actual plant operating between these extreme states would have several extractions; two serve to demonstrate the method.)

Ans. **(a)** 0.273, 0.090, **(b)** 50.9%, **(c)** 2.92×10^6 lb/hr, **(d)** 6700 Btu/kW-hr.

9.25 A supercritical reheat-regenerative cycle has been designed in which the steam generator receives feedwater at 5500 psia, and the turbine receives steam at 4500 psia and 1150°F (state 1). Between the throttle and the first reheat point, 10% of the throttle steam is extracted for feedwater heating; the remainder is withdrawn at 460 psia and 550°F and is reheated to 440 psia and 1000°F. The steam reenters the engine, and another 20% of the throttle steam is extracted for feedwater heating during the subsequent expansion to the second reheat point. The remaining 70% of the throttle steam is withdrawn at 55 psia and 500°F and reheated to 50 psia and 1000°F; final expansion occurs to the condenser pressure of 1 psia. The estimated combined heat rate for the 120,000-kW engine is 8300 Btu/kW-hr. At state 1, $h_1 = 1504.5$ and $s_1 = 1.5011$; for the feedwater, $h_f = 415$. Estimate the hourly flow of steam to the throttle of the engine.

Ans. 668,400 lb/hr.

9.26 In an ideal mercury-steam cycle, saturated mercury vapor enters the mercury turbine at 100 psia and is exhausted into the condenser-boiler at 1 psia ($h_f = 14$). Saturated steam at 360 psia is generated in the condenser-boiler, and the steam turbine exhausts at 1.50 in. Hg abs. ($h_f = 60$). Sketch the Ts diagram, use Items B 16 and B 34, and find **(a)** the work of the mercury turbine for 1 lb Hg, **(b)** the work of the steam turbine for 1 lb of steam, **(c)** the gross work of the two turbines per 100,000 lb/hr of steam generated, and **(d)** the thermal efficiency of the binary-vapor system.

Ans. **(a)** 44 Btu/lb Hg, **(b)** 381 Btu/lb, **(c)** 26,700 kW, **(d)** 54.4%.

9.27 A pressurized-water nuclear power plant consists of two independent loops (see figure). The water in the primary loop receives heat from the reactor and delivers it to the boiler; the secondary loop (solid lines) is through the turbine for power generation. At full load the turbine is producing 25,000 hp; pressures and temperatures are indicated on the diagram. **(a)** Determine the ideal thermal efficiency for the secondary loop. If the temperature rise in the primary loop is limited to 30°F, what will be the overall thermal efficiency of the H_2O systems? **(b)** Regenerative feedwater heating with steam extracted at 260°F is added to the foregoing cycle (see dotted addition). Compute the thermal efficiencies as in **(a)**. Is there an improvement? **(c)** Reheat (connection not shown) is provided at the extraction point in **(b)** and the steam that does not go to the feedwater heater is heated to 500°F at 30 psia. Other data are as already given. Is the thermal efficiency with this arrangement better than in **(b)** or **(a)**? Discuss.

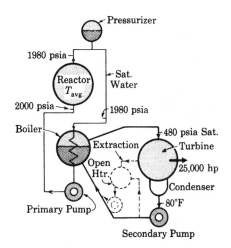

Problem 9.27

Ans. **(a)** 35.8%, 35.4%, **(b)** 38.2%, 37.5%, **(c)** 36.1%, 35.7%.

9.28 A steam cycle is required to supply 50,000 lb/hr of process steam at 20 psia, dry and saturated, and to generate 5000 kW of by-product power. The process steam is not returned to the system. The turbine receives

74,100 lb/hr of throttle steam with $h_1 = 1366$ Btu/lb and expands it to the extraction state at 20 psia. There are bled 59,900 lb/hr (50,000 lb/hr to process; 9,900 lb/hr to an open feedwater heater), and the remainder expands to the condenser pressure of 1 in. Hg abs. Makeup water at 70°F to re- place the process steam joins the saturated condensate from the condenser and thence flows to the open heater. A boiler-feed pump closes the cycle from the heater to the boiler. Sketch the energy diagram and determine the temperature of the feedwater to the boiler.

10

IMPERFECT GASES

INTRODUCTION   10.1

Unhappily, most substances do not act as ideal gases, nor do we have tables, such as the steam tables or ammonia tables, of their basic properties. For example, such an ordinarily well-behaved gas as H_2 has, at 10 psia and 60°R, an $h = 371.75$ Btu/lb; at 1500 psia and 60°R, $h = 208.02$ Btu/lb, whereas for an ideal gas the enthalpy does not vary with pressure. At 500°R, it acts better; $h = 1680.83$ at 10 psia versus $h = 1699.5$ at 1500 psia, but the deviation from ideal gas behavior may be significant.

If the answer to an engineering problem is an important one, the first thing to do is to search the literature for the properties needed to obtain the desired results (for *TABLES* example, see references numbered 0.1–0.32). If such tabulated properties cannot be found, there are different ways of proceeding, some aspects of which are discussed in this and the next chapter. We should keep in mind that, since thermodynamics produces absolutely no numerical answers without some numerical knowledge of some properties, a certain amount of experimental data is essential. In dire situations, experiments may be necessary to develop needed data.

EQUATIONS OF STATE   10.2

Equations of state, designed to fit known experimental values, define some portion of a thermodynamic surface, Figs. 3/5 and 3/6. Any three properties may be used, but the usual ones for algebraic expressions are p, v, and T, because they are measurable. Such equations may be used in ways indicated later to compute changes of other properties and they serve as interpolation formulas to compute values for tables of properties. Since equations of state may be programmed for computers, they have achieved renewed interest for studies that would likely have previously been avoided because of the tedious efforts required.

Over the years, many equations of state have been proposed, some of which attempt to care for all gases (none is universally applicable), and some are designed to represent accurately the relation between p, v, and T for a particular substance over a limited range. Among the most useful are the so-called virial equations, *VIRIAL EQUATIONS* sometimes purely empirical, but whose coefficients may be representative of the

255

effect of intermolecular forces, quantitatively evaluated by statistical mechanics,[4.4]

(10-1) $$\frac{pv}{RT} = 1 + \frac{B(T)}{v} + \frac{C(T)}{v^2} + \frac{D(T)}{v^3} + \cdots$$ [EXPLICIT IN p]

(10-2) $$\frac{pv}{RT} = 1 + B'(T)p + C'(T)p^2 + D'(T)p^3 + \cdots$$ [EXPLICIT IN v]

where the *virial coefficients* B, C, D, B', C', and D' are functions of temperature; B, C, and so on are called, respectively, the second, third, and so on, virial coefficients.

There are numerous equations of state in use for one purpose or another, a few of which are given in *Problems*. Many of them contain more than two constants (at least one, as many as nine), but Redlich and Kwong[10.18] have proposed a largely empirical, two-constant equation that often fits experimental states better than the more complex equations.[10.17] It is

(10-3) $$\left[p + \frac{a}{T^{0.5}v(v + b)} \right](v - b) = RT$$

(a) $$a = 0.4278\frac{R^2 T_c^{2.5}}{p_c} \quad \text{and} \quad b = 0.0867\frac{RT_c}{p_c}$$

for T K, p atm, and $\bar{v}$ cm^3/gmole. At high pressures, the volume of all gases (any temperature) approaches the value of $0.26\bar{v}_c$[10.17]; equation (10-3) satisfies the condition that $b = 0.26\bar{v}_c$.

10.3   VAN DER WAALS' EQUATION

One of the earliest equations of state for real gases (1873), still often used because of its relative simplicity, is one proposed by a Dutch physicist, van der Waals; in different forms, it is

(10-4A) $$\left(p + \frac{a}{v^2}\right)(v - b) = RT$$

(10-4B) $$p = \frac{RT}{v - b} - \frac{a}{v^2}$$

(10-4C) $$pv^3 - (pb + RT)v^2 + av - ab = 0$$

which is seen to be a cubic equation in terms of v. To put it into the virial form (10-1), arrange (10-4B) in the form

(a) $$pv = RT\left(1 - \frac{b}{v}\right)^{-1} - \frac{a}{v}$$

and apply the binomial theorem to $(1 - b/v)^{-1}$ to get

(b) $$\left(1 - \frac{b}{v}\right)^{-1} = 1 + \frac{b}{v} + \frac{b^2}{v^2} + \frac{b^3}{v^3} + \cdots$$

Substitution in **(a)** gives

(c) $$pv = RT + \frac{RTb - a}{v} + \frac{RTb^2}{v^2} + \frac{RTb^3}{v^3} + \cdots$$

where the v^{-3} term is likely to be negligible.

For best agreement for a limited range of properties, a and b can be found for known p, v, and T at two states. However, for what we call a "van der Waals (vdW) gas," a and b are as found below. The constant b, first introduced by Clausius, is intended to be proportional to the volume of the molecules. The molecules of gases that are acting nearly ideally are so far apart, relative to their size, that they occupy a negligible proportion of the volume. Observe in equation (10-4B) that the effect of larger b is to increase the computed p. Since the molecules at usual spacing have an attraction for each other, which is some function of their masses and of their distance apart, the pressure that the gas exerts on the walls of the container is somewhat less than it would be if there were no intermolecular forces; hence, the correction a/v^2 in (10-4B), proportional to the square of the density, was intended to account for the molecular attractive forces.

If various temperatures are assumed in equation (10-4) and a series of isothermal curves are plotted on the pv plane for a particular substance, the result will be about as pictured in Fig. 10/1. Because the equation is a cubic in v, it has three roots. Above curve C, Fig. 10/1, where the equation is quite useful, there is one real root and two imaginary ones; below C, there are three real roots; at C, which is the critical point, the three roots are equal. The part $adeb$ of a curve such as A does not represent actual behavior, except as a metastable state may occur. Let the dotted curve aC represent the saturated vapor line. We know that, for equilibrium states, the actual isotherm in the two-phase region remains at constant pressure afb, Fig. 10/1. Outside of the two-phase region, the van der Waals equation usually gives more accurate results than $pv = RT$, but there are states where the van der Waals equation with constants as given in Item B 18, computed from critical data as described below, gives answers seriously in error. Hence, this equation should be used only when it is known to apply with the required accuracy.

One of the requirements for a general equation of state is that the critical isotherm should have a point of inflection at the critical point, the mathematical requirements being that the curve is not only horizontal at that point, Fig. 10/1 $(\partial p/\partial v)_T = 0$, but

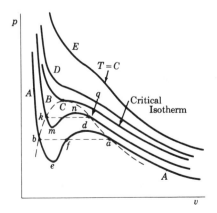

Van der Waal's Isotherms. Also typical of similar equations of state. The part *ad* can represent supersaturated vapor (§ 18.11); *be* represents states of overexpansion of the liquid.[2.4]

Fig. 10/1

also that the rate of change of slope be zero, $(\partial^2 p/\partial v^2)_T = 0$. These two conditions define the critical point of an equation of state. Making the differentiations of equation (10-4B), we get

(d)
$$\left(\frac{\partial p}{\partial v}\right)_T = -\frac{RT}{(v-b)^2} + \frac{2a}{v^3} = 0$$

(e)
$$\left(\frac{\partial^2 p}{\partial v^2}\right)_T = \frac{2RT}{(v-b)^3} - \frac{6a}{v^4} = 0$$

Since these are the conditions at the critical point, let $v = v_c$, the critical volume, and $T = T_c$, the critical temperature. Then solving equations **(d)** and **(e)** simultaneously for v_c, we get

(f)
$$v_c = 3b \quad \text{or} \quad b = \frac{v_c}{3}$$

Using this value of v_c in equation **(d)**, we get

(g)
$$T_c = \frac{8a}{27Rb}$$

Substitute the values of v_c and T_c from **(f)** and **(g)** into (10-4b) and get

(h)
$$p_c = \frac{a}{27b^2}$$

Since critical pressure and temperature data are more common and generally more accurate than critical volume data, equations **(g)** and **(h)** are the ones usually used to compute a and b. A check shows that b obtained from equation **(f)** is different from the values in Item B 16, which is to say that the van der Waals equation is not consistent with actual measurements of critical point properties. Therefore, unless the constants are adjusted, it should be expected to give inaccurate answers in this vicinity.

Note the units of the constants a and b in Item B 18 and make other units consistent; ft^3/pmole volumes, and the universal gas constant is $R = 0.73$ atm-ft^3/pmole-°R; conversion, 2.72 Btu/atm-ft^3.

10.4   USE OF AN EQUATION OF STATE FOR PROPERTY AND ENERGY CHANGES

In the next chapter, general methods for computing changes of internal energy, enthalpy, entropy, and other details are developed, methods that may be used with equations of state (or with raw data). At this point, we wish to repeat that basic energy relations remain valid: $dQ = c\,dT$ with the appropriate specific heat; $dQ = T\,ds$; $p\,dV$ and $-v\,dp$ have the meanings previously defined; and so on. For example, suppose a van der Waals gas undergoes an isothermal process; we use equation (10-4B) and get

(a)

$$\int p \, dv = RT \int_1^2 \frac{dv}{v-b} - a \int_1^2 \frac{dv}{v^2}$$

$$= RT \ln \frac{v_2 - b}{v_1 - b} + a \left(\frac{1}{v_2} - \frac{1}{v_1} \right)$$

(b)

$$-\int v \, dp = \int p \, dv - \Delta pv$$

where p_1, p_2, v_1, and v_2 conform to the van der Waals equation. Equation **(a)** gives the work of a nonflow isothermal process of a vdW gas; if the $\int p \, dv$ from **(a)** is used, equation **(b)** evaluates $W_{sf} + \Delta K + \Delta P$ for a steady-flow isothermal. The van der Waals equation (10-4B) is explicit in p; if the equation of state is explicit in v, the $-\int v \, dp$ would be easier to obtain for $T = C$.

REDUCED COORDINATES AND COMPRESSIBILITY FACTOR 10.5

A reduced property is its actual value divided by its value at the critical point;

(a)

$$p_R \equiv \frac{p}{p_c} \qquad v_R \equiv \frac{v}{v_c} \qquad T_R \equiv \frac{T}{T_c}$$

$$p = p_R p_c \qquad v = v_R v_c \qquad T = T_R T_c$$

where the capital subscripts R are used to avoid confusion with the *relative* values (p_r, v_r) in the *Gas Tables*; p_R is the **reduced pressure,** v_R the **reduced volume**, and T_R the **reduced temperature**.

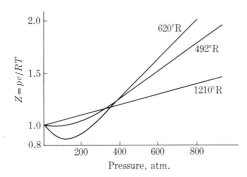

Compressibility Factor, Nitrogen (after **Fig. 10/2** Dodge[1.16]). The variables p and T could be changed to p_R and T_R.

For ideal gases $pv/(RT) = 1$; for all actual gases (see Fig. 10/2),

(b)

$$\frac{pv}{RT} \to 1 \quad \text{as} \quad p \to 0$$

Thus, $pv/(RT)$ is a variable for actual gases, called the *compressibility factor Z*,

(10-5) $Z \equiv \dfrac{pv}{RT}$ $pv = ZRT$ $pV = nZ\bar{R}T$ $p\bar{v} = Z\bar{R}T$

                                                       [1 LB]   [n MOLES]   [1 MOLE]

Values of this factor may be given in graphical form, illustrated by Fig. 10/2, and spot values of Z can be used to compute an unknown p, v, or T. Also such curves can be used in determining changes of other properties (h, u, s, and so on, Chapter 11).

As one can see, a large quantity of data is required to plot a Zp chart for a particular substance. As a consequence, another way of taking advantage of compressibility factors has been developed, called the *principle of corresponding states*. This principle can be used in connection with equations of state and was in fact first proposed by van der Waals in 1881.

Two fluids (this idea is applied also to liquids) are said to be in corresponding states when they have two reduced coordinates in common, say p_R and T_R; that is, the principle of corresponding states is expressed mathematically as $Z = f(p_R, T_R)$. The principle presumes that the thermodynamic behavior of all fluids is the same in terms of reduced coordinates and that each property change for all fluids can be obtained from the same equation [for example, equation (11-21)]. For the principle to be universally true, the substances, with a particular p_R and T_R, must have the same value of the reduced volume and of the compressibility factor; in short, the following equation should hold,

(c) $Z = \dfrac{pv}{RT} = \dfrac{p_R p_c v_R v_c}{RT_R T_c} = \left(\dfrac{p_R v_R}{T_R}\right)\left(\dfrac{p_c v_c}{RT_c}\right) = Z_c \dfrac{p_R v_R}{T_R}$

where $Z_c = p_c v_c/(RT_c)$. A check of Item B 18 shows that the compressibility factor Z_c at the critical point is *not* the same for all gases. Therefore the principle of corresponding states does *not* hold as stated, but may be modified to $Z = f(Z_c, p_R, T_R)$. As a matter of fact the principle in its application has been modified in various ways.[1.13,4.4] A four-parameter approach with improved accuracy has been formulated by Hsieh.[10.16]

This principle fails significantly to give good results for some substances. The gases that have nearly spherical molecules (examples, the monatomics, H_2, CH_4) have a value of Z_c in the vicinity of 0.3; for polar molecules, Z_c for many will be about 0.23; for hydrocarbons, more nearly about 0.27, which is close to an average for all gases. These variations can be related to the structure of the molecule.[4.4] It has been found that if the principle of corresponding states is applied to groups of substances with similar values of Z_c, say, three different Z_c values, the results are more accurate.

10.6   GENERALIZED COMPRESSIBILITY CHART

The general appearance of a generalized compressibility chart (in terms of reduced properties) is shown in Fig. 10/3, which is an extract of one of the working charts in Appendix B, Items B 19, B 20, and B 21. These working charts are made up from data on 26 gases and have a correlation with actual properties within about 2.5% with a few exceptions, particularly water and ammonia. For hydrogen, helium, neon, ammonia, and water, better results are obtained by using pseudo-critical-point data,[1.16] T_{pc} and p_{pc};

(a) $$T_{pc} = T_c + 14.4°R \quad \text{and} \quad T_R = \frac{T}{T_c + 14.4}$$

(b) $$p_{pc} = p_c + 8 \text{ atm} \quad \text{and} \quad p_R = \frac{p}{p_c + 8}$$

Even with these adjustments, the results may be significantly in error unless $T_R > 2.5$.[10.5] With or without such modifications and as with any empirical approach, one must be highly conscious of the possibility of grievous departures from truth.

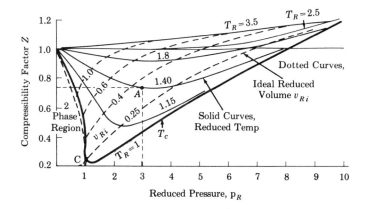

Reduced Pressure, p_R

Generalized Compress- **Fig. 10/3** *ibility Chart. At $p_R = 3$ and $T_R = 1.40$, locate A and find $Z \approx 0.73$; also, $0.4 > v_{ri} > 0.25$ (not enough lines here for interpolations). (Not exactly to scale.)*

In Fig. 10/3, the first thing to observe is that from some lower pressure, say $p_R = 1$, all gases approach ideal ($Z \to 1$) as the pressure approaches zero ($p \to 0$). Next, the deviation from ideal-gas behavior is extreme in the vicinity of the critical point, $p_R = 1$ and $T_R = 1$, being almost 80% off here. Not shown in Fig. 10/3 but evident in Item B 21, the extreme departure at very high pressures ($p_R \approx 40$) is of the same order of magnitude. If $p_R < 5$ and $T_R > 2.5$, or thereabouts, the value of Z does not vary greatly from unity.

Shown dotted in Fig. 10/3 and the working charts are curves of constant v_{Ri}, called the **ideal reduced volume**, proposed by Su,[10.3] and defined by

(10-6) $$v_{Ri} \equiv \frac{v}{v_{ci}} = \frac{vp_c}{RT_c} = v_R Z_c = \frac{T_R}{p_R} Z \quad \text{and} \quad Z = \frac{p_R v_{Ri}}{T_R} Z_c$$

where v is the actual volume and v_{ci} is the "ideal" critical volume, because it is computed from $v_{ci} = RT_c/p_c$. The other transformations are made with help from equation **(b)**, § 10.5. Since $v_{Ri} = f(T_R, p_R)$ for a particular Z (not true of v_R), v_{Ri} can be plotted on the Z, p_R coordinates, giving curves that are of considerable assistance, via equation (10-6), especially when a volume is a given property.

Since Z may be either larger, or smaller, than unity, the actual volume at a particular p and T may be either larger or smaller than that of an ideal gas $(v = ZRT/p)$. It is larger when its temperature is above the so-called Boyle isotherm (Boyle $T_R \approx 2.5$) and smaller when it is below.

10.7 Example—Volume Given

A 12.5-ft^3 cylinder contains 150 lb of CO_2 at 140°F. Compute the pressure, using a generalized compressibility chart.

Solution. From Item B 18: $\bar{v}_c = 1.51$ ft^3/pmole, $Z_c = 0.276$, $T_c = 548$, $M = 44$ lb/pmole.

$$T_R = \frac{600}{548} = 1.094 \qquad v_{Ri} = v_R Z_c = \left[\frac{(12.5/150)(44)}{1.51}\right](0.276) = 0.668$$

From Item B 20, read $Z = 0.65$. Then the pressure should be approximately

$$p = \frac{ZRT}{v} = \frac{(0.65)(1545)(600)}{(12.5/150)(44)(144)} = 1140 \text{ psia} \quad \text{or} \quad 77.6 \text{ atm}$$

By a table of properties not in the text,[0.18] $p = 79.6$ atm; a fair check. The ideal gas equation gives about 120 atm.

10.8 GENERALIZED EQUATIONS OF STATE

The van der Waals equation, and others, can be transformed into reduced coordinates. Substitute the values of p_c, v_c, and T_c from equations (f), (g), and (h), § 10.3, into (a), § 10.5, and then substitute the corresponding values of p, v, and T into equation (10-4). For example,

(a) $$p_c = \frac{a}{27b^2} \qquad p_R = \frac{p}{a/(27b^2)} \qquad p = \frac{ap_R}{27b^2}$$

The reduced van der Waals equation, from equation (10-4), becomes

(10-7) $$\left(p_R + \frac{3}{v_R^2}\right)(3v_R - 1) = 8T_R$$

Notice that there are no unknown or experimental constants in this equation except as the critical coordinates p_c, v_c, and T_c are involved in the reduced coordinates. To get the value of Z_c for a van der Waals gas, use (f), (g), and (h), § 10.3, in

(b) $$Z_c = \frac{p_c v_c}{RT_c} \qquad \text{and get} \qquad Z_c = \frac{3}{8} = 0.375$$

Observation of the disagreement between this value and those for actual gases in Item B 18 suggests that the van der Waals reduced equation is inaccurate in the vicinity of the critical-point state. To obtain better accuracy, Su and Chang[10.2] modified the van der Waals reduced equation by the use of the ideal reduced volume. From equation (10-6), $v_R = v_{Ri}/Z_c = v_{Ri}/0.375$; this value into (10-7) gives

(10-8)
$$p_R = \frac{T_R}{v_{Ri} - 0.125} - \frac{0.422}{v_{Ri}^2}$$

which serves a wider range more satisfactorily than (10-7).

Example   10.9

What mass of nitrogen is contained in a 10-ft^3 vessel at a pressure of 840 atm and 820°R? Make the computation by using (a) the ideal gas equation, (b) the compressibility factor, (c) the van der Waals equation, (d) the reduced van der Waals equation, and (e) the modified reduced van der Waals equation.

Solution.   (a) Using $R = 55.1$ ft-lb/lb-°R from Item B 1, the ideal gas equation yields

(a)
$$m = \frac{pV}{RT} = \frac{(840)(14.7)(144)(10)}{(55.1)(820)} = 394 \text{ lb}$$

(b) From Item B 21 for $p_R = 840/33.5 = 25.1$ and $T_R = 820/227 = 3.61$, we read $Z = 1.6$, where $p_c = 33.5$ and $T_c = 227$ are taken from Item B 18. Using $R = 0.73$ atm-ft^3/pmole-°R and equation (10-5), we find

(b)
$$n = \frac{pV}{ZRT} = \frac{(840)(10)}{(1.6)(0.73)(820)} = 8.78 \text{ moles}$$

or $m = nM = (8.78)(28) = 246$ lb.

(c) For the van der Waals equation (10-4B), use $a = 346$ and $b = 0.618$ from Item B 18, and note that the volume $v = \bar{v} = 10/n$ ft^3/pmoles for n pmoles and $R = 0.73$.

$$p = \frac{RT}{\bar{v} - b} - \frac{a}{\bar{v}^2}$$

(c)
$$840 = \frac{(0.73)(820)}{(10/n) - 0.618} - \frac{346}{(10/n)^2}$$

After several trials, assuming various values of n, we find that $n = 8.6$ pmoles satisfies the equation as closely as warranted by the expected accuracy. Therefore

(d)
$$m = nM = (8.6)(28) = 241 \text{ lb}$$

(d) We found the reduced coordinates $p_R = 25.1$ and $T_R = 3.61$ in (b), so we now solve equation (10-7) by trial and error for v_R. After several trials, assuming various values of v_R, and an interpolation, we check $v_R = 0.628$ and find

$$p_R = \frac{8T_R}{3v_R - 1} = \frac{3}{v_R^2}$$

(e)
$$25.1 \approx \frac{(8)(3.61)}{(3)(0.628) - 1} - \frac{3}{(0.628)^2} \approx 25.1$$

close enough. For $\bar{v}$ ft^3/pmole, the number of moles is $n = 10/\bar{v} = 10/(\bar{v}_c v_R)$; and $m = nM$.

For $\bar{v}_c = 1.44$, Item B 18, we have

(f) $n = \dfrac{10}{(1.44)v_R}$ and $m = nM = \dfrac{(10)(28)}{(1.44)(0.628)} = 310 \text{ lb}$

This is probably not a good answer, but it is better than that from $pV = mRT$

(e) Using the modified reduced properties equation (10-8) and solving for v_{Ri} by trial, we have for $v_{Ri} = 0.235$,

(g) $p_R = \dfrac{T_R}{v_{Ri} - 0.125} - \dfrac{0.422}{v_{Ri}^2} \approx \dfrac{3.61}{0.235 - 0.125} - \dfrac{0.422}{(0.235)^2} \approx 25.15$

versus $p_R = 25.1$. Since $v_{Ri} = 0.235 = \bar{v}/(RT_c/p_c)$, we have

(h) $\bar{v} = \dfrac{(0.235)(0.73)(227)}{33.5} = 1.161 \text{ ft}^3/\text{pmole}$

For $n = V/\bar{v} = 10/1.161$ pmoles, we have

(i) $m = nM = \left(\dfrac{10}{1.161}\right)(28) = 241 \text{ lb}$

By a table of properties of N_2,[0.18] we find $m = 250$ lb. For this particular case, the conclusion is that the compressibility factor method gives the best answer (246); van der Waals equations (10-4) and (10-8) do fairly well (241); but the others are seriously in error.

10.10   ENTHALPY DEVIATION

Consider a substance to be in an ideal-gas state ($p \to 0$), say $p°$ and $T°$. Let its state be changed at $T = C$ until the pressure is p. The difference between its actual enthalpy h and the enthalpy h^* it would have *if it had remained an ideal gas* is called the **enthalpy deviation** (also *enthalpy departure, residual enthalpy*). In Fig. 10/4, let the substance be in state $1°$; the *actual* isothermal line is as shown, say, to point 1; but if it were an ideal gas, the enthalpy change would have been zero $[h = f(T)]$ and at the end state at the pressure p_1 would be 1^*. The enthalpy deviation is $\Delta h_{d1} = [h_1^* - h_1]_T$. Note the symbolization; $1°$, $2°$ represent reference ideal gas states; $h_1°$, $s_1°$, for example, represent these properties in this state; 1^*, 2^* represent states of a corresponding ideal gas in other than a reference state; h_1^*, s_1^*, and so on, represent properties of this hypothetical gas at p_1 and T_1; $h_1° = h_1^*$ and $T_1° = T_1^*$. Given an imperfect gas that changes from state 1 to state 2, Fig. 10/4, it is desired to know $\Delta \bar{h} = \bar{h}_2 - \bar{h}_1$. This $\Delta \bar{h}$ may be determined by (1) finding $\Delta \bar{h}_{d1}$ in tables or charts (moving from an actual state to an ideal gas state); (2) determining $\Delta \bar{h}^*$, the enthalpy change of the ideal gas between the specified states; and (3) finding $\Delta \bar{h}_{d2}$ (moving from an ideal gas state at 2 to the actual gas in state 2). These quantities are then handled as follows, along the path shown in Fig. 10/4:

$$\Delta \bar{h} = \bar{h}_2 - \bar{h}_1 = (\bar{h}_1^* - \bar{h}_1) + (\bar{h}_2° - \bar{h}_1°) + (\bar{h}_2 - \bar{h}_2^*)$$

(10-9)

$$= \Delta \bar{h}_{d1} + \int C_p^* \, dT - \Delta \bar{h}_{d2} = \Delta \bar{h}_{d1} + \Delta \bar{h}^* - \Delta \bar{h}_{d2}$$

$$\quad\quad\quad\quad [\text{CHART}] \quad [\text{IDEAL} \quad [\text{CHART}]$$
$$\quad\quad\quad\quad\quad\quad\quad\quad \text{GAS}]$$

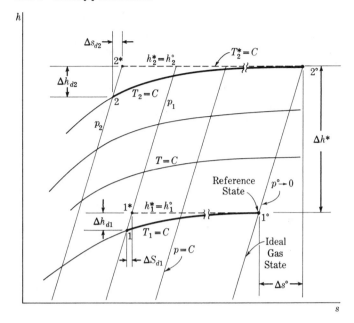

Gas with Positive Joule–   **Fig. 10/4**
Thomson Coefficient.

where $\bar{h}° = \bar{h}*$ may be taken from the *Gas Tables* if appropriate, or the $\int C_p^* \, dT$ may be evaluated for some average C_p^* suitable to the temperature range or by means of a function of T, as in Table I, p. 53;

(a) $\Delta\bar{h}_d \equiv \bar{h}* - \bar{h} = (T_c) \times$ (chart reading), obtained from Items B 22 and B 23, Btu/pmole.

Even though the principle of corresponding states has serious shortcomings, it can be advantageously applied to obtaining deviations of actual properties from the corresponding ideal-gas properties, the theory of which is given in Chapter 11 (§§ 11.13 and 11.18). Items B 22 and B 23 give enthalpy deviations $\Delta\bar{h}_d$ in terms of p_R, T_R for $Z_c = 0.27$. It is to be expected that the error from these charts will increase as the variation of Z_c from 0.27 increases.

References [1.13] and [10.8] give correction factors for gases whose value of Z_c is significantly different from 0.27, and those desiring somewhat better answers to engineering problems (by this approach) should consult one of these references.

ENTROPY DEVIATION   10.11

Mostly, the same remarks apply to entropy as to enthalpy regarding the use of an ideal gas state and generalized properties to get the actual entropy change. Thus by comparison with Fig. 10/4 we can write [see equation (11-21)] for 1 mole,

(a) $\Delta\bar{s} = \bar{s}_2 - \bar{s}_1 = (\bar{s}_1^* - \bar{s}_1) + (\bar{s}_1° - \bar{s}_1^*)_T + (\bar{s}_2° - \bar{s}_1°)_p + (\bar{s}_2^* - \bar{s}_2°)_T + (\bar{s}_2 - \bar{s}_2^*)$

$$= \Delta\bar{s}_{d1} - \bar{R}\ln\frac{p°}{p_1} + \int_{1°}^{2°} C_p^* \frac{dT}{T} - \bar{R}\ln\frac{p_2}{p°} - \Delta\bar{s}_{d2}$$

(10-10) $\Delta\bar{s} = \Delta\bar{s}_{d1} + \bar{\phi}_2 - \bar{\phi}_1 - \bar{R}\ln\dfrac{p_2}{p_1} - \Delta\bar{s}_{d2} = \Delta\bar{s}_{d1} + \Delta\bar{s}^* - \Delta\bar{s}_{d2}$

   [CHART]   [IDEAL GAS]   [CHART]

where we have used equation (6-13) for the entropy change of the ideal gas; $\Delta\bar{\phi} = \int C_p\, dT/T$ may be found in the *Gas Tables* for some gases, or the integral may be made with a suitable average C_p or with C_p as a function of T. Values of $\Delta\bar{s}_d \equiv s^* - s$ are given in Items B 24 and B 25 for $Z_c = 0.27$.

Generalized charts (or tables) have been constructed for other properties, including internal energy, density, specific heat, and fugacity; and, of course, they can be drawn for any property. We shall let the ones for enthalpy and entropy be our examples of this approach. If the absolute entropy of an imperfect gas is desired, start with a known value, as in Item B 11, and use equation (6-15).

10.12   Example—Isothermal Process

In one stage during the compression of air, the air enters at $p_1 = 40$ atm and $T_1 = 520°R$. If it is compressed isothermally to 220 atm in a steady-flow reversible process, what work is done, Btu/lb ($\Delta K = 0$)? What is the heat? See Fig. 10/5.

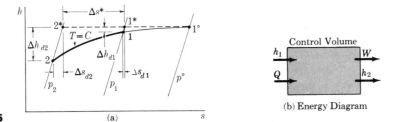

Fig. 10/5   (a)

(b) Energy Diagram

Solution.   From equations (4-16) and (4-15), which is $T\,ds = dh - v\,dp$, we have $W = -\int v\,dp$ and

(a)
$$W = -\int v\,dp = \int(-dh + T\,ds) = h_1 - h_2 + T(s_2 - s_1)$$

which incidentally is equal to $-\Delta G$ (§ 5.24); equation **(a)** also comes directly from the energy diagram. Thus, we need only to evaluate the properties shown in **(a)** to compute the work. From Item B 18 for air: $T_c = 239°R$, $p_c = 37.2$ atm, $v_c = 1.33$ ft^3/pmole, and $Z_c = 0.284$.

(b)
$$p_{R1} = \frac{40}{37.2} = 1.075 \qquad p_{R2} = \frac{220}{37.2} = 5.92 \qquad T_R = \frac{520}{239} = 2.175$$

Using Items B 23 and B 25,

(c)
$$\Delta\bar{h}_{d1} = T_c(0.4) = (239)(0.4) = 96 \text{ Btu/pmole}$$

(d)
$$\Delta\bar{h}_{d2} = (239)(2) = 478 \text{ Btu/mole}$$

(e)
$$\Delta\bar{s}_{d1} = 0.16; \qquad \Delta\bar{s}_{d2} = 0.85 \text{ Btu/pmole-°R}$$

(f)
$$\Delta\bar{h}^* = 0, \text{ because } T = C \qquad\qquad\qquad \text{[Fig. 10/5]}$$

(g)
$$\Delta\bar{s}^* = \Delta\bar{\phi} - \bar{R}\ln\frac{p_2}{p_1} = 0 - 1.986\ln\frac{220}{40} = -3.3854 \text{ Btu/pmole-°R}$$

(h)
$$\Delta \bar{h} = \Delta \bar{h}_{d1} - \Delta \bar{h}_{d2} = 96 - 478 = -382 \text{ Btu/pmole}$$

(i)
$$T\,\Delta \bar{s} = T(\Delta \bar{s}_{d1} + \Delta \bar{s}^* - \Delta \bar{s}_{d2}) = (520)(-4.07) = -2116 \text{ Btu/pmole}$$

(j)
$$W = -\Delta h + T\,\Delta s = \frac{+382 - 2116}{29} = -59.8 \text{ Btu/lb}$$

where, for air, the equivalent molecular mass is $M \approx 29$ lb/pmole. The heat is $T\,\Delta s$, or $Q = -2116/29 = -73$ Btu/lb. (For an ideal gas at $T = C$, $W = Q = T\,\Delta s$.) As a matter of interest, we find by table,[0.18] $\Delta \bar{s} = -4.09$ Btu/pmole-°R and $\Delta \bar{h} = -382$ Btu/pmole, an elegant, coincidental, but in this case, honest agreement with the chart results. If it is assumed that the van der Waals equation is suitable, W could be obtained from equation **(a)**, § 10.4, minus Δpv. Also, if the given process had been a reversible *nonflow* isothermal process, $W = \int(T\,ds - du)$; and Δu could have been obtained from $\Delta u = \Delta h - \Delta pv$, where $\Delta pv = R\,\Delta ZT$ is obtained with help from the compressibility charts.

Example—Throttling Process 10.13

Methane is throttled from state 1 at 100 atm and 520°R to $p_2 = 10$ atm. (See § 11.19.) Compute its final temperature and the change of entropy. For this temperature range, let $C_p = 7.95 + 6.4T^4/10^{12}$ Btu/pmole-°R at low pressure.

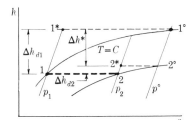

When an isenthalpic curve on the Tp plane has negative slope, temperature drops during throttling; see Fig. 11/4. **Fig. 10/6**

Solution. The process is shown 1-2 on Fig. 10/6. The basic method of solution is to assume different values of T_2 until the change Δh, equation (10-9), is zero; or we may be able to come close enough that an interpolated value will be adequate. From Item B 18: $T_c = 344$°R, $p_c = 45.8$ atm, $Z_c = 0.29$; therefore $T_{R1} = 520/344 = 1.51$, $p_{R1} = 100/45.8 = 2.18$; $\Delta h_{d1} = T_c(2.08) = (344)(2.08) = 716$ Btu/pmole; $p_{R2} = 10/45.8 = 0.218$.

Assume $T_2 = 445$°R; then $T_{R2} = 445/344 = 1.29$; from Item B 22, $\Delta h_{d2} = T_c(0.25) = (344)(0.25) = 86$ Btu/pmole;

(a) $\Delta h^* = \displaystyle\int C_p\,dT = \int \left(7.95 + \frac{6.4T^4}{10^{12}}\right)dT = \left[7.95T + \frac{6.4T^5}{5 \times 10^{12}}\right]_{520}^{445} = -622.4 \text{ Btu/pmole}$

Substituting into equation (10-9), we get

(b) $\Delta h = \Delta h_{d1} + \Delta h^* - \Delta h_{d2} = 716 - 622.4 - 86 = +7.6 \text{ Btu/pmole}$

Considering the relative sizes of these numbers, we have (by coincidence and by virtue of the impossibility of finding small changes in the deviations accurately from the charts) probably arrived at as good an answer as can be obtained with the means at hand. However, to illustrate the procedure, we continue. Assume $T_2 = 440$°R; then $T_{R2} = 440/344 = 1.28$; $\Delta \bar{h}_{d2} = T_c(0.26) = (344)(0.26) = 89.5$ Btu/pmole.

(c)
$$\Delta \bar{h}^* = \left[7.95T + \frac{6.4T^5}{5 \times 10^{12}} \right]_{520}^{440} = -663.5 \, \text{Btu/pmole}$$

(d)
$$\Delta \bar{h} = \Delta \bar{h}_{d1} + \Delta \bar{h}^* - \Delta \bar{h}_{d2} = 716 - 663.5 - 89.5 = -37 \, \text{Btu/pmole}$$

By interpolation, we arrive at $T_2 = 444°$R, or $-16°$F, to the nearest whole degree. For entropy, from Items B 24 and B 25: for $T_{R2} = 444/344 = 1.29$, $\Delta \bar{s}_{d2} = 0.16$; $\Delta \bar{s}_{d1} = 1.13 \, \text{Btu/pmole-°R}$.

(e)
$$\Delta \bar{s}^* = \int C_p^* \frac{dT}{T} - \bar{R} \ln \frac{p_2}{p_1} = \int \left(7.95 + \frac{6.4T^4}{10^{12}} \right) \frac{dT}{T} - \bar{R} \ln \frac{10}{100}$$

$$= \left[7.95 \ln T + \frac{6.4T^4}{4 \times 10^{12}} \right]_{520}^{444} + 4.573 = +3.19 \, \text{Btu/pmole-°R}$$

Using equation (10-10) for $\Delta \bar{s}$ and $M = 16.03 \, \text{lb/pmole}$, we get

(f)
$$\Delta s = \frac{1}{16.03}(1.13 + 3.19 - 0.16) = 0.259 \, \text{Btu/lb-°R}$$

By table,[0.18] we find $T_2 = 437°$R to the nearest whole degree.

10.14   Example—Isentropic Process

It is decided to investigate a low temperature situation in connection with liquefying gases. Air at 300 atm and 450°R enters a turbine and expands to $p_2 = 30$ atm. Assuming an isentropic process, compute the final temperature and the work done ($\Delta K = 0$). See Fig. 10/7.

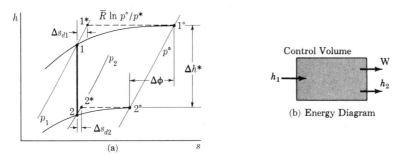

Fig. 10/7

(a)

(b) Energy Diagram

Solution.   The general procedure is to assume temperature T_2 until it is found that Δs from (10-10) is zero; $\Delta \bar{s}_{d1}$ and the log term can be evaluated at the outset.

(a)
$$-\bar{R} \ln \frac{p_2}{p_1} = 1.986 \ln \frac{p_1}{p_2} = 1.986 \ln \frac{300}{30} = 4.573 \, \text{Btu/pmole-°R}$$

For $p_c = 37.2$ atm and $T_c = 239°$R, we find

(b)
$$p_{R1} = \frac{300}{37.2} = 8.07 \quad \text{and} \quad T_{R1} = \frac{450}{239} = 1.88$$

These values and Item B 25 yield $\Delta \bar{s}_{d1} = 1.32 \, \text{Btu/pmole-°R}$. Assume $T_2 = 235°$R; then $T_{R2} = 235/239 = 0.984$; $p_{R2} = 30/37.2 = 0.807$. From Item B 24, we find that the desired point lies between the $T_R = 1$ curve and the saturated vapor curve. After some search of the literature, we find the saturation temperature of air at 30 atm as about 230°R; the corresponding $T_{Rs} = 230/239 = 0.964$, which is the value of the reduced temperature at the saturation curve and $p_R = 0.807$, Item B 24. This knowledge permits a graphical location of the point where $T_{R2} = 0.984$; from which $\Delta s_{d2} = 2.3$ Btu/pmole-°R.

From the air table, we find $\Delta\phi = 0.40163 - 0.5571 = -0.1555$ Btu/lb-°R for 450°R to 235°R, or $-(29)(0.1555) = -4.51$ Btu/pmole-°R. Substituting the computed values into equation (10-12), we get

(c) $\Delta\bar{s} = 1.32 + (-4.51) + 4.573 - 2.3 = -0.91$ Btu/pmole-°R

the negative sign indicating that 235°R is too low. After two more computations as above and an interpolation, we have $T_2 = 247$°R, not much above $T_c = 239$°R; for which $T_{R2} = 247/239 = 1.032$. Now enter Items B 22 and B 23, use Δh^* from the air table, and find $W = -\Delta h$ as follows. By equation (10-9) in Btu/lb,

(d) $\Delta h = \Delta h_{d1} + \Delta h^* - \Delta h_{d2}$

$$= \frac{(2.72)(239)}{29} + (58.92 - 107.5) - \frac{(2.3)(239)}{29} = -45.1 \text{ Btu/lb}$$

or $W = +45.1$ Btu/lb. The net deviation of enthalpy is about 3.5 Btu/lb in this case; but with the correction, the final result checks fairly well with Δh read from a Ts chart for air.

FUGACITY AND THE GENERALIZED FUGACITY COEFFICIENT   10.15
CHART

The time has come to introduce another thermodynamic property, fugacity f. Later (see Chapter 13) it will assume an important role when we will be discussing mixtures other than ideal gases that might undergo a chemical process—for example, a combustible mixture of gasoline vapor and air.

Fugacity is by definition a pseudopressure that accounts for the nonideality of a gas; it has the units of pressure. An equation that is normal for ideal gases may be used for real gases through the substitution of fugacity f for pressure p.

To develop the concept of fugacity, we use the Gibbs function $G = h - Ts$ in differential form

(a) $dG = dh - T\,ds - s\,dT$

$$= du + p\,dv + v\,dp - T\,ds - s\,dT$$

$$= v\,dp - s\,dT \qquad \text{where } T\,ds = du + p\,dv$$

At constant temperature, equation (a) becomes

(b) $dG_T = vdp_T$

For an ideal gas ($pv = RT$), equation **(b)** becomes

(c) $dG_T = \frac{RT}{p}dp_T = RT\,d(\ln p)_T$

For a real gas with $pv = ZRT$

(d)
$$dG_T = \frac{ZRT}{p} dp_T = ZRT d(\ln p)_T$$

We can also account for the real gas effect by substituting fugacity f for pressure p in equation **(c)** and obtain

(e)
$$dG_T = RT d(\ln f)_T$$

where it should be noted that

(f)
$$\lim_{p \to 0} (f/p) = 1$$

Now consider a gas undergoing an isothermal process from a low pressure $pv = RT$) to a high pressure $(pv = ZRT)$; let us view the change in Gibbs function for this process.

(g)
$$dG_T = ZRT d(\ln p)_T = RT d(\ln f)_T$$

or
$$Z d(\ln p)_T = d(\ln f)_T$$

Further, we note that

(h)
$$dp/p = dp_R/p_R \qquad \text{where} \quad p_R = p/p_c$$

and
$$d(\ln p) = d(\ln p_R)$$

from which it follows that

(i)
$$Z d(\ln p_R)_T - d(\ln p_R)_T = d(\ln f)_T - d(\ln p)_T$$

$$(Z - 1) d(\ln p_R)_T = d(\ln f/p)_T$$

and
$$\int d(\ln f/p)_T = \int (Z - 1)d(\ln p_R)_T$$

or

(j)
$$\ln (f/p)_T = [(Z - 1)d(\ln p_R)_T$$

Here we see that as $p \to 0$, $Z \to 1$ and the right side of equation **(j)** goes to zero; thus, $f \to p$. By integrating between $p = 0$ (ideal gas) and a known p, we can obtain the value of f/p, the fugacity coefficient. The compressibility factor Z may be obtained graphically for any temperature from the generalized compressibility chart. From equation **(j)**, the generalized fugacity chart results; see Fig. 10/8 and Item B 37.

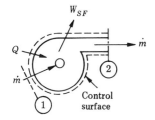

Illustration of Generalized Fugacity Coefficient Chart.   **Fig. 10/8**

Example—Isothermal Process   **10.16**

Methane is compressed reversibly and isothermally in a steady-flow steady-state rotary compressor from 101 kPa abs, 277 K to 23, 210 kPa abs. For the process, $\Delta P = 0$, $\Delta K = 0$. Find the work W_{SF} per kgmole of flowing methane. See Fig. 10/9.

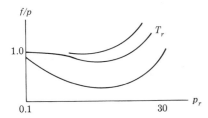 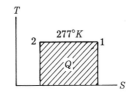

Rotary Pump and Ts Diagram.   **Fig. 10/9**

Solution.   For the process, $P_1 + K_1 + h_1 + Q = P_2 + K_2 + h_2 + W_{SF}$

$$W_{SF} = h_1 - h_2 + Q$$

$$= h_1 - h_2 + T(s_2 - s_1)$$

$$= (h_1 - T_1 s_1) - (h_2 - T_2 s_2)$$

$$= G_1 - G_2 = -\Delta G_T$$

$$p_{R_1} = p_1/p_c = 101/4642 = 0.022$$

where from Item B 18, $p_c = 45.8$ atm $= 4642$ kPa and $T_c = 344°R = 191$ K

$$p_{R_2} = 23,210/4642 = 5$$

$$T_{R_1} = T_{R_2} = 277/191 = 1.45$$

From Item B 37, $f_1/p_1 = 1.0$, $f_2/p_2 = 0.7$

and

$$f_1 = (1)(101) = 101 \text{ kPa abs}$$

$$f_2 = (0.7)(23, 210) = 16,247 \text{ kPa abs}$$

$$W_{SF} = (8.3143)(277) \ln 101/16,247 = -11,701 \text{ kJ/kgmole}$$

where

$$W_{sF_T} = -\Delta G_T = \bar{R}T \ln f_1/f_2$$

10.17   CLOSURE

The examples of this chapter and many of the problems in *Problems* relate to substances for which tabular properties are available. There are two reasons for such choices. First, even though tabulated properties are often somewhere in the literature, they may not be readily available, or they might not be found by the engineer. Second, and perhaps more important for pedagogical purposes, the opportunity to compare answers by the procedures of this chapter with values directly derived from experiment will help to build confidence or a lack thereof in a sample of situations, the first step in achieving the background knowledge needed for engineering judgment.

It is unnecessary to warn against jumping to general conclusions. For example, a common notion is that, at $p = C$, gases tend to act more nearly ideal as the temperature increases; while the idea is all right in the majority of actual cases, an examination of Item B 19 shows that it is not always true.

PROBLEMS

SI UNITS

10.1 Two constants, a and b, are included in the van der Waals equation of state. Discuss the significance of each.

10.2 The van der Waals equation of state is

$$\left(p + \frac{a}{v^2}\right)(v - b) = RT$$

where a and b are constants. Its critical isotherm has a point of inflection at the critical point. **(a)** Show that these expressions for the constants must obtain: $a = 9RT_cv_c/8 = 3p_cv_c^2$, $b = v_c/3$. **(b)** Compute the approximate values of the constants a and b for the fuel octane C_8H_{18} and compare results with values of a and b found in Item B 18.

10.3 The same as **10.2** except the equation of state is that of Redlich-Kwong:

$$\left[p + \frac{a}{T^{0.5}v(v + b)}\right](v - b) = RT$$

the constants are $a = 0.4278\,R^2T_c^{2.5}/p_c$, $b = 0.0867RT_c/p_c$, and the fuel is butane C_4H_{10}.

10.4 By manipulating the expressions for each of the respective constants a and b,

determine the compressibility factor at the critical point $Z_c = p_cv_c/RT_c$ for a gas whose equation of state is **(a)** van der Waals, **(b)** Redlich-Kwong. **(c)** Calculate each Z_c of these two gases by substituting values of the critical properties of a given gas, say air, directly into the expression; see Item B 18 for values. **(d)** Compare all results with the values of Z_c given in Item B 18.

10.5 The pressure and temperature of a given fluid are 100 atm and 633 K, respectively. Determine the compressibility factor and the density if the fluid is **(a)** nitrogen, **(b)** ammonia, **(c)** heptane. **(d)** Compare each density to that obtained assuming the gas to be ideal.

10.6 There are required at least 20 kg of steam at 600 bar, 750°C to conduct an experiment; a 140-ℓ heavy duty tank is available for storage. Predict if this is adequate storage capacity using **(a)** the ideal gas theory, **(b)** the compressibility factor, **(c)** the van der Waal's equation, **(d)** the Mollier chart, Item B 16 (SI).

10.7 The Berthelot equation of state is

$$p = \frac{RT}{v - b} - \frac{a}{Tv^2}$$

where a and b are constants. (See **10.9**, **10.27**.) Show that to meet the requirement that the critical isotherm should have a point of inflection at the critical point,

$$a = \frac{27}{64} \frac{R^2 T_c^2}{p_c} \quad \text{and} \quad b = \frac{RT_c}{8p_c}$$

Show all steps required in arranging it in the virial form

$$pv/(RT) = 1 + (b - a/T^2)/v + b^2/v^2 + b^3/v^3 + \cdots$$

10.8 Nitrogen is stored in a $0.425 \, \text{m}^3$ tank at 560 atm, 456 K. Determine the mass of nitrogen in the tank using **(a)** ideal gas theory, **(b)** the compressibility factor, **(c)** van der Waals equation.

Ans. **(a)** 178, **(b)** 132, **(c)** 132.3 kg.

10.9 A hospital requires that 225 kg of medicinal oxygen be on hand at any one time. The oxygen is stored in 100-ℓ drums at 272 atm, 300 K. Determine the number of storage drums required based on **(a)** $pv = RT$, **(b)** $pv = ZRT$, **(c)** van der Waals equation, **(d)** Berthelot's equation, **(e)** Redlich-Kwong equation. Which answer would you select as being the more accurate?

10.10 Find the density of air at 13.79 MPaa and 172.8 K, using **(a)** $pv = RT$, **(b)** $pv = ZRT$, **(c)** van der Waals equation. **(d)** Which of the three answers do you believe is more accurate?

Ans. **(a)** 278.2, **(b)** 415.8 **(c)** 398 kg/m³.

10.11 Nitrogen is stored at 200 atm, 335 K. Determine its density using **(a)** the reduced Dieterici equation, **(b)** the modified reduced Dieterici equation, **(c)** the ideal gas theory, **(d)** the compressibility factor. Which of the foregoing answers would you deem more accurate? Dieterici equation: $p(v - b) \cdot e^{a/RTv} = RT$.

10.12 **(a)** Integrate the expression $\int p \, dv$ for a van der Waals fluid undergoing an isobaric nonflow process. **(b)** Let the fluid have properties similar to those of butane C_4H_{10} and determine the nonflow work of 1 kg-mole as it expands isobarically and reversibly from 75 atm, 850 K to 2778 K.

10.13 An ideal turbine receives nitrogen at 300 atm, 500 K and expands it in a steady flow manner to 5 atm; the flow rate is

2 kg/sec. Assume the nitrogen obeys the Redlich-Kwong equation of state and calculate the power output of the turbine. See Table I for the specific heat c_p for the nitrogen.

10.14 **(a)** Transform the van der Waals equation of state to reduced coordinates. See **10.2**. **(b)** Modify this reduced equation by introducing the ideal critical volume $v_{ci} = RT_c/p_c$ and the ideal reduced volume $v_{Ri} = v/v_{ci}$. **(c)** Develop an expression for the compressibility factor $Z = pv/RT$ based on the reduced equation of van der Waals.

10.15 The same as **10.14** except that the Berthelot equation is to be transformed and then modified. See **10.7**.

10.16 Saturated steam at 200 bar is to be heated at constant pressure to 650°C. Using appropriate deviations, determine Δh, the heat added, Δs, and the specific volume at state 2. Compare results with values read from the Mollier chart, Item B 16 (SI).

10.17 There are throttled 2.27 kg/s of nitrogen from 200 atm, 200 K to 15 atm; sink temperature $t_0 = 15.6°C$. Compute **(a)** the final temperature, **(b)** ΔS, **(c)** the change of availability. For this temperature range, use c_p from Item B 1.

Ans. (approx.) **(a)** 139 K, **(b)** 0.5448 kJ/K-s, **(c)** −356 kJ/s.

10.18 Show that the fugacity of a van der Waals gas is given by the expression

$$\ln f = -\ln (v - b) - \frac{2a}{RTv} + \frac{b}{v - b}$$

10.19 Chilled drinking water, say at 1 atm, 283 K represents the state of a compressed liquid. Show that the fugacity of this compressed liquid is approximately equal to the fugacity of the liquid in its saturated state at the same temperature.

10.20 A booster station compresses 50 kg/sec of methane from 14 atm, 285 K to 92 atm in a reversible and isothermal steady-state steady-flow manner. Use the generalized fugacity coefficient chart (Item B 37) and calculate the compression work and the heat transfer. Now solve using ideal gas theory and compare answers.

10.21 Write a computer program for a van der Waals gas using its virial form and let enthalpy h and entropy s be calculated at selected pressures and temperatures. Account for the variation of the specific heat $c_p = \alpha + \beta T + \gamma T^2$.

ENGLISH UNITS

10.22 The property relation $(pv/RT) = Z$ is known as the compressibility factor for a given substance. For an ideal gas, $Z \equiv 1$ at any state. **(a)** Find the compressibility factor Z_c for a van der Waals gas at the critical point. **(b)** Using this value of Z_c, rewrite the expressions for the constants a and b excluding the property v_c. See **10.2**.

10.23 The same as **10.22** except use the Redlich-Kwong gas. See **10.3**.

10.24 A study of the compressibility charts, Items B 19 through B 21, will reveal that certain combinations of p_R and T_R will produce compressibility factors of $Z \equiv 1$, as if the gas were ideal. **(a)** List several of these combinations from Item B 20. **(b)** What will be the state (p psia, $t°F$) of a given fluid (other than ideal) to produce $Z \equiv 1$ if that fluid is air? Ammonia? Helium?

10.25 In a state where the pressure is 30 atm, a compressibility factor of 0.8 is noted. Determine the temperature and the mole volume if the substance is **(a)** air, **(b)** hydrogen, **(c)** ethylene.

Ans. **(a)** $-187.5°F$, 5.35 ft³/mol.

10.26 Refer to § 10.3 and derive the virial form of the Redlich-Kwong equation of state.

10.27 Find the density of air at 4000 psia and 1040°F **(a)** as an ideal gas, **(b)** as a Berthelot gas (see **10.7**), **(c)** using $pv = ZRT$.

Ans. **(a)** 7.20, **(b)** 6.51, **(c)** 6.55 lb/ft³.

10.28 Superheated vapor tables show that at 5500 psia, 1200°F the density of water vapor is 6.6 lb/ft³. Determine the percentage error involved if the density is calculated by using **(a)** the modified reduced van der Waals equation, **(b)** the ideal gas theory.

Ans. **(a)** 7.73%, **(b)** 15.60%.

10.29 Compute the pressure of 5 lb of carbon monoxide contained in 0.5 ft³ at 40°F, using **(a)** $pv = RT$, **(b)** $pv = ZRT$, and **(c)** van der Waals equation.

Ans. **(a)** 1915, **(b)** 1825, **(c)** 1780 psia.

10.30 Methane at 600 psia, 100°F, is flowing in a pipe line at the rate of 10,000 cfm. Determine the mass flow rate using **(a)** ideal gas theory, **(b)** compressibility factor, **(c)** Redlich-Kwong equation.

10.31 If 5 lb of ethane are contained in a 1-ft³ tank at 80°F, determine the pressure by means of the reduced Berthelot equation and compare this value to that obtained through the ideal gas theory and the compressibility factor.

10.32 Oxygen initially at 1000 psia and 0.4 ft³/lb undergoes an isothermal process to $p_2 = 200$ psia. If the process is steady flow with $\Delta K = -2$ Btu/lb and performs according to the Berthelot equation (**10.7**), determine a, b, $-\int v \, dp$, and the work W. The mass flow rate is 16 lb/sec.

Ans. 96,400, 0.506, 1370 atm-ft³/pmole, 2000 kW.

10.33 **(a)** What are the enthalpy and internal energy of N_2 at 360°R and 70 atm measured from an ideal gas base of zero at 0°R and 1 atm? **(b)** What is the absolute entropy of N_2 at 360°R and 70 atm? Use the deviation charts.

10.34 A gas turbine receives 1 lb/sec air at 300 atm, 450°R and expands it reversibly to 30 atm in an adiabatic steady flow manner; $\Delta P = 0$; $\Delta K = 0$. **(a)** Use the deviation charts and find the exit air temperature and the work. **(b)** Solve using ideal gas theory and compare with foregoing answers.

Ans. **(a)** 247°R, 45.1 Btu/sec, **(b)** 233°R, 52.1 Btu/sec.

10.35 If 1 mole of oxygen is confined in a rigid tank while being refrigerated from 1500 psia, 450°R, to 300°R, determine **(a)** the final pressure, **(b)** the heat, **(c)** the change of entropy. Use the deviation charts.

Ans. (approx.) **(a)** 50.5 atm, **(b)** -945 Btu, **(c)** -2.57 Btu/°R.

10.36 A manufacturer of pressurized gas containers desires to obtain a plot of pressure versus volume at selected temperatures. Use one of these three equations of state on the reduced basis, van der Waals, Redlich-Kwong, Berthelot, and write the computer program.

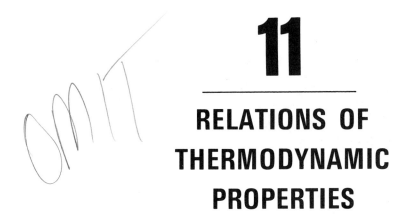

11

RELATIONS OF THERMODYNAMIC PROPERTIES

The science of thermodynamics has evolved from the basic laws of thermodynamics largely via the mathematical manipulation of properties (point functions). The numerical knowledge found in tables of properties is dependent on various of these property relations and, in extreme cases, on only a small amount of experimental data. At the outset, think of the system as being a pure substance whose state is defined by p, v, and T, the kind of system we have concentrated on so far. Since the mathematical approach is general, the methods used should be thought of in a general sense for their possible future use in different kinds of systems. There are so many interrelations of the mathematics and thermodynamics that there are usually several ways in which any particular equation can be obtained. On this account, the reader would gain greater comprehension if he devised derivations other than those given.

Most of the discussions of this chapter are related to a pure substance, though the mathematical approach is applicable to other systems, as shown in the latter part of the chapter. Magnetic, electric, surface tension, and other effects are absent or negligible unless specified. For the pure-substance systems, the most available measured data are pressures, temperatures, and volumes, especially p and T. Therefore, it is appropriate whenever possible to express changes of other significant properties in terms of p, T, v.

Let us first review briefly certain useful mathematical relations. Let z be a function of two independent properties, say $z = f(x, y)$, where x, y, and z here represent *any* three properties of a pure substance; then the total differential dz is

$$(11\text{-}1) \qquad dz = \left(\frac{\partial z}{\partial x}\right)_y dx + \left(\frac{\partial z}{\partial y}\right)_x dy$$

which in words says that a change in the property z can be considered as occurring in two steps, one step while y is constant, $(\partial z/\partial x)_y\, dx$, and another while x is

275

constant, $(\partial z/\partial y)_x \, dy$. Equation (11-1) can be extended to any number of independent variables. Since there are so many properties, it is expedient to indicate the variable that is being held constant by a subscript, as x in $(\partial z/\partial y)_x$. The partial notation $(\partial z/\partial y)_x$ indicates in itself that z is a function of y and x, and that this is the derivative of z with a respect to y while x is held constant. To illustrate with thermodynamic properties, consider $s = s(p, T)$ and write the differential of s according to equation (11-1);

(a) $$ds = \left(\frac{\partial s}{\partial T}\right)_p dT + \left(\frac{\partial s}{\partial p}\right)_T dp.$$

Think of a change of entropy with respect to a thermodynamic surface, Fig. 3/5, and let the entropy change along reversible paths, say first at constant pressure and then at constant temperature, between any two equilibrium states 1 and 2 on this surface; this thought process gives a physical meaning to ds in equation **(a)**.

For equation (11-1) to be valid for our purposes, dz must be an exact differential (that is, a point function as opposed to a path function such as dQ and dW), and an equation relating point functions must meet the test for exactness: from the calculus for a continuous function, the derivatives of the coefficients of dx and dy in (11-1) are equal;

(b) $$\frac{\partial}{\partial y}\left(\frac{\partial z}{\partial x}\right) = \frac{\partial}{\partial x}\left(\frac{\partial z}{\partial y}\right) \quad \text{or} \quad \frac{\partial^2 z}{\partial x \, \partial y} = \frac{\partial^2 z}{\partial y \, \partial x}$$

To symbolize the principle differently, write (11-1) as

(c) $$dz = M \, dx + N \, dy$$

where

(11-2) $$\left(\frac{\partial z}{\partial y}\right)_x = N \quad \text{and} \quad \left(\frac{\partial z}{\partial x}\right)_y = M$$

Then, since the order of differentiation does not matter, we have Euler's theorem,

(11-3) $$\left(\frac{\partial M}{\partial y}\right)_x = \left(\frac{\partial N}{\partial x}\right)_y \qquad \text{[EXACTNESS TEST]}$$

where M and N are basically functions of x and y; $M(x, y)$, $N(x, y)$. If it is unknown whether or not a differential is exact (when not exact, it is technically called a Pfaffian), the test of (11-3) supplies the answer, because it is a necessary and sufficient one. When it is known that a differential is exact, as we know for any thermodynamic property, the test for exactness often produces a useful relation.

To get a relation of the partial derivatives, let $dz = 0$ in equation (11-1), and find

(d) $$\left(\frac{\partial z}{\partial x}\right)_y\left(\frac{\partial x}{\partial y}\right)_z + \left(\frac{\partial z}{\partial y}\right)_x = 0$$

perhaps the easiest form to recall is

(11-4) $$\left(\frac{\partial z}{\partial x}\right)_y\left(\frac{\partial x}{\partial y}\right)_z\left(\frac{\partial y}{\partial z}\right)_x = -1$$

for two independent variables. If instead of x, y, and z, the coordinates were thermodynamic properties, this equation could be useful in checking the compatibility of experimental data, perhaps of different origins.

To introduce a fourth variable v (standing for *any* property), we may first write the differential of $x(v, y)$;

(e)
$$dx = \left(\frac{\partial x}{\partial v}\right)_y dv + \left(\frac{\partial x}{\partial y}\right)_v dy$$

and substitute this value of dx into (11-1). This gives

(f)
$$dz = \left(\frac{\partial z}{\partial x}\right)_y\left[\left(\frac{\partial x}{\partial v}\right)_y dv + \left(\frac{\partial x}{\partial y}\right)_v dy\right] + \left(\frac{\partial z}{\partial y}\right)_x dy$$

$$= \left(\frac{\partial z}{\partial x}\right)_y\left(\frac{\partial x}{\partial v}\right)_y dv + \left[\left(\frac{\partial z}{\partial x}\right)_y\left(\frac{\partial x}{\partial y}\right)_v + \left(\frac{\partial z}{\partial y}\right)_x\right] dy$$

Next write the differential of $z(v, y)$;

(g)
$$dz = \left(\frac{\partial z}{\partial v}\right)_y dv + \left(\frac{\partial z}{\partial y}\right)_v dy$$

Since the coefficients of dv in **(f)** and **(g)** must be equal, we find

(11-5)
$$\left(\frac{\partial z}{\partial x}\right)_y\left(\frac{\partial x}{\partial v}\right)_y\left(\frac{\partial v}{\partial z}\right)_y = 1$$

a valid relationship, useful in changing a variable, for any four properties; that is, the system has 3 independent variables. In these various equations, any two of the properties x, y, z, or v may interchange positions, but the patterns shown must be maintained.

Another means of obtaining a possibly useful relation is to let $dz = 0$ in equation (11-1) or **(c)**; this gives

(11-6)
$$\left(\frac{\partial y}{\partial x}\right)_z = -\frac{M}{N} \quad \text{or} \quad \left(\frac{\partial x}{\partial y}\right)_z = -\frac{N}{M}$$

MAXWELL RELATIONS* 11.3

Significant property relations, known as the *Maxwell relations*, are found by applying the exactness test to the functions for the properties internal energy u, enthalpy h, Helmholtz function A, and Gibbs function G (§ 5.24). The results apply

* James Clerk Maxwell (1831–1879), another of the colossal minds of the nineteenth century, was born of wealthy parents near Edinburgh. At the age of fifteen, he presented a paper to the Royal Society of Edinburgh on the calculation of the refractive index of a material. He graduated from Cambridge, and by the time he was twenty-nine, he was a professor of natural philosophy at Kings College, London. He wrote voluminously on scientific matters, his greatest contributions being in electromagnetic theory. In thermodynamics, he contributed the Maxwell relations, which are mathematical relationships essential to advanced study of properties. He aided significantly in promulgating kinetic theory and the new science of thermodynamics.

to a pure substance (no chemical reaction) and the symbols will be as for unit mass, but the relations hold as well for 1 mole. From the familiar $T\,ds = du + p\,dv$, we have

(4-14)
$$du = T\,ds - p\,dv$$

where $M = T$ and $N = -p$. Application of the exactness test, equation (11-2), gives

(a)
$$\left(\frac{\partial T}{\partial v}\right)_s = -\left(\frac{\partial p}{\partial s}\right)_v$$

the Maxwell relation I, Table V. In equation (4-14), let $v = C$ $(dv = 0)$, or use equation (11-2), and get

(b)
$$T = \left(\frac{\partial u}{\partial s}\right)_v$$

an expression often considered to be the definition of thermodynamic temperature.

TABLE V   Maxwell Relations and Others, Pure Substance

Function: $du = T\,ds - p\,dv$;

By (11-2): $\left(\dfrac{\partial u}{\partial v}\right)_s = -p = \left(\dfrac{\partial A}{\partial v}\right)_T$ and

By (11-6): $\left(\dfrac{\partial s}{\partial v}\right)_u = \dfrac{p}{T}$

Maxwell relation I: $\left(\dfrac{\partial T}{\partial v}\right)_s = -\left(\dfrac{\partial p}{\partial s}\right)_v$

$\left(\dfrac{\partial u}{\partial s}\right)_v = T = \left(\dfrac{\partial h}{\partial s}\right)_p$

Basic: $du = \left(\dfrac{\partial u}{\partial s}\right)_v ds + \left(\dfrac{\partial u}{\partial v}\right)_s dv$

Function: $dh = T\,ds + v\,dp$;

By (11-2): $\left(\dfrac{\partial h}{\partial s}\right)_p = T = \left(\dfrac{\partial u}{\partial s}\right)_v$ and

By (11-6): $\left(\dfrac{\partial s}{\partial p}\right)_h = -\dfrac{v}{T}$

Maxwell relation II: $\left(\dfrac{\partial T}{\partial p}\right)_s = \left(\dfrac{\partial v}{\partial s}\right)_p$

$\left(\dfrac{\partial h}{\partial p}\right)_s = v = \left(\dfrac{\partial G}{\partial p}\right)_T$

Basic: $dh = \left(\dfrac{\partial h}{\partial s}\right)_p ds + \left(\dfrac{\partial h}{\partial p}\right)_s dp$

Function: $dA = -p\,dv - s\,dT$;

By (11-2): $\left(\dfrac{\partial A}{\partial T}\right)_v = -s = \left(\dfrac{\partial G}{\partial T}\right)_p$ and

By (11-6): $\left(\dfrac{\partial v}{\partial T}\right)_A = -\dfrac{s}{p}$

Maxwell relation III: $\left(\dfrac{\partial p}{\partial T}\right)_v = \left(\dfrac{\partial s}{\partial v}\right)_T$

$\left(\dfrac{\partial A}{\partial v}\right)_T = -p = \left(\dfrac{\partial u}{\partial v}\right)_s$

Basic: $dA = \left(\dfrac{\partial A}{\partial v}\right)_T dv + \left(\dfrac{\partial A}{\partial T}\right)_v dT$

Function: $dG = v\,dp - s\,dT$;

By (11-2): $\left(\dfrac{\partial G}{\partial T}\right)_p = -s = \left(\dfrac{\partial A}{\partial T}\right)_v$ and

By (11-6): $\left(\dfrac{\partial p}{\partial T}\right)_G = \dfrac{s}{v}$

Maxwell relation IV: $\left(\dfrac{\partial v}{\partial T}\right)_p = -\left(\dfrac{\partial s}{\partial p}\right)_T$

$\left(\dfrac{\partial G}{\partial p}\right)_T = v = \left(\dfrac{\partial h}{\partial p}\right)_s$

Basic: $dG = \left(\dfrac{\partial G}{\partial p}\right)_T dp + \left(\dfrac{\partial G}{\partial T}\right)_p dT$

Similarly, with $s = C$ or by equation (11-2),

(c) $$\left(\frac{\partial u}{\partial v}\right)_s = -p$$

By equation (11-6), or by letting $u = C$ in (4-14), we have

(d) $$\left(\frac{\partial s}{\partial v}\right)_u = \frac{p}{T}$$

The other functions are handled first as follows and then with the same mathematical procedures to get the relations summarized in Table V.

DEFINITION: $\quad h = u + pv \quad\quad dh = du + p\,dv + v\,dp$

(4-15) $\quad\quad dh = T\,ds + v\,dp, \quad$ because $\quad du + p\,dv = T\,ds$

DEFINITION: $\quad A = u - Ts \quad\quad dA = du - T\,ds - s\,dT$

(11-7) $\quad\quad dA = -p\,dv - s\,dT \quad$ because $\quad du - T\,ds = -p\,dv$

DEFINITION: $\quad G = h - Ts \quad\quad dG = dh - T\,ds - s\,dT$

(11-8) $\quad\quad dG = v\,dp - s\,dT \quad$ because $\quad dh - T\,ds = v\,dp$

Since there are many other partial derivatives (Hougen, Watson, and Ragatz[1.13] say there are 168 partial derivatives involving p, v, T, s, u, h, A, and G), space forbids an extensive development here.[11.1] For convenience, we shall refer to the Maxwell relations by roman numerals shown in Table V.

Consider Maxwell relations III and IV. By the third law, $s \to 0$ as $T \to 0$. From these relations, we can then say

$$\lim_{T \to 0} \left(\frac{\partial v}{\partial T}\right)_p = 0 \quad\text{and}\quad \lim_{T \to 0} \left(\frac{\partial p}{\partial T}\right)_v = 0$$

which is to say that the $p = C$ curve on the vT plane and the $v = C$ curve on the pT plane have zero slope at $0°R$ (not as pictured in Fig. 6/2 for an ideal gas).

Example—Change of Entropy, T = C, by Equation of State 11.4

Derive an equation giving the change of entropy of a gas that follows the Clausius equation of state, which is

(a) $$p = \frac{RT}{v - b}$$

Solution. Since either of the Maxwell relations III or IV involves the change of entropy at $T = C$, either may be used. Using Maxwell relation III, differentiating **(a)** above with respect to T with $v = C$, we get

(b) $$\left(\frac{\partial p}{\partial T}\right)_v = \frac{R}{v - b} = \left(\frac{\partial s}{\partial v}\right)_T$$

Drop the partial symbols and integrate;

(c)
$$\int_1^2 ds = s_2 - s_1 = R \int_1^2 \frac{dv}{v - b} = R \ln \frac{v_2 - b}{v_1 - b}$$
q.e.d.

11.5   CLAUSIUS-CLAPEYRON EQUATION

As an immediate application of Maxwell's relations, we shall derive the Clausius-Clapeyron equation, also often called the *Clapeyron equation.* One of several ways to derive it is to multiply and divide one side of the Maxwell relation III by T:

(a)
$$\left(\frac{\partial p}{\partial T}\right)_v = \frac{T}{T}\left(\frac{\partial s}{\partial v}\right)_T$$

and think of it with respect to a phase change of a pure substance under equilibrium conditions. For such a process, $T = C$, and the pressure and temperature are not a function of volume (a univariant system). For this application then, the subscripts are superfluous. During a phase change, all extensive properties change proportionally: that is, at an intermediate state, $xh_{fg}/(xv_{fg}) = h_{fg}/v_{fg}$. Therefore we may integrate the right-hand side of **(a)** from saturated liquid to saturated vapor at a particular p, T, and, noting that $T\Delta s = h_{fg}$, we get

(b)
$$\frac{dp}{dT} = \frac{T(s_2 - s_1)}{T(v_2 - v_1)} = \frac{Ts_{fg}}{Tv_{fg}} = \frac{h_{fg}}{Tv_{fg}}$$

where dp/dT is the slope of the curve separating the liquid and vapor phases on the pT plane, Fig. 11/1. This important equation is useful: (1) for checking the consistency of experimental data, (2) especially for determining the latent heat of evaporation h_{fg} from only p, v, T data, (3) for finding any one of the terms in the equation when all others are known, and (4) for a possible definition of absolute temperature, $T = (h_{fg}/v_{fg})(dT/dp)$.

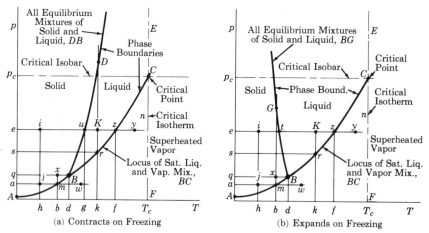

(a) Contracts on Freezing   (b) Expands on Freezing

Fig. 11/1   *Phase Diagram of Fig. 3/6(a) repeated.* The slope of the liquid-solid phase boundary *BD* is positive when the volume change v_{if} increases during melting. If this change is a decrease, as for H_2O, the slope becomes negative.

The same idea applied to the two-phase solid-liquid system and the two-phase solid-vapor system, gives

(c)
$$\frac{dp}{dT} = \frac{h_{if}}{Tv_{if}} \quad \text{and} \quad \frac{dp}{dT} = \frac{h_{ig}}{Tv_{ig}}$$

where subscript i represents the solid (ice) phase, § 3.3, and, for example, v_{ig} is the change of volume during sublimation. Equations **(b)** and **(c)** may be generalized as follows:

(11-9)
$$\frac{dp}{dT} = \frac{\Delta h \text{ (for phase change, heat added)}}{T\Delta v \text{ (for phase change, heat added)}}$$

Thinking of phase changes from liquid to vapor or from solid to vapor, we notice that Δh is always positive and the volume change Δv is always positive; for the reverse change of phase, both changes are always negative. Therefore the slope of phase curves AB and BC of Fig. 11/1 is always positive. For most substances v_{if} is positive during melting, giving a positive slope to BD as pictured in Fig. 11/1.

If the pressure is low enough that the vapor acts nearly as an ideal gas at saturation states and if $v_g \gg v_f$, then $v_{fg} \approx v = RT/p$ and $v_{ig} \approx v = RT/p$, which, substituted into (11-9), gives

(11-10)
$$h_{fg} = \frac{RT^2}{p}\frac{dp}{dT} \quad \text{or} \quad \frac{dp}{p} = \frac{h_{fg}}{R}\frac{dT}{T^2} \qquad [Z \approx 1]$$

also called the Clapeyron equation, an expression for which no volume data are needed. Equation (11-10) is useful at pressures below those found in tables of properties of vapors and elsewhere. For any of the forms involving vapor as one phase, it should be observed (vapor tables) that for small changes in temperature (or p) the heat of vaporization changes little at low pressures, and therefore can often be handled as a constant. There are in the literature, however, empirical equations that give p, h_{fg}, and/or v_{fg} as functions of temperature, which also make it possible to integrate (11-9).

Example—Clapeyron Equation   11.6

With the information at hand, determine the partial pressure of the vapor in atmospheric air at $-10°F$ and 1 atm if the relative humidity is $\phi = 60\%$.

Solution. At $32°F$, we find $p_{\text{sat}} = 0.08859$ psia and $h_{fg} = 1075.5$ Btu/lb. Since the heat of fusion of H_2O is closely 144 Btu/lb, the enthalpy of sublimation at $32°F$ is $h_{ig} = h_{if} + h_{fg} \approx 144 + 1075.5 = 1219.5$ Btu/lb. (Since the problem necessarily involves an approximation, greater precision here is unjustified.) Having already observed that the heat of a change of phase does not change rapidly at low pressure, we assume h_{ig} constant and use equation (11-10); $T_1 = 492°R$, $T_2 = 450°R$, and $R = 85.7$ ft-lb/lb-°R for H_2O. Converting 1219.5 to ft-lb for a consistent energy unit, we have

(a)
$$\int_1^2 \frac{dp}{p} = \frac{h_{ig}}{R}\int_1^2 \frac{dT}{T^2} \quad \text{or} \quad \ln\frac{p_{v2}}{p_{v1}} = -\frac{h_{ig}}{R}\left(\frac{1}{T_2} - \frac{1}{T_1}\right)$$

(b)
$$\ln\frac{p_{v1}}{p_{v2}} = \frac{(1219.5)(778)}{85.7}\left(\frac{1}{450} - \frac{1}{492}\right) = 2.1$$

from which $p_{v1}/p_{v2} = 8.17$; or $p_{v2} = 0.08854/8.17 = 0.01084$ psia (vs. 0.0108 psia in **K** and **K**,[0.7] a good check. Such tabular data as in reference [0.7] are not too common). For $\phi = 0.60$, $p_v(-10°F) = 0.6p_{v2} = (0.6)(0.01064) = 0.00638$ psia.

11.7   COEFFICIENTS OF EXPANSIONS AND COMPRESSIBILITIES; MODULI

There are a number of experimentally determined intensive properties that are particularly useful in dealing with solids and liquids. *These various coefficients are not constant* and must therefore be looked upon as an instantaneous value or an appropriate average value, if a single number is used. One with which the reader is familiar is the (thermal) *coefficient of linear expansion* α, applicable to solids,

$$(11\text{-}11) \qquad \alpha = \frac{1}{L}\left(\frac{\partial L}{\partial T}\right)_p$$

which is the change in length per unit length for one-degree change of temperature, the pressure remaining constant.

Analogous to this concept for a solid, liquid, or gas is the *expansion coefficient* β, also called the *coefficient of volume expansion* and *expansivity*,[1.1] defined by

$$(11\text{-}12) \qquad \beta = \frac{1}{v}\left(\frac{\partial v}{\partial T}\right)_p$$

which is the change in volume per unit volume for a unit change of temperature at constant pressure; the unit is 1/(temperature unit), say $°R^{-1}$. When β is positive, an isothermal increase in pressure results in rejected heat; when β is negative, an isothermal increase in pressure is accompanied by heat added.

The *compressibility* κ or *coefficient of compressibility* (not to be confused with the *compressibility factor*) is the decrease of volume per unit volume for unit increase in pressure, and there are two values in use: one when the entropy remains constant and the other when the temperature remains constant. These properties are defined by

$$(11\text{-}13) \qquad \text{adiabatic compressibility, } \kappa_s = -\frac{1}{v}\left(\frac{\partial v}{\partial p}\right)_s$$

$$(11\text{-}14) \qquad \text{isothermal compressibility, } \kappa_T = -\frac{1}{v}\left(\frac{\partial v}{\partial p}\right)_T$$

for which the unit is 1/(pressure unit). Since $v = 1/\rho$, these properties can be readily expressed in terms of densities; $dv = d(1/\rho) = -d\rho/\rho^2$. Both κ_s and κ_T are positive; the negative signs in (11-13) and (11-14) recognize that the volume decreases as the pressure increases.

The reciprocals of the compressibilities are called *bulk moduli* ζ;

$$(11\text{-}15) \qquad \text{adiabatic bulk modulus, } \zeta_s = -v\left(\frac{\partial p}{\partial v}\right)_s = \rho\left(\frac{\partial p}{\partial \rho}\right)_s$$

(11-16)   isothermal bulk modulus, $\zeta_T = -v\left(\dfrac{\partial p}{\partial v}\right)_T = \rho\left(\dfrac{\partial p}{\partial \rho}\right)_T$

for which the units are the same as the pressure units.

We may obtain a relation involving β and κ_T by writing the p, v, T equivalent of equation (11-4);

(a)
$$\left(\frac{\partial v}{\partial p}\right)_T\left(\frac{\partial T}{\partial v}\right)_p\left(\frac{\partial p}{\partial T}\right)_v = -1$$

Substitute $(\partial v/\partial p)_T = -v\kappa_T$, $(\partial v/\partial T)_p = v\beta$, and, with Maxwell relation III, Table V, get

(b)
$$\left(\frac{\partial p}{\partial T}\right)_v = \frac{\beta}{\kappa_T} = \left(\frac{\partial s}{\partial v}\right)_T$$

Sample value: *methyl alcohol*, $\beta = 1.259 \times 10^{-3}\,\mathrm{K}^{-1}$ at 20°C, $\kappa_T = 79 \times 10^{-6}\,\mathrm{atm}^{-1}$ at 0°C and 1–500 atm, and $\kappa_T = 29 \times 10^{-6}\,\mathrm{atm}^{-1}$ at 0°C and 2500–3000 atm; *gold*, $\beta = 4.41 \times 10^{-9}$ at 0–100°C; *lead*, $\beta = 8.4 \times 10^{-9}$ at 0–100°C; *benzene*, $\kappa_T = 9 \times 10^{-7}$ at 16°C and 8 to 37 atm; CO_2, $\kappa_T = 1.74 \times 10^{-8}$ at 13°C and 60 atm, $\kappa_T = 4.4 \times 10^{-8}$ at 13°C and 90 atm.

Example—Variation of Internal Energy with Pressure   11.8

Find an equation for the change of internal energy with pressure while the temperature is constant.

Solution. What the problem says in mathematical notation is: find $(\partial u/\partial p)_T$. This suggests writing the differential of $u(p, T)$, considering this result with another equation with du in it, as for example, $du = T\,ds - p\,dv$. By equation (11-1),

(a)
$$du = \left(\frac{\partial u}{\partial p}\right)_T dp + \left(\frac{\partial u}{\partial T}\right)_p dT = T\,ds - p\,dv$$

(b)
$$\left(\frac{\partial u}{\partial p}\right)_T = T\frac{ds}{dp} - p\frac{dv}{dp} - \left(\frac{\partial u}{\partial T}\right)_p \frac{dT}{dp}$$

If Δu is to be evaluated at $T = C$, the last term in this equation is zero ($dT = 0$), and we have

(c)
$$\left(\frac{\partial u}{\partial p}\right)_T = T\left(\frac{\partial s}{\partial p}\right)_T - p\left(\frac{\partial v}{\partial p}\right)_T$$

(d)
$$= -T\left(\frac{\partial v}{\partial T}\right)_p + vp\kappa_T$$

(e)
$$= -Tv\beta + vp\kappa_T$$

where Maxwell relation IV and equation (11-14) were used to get (d), and where equation (11-12) permits the final form. With information concerning the properties on the right-hand side, Δu may be found by integration.

11.9   Example—Isothermal Work and Heat for a Solid

To emphasize the generality of the foregoing approach, let a piece of copper be compressed at a constant temperature of 540°R from 100 psia to 20,000 psia. For 1 lb, compute the work, the entropy change, the heat, and the change of internal energy. From the literature,[1.1] we find $\kappa_T = 0.776 \times 10^{-12}$ cm²/dyne and $\beta = 49.2 \times 10^{-6}$ K⁻¹, which are assumed to be average values; also $\rho = 1/v = 555$ lb/ft³.

Solution.   Converting units:

$$\kappa_T = \left(\frac{0.776 \text{ cm}^2}{10^{12} \text{ dyne}}\right)\left(444,800 \frac{\text{dyne}}{\text{lb}}\right)\left(\frac{1}{929} \frac{\text{ft}^2}{\text{cm}^2}\right) = 3.72 \times 10^{-10} \text{ ft}^2/\text{lb}$$

$$\beta = \left(\frac{49.2}{10^6} \frac{1}{\text{K}}\right)\left(\frac{5}{9} \frac{\text{K}}{°\text{R}}\right) = 2.74 \times 10^{-5} °\text{R}^{-1}$$

$$p_1 = (100)(144) = 1.44 \times 10^4 \text{ psf} \qquad p_2 = (20,000)(144) = 288 \times 10^4 \text{ psf}$$

The work of a reversible nonflow process is $\int p\, dv$. After equation (11-1), we write

(a)
$$dv = \left(\frac{\partial v}{\partial p}\right)_T dp + \left(\frac{\partial v}{\partial T}\right)_p dT = -\kappa_T v\, dp + \beta v\, dT$$

where we have used equations (11-14) and (11-12). For points on an isotherm, this equation becomes

(b)
$$dv = \left(\frac{\partial v}{\partial p}\right)_T dp = -\kappa_T v\, dp \qquad\qquad [T = C]$$

Drop the partial notation, but keep in mind the isothermal restraint. Hence, with κ_T and v taken as constant, as they very nearly are, the reversible work becomes

(c)
$$W = \int p\, dv = -\kappa_T v \int p\, dp = -\frac{\kappa_T v}{2}(p_2^2 - p_1^2)$$

$$= -\frac{1}{2}\left(\frac{3.72}{10^{10}}\right)\left(\frac{1}{555}\right)(10^8)(288^2 - 1.44^2) = -2.78 \text{ ft-lb/lb}$$

We shall derive general equations for Δs changes later, but for this application we find that Maxwell relation IV, Table V, involves a change of entropy at constant T, which, together with (11-12) yields

(d)
$$ds = -\left(\frac{\partial v}{\partial T}\right)_p dp = -\beta v\, dp \qquad\qquad [T = C]$$

$$\Delta s = -\beta v(p_2 - p_1) = -\left(\frac{2.74}{10^5}\right)\left(\frac{1}{555}\right)(10^4)(288 - 1.44) = -0.141 \text{ ft-lb/lb-°R}$$

for constant β, v, and T. For constant T and a reversible process.

(e)
$$Q = T\,\Delta s = -T\int \beta v\, dp = (540)(-0.141) = -76.1 \text{ ft-lb/lb}$$

(f)
$$\Delta u = Q - W = -76.1 - (-2.78) = -73.3 \text{ ft-lb/lb}$$

Observe the large decrease of stored energy accompanying this increase of pressure.

IDEAL-GAS EQUATION OF STATE AND THE COEFFICIENTS 11.10

As a step in gaining familiarity with the equations to be developed giving property changes, the reader may apply them to an ideal gas. Differentiating $pv = RT$, we get the various partial derivatives as follows:

(a)
$$p\,dv + v\,dp = R\,dT$$

(b)
$$\left(\frac{\partial v}{\partial p}\right)_T = -\frac{v}{p} \qquad \left(\frac{\partial v}{\partial T}\right)_p = \frac{R}{p} \qquad \left(\frac{\partial p}{\partial T}\right)_v = \frac{R}{v}$$

These expressions may be manipulated in any rational manner; for example, using $p = RT/v$, we get $(\partial v/\partial p)_T = -v^2/(RT)$. Using these partials in the definitions of the coefficients of expansion and compressibility, we find

(c)
$$\beta = \frac{1}{v}\left(\frac{\partial v}{\partial T}\right)_p = \frac{1}{v}\frac{R}{p} = \frac{1}{T} \qquad\qquad [pv = RT]$$

(d)
$$\kappa_T = -\frac{1}{v}\left(\frac{\partial v}{\partial p}\right)_T = -\frac{1}{v}\left(-\frac{v}{p}\right) = \frac{1}{p} \qquad\qquad [pv = RT]$$

If $s = C$, then $pv^k = C$ for constant k; differentiation of the latter equation provides a value of $(\partial v/\partial p)_s$;

(e)
$$[pkv^{k-1}\,dv + v^k\,dp = 0]_s \qquad \text{or} \qquad \left(\frac{\partial v}{\partial p}\right)_s = -\frac{v}{pk}$$

(f)
$$\kappa_s = -\frac{1}{v}\left(\frac{\partial v}{\partial p}\right)_s = \frac{1}{pk} \qquad\qquad [pv = RT]$$

Gas with Compressibility Factor Z. If $Z \neq 1$, the procedure may be quite similar. For example, differentiating $pv = ZRT$ with respect to T and with $p = C$ gives a result applicable to imperfect gases:

(g)
$$\left(\frac{\partial v}{\partial T}\right)_p = \frac{ZR}{p} + \frac{RT}{p}\left(\frac{\partial Z}{\partial T}\right)_p = \frac{R}{p}\left[Z + T\left(\frac{\partial Z}{\partial T}\right)_p\right] \qquad [pv = ZRT]$$

GENERAL EQUATIONS FOR CHANGE OF ENTROPY 11.11

To find an equation for ds in terms of, say, v and T, write [after (11-1); $s = s(v, T)$]

(a)
$$ds = \left(\frac{\partial s}{\partial T}\right)_v dT + \left(\frac{\partial s}{\partial v}\right)_T dv$$

Using equation (2-21), $c_v \equiv (\partial u/\partial T)_v \equiv [(\partial u/\partial s)(\partial s/\partial T)]_v = T(\partial s/\partial T)_v$, where $(\partial u/\partial s)_v = T$ from Table V, we find the coefficient of dT in **(a)** as

(b)
$$\left(\frac{\partial s}{\partial T}\right)_v = \frac{1}{T}\left(\frac{\partial u}{\partial T}\right)_v = \frac{c_v}{T}$$

Substituting $(\partial s/\partial v)_T = (\partial p/\partial T)_v$ from Maxwell relation III, we obtain

$$(11\text{-}17) \qquad ds = \frac{c_v}{T}\,dT + \left(\frac{\partial p}{\partial T}\right)_v dv$$

From this equation, we see that the variation of entropy with volume while T remains constant is

(c) $\qquad\qquad\qquad\qquad \Delta s = \int \left(\frac{\partial p}{\partial T}\right)_v dv$

Check for an ideal gas and compare with equation (6-12).

For the usual reason, an equation for ds in terms of p and T is more useful. To get this, write the differential of $s = s(p, T)$;

(d) $\qquad\qquad\qquad ds = \left(\frac{\partial s}{\partial T}\right)_p dT + \left(\frac{\partial s}{\partial p}\right)_T dp$

Since $c_p \equiv (\partial h/\partial T)_p = [(\partial h/\partial s)(\partial s/\partial T)]_p = T(\partial s/\partial T)_p$, where $(\partial h/\partial s)_p = T$ from Table V, the coefficient of dT in (d) becomes

(e) $\qquad\qquad\qquad \left(\frac{\partial s}{\partial T}\right)_p = \frac{1}{T}\left(\frac{\partial h}{\partial T}\right)_p = \frac{c_p}{T}$

Using also the Maxwell relation IV, $(\partial s/\partial p)_T = -(\partial v/\partial T)_p$, transform equation (d) to

$$(11\text{-}18A) \qquad ds = \frac{c_p}{T}\,dT - \left(\frac{\partial v}{\partial T}\right)_p dp$$

Observe that the partials in either equation (11-17) or (11-18) can be eliminated with suitable equations of state. Compare (11-18A) with equation (11-12) and note that $(\partial v/\partial T)_p = v\beta$; then

$$(11\text{-}18B) \qquad ds = \frac{c_p}{T}\,dT - \beta v\,dp$$

If $T = C$, we have $ds = -[Bv\,dp]_T$, which could be used, for example, for a liquid or solid whose β is known, as in § 11.8. If the entropy $s°$ at some reference state is known, s at any state p, T is then, from (11-18A),

$$(11\text{-}19) \qquad s = s° + \int_{T°}^{T} c_p \frac{dT}{T} - \int_{p°}^{p} \left(\frac{\partial v}{\partial T}\right)_p dp$$

11.12   ENTROPY DEVIATION

It is convenient now, at times, to distinguish not only between an ideal gas and a real gas, but also between an ideal gas and a real gas that is behaving like an ideal gas. To do this, let the properties of the real gas be represented by T, s, h, and so on, the properties of an ideal gas by T^*, s^*, h^*, and so on, and the properties of a real gas behaving (closely) as an ideal gas by $T°$, $p°$, $s°$, $h°$, and so on, which is also a convenient reference state. (Note that the substance may be a liquid or solid, and reference states may be in these phases. We are simply paying particular attention to imperfect gases.) All gases approach the ideal gas as their pressure approaches zero; hence $p° \to 0$ represents some low pressure, where $c_p°$, $c_v°$ are known, that results in nearly ideal-gas behavior, and $s°$ is the entropy of the real gas in a low-pressure region.

Given a knowledge of entropy values at a certain reference state $p°$, $T°$, the second term of (11-18A) could be integrated from state 1 to this pressure at constant $T_1 = C$, that is, 1 to 1° in Fig. 11/2. Then the first term could be integrated at $p° = C$ from 1° to 2° (if $c_p°$ is a known function of T at this pressure—Table I, p. 53), and finally the second term is again integrated (constant T_2) from the low pressure state to the final pressure p_2, 2° to 2. This procedure as depicted in Fig. 11/2 provides an internally reversible path from state 1 to 2. The integrations can be made graphically if sufficient p, v, T data are available.

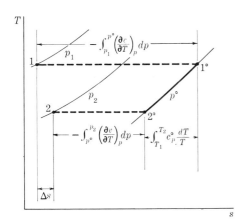

The change of entropy of a real gas via a state $p°$ at 1° and 2° in which the real gas acts nearly as an ideal gas $(c_p° = c_p^*)$.   **Fig. 11/2**

A practical way for gaseous substances involves the change of entropy of an ideal gas that has all the attributes of the real gas except that it acts as an ideal gas should. Use equation (6-13), with the superscript * to differentiate an ideal gas from an actual one, as follows:

(a)
$$ds^* = c_p^* \frac{dT}{T} - R\frac{dp}{p} = d\phi - R\frac{dp}{p}$$

The difference between the entropy of the actual gas, say, state 2, Fig. 11/3, and that of the corresponding ideal gas measured from the same reference state to the same T and p, state 2*, Fig. 11/3, is a deviation from ideal behavior. Thus, at $T = C$ from a known state 2°, to a higher-pressure state, equations (11-18A) and **(a)** become

(b)
$$s - s° = -\int_{p°}^{p} \left(\frac{\partial v}{\partial T}\right)_p dp \quad \text{and} \quad s^* - s° = -R\int_{p°}^{p} \frac{dp}{p}$$

[ACTUAL SUBSTANCE]   [CORRESPONDING IDEAL] where s is the actual entropy, as s_2 in Fig. 11/3; s^* is the entropy of an ideal gas after it has been compressed from a low pressure to the same state (p, T) as the actual gas; and $s°$ is the actual entropy at some known state. By subtraction of the foregoing equations, $s°$ is eliminated, and we get

$$(11\text{-}20\text{A}) \quad \Delta s_d = (s^* - s)_T = \int_{p°}^{p}\left[\left(\frac{\partial v}{\partial T}\right)_p dp - R\frac{dp}{p}\right] = \int_{p°}^{p}\left[\left(\frac{\partial v}{\partial T}\right)_p - \frac{R}{p}\right]dp$$

$$(11\text{-}20\text{B}) \qquad\qquad d(\Delta s_d) = \left[\left(\frac{\partial v}{\partial T}\right)_p - \frac{R}{p}\right]dp$$

the basic definition of **entropy deviation**, § 10.11. An isothermal for an ideal gas is also a change at $h = C$ because $h = f(T)$; thus, between pressures $p°$ and p_2, the ideal gas would change from 2° to 2*, but the actual isothermal results in an actual change from 2° to 2, Fig. 11/3.

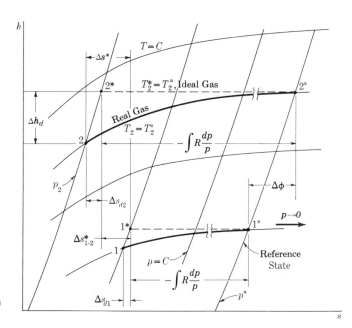

Fig. 11/3   Temperature decreases with throttling.

11.13 ENTROPY DEVIATION BY CORRESPONDING STATES

To convert (11-20) for use by the principle of corresponding states, first introduce the compressibility *factor* by differentiating $pv = ZRT$ with respect to T with $p = C$; this gives equation **(g)** in § 11.10 whose value of $(\partial v/\partial T)_p$ is substituted in (11-20). This gives

$$\textbf{(a)} \qquad (s^* - s)_T = \int_{p°}^{p}\left[\frac{RZ}{p} + \frac{RT}{p}\left(\frac{\partial Z}{\partial T}\right)_p - \frac{R}{p}\right]dp$$

$$= R\int_{p°}^{p}\left[Z - 1 + T\left(\frac{\partial Z}{\partial T}\right)_p\right]\frac{dp}{p}$$

an equation that can be integrated for a particular gaseous substance, given its compressibility factors. Note that s has the same units as R, since the other units cancel on the right side. Introducing reduced coordinates, $T = T_c T_R$, $p = p_c p_R$, and noting that, for example,

(b) $$\frac{dp}{p} = \frac{d(p_c p_R)}{(p_c p_R)} = \frac{dp_R}{p_R} = d(\ln p_R) = d(\ln p)$$

we get

(11-21) $$(s^* - s)_T = \Delta s_d = R \int_{p_R^0}^{p_R} \left[Z - 1 + T_R \left(\frac{\partial Z}{\partial T_R} \right)_p \right] \frac{dp_R}{p_R}$$

which is the value plotted in Items B 22 and B 23. The integration is made graphically from generalized compressibility data (Fig. 10/3).

Referring to Fig. 11/3, we see that the path for computing Δs_{1-2} is 1-1*-2*-2; that is,

(a) $$\Delta s = s_2 - s_1 = (s_2 - s_2^*) + (s_2^* - s_1^*) - (s_1 - s_1^*)$$

(11-22) $$\Delta s = -\Delta s_{d2} + \left(\int c_p^* \frac{dT}{T} - R \ln \frac{p_2}{p_1} \right) + \Delta s_{d1}$$

which is the same as equation (10-10) and in which s, c_p^*, and R have the same units; $s_2^* - s_1^* = \Delta s^*$ is given by equation (6-13). If the entropy in state 2 is desired with respect to a known entropy, the reference state could be state r, and the foregoing equation becomes

(11-23) $$s_2 = s_r + (s_2 - s_2^*) + (s_2^* - s^\circ) - (s_r - s^\circ)$$

$$= s_r - \Delta s_{d2} + \Delta s^* + \Delta s_{dr}$$

if the reference state is an ideal gas state, $\Delta s_{dr} = 0$. If s_r is the absolute entropy, then (11-23) gives s_2 absolute.

Example—Entropy of Compressed Water **11.14**

(a) Compute the approximate change of entropy of 1 lb of water when its state is changed from saturated water at 100°F to a pressure of 3000 psia with $T = C$, using steam table data. (b) Also compute Δs for the same change, using the experimental coefficient $\beta = 205 \times 10^{-6} \, ^\circ R^{-1}$, assumed constant (which it nearly is for this range of Δp).

Solution. (a) Any calculation made on the basis of large increments will be approximate. As a first approximation, consider the liquid to be incompressible, $\kappa_T = -(\partial v/\partial p)_T/v = 0$, and use equation (11-18A). Find the approximate value of $(\partial v/\partial T)_p$ from Item B 13, as follows:

t	v_f	Δv_f	$\dfrac{\Delta v}{\Delta T}$
120°F	0.016204		
		0.000132	$\dfrac{0.000132}{40} = 3.3 \times 10^{-6}$
80°F	0.016072		

Since the saturation pressure changes little in this range of temperature, the value shown is virtually at constant pressure. There are not always enough decimals in vapor table values for the difference of volumes to be accurate; a large Δt will average the error, although the $40°$ interval is not needed in this case. Ignore $p_1 = 0.95$ psia ($\ll 3000$), and get

$$\Delta s_T = -\int \left(\frac{\partial v}{\partial T}\right)_p dp = -\left(\frac{\Delta v}{\Delta T}\right)_p \Delta p$$

$$= -(3.3 \times 10^{-6})\left(\frac{3000 \times 144}{778}\right) = -1.841 \times 10^{-3}\,\text{Btu/lb-°R}$$

where 778 converts the foot-pound energy unit to Btu. Using Table II, p. 70, for compressed water and Item B 13, we find $\Delta s_T = 0.1277 - 0.1295 = -0.0018$ Btu/lb-°R, a more accurate value. This calculation suggests that a helpful correction can be made for the entropy of a compressed liquid in the manner shown.

(b) If the coefficient of volume expansion is available, a correction for Δs of a compressed liquid can be made with equation (11-18B), $T = C$;

$$\Delta s_T = -v\beta\,\Delta p = -\frac{(0.01613)(205 \times 10^{-6})(3000 - 0.95)(144)}{778}$$

$$= -1.77 \times 10^{-3}\,\text{Btu/lb-°R}$$

where v is from the steam tables. Observe that β changes significantly with temperature. Also, it is significant that when β is positive, an increase in pressure results in a decrease of entropy (true of liquid water above about 39°F); when β is negative, an increase in pressure results in an increase of entropy (true of water below about 39°F); when β is zero, a change in pressure produces no change in entropy (true of water at about 39°F).

11.15   CHANGE OF ENTROPY WITH VOLUME, $T = C$

In equation (11-17), we have the coefficient $(\partial p/\partial T)_v$. Write equation (11-4) for p, v, t and get

(a)
$$\left(\frac{\partial p}{\partial T}\right)_v = -\left(\frac{\partial v}{\partial T}\right)_p\left(\frac{\partial p}{\partial v}\right)_T = \frac{-\beta v}{-\kappa_T v} = \frac{\beta}{\kappa_T}$$

Substitute this equivalent into (11-17) and find

(b)
$$ds = \frac{c_v}{T}\,dT + \frac{\beta}{\kappa_T}\,dv$$

and for $T = C$,

(c)
$$ds_T = \left(\frac{\partial p}{\partial T}\right)_v dv = \frac{\beta}{\kappa_T}\,dv$$

Also, we note that $(\partial p/\partial T)_v$ can be obtained from an equation of state.

GENERAL EQUATION FOR CHANGE OF INTERNAL ENERGY   11.16

Substitute the value of ds from equation (11-17) into $du = T\,ds - p\,dv$ and get

(a)
$$du = T\left[c_v\frac{dT}{T} + \left(\frac{\partial p}{\partial T}\right)_v dv\right] - p\,dv$$

(11-24)
$$du = c_v\,dT + \left[T\left(\frac{\partial p}{\partial T}\right)_v - p\right]dv$$

which gives the internal change of a pure substance in terms of p, v, T, and c_v. This equation for du in terms of the independent variables T and v may be integrated with an appropriate equation of state, or we may use the equation **(a)** in § 11.15 and obtain

(b)
$$du = c_v\,dT + \left[\frac{T\beta}{\kappa_T} - p\right]dv$$

from which we see that the change of internal energy at constant temperature is given by

(c)
$$\left(\frac{\partial u}{\partial v}\right)_T = \frac{T\beta}{\kappa_T} - p$$

If du in terms of p and T as the independent variables is desired, substitute ds from equation (11-18A) into $du = T\,ds - p\,dv$ and get

(d)
$$du = T\left[\frac{c_p\,dT}{T} - \left(\frac{\partial v}{\partial T}\right)_p dp\right] - p\,dv$$

Eliminate dv with the differential of $v(T, p)$.

(e)
$$dv = \left(\frac{\partial v}{\partial T}\right)_p dT + \left(\frac{\partial v}{\partial p}\right)_T dp$$

(11-25)
$$du = c_p\,dT - T\left(\frac{\partial v}{\partial T}\right)_p dp - p\left(\frac{\partial v}{\partial T}\right)_p dT - p\left(\frac{\partial v}{\partial p}\right)_T dp$$

$$= \left[c_p - p\left(\frac{\partial v}{\partial T}\right)_p\right]dT + \left[-p\left(\frac{\partial v}{\partial p}\right)_T - T\left(\frac{\partial v}{\partial T}\right)_p\right]dp$$

(f)
$$du = (c_p - pv\beta)\,dT + (pv\kappa_T - Tv\beta)\,dp$$

where we have used (11-12), $\beta = (\partial v/\partial T)_p/v$, and (11-14), $\kappa_T = -(\partial v/\partial p)_T/v$. From equation (11-25), we may get equation **(e)**, § 11.8, for $(\partial u/\partial p)_T$.

The change of internal energy may be set up in terms of reduced coordinates, as is enthalpy below, but we shall assume that the detail given for entropy and enthalpy is sufficient. In this connection, we might recall that

(g) $$u = h - pv = h - ZRT$$

11.17   GENERAL EQUATION FOR ENTHALPY CHANGE

With general equations for ds at hand, (11-17) and (11-18), we may substitute each of these in our familiar equation (4-15), $dh = T\,ds + v\,dp$. The usual form for dh is obtained with equation (11-18); to wit,

(a) $$dh = T\left[c_p\frac{dT}{T} - \left(\frac{\partial v}{\partial T}\right)_p dp\right] + v\,dp$$

(11-26) $$dh = c_p\,dT + \left[v - T\left(\frac{\partial v}{\partial T}\right)_p\right]dp$$

the enthalpy change of any pure substance in terms of p, v, T, and c_p; $(\partial v/\partial T)_p = v\beta$, equation (11-12).

11.18   ENTHALPY DEVIATION

Obtaining the equation of the entropy deviation was somewhat complicated by the fact that the entropy of an ideal gas *is* a function of pressure. Since the ideal gas enthalpy is a function of temperature only, we have $h^* = h°$, Fig. 11/3. The enthalpy deviation is defined as the ideal gas enthalpy minus the actual enthalpy, § 10.10, both measured at the same temperature;

(a) $$d(\Delta h_d) = \left(\frac{\partial h}{\partial p}\right)_T dp$$

That is, since $h^* = h°$, Δh as integrated for $T = C$ from a low pressure ("ideal gas") state is Δh_d. From (11-26), we have

(b) $$d(\Delta h_d) = dh_T = \left[v - T\left(\frac{\partial v}{\partial T}\right)_p\right]dp_T$$

where the subscript T is a reminder of this constraint. Let the enthalpy in the reference state be represented by $h°$. Then, using $pv = ZRT$ and the consequent differential with $p = C$, we get equation **(g)**, § 11.10, to wit,

(c) $$\left(\frac{\partial v}{\partial T}\right)_p = \frac{ZR}{p} + \frac{RT}{p}\left(\frac{\partial Z}{\partial T}\right)_p$$

Using this in equation **(b)**, we find

(d) $$(h - h°)_T = \int_{p°}^{p}\left[v - \frac{ZRT}{p} - \frac{RT^2}{p}\left(\frac{\partial Z}{\partial T}\right)_p\right]dp$$

Since $v = ZRT/p$ and $h° = h^*$, we have, by changing signs,

$$(11\text{-}27)\quad \Delta h_d = (h^* - h)_T = RT^2 \int_{p°}^{p} \left(\frac{\partial Z}{\partial T}\right)_p \frac{dp}{p} = RT^2 \int_{p°}^{p} \left(\frac{\partial Z}{\partial T}\right)_p d(\ln p)$$

an equation that is Δh_d in general in terms of Z for $T = C$ and one that could be graphically integrated with values of Z for a particular substance at hand. In terms of generalized coordinates, $T = T_R T_c$ and $p = p_R p_c$, (11-27) is transformed to

$$(11\text{-}28)\qquad \frac{(h^* - h)_T}{T_c} = \frac{\Delta h_d}{T_c} = RT_R^2 \int_0^{p_R} \left(\frac{\partial Z}{\partial T_R}\right)_p \frac{dp_R}{p_R}$$

$$= RT_R^2 \int_0^{p_R} \left(\frac{\partial Z}{\partial T_R}\right)_p d(\ln p_R)$$

which is the term plotted in Items B 20 and B 21, a term whose units are the same as those for R. The integration can be made graphically with the aid of a generalized compressibility chart, Fig. 10/3.* The results obtained are compatible with true values only insofar as the principle of corresponding states is valid. If the substance passes into or through a two-phase state, the change of its enthalpy may be found along a $p = C$ line involving its latent heat.

The change of enthalpy between any two states is given in terms of enthalpy deviation, $\Delta h_d = h^* - h$, as follows:

$$(e)\qquad h_2 - h_1 = (h_1^* - h_1) + (h_2° - h_1°) + (h_2 - h_2^*)$$

$$(10\text{-}9)\qquad h_2 - h_1 = \Delta h_{d1} + \Delta h^* - \Delta h_{d2}$$

as used in Chapter 10; see Fig. 11/3.

JOULE–THOMSON COEFFICIENT 11.19

The experiment is described in § 6.11, and the Joule-Thomson coefficient $\mu = (\partial T/\partial p)_h$ is readily obtained from equation (11-26) by letting $dh = 0$ and dividing through by dp;

$$(11\text{-}29)\qquad \mu \equiv \left(\frac{\partial T}{\partial p}\right)_h = \frac{1}{c_p}\left[T\left(\frac{\partial v}{\partial T}\right)_p - v\right] = \frac{1}{c_p}[Tv\beta - v] = -\frac{1}{c_p}\left(\frac{\partial h}{\partial p}\right)_T$$

where we have used $(\partial v/\partial T)_p = v\beta$ and where we see by (11-26) that the bracketed term is equal to $(\partial h/\partial p)_T$.

As mentioned in Chapter 6, $\mu = 0$ for an ideal gas (check it), but μ can be zero when the gas is not ideal (even when a change of state occurs). To explain, we consider an isenthalpic curve, Fig. 11/4, which is obtained by starting, say, at a constant p_5 and T_5 upstream from the porous plug (equivalent to p_1 and T_1, Fig.

---

* A procedure: Plot $p_R = C$ lines with Z as ordinate and T_R as abscissa; measure slopes $(\partial Z/\partial T_R)_p$ and plot these as ordinates against $\ln p_R$, a curve for each T_R; the area under this last curve is the integral in (11-2B) between desired limits of p_R.

11/5), and regulating the flow, perhaps by valves, so that the pressure on the downstream side is constant at p_6, p_7, and so on. The corresponding temperatures are read, from which an isenthalpic curve 5-10-6 can be drawn. This procedure is repeated, starting at some other p and T, say p_{12} and T_{12}, until a family of such curves is obtained, if desired. The highest point on each of these $h = C$ curves is where its slope is zero, $(\partial T/\partial p)_h = 0$. A smooth curve drawn through these peaks is the locus of all such points, where $\mu = 0$, and is called the **inversion curve**. We note that the temperature rises during a throttling process that occurs entirely outside the inversion curve, such as 5-11, 5-10, 12-13; the slope μ of the curve is negative. For all isenthalpic state changes entirely within the inversion curve, such as 1-2, 8-6, the temperature decreases upon throttling. A process that proceeds from a state outside the inversion curve to a state inside may experience either a temperature increase (as 5-8), a decrease (as 5-6), or coincidentally have no change of temperature. To cool a gas by throttling then, the initial and final states must be properly related. Generally, the states are such that they fall inside the inversion curve. For small state changes in the vicinity of the inversion curve, the consequences of $\mu = 0$ are almost valid; that is, $T/v = C$ as obtained by setting the first bracket in equation (11-29) equal to zero

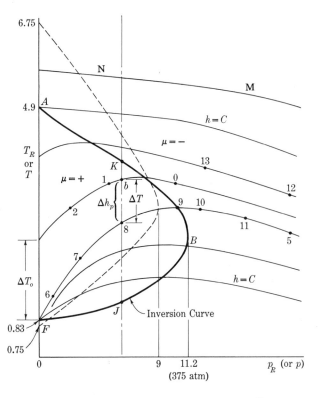

Fig. 11/4   *Inversion Curve.* The solid curves show roughly these characteristics of N_2; the curvature of the $h = C$ lines is exaggerated in parts for emphasis. The dotted inversion curve is about as predicted by the van der Waals reduced equation of state, equation (10-7), § 10.8. Observe that if the temperature is high enough, anywhere on curve *MN*, for example, the throttling process is accompanied by a temperature increase for any pressure range; μ is always positive. Actual properties for N_2 are: $T_A = 1120°R$, $T_F = 186°R$, $T_B = 554°R$, $p_B = 376$ atm. The numbers shown are reduced coordinates. Note that Δp is always negative: μ is positive when ΔT is negative.

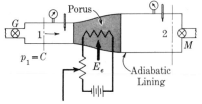

Fig. 11/5   *Modified Joule–Thomson Experiment.* A schematic representation.

and integrating. Sample values of the maximum inversion temperatures at A are: air, $1085°R$; argon, $1300°R$; helium, $\sim 72°R$; hydrogen, $363°R$.

Other charts may be plotted from the data of the experiment; for example μ versus p. In Fig. 11/4, one could measure the slope $\mu = (\partial T/\partial p)_h$ at a number of points, such as 6, 8, 11, and so on, along an $h = C$ line and then plot these slopes against the corresponding pressures to get a curve on the μp plane. The area under such a curve is $\Delta T = \int \mu\, dp]_h$. Plots with coordinates of pressure or temperature against enthalpy may also be made. From the measurements of p and T as described above, μ at any state can be found; then an instantaneous c_p could be computed from equation (11-29), with volume data at hand; but see equations (11-31) and (11-35).

With c_p and μ known, specific or mole volumes can be computed as follows. From equation (11-29), multiply

(a)
$$c_p\mu = T\left(\frac{\partial v}{\partial T}\right)_p - v$$

by dT/T^2, keeping in mind the constant p constraint, and get

(b)
$$\frac{c_p\mu\, dT}{T^2} = \left[\frac{T\, dv - v\, dT}{T^2}\right]_p = d\left(\frac{v}{T}\right)_p$$

from which, with integration from a reference state $v°/T°$ to v/T,

(c)
$$\frac{v}{T} = \frac{v°}{T°} + \int\left(\frac{c_p\mu}{T^2}\, dT\right)_p$$

Integration of **(c)** from $T°$ to T makes possible the computation of v. Also a form of equation **(c)** is used to correct the temperature reading from a constant pressure gas thermometer.

A modified experiment is indicated in Fig. 11/5, wherein there is provided a means of heating the gas passing through the porous plug. If the flow of energy E_c into the plug is regulated just right, the temperature at 2 may be maintained the same as at 1. If data are collected for a number of different final pressures (all at same T), an isotherm can be plotted on the hp plane; other isotherms in a similar manner. The slope of such an isotherm is $(\partial h/\partial p)_T$ $(= \mu_T)$, which can therefore be evaluated. Writing equation (11-4) for h, T, p, we get

(d)
$$\mu_T \equiv \left(\frac{\partial h}{\partial p}\right)_T = -\left(\frac{\partial h}{\partial T}\right)_p\left(\frac{\partial T}{\partial p}\right)_h = -c_p\mu$$

Thus, data from the two experiments permit a computation of c_p. This approach is useful for high pressures where calorimetric measurements are more difficult (but spectroscopic procedures at low pressure are most accurate).

Using $pv = ZRT$ with equation (11-29), we obtain an expression from which a change of Z can be computed by a graphical integration, with values of μ known; or μ can be computed from known Z's. Evidently an equation of state can also be used to evaluate μ from (11-29). The *dotted* inversion curve of Fig. 11/4 is the reduced van der Waals equation, and it shows that this method of determining the inversion temperature is subject to error. The equation of this curve is obtained by getting the partial from the reduced van der Waals equation into (11-29) and equating $\mu = 0$.

In equation (a), § 11.18, we find the partial that defines μ_T; therefore

(e)
$$\Delta h_d = \int \mu_T \, dp = -\int c_p \mu \, dp$$

Reflecting upon the brief discussion above, we conclude that a surprising amount of information can be obtained from these simple flow experiments.

11.20   THE SPECIFIC HEATS

An equation for the specific heat c_v in terms of p, v, T is obtained directly from (11-17) by letting $ds = 0$;

(11-30)
$$c_v = -T\left(\frac{\partial p}{\partial T}\right)_v \left(\frac{\partial v}{\partial T}\right)_s = \frac{\beta T}{\kappa_T}\left(\frac{\partial v}{\partial T}\right)_s$$

the final form by using equation (b), § 11.7. Similarly, if $ds = 0$ in equation (11-18), we get

(11-31)
$$c_p = T\left(\frac{\partial v}{\partial T}\right)_p \left(\frac{\partial p}{\partial T}\right)_s = T\beta_v \left(\frac{\partial p}{\partial T}\right)_s$$

see equation (b) below. The ratio of specific heats is then

(a)
$$k = \frac{c_p}{c_v} = -\frac{(\partial v/\partial T)_p (\partial p/\partial T)_s}{(\partial p/\partial T)_v (\partial v/\partial T)_s} = -\frac{(\partial v/\partial T)_p (\partial p/\partial v)_s}{(\partial p/\partial T)_v}$$

Entropy cannot be deleted entirely from the expression for k. Applying equation (11-4) to the properties p, v, T, we find

(b)
$$\left(\frac{\partial v}{\partial T}\right)_p \left(\frac{\partial T}{\partial p}\right)_v = -\left(\frac{\partial v}{\partial p}\right)_T$$

which, used in equation (a), gives

(11-32)
$$k = \frac{c_p}{c_v} = \frac{(\partial v/\partial p)_T}{(\partial v/\partial p)_s} = \frac{\kappa_T}{\kappa_s}$$

where the isothermal and isentropic compressibilities, § 11.7, have been used. We see that measurements of these compressibilities result in a value of k independent of calorimetric data, the kind of check that experimentalists welcome. For *gases* not too far removed from saturation states, c_p increases rapidly with pressure, and at the higher pressures, ones greater than $p_R = 2$, it may decrease somewhat, but it remains at a much higher value than at 1-atm pressure. Since the value of c_v does not increase so rapidly with pressure as c_p, the ratio k increases materially with pressure—with temperature constant. These observations may be deduced qualitatively from equations (11-30) and (11-31).

To get an equation for the difference of specific heats, one could simply subtract equations (11-31) and (11-30). However, a shorter form is more directly obtained in another way. Consider that equation (11-17) is applied to a constant pressure process;

(c)
$$ds_p = \left[\frac{c_v}{T} dT + \left(\frac{\partial p}{\partial T} \right)_v dv \right]_p$$

From equation (11-18), for $p = C$, we get

(d)
$$ds_p = \left(\frac{c_p}{T} dT \right)_p$$

Since **(c)** and **(d)** both give ds_p, equate them and multiply through by T/dT to get

(11-33)
$$c_p - c_v = T \left(\frac{\partial v}{\partial T} \right)_p \left(\frac{\partial p}{\partial T} \right)_v$$

From equation **(b)** above and using equation **(b)**, § 11.7, we have

(e)
$$\left(\frac{\partial p}{\partial T} \right)_v = - \left(\frac{\partial v}{\partial T} \right)_p \left(\frac{\partial p}{\partial v} \right)_T = \frac{\beta}{\kappa_T}$$

which, substituted into (11-33), gives

(11-34)
$$c_p - c_v = -T \frac{(\partial v / \partial T)_p^2}{(\partial v / \partial p)_T} = -T \frac{(\partial p / \partial T)_v^2}{(\partial p / \partial v)_T} = \frac{T v \beta^2}{\kappa_T}$$

Test this equation (11-33) for an ideal gas. Since v, T, and the isothermal compressibility κ_T from equation (11-14) must be positive, equation (11-34) shows that $c_p - c_v$ is always positive, except when $(\partial v / \partial T)_p = 0$, where $c_p = c_v$ (true for liquid H_2O at maximum density, 39°F). The specific heats are also equal at 0°R; $c_p - c_v \to 0$, as $T \to 0$, as concluded from equation (11-34). Recall that $c_p - c_v = R$, the ideal gas constant for a gas that follows $pv = RT$.

For a low pressure (ideal) gas, the specific heats are functions of temperature only. Since this is so and since the specific heats of many substances at low pressure are available, expressions that show the variations of c_p and c_v along an isothermal are useful. Apply the exactness test (11-3) to equation (11-26) for dh, which is known to be an exact differential;

(11-26)
$$dh = c_p dT + \left[v - T \left(\frac{\partial v}{\partial T} \right)_p \right] dp$$
$$\quad\quad [M] \quad\quad\quad [N]$$

(f)
$$\left(\frac{\partial c_p}{\partial p} \right)_T = \left[\left(\frac{\partial v}{\partial T} \right)_p - T \left(\frac{\partial^2 v}{\partial T^2} \right)_p - \left(\frac{\partial v}{\partial T} \right)_p \right] = -T \left(\frac{\partial^2 v}{\partial T^2} \right)_p$$

Transposing ∂p and integrating the left side, we get

$$(11\text{-}35) \qquad (c_p - c_p^\circ)_T = (\Delta c_p)_T = -T \int_{p^\circ}^{p} \left(\frac{\partial^2 v}{\partial T^2}\right)_p dp$$

where c_p is the value at some pressure p and c_p° is a known value at some reference state at the same temperature and the integral is with $T = C$. Applying equation (11-35) to an ideal gas, we find that $\Delta c_p = 0$; that is, that c_p is not a function of the pressure of an ideal gas. Transform $(\partial^2 v/\partial T^2)_p = [\partial(\partial v/\partial T)/\partial T]_p = [\partial(v\beta)/\partial T]_p = v\beta^2$, where $(\partial v/\partial T)_p = v\beta$; the final form is good only if β is constant. This gives

(g) $\qquad\qquad\qquad (dc_p)_T = -Tv\beta^2 dp \qquad\qquad [\text{LIQUID, SOLID, } \beta = C]$

from which, because β^2, v, and T are always positive, we conclude that c_p for a liquid or solid decreases as the pressure increases.

To get the variation of the constant volume specific heat with volume, apply the exactness test to equation (11-24) for du;

(h) $\qquad\qquad\qquad \left(\frac{\partial c_v}{\partial v}\right)_T = T\left(\frac{\partial^2 p}{\partial T^2}\right)_v$

$$(11\text{-}36) \qquad (c_v - c_v^\circ)_T = (\Delta c_v)_T = T \int_{v^\circ}^{v} \left(\frac{\partial^2 p}{\partial T^2}\right)_v dv$$

where v° is the volume at low pressure p°. Equations (11-35) and (11-36) are recognized as ones that can be interpreted as the *specific heat deviations*, when the change is evaluated from a low pressure p° to some actual higher p, and they can be converted to generalized properties as we have done for Δs_d and Δh_d.

11.21   Example—Difference of Specific Heats for a Solid

For the copper in the example of § 11.9, determine the difference of specific heats and the specific heat at constant volume. From § 11.9, we have $\kappa_T = 3.72 \times 10^{-10} \text{ ft}^2/\text{lb}$, $\beta = 2.74 \times 10^{-5} \text{ °R}^{-1}$, $T = 540°R$, and $\rho = 555 \text{ lb/ft}^3$.

Solution. By the last form of equation (11-34), we have

(a) $\qquad c_p - c_v = \dfrac{Tv\beta^2}{\kappa_T} = \dfrac{T\beta^2}{\rho\kappa_T} = \dfrac{(540)(2.74)^2(10^{-10})}{(555)(3.72)(10^{-10})(778)} = 0.002546 \text{ Btu/lb-°R}$

Observe that the pound unit in κ_T is a force pound (from pressure), and a mass pound in ρ; without 778, the units are therefore ft-lb/lb-°R. From Marks' *Handbook*, we find, for copper, $c = 0.092$ Btu/lb-°R "at room temperature." Considering the difficulty of maintaining a constant volume temperature change, it is quite safe to assume that the given c is c_p. Consequently $c_v = 0.092 - 0.0025 = 0.0895$, a value that substantiates previous remarks about the difference $c_p - c_v$ for solids and liquids often being small. However, one should not draw sweeping conclusions if one is faced with the necessity of using c_v of solids or liquids, because the variation is often greater than 10%; besides, the difference increases with temperature.

In § 11.9, we found the equation for the $p\,dv$ work of a solid [see equation **(c)**] and the equation for the change of entropy for an isothermal process. For an isentropic process, we naturally think of letting $ds = 0$; from equation (11-18A), we get

(a)
$$c_p \frac{dT}{T} = \left(\frac{\partial v}{\partial T}\right)_p dp \quad \text{or} \quad dT = \frac{T_m \beta v}{c_p} dp \qquad [s = C]$$

where one may assume that v and β do not change significantly during the process and where the temperature change is so small that integration with an average T_m, or T_1, is valid. Since the nonflow reversible work is $dW = p\,dv$, we first write the differential of $v(s, p)$:

(b)
$$dv = \left(\frac{\partial v}{\partial s}\right)_p ds + \left(\frac{\partial v}{\partial p}\right)_s dp = -\kappa_s v\,dp \qquad [s = C]$$

in which we let $ds = 0$ and $(\partial v/\partial p)_s = -\kappa_s v$ from equation (11-13). This value of dv into $p\,dv$ gives

(c)
$$W = -\kappa_s v \int_1^2 p\,dp = -\kappa_s v \left(\frac{p_2^2 - p_1^2}{2}\right)$$

For the isentropic $Q = 0$ and $\Delta U = -W$.

For a steady-flow process, $-v\,dp = dK + dP + dW$, equation (4-16), in which $-\int v\,dp$ may be evaluated from $\int p\,dv - \Delta pv$. Also observe that $dh = T\,ds - v\,dp$ reduces to $dh = -v\,dp$ for the isentropic, and that the pressure increase may be large enough to cause a change of volume that should not be overlooked. From the compressibilities, § 11.7, we may write

(d)
$$dv = -\kappa_T v\,dp \qquad \text{and} \qquad dv = -\kappa_s v\,dp$$
$$[T = C] \qquad\qquad\qquad [s = C]$$

one of which was used in equation **(c)**. We have already pointed out the approximation of letting $v = C$ in $-\int v\,dp$ to get $-v\int dp$.

The change of enthalpy for an isothermal process is obtained from equation (11-26) by setting $dT = 0$. Using $(\partial v/\partial T)_p = \beta v$, we have

(e)
$$[dh]_T = v\,dp - \beta Tv\,dp = (1 - \beta T)v\,dp$$

which is especially easy to integrate for those situations where β and v, as well as T, are nearly enough constant that an average value is suitable.

SPECIFIC HEATS OF SOLIDS   11.23

About 150 years ago Dulong and Petit observed the specific heat c of many solid crystalline *elements* to be such that $cM = 6$, which is the molal specific heat C for $M =$ the *atomic* mass; $C = 6\,\text{Btu/lb atom-°R}$ (or cal/gatom-K for elements).

Assume the solid crystal to have the same properties in all directions (isotropic), its atoms to have certain mean locations, about which they vibrate; they cannot move about as gas molecules do. Their vibrations can occur in any direction so that the vibrational energy corresponds to three degrees of freedom, § 2.22. Assuming harmonic vibrations, the average potential energy is the same as the average kinetic energy, § 2.11. Then in accordance with the principle of equipartition of energy, $f = 6$ in equation (2.29), § 2.22, and the energy of the atom is $f\kappa T/2 = 3\kappa T$. On the atomic mass basis, $\bar{u} = f\bar{R}T/2 = 3\bar{R}T$, or $C_v = (\partial \bar{u}/\partial T)_v = 3\bar{R} = 3 \times 1.986 = 5.958$ Btu/pmole-°R (or cal/gatom-K). The gas constant appears in the specific heat for a solid because of Boltzmann's constant κ for a molecule. The agreement with Dulong and Petit was encouraging, but too many monatomic solids possess specific heats that differ materially from 6 at ordinary temperatures, and all of them decrease as the temperature decreases. It was not until the advent of quantum mechanics that an explanation was found, first by Einstein, who derived the quantum theory value of the average energy per degree of freedom, instead of using κT. In Einstein's model, each atom had the same frequency ν, vibrating independently of its neighbors. The value of the frequency was determined so as to make Einstein's equation fit experimental data through a fairly wide temperature range, but it failed at very low temperatures.[5.2]

Debye's model gave a striking improvement at quite low temperatures. He assumed that the vibrations were standing waves not independent of each other, a reasonable assumption considering that the atoms in a crystal are quite close enough to have force interactions, but he did assume that there was a limiting maximum frequency. It is this maximum frequency that is determined so as to make the quantum theory equation match measured values, and it is a characteristic number of the substance.

The Dulong and Petit value serves fairly well at "ordinary" temperatures for monatomic solids whose molecular weight is greater than 40. For unusual states, refer to advanced physics texts and those on statistical thermodynamics.[5.2]

11.24   THE PRIMITIVE SURFACE OF GIBBS

In Chapter 3 we presented several thermodynamic surfaces and phase diagrams relating to the pure substance. Now we will introduce a set of state points that describe the pure substance in the E-V-S region and name the resulting surface the Gibbs primitive surface; details follow.

To define fully this primitive surface we need a relation between the five properties E, V, S, p, and T; this relation already exists in the form of the first law as we apply it to a closed system

(a)
$$dQ = dE + dW_n$$

$$T\,dS = dE + p\,dV$$

(b)
$$dE = T\,dS - p\,dV$$

Now since E, V, and S are properties, it follows that

(c)
$$E = f(S, V) \quad \text{or} \quad f(E, S, V) = 0$$

and we may write

(d) $$dE = (\partial E/\partial S)_V\, dS + (\partial E/\partial V)_S\, dV$$

By comparing equations **(b)** and **(d)**, we note

(e) $$(\partial E/\partial S)_V = T$$

(f) $$(\partial E/\partial V)_S = -p$$

Equations **(e)** and **(f)** may be interpreted geometrically as defining a plane which is tangent to the primitive surface described by equation **(c)**. This is depicted in Fig. 11/6.

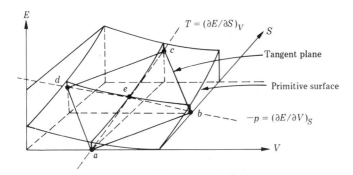

Gibbs Primitive Surface with Tangent Plane.   **Fig. 11/6**

Observe that the plane a-b-c-d is tangent to the primitive surface at point E and lies beneath the surface—it always will be beneath because the surface will be ever concave upwards.

We may now define the Gibbs primitive surface as being a set of state points that represents all homogeneous equilibrium states of a pure substance at rest in the E-V-S region; the pressure p and temperature T are required to describe the degree of curvature of the surface at each point in the set—that is, they describe each of the innumerable planes that are tangent to the surface.

THE DERIVED SURFACE OF GIBBS   11.25

There are a great number of surfaces that may be derived from the Gibbs' primitive surface. For instance, let us relax the requirement of the substance being at rest, that is, all particles are now in motion with a common velocity v. We note then that

$$E = U + v^2/2 \quad \text{or} \quad \Delta E = v^2/2$$

and all other properties remain unchanged as

$$\Delta S = 0 \quad \Delta V = 0.$$

Thus, this derived surface is one that is parallel (so to speak) with the primitive surface but is displaced vertically upward by a value of ΔE to account for the motion.

It may also be noted that if the substance is in more than one phase as solid S, liquid L, and/or vapor V, we may then have a surface like Fig. 11/7.

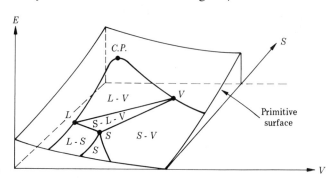

Fig. 11/7   *Primitive and Derived Surfaces.*

11.26   THERMODYNAMIC RELATIONS FROM THE GIBBS PRIMITIVE SURFACE

Now let us develop a relation or two from the primitive surface. Fig. 11/7 reveals the three phases of a pure substance; the triple point is the triangle with the vertices S, L, V.

Consider the liquid-vapor line L-V and note that the tangent plane must touch all along this line since p and T are the same at all of its points. If we construct a plane of constant V through point L and one of constant S through point V, the two planes will intersect each other along a line AB that is parallel to the axis E; see Fig. 11/8(a).

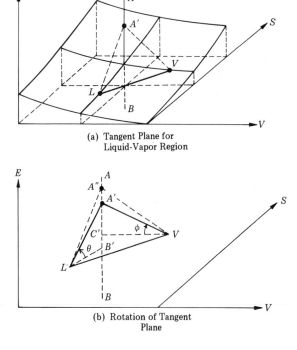

(a) Tangent Plane for
Liquid-Vapor Region

(b) Rotation of Tangent
Plane

Fig. 11/8

From the geometry of Fig. 11/8(b),

$$\tan \theta = \frac{A'B'}{LB'} = T \quad \text{and} \quad \tan \phi = \frac{A'C'}{VC'} = -p$$

With a small rotation of the tangent plane,

$$\tan (\theta + d\theta) = \frac{A'B' + A'A''}{LB'}$$

$$= \frac{A'B'}{LB'} + \frac{A'A''}{LB'} = T + dT$$

and

$$\frac{A'A''}{LB'} = dT$$

Likewise,

$$\tan (\phi + d\phi) = \frac{A'C' + A'A''}{VC'}$$

$$= \frac{A'C'}{VC'} + \frac{A'A''}{VC'} = |-p| + |-dp|$$

and

$$\frac{A'A''}{VC'} = |-dp|$$

But $A'A'' = dT(LB') = dp(VC')$

$$\frac{dT}{dp} = \frac{VC'}{LB'}$$

Now VC' is $v_v - v_L = v_{fg}$

and LB' is $s_v - s_L = s_{fg}$.

Finally,

$$\frac{dT}{dp} = \frac{v_{fg}}{s_{fg}} = \frac{v_{fg}}{h_{fg}/T}$$

or

$$h_{fg} = Tv_{fg} \frac{dp}{dT},$$

the Clausius-Clapeyron relation.

Again note that

$$\frac{A'A''}{LB'} = (dT)_v = (dT)_s$$

and

$$\frac{A'A''}{VC'} = -(dp)_s = -(dp)_v$$

Now

$$\frac{A'A''}{(VC')(LB')} = \frac{A'A''}{(LB')(VC')}$$

$$\frac{(dt)_v}{VC'} = \frac{-(dp)_s}{LB'}$$

Rearranging, we get

$$\frac{(dT)_v}{-(dp)_s} = \frac{VC'}{LB'}$$

But $VC' = (v_{fg})_s$ and $LB' = (s_{fg})_v$

or

$$\frac{(v_{fg})_s}{(s_{fg})_v} = \frac{(dv)_s}{(ds)_v}$$

Therefore,

$$\frac{(dT)_s}{-(dp)_v} = \frac{(dv)_s}{(ds)_v}$$

or $(\partial T/\partial v)_s = -(\partial p/\partial s)_v$, one of the Maxwell relations.

Following with other geometric manipulations, the remaining three Maxwell relations may be developed.

11.27   MIXTURES OF IMPERFECT GASES

A mixture of *ideal* gases (Chapter 12) presents no problem because the various approaches suggested below lead to the same result; that, for example, Dalton's law of partial pressures (§ 6.10) and Amagat's law of partial volumes (§ 12.2) give the same answers. It has been shown that a component ideal gas i in a mixture of gases exerts a partial pressure $p_i = X_i p_m$, where X_i is the mole fraction or volumetric fraction of i, but a check shows that it was assumed that the compressibility factor $Z = 1$ (or is the same for all the components). For gases with different values of Z and for n_i moles of each of i components,

(a) $$p_m = \frac{n_m Z_m \bar{R} T_m}{V_m} = \frac{n_a Z_a \bar{R} T_m}{V_m} + \frac{n_b Z_b \bar{R} T_m}{V_m} + \cdots = \frac{\bar{R} T_m}{V_m} \sum_i n_i Z_i$$

for any number of components, $A, B, ...$, in which the individual terms on the right could be thought of as the pressure exerted by the respective individual components if each should occupy V_m at T_m alone. Since it is customary to *define* the *partial pressure* of an actual gas by $p_i = X_i p_m$, and so on, the terms of equation **(a)** are *not* called *partial* pressures. For a particular definition of Z as given below, the equation can be said to be Dalton's law of *additive pressures* in terms of the compressibility factor Z. The principle of additive pressures can also be expressed in terms of any equation of state, but preferably one explicit in pressure,[1.16] say, van der Waals. For one component i, van der Waals' equation becomes,

(b)
$$p_i = \frac{n_i \bar{R} T_m}{V_m - n_i b} - \frac{n_i^2 a}{V_m^2}$$

obtained by letting $v_i = V_m/n_i$ ft^3/mole of i in equation (10-4); the total pressure is assumed to be $p_m = p_a + p_b + \cdots = \Sigma p_i$. Besides the possibility of using any one of several other equations of state instead of van der Waals, there are other ways of using the additive pressure principle, which we shall not go into here.[1.16]

Amagat's principle of *additive volumes* is $V_m = V_a + V_b + \cdots = \Sigma V_i$, where V_i represents the volume that would be occupied by component i, at p_m and T_m. As for additive pressures, this principle can be used in several ways, including the use of compressibility factors and equations of state, preferably ones explicit in v. These various approaches in general give different answers. Sometimes one approach is better; sometimes, another.

Coming to grips with particular plans, note first that the manner of choosing the individual compressibility factors in equation **(a)** is not defined. In **(a)**, let $V_m = n_m \bar{v}_m$, where n_m is the moles of mixture; then $n_i/n_m = X_i$, the mole fraction of gas i, and so on. Now, considering that Z_m is the compressibility factor of the mixture, we may obtain an estimate of it from **(a)**, which reduces to the following (by dividing through by n_m; $X_i = n_i/n_m$);

(c)
$$Z_m = X_a Z_a + X_b Z_b + \cdots = \sum_i X_i Z_i$$

in which the individual component Z's can be defined in either one of the following two ways (if the volume of 1 mole of constituent i is $\bar{v}_i$, its volume for an X_i mole fraction at its partial pressure is $X_i \bar{v}_i$, which is the volume of 1 mole of mixture $\bar{v}_m$; $X_i \bar{v}_i = \bar{v}_m$):

(d)
$$Z_i = \frac{p_i \bar{v}_i}{\bar{R} T_m} = \frac{p_i \bar{v}_m}{X_i \bar{R} T_m} \qquad \text{[ADDITIVE PRESSURES]}$$

(e)
$$Z_i = \frac{p_m \bar{v}_i}{\bar{R} T_m} \qquad \text{[ADDITIVE VOLUMES]}$$

where there is 1 mole of mixture in **(d)**, and **(e)** is written for 1 mole of i at p_m and T_m. If the values of the individual Z's from **(d)** are substituted in **(c)**, we have

(f)
$$Z_m = \frac{p_m \bar{v}_m}{\bar{R} T_m} = X_a \frac{p_a \bar{v}_m}{X_a \bar{R} T_m} + X_b \frac{p_b \bar{n}_m}{X_b \bar{R} T_m} + \cdots = \frac{\bar{v}_m}{\bar{R} T_m} \sum_i p_i$$

the value of Z_m by additive pressures. Next, substituting Z's from (e) into (c), we find

(g) $$Z_m = \frac{p_m \bar{v}_m}{\bar{R} T_m} = X_a \frac{p_m \bar{v}_a}{\bar{R} T_m} + X_b \frac{p_m \bar{v}_b}{\bar{R} T_m} + \cdots = \frac{p_m}{\bar{R} T_m} \sum_i X_i \bar{v}_i \qquad \begin{array}{l}\text{[ADDITIVE}\\\text{VOLUMES]}\end{array}$$

when the components are at the same p_m and T_m. Equation (g) defines Z in accordance with the additive volume principle.

A procedure in the solution of problems by the foregoing method is: compute p_R, T_R, and/or v_{Ri} for each gas by either or both principles, equation (d) or (e); utilize a chart of compressibility factors; and compute Z_m from (c). One unknown can be determined from $p_m \bar{v}_m = Z_m \bar{R} T_m$. Perhaps a different attack is necessitated by the known data, say a trial solution.

Actually, we have no vindicated theory for handling mixtures of imperfect gases. Without daring to draw a line, we might say that additive volumes tend to give better answers at high reduced pressures; additive pressures, at low reduced pressures. Hougen[1.13] points out that Amagat's rule gives better results for mixtures of N_2 and H_2 but that Dalton's rule is better for mixtures of argon and ethylene. And so it goes. The search for a surer method has gone on unceasingly, with one of the simplest suggestions being Kay's rule[0.8]:

(h) $$p_{cm} = X_a p_{ca} + X_b p_{cb} + \cdots = \sum_i X_i p_{ci}$$

(i) $$T_{cm} = X_a T_{ca} + X_b T_{cb} + \cdots = \sum_i X_i T_{ci}$$

where p_{cm} and T_{cm} are the so-called pseudocritical pressure and temperature of the mixture, X_i is the mole fraction of a constituent, and p_{ci} and T_{ci} are critical values for constituent i. With reduced p_R and T_R for p_{cm} and T_{cm}, a value of Z_m for the mixture can be determined from a generalized compressibility chart. Also a pseudo reduced volume v_{Ri} can be computed from p_{cm} and T_{cm}. Then the mixture is handled as a single pure substance (no changes of phase). Kay's rule, besides being easier to use, usually gives better answers overall than either Amagat's or Dalton's rules applied to the compressibility factor. There are other more involved rules that give somewhat better agreement with experiment, at least in the area tested.[1.5,1.13]

11.28   ELASTIC SYSTEM

We should emphasize the generality of the mathematical approach by looking at systems that are not defined by p, v, T. Considering an elastic system, say a metal rod, with which the reader is probably familiar from a course in strength of materials, we found in § 2.11, equation (f), that $\sigma \varepsilon / 2$ is the stored energy per unit volume that comes from work done on the elastic body; $\sigma = F/A$ is the uniform normal stress, $d\varepsilon = dL/L$, where ε is the unit deformation. In general differential form, this work is $dW = |\sigma\, d\varepsilon|$; if σ is a tensile stress, it is positive, $\Delta \varepsilon$ is an increase, but the work is done *on* the body. Hence, if the elastic member is the system and if we maintain the same sign convention, we say $dW = -\sigma\, d\varepsilon$. Remember that this is work per unit volume, § 2.11.

In § 2.15, we pointed out that dW in the energy equations could be interpreted to be any "kind" of work, equation (2-19). In the elastic system, the only work is that done to change the deformation, as given above ($p\,dv$ work is quite negligible). Therefore, the nonflow energy equation (4-7) becomes

(a) $$dQ = dU - \sigma\,d\varepsilon$$

For reversible processes, $dQ = T\,dS$; hence

(11-38) $$dU = T\,dS + \sigma\,d\varepsilon$$

To get various property relations, handle this equation as was done in Table V. The exactness test gives the equivalent Maxwell relation:

(b) $$\left(\frac{\partial T}{\partial \varepsilon}\right)_s = \left(\frac{\partial \sigma}{\partial S}\right)_\varepsilon$$

(c) $$\left(\frac{\partial U}{\partial S}\right)_\varepsilon = T \qquad\qquad [\text{FROM (11-38)}, \varepsilon = C]$$

(d) $$\left(\frac{\partial U}{\partial \varepsilon}\right)_s = \sigma \qquad\qquad [\text{FROM (11-38)}, S = C]$$

(e) $$\left(\frac{\partial S}{\partial \varepsilon}\right)_U = -\frac{\sigma}{T} \qquad\qquad [\text{FROM (11-38)}, U = C]$$

Reviewing definitions to be used later, we note that the coefficient of linear expansion at constant pressure, § 11.7, becomes (at constant stress in this system— $d\varepsilon = dL/L$)

(f) $$\alpha = \left(\frac{\partial \varepsilon}{\partial T}\right)_\sigma$$

which varies with temperature, but not much for a small ΔT. The familiar Young's modulus (note: at $T = C$) is constant for metals within the elastic limit and at normal temperatures, $E = \sigma/\varepsilon$, but not for all situations. In general then,

(g) $$E = \left(\frac{\partial \sigma}{\partial \varepsilon}\right)_T \quad [E \text{ IS NOT ENERGY IN THIS SECTION}]$$

Write the differential of $\varepsilon(T, \sigma)$ and use these definitions;

(h) $$d\varepsilon = \left(\frac{\partial \varepsilon}{\partial T}\right)_\sigma dT + \left(\frac{\partial \varepsilon}{\partial \sigma}\right)_T d\sigma = \alpha\,dT + \frac{1}{E}\,d\sigma$$

which says that the total change of deformation is that due to temperature change plus that due to stress change. From equation **(h)** for $d\varepsilon = 0$, we find

(i) $$\left(\frac{\partial \sigma}{\partial T}\right)_\varepsilon = -\alpha E$$

which is the case of a member not allowed to change its length while undergoing a temperature change. Since $\alpha = \alpha(T)$, equation **(i)** is easily integrated for $E = C$, $\int d\sigma = -E \int \alpha(T) \, dT$, to find the change in σ, say for a specified change in temperature. For a small ΔT, use an average value of α, constant E, and find $\sigma_2 - \sigma_1 = -\alpha E \, \Delta T$, where a compressive stress is negative. The exactness test applied to the second form of equation **(h)** produces

(j)
$$\left(\frac{\partial \alpha}{\partial \sigma}\right)_T = \left[\frac{\partial(1/E)}{\partial T}\right]_\sigma$$

which gives the variation of α with stress at $T = C$. For metals with E constant, which is virtually true for a small temperature range, we readily conclude that α is not affected by a $\Delta\sigma$. However, at higher temperatures, E begins to decrease at an increasing rate with temperature. For example, with information on α and σ so that the slope $\partial\alpha/\partial\sigma$ could be determined at various temperatures, this slope as ordinate could be plotted against temperature, and the area under the curve between any two temperatures would represent the change of $1/E$ with $\sigma = C$. The same units of force and area in σ and E and temperature in α and T must be used. For most materials, α is positive, but for rubber and elastomers it is negative; E is always positive.

To get an equation for the change of entropy, first write the differential of $S(T, \varepsilon)$;

(k)
$$dS = \left(\frac{\partial S}{\partial T}\right)_\varepsilon dT + \left(\frac{\partial S}{\partial \varepsilon}\right)_T d\varepsilon$$

Differentiate the Helmholtz function, $A = U - TS$, use equation (11-37), and apply the exactness test;

(l)
$$dA = dU - T \, dS - S \, dT = \sigma \, d\varepsilon - S \, dT$$

(m)
$$\left(\frac{\partial \sigma}{\partial T}\right)_\varepsilon = -\left(\frac{\partial S}{\partial \varepsilon}\right)_T$$

Write equation (11-5) for U, T, S with $\varepsilon = C$;

(n)
$$\left(\frac{\partial U}{\partial T}\right)_\varepsilon \left(\frac{\partial T}{\partial S}\right)_\varepsilon \left(\frac{\partial S}{\partial U}\right)_\varepsilon = 1$$

(o)
$$\left(\frac{\partial S}{\partial T}\right)_\varepsilon = \frac{1}{T}\left(\frac{\partial U}{\partial T}\right)_\varepsilon$$

where equation **(c)** was used to get the T. Substitute the equivalents found in **(m)** and **(o)** into **(k)** and get

(p)
$$dS = \frac{1}{T}\left(\frac{\partial U}{\partial T}\right)_\varepsilon dT - \left(\frac{\partial \sigma}{\partial T}\right)_\varepsilon d\varepsilon$$

For an isentropic process, $dS = 0$ and

(q)
$$\left(\frac{\partial T}{\partial \varepsilon}\right)_S = T\left(\frac{\partial T}{\partial U}\right)_\varepsilon \left(\frac{\partial \sigma}{\partial T}\right)_\varepsilon$$

which is the change of temperature with deformation at constant entropy. The ∂U is related to $C_\varepsilon \equiv (\partial U/\partial T)_\varepsilon$, which is closely the constant volume specific heat, a number that is always positive. Also, T is always positive. For metals and most materials, the stress decreases with temperature increase while $\varepsilon = C$; that is, $(\partial \sigma/\partial T)_\varepsilon$ is negative. (This is algebraic. Suppose the initial stress is $\sigma_1 = -10$ ksi, a compressive stress. Let the temperature increase until $\sigma_2 = -20$ ksi. This means a decrease in σ.) On the other hand, rubber has a positive value of $(\partial \sigma/\partial T)_\varepsilon$. It follows that if a rubber band is quickly stretched, $\Delta \varepsilon$ positive, its temperature rises.

If more relations are needed, one may set up a property analogous to enthalpy, $\zeta \equiv U - \sigma \varepsilon$, and then the analogous Gibbs function, $\zeta - TS$; manipulate them as explained for Table V.[1.10]

PARAMAGNETIC SYSTEM 11.29

A material that reacts strongly with a magnet is called a *ferromagnetic material*; examples: iron, nickel, cobalt, liquid oxygen, and a number of metal alloys. Their permeabilities μ are much greater than μ_0, the permeability of free space, § 2.15. Most materials react very weakly in a magnetic field, and are *diamagnetic*, meaning that a free needle of the material in a field assumes a position across the field; its permeability $\mu < \mu_0$. A material whose permeability μ is slightly greater than μ_0 is said to be *paramagnetic*, and it may undergo approximately a reversible magnetizing process. A free needle made of it becomes aligned with the field.[1.27]

Considering the paramagnetic material, assume, as for the elastic system, that $p\,dV$ work is negligible and that the system is the material itself. From § 2.15, we write the work per *unit volume* to change the magnetism as $dW = -\mu_0 \mathbf{H} \cdot d\mathbf{M}$, the negative sign because a work input is required to increase the induced magnetism. The vector $\mathbf{H}$ is the magnetic field strength, and vector $\mathbf{M}$ the magnetic moment; both are properties. (The dot product for collinear vectors is the same as the ordinary product.) The first law for a reversible process then yields

(a)
$$T\,dS = dU - \mu_0 \mathbf{H} \cdot d\mathbf{M}$$

comparable to equation (11-37) for the elastic system. One may define the analogue of enthalpy as $\zeta = U - \mu_0 \mathbf{H} \cdot \mathbf{M}$ and from this an equivalent Gibbs function; also use the Helmholtz function and find numerous property relations by the approach more or less detailed in previous discussions. A simple alternative is to set up the equivalent symbols for the two systems (for example, $p \to -\mu_0 \mathbf{H}$, $V \to \mathbf{M}$, $h \to \zeta$, $A \to A$, $\xi = \zeta - TS \to G$), and then substitute these equivalent symbols in place of those in Table V relations.

In such an analysis, we find that the temperature increases as the material is magnetized, decreases as it is demagnetized, a characteristic that is used to obtain the lowest temperatures, approaching $0°R$. To do this, a paramagnetic salt in a strong magnetic field is cooled to as low a temperature as possible by liquid helium; then, on the removal of the magnetic field, the material's temperature drops to a fraction of a degree. Using cerium-magnesium nitrate, a temperature of 0.006 K can be obtained.[1.11]

11.30   REVERSIBLE CELL

There are a number of reversible cells, a common one being the lead storage battery. If the opposing emf is regulated to be slightly lower or higher than the emf of the cell, current can be made to flow either way and the chemical reaction in the cell may proceed in either direction, depending on the relative magnitude of the voltage applied to the cell. The amount of work that can be obtained from the reaction depends on the number of electrons released during the reaction (each one has the electronic charge) called its *valence*, N_e. For 1 mole, the charge is a Faraday $\mathscr{F} = 96,487$ coulombs for unit valence (1 available electron per molecule), or $N_e\mathscr{F} = \mathcal{Q}$ coulombs per mole for N_e electrons per molecule. This gives the work of the cell from equation (2-17) as,

(11-39) $$W = \mathcal{Q}\mathscr{E} = N_e\mathscr{F}\mathscr{E} \text{ J/gmole.}$$

To get property relations for this system, we have the electrical work by equation (2-17), $dW_e = \mathscr{E}\,d\mathcal{Q}$, where $\mathscr{E}$ is the emf. If the cell gives off a gas, there is $p\,dV$ boundary work (imagine a boundary moving against the surrounding's pressure as the gas is generated); so the system's total work [dW in equation (4-7)] is the sum of these. The first law balance for a reversible process, with $dQ = T\,dS$, is then

(a) $$T\,dS = dE_s + p\,dV - \mathscr{E}\,d\mathcal{Q}$$

where the negative sign is used because the charge decreases as electrical work is delivered. However, the $p\,dV$ work is ordinarily small compared to other energy quantities and is generally dropped, equivalent to assuming $V = C$. Equation (a) then becomes

(b) $$dE_s = T\,dS + \mathscr{E}\,d\mathcal{Q}$$

The analogous symbols here are $-p \to \mathscr{E}$ and $V \to \mathcal{Q}$. Corresponding to Maxwell relations I and II, Table V, we obtain

(c) $$\left(\frac{\partial T}{\partial \mathcal{Q}}\right)_S = \left(\frac{\partial \mathscr{E}}{\partial S}\right)_\mathcal{Q} \quad \text{and} \quad \left(\frac{\partial T}{\partial \mathscr{E}}\right)_S = -\left(\frac{\partial \mathcal{Q}}{\partial S}\right)_\mathscr{E}$$

The second equation (c) would be obtained via the analogous enthalpy $\zeta = E_s - \mathscr{E}\mathcal{Q}$. Let the specific heat at constant charge $\mathcal{Q}$ be $c_\mathcal{Q} \equiv (\partial E_s/\partial T)_\mathcal{Q}$. Then the equation analogous to (11-17) is

(d) $$dS = c_\mathcal{Q}\frac{dT}{T} - \left(\frac{\partial \mathscr{E}}{\partial T}\right)_\mathcal{Q} d\mathcal{Q}$$

If we observe that when p and V are constant, as is true if there is no volume change during the reversible chemical process, then $\Delta E_s = \Delta H$, where E_s is the total stored energy, including chemical. For $G = H - TS$ with $T = C$, differentiation gives

(e) $$dG = dH - T\,dS = T\,dS + \mathcal{E}\,d\mathcal{Q} - T\,dS = \mathcal{E}\,d\mathcal{Q}$$

where equation **(b)** was used to replace dH. For this reversible reaction, the work is $-\Delta G$, as we learned in Chapter 7 by a different argument.

Return to equation **(b)**, let $dE_s = dH$, for $p = C$, $V = C$, and use equation **(d)** to get the enthalpy change at $T = C$;

(f) $$dH = -T\left(\frac{\partial \mathcal{E}}{\partial T}\right)_{\mathcal{Q}} d\mathcal{Q} + \mathcal{E}\,d\mathcal{Q}$$

(g) $$\Delta H = N_e\mathcal{F}\left[\mathcal{E} - T\left(\frac{\partial \mathcal{E}}{\partial T}\right)_{\mathcal{Q}}\right]$$

the integrated form, where $N_e\mathcal{F}$ comes from equation (11-39), given previously. This equation permits the computation of the enthalpy of a reaction from measured electrical data, if the reaction can be made to happen in a cell.

CLOSURE 11.31

The mathematical background of thermodynamics is impeccable, and it can, of course, be carried much further than is done here. As better data are obtained, we can, as you see, be more sure about all related properties.

Since digital computers have become generally available, the mathematical relations and equations of state have acquired new interest, because problems that were previously so forbidding as to discourage attack now become inviting. It is likely true that much of the newer tabulated properties would not be available if it were not for this new tool.

NOTE TO READER. Subsequent chapters may be covered in any desired sequence.

PROBLEMS

SI UNITS

11.1 Test the following expressions and state which of them have characteristics of a property, that is, are point functions: **(a)** $p\,dv$, **(b)** $(p\,dv + v\,dp)$, **(c)** $(3v^2/p^2)\,dv - (2v^3/p)\,dp$, **(d)** $(2x^2y + 5)\,dx + (2/3x^3 + 7)\,dy$, **(e)** $x^3y + 3xy$ where x and y are general properties.

11.2 Prove that work W is not a property. Apply the exactness test to a closed gaseous system undergoing a reversible adiabatic process.

11.3 Mathematics dictates that four point-function variables x, y, z, v are related by the expression $(\partial z/\partial x)_y(\partial x/\partial v)_y(\partial v/\partial z)_y = 1$. Demonstrate the validity of this statement using water properties in the region of 20 bar, 300°C. Replace x, y, z, v with t, p, s, h.

11.4 During the analytical development of the Maxwell relations, four other pairs of equalities are noted and equal respectively to T, v, $-p$, $-s$. Derive these relations.

11.5 Employing Maxwell relation I, $(\partial T/\partial v)_s = -(\partial p/\partial s)_v$, and the mathematical relation in **11.27**, develop the remaining three Maxwell relations.

11.6 Demonstrate the validity of each of the four Maxwell relations by evaluating the eight derivatives for steam in the region of 20 bar 400°C, using finite intervals (Δp) for differentials (dp, ∂p).

11.7 Compute the Gibbs function for about four states of superheated steam at 300°C, plot the Gp curve at constant T, and show that $(\partial G/\partial p)_T = v$, numerically (approximating the differentials by finite intervals). Use 1 bar pressure intervals for the plot.

11.8 Express the coefficients of expansion and compressibility in terms of density instead of specific volume.

11.9 Starting with the basic relation $\beta = (1/v)(\partial v/\partial T)_p$ determine the coefficient of expansion for a gas whose equation of state is

(a) $$pv = RT$$

(b) $$\left(p + \frac{a}{v^2}\right)(v - b) = RT$$

(c) Now develop the expressions for the two coefficients of compressibility, isothermal and isentropic, for these two gases. Assume constant specific heats.

11.10 Demonstrate that the equation of state for a pure substance may be written in the differential form: $dv/v = \beta\, dT - K_T dp$. Remember, $v = f(T, p)$. Now let $\beta = 1/T$ and $K_T = 1/p$ and find the equation of state.

11.11 Show that the difference between the two compressibility coefficients is given by the expression $\kappa_T - \kappa_s = T\beta^2 v/c_p$.

11.12 **(a)** Write the differential of v for $v = f(p, T)$ and express the volume change in terms of the isothermal compressibility and the expansion coefficient. **(b)** Compute the volume change for a unit mass of copper for which $K_T = 0.776 \times 10^{-12}\ \text{cm}^2/\text{dyne}$, $\beta = 49.2 \times 10^{-6}/\text{K}$ and density is $\rho = 8888\ \text{kg/m}^3$; the pressure is increased from 690 kPaa to 137.9 MPaa while the temperature remains constant at 300 K.
Ans. $-1.20 \times 10^{-7}\ \text{m}^3/\text{kg}$.

11.13 **(a)** Start with the entropy function $s = f(T, v)$ and derive the general expression for the entropy change

$$ds = c_v\frac{dT}{T} + \left(\frac{\partial p}{\partial T}\right)_v dv$$

(b) Now use the function $s = f(T, p)$ and obtain the general expression

$$ds = c_p\frac{dT}{T} - \left(\frac{\partial v}{\partial T}\right)_p dp$$

11.14 Using the expression for ds in **11.13(a)**, show that the entropy change along an isotherm takes the form of $ds_T = \beta\, dv/K_T$ for an ideal gas.

11.15 Evaluate the change of entropy along an isotherm for 1 kg of steam in the region of 50 bar, 400°C using the two general expressions in **11.13**. Use finite differences and compare the results of each to the values as taken directly from Item B 16 (SI).

11.16 **(a)** Apply each of the two general entropy equations (see **11.13**), to a van der Waals gas undergoing a process and derive an equation for the change of entropy. **(b)** Now apply the general equations (11-24) and (11-26) to the gas and determine the changes of internal energy and enthalpy for the process.
Ans. **(a)** $ds = c_v dT/T + R\, dv/(v - b)$, **(b)** $du = c_v\, dT + a\, dv/v^2$.

11.17 Find the changes of entropy, internal energy, and enthalpy for 2 kg acetylene undergoing a state change from 15 atm, 278 K, to 30 atm, 556 K, assuming that the acetylene conforms to van der Waals' equation of state. Use the general expressions developed in **11.16** and compare with ideal gas theory.

11.18 Steam at 400°C is compressed isothermally from 35 bar to 110 bar. Determine the changes of enthalpy and entropy by each of these methods and compare: **(a)** extract the enthalpy and entropy values directly from the Mollier chart, Item B 16(SI); **(b)** use the deviation charts.

11.19 Gaseous CO_2 is compressed isentropically from 110 atm, 304 K to 729 atm for storage purposes; the flow rate is 1 kg/min. Use the deviation charts and find **(a)** the final temperature and **(b)** the work required. **(c)** If the gas behaved in an ideal manner, what would be the final temperature and work? Compare answers.

11.20 Steam at 700 bar, 650°C, is throttled slowly through an adiabatic plug to 40 bar; the sink temperature is 38°C. Determine **(a)** the average Joule-Thomson coefficient, **(b)** the change in availability of the flowing steam.
Ans. **(a)** +0.333 K/bar, **(b)** −395 kJ/kg.

11.21 In this exercise, c_p, c_v, and k for steam will be computed. **(a)** Locate these two state points on Item B 16 (SI): 20 bar, 400°C and 30 bar, 450°C. Obtain data and compute $c_v = T(\partial s/\partial T)_v$ using finite differences. **(b)** Determine $c_p = (\partial h/\partial T)_p$ in this region say between 400°C and 450°C at 25 bar. **(c)** Now determine $k = c_p/c_v$ from these two specific heat values.

11.22 Relevant to the Mollier (hs) diagram, see Item B 16 (SI), prove that the constant pressure lines thereon are parallel to each other at their points of intersection with a common isotherm.

11.23 A force B acts on a tight rubber band of length L and does a differential amount of work $dW = B\,dL$ in a reversible process. By manipulation of fundamental mathematical relations, show that $(\partial u/\partial L)_T = B - T(\partial B/\partial T)_L$.

11.24 A gas obeys the equation of state $p(v - b) = RT$. **(a)** Is the difference $c_p - c_v$ variable or constant? Increase or decrease? **(b)** Is the internal energy change or the enthalpy change dependent in any way on the pressure change? **(c)** Will the temperature of this gas change in a throttling process? Increase or decrease? **(d)** Find equations for: the inversion temperature, the expansion coefficient, the isothermal compressibility; **(e)** the Tv relation for an isentropic process, with specific heats constant; **(f)** the isentropic compressibility coefficient.

11.25 An isothermal trace, less than that of critical, is to be plotted on the pV-plane for a van der Waals gas. Select the constants for a given gas from Item B 18, write the virial form of the van der Waals equation, use a pressure range of 1–50 atm, and calculate the various volumes at each pressure and for the selected temperature. Write this computer program.

ENGLISH UNITS

11.26 Employ the exactness test and determine if the differential $dz = (x + 3y)\,dx + (x + 3y)\,dy$ is exact. Integrate it clockwise around the square whose corners are $(0, 0)$, $(0, 2)$, $(2, 2)$, $(2, 0)$. Assuming that dz is an exact differential, what would be the value of this cyclic integration?

11.27 Demonstrate the validity of equation (11-4), $(\partial x/\partial y)_z(\partial y/\partial z)_x(\partial z/\partial x)_y = -1$, by replacing the x, y, z functions with p, v, T properties and then evaluating the three

derivatives from **(a)** the ideal gas equation of state, **(b)** the Redlich-Kwong equation of state given in **10.3**.

11.28 Use the exactness test and demonstrate that heat Q is not a property. Suggest use a closed gaseous system and let it undergo a reversible process.

11.29 Replace the x, y, z functions in the mathematical relation in **11.27** with p, T, S; thence evaluate each of the partial derivatives for steam in the region of 100 psia, 500°F, using appropriate finite intervals (Δp) for differentials ($dp, \partial p$). Is the product of these three ratios in keeping with the equality?

11.30 Develop expressions for the four exact total derivatives which in turn serve as the base for the analytical development of the four Maxwell relations.

11.31 Using the Gibbs primitive surface as described in the EVS-region (see § 11.24) develop the Maxwell relations through trigonometric and descriptive geometric manipulations.

11.32 Demonstrate the validity of each of the four Maxwell relations by evaluating the eight derivatives for Freon F12 in the region of 100 psia, 160°F, using finite differences (as Δp for dp or ∂p); use Item B 35.

11.33 For a vapor obeying the ideal gas equation, show that the change in the saturation vapor pressure is given by $dp/p = h_{fg}\,dT/(RT^2)$ if the volume of the saturated liquid is assumed to be negligible.

11.34 Using the Clapeyron relation, check the latent heat h_{fg} for steam at 140°F using pressure, temperature, and volume data taken from tables.

11.35 Plot the saturated liquid-vapor curve on the pT plane for steam and estimate the slope at a temperature of 50°F. Then on the assumption that steam is an ideal gas, compute the latent heat of vaporization. Discuss any deviation from the table value of h_{fg} from the viewpoint of error in measurement of the plotted slope or error in assumption of ideal gas.

11.36 Compute the approximate pressure needed to maintain an ice-water mixture in equilibrium at 31°F. Assume that Δv and Δh are independent of pressure. From Keenan and Keyes *Steam Tables* we find $v_f = 0.01602$, $v_i = 0.01747\ \text{ft}^3/\text{lb}$, and $h_i =$

−143.8; assume $h_f = -1$ Btu/lb; the initial $p_1 = 0.0847$ psia.   *Ans.* 1084 psia.

11.37 Find the equilibrium temperature of a mixture of ice and liquid water when the pressure is changed from that at the triple point value to that of one atmosphere. See **11.36** for pertinent data.

11.38 Show for an ideal gas that κ_T is k times the value of κ_s where $k = c_p/c_v$.

11.39 Prove that the slope of an isotherm on the βp-plane is equal to and opposite (in signs) the slope of an isobar on the $K_T T$-plane; state any required assumptions.

11.40 (a) For a van der Waals fluid, verify each of the following coefficients:

$$\beta = \frac{Rv^2(v - b)}{RTv^3 - 2a(v - b)^2}$$

$$\kappa_T = \frac{v^2(v - b)^2}{RTv^3 - 2a(v - b)^2}$$

(b) What is the value of κ_T/β expressed in its simplest form? (c) What do the above relations become when $a = 0$, $b = 0$ (ideal gas)?

11.41 Develop the general equations for du and dh employing the expressions given in **11.13**.

11.42 Estimate the change of enthalpy for 1 lb of steam at 500°F undergoing an isothermal process between 50 psia and 400 psia; use only pvT data from the vapor tables and the general equation (11-26) for dh. HINT: plot $p = C$ curves on the vT plane; obtain the slope at the desired T and then find the bracketed part in the general equation between the desired pressures at suitable pressure intervals; finally plot a curve of the bracketed value vs pressure—the area under this curve is the graphical integration of $\int dh_T$.

11.43 (a) Develop the expression for the Joule-Thomson coefficient as follows:

$$\mu = \left(\frac{\partial T}{\partial p}\right)_h = \left(\frac{v}{c_p}\right)[T\beta - 1]$$

(b) In the case of the Joule-Thomson coefficient being constant ($\mu = c$) for a given fluid, prove that the following expression must hold: $h = f(T - \mu p)$.

11.44 Determine the Joule-Thomson coefficient for a gas whose equation of state is (a) $pv = RT$, (b) $p(v - b) = RT$, (c) $(p + a/v^2)(v - b) = RT$.

11.45 Study Item B 34 and state whether the Joule-Thomson coefficient for mercury is positive or negative. Give reasons for your answer.

11.46 Estimate the inversion temperature for steam and hydrogen, each at zero pressure, assuming that van der Waals equation of state governs each fluid.

11.47 Derive the relation for the difference of specific heats which states that $c_p - c_v = vT\beta^2/\kappa_T$.

11.48 Derive the relation for the ratio of specific heats, $c_p/c_v = k = \kappa_T/\kappa_s$ where κ_T and κ_s are the compressibility factors, isothermal and isentropic, respectively.

11.49 Show that for a van der Waals gas,

$$c_p - c_v = \frac{TR^2}{[TR - 2a(v - b)^2/v^3]}$$

11.50 (a) Using the two state points for steam 2600 psia, 900°F and 2900 psia, 1000°F, compute $c_v = T(\partial s/\partial T)_v$. (b) Determine c_p in this region, say between 900°F and 1000°F at 2700 psia. (c) Determine k from these two specific heats.

          Ans. (a) 0.474, (b) 0.732, (c) 1.545.

11.51 Start with the basic definition of c_v and develop the expression which shows that the rate of change of c_v with respect to volume along an isotherm is $(\partial c_v/\partial v)_T = T(\partial^2 p/\partial T^2)_v$.

11.52 The same as **11.51** except that the rate of change of c_p with respect to pressure along an isotherm is shown to be $(\partial c_p/\partial p)_T = -T(\partial^2 v/\partial T^2)_p = -Tv\beta^2$.

11.53 This is an exercise with compressibilities and the expansion coefficient from finite differences. Let the pressure on saturated water at 200°F be increased to 1000 psia. (a) Using the compressed liquid table for H_2O, compute the average isothermal compressibility. (b) Using saturation values at 190°F and 210°F, estimate the expansion coefficient at 200°F. (c) Show that $\kappa_T - \kappa_s = T\beta^2 v/c_p$ and compute κ_s. Estimate c_p from $(\Delta h/\Delta T)_p$ at 190°F and 210°F. (d) If the increase in pressure is brought about nonflow, show that $dW_n = -\kappa_s vp\,dp$, and compute the nonflow work W_n. **(e)** Compute the isentropic rise in temperature. **(f)** Show that the steady flow work $dW_{sf} = dW_n - d(pv)$, and compute W_{sf}.

11.54 Assume 10 lb of liquid water at 200°F are isothermally decompressed to a saturated state; the process is accompanied with a decrease in enthalpy of 30 Btu. Estimate the initial pressure. Compare results with the value found in Table II.

11.55 The system is an elastic wire of cross-sectional area A and length L subjected to a tensile force F. **(a)** Show that

$$\alpha = -\frac{1}{AE_y}\left(\frac{\partial F}{\partial T}\right)_\varepsilon$$

(b) The equation of state for the wire is (K = constant)

$$\sigma = \frac{KT}{A}\left[\frac{L}{L_0} - \left(\frac{L_0}{L}\right)^2\right]$$

where L_0 is the free (unloaded) length. Show that at constant temperature Young's modulus is

$$E_y = \sigma + \frac{3KT}{A}\left(\frac{L_0}{L}\right)^2$$

(c) Using dS from equation **(p)**, § 11.28, and equation (11-38), show that

$$dU = \left(\frac{\partial U}{\partial T}\right)_\varepsilon dT + [\sigma + T\alpha E]\, d\varepsilon.$$

11.56 Find the expression for the partial derivative $(\partial u/\partial p)_T$ for a fluid whose equation of state is **(a)** $pv = RT$, **(b)** $pv = RT + Ap$, where A is a function of temperature only.

12

MIXTURES OF
GASES AND VAPORS

12.1 INTRODUCTION

Having already solved problems with air as the working substance, we know how to handle mixtures of ideal gases, provided we have certain characteristics of the mixture (c_p, c_v, R). So, first we shall investigate how to get such needed properties of ideal gas mixtures, and then look at the more involved problem of a mixture that contains a component that undergoes a phase change during the process.

12.2 DESCRIPTION OF MIXTURES

A **gravimetric analysis** gives the mass fraction of each component in the mixture, applicable to solids, liquids, or gases. For example, the gravimetric fraction of the i component, also called the **mass fraction f_m**, in a mixture is

$$(12\text{-}1) \qquad f_{mi} = \frac{\text{mass of } i \text{ component}}{\text{mass of mixture}} = \frac{m_i}{\sum\limits_{i} m_i} = \frac{m_i}{m_m} \rightarrow \frac{\text{lb}_i}{\text{lb}_m}$$

where f_{mi} is the mass fraction of component i, and $\Sigma\, m_i$ represents the sum of the masses of the i components in the mixture, simply designated m_m; $\Sigma f_{mi} = 1$.

A **volumetric analysis** is frequently used to describe mixtures of gases. **Amagat's law** (also called the *additive volume law*) for ideal gases is: *The total volume of a mixture of gases is equal to the sum of the volumes that would be occupied by the various components each at the pressure p_m and temperature T_m of the mixture.* For example, suppose that three gases X, Y, Z have been separated into compartments, Fig. 12/1, each at p_m and T_m; the corresponding volume of each gas is its *partial volume*; and if the basis of computations is 1 mole of mixture, these partial volumes are also the *mole fractions*, as shown by the following illustration. Apply $pV = n\bar{R}T$ to any one component and to the mixture, and divide;

$$\textbf{(a)} \qquad \frac{p_m V_a}{p_m V_m} = \frac{n_a \bar{R} T_m}{n_m \bar{R} T_m} \qquad \text{or} \qquad \frac{V_a}{V_m} = \frac{n_a}{n_m} = X_a$$

where $n_m = \Sigma\, n_i$. In general form, the mole fraction of the i component of a gas mixture in thermal equilibrium is

(12-2)
$$X_i = \frac{n_i}{\sum_i n_i} = \frac{V_i}{\sum_i V_i} \qquad [\Sigma\, X_i = 1]$$

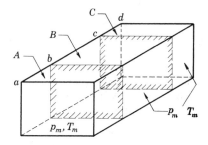

Total volume is V_m. When $V_m = 1\ \mathrm{ft}^3$, the volume of gas   **Fig. 12/1**
A is x_a (or x_A, as desired); of gas B, x_b, of gas C, x_c.

If the mole fraction of the i component in a mixture is X_i, then (for pmoles)

(b)
$$\left(X_i \frac{\mathrm{pmole}_i}{\mathrm{pmole}_m}\right)\left(M_i \frac{\mathrm{lb}_i}{\mathrm{pmole}_i}\right) = X_i M_i \text{ lb of } i \text{ per pmole of mixture}$$

where M_i is the molecular mass of component i. It follows that the *equivalent molecular mass* of the mixture (which is basically defined as the total mass m_m divided by the number of moles of mixture n_m—$M_m = m_m/n_m$) is

(12-3)
$$M_m = X_a M_a + X_b M_b + \cdots = \sum_i X_i M_i$$

for i components; units are, say, lb/pmole of mixture. If the specific gas constant of the mixture is desired, we have

(c)
$$R_m = \frac{\bar{R}}{M_m} \qquad \text{say} \qquad \frac{1545}{M_m}\,\text{ft-lb/lb-}^\circ\text{R}$$

Since (12-3) gives the mass of the mixture and equation **(b)** the mass of a component, the mass fraction of the i component is

(12-4)
$$f_{mi} = \frac{X_i M_i}{\sum_i X_i M_i} \qquad \text{or} \qquad f_{mi} = \frac{n_i M_i}{\sum_i n_i M_i} \qquad [\text{IDEAL GASES}]$$

the means of converting from volumetric analysis to gravimetric analysis. In converting from gravimetric to volumetric fractions, the basic thought process is, say,

(d)
$$\left(f_{mi}\frac{\mathrm{lb}_i}{\mathrm{lb}_m}\right)\left(\frac{\mathrm{pmole}_i}{M_i\,\mathrm{lb}_i}\right) = \frac{f_{mi}}{M_i}\frac{\text{pmoles of } i}{\text{lb of mixture}}$$

From this, we conclude that the total number of moles of mixture per pound of mixture is $\Sigma f_{mi}/M_i$. Therefore, the mole fraction of each i component is

$$(12\text{-}5) \qquad\qquad X_i = \frac{f_{mi}/M_i}{\Sigma\,(f_{mi}/M_i)}$$

In this context, we should recall Dalton's law of additive pressures, §6.10, and a consequence;

$$(6\text{-}11) \qquad\qquad p_m = \sum_i p_i \quad \text{and} \quad X_i = \frac{p_i}{\displaystyle\sum_i p_i} = \frac{p_i}{p_m}$$

To get an expression for the density of a mixture, write $p = \rho RT$, equation (6-4), for each component and add densities for i components as follows, using equations (c) and (6-11) above:

$$(e) \qquad \sum_i \rho_i = \sum_i \frac{p_i}{R_i T_m} = \frac{p_m}{1545\,T_m}\sum_i X_i M_i = \frac{M_m p_m}{1545\,T_m} = \frac{p_m}{R_m T_m} = \rho_m$$

which is to say that the density of the ideal-gas mixture is the sum of the densities of its components, each at its partial pressure.

12.3   Example

A mixture of gases contains 2 moles of O_2, 3 moles of CO, and 5 moles of H_2 at 40°F and 200 psia. For the mixture, determine (a) the volumetric and gravimetric analyses, the equivalent molecular mass, and the specific gas constant; (b) the partial pressure of the O_2; and (c) the volume occupied V_m.

Solution. (a) We shall use rounded off values of the molecular masses because that is satisfactory for many engineering purposes. (See Item B 1.) A tabulation keeps things orderly.

The total number of moles is $2 + 3 + 5 = 10$; therefore the mole fractions are $X_O = 2/10 = 0.2$, $X_{CO} = 3/10 = 0.3$, and $X_H = 5/10 = 0.5$. See column (3) in the tabulation of this example. Now multiply $(X$ moles) $(M$ lb/mole); that is, the values in column (2) by the corresponding values in (3) to get column (4). These numbers in column (4) are the masses of each component per mole of mixture; the sum is therefore the mass of the mixture per mole of mixture or $M_m = 15.8$. (*Ans.*)

(1)	(2)	(3)	(4)	(5)
Gas	M_i	X_i	(3) × (2) (Mass)	f_m $X_iM_i/15.8$
O_2	32	0.2	6.4	0.405
CO	28	0.3	8.4	0.532
H_2	2	0.5	1.0	0.063
Total		1.0	$\Sigma X_i M_i = 15.8$	1.000

The gravimetric analysis in column (5) is obtained by dividing the values in column (4) by the total of column (4); to illustrate, $6.4/15.8 = 0.405$.

$$(a) \qquad\qquad R_m = \frac{1545}{15.8} = 97.9 \text{ ft-lb/lb-}°\text{R}$$

(b) For the partial pressure of O_2, we have $p_O = X_O p_m = (0.2)(200) = 40$ psia.

(c) The total volume is the volume of any constituent, say, O_2;

(b)
$$V_m = \frac{1545 n_O T_m}{p_O} = \frac{(1545)(2)(500)}{(40)(144)} = 268 \text{ ft}^3$$

Just for instructive purposes,

(c)
$$V_m = \frac{1545 n_m T_m}{p_m} = \frac{(1545)(10)(500)}{(200)(144)} = 268 \text{ ft}^3$$

PROPERTIES OF A MIXTURE 12.4

By the Gibbs-Dalton principle, the value of any extensive property of an ideal gas mixture is the sum of those particular properties that each individual component would have if it existed alone in the particular volume. First, consider a unit mass of mixture with f_{mi} as the mass fraction of a component. The specific enthalpy is

(a)
$$h_m = f_{ma} h_a + f_{mb} h_b + f_{mc} h_c + \cdots = \sum_i f_{mi} h_i$$

For a mass m_i of each i component, the total enthalpy of a mixture is $H_m = \sum m_i h_i$. The specific internal energy and the total internal energy are

(b)
$$u_m = f_{ma} u_a + f_{mb} u_b + f_{mc} u_c + \cdots = \sum_i f_{mi} u_i \qquad U_m = \sum_i m_i u_i$$

If we substitute $c_p T$ for h's in equation **(a)** and $c_v T$ for u's in equation **(b)**, and cancel the T's, we get

(c)
$$c_{pm} = f_{ma} c_{pa} + f_{mb} c_{pb} + f_{mc} c_{pc} + \cdots = \sum_i f_{mi} c_{pi}.$$

(d)
$$c_{vm} = f_{ma} c_{va} + f_{mb} c_{vb} + f_{mc} c_{vc} + \cdots = \sum_i f_{mi} c_{vi}.$$

Next consider 1 mole of mixture with a mole fraction x_i of the i component. The molal enthalpy and the total enthalpy are

(e)
$$\bar{h}_m = X_a \bar{h}_a + X_b \bar{h}_b + X_c \bar{h}_c + \cdots = \sum_i X_i \bar{h}_i \qquad H_m = \sum_i n_i \bar{h}_i$$

The molal internal energy and the total internal energy are

(f)
$$\bar{u}_m = X_a \bar{u}_a + X_b \bar{u}_b + X_c \bar{u}_c + \cdots = \sum_i X_i \bar{u}_i \qquad U_m = \sum_i n_i \bar{u}_i$$

Again, substituting $C_p T$ and $C_v T$ into **(e)** and **(f)** and canceling the T's, we get the molal specific heats of a gas mixture as

(g)
$$C_{pm} = X_a C_{pa} + X_b C_{pb} + X_c C_{pc} + \cdots = \sum_i X_i C_{pi}$$

(h)
$$C_{vm} = X_a C_{va} + X_b C_{vb} + X_c C_{vc} + \cdots = \sum_i X_i C_{vi}$$

It is also true that $C_{vm} = C_{pm} - \bar{R}$, equation (6-8); $k_m = c_{pm}/c_{vm}$; and $\bar{u}_m = \bar{h}_m - \bar{R}T$ for an ideal gas. The foregoing relations presume that internal energy is not dependent on the pressure (Joule, § 6.7). However, the entropy is. Equation (6-15) gives the absolute entropy of an ideal gas. By the Gibbs-Dalton principle, the entropy of a mixture of i components is

(i)
$$S_m = \sum_i \left[n_i \left(\bar{s}_i^\circ + \int_{T^\circ}^{T_m} \frac{C_p \, dT}{T} - \bar{R} \ln \frac{p_i}{p^\circ} \right) \right]$$

for n_i moles of each i component. The entropy of 1 mole of the mixture at the *standard state temperature* is

(j)
$$\bar{s}_m = \sum_i \left[X_i \left(\bar{s}_i^\circ - \bar{R} \ln \frac{p_i}{p^\circ} \right) \right]$$

for X_i mole fraction; either n_i or X_i is used as convenient. If p_i is in atmospheres, $p^\circ = 1$. If the mixture is at 1 atm, $p_m = p^\circ = 1$, and $p_i/p_m = p_i/p^\circ = X_i$, equation (6-11), § 12.2. Values of s° are found in Table III, § 6.12, and Item B 11.

If the i components of an ideal gas mixture are initially in individual equilibrium states defined by p_{i1} and T_{i1}, the change of entropy during their adiabatic mixing to an equilibrium state is

(k) $\Delta S = \sum_i \left[n_i \left(\bar{\phi}_{i2} - \bar{\phi}_{i1} - \bar{R} \ln \frac{p_{i2}}{p_{i1}} \right) \right]$ and $\Delta \bar{s} = \sum_i \left[X_i \left(\bar{\phi}_{i2} - \bar{\phi}_{i1} - \bar{R} \ln \frac{p_{i2}}{p_{i1}} \right) \right]$

[TOTAL AMOUNT]   [PER MOLE MIXTURE]

where p_{i2} is the partial pressure of the i component in the mixture and $\bar{\phi}_{i2}$ is for the temperature T_m of the mixture; for constant C_p, we have $\Delta \bar{\phi}_i = C_p \ln (T_m/T_{i1})$. If the components are at the same temperature before and after adiabatic mixing, $\Delta \bar{\phi}_i = 0$, and the ΔS is that owing to the mixing only, called the *entropy of mixing*. If the components are also at 1 atm pressure before mixing and p_i atm afterwards,

(l) $\Delta S = \sum_i \left[n_i(-\bar{R} \ln p_i) \right]$ and $\Delta \bar{s} = \sum_i \left[X_i(-\bar{R} \ln p_i) \right]$

[TOTAL AMOUNT]   [PER MOLE MIXTURE]

The absolute entropy $\bar{s}_m^\circ$ of a mole of mixed gases in the standard state ($p_m = p^\circ = 1$ atm), for $X_i = p_i/p_m$, is

(m)
$$\bar{s}_m^\circ = \sum_i (X_i \bar{s}_i^\circ - X_i \bar{R} \ln X_i)$$

Let us note in passing that when two gases at different temperatures diffuse, we can say that one loses heat to the other, but what happens is that the more energetic molecules transfer molecular energy by impacts with the less energetic ones, until macroscopically we observe only one average molecular translational energy (uniform temperature).

MIXTURES OF GASES WITH A SUBSTANCE UNDERGOING PHASE CHANGE   12.5

There are situations where one of the gases is a vapor close to its saturation state with some change of phase during a process, or where gases are in contact with a liquid (or saturated solid) with a change of phase of some of the liquid (solid) during a process. Examples of everyday experience include the mixture of air and gasoline for an automobile engine, and normal atmospheric air that invariably contains H_2O vapor, usually somewhat superheated. Other components of the air condense if the temperature is sufficiently low.

In these gas-vapor mixtures, the condensable component shall be called the *vapor*, abbreviated v; the gases are differentiated as a class as the *dry gases*, abbreviated dg—or as *dry air*, abbreviated da, when we specifically deal with air and some condensable component. First, there are a few technical terms pertaining to this subject that must be known.

DEW POINT   12.6

Imagine a superheated vapor in some state 1, Fig. 12/2(a), to be contained in a constant pressure system, Fig. 12/2(b), and let it cool isobarically, 1-c. When its state becomes saturated at c, further heat loss will cause some condensation; therefore state c is called the **dew point**. Although any superheated vapor has such a dew point, this term is used only with respect to gas-vapor mixtures. If the temperature at c is below the triple-point temperature, condensation to the solid phase occurs. When the pressure of the vapor at 1, p_{v1}, is known, the dew point is the saturation temperature corresponding to p_{v1}.

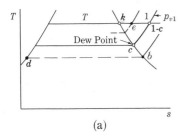

(a)   (b)

Dew Point and Relative Humidity. The   **Fig. 12/2**
Ts diagram depicts the state of the
vapor only.

A superheated vapor may cool to the saturated vapor state by processes other than $p = C$, but in these cases, the process by which the condensation point is reached is spelled out, as, for example, the *dew point at constant volume*.

For a gas-vapor mixture, an isobaric process is one of constant total pressure p_m. In this event, when the gas is cooled below the dew point of its vapor component, the mixture is in thermal equilibrium. If, for example, it goes to t_b, Fig. 12/2(a), some of the vapor must condense and the vapor pressure must now be the saturation of the vapor must condense and the vapor pressure must now be the saturation pressure corresponding to temperature t_b. In most problems, when this occurs, we assume that the mixture contains saturated vapor in state b, and that the saturated liquid in state d has settled out. (You have noticed drops of water on the cold windowpane in the winter because adjacent air has been cooled below the dew point.)

If the pressure of the mixture p_m remains constant and the vapor content decreases or increases, the partial pressures, by Dalton's law, are related as follows, $p_{dg} = p_m - p_v$, where p_{dg} is the partial pressure of the dry gases.

12.7   RELATIVE HUMIDITY

The *relative humidity* ϕ is defined as the ratio of the actual partial pressure of the vapor, say p_{v1}, Fig. 12/2(a), divided by the saturation pressure corresponding to the actual temperature of the mixture (vapor), p_{vk}. When the vapor component is performing closely as an ideal gas and $p_v \approx \rho_v R_v T$, we see that the vapor pressure is (closely) directly proportional to the density ($t_1 = t_k$). We then write*

(12-6)
$$\phi \equiv \frac{\text{actual } p_v \text{ at } T}{\text{saturation } p_v \text{ at } T} = \frac{p_{v1}}{p_{vk}} \approx \frac{\rho_{v1}}{\rho_{vk}} = \frac{v_{vk}}{v_{v1}}$$

where the subscripts refer to Fig. 12/2. Note that relative humidity is a property of a superheated vapor, but one used only for gas-vapor mixtures. In many such mixtures the partial pressure of the vapor is so low that the vapor molecules are widely enough separated that their interaction forces are relatively small (zero for an ideal gas), and the ideal gas laws give reasonably good answers. For example, the partial pressure of the H_2O in the atmosphere is generally well below 1 psia. Partial pressures of steam below 32°F are given in Keenan and Keyes.[0.7]

We commonly speak of a saturated gas (or *saturated air*) meaning that the vapor component in the gas (H_2O in the air) is *saturated vapor*. In this state at a particular temperature, the gas (air) can hold no more vapor (steam); if the temperature of the mixture is raised, more vapor can be added; if the temperature is decreased, some vapor must condense.

12.8   HUMIDITY RATIO

The *humidity ratio* ω (also called *absolute humidity* and *specific humidity*) is the mass of vapor per pound of dry gas;

(12-7)
$$\omega \equiv \frac{m_v}{m_{dg}} = \frac{\rho_v}{\rho_{dg}} \to \frac{\text{mass}_v/\text{ft}^3}{\text{mass}_{dg}/\text{ft}^3} \to \frac{\text{lb v}}{\text{lb dg}}$$

in which $\rho_{dg} = (p/RT)_{dg}$, where p_{dg} is the partial pressure of the dry gas; $\rho_v = 1/v$, in which v_v or ρ_v may be obtained with help from equation (12-6), but preferably from vapor tables if possible. When the vapor approximates an ideal gas, use $\rho = p/(RT)$ in equation (12-7) to get

(a)
$$\omega \equiv \frac{\rho_v}{\rho_{dg}} = \frac{(p_v)(R_{dg}T)}{(R_vT)(p_{dg})} = \frac{p_vR_{dg}}{p_{dg}R_v} = \frac{p_vR_{dg}}{R_v(p_m - p_v)} \frac{\text{lb v}}{\text{lb dg}}$$

Solve for p_v and find

(12-8)
$$p_v = \frac{R_vp_m\omega}{R_{dg} + \omega R_v}$$

* No relation to ϕ in the *Gas Tables.*

applicable at a particular state. If equations **(a)** and (12-8) are applied to an air-steam mixture $R_{dg} = R_a = 53.3$ and $R_v = 1545/18.016 = 85.7$ for H_2O; and we find

(12-9)
$$\omega = \frac{53.3 p_v}{85.7 p_a} = \frac{0.622 p_v}{p_m - p_v} \quad \text{or} \quad p_v = \frac{p_m \omega}{0.622 + \omega}$$

[AIR-STEAM MIXTURE ONLY]

Observe that as the mixture undergoes a process during which the mass of vapor changes, the mass of dry gas remains constant (for a constant mass system), which makes 1 lb of dry gas a rational basis for computations. The corresponding mass of the mixture is $1 + \omega$ lb. The volumetric fraction of steam vapor in air, by equation (12-5), § 12.2, is

(b)
$$X_v = \frac{p_v}{p_m} = \frac{\omega/18}{\omega/18 + 1/29} = \frac{\omega}{0.622 + \omega}$$

only for
H_2O vapor
and air
$\dfrac{M_v}{M_a} = 0.622 = \dfrac{18}{29}$

where $M_a \approx 29$ and $M_v \approx 18$.

ADIABATIC SATURATION PROCESS 12.9

An *adiabatic saturation process* is a steady-flow process at total constant pressure through a control volume for which there is no heat, and in which a liquid and the gas flowing through (presumably) arrive at a local equilibrium state containing saturated vapor before the exit section is reached. In Fig. 12/3, any gas that passes over and/or through the liquid causes the liquid to evaporate until the gas contains the liquid's vapor in a saturated state. Call the entrance section d and the exit section w; at steady-state operation, the input liquid is at temperature t_w, called the adiabatic saturation temperature. Inside, there results an energy interaction between the incoming gas and the liquid, the latent heat of evaporation coming from the gas to evaporate the liquid; temperature $t_w < t_d$. Applying the idea to air-water mixtures, we note that atmospheric air entering the process already has an amount ω_d of H_2O in it. Since this process is irreversible, it is shown dotted in Fig. 12/4, process 1-w ($1 = d$). Since $\Delta K \approx 0$ and since no work is done and $Q = 0$, we write the energy balance for the energy diagram of Fig. 12/3:

(a)
$$h_{ad} + \omega_d h_{vd} + (\omega_w - \omega_d)h_{fw} = h_{aw} + \omega_w h_{vw} \text{ Btu/lb da}$$

Using the enthalpy of evaporation symbol $h_{fgw} = h_{vw} - h_{fw} = h_{gw} - h_{fw}$ at the exit

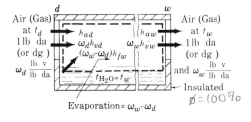

Evaporation = $\omega_w - \omega_d$

Adiabatic Saturation. The vessel is long enough **Fig. 12/3** that there is ample time for the air and water to reach equilibrium with respect to one another (at the adiabatic saturation temperature t_w). The process may be speeded by pumping and spraying the water into the air stream, with only slight deviation from the adiabatic saturation process.

state, solve for ω_d, the humidity ratio of the entering air (gas);

$$(12\text{-}10) \qquad \omega_d = \frac{\omega_w h_{fgw} + h_{aw} - h_{ad}}{h_{vd} - h_{fw}} \text{ lb v/lb da} \qquad \left[\omega_d = \frac{p_{vd}}{p_{ad}}\right]$$

where $h_{aw} - h_{ad} = c_p \, \Delta t_a = -0.24(t_d - t_w)$, the change of specific enthalpy for dry air (when $c_p = 0.24$). The adiabatic saturation temperature depends on the entering pressure, temperature, and humidity ratio, and this saturation state at section w, Fig. 12/3, can be used experimentally to determine the initial state. After solving for ω_d from (12-10), use (12-8) or (12-9) to compute p_v at the initial state. A semiempirical equation was developed by Carrier[12.2] that simplifies the computation of the partial pressure of the vapor for ordinary atmospheric air only:

$$(\textbf{b}) \qquad p_v = p_{vw} - \frac{(p_m - p_{vw})(t_d - t_w)}{2830 - 1.44 t_w} \qquad [\text{CARRIER'S EQUATION}]$$

which is different from Carrier's original equation only in the constants.*

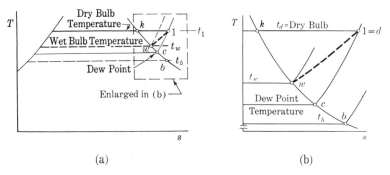

(a)   (b)

Fig. 12/4   *Adiabatic Saturation on Ts Plane.* For some other substances in such a process, ω_d may be zero. The adiabatic saturation temperature t_w is closely approximated by the so-called wet-bulb temperature for atmospheric air (§ 12.10).

12.10   WET-BULB TEMPERATURE

The temperature of a gas-vapor mixture read with a dry thermometer is called the **dry-bulb temperature.** If the bulb of the thermometer is surrounded by a wet gauze, liquid is evaporated as air (gas) is blown over it, Fig. 12/5, causing the temperature to drop in a process similar to the adiabatic saturation process. This lower reading, when properly taken, is called the **wet-bulb temperature.** One of the commonest instruments is a sling **psychrometer**, which consists of two thermometers, one dry bulb and one wet bulb, attached to a handle so that they may be easily whirled about the axis of the handle. The amount by which the moisture on the gauze is cooled, $t_d - t_w$, is called the *wet bulb depression.*

The amount of this depression and the rate of evaporation of the water depends in part on the amount of steam already in the air. If the air is saturated, none of the water on the gauze evaporates because the air is already a saturated mixture with

* Equation **(b)** is derived in some detail in the first and second editions of this book under the title of *Applied Thermodynamics.*

respect to the steam; the wet-bulb and dry-bulb temperatures are the same. The less "moisture" (steam) carried by the air, the more that must be evaporated in order to result in saturation; hence the rate of evaporation being greater, the wet-bulb temperature will be lower.

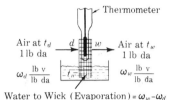

Air at t_d
1 lb da
$\omega_d \dfrac{\text{lb v}}{\text{lb da}}$

Air at t_w
1 lb da
$\omega_w \dfrac{\text{lb v}}{\text{lb da}}$

Water to Wick (Evaporation) = $\omega_w - \omega_d$

Wet-Bulb Thermometer. Approximates adiabatic saturation.   **Fig. 12/5**

The wet·bulb reading with effective equilibrium at the liquid-air interface is affected by: radiation to the thermometer, the velocity of air relative to the thermometer, the design of the instrument, the rate of diffusion of the evaporated water into the air stream, the temperature of the water on the wick, and the heat transfer rates.[12.5] It is fortunate that for the usual atmospheric conditions, these various effects tend to equalize in a way such that the wet-bulb reading is a good approximation of the adiabatic saturation temperature.

PSYCHROMETRIC CHART   12.11

With the wet-bulb and dry-bulb temperatures, the psychrometric chart, Fig. 12/6, on which any point defines a particular state, provides quick answers for many practical problems. See Item B 17 for a working chart. On it are plotted data for finding the amount of steam in the air, the relative humidity, and other useful information. Such charts are designed for a particular total pressure p_m, usually, of course, 1 atm. If the atmospheric pressure is much different from standard, one must return to fundamental theory, say, equations (12-9) and (12-10), or use correction factors[12.1] (which are given on some charts). Since the embryo engineer is interested in learning principles, he will use the chart at this time only for checking purposes.

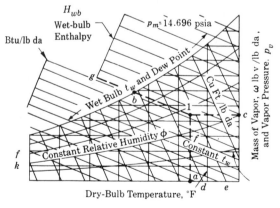

Dry-Bulb Temperature, °F

Form of Psychrometric Chart. The **Fig. 12/6** wet-bulb enthalpy H_{wb} is the enthalpy of the saturated mixture at the wet-bulb temperature. This chart may be entered with various data. Suppose the wet- and dry-bulb temperatures are known; find the wet-bulb temperature at b, the dry-bulb at a, and follow constant temperature lines until they meet at state 1. Move to right or left to c and read values of ω and p_{v1}, estimate the volume between the lines d and e; estimate ϕ between lines f and k; follow constant wet bulb to g and read value of the wet-bulb enthalpy (§ 12.15).

ENTHALPY AND ENTROPY OF SUPERHEATED VAPOR AT LOW PRESSURE   12.12

The ASME S.T. gives properties of superheated steam to pressures beginning at 0.12 psia. In the absence of the detail of these tables, we may assume that superheated steam at pressures below 1 psia is an ideal gas (arbitrary dividing line), which is to say that its enthalpy is a function of temperature only (§ 6.7). Therefore, an approximation that may be convenient for our purposes is (Fig. 12/4 for definition of subscripts)

(12-11) $h_{v1} \approx h_g$ at temperature $t_1 = h_{gk}$ Btu/lb v   [$p \leq 1$ psia]

an idea that may be applied to other vapors when the vapor pressure is "low." In situations where the temperature of a mixture is unknown, it may be a convenience to express the enthalpy of the vapor in terms of temperature to avoid an iterative solution for the unknown temperature; closely, for a t_d°F dry bulb,

(a) $h_v \approx 1061 + 0.445t_d$ Btu/lb v   [SUPERHEATED H_2O VAPOR, $p_v < 1$ psia]

Equation (a) results from plotting h_v (Btu/lb) versus t_d (°F) in rectangular coordinates; a straight line results with 1061 as the vertical intercept and 0.445 the slope.

The entropy of a vapor at low pressure, in the absence of tabular values, say, in state 1, Fig. 12/4, may be estimated from states c or k on the assumption of ideal gas action. Applying equation (6-13) to k-1, we have (entropy $\Delta\phi_{k-1} = 0$, R_v Btu/lb-°R)

(b) $s_1 = s_{gk} - R_v \ln \dfrac{p_{v1}}{p_{vk}} = s_{gk} + R_v \ln \dfrac{p_{vk}}{p_{v1}}$ Btu/°R-lb v

where s_{gk} is the entropy of saturated vapor at state k.

12.13   Example

Water vapor is at 0.5 psia, 100°F. Find its enthalpy using Item B 13 and considering the vapor to respond as an ideal gas; compare value with that obtained using equation (a), § 12.12.

Solution.   From Item B 13, $h_v = 1105.1$ Btu/lb. Using equation (a), § 12.12,

$$h_v = 1061 + (0.445)(100)$$

$$= 1105.5 \text{ Btu/lb}$$

Results compare favorably.

12.14   Example—Properties of Atmospheric Air

The dry- and wet-bulb temperatures of air are found to be $t_d = 83$°F and $t_w = 68$°F. The barometer is $p_m = 29.4$ in. Hg. Determine (a) the humidity ratio, (b) the partial pressure of the vapor (from theory and from Carrier's equation), (c) the relative humidity, (d) the dew point, (e) the density of the air, (f) the density of dry air at the same p_m and t_d, and (g) the entropy of the atmospheric air.

Solution.   (a) *Humidity Ratio.* Let the state be represented by 1, Fig. 12/7. First, obtain values needed for the solution of equation (12-10).

At $t_w = 68$°F $h_{fgw} = 1055.2$ $h_{fw} = 36.05$ Btu/lb

At $t_d = 83$°F $h_{gb} \approx h_{v1} = 1097.7$ Btu/lb

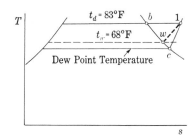

Fig. 12/7

To get ω_w, find ρ_{vw} and ρ_{aw}. At $t_w = 68°F$, the partial pressure of the steam is $p_{vw} = 0.339$ psia (from steam tables) and $v_g = 926.5$ ft^3/lb; or $\rho_{vw} = 1/926.5$ lb/ft^3. The pressure of the dry air at saturation, state w, Fig. 12/7, for $p_m = (0.491)(29.4) = 14.45$ psia, is

(a)
$$p_{aw} = p_m - p_{vw} = 14.45 - 0.339 = 14.111 \text{ psia}$$

(b)
$$\rho_{aw} = \frac{p_{aw}}{R_a T_a} = \frac{(14.11)(144)}{(53.3)(528)} = 0.0721 \text{ lb/ft}^3$$

(c)
$$\omega_w = \frac{\rho_{vw}}{\rho_{aw}} = \frac{1}{(926.5)(0.0721)} = 0.015 \text{ lb v/lb da}$$

which is the humidity ratio *after* adiabatic saturation, state w. Using these various values in equation (12-10), we get the humidity ratio of the original air as

(d)
$$\omega_1 = \frac{(0.015)(1055.2) - (0.24)(83 - 68)}{1097.7 - 36.05} = 0.0115 \text{ lb v/lb da}$$

or $(0.0115)(7000 \text{ grains/lb}) = 80.5$ grains v/lb da. (*Gas Tables* may be used for h_{air}.)

(b) ***Partial Pressure of Vapor.*** From equation (12-9), the vapor pressure p_{v1}, Fig. 12/7, is

(e)
$$p_{v1} = \frac{\omega_1 p_m}{0.622 + \omega_1} = \frac{(0.0115)(14.45)}{0.622 + 0.0115} = 0.262 \text{ psia}$$

From Carrier's equation **(b)**, § 12.9, we have

$$p_{v1} = p_{vw} - \frac{(p_m - p_{vw})(t_d - t_w)}{2830 - 1.44 t_w}$$

(f)
$$= 0.339 - \frac{(14.11)(15)}{2830 - (1.44)(68)} = 0.2615 \text{ psia}$$

(c) ***Relative Humidity.*** By equation (12-6), the relative humidity is

(g)
$$\phi = \frac{p_{v1}}{p_{vb}} = \frac{0.262}{0.5587} = 46.9\%$$

where $p_{vb} = 0.5587$ psia corresponds to the dry-bulb temperature $t_b = 83°F$.

(d) ***Dew Point.*** At c, the vapor pressure is $p_{vc} = p_{v1} = 0.262$ psia. The saturation temperature corresponding to this pressure is found in the steam tables as $60.64°F$, the dew point.

(e) **Density of mixture.** The density of the mixture is the sum of the densities of the constituents; $\rho_m = \rho_a + \rho_v$. For the dry air in state 1, Fig. 12/7,

(h) $p_{a1} = p_m - p_{v1} = 14.45 - 0.262 = 14.188$ psia

(i) $\rho_{a1} = \dfrac{p_{a1}}{R_a T_a} = \dfrac{(14.188)(144)}{(53.3)(543)} = 0.0706$ lb/ft^3

From equation (12-6), we get the density of the vapor at 1 for $v_{gb} = 577.6$ ft^3/lb;

(j) $\rho_{v1} = \phi \rho_{vb} = \dfrac{\phi}{v_{gb}} = \dfrac{0.469}{577.6} = 0.00081$ lb/ft^3

(k) $\rho_m = \rho_{a1} + \rho_{v1} = 0.0706 + 0.00081 = 0.07141$ lb/ft^3

(f) **Density of Dry Air.** For dry air at 29.4 in. Hg $= 14.45$ psia and 83°F,

(l) $\rho_a = \dfrac{p_a}{R_a T_a} = \dfrac{(14.45)(144)}{(53.3)(543)} = 0.0719$ lb/ft^3

It is interesting to note that dry air at a particular temperature and pressure is heavier than atmospheric air at the same temperature and pressure, $0.0719 > 0.07141$. For practice, *check the foregoing answers by a psychrometric chart*, noting that the chart is constructed for a $p_m = 14.696$ psia.

(g) **Entropy.** At 83°F, $s_{gb} = 2.0275$. By equation (b), § 12.12,

(m) $s_{v1} = 2.0275 + 0.11023 \ln \dfrac{0.5587}{0.262} = 2.1109$ Btu/°R-lb v

where, for H_2O, $R_v = 0.11023$ Btu/lb-°R. Using the air table for the entropy of dry air at 543°R ($\sim$83°F) and 29.4 in. Hg, we get

(n) $s_{da} = \phi - R_{da} \ln \dfrac{p}{p^\circ}\Big|_{da} = 0.6021 + R_{da} \ln \dfrac{29.92}{29.4} = 0.6033$ Btu/°R-lb da

For $\omega_1 = 0.0115$ lb v/lb da, the entropy of the mixture (for purposes of *changes* of entropy) is taken as

(o) $S_{m1} = 0.6033 + (0.0115)(2.1109) = 0.6275$ Btu/°R-lb da

As long as the basis is per unit mass of dry gas and the mass of vapor (plus its liquid, if two phases exist at any state considered) is the constant, different datums for dry gas and vapor are all right. Especially at low pressures, the entropy of the liquid may be taken as independent of pressure (use s_f as usual).

12.15   ENTHALPY, INTERNAL ENERGY, AND WET-BULB ENTHALPY
OF A GAS-VAPOR MIXTURE

It is the practice in air conditioning computations to reckon the enthalpy of dry air from 0°F and the enthalpy of steam from the table's datum. Thus the enthalpy of a gas-vapor mixture in general may be taken as

(12-12) $H_m = h_a + \omega h_v = c_p t + \omega h_v$ Btu/lb dg where ω is the humidity ratio at $t°F$, and h_v is obtained directly from tables or in accordance with § 12.12.

By § 12.4 and for the same datum states as above, the internal energy of the mixture is

(a) $$U_m = u_{dg} + \omega u_v = c_v t + \omega(h_v - p_v v_v)$$

Gas table values for h_{dg}, h_a, u_{dg}, u_a are also appropriate.

A property commonly used in commercial computations in air conditioning is the enthalpy of the mixture at the wet-bulb H_{wb}, called the wet-bulb enthalpy;

(b) $$H_{wb} = h_{aw} + \omega_w h_{gw} = 0.24t_w + \omega_w h_{gw} \text{ Btu/lb da}$$

where h_{gw} is the enthalpy of saturated steam and ω_w is the humidity ratio, both at the wet-bulb temperature t_w. Since *wet-bulb enthalpy is a function of the wet-bulb temperature only*, it can be included on a psychrometric chart for a particular p_m without complication, Fig. 12/6. Also, inasmuch as it is nearly equal to the enthalpy of a mixture, it is sometimes used to compute $Q \approx \Delta H_{wb}$ instead of $Q = \Delta h$ in steady-flow processes where $W = 0$ and $\Delta K = 0$. Observe that the right-hand side of equation **(a)**, § 12.9, is H_{wb} and that the enthalpy of the given mixture on the left-hand side is $H_m = h_{ad} + \omega_d h_{vd}$, where the subscripts refer to Fig. 12/3. Thus, wet-bulb enthalpy may be corrected to enthalpy of an unsaturated mixture by subtracting the term $(\omega_w - \omega_d)h_{fw}$;

(c) $$H_m = H_{wb} - (\omega_w - \omega_d)h_{fw} = h_{aw} + \omega_w h_{vw} - (\omega_w - \omega_d)h_{fw} \text{ Btu/lb dg}$$

Example 12.16

For the mixture in the example of § 12.14 ($t_d = 83°F$, $t_w = 68°F$), determine the enthalpy and the wet-bulb enthalpy, for comparison.

Solution. For equation (12-12), we find $h_{v1} = 1097.9$ by a double interpolation in the superheat table (ASME). By the approximation of equation (12-11), we find $h_{v1} \approx h_{g1} = 1097.7$. For the ordinary engineering problem, the difference is not significant. Using $h_{v1} = 1097.9$ as long as we have it, we get

(a) $$H_m = (0.24)(83) + (0.0115)(1097.9) = 32.55 \text{ Btu/lb da}$$

(b) $$H_m = 32.55/1.0115 = 32.2 \text{ Btu/lb mixture}$$

where $1097.9 = h_g \approx h_{vb}$, Fig. 12/7, from the steam tables; $h_{v1} = h_{vb}$. The wet-bulb enthalpy of the mixture is

(c) $$H_{wb} = (0.24)(68) + (0.015)(1091.2) = 32.69 \text{ Btu/lb da}$$

where $1091.2 = h_g$ at $t_w = 68°F$ and $\omega_w = 0.015$ from § 12.14. Check this answer against a psychrometric chart and note that it is somewhat different from the computed value of H_m.

MIXTURES OTHER THAN AIR-STEAM 12.17

Vapor properties for other gas-vapor mixtures are not always at hand. Chemical handbooks will contain enough data for certain purposes, concerning latent heats, specific heats, saturation pressures, and temperatures. The curves of Fig. 12/8 are typical of available data. The use of this information is illustrated by the following example.

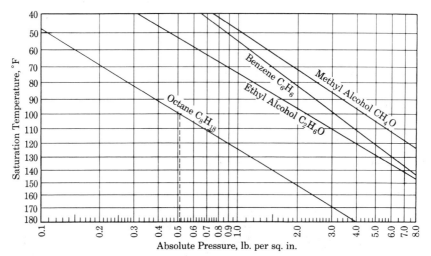

Fig. 12/8   *Pressures and Temperatures of Saturated Vapors.* To use: Enter chart for, say a known temperature of 100°F. Move toward the right along a constant temperature line until the desired curve is reached. Move down to read the corresponding saturated vapor pressure. Following the dotted line for octane, find a pressure of 0.51 psia. The curves are approximately straight lines only on special coordinates.

12.18   Example—Fuel-Air Mixture

A fuel-air mixture in an intake manifold, consisting of octane (C_8H_{18}) and air, is at 122°F and a total pressure of $p_m = 12$ psia. Determine the pounds of air per pound of fuel if (a) the mixture is saturated with respect to the octane, and (b) the relative humidity of the octane is 21%; but first determine the dew point at this humidity.

Solution.   (a) From Fig. 12/8, the partial pressure of the saturated vapor of octane at 122°F is found to be $p_{v3} = 0.95$ psia, Fig. 12/9. The molecular weight of octane is 114.22, from which its ideal gas constant is

(a)
$$R_{\text{octane}} = \frac{1545}{114.22} = 13.5$$

From $p = \rho/RT$, we find the approximate density of octane vapor to be

(b)
$$\rho_{v3} = \frac{p}{RT} = \frac{(0.95)(144)}{(13.5)(582)} = 0.0174 \text{ lb/ft}^3$$

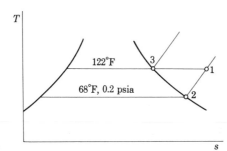

Fig. 12/9

The partial pressure of the air at state 3 of the vapor is $12 - 0.95 = 11.05$ psia, from which the density of the air is

(c) $$\rho_{a3} = \frac{p}{RT} = \frac{(11.05)(144)}{(53.3)(582)} = 0.0513 \text{ lb/ft}^3$$

Then the mass of octane vapor per pound of air (or the humidity ratio) is

(d) $$\omega_3 = \frac{\rho_{v3}}{\rho_{a3}} = \frac{0.0174}{0.0513} = 0.339 \text{ lb v/lb da}$$

This ratio is equivalent to $1/0.339 = 2.95$ lb air/lb fuel. (Such a mixture is not likely in ICE practice. See § 13.6.)

(b) Find the partial pressure of the octane from its relative humidity;

(e) $$\phi = \frac{p_{v1}}{p_{v3}} = 0.21 = \frac{p_{v1}}{0.95}$$

or $p_{v1} = 0.2$ psia. From Fig. 12/8, the corresponding saturation temperature is $t_2 = 68°F$, the dew point. The density of saturated vapor at 122°F has been found as $\rho_{v3} = 0.0174 \text{ lb/ft}^3$. Using the approximation of equation (12-6), $\phi = \rho_{v1}/\rho_{v3}$, we have

(f) $$\rho_{v1} = \phi \rho_{v3} = (0.21)(0.0174) = 0.00363 \text{ lb/ft}^3$$

The partial pressure of the air is $p_a = p_m - p_{v1} = 12 - 0.2 = 11.8$ psia; from which

(g) $$\rho_{a1} = \frac{(11.8)(144)}{(53.3)(582)} = 0.0548 \text{ lb/ft}^3$$

(h) $$\frac{1}{\omega_1} = \frac{\rho_{a1}}{\rho_{v1}} = \frac{0.0548}{0.00363} = 15.1 \text{ lb air/lb v (fuel)}$$

Compare this answer with $r_{a/f}$ in § 13.6.

Example—Air Conditioning Process   12.19

Atmospheric air in state 1 has the following properties: $p_m = 29.92$ in. Hg, $t_d = 100°F$, and $\omega_1 = 0.0295$ lb v/lb da; the rate of flow is 100,000 cfm. This air is cooled in a steady-flow, constant-total-pressure process to 60°F, state 2, Fig. 12/10, and then heated to 85°F, state 3. (a) Determine the relative humidity of the original mixture. (b) How much moisture is deposited during the cooling to 60°F? (c) How much heat is removed during cooling? (d) How many tons of refrigeration are required? (e) What is the volume of the original 100,000 cfm after cooling? (f) What is the relative humidity of the air in state 3, Fig. 12/10? (Except that t_2 would probably be somewhat lower, this problem exemplifies the elementary summer air conditioning process. Use the psychrometric chart to check results.)

Solution. (a) First find the partial pressure p_{v1}, equation (12-9);

(a) $$p_{v1} = \frac{\omega_1 p_m}{0.622 + \omega_1} = \frac{(0.0295)(14.7)}{0.622 + 0.0295} = 0.666 \text{ psia}$$

From the steam tables, we find $p_{vb} = 0.9492$ psia for 100°F (Item B 13) and then

(b) $$\phi_1 = \frac{p_{v1}}{p_{vb}} = \frac{0.666}{0.9492} = 70.2\%$$

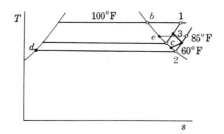

(b) The condensed H_2O is obtained by subtracting ω_2 from ω_1. To get an idea of where state 2 for the H_2O is, we enter the steam tables and find the dew point c, Fig. 12/10, corresponding to $p_{v1} = 0.666$ as $t_c = 88.5°F$. Thus at 60°F, the air has been cooled below the dew point and H_2O has been condensed. At 60°F, $p_{v2} = 0.2561$ psia is taken from the tables; then from equation (12-9), we get

(c)
$$\omega_2 = \frac{0.622 p_{v2}}{p_m - p_{v2}} = \frac{(0.622)(0.2561)}{14.7 - 0.2561} = 0.01103 \text{ lb v/lb da}$$

(d)
$$\omega_1 - \omega_2 = 0.0295 - 0.01103 = 0.01847 \text{ lb v/lb da}$$

For the $\dot{V}_1 = 100,000$ cfm, the mass of dry air is

(e)
$$\dot{m}_a = \frac{p_{a1}\dot{V}_{a1}}{R_a T_1} = \frac{(14.7 - 0.666)(144)(100,000)}{(53.3)(560)} = 6770 \text{ lb/min}$$

The condensation then amounts to

(f)
$$\frac{(6770)(0.01847)}{8.33 \text{ lb/gal } H_2O} = 15 \text{ gpm}$$

which is equivalent to 900 gal/hr.

(c) The heat transferred in a steady-flow process where $W = 0$ and $\Delta K = 0$ is $Q = \Delta H$. The enthalpy of the vapor in the original air is $h_{v1} \approx h_{gb} = 1105.1$ Btu/lb v at 100°F; or

(g)
$$H_{v1} = \omega_1 h_{v1} = (0.0295)(1105.1) = 32.6 \text{ Btu/lb da}$$

At state 2, the enthalpy of the H_2O should account for the condensed steam; that is,

(h)
$$H_{v2} = \omega_2 h_{g2} + (\omega_1 - \omega_2)h_{f2}$$
$$= (0.01103)(1087.7) + (0.01847)(28.06) = 12.515 \text{ Btu/lb da}$$

where h_{g2} and h_{f2} correspond to a temperature of $t_2 = 60°F$. By equation (12-12),

(i)
$$Q = \Delta H_m = c_p(t_2 - t_1) + H_{v2} - H_{v1}$$
$$= 0.24(60 - 100) + 12.515 - 32.6 = -29.7 \text{ Btu/lb da}$$

The heat removed per pound of original mixture is $29.7/1.0295 = 28.85$ Btu/lb mixture, closely approximated by ΔH_{wb}, which differs by about 4%.

(d) The total amount of heat transferred is therefore

(j)
$$m_a \, \Delta H_m = (6770)(29.7) = 201,000 \text{ Btu/min}$$

One ton of refrigeration is defined as cooling at the rate of 200 Btu/min (§ 18.4); therefore the ideal refrigeration capacity required is $201,000/200 = 1005$ tons.

(e) In 100,000 cfm of atmospheric air, there are also 100,000 cfm of dry air and of vapor, and so on; hence the volume of the mixture after it is cooled is

(k) $$V_{m2} = V_{a2} = \frac{m_a R_a T_2}{p_{a2}} = \frac{(6770)(53.3)(520)}{(14.44)(144)} = 90,300 \text{ cfm}$$

The foregoing computation can be checked using the steam: $(6770\omega_2 = \text{lb v})$

(l) $$(m_a \text{ lb da})(\omega_2 v_{g2} \text{ ft}^3 \text{ v/lb da}) = (6770)(0.01103)(1207.6) = 90,200 \text{ cfm}$$

(f) The partial pressure of the steam remains constant during the heating from 2 to 3; so $p_{v3} = p_{v2} = 0.2561$ psia. Saturated steam at 85°F has a pressure of $p_{ve} = 0.5958$ psia, Item B 13. The relative humidity at 3 is

(m) $$\phi_3 = \frac{p_{v3}}{p_{ve}} = \frac{0.2561}{0.5958} = 43\%$$

Example—Constant Volume Process **12.20**

A 10-ft^3 tank contains atmospheric air compressed to $p_{m1} = 100$ psia, $t_1 = 300°F$, and with $\phi_1 = 5\%$. This mixture cools to 80°F. Determine (a) the initial pressure of the dry air and vapor, (b) the state at which condensation begins, (c) the partial pressures after the cooling, (d) the amount of H_2O condensed, and (e) the heat transferred.

Solution. (a) From Item B 13, we use the saturation pressure of $p_{vb} = 67.01$ psia corresponding to 300°F and find, Fig. 12/11, $p_{v1} = \phi p_{vb} = (0.05)(67.01) = 3.35$ psia; then $p_{a1} = p_{m1} - p_{v1} = 100 - 3.35 = 96.65$ psia.

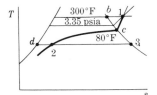

Process for the Vapor, V = C. **Fig. 12/11**

(b) Since the steam pressure is 3.35 psia, the vapor will depart more from ideal gas behavior than in previous examples. Using the complete steam tables, we find by interpolation

(a) $$v_{v1} = 134.8 \text{ ft}^3/\text{lb at } 3.35 \text{ psia and } 300°F$$

Since the volume and mass remain constant, the specific volume remains constant, and condensation is on the point of beginning when the volume of *saturated* vapor $v_g = 134.8$ ft^3/lb. Entering the saturated steam table (temperature), move along the column headed v_g, looking for 134.8. Interpolation gives a temperature of $t_c = 136.2°F$, Fig. 12/11, the dew-point temperature when the steam is cooled at *constant volume*. (In the absence of these tables, especially for pedagogical purposes, use $\phi \approx \rho_{v1}/\rho_{vb} = v_{vb}/v_{v1} = 6.4658/v_{v1} = 0.05$, and get $v_{v1} = 129.3$ ft^3/lb, by the ideal gas approach. The dew point for this volume $v_{gc} = v_{v1}$ is a little greater than 138°F.)

(c) Since each constituent changes state at constant volume, the dry air follows Charles' law. Thus, the pressure of the dry air at 80°F is

(b) $$p_{a2} = \frac{T_2}{T_1} p_{a1} = \frac{(540)(96.65)}{760} = 68.7 \text{ psia}$$

Because the final temperature $t_2 = 80°F$ is less than the dew point $t_c = 136.2°F$, some of the original vapor has condensed; therefore the vapor that remains is saturated in state 3, Fig. 12/11, and it must be at the saturation pressure corresponding to 80°F, or $p_{v2} = p_{v3} = 0.5068$ psia, Item B 13.

(d) To find the amount of condensation between states 1 and 2, find ω_1 and ω_2 and subtract. Using $\omega = v_a/v_v$, we first find

(c)
$$v_{a1} = \frac{R_a T_1}{p_{a1}} = \frac{(53.3)(760)}{(96.65)(144)} = 2.91 \text{ ft}^3/\text{lb da}$$

Having found $v_{v1} = 134.8$ from the steam tables, we get

(d)
$$\omega_1 = \frac{v_{a1}}{v_{v1}} = \frac{2.91}{134.8} = 0.02159 \text{ lb v/lb da}$$

At 80°F, $v_g = 633.3 = v_{v3}$, Fig. 12/11. Since the quantity of air does not change, its specific volume remains unchanged: $v_{a2} = v_{a1} = v_{a3} = 2.91 \text{ ft}^3/\text{lb da}$. Therefore,

(e)
$$\omega_2 = \frac{2.91}{633.3} = 0.00459 \text{ lb v/lb da}$$

The amount condensed is $\omega_1 - \omega_2 = 0.02159 - 0.00459 = 0.017 \text{ lb H}_2\text{O/lb da}$. For $m_a = 10/2.91 = 3.436$ lb in the tank, the total condensation is $(3.436)(0.017) = 0.0584 \text{ lb H}_2\text{O}$. This liquid is saturated, its state being represented by d, Fig. 12/11.

(e) For $V = C$, $Q = \Delta U$. The heat from the air is

(f)
$$Q_a = \Delta U_a = m_a c_v \Delta T = (3.436)(0.1714)(80 - 300) = -129.6 \text{ Btu}$$

The quality of the H_2O (gravimetric fraction of vapor per lb H_2O) at 2 is

(g)
$$x_2 = \frac{0.00459 \text{ lb v/lb da}}{0.02159 \text{ lb H}_2\text{O/lb da}} = 21.26\%$$

(h)
$$h_{v2} = (h_f + x_2 h_{fg})_2 = 48.037 + (0.2126)(1048.4) = 270.9 \text{ Btu/lb v}$$

(i)
$$v_{v2} = v_{v1} = 134.8 \text{ ft}^3/\text{lb v (no change in total H}_2\text{O)}$$

(j)
$$u_{v2} = h_{v2} - \frac{p_{v2} v_{v2}}{J} = 270.9 - \frac{(0.5068)(144)(134.8)}{778} = 258.2 \text{ Btu/lb v}$$

where all values are for both phases of H_2O at 80°F. The value of $h_{v1} = 1195.2 \text{ Btu/lb v}$, as obtained from the superheat table for $p_{v1} = 3.35$ psia and $t_1 = 300°F$. (See also the Mollier chart, Item B 16.) Using $v_{v1} = 134.8$ as found in part (b), we have

(k)
$$u_{v1} = h_1 - \frac{p_1 v_1}{J} = 1195.2 - \frac{(3.35)(144)(134.8)}{778} = 1111.2 \text{ Btu/lb v}$$

For $m_{H_2O} = 10/v_{v1} = 10/134.8 = 0.07418$ lb, the heat rejected by the H_2O during the process is

(l)
$$Q_{H_2O} = m_{H_2O}(u_{v2} - u_{v1}) = 0.07418(258.2 - 1111.2) = -63.3 \text{ Btu}$$

(m)
$$Q = Q_a + Q_{H_2O} = -129.6 - 63.3 = -192.9 \text{ Btu}$$

the rejected heat. If the entire mixture has been taken as *air alone* (heat of condensation neglected), the computed value of the heat would have been

(n) $mc_v \, \Delta T = (3.5102)(0.1714)(-220) = -132.4 \text{ Btu}$

which is badly in error.

MIXING STREAMS   12.21

Sometimes in air conditioning systems, the refrigerated air is tempered by mixing with recirculated air or fresh atmospheric air. In cases where mixing streams contain a certain component that we wish to keep account of, as H_2O in conditioned air, mass balances involving that particular substance are generally required. Also, in computing energy exchanges, total quantities of each stream (say, per minute) must be used.

Other examples of multiple streams include: (1) cooling towers, Fig. 12/12, wherein atmospheric air mixes with a warm water input and reduces its temperature, in a phenomenon much like adiabatic saturation, § 12.9; (2) air conditioning systems that use sprays of cold water to cool the air below its dew point.

Induced-Draft Cooling Tower. Cooling towers, used to cool condenser water, depend on natural draft, or are provided with forced-draft or induced-draft fans for increased capacity. If fans are used, the rate of cooling may be governed to suit the load by regulating the flow of air. The power required for induced draft is greater than for forced draft; however, the recirculation of the discharged hot air is more likely to occur under unfavorable conditions with forced draft, an action that reduces the capacity of the tower. (*Courtesy Foster Wheeler Corp., New York.*)   **Fig. 12/12**

In the energy diagram of Fig. 12/13, the air enters at boundary 1 (with its moisture ω_1) and departs at 2. The water enters at boundary A and departs at B. There is assumed to be no change of mass or stored energy within the boundaries of the control volume. For m lb H_2O/lb da, the mass balance for the H_2O is

(a) $\omega_1 + m_A = \omega_2 + m_B$

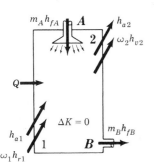

Energy Diagram for Mixing Streams of Gas and Liquid. The symbols are for air and water, the commonest problem. All masses are for a flow of 1 lb of dry gas (air).   **Fig. 12/13**

The energy balance is

(b) $Q + h_{a1} + \omega_1 h_{v1} + m_A h_{fA} = h_{a2} + \omega_2 h_{v2} + m_B h_{fB}$ Btu/lb da

which, of course, can be rearranged for the purpose of a particular problem. Usually, Q is assumed to be zero. Using the mass balance, equation **(a)**, either m_B or m_A can be eliminated in **(b)**; $\Delta h_a = c_p \Delta t$, or gas table values may be used.

In cooling towers, the smaller ω_1, the greater the cooling effect of the air on the water. The relative humidity at 2 should be greater than 90%. Makeup water in the condenser cooling system in the amount of $\omega_2 - \omega_1$ is needed. When this idea is used in air conditioning, the water A, Fig. 12/13, is cold enough to cool the air well below its dew point to remove H_2O from the air, a principal objective in the usual air conditioning system so that the final relative humidity of the room air is at a comfortable level. The events of the example in § 12.19 may be brought about in this manner.

12.22   CLOSURE

In several instances in this chapter and elsewhere in the text, the enthalpy of saturated liquid, h_f, is used where the liquid is actually subcooled, the reason being that *the sub-cooling is so small as to be negligible.*

If no condensation or evaporation of the vapor occurs, its partial pressure remains constant as long as the total pressure is constant (ideal gas law). If the air is saturated, the dew-point, wet-bulb, and dry-bulb temperatures are the same. See § 11.27 for more on mixtures of imperfect gases. If the gases in the mixture should vary markedly from ideal, the answers obtained by the methods of this chapter may be unsatisfactory.

PROBLEMS

SI UNITS

12.1 The gravimetric analysis of dry air is approximately $O_2 = 23.1\%$ and $N_2 = 76.9\%$. Calculate **(a)** the volumetric analysis, **(b)** the gas constant, **(c)** the respective partial pressures, and **(d)** the specific volume and density at 1 atm, 15.6°C. **(e)** How many kilograms of O_2 must be added to 2.27 kg air to produce a mixture which is 50% O_2 by volume?
Ans. **(a)** 20.81% O_2, **(b)** 288.2 J/kg-K, **(c)** 21.08 kPaa for O_2, **(d)** 0.84 m³/kg, **(e)** 1.471 kg.

12.2 **(a)** How many kilograms of nitrogen must be mixed with 3.60 kg of carbon dioxide in order to produce a gaseous mixture that is 50% by volume of each constituent? **(b)** For the resulting mixture, determine M_m, R_m, and

the partial pressure of the N_2 if that of the CO_2 is 138 kPaa.

12.3 A 283-ℓ drum contains a gaseous mixture at 689.48 kPaa and 37.8°C whose volumetric composition is 30% O_2 and 70% CH_4. How many kilograms of mixture must be bled and what mass of O_2 added in order to produce at the original pressure and temperature a mixture whose new volumetric composition is 70% O_2 and 30% CH_4?
Ans. Bleed 0.90 kg, add 1.381 kg O_2.

12.4 A gaseous mixture of CH_4, N_2, CO, and O_2 occupies a vessel at the respective partial pressures of 140, 55, 70, and 15 kPaa. Find **(a)** the volumetric and gravimetric analyses, **(b)** M_m and R_m, **(c)** the specific heats c_p and c_v for the mixture, and **(d)** the volume occupied by 45 kg of the mixture at p_m and 32°C.

12.5 The humidity ratio of atmospheric air at 1 atm, 26.7°C is 0.016 kg v/kg da. Find **(a)** p_v, **(b)** ϕ, **(c)** the dew point, **(d)** ρ_m, **(e)** the volumetric and gravimetric percentages of the H_2O, and **(f)** the mass of H_2O in 28.32 m³ of the atmospheric air.

Ans. **(a)** 2.54 kPaa, **(b)** 72.6%. **(c)** 21.3°C, **(d)** 1.166 kg/m³, **(e)** 2.51% by vol, **(f)** 0.52 kg.

12.6 The temperature in a steam condenser is 32.2°C, and there are 0.075 kg da/kg H_2O present. If the H_2O is saturated vapor, find **(a)** ω, **(b)** the condenser vacuum for a 1-bar barometer, **(c)** ρ_m, **(d)** the percentage by volume and the percentage by mass of the air.

Ans. **(a)** 13.31 kg v/kg da, **(b)** 96.29 kPa vac, **(c)** 0.0368 kg/m³, **(d)** 4.46%, 6.98%.

12.7 The enthalpy for steam vapor at low pressures, and within the temperature range 10–60°C, may be obtained from the expression $h_v = 2500 + 1.817\, t_v$, kJ/kg, where t_v is in °C. Demonstrate that this equation results from plotting h_g for steam on rectangular coordinate paper and noting that a straight line results. See Item B 16 (SI) for values.

12.8 An adiabatic evaporative process occurs as follows: 6.8 kg/s da at 1 bar, 104.4°C and 0.454 kg/s wet steam ($y = 10\%$) at 1 bar each enter an adiabatic chamber and leave after mixing as a gas-vapor mixture at 1 bar; all liquid initially in the steam evaporates during mixing to form added vapor. Sketch an energy diagram of the steady flow system and for the exit mixture find **(a)** the temperature, **(b)** the humidity ratio, **(c)** the relative humidity. **(d)** What is the entropy production and the irreversibility for the mixing process if $t_0 = 15.6°C$?

12.9 A gaseous mixture at 137.9 kPaa, 93.3°C has the following volumetric analysis: 35% CO_2, 50% N_2, 15% H_2O. With respect to the H_2O, find **(a)** the humidity ratio, **(b)** the dew point, **(c)** the relative humidity, and **(d)** the degree of saturation. **(e)** What is the density of the mixture?

Ans. **(a)** 0.092 kg v/kg dg, **(b)** 60.8°C, **(c)** 26%, **(d)** 0.13, **(e)** 1.45 kg/m³.

12.10 The volumetric percentages of a gas-vapor mixture are 90% H_2 and 10% H_2O when the temperature and total pressure are 65.6°C and 137.9 kPaa. This mixture is cooled with constant total pressure at the rate of 28.32 m³/min, measured at the initial state, to 15.6°C. Determine the initial relative humidity and the rate at which condensation occurs.

Ans. 53.8%, 1.122 kg/min.

12.11 There are heated 2 m³ of air-steam mixture, initially saturated at 1 bar, 20°C, at constant volume. If the final temperature is 65°C, find **(a)** the masses of dry air and water vapor, **(b)** the relative humidity and partial pressures for the final state, **(c)** the heat.

12.12 To what temperature must saturated air at 1 bar and 30°C be heated at constant volume in order to quarter its initial relative humidity?

12.13 An air-steam mixture, 4.535 kg and initially saturated at 517.11 kPaa, 32.2°C, is expanded isothermally until the partial pressure of the vapor becomes 1.227 kPaa. For the final state, find **(a)** the partial pressure of the dry air, **(b)** the relative humidity, **(c)** the dew point, **(d)** the heat.

Ans. **(a)** 130.52 kPaa, **(b)** 25.5%, **(c)** 10°C, **(d)** 544 kJ.

12.14 With 150 m³/min of saturated air at 1 bar, 10°C, there are mixed 100 m³/min of air at 1 bar, 40°C db, 60% relative humidity in a steady flow adiabatic process. For the resulting mixture determine **(a)** ω, **(b)** t_d, **(c)** ϕ, **(d)** dewpoint, and **(e)** ρ.

12.15 Show that if atmospheric air be expanded (or compressed) isentropically without condensation of the water vapor (ideal gas components), the respective variations in entropy of the two components, dry air and water vapor, will be other than zero—but equal and opposite such that the entropy of the mixture remains constant.

12.16 In a divided adiabatic chamber (see figure), there are 11.338 kg of dry air on one side at 1 bar, 99.6°C and 0.454 kg of saturated water vapor on the other side at 1 bar. The partition is removed and mixing occurs to equilibrium. For the final state,

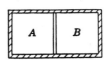

Problem 12.16

determine **(a)** the partial pressure of the water vapor, **(b)** the relative humidity. **(c)** For the system as a whole, find the change of internal energy and of entropy caused by the mixing process.

ENGLISH UNITS

12.17 A mixture of 1 lb of CO and 3 lb of N_2 is in a tank at 30 psia and 80°F. Determine **(a)** the partial pressure of each gas, **(b)** the volume of the tank, and **(c)** the volume that each gas would occupy if it were separated from the other gases but at 30 psia and 80°F. **(d)** If these gases were originally separated by a partition as in **(c)** and the partition is removed, what is the net change in entropy after the gases have diffused to equilibrium?

Ans. **(a)** 7.5, 22.5 psia; **(b)** 27.5 ft^3; **(c)** 6.88, 20.62 ft^3; **(d)** +0.159 Btu/°R.

12.18 A rigid tank containing 3 lb of a gaseous mixture of nitrogen and carbon dioxide (each 50% by volume) at 40 psia and 150°F receives 1 lb more of nitrogen with the temperature remaining at 150°F. For the resulting 4 lb, determine **(a)** the gravimetric and volumetric analyses, **(b)** the pressure, and **(c)** the molecular weight.

Ans. **(a)** For N_2: $f = 54.2\%$, $X = 65\%$; **(b)** 57.1 psia; **(c)** 33.6 lb/mol.

12.19 Atmospheric air is at 14.65 psia, $t_d = 94°F$, and $t_w = 68°F$. Find **(a)** the partial pressure of the water vapor, **(b)** the humidity ratio, **(c)** the dew point, **(d)** the relative humidity, **(e)** the volumetric and gravimetric percentages of the H_2O, and **(f)** the volume occupied by the H_2O in 1 lb da.

Ans. **(a)** 0.2027 psia, **(b)** 0.00873 lb v/lb da, **(c)** 53.5°F, **(d)** 25.6%, **(e)** 1.38% by vol, **(f)** 14.20 ft^3.

12.20 A low-pressure, air–steam mixture at 150°F has a dew point of 80°F. If the steam/air mass ratio is 4/100, find **(a)** ϕ, **(b)** p_m, **(c)** ρ_m, **(d)** the volumetric percentage of the H_2O, and **(e)** the volume occupied by the dry air in 1 lb vapor.

Ans. **(a)** 13.65%, **(b)** 17.072 in. Hg abs, **(c)** 0.0363 lb/ft^3, **(d)** 6.05%, **(e)** 711 ft^3.

12.21 For a given state, the atmospheric conditions are $p_m = 14.7$ psia, $t_d = 89°F$, $t_w = 74°F$. Making the solution from the psychrometric chart only, find **(a)** the dew point, **(b)** ω, **(c)** p_v, **(d)** ϕ, and **(e)** the specific volume. **(f)** How many pounds of water vapor are present in 15,000 ft^3 of this air? **(g)** How many pounds of water are required to saturate the 15,000 ft^3 of air at 74°F if t_w remains constant? If t_d remains constant? Assume that the total pressure remains unchanged.

Ans. **(a)** 68°F, **(b)** 0.01458 lb v/lb da, **(c)** 0.3035 psia, **(d)** 50%, **(e)** 14.15 ft^3/lb da, **(f)** 15.46 lb, **(g)** 3.63 lb for $t_w = C$.

12.22 Saturated atmospheric air at 14.7 psia, $t_d = 54°F$ undergoes a constant pressure humidifying process until its relative humidity is halved while its humidity ratio is doubled. Make a psychrometric chart sketch and find **(a)** final t_d, t_w, and dew-point temperatures, **(b)** moisture added to 2000 cfm of initial atmospheric air, lb/min, **(c)** heat required. Btu/min.

12.23 There are heated 1350 cfm of atmospheric air initially saturated at 14.70 psia, 65°F db until the final dry-bulb temperature is $t_{d2} = 90°F$; the dew-point temperature remains constant during this heating process. Use the psychrometric chart only (show sketch) and find **(a)** ϕ_2, **(b)** initial mass of atmospheric air, lb/min, **(c)** heat required, Btu/min.

12.24 An air conditioned space requires 5000 cfm of free air (§ 14.5) as measured at 100°F db and 13 psia (barometer); $t_w = 84°F$ (state 1). This air is processed through a cold water spray until the dew point of the air to be delivered to the space reaches the required value. After reaching the space, the air (with no H_2O liquid) is warmed by heat flow into the building to 70°F and 49% relative humidity (state 2) without a change of H_2O content. Find **(a)** p_{v1} from Carrier's equation and the dew point of the free air, **(b)** the dew point at state 2 and the H_2O condensed during cooling, and **(c)** the refrigeration done during processing expressed in tons of refrigeration (1 ton of refrigeration ≡ 200 Btu/min rate of cooling). **(d)** During cooling, cold water at 36°F is sprayed into the air and is warmed to 48°F at exit. Sketch an energy diagram and find the amount of 36°F water needed.

Ans. **(a)** d.p. ≈ 80°F; **(b)** 50°F, 303 lb/hr; **(c)** 46.8 tons; **(d)** 93.2 gpm.

12.25 Demonstrate that if h_g for steam

be plotted against temperature using a range of say 50–150°F, a straight line will obtain on rectangular coordinates with the resulting equation $h_v = 1061 + 0.445\, t_v$.

12.26 Atmospheric air at 14.55 psia and 80°F db undergoes an adiabatic saturation process to a final temperature of 58°F. For the state of the air just prior to the adiabatic process, find **(a)** t_w, **(b)** p_v, **(c)** ϕ, **(d)** ω, and **(e)** ρ_m. **(f)** For the state after the process, what are p_v, t_w, ω, and ρ_m?

12.27 The state of the atmospheric air is $p_m = 29.60$ in. Hg abs, $t_d = 80°F$, $t_w = 60°F$. **(a)** Find ω, p_v, ϕ, and the dew point, using the principle of adiabatic saturation. **(b)** Check p_v by Carrier's equation. **(c)** Solve **(a)**, using the psychrometric chart (show all chart readings on a sketch).
Ans. 0.00656 lb v/lb da, 0.31 in. Hg abs, 30%, 45.8°F.

12.28 If an open pan of water be placed in a $16 \times 20 \times 9$-ft room where $p_m = 14.50$ psia, $t_d = 78°F$, $t_w = 54°F$, approximate the maximum loss of water from the pan that may occur. Neglect infiltration and exfiltration losses, except assume that the total pressure and the dry bulb temperature remain constant. *Ans.* 3.65 lb.

12.29 There are added (by mass) 2 parts of benzene vapor to 13 parts of dry air to form a gas-vapor mixture at 15 psia, 110°F. Find the relative humidity ϕ based upon the benzene vapor.

12.30 Mix 6 lb of ethyl alcohol vapor with 18 lb oxygen to form a gas-vapor mixture at 20 psia, 140°F. Find **(a)** gravimetric analysis, **(b)** volumetric analysis, **(c)** relative humidity, **(d)** dew point temperature.
Ans. **(a)** 75% O_2, **(b)** 81.2% O_2, **(c)** 57%, **(d)** 119°F.

12.31 Through the manifold of an automobile engine there moves 15 lb/min dry air, 1 lb/min octane vapor, and 0.20 lb/min water vapor, all mixed at 12 psia and 130°F. Find the relative humidity and the dew point of the mixture **(a)** with respect to the octane vapor and **(b)** with respect to the water vapor.
Ans. **(a)** 17.21%, 68°F, **(b)** 11.21%, 59.1°F.

12.32 For summer comfort, atmospheric air in state 1, at $p_m = 24.45$ in. Hg, 80°F db, and 70% relative humidity, is to be cooled at

constant total pressure to 50°F (state 2). Then the air-vapor mixture at 2 is heated to state 3 at 70°F (by heat flux into the space). For a flow of 1000 cfm at 1, compute: **(a)** the lb/min of dry air, **(b)** the humidity ratios at 1, 2, and 3, **(c)** the lb/min of H_2O condensed, **(d)** the dew-point temperatures at 1, 2, and 3, **(e)** the heat removed from state 1 to 2, Btu/min, **(f)** the relative humidity at state 3. **(g)** Solve the problem by psychrometric chart, specifying also the wet-bulb temperatures at 1, 2, and 3. Show solution on a large sketch. Is there a theoretical reason for the answers by chart not being the same as those computed? Discuss.
Ans. **(a)** 58.2 lb/min, **(b)** 0.019, 0.00936, **(c)** 0.56 lb/min, **(d)** 69.2°F, 50°F, **(e)** 1032 Btu/min, **(f)** 49.1%.

12.33 Air at 110°F, $\phi = 10\%$ in a desert location ($p_m = 13$ psia) is cooled by contact with water (steady flow through wet screens). The air leaves the cooler at 76°F, $\phi = 80\%$. **(a)** If the water enters at 80°F and cools to 70°F as it leaves the unit, how much is supplied to the system (considered adiabatic) per pound of incoming air? **(b)** How much make-up water is needed during 1 hr for 1000 cfm of incoming air? Sketch an energy balance.
Ans. **(a)** 0.369 lb, **(b)** 41.5 lb/hr.

12.34 Assume 20 ft^3 of an air-steam mixture are initially at 14.70 psia, 96°F db, and 100% relative humidity. The air is cooled at constant volume to 40°F. Find **(a)** ω_1, **(b)** the partial pressures at the final state, **(c)** ω_2, **(d)** the temperature at which condensation started, **(e)** the total amount of vapor condensed, and **(f)** the heat.
Ans. **(a)** 0.03775 lb v/lb da, **(b)** 12.45, 0.1217 psia, **(c)** 0.006075 lb v/lb da, **(d)** 96°F, **(e)** 0.0427 lb, **(f)** −63 Btu.

12.35 Atmospheric air, initially saturated at 15 psia, 110°F, is compressed isothermally until the final pressure is $p_{m2} = 45$ psia. Sketch the Ts diagram and for 20 lb atmospheric air find **(a)** the final relative humidity, **(b)** the amount of moisture condensed, lb.

12.36 There are mixed 8500 cfm of saturated air at 14.60 psia, 48°F db, with 6600 cfm of air at 14.60 psia, 92°F db, 60% relative humidity, in a steady flow adiabatic process. For the resulting mixture, determine

(a) ω, (b) t_d, (c) ϕ, (d) the dew point, (e) ρ, and (f) μ.

Ans. (a) 0.0123 lb v/lb da, (b) 66.3°F, (c) 88.7%, (d) 62.9°F.

12.37 In an air-conditioning system, there are mixed adiabatically two steady flow streams; $\Delta P = 0$, $\Delta K = 0$. One stream A is atmospheric air at 29.92 in. Hg abs, 90°F db, $\phi = 38\%$, 500 lb/min; the other stream B is recirculated air at 29.92 in. Hg abs, 70°F db, $\phi = 70\%$, 400 lb/min. Determine (a) the humidity ratio of each stream A, B, C (the resulting mixture), (b) the dry bulb temperature and dew point of stream C, (c) the entropy production of the mixing process.

Ans. (a) 0.0114, 0.01094, 0.0112 lb v/lb da, (b) 82°F, (c) 374 Btu/°R-min.

12.38 Water enters a *cooling tower* at 110°F and leaves at 76°F. The tower receives 500,000 cfm of atmospheric air at 29.75 in. Hg abs, 70°F db, 40% relative humidity; the air leaves the tower at 95°F db, $\phi = 95\%$. Find (a) the mass per min of dry air passing through the tower, (b) the volume of incoming water cooled, gpm, and (c) the amount of water evaporated each hour.

12.39 A mechanical draft *cooling tower* receives 250,000 cfm of atmospheric air at 29.60 in. Hg abs, 84°F db, 45% relative humidity and discharges the air saturated at 98°F. If the tower receives 3500 gpm of water at 104°F, what will be the exit temperature of the cooled water?   *Ans.* 83°F.

12.40 Program this problem. For a given set (or sets) of psychrometric data (t_d, t_w, p_m) relating to atmospheric air, employ Carriers equation and find the relative humidity ϕ, the humidity ratio ω, and the dew point temperature t_{dp}.

13

REACTIVE SYSTEMS

INTRODUCTION 13.1

It was not too long ago that the chemical processes were left almost entirely to the chemist and the chemical engineer. Then the energy crisis came with the stark realization that all earthly fuels are finite. Suddenly everybody became involved with energy regardless of their grasp of the subject matter contained in this chapter. Although most people will never comprehend combustion, it is expected that all engineers have a working knowledge of fossil fuels and their rapid oxidation in air.

The controlled burning of a fuel in air is an exothermic reaction accompanied by a substantial increase in temperature. Like all working thermodynamic processes, it should be an efficient one. Four factors (among others) must be correct if a combustion process merits any consideration—these follow and spell MATT. M for good mixing of fuel and oxidizer; A for adequate air to supply the oxygen; T for ignition temperature to start and maintain the combustion process; T for time to permit complete combustion.

Perhaps this chapter should prove interesting since you are so acutely aware of the role that combustion plays with our nation in maintaining its image and status among the other countries of this world.

FUELS 13.2

Fossil fuels may be classified conveniently as (1) solid, (2) liquid, and (3) gaseous. Coal, in many different grades, is the most common solid fuel. The most widely used liquid fuels are hydrocarbon derivatives, of the chemical form $C_x H_y$, where the subscripts x and y have many different values. The alcohols, of the chemical form $C_x H_y O_z$, have application where a price disadvantage is outweighed by other advantages. The gaseous fuels are generally the cleanest, and where natural gas is plentiful, probably the cheapest. Manufactured and byproduct gases are also used for heating and power production. The space age has brought about the development of "fuels," particularly reactants that combine rapidly and release large amounts of energy, as needed for rocket propulsion.

13.3   ANALYSIS OF FUELS

The *ultimate analysis* is one that specifies the various percentages by mass (or the mass fractions f_m) of the elements in the fuel, typical for solid and liquid fuels.

A gaseous fuel may be defined either by a gravimetric or a volumetric (molal) analysis (§ 12.2). The combustible elements are mostly carbon and hydrogen.

If the reacting fuel is a single molecular type, we may determine a gravimetric analysis of the different atoms from the chemical formula. For the hydrocarbon octane C_8H_{18}, which is close to the average molecular form in gasoline (a mixture of numerous different molecules), we have

(a) $8 \times 12 + 18 \times 1 \approx 114$ lb/pmole of octane

where, in a molecule of fuel, there are 8 atoms of carbon of atomic mass 12 and 18 atoms of hydrogen of approximate atomic mass 1. (In most combustion problems, the whole number nearest the atomic or molecular mass is sufficiently accurate.) Therefore, in 1 mole of octane, there are $8 \times 12 = 96$ lb C and 18 lb H_2. The gravimetric fractions are $f_{mC} = 96/114 = 0.842$ C, $f_{mH_2} = 18/114 = 0.158$ H_2.

In the opposite direction, suppose a hydrocarbon has been found to contain 80% C and 20% H_2 by mass ($f_{mC} = 0.80$, $f_{mH_2} = 0.20$). The corresponding number of atoms is $80/12 = 6.67$ and $20/1 = 20$, which numbers are proportional to the number of atoms of C and H_2, respectively. Equivalent hydrocarbons would be

$$C_{6.67}H_{20} \quad \text{or} \quad CH_3 \quad \text{or} \quad C_2H_6$$

or any one where the number of H_2 *atoms* is $20/6.67 = 3$ times the number of C atoms; designated $(CH_3)_x$. Notice that the equivalent molecular mass of $C_{6.67}H_{20}$ is 100.

13.4   COMPOSITION OF AIR

Atmospheric air has a volumetric composition of 20.99% oxygen, 78.03% nitrogen, somewhat less than 1% argon, with small quantities of several inert gases such as water vapor, carbon dioxide, helium, hydrogen, and neon. For most engineering calculations, it is usually accurate enough to include all inert gases as nitrogen and to use the analysis: 21% oxygen and 79% "atmospheric" nitrogen by volume. Thus, in 100 moles of air, there are approximately 21 moles of O_2 and 79 moles of N_2, or

(a) $\dfrac{79}{21} = 3.76 \dfrac{\text{moles } N_2}{\text{mole } O_2}$ or $3.76 \dfrac{\text{ft}^3 N_2}{\text{ft}^3 O_2}$

a useful number for the current study. The approximate gravimetric composition of air is 23.1% O_2, 76.9% N_2, or there are $76.9/23.1 = 3.32$ lb N_2/lb O_2.

13.5   AIR-FUEL RATIOS

If the proportions of the constituents of the reactants are such that there are exactly enough molecules of oxidizer to bring about a complete reaction to stable molecular forms in the products, these are said to be the *stoichiometric* proportions. For these proportions, there is an *ideal* amount of oxidizer; for fuels reacting with pure O_2, we speak of "100% air" (or 100% O_2).

In practice, to assure a complete reaction of the fuel (complete combustion), an excess of air or of oxygen is supplied. In this case, for example, we may speak of 120% air, 200% air, and so on; or 20% or 100% *excess air*, and so on. Any air over and above the amount practically required for the particular situation results in more energy leaving the system as stored (molecular) energy, inasmuch as the temperature to which the products are cooled will probably be as low as economically feasible anyway. Experience suggests the following amount of *excess* air for steam power plant furnaces: pulverized coal, 15–20%. fuel oil, 5–20%; natural gas, 5–12%.

There are occasions when there is a deficiency of oxidizer, and for fuels, we speak of, for example, "80% air" (or 80% O_2). In general, if there is not some excess, the loss from unburned fuel, incomplete combustion, is uneconomically large.

Inasmuch as the various principles are familiar, a few examples will serve to explain their application. As you recall, the atomic species are conserved in a chemical process. If one will therefore methodically write out the atom balance, equivalent to a mass balance of that element, for each atom involved, one is more likely to obtain correct chemical equations.

Example—Combustion of Octane 13.6

If gaseous octane $C_8H_{18}(g)$ is burned in ideal air, what volume of air at 140°F and 14 psia is required? Determine the relative volumes and masses of the constituents of the reactants and of the products when the H_2O is liquid $H_2O(l)$, and compute the air-fuel ratio. Compute an approximate value of the equivalent molecular weight for the products when the H_2O is gaseous, $H_2O(g)$.

Solution. For stoichiometric combustion, the products are H_2O, CO_2, and N_2 (the N_2 is considered to be inert, passing through the reaction chemically unchanged). Depending on the final temperature, the H_2O may be liquid or gas (vapor). If the products return to approximately atmospheric pressure, we assume that all the H_2O formed during combustion has condensed. In this example, both phases are considered. We recall that the coefficients of the chemical symbols can represent either the number of molecules in the reaction or, preferably, moles of each substance. Let 1 mole of fuel be the basis of the equation and write the chemical equations with unknown coefficients represented by letters, say, a, b, c.

(a)
$$C_8H_{18}(g) + aO_2 + 3.76aN_2 \rightarrow bH_2O + cCO_2 + 3.76aN_2$$

For 3.76 moles of N_2 per mole of O_2 in air, § 13.4, the coefficient of atmospheric N_2 is $3.76a$, for a moles of atmospheric O_2. The atom balances give

$$C: \quad 8 = c \quad \text{or} \quad c = 8$$

$$H_2: \quad 18 = 2b \quad \text{or} \quad b = 9$$

$$O_2: \quad 2a = b + 2c \quad \text{or} \quad a = 12.5 \quad 3.76a = 47$$

Then, if the final temperature is low enough to have condensed the H_2O, we get

(b)

	Reactants			Products	
	$C_8H_{18}(g) + 12.5O_2 + 47N_2 \rightarrow 9H_2O(l) + 8CO_2 + 47N_2$				
Moles:	1	+ 12.5	+ 47	→ 9	+ 8	+ 47
Relative volume:	1	+ 12.5	+ 47	→ 0	+ 8	+ 47
Relative mass:	114	+ 400	+ 1316 = 162	+ 352	+ 1316
Mass/lb fuel	1	+ 3.51	+ 11.54 = 1.42	+ 3.09	+ 11.54

The first line below the equation is simply the number of moles of each component for complete combustion in stoichiometric air. The "Relative volume" line shows relative volumes when reactants and products are at the *same temperature and pressure* (Avogadro); liquid H_2O is shown with zero volume (it is so small compared to gas volumes). Sample interpretation: 1 ft^3 fuel requires 12.5 ft^3 O_2 and results in 8 ft^3 CO_2 (same p and T). The volumetric analyses of the reactants and of the products can be obtained by dividing the volume of one constituent by the total volume: for reactants, suppose 12.5/60.5 is the volumetric fraction of O_2; for the products, 8/55 is the volumetric fraction of CO_2.

The "Relative mass" line is obtained by multiplying the number of moles of each component by its molecular mass M lb/mole; for O_2, this is $(12.5)(32) = 400$. The "Mass/lb fuel" line is obtained by dividing each term of the preceding line by the mass of fuel (114). Interpretation of the last line: it takes 3.51 lb O_2/lb fuel; there results 1.42 lb H_2O/lb f, in addition to H_2O that was already in atmospheric air; f stands for fuel.

One of the required answers is $r_{a/f} = 3.51 + 11.54 = 15.05$ lb air/lb f, which is the air-fuel ratio ($r_{a/f} = 15.1$ when the decimal parts of the molecular weights are used); the fuel-air ratio $r_{f/a} = 1/15.05 = 0.0664$ lb f/lb air. If the relative masses of the reactants are not needed, a quicker way to compute the air is to use the equivalent molecular weight of air, say 29, and the total moles, $12.5 + 47 = 59.5$ moles of air. Then

(c)
$$r_{a/f} = \frac{\text{lb air/pmole f}}{\text{lb fuel/pmole}} = \frac{(29)(59.5)}{114} = 15.1 \text{ lb air/lb fuel}$$

Small differences in answers by different methods are to be expected when approximate atomic weights are used.

At this stage, the volume of air may be computed in several different ways. For 59.5 moles of air per mole of fuel, equation (6-4) gives

(d)
$$V_a = \frac{1545nT}{p} = \frac{(1545)(59.5)(600)}{(14)(144)} = 27,380 \text{ ft}^3/\text{mole fuel}$$

or $27,380/114 = 240$ ft^3/lb fuel.

The total mass of products per mole of fuel is $162 + 352 + 1316 = 1830$ lb. The total moles of gaseous products with $H_2O(g)$ per mole of fuel is $9 + 8 + 47 = 64$. The molecular mass is then $M = 1830/64 = 28.6$ lb/mole of products.

13.7   Example—Volume of Products

For the products of combustion found in § 13.6, determine the volume at 240°F and 14 psia in ft^3/lb fuel.

Solution.   There are also several ways in which this problem may be solved at this stage. Since at the stated temperature, the H_2O is gaseous, there are $9 + 8 + 47 = 64$ moles of products per mole of fuel. Use the ideal gas equation for 114 lb fuel and get

(a)
$$V = \frac{1545nT}{114p} = \frac{(1545)(64)(700)}{(114)(14)(144)} = 301 \text{ ft}^3/\text{lb fuel}$$

Also, the volume can be computed by using any component of the products at its partial pressure. Since this idea is often useful, let us check the above computation. Compute the partial pressure of the 47 moles of nitrogen as

(b)
$$p_N = p_x = X_x p_m = (47/64)14 = 10.3 \text{ psia}$$

Use the 11.54 lb N_2/lb fuel obtained from § 13.6, and get

(c)
$$V = \frac{mRT}{p} = \frac{(11.54)(1545/28)(700)}{(10.3)(144)} = 301 \text{ ft}^3/\text{lb fuel}$$

the volume of the *mixture*. This sort of computation is preferably made with that component of the mixture that is most nearly an ideal gas and whose volumetric percentage is most accurately known (if there is a difference).

COMBUSTION IN EXCESS AND DEFICIENT AIR   13.8

Excess air is presumed to pass through the reaction chemically unchanged. Suppose it is desired to have 100% excess air ("200% air"). First, the combustion equation must be balanced for stoichiometric air so that the ideal amount is known; then the amount of air corresponding to a given percentage of excess (or deficiency) can be determined. For 100% excess air in equation **(b)**, § 13.6, there would be 25 moles O_2 and 94 moles N_2. The balanced equation becomes

(a) $C_8H_{18}(g) + 25O_2 + 94N_2 \rightarrow 9H_2O(l) + 8CO_2 + 12.5O_2 + 94N_2$

With *deficient air*, one must make assumptions regarding the composition of the products at this stage. We shall assume that all the H_2 goes into H_2O (because there is a strong affinity between H_2 and O_2) and that the incompleteness of combustion results in the appearance of some CO as well as CO_2. For this assumption, there are not too many unknowns. As an example, suppose that 80% of stoichiometric air is supplied for the octane in equation **(b)**, § 13.6; then the O_2 in the reactants is $(0.80)(12.5) = 10$ moles, and the corresponding N_2 is $(3.76)(10) = 37.6$ moles. The chemical equation becomes

(b) $C_8H_{18}(g) + 10O_2 + 37.6N_2 = aCO + bCO_2 + 9H_2O + 37.6N_2$

From the carbon and oxygen atom balances,

$$C: \quad 8 = a + b \quad \text{or} \quad a = 8 - b$$

$$O_2: \quad 20 = a + 2b + 9$$

A simultaneous solution gives $b = 3$ moles and $a = 5$ moles. The balanced equation can now be written and other computations made as illustrated in §§ 13.6 and 13.7.

Example—Dew Point of H₂O in Products   13.9

(a) *Stoichiometric dry air.* Let the reaction be for octane, as per equation **(b)**, § 13.6. For a total pressure $p_m = 14$ psia, determine the dew point of the H_2O.

Solution. For 9 moles of gaseous H_2O and 64 moles of products, the partial pressure of the H_2O by Dalton's law is

(a) $$p_v = X_v P_m = \left(\frac{9}{64}\right)14 = 1.97 \text{ psia}$$

The corresponding saturation temperature is 125.5°F, Item B 13, almost the dew point (§ 12.6).

(b) *Humid Air, 20% excess.* Let there be 20% excess air in the combustion of C_8H_{18}. The humidity ratio of the air supplied is $\omega = 0.015$ lb v/lb da; the total pressure is 14 psia. Find the dew point of the H_2O.

Solution. The amounts of O_2 and N_2 are obtained by multiplying their coefficients by 1.20 (for the 20% excess); moles of $N_2 = (1.2)(47) = 56.4$. The moles of H_2O ($= 9$) and of CO_2

$(= 8)$ are unchanged. But since there are now $(1.20)(12.5) = 15$ moles of O_2 and only 12.5 are needed for the reaction, the products will contain $15 - 12.5 = 2.5$ moles of O_2. This gives the total number of moles of products as

(b) $$n_p = \sum_i n_i = 9 + 8 + 2.5 + 56.4 = 75.9 \text{ moles}$$

The amount of dry air accounted for in the reaction equation is $15 + 56.4 = 71.4$ moles, or $(71.4)(29)$ lb da/pmole f, where $M_{da} = 29$. For $\omega = 0.015$ lb v/lb da, the vapor entering the reaction per mole of fuel is $m_v = (0.015)(71.4)(29)$ lb, and the moles of entering vapor are

(c) $$n_{v\text{in}} = \frac{m_v n_{da} M_{da}}{M_v} = \frac{(0.015)(71.4)(29)}{18} = 1.725 \text{ pmoles}$$

Total vapor in products $n_v = 9 + 1.725 = 10.725$ pmoles and

(d) $$p_v = \frac{n_v}{n_p} p_m = \frac{10.725}{77.63}(14) = 1.934 \text{ psia}$$

from which the saturation temperature (dew point) is 124.8°F. It is coincidental that the answers in parts (a) and (b) are virtually the same. In both cases, there is a small error because the H_2O is not acting ideally at its partial pressure.

13.10   **Example—Finding Air Required and Products from Combustion, Given the Gravimetric Analysis of Fuel**

    Whenever a gravimetric analysis is given, a common practice with coal and other solid fuels, *convert to moles of each element*; that is, if f_m is a *percentage* number,

(a) $$\frac{(f_m \text{ lb}/100 \text{ lb fuel})}{(M_i \text{ lb/mole})} = \frac{f_{mi}}{M_i} \text{ moles}/100 \text{ lb fuel} \qquad [\sum f_{mi} = 100]$$

Given the ultimate analysis of an Illinois coal[13.3] as follows, *dry basis*:

f_m	67.34% C	4.67% H_2	8.47% O_2	1.25% N_2	4.77% S	13.5% ash
f_m/M	5.61 C	2.33 H_2	0.265 O_2	0.045 N_2	0.149 S	———

The second line is obtained by dividing each f_m by the corresponding M of the element and the numbers may now be taken as *moles per* 100 *lb of dry fuel*. Since the coefficients in the chemical equation are moles, we write

$$[5.61C + 2.33H_2 + 0.265O_2 + 0.045N_2 + 0.149S] + aO_2 + 3.76aN_2 \rightarrow$$

(b)

$$bCO_2 + cH_2O + dSO_2 + (3.76a + 0.045)N_2$$

where the brackets are used only to keep the elements of the fuel clearly identified. The material balances are

$$C: \qquad 5.61 = b \qquad\qquad\qquad b = 5.61$$

$$H_2: \qquad 2.33 \times 2 = 2c \qquad\qquad c = 2.33$$

$$S: \qquad 0.149 = d \qquad\qquad\qquad d = 0.149$$

$$O_2: \qquad (0.265 \times 2) + 2a = 2b + c + 2d \qquad a = 6.66$$

The chemical equation for stoichiometric air is

(c)
$$[5.61C + 2.33H_2 + 0.265O_2 + 0.045N_2 + 0.149S] + 6.66O_2 + 25.04N_2 \rightarrow$$

$$5.61CO_2 + 2.33H_2O + 0.149SO_2 + 25.09N_2$$

The moles of air are $6.66 + 25.04 = 31.7$ moles/100 lb dry fuel. At 29 lb/mole, we get

(d) $3.17 \times 29 = 919.3$ lb air/100 lb dry fuel   or 9.19 lb air/lb dry fuel

With the chemical equation, almost any desired ratio is easily found. Common ratios found are: mass of dry products (no H_2O) per pound of carbon in the fuel; mass of dry products per pound of dry coal; volumetric percentage of CO_2 in the products; partial pressure of the H_2O in the hot products; and the volumetric and gravimetric analyses of the products. In furnace work, the products are called *flue gas*.

If there is considerable moisture in the coal, as in many low-grade coals, some of the heating value of the combustible elements is necessarily used to evaporate the H_2O and the latent heat of evaporation goes out the stack as a loss. Naturally, we shall be interested in the air-fuel ratio for coal *as fired*. As an example, let the moisture content of the coal be 12%. Then, there is $100 - 12 = 88\%$ of dry coal, or 0.88 lb dry fuel/lb wet fuel (as fired—abbreviated f.a.f.). Thus,

(e) $$r_{a/f} = \left(9.19\frac{\text{lb air}}{\text{lb dry fuel}}\right)\left(0.88\frac{\text{lb dry fuel}}{\text{lb as fired}}\right) = 8.1 \text{ lb air /lb f.a.f.}$$

The sulfur content is frequently neglected in computations such as these. Moreover, the nitrogen in the fuel is usually a small quantity, and as such it would be little missed in sliderule calculations. On the other hand, the practice is to include the oxygen, letting it count toward that required for combustion, as in this example.

ANALYZING THE PRODUCTS OF COMBUSTION 13.11

The products of combustion are analyzed in order to learn about the combustion process, say, the amount of air supplied. The Orsat apparatus, Fig. 13/1, determines the percentages of CO_2, CO, and O_2 in the dry products (although the test sample is saturated with H_2O). Since the products from an internal combustion engine contain hydrogen H_2 and hydrocarbons, principally methane, CH_4, these constituents can be: (1) ignored, (2) determined by more complete test, or (3) estimated. Experience suggests that the amount of CH_4 is about 0.22% of the volume of dry exhaust gas, and that the amount of free H_2 by volume is about half of the volumetric percentage of CO.[0.24]

From the product analysis, we can construct the corresponding theoretical chemical equation or obtain part of it, depending on what is known about the fuel. There are inevitable sources of error, such as unburned fuel in the ash and smoke. The analyses of the fuel and of the products may be complete, in which case several material balances may be made. However, we can get a good approximation of the air supplied (if the free O_2, H_2, and N_2 contents are small), from a carbon balance, knowing only the amount of carbon burned per unit mass of fuel. Detail is best covered by examples.

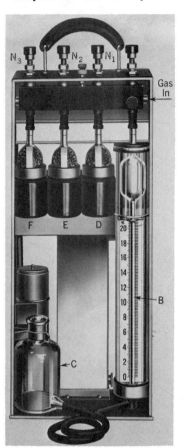

Fig. 13/1   *Flue-Gas Analyzer.* The bottle *C*, containing water, is connected to the burette *B* by a rubber tube. Raising and
lowering the bottle causes the water to flow into and out of
the burette. In operation, the air in the burette and the
adjoining passages is first driven out by displacing it with
the water in *C*. A sample of the products of combustion is
then drawn into the burette *B*. During these preliminary
operations, the needle valves N_1, N_2, and N_3 to the solution
containers *D*, *E*, and *F* have been closed. With valve N_1
open, the sample of flue gas in the burette is forced into
container *D* by raising bottle *C*. A solution of potassium
hydroxide in *D* absorbs the carbon dioxide of the flue gas,
leaving the other constituents unaffected. Remaining gas is
returned to the burette by lowering *C*, and the loss of
volume is noted. Successively, the gas is then forced into *E*,
where a solution of pyrogallic acid in a solution of potassium hydroxide absorbs the oxygen, and into *F*, where a
solution of cuprous chloride in ammonia absorbs the carbon monoxide. The remaining gas is assumed to be all
nitrogen, or an estimation of other products is made.
(*Courtesy Ellison-Draft Gage Co., Chicago*)

13.12   Example—Actual Air-Fuel Ratio

A flue-gas analysis of the dry products of the coal in the example of § 13.10 is made and
found to be 15% CO_2, 3.5% O_2, and 0.2% CO, with the remainder assumed to be nitrogen,
81.3% N_2; all by volume of dry gases. What excess or deficiency of air is supplied?

Approximate Solution, Assuming Carbon Only Is Known.   For the coal of §13.10, we
know that the carbon is 0.6734 lb C/lb dry fuel or 5.61 moles per 100 lb of dry fuel. A partial
chemical equation is

(a) $a[5.61C + \cdots] + \cdots \rightarrow 15CO_2 + 0.2CO + 3.5O_2 + 81.3N_2 + \cdots$

The dry-product analysis is written as given, the basis of the equation being 100 moles of
products. The coefficient *a* is obtained from a carbon balance;

$$\text{C:}\quad 5.61a = 15 + 0.2 \quad \text{or} \quad a = 2.71 \quad 5.61a = 15.2$$

(b) $15.2C + \cdots \rightarrow 15CO_2 + 0.2CO + 3.5O_2 + 81.3N_2 + \cdots$

In this approach, we assume that the nitrogen measures the amount of air, a good assumption
if no N_2 is in the fuel, since it passes through the ideal reaction unaffected. The moles of O_2
that accompanied 81.3 moles N_2 in the air are $81.3/3.76 = 21.6$ moles O_2, from which moles of
air are $81.3 + 21.6 = 102.9$. The mass of air per pound of carbon, which is based on *all the*

carbon being burned, is

(c) $\dfrac{(102.9)(29)\,\text{lb air}}{(15.2)(12)\,\text{lb C}} = 16.36\ \text{lb air/lb C burned}$

Using the known carbon content (0.6734) and moisture content (0.12 or 0.88 lb dry fuel/lb f.a.f.), we have

(d) $r_{a/f} = \left(16.36\dfrac{\text{lb air}}{\text{lb C}}\right)\left(0.6734\dfrac{\text{lb C}}{\text{lb dry f}}\right)\left(0.88\dfrac{\text{lb dry f}}{\text{lb f.a.f.}}\right) = 9.7\dfrac{\text{lb air}}{\text{lb f.a.f.}}$

versus 8.1 lb air for stoichiometric combustion. The excess is $(9.7 - 8.1)/8.1 = 19.8\%$.

Solution From Ultimate Analysis. When more is known of the fuel analysis, a somewhat more accurate computation of the air can be made. Except in extreme cases, the sulfur is ignored (the SO_2 is in the CO_2 percentage as determined by the Orsat because potassium hydroxide absorbs it too). Using mole values from § 13.10, we have

(e) $a[5.61C + 2.33H_2 + 0.265O_2 + 0.045N_2] + bO_2 + 3.76bN_2 \rightarrow$

$15CO_2 + 0.2CO + 3.5O_2 + (0.045a + 3.76b)N_2 + dH_2O$

If there is hydrogen in the fuel, there is bound to be some H_2O in the products. The coefficient a is determined from the carbon balance and was found above; $a = 2.71$. From other material balances, we have,

H_2: $(2.71)(2.33)(2) = 2d$ $d = 6.31$

O_2: $(2.71)(0.265)(2) + 2b = 30 + 0.2 + 7 + 6.31$ $b = 21.03$

N_2: $(0.045a + 3.76b)$ $= 79.19$

The balanced equation is $(3.76b = 79.07)$

(f) $[15.2C + 6.31H_2 + 0.72O_2 + 0.122N_2] + 21.03O_2 + 79.07N_2 \rightarrow$

$15CO_2 + 0.2CO + 3.5O_2 + 79.19N_2 + 6.31H_2O$

The air supplied is the O_2 and N_2 on the left-hand side, outside the brackets that enclose the fuel; $21.03 + 79.07 = 100.1$ moles of air. As before,

(g) $r_{a/f} = \dfrac{(100.1)(29)(0.6734)(0.88)}{(15.2)(12)} = 9.43\ \text{lb air/lb f.a.f.}$

Compare with the preceding answer. Besides different procedures, an experimental error is no doubt involved in taking readings. The advantage of the second method is that any desired ratio is quickly obtained with the balanced equation. Perhaps you may notice, though it is not essential in this example to do so, that the amount of fuel represented within the brackets is $100a = 271$ lb/100 moles dry products.

Example—Air for Hydrocarbon of Unknown Composition   13.13

Since commercial petroleum fuels are mixtures of numerous hydrocarbons, it is often advantageous to get an estimate of the air-fuel ratio without the expense of an analysis of the fuel (which no doubt varies from delivery to delivery). In the following approach, the fuel is assumed to be composed of carbon and hydrogen only in the form C_xH_y, and it should therefore contain only small amounts of O_2, N_2, and S in order to avoid significant error.

The *dry* exhaust from an automotive engine at 1-atm pressure has a volumetric analysis as follows: 12.5% CO_2, 3.1% O_2, 0.3% CO, which is the information obtained from the Orsat. (There should also be about 0.22% CH_4 and 0.15% H_2. See § 13.11. These quantities may be included if desired, but we shall omit them to shorten the presentation.) Assume that the remainder of the exhaust is N_2 = 84.1%. (a) Set up the theoretical combustion equation, finding values of x and y in C_xH_y. (b) Determine the air-fuel ratio.

Solution.   (a) The products analysis shows no H_2O, but *do not forget to include it* because it is sure to be there when the fuel contains hydrogen. Since there is assumed to be no O_2 in the fuel, the O_2 on the left-hand side of the chemical equation is that which accompanies the N_2 in the air; 84.1/3.76 = 22.4 moles O_2.

$$C_xH_y + 22.4O_2 + 84.1N_2 \rightarrow 12.5CO_2 + 3.1O_2 + 0.3CO + 84.1N_2 + aH_2O$$

The material balances:

C:	$x = 12.5 + 0.3$	$x = 12.8$
O_2:	$(2)(22.4) = (2)(12.5) + (2)(3.1) + 0.3 + a$	$a = 13.3$
H_2:	$y = 2a = (2)(13.3)$	$y = 26.6$

Thus, *provided* all the carbon and hydrogen burned, the fuel is a mixture of hydrocarbons for which the average molecule is $C_{12.8}H_{26.6}$, or for which the ratio $y/x = 26.6/12.8$. The chemical equation tells no more. The balanced equation then is

$$C_{12.8}H_{26.6} + 22.4O_2 + 84.1N_2 \rightarrow 12.5CO_2 + 3.1O_2 + 0.3CO + 84.1N_2 + 13.3H_2O$$

(b) The "molecular mass" of $C_{12.8}H_{26.6}$ is $(12)(12.8) + 26.6 = 180.2$ lb/mole. Since the moles of air per mole of fuel are $22.4 + 84.1 = 106.5$, we have

$$r_{a/f} = \frac{(106.5)(29)}{180.2} = 17.1 \text{ lb air/lb fuel}$$

If other information is desired, such as pounds of H_2O in exhaust per pound of fuel, the dew point of the H_2O in the products, and the estimated gravimetric analysis of the fuel, follow procedures previously explained. (If the reader desires to do these as an exercise, the answers are, respectively, 1.33, 121°F, 0.854C, 0.146H.)

13.14   HEATS OF REACTIONS

For fuels, the heat of the chemical reaction is called *heating value*, *heat of combustion*, *enthalpy of reaction* h_{rp}, or *internal energy of reaction* u_{rp}. *Heating value* is a general name, without specific meaning unless the way in which it is measured is defined (or implied) by the context. If the heating value is measured by a bomb calorimeter (ideally a rigid constant volume chamber), typical for solid and liquid fuels, the heating value obtained is that at constant volume q_v, the same as the internal energy of the reaction $-u_{rp}$. The heating value of gaseous fuels is generally found in a steady-flow, essentially constant pressure calorimeter, from which is obtained the constant pressure heating value q_p, the same numerically as the enthalpy of reaction $|h_{rp}|$. In each case for fuels that contain hydrogen, there is a *higher heating value* q_{vh}, q_{ph}, when the H_2O in the products has condensed [$H_2O(l)$], and a *lower heating value* q_{vl}, q_{pl}, when the H_2O has not condensed [$H_2O(g)$]. The lower heating value q_1 is equal to the higher heating value minus the latent heat of the condensed water vapor. In test work the heating value is converted to a standard state, symbolized as q_p°, q_v°, and so on, and with sufficient data any heating value can be computed from another, as explained later. The enthalpy and the internal energy of the reaction, from the chemistry point of view, are negative when the reaction is exothermic. On the other hand, engineers customarily think of *heating value* as positive; hence, $q_p^\circ = -h_{rp}^\circ$ and $q_v^\circ = -u_{rp}^\circ$.

ENTHALPY OF COMBUSTION 13.15

Consideration of equation (4-14) in the form $dQ = dH - V\,dp$ makes it evident that the heat of a $p = C$ process is ΔH, whether the process occurs in an open steady-flow system or a closed system. For a steady-flow calorimeter, we first imagine the process to be at standard temperature (77°F, 25°C in our tables) and pressure (1 atm), and set up the energy diagram of Fig. 13/2. The energy of the reactants, say fuel and air (saturated with H_2O), consists of the sensible internal energy of all the reactants U_r° and the chemical energy of the fuel E_{ch}, giving a total energy content of $E_t = U_r^\circ + E_{ch}$. Also crossing the boundary of the system is the flow work energy pV_r°. Departing the system are the internal energy of the products U_p° and flow work pV_p°, plus heat Q. This particular heat is the enthalpy of the reaction h_{rp}°. When the fuel is oxidized by a molecular oxygen, engineers call it the *heating value*; $q = -h_{rp}^\circ$.

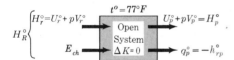

Control Volume for Open-System Calorimeter. The energy balance presumes that at the entrance and exit sections there is internal local thermal equilibrium, with negligible change of kinetic energy; $p = C$, $T_2 = T_1$. **Fig. 13/2**

The *total enthalpy* H_R° of the reactants in the standard state is the total energy stored in the reactants $U_r^\circ + E_{ch}$ plus the flow work pV_r° (no nuclear); Fig. 13/2. An energy balance for Fig. 13/2 gives

(13-1A)
$$H_R^\circ = H_r^\circ + E_{ch} = H_p^\circ + (-h_{rp}^\circ)$$
$$\text{[IN]} \qquad \text{[OUT]}$$

(13-1B)
$$E_{ch} = H_p^\circ - H_r^\circ + (-h_{rp}^\circ)$$

(13-1C)
$$h_{rp}^\circ = H_p^\circ - H_R^\circ$$

This last equation is a definition of **enthalpy of reaction**, which is to say that the total enthalpy of the reactants at T°, p° is greater than the enthalpy of the products at T°, p° by the amount $-h_{rp}^\circ$.

As the test of Fig. 13/2 was defined (entering air saturated with H_2O), the corresponding heating value is the higher one q_h° at constant pressure, because, since the air is already saturated, practically all the H_2O formed by combustion is necessarily condensed when the original temperature is reached; part of the heat rejected q_p° is the latent heat of evaporation of this H_2O. We may compute the lower heating value from

(13-2) $$q_l^\circ = q_h^\circ - m_w h_{fg}^\circ = q_h^\circ - 1050.1\, m_w \text{ Btu/lb fuel}$$

where m_w is pounds of H_2O formed per pound of fuel, $h_{fg} = 1050.1$ for H_2O at $77°F$, and $1050.1\, m_w$ is the latent heat of the water formed.

13.16   SOME REMARKS ON HEATING VALUES

The energy derived from combustion in support of a working (and heating) process is the base for determining the efficiency of that thermodynamic process. The measure is taken as the product of the amount of fuel consumed, during the process, and its heating value. A choice of which heating value (lower or higher) to use poses a problem. The solution is somewhat simple.

If the thermodynamic system can utilize the energy contained in the water vapor derived from burning the hydrogen in the fuel, the higher heating value is used. A case in question is the electric generating plant using fossil fuel. It is quite easy to reduce the temperature of the exit products to below their dewpoint (which we don't do for maintenance reasons).

On the other hand, it is expected that the exhaust products from an automobile engine will never approach their dewpoint temperature or utilize the energy in the water vapor—thus, the lower heating value is used here.

13.17   ENTHALPY CHANGE DURING REACTION

Since there is no common datum for measuring the energy of different substances, we must have a known "crossover" (a known enthalpy of the reaction) to get the change of enthalpy between the reactants in any state 1 and the products in any state 2. This known enthalpy of reaction may be at any p, T, but the available one is usually for the standard state p°, T°. Referring to Fig. 13/3, we see that the enthalpy change from state 1 to 2 is the algebraic sum of the changes 1 to d, d to c, and c to 2; therefore $\Delta H = H_{rp}$ is

(a) $$H_{p2} - H_{r1} = (H_{p2} - H_{pc}^\circ) + (H_{pc}^\circ - H_{rd}^\circ) + (H_{rd}^\circ - H_{r1})$$

Fig. 13/3   *Enthalpy Change, Reactants to Products.* Since curves *bc*-2 and *ad*-1 are not parallel, $-h_{rp}$ along *mn* at temperature T is in general not the same as at $T°$. Process 1-*g*, where $Q = 0$ and $W = 0$, is an adiabatic combustion process, equation (13-5); $S = C$.

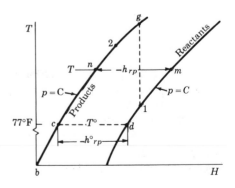

where

$H^\circ_{pc} - H^\circ_{rd} = h^\circ_{rp}$ Btu/lb (or $\bar{h}^\circ_{rp}$ Btu/pmole)
$H_{p2} - H^\circ_{pc}$ is ΔH_p for the products from the crossover dc to the actual state 2,
$H^\circ_{rd} - H_{r1}$ is ΔH_r for all the reactants from the actual state 1 to the crossover dc.

All the foregoing equations may be in terms of either moles n or mass m. Since the coefficients in the chemical equation give the moles of each constituent, the mole basis is usually the more convenient. If the constant pressure specific heats C_m of the mixtures are known,

(b)
$$H_{p2} - H^\circ_{pc} = n_p \int_{T^\circ}^{T_2} C_{pm} \, dT$$

(c)
$$H^\circ_{rd} - H_{r1} = n_r \int_{T_1}^{T^\circ} C_{rm} \, dT$$

However, we commonly compute sensible enthalpy changes of the individual constituents and add them; thus,

(13-3A)
$$H_{p2} - H^\circ_p = \sum_p n_j \int_{T^\circ}^{T_2} C_{pj} \, dT = \sum_p [n_j(h_{j2} - h^\circ_j)]_p$$

(13-3B)
$$H^\circ_r - H_{r1} = \sum_r n_i \int_{T_1}^{T^\circ} C_{ri} \, dT = \sum_r [n_i(h^\circ_i - h_{i1})]_r$$

where subscripts p and r stand for products and reactants; n_j, n_i are the number of moles of j, i constituents; and C_{pj}, C_{ri} are constant pressure, molal specific heats of constituents j, i in products and reactants, respectively. For the more common reactions of fuels, one may have at hand tabular values of sensible enthalpies, as in the *Gas Tables*, Items B 2 through B 10. The foregoing discussion presumes that the fuel is completely oxidized to stable forms. If this is not true, the use of the enthalpy of formation, § 13.29, is a handier procedure in obtaining the ΔH of the reaction.

COMBUSTION PROCESS, RANDOM STATES 13.18

To get an equation for a reaction between any two states 1 and 2 for the case of complete oxidation, first set up an energy balance. In Fig. 13/4, let the reactants pass from a local equilibrium state 1 at the entrance boundary of the control volume to a definable state 2 at the exit. We have

(a)
$$-Q = H_{r1} - H_{p2} + E_{ch}$$

Use the value of E_{ch} from equation (13-1B) in **(a)** and get

(13-4A)
$$H_{p2} - H^\circ_p = H_{r1} - H^\circ_r + q^\circ_p + Q$$

Open System, Steady Flow $\Delta K \approx 0$. **Fig. 13/4**

where $q_p^\circ = -h_{rp}^\circ$ may be either the lower or higher value, depending on the final state 2. At times it will be convenient to let $H_{p2} - H_p^\circ = H_{p2}^\circ$, which is the enthalpy of the products in state 2 measured from the standard state. Similarly, let $H_{R1}^\circ = H_{r1} - H_r^\circ + q_p^\circ$, where H_{R1}° is the total enthalpy at state 1 measured from the standard state. With these symbols, equation (13-4A) becomes

$$(13\text{-}4\text{B}) \qquad\qquad H_{p2}^\circ = H_{R1}^\circ + Q$$

Obtain $H_{p2} - H_p^\circ$ and $H_{r1} - H_r^\circ$ in accordance with equation (13-3).

For **adiabatic combustion**, $Q = 0$ and equation (13-4) becomes

$$(13\text{-}5) \qquad\qquad H_{p2} - H_p^\circ = H_{r1} - H_r^\circ + q_l^\circ$$

$$[\text{ADIABATIC COMBUSTION}]$$

where the units of each term match those of $q_l^\circ = -h_{rp}^\circ$, Btu/lb fuel or Btu/pmole fuel. Equation (13-5) can be used to determine the enthalpy of the products H_{p2} and the temperature at state 2. This temperature T_2 is called the **adiabatic flame temperature** (with dissociation neglected here—see §§ 13.34 and 13.35).

In most problems *related to combustion*, the variation of enthalpy with pressure can be neglected. Remember to observe the phase of each substance in the reaction, especially whether the fuel is liquid, gaseous, or solid (heating value q is different for each phase) and whether the H_2O is liquid or gaseous.

To get the entropy change during a reaction, first compute the absolute entropy of each constituent of the reactants at its temperature and partial pressure; the sum of these values is the entropy of the reactants. The absolute entropy of the products is determined similarly. In mathematical shorthand,

$$(13\text{-}6) \qquad\qquad \Delta S = S_{rp} = \sum_p n_j \bar{s}_j^a - \sum_r n_i \bar{s}_i^a$$

where there are j constituents of products and i constituents of reactants; $\bar{s}_i^a$ and $\bar{s}_j^a$ are in accordance with $\bar{s}^a$ of equation (6-15).

$$(6\text{-}15) \qquad \bar{s}^a = \bar{s}^\circ + \int_{T^\circ}^{T} \frac{C_p\, dT}{T} - \bar{R} \ln \frac{p}{p^\circ} = \bar{s}^\circ + \bar{\phi} - \bar{\phi}^\circ - \bar{R} \ln p$$

for p atm. Since $\bar{\phi}^\circ$ in the gas tables in Appendix B is approximately the same as $\bar{s}^\circ$, Item B 11, $\bar{s} = \bar{\phi} - \bar{R}\ln p$ for this case.

13.19 SENSIBLE ENTHALPY OF LIQUID OCTANE

With respect to a zero datum for octane *vapor* at $0°R$, the enthalpy of *liquid* octane at temperature $T°R$ may be taken as [0.6]

$$(13\text{-}7) \qquad\qquad h_f = 0.5T - 287 \text{ Btu/lb fuel}$$

Since the amount of energy is relatively small and since the sensible enthalpy of other hydrocarbons is not too different, equation (13-7) may be used with small error for any hydrocarbon fuel.

Example—Temperature After Combustion, Ideal Diesel Engine **13.20**

Let the constant pressure combustion process be adiabatic from state 2 to state 3, Fig. 13/5; the temperature at the end of compression $T_2 = 1510°R$; and let there be 200% stoichiometric air (100% excess air), compression ratio $r_k = 13.5$. (a) Compute the temperature T_3 at the end of combustion if liquid octane fuel, C_8H_{18}, is injected at 80°F. (Octane is not a Diesel fuel, but we already have the combustion equation for it and the results would not be much different for the actual fuel.)

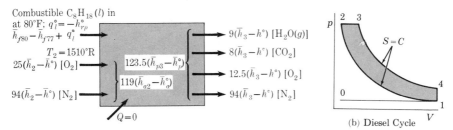

(a) Enthalpy Balance, States 2-3

(b) Diesel Cycle

Enthalpies are measured from $t° = 77°F$; fuel and air have different entering temperatures. **Fig. 13/5**

Solution Using Products Table, Item B9. [The "Products" tables are strictly for a hydrocarbon of the form $(CH_2)_n$. The error in using them for the usual hydrocarbons is expected to be small.] If there is 200% air, the number of moles of air is doubled and the chemical equation from equation **(b)**, § 13.6, is

(a) $$C_8H_{18}(l) + 25O_2 + 94N_2 \rightarrow 9H_2O(g) + 8CO_2 + 12.5O_2 + 94N_2$$

Since we shall use the standard enthalpy of reaction (at 77°F), all sensible enthalpies are correctly measured from this temperature. Knowing that $Q = \Delta H$ for a $p = C$ process, we may use an "enthalpy diagram" to get the energy balance. The enthalpy at state 2, Fig. 13/5, is that of the air plus that of the fuel. After equation (13-3), for 1 pmole of fuel, the sensible enthalpy of the reactants measured from the 77°F datum is

$$H_{r2} - H_r° = \sum_r [n_i(h_{i2} - h_i°)]$$

(b)
$$= M_f(0.5)(T_2 - T°) + 119M_a(h_{a2} - h_a°)$$

$$= (114.2)(1.5) + (119)(29)(371.82 - 128.34) = 840,420 \text{ Btu/pmole fuel}$$

where M_f is the molecular mass of octane, used to convert equation (13-7) to a mole basis; M_a is the molecular mass of air to convert the unit mass values of h to a mole basis: $119 = $ moles of air from equation **(a)**; and h_{a2}, $h_a°$ are from Item B 2 for the specified temperatures.

Inasmuch as Item B 9 gives the enthalpy of the products for "200% air" directly, we have the enthalpy H_{p3} measured from the 77°F datum as

(c) $$H_{p3}° = n_p(\bar{h}_{p3} - \bar{h}_p°) = 123.5(\bar{h}_{p3} - 3774.9)\text{ Btu}$$

For $H_{R2} = H_{r2} - H_r° + (-\bar{h}_{rp}°)$, for $Q = H_{p3}° - H_{R2}° = 0$, and with $H_{p3}°$ from **(c)**, we get

(d)
$$\bar{h}_{p3} = 3774.9 + \frac{840,420}{123.5} + \frac{(114.2)(19,100)}{123.5} = 28,242 \text{ Btu/pmole f}$$

where $-\bar{h}_{rp}^\circ = (114.2)(19,100)$, from Item B 12, for a pmole of *liquid* octane. (Note that this value of $-\bar{h}_{rp}^\circ = \bar{q}_i^\circ$ accounts for the heat to vaporize the liquid fuel at 77°F; that is, the difference between the columns headed "Gaseous Fuel" and "Liquid" in Item B 12 is the latent heat of evaporation.) With $h_{p3} = 28,242$, enter Item B 9, interpolate to the nearest whole degree, and find $T_3 = 3466°R$, the *adiabatic flame temperature*.

Solution Using Individual Products. If the mixture is one that is neither exactly "200% air" nor "400% air," interpolation can be made as explained in the Keenan and Kaye *Gas Tables* for a hydrocarbon fuel. But in the general case, one does not have the properties of the product mix of reaction, and it may be necessary to consider each component separately (see also § 13.28). In this event, a plan of tabulation somewhat as follows can be used. Refer to Fig. 13/5.

$$
\text{Enthalpy at 2}
\begin{cases}
\text{C}_8\text{H}_{18}: \bar{h}_{f80} - \bar{h}_{f77} - \bar{h}_{rp}^\circ = 114.2(1.5 + 19,100) = 2{,}182{,}390 \\[1mm]
\text{O}_2: \quad 25(11{,}098.3 - 3725.1) \qquad\qquad\quad = \quad 184{,}330 \\[1mm]
\text{N}_2: \quad 94(10{,}724.6 - 3729.5) \qquad\qquad\quad = \quad 657{,}540 \\[1mm]
\hline
\text{Total enthalpy (measured from 77°F)}, H_{R2}^\circ \quad = 3{,}024{,}260
\end{cases}
$$

$$
\text{Enthalpy at 3}
\begin{cases}
\text{H}_2\text{O}: 9(\bar{h}_3 - 4258) & \text{O}_2: 12.5(\bar{h}_3 - 3725.1) \\[1mm]
\text{CO}_2: 8(\bar{h}_3 - 4030.2) & \text{N}_2: 94(\bar{h}_3 - 3729.5)
\end{cases} H_{p3}^\circ
$$

For $H_{p3}^\circ - H_{R2}^\circ = 0$, we have

$$3{,}024{,}230 = 9(\bar{h}_3 - 4258) + 8(\bar{h}_3 - 4030.2) + 12.5(\bar{h}_3 - 3725.1) + 94(\bar{h}_3 - 3729.5)$$

$$\quad\;\; [\text{H}_2\text{O}] \qquad\qquad [\text{CO}_2] \qquad\qquad\; [\text{O}_2] \qquad\qquad\;\; [\text{N}_2]$$

The procedure from here is: assume some temperature T_3 and compute the enthalpies of each of the components of the products, continuing to do so until the equation balances; $T_3 = 3466°R$ as found from the Products table should result in a virtual balance.

13.21   Example—Availability and Irreversibility for Combustion

Let the events of this example be an adiabatic reaction of $\text{C}_8\text{H}_{18}(\text{l})$ with air, as in § 13.19, at a constant pressure of 254 psia from a temperature of 1510 to 3466°R, followed by a steady-flow process to the dead state. The combustion is represented by process 2-3 of the previous example, Fig. 13/5(b). Find (a) the steady-flow availability of the products in state 3, Fig. 13/5(b); (b) the change of entropy and the irreversibility of the combustion process; and (c) the change of availability of the combustion process. Assume a sink at $t_0 = t^\circ = 77°F$ and $p_0 = 14.4$ psia. The chemical equation for convenient reference is

(a) $\text{C}_8\text{H}_{18}(\text{l}) + 25\text{O}_2 + 94\text{N}_2 \rightarrow 9\text{H}_2\text{O}(\text{g}) + 8\text{CO}_2 + 12.5\text{O}_2 + 94\text{N}_2$

Solution. (a) Assume that the properties of the products are as given in Item B 9, closely true. The availability of the 123.5 moles of products at 3466°R and 254 psia is, equation (5-8),

(b) $-\Delta\mathscr{A}_{f/0} = -\Delta H + T_0\,\Delta S = n_p[(\bar{h}_3 - \bar{h}_0) - T_0(\bar{s}_3 - \bar{s}_0)]_p$

$$= 123.5\left[28{,}242 - 3774.9 - 537\left(61.24 - 46.3 - \bar{R}\ln\frac{254}{14.4}\right)\right]$$

$$= 2{,}408{,}890 \text{ Btu/pmole } \text{C}_8\text{H}_{18}, \text{ or } \quad 21{,}094 \text{ Btu/lb f}$$

where

$$\bar{s}_3^a - \bar{s}_0 = \bar{\phi}_3 - \bar{\phi}_0 - \bar{R} \ln \frac{p_3}{p_0}$$

with properties taken from Item B 9; $h_0 = h°$ because the sink temperature is taken as 77°F. The value from **(b)** is the work that would be done as the products pass reversibly to the defined dead state. [NOTE: The reader may be bothered by the fact that the availability from equation **(b)** is greater than the enthalpy of the reaction which, at 1510°R, is about 18,500 Btu/lb. But about 30% of this availability represents an input of availability in getting the air from the dead state to 1510°R and 254 psia. Also, if the combustion occurred in a Diesel engine, the "practical" ideal engine would be able to utilize availability only in an isentropic expansion to the pressure of the surroundings, leaving in the working substance at discharge some 25% of the availability in **(b)**. The compression and expansion in a Diesel engine are not steady flow, but the order of magnitude of the numbers is the same. More detail on the Diesel cycle is in Chapter 15.]

(b) To get the change of entropy during 2-3, Fig. 13/5(b), we first find the entropy of each reactant in state 2 and then of each product in state 3. Let the specific heat of liquid octane be taken as 0.5 Btu/lb-°R or $(0.5)(114.2) = 57.1$ Btu/pmole-°R. The absolute entropy of the fuel as it enters the pump that delivers it to the cylinder at 80°F (540°R) is closely

(c) $$C_8H_{18}(l) \text{ at 1 atm:} \quad \bar{s}^a = \bar{s}° + \int \frac{C\,dT}{T} = 85.5 + 57.1 \ln \frac{540}{537}$$

or $\bar{s}^a = 85.8$ Btu/pmole-°R; all $\bar{s}°$ values are from Item B 11. Its entropy at the beginning of the reaction is its value after the pump work has been done. Taking $\beta = 0.43 \times 10^{-3}\,°R^{-1}$, $\rho = 45$ lb/ft³ from a convenient handbook,[14.7] assuming that the pumping to 254 psia is isothermal, and following equation **(d)** in the example of § 11.9, we get the change of entropy as

(d) $$\Delta \bar{s} = -\beta v(p_2 - p_1) = -\frac{(0.43)(254 - 14.4)(144)(114)}{(10^3)(45)(778)} = -0.0483 \text{ Btu/pmole-°R}$$

which compared to $\bar{s}^a$ above, is seen to be rather negligible. So, we say $\bar{s}_f^a = 85.8$ Btu/pmole-°R. For isentropic pumping, $\Delta s = 0$.

Using mole values from equation **(a)**, we find the partial pressures of the various gaseous constituents:

$$p_2[O_2] = \left(\frac{25}{119}\right) 17.28 = 3.63 \text{ atm} \qquad p_3[H_2O] = \left(\frac{9}{123.5}\right) 17.28 = 1.26 \text{ atm}$$

$$p_2[N_2] = \left(\frac{94}{119}\right) 17.28 = 13.65 \text{ atm} \qquad p_3[CO_2] = \left(\frac{8}{123.5}\right) 17.28 = 1.12 \text{ atm}$$

$$p_3[N_2] = \left(\frac{94}{123.5}\right) 17.28 = 13.15 \text{ atm} \qquad p_3[O_2] = \left(\frac{12.5}{123.5}\right) 17.28 = 1.75 \text{ atm}$$

Since we are assuming that the N_2 passes through the reaction with its molecular structure unchanged, its change of entropy must be computed directly by equation (6-13); from 1510 to 3466°R and with partial pressures as just computed,

$$\Delta S[N_2] = n_i\left(\Delta\bar{\phi}_i - \bar{R} \ln \frac{p_3}{p_2}\right)$$

(e)

$$= 94\left(59.861 - 53.121 - \bar{R} \ln \frac{13.15}{13.65}\right) = 640.5 \text{ Btu/°R}$$

Next, compute the absolute entropy S^a of the other constituents by equation (6-15) or (6-16);

(f) $S_2^a[O_2] = n_i(\bar{\phi}_i - \bar{R} \ln p_i)_2$

$= 25(56.709 - \bar{R} \ln 3.63) = 1353.7 \text{ Btu/}°\text{R}$

(g) $S_3^a[H_2O] = 9(62.7587 - \bar{R} \ln 1.26) = 560.7 \text{ Btu/}°\text{R}$

(h) $S_3^a[CO_2] = 8(73.3219 - \bar{R} \ln 1.12) = 585 \text{ Btu/}°\text{R}$

(i) $S_3^a[O_2] = 12.5(63.8263 - \bar{R} \ln 1.75) = 783.9 \text{ Btu/}°\text{R}$

If desired, ΔS for 12.5 moles of O_2 could be computed, as is done for N_2; then equation **(f)** would have been for 12.5 moles, or half the value computed. The difference of the absolute entropies computed, state 3 minus state 2, is

(j) $560.7 + 585 + 783.9 - 1353.7 - 85.8 = 490.1 \text{ Btu/}°\text{R}$

Including the change for the N_2, the change of entropy for the reaction is

(k) $S_{rp} = 640.5 + 490.1 = 1130.6 \text{ Btu/}°\text{R-pmole } C_8H_{18}$

Since the closed-system combustion process is adiabatic, there is no interaction with the surroundings except for the work done on the boundary (piston); therefore the irreversibility is

(l) $I = T_0 S_{rp} = (537)(1130.6) = 607,132 \text{ Btu per mole of fuel or } 5316 \text{ Btu/lb f}$

(c) The change of availability during the combustion process is not the same as the availability of the products. Since $H_{p3}° = H_{R2}°$,

(m) $\Delta \mathscr{A}_f = H_{p3}° - H_{R2}° - T_0(S_{p3} - S_{r2}) = -T_0 S_{rp} = -607,132 \text{ Btu}$

the negative of the irreversibility (in this case).

13.22   Example—Fuel Cell Reaction

To complete the picture and emphasize the comparison of a reversible reaction to the above calculations, compute the work that would be done if $C_8H_{18}(l)$ reacted reversibly with O_2 in a fuel cell at $600°\text{R}$ and 5 atm pressure.

Solution.   Let the reaction be

(a) $C_8H_{18}(l) + 12.5O_2 \rightarrow 9H_2O(g) + 8CO_2$

Concerning entropy computations, we found above that the entropy of the liquid C_8H_{18} changed very little for a much larger increase in pressure; hence, we shall ignore this correction, in view of the size of other pertinent entropy values. For 17 moles of gaseous products, the partial pressures (assuming Dalton's law is not too much in error) are $p(H_2O) = (9/17)5 = 2.65$ atm and $p(CO_2) = (8/17)5 = 2.35$ atm. For $C_8H_{18}(l)$, use $c_p = 0.5$ Btu/lb-°R, or 57.1 Btu/pmole-°R. At $600°\text{R}$, we have

(b) $\bar{s}^a[C_8H_{18}(l)] = \bar{s}° + \int_{537}^{600} \frac{C \, dT}{T} = 85.5 + 57.1 \ln \frac{600}{537} = 94.61 \text{ Btu/}°\text{R-pmole}$

(c) $S^a[O_2] = n_i(\bar{\phi}_i - \bar{R} \ln p_i) = 12.5(49.762 - \bar{R} \ln 5) = 582.075 \text{ Btu/}°\text{R}$

(d) $S^a[H_2O(g)] = 9(45.97 - \bar{R} \ln 2.65) = 396.3123$ Btu/°R

(e) $S^a[CO_2] = 8(52.038 - \bar{R} \ln 2.35) = 402.7296$ Btu/°R

(f) $S_{rp} = 396.31 + 402.73 - 582.07 - 94.61 = 122.36$ Btu/°R

for 1 pmole of C_8H_{18}. Obtain enthalpies, measured from 77°F, and use the lower heating value:

(g)
$$H[C_8H_{18}(l)] = C(T - T°) + (-\bar{h}^\circ_{rp})$$
$$= 57.1(600 - 537) + (19,100)(114.2) = 2,184,817 \text{ Btu/pmole}$$

(h) $H[O_2] = n(\bar{h} - \bar{h}^\circ) = 12.5(4168.3 - 3725.1) = 5540$ Btu

(i) $H[H_2O(g)] = 9(4764.7 - 4258.0) = 4560$ Btu

(j) $H[CO_2] = 8(4600.9 - 4030.2) = 4566$ Btu

(k) $\Delta H = H_{rp} = 4560 + 4566 - 2,184,817 - 5540 = -2,181,231$ Btu/pmole C_8H_{18}

For the constant T, p process, the reversible work is $W = -\Delta G$, or

(l)
$$W = -(\Delta H - T\,\Delta S) = 2,181,231 - (600)(122.36)$$
$$= 2,107,815 \text{ Btu/pmole } C_8H_{18} \quad \text{or} \quad 18,457 \text{ Btu/lb}$$

corresponding to an efficiency based on the standard enthalpy of reaction of $W/\bar{q}^\circ_i = 18,457/19,100 = 96.6\%$.

CONSTANT VOLUME COMBUSTION   13.23

In the closed, constant volume system (bomb calorimeter), place a small amount of water in order to saturate the reactants with H_2O. Then, when the products have cooled back to the standard temperature of 77°F, the H_2O formed by combustion condenses and the higher heating value is obtained. The total energy stored initially is the sensible internal energy of the reactants U°_r plus the chemical energy E_{ch}, whose magnitude is unknown and unneeded. After combustion and cooling to the standard temperature, the energy stored in the system is U°_p, the internal energy of the products. There is no work, so that the energy crossing the boundary consists only of heat Q, a particular value represented by the symbol $q^\circ_v = -u^\circ_{rp}$. The decrease in stored energy is equal to the energy leaving the system, from $Q = \Delta E_t + W$, equation (4-6);

(a) $E^\circ_{t1} - E^\circ_{t2} = -Q = q^\circ_v$

The magnitude of q°_v is all that is learned from the test. Substitute $E^\circ_{t1} = U^\circ_r + E_{ch}$ and $E^\circ_{t2} = U^\circ_p$, Fig. 13/6, and find

(b) $E_{ch} = U^\circ_p - U^\circ_r + q^\circ_v$ Btu/lb fuel,   or Btu/pmole fuel

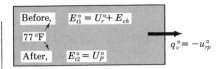

Fig. 13/6   *Closed-System Calorimeter, $V = C$, $t_2 = t_1 = 77°F$.*

Now consider a constant volume reaction between *any* two states 1 and 2, the energy diagram as in Fig. 13/7. By the first law,

(c) $Q = E_{t2} - E_{t1} = U_{p2} - U_{r1} - E_{ch}$

where E_t is the total stored energy. The value of E_{ch} in equation **(b)** substituted into this one gives

(13-8) $U_{p2} - U_p^\circ = U_{r1} - U_r^\circ + q_v^\circ + Q = U_{r1} - U_r^\circ - u_{rp}^\circ + Q$

with Q of conventional sign [heat added $(+)$, rejected $(-)$]. When $Q = 0$,

(13-9) $U_{p2} - U_p^\circ = U_{r1} - U_r^\circ + q_v^\circ = U_{r1} - U_r^\circ - u_{rp}^\circ$

[ADIABATIC COMBUSTION—CONSTANT VOLUME]

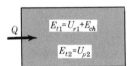

Fig. 13/7   *For Adiabatic Combustion, $Q = 0$.*

If the H_2O is condensed after combustion, the *higher heating value at constant volume* q_{hv}° applies. If the H_2O is not condensed, as it would *not* be for adiabatic combustion, use the *lower heating value at constant volume*, which is obtained by subtracting the internal energy change during vaporization of the H_2O at the test temperature. Let m_w lb/lb fuel be the mass of H_2O formed during combustion; then, for $u_{fg}^\circ = 990.8$ Btu/lb at 77°F,

(13-10) $q_{lv}^\circ = q_{hv}^\circ - m_w u_{fg}^\circ = q_{hv}^\circ - 990.8\, m_w$ Btu/lb fuel

13.24   RELATION OF CONSTANT PRESSURE AND CONSTANT VOLUME HEATING VALUES

Since we are assuming that the gaseous constituents are ideal, let $H = U + pV = U + n\bar{R}T$, for n moles of ideal *gaseous* constituents. Then by equations (13-1) and **(b)**, § 13.21, we have $(-h_{rp}^\circ = q_p^\circ)$

(a) $\bar{q}_p^\circ = H_r^\circ - H_p^\circ + E_{ch} = U_r^\circ + n_r\bar{R}T^\circ - (U_p^\circ + n_p\bar{R}T^\circ) + E_{ch}$

(b) $\bar{q}_v^\circ = U_r^\circ - U_p^\circ + E_{ch}$ Btu/pmole fuel

By subtraction

(c) $$\bar{q}_p^\circ - \bar{q}_v^\circ = n_r \bar{R} T^\circ - n_p \bar{R} T^\circ = \bar{R} T^\circ (n_r - n_p)\ \text{Btu/pmole}$$

(13-11) $$q_p^\circ - q_v^\circ = \frac{RT^\circ}{M_f}(n_r - n_p)\ \text{Btu/lb fuel}$$

where M_f is the molecular weight of the fuel; $\bar{R} = 1.986$. The moles of *gaseous* products n_p may be greater than, equal to, or less than the moles of *gaseous* reactants n_r; thus, either $q_p^\circ < q_v^\circ$, or $q_p^\circ > q_v^\circ$, where both are either higher heating values or lower heating values. Since the moles of inert gases appear on each side of the chemical equation, they may be ignored in equations **(c)** and (13-11).

Example 13.25

Relatively compare the heating values q_p° and q_v°, higher and lower, for propane C_3H_8.
Solution. The stoichiometric equation is

$$1C_3H_8 + 5O_2 \rightarrow 3CO_2 + 4H_2O$$

For the higher heating value, the water condenses and $n_p = 3$; $n_r = 6$. Thus,

$$q_p^\circ - q_v^\circ = \frac{RT^\circ}{m_f}(n_r - n_p)$$

$$q_p^\circ - q_v^\circ > 0$$

$$q_{ph}^\circ > q_{vh}^\circ$$

For the lower heating value, $n_p = 7$ and $n_r = 6$. Here

$$q_p^\circ - q_v^\circ < 0$$

or

$$q_{p1}^\circ < q_{v1}^\circ$$

Example—Computing Heating Values at Constant Volume 13.26

What are the lower and higher heating values of octane C_8H_{18} at constant volume, liquid fuel?

Solution. Although the bomb calorimeter in general gives more accurate results than the steady-flow calorimeter, most reports in the literature are for $-h_{rp}^\circ$. From Item B 12, we have $q_{hp}^\circ = 20{,}591$ for $C_8H_{18}(l)$. Ignore the quantity of inert gases in the difference $n_r - n_p$. From § 13.6,

(a) $$C_8H_{18}(l) + 12.5O_2 + \cdots \rightarrow 9H_2O + 8CO_2 + \cdots$$

for $H_2O(l)$, $n_r = 12.5$ and $n_p = 8$. Therefore, from (13-11), we get

$$q_{hv}^\circ = q_{hp}^\circ + \frac{\bar{R}T}{M_f}(n_p - n_r) = 20{,}591 + \frac{(1.986)(537)}{114.2}(-4.5) = 20{,}549\ \text{Btu/lb}$$

the higher heating value at $V = C$. From (13-10),

(b) $q_{lv}^{\circ} = q_{hv}^{\circ} - 990.8\,m_w = 20{,}549 - (990.8)(1.42) = 19{,}140\ \text{Btu/lb}$

where $m_w = (9)(18)/114 = 1.42$ lb H_2O/lb f is taken from §13.6. The lower heating value at constant volume can also be obtained directly from the lower heating value at constant pressure, $-h_{rp}^{\circ}\,[C_8H_{18}(l)] = 19{,}100\ \text{Btu/lb}\ [H_2O(g)]$. The reaction equation above shows that for H_2O gaseous, $n_r = 12.5$ and $n_p = 17$; thus,

(c) $q_{lv}^{\circ} = q_{lp}^{\circ} + \dfrac{\bar{R}T^{\circ}}{M_f}(n_p - n_r) = 19{,}100 + \dfrac{(1.986)(537)}{114.2}(+4.5) = 19{,}142\ \text{Btu/lb}$

From the test value of the bomb calorimeter, we may similarly compute $-h_{rp}^{\circ}$.

13.27   HEATING VALUE AT OTHER THAN STANDARD TEMPERATURE

If it is desired to convert the heating value at standard temperature to that at some other temperature, imagine the calorimeter test to be run from and back to the desired temperature, say, state 2. Then the heating value at temperature 2 is

(a) $q_{p2} = H_{r2} - H_{p2} + E_{ch}$ $[T_2 = C]$

after Fig. 13/2. Equation (13-1) may be written

(13-1) $q_p^{\circ} = H_r^{\circ} - H_p^{\circ} + E_{ch}$

Subtract corresponding sides of this equation from equation **(a)** and solve for q_{p2};

(b) $q_{p2} = q_p^{\circ} + (H_{r2} - H_r^{\circ}) - (H_{p2} - H_p^{\circ})$

[PRESSURE CONSTANT]

the heating value q_{p2} at any temperature 2, each term in the same units, and the parenthetic terms are obtained in accordance with equation (13-4). The nitrogen in the reactants is the same as in the products and may be ignored in the foregoing equation with some saving in time. A similar treatment of the constant volume combustion yields (q_{v2} at T_2):

(c) $q_{v2} = q_v^{\circ} + (U_{r2} - U_r^{\circ}) - (U_{p2} - U_p^{\circ})$

[VOLUME CONSTANT]

The hypothetical heating value at $0°R$ is obtained by letting the sensible enthalpies H_{r2} and H_{p2} be zero; thus,

(d) $-h_{rp0} = q_{p0} = q_p^{\circ} + H_p^{\circ} - H_r^{\circ} = -h_{rp}^{\circ} + H_p^{\circ} - H_r^{\circ}$

If tabulated values of q_{p0} were available, the arithmetic of some solutions would be much reduced. For example, substituting the value of q_p° from the foregoing equation into (13-5), we get

(e)
$$H_{p2} = H_{r1} + q_{p0} + Q$$

where H_{p2} and H_{r1} are measured from $0°R$; thus the gas value tables could be used without a subtraction involving the standard state $(h°)$.

Example—Enthalpy of Reaction at $0°R$ Temperature   13.28

Compute the lower heating value of gaseous octane at $0°R$; all components in the reaction are gaseous with their enthalpies taken as zero at $0°R$.

Solution.   The reaction equation is $C_8H_{18}(g) + 12.5O_2 → 8CO_2 + 9H_2O(g)$. The enthalpy changes of N_2 and other diluting gases are zero and can be ignored. From Item B 12, we find $q_l = -h_{rp}° = 19,256$ Btu/lb $C_8H_{18}(g)$, the lower heating value $[H_2O(g)]$. In equation **(b)**, § 13.24, the second state is at $0°R$, designated as H_0; $H_2 - H° = H_0 - H°$. Since the partial pressure of the octane is low if the reaction occurs at 1-atm pressure, we could estimate its sensible enthalpy as c_pT, but we recall that this value is usually high if c_p is the value at $T°R$. If we use equation (13-7), the results should be somewhat more accurate; from this, we get $h_f = 0.5T - 287 = -18.5$ Btu/lb. From Item B 11, we have $h_{fg} = 17,856/114.22 = 156.3$ Btu/lb for C_8H_{18} at $77°F$; from Item B 12, we get $h_{fg} = q_i°(g) - q_i°(l) = 19,256 - 19,100 = 156$ Btu/lb; take your choice. Then, as measured from an ideal gas state at $0°R$, the sensible enthalpy of the octane is

(a)
$$h(C_8H_{18}) = -18.5 + 156.3 = 137.8 \text{ Btu/lb}$$

Summarizing according to equation **(b)**, § 13.27, for $537°R$, we have

$$C_8H_{18}: \quad H_0 - H° = 0 - 137.8 = -137.8 \text{ Btu/lb}$$

$$O_2: \quad H_0 - H° = 0 - \frac{(12.5)(3725.1)}{114.22} = -407.7 \text{ Btu/lb fuel}$$

$$\text{Reactants:} \quad H_{r0} - H_r° = -137.8 - 407.7 = -545.5 \text{ Btu/lb fuel}$$

$$CO_2: \quad H_0 - H° = 0 - \frac{(8)(4030.2)}{114.22} = -282.3 \text{ Btu/lb fuel}$$

$$H_2O: \quad H_0 - H° = 0 - \frac{(9)(4258)}{114.22} = -335.5 \text{ Btu/lb fuel}$$

$$\text{Products:} \quad H_{p0} - H_p° = -282.3 - 335.5 = -617.8 \text{ Btu/lb fuel}$$

Then from equation **(b)**, § 13.27, we get the heating value at $0°R$ as

$$-h_{rp0} = q_{l0} = 19,256 - 545.5 - (-617.8) = 19,328.3 \text{ Btu/lb fuel}$$

This is, of course, a figment of the imagination, but since enthalpy is a point function, it does not matter by what path the change of enthalpy between two particular states is obtained.

ENTHALPY OF FORMATION   13.29

The heat involved (exothermic or endothermic) when a compound is formed from its elements at constant temperature and pressure is called the *heat of formation* or the *enthalpy of formation* Δh_f. When the reaction occurs (or is imagined to occur) at a standard state, usually 77°F and 1 atm for each constituent as before, the heat is the *standard* enthalpy of formation Δh_f°. According to this plan, the enthalpy of formation is the enthalpy of the compound as measured from a datum at which the *elements* (most stable isotope) have zero enthalpy. Let CO_2 be formed by a reaction of C with O_2 at 77°F;

(a) $$C(s) + O_2 \rightarrow CO_2(g)$$

We could measure the heat Q of this reaction and this would be the enthalpy of formation of CO_2 because this reaction is forming CO_2 from its elements; $\Delta h_f^\circ = -169{,}293$ Btu/pmole of CO_2, Item B 11, naturally a negative number for this reaction because energy leaves the system when $T = C$. In this case, this heat is also the heating value of 1 mole of carbon, which is 14,097 Btu/lb C, Item B 12, when the C is oxidized by O_2. The enthalpy of combustion and the enthalpy of formation are the same *only* when an *element* is oxidized by an element (O_2). The same heat is involved, except for sign, if the reaction is performed in the opposite direction, a compound (CO_2) converted into its elements (C and O_2). From this observation, Hess formulated the law that the net enthalpy change ($p = C$) in chemical processes is independent of the sequence of the processes. From this principle, it follows that the change of enthalpy of formation during a constant p, T chemical reaction is equal to the enthalpy of the reaction; at the standard state,

(13-12) $$-H_{rp}^\circ = (\Delta H_f^\circ)_r - (\Delta H_f^\circ)_p$$

where

$$(\Delta H_f^\circ)_r = \sum_r n_i (\Delta h_f^\circ)_i \qquad \text{the sum for } i \text{ constituents of reactants}$$

$$(\Delta H_f^\circ)_p = \sum_p n_j (\Delta h_f^\circ)_j \qquad \text{the sum for } j \text{ constituents of products}$$

In making an enthalpy balance as we did in § 13.20, with the enthalpies measured from the defined 77°F datum, we use the enthalpies of formation for which the datum is *zero enthalpy for elements at* 1 *atm and* 77°F, the standard state. At some temperature other than standard, the *total* enthalpy h_t measured from the base just defined is

(b) $$h_t = \Delta h_f^\circ + (h - h^\circ)$$

for either unit mass or 1 mole, where h is the sensible enthalpy at a specified temperature. For n_i moles of the ith constituent, the enthalpy of a mixture at any temperature T is then

(13-13) $$H_m = \sum_i [n_i (\Delta h_f^\circ)_i + n_i (\bar{h}_i - \bar{h}_i^\circ)]$$

applicable to reactants H_R or products H_p; see equation (13-4). This approach is particularly advantageous when a reaction is incomplete and when significant dissociation (§ 13.35) is to be accounted for. The energy balance for a steady-flow system with $W = 0$ and $\Delta P = 0$ is

(c)
$$Q + H_R + K_R = H_p + K_p$$

where H_R and H_p are computed according to equation (13-13) and $K = mv^2/(2\mathbf{k})$, say, for m lb/pmole of fuel. (Normally during a reaction, no mass crosses the boundary of the control volume except the main streams.) Formalizing, the change of enthalpy in terms of enthalpies of formation from state 1 to state 2 is, for j components of products and i components of reactants,

(d)
$$\Delta H = \sum_p n_j(\bar{h}_2 - \bar{h}^\circ)_j + \sum_p n_j(\Delta \bar{h}_f^\circ)_j - \sum_r n_i(\Delta \bar{h}_f^\circ)_i - \sum_r n_i(\bar{h}_1 - \bar{h}^\circ)_i$$

An example will help with the detail.

Example—Temperature After Combustion, Diesel Engine   13.30

Let the data be the same as in example of § 13.20, except that enthalpies of formation of each compound are to be used.

Solution. Use the results of § 13.20 as far as possible: the moles of each component; the value of $\bar{h}_{f80} - \bar{h}_{f77} = 171.3$ Btu/pmole for the fuel (all combustible). The standard enthalpy of formation of $C_8H_{18}(l)$ is $\Delta \bar{h}_f^\circ = -107,532$ Btu/mole, Item B 11. Compute the enthalpies in accordance with equation (13-13), and as set up in Fig. 13/8, letting $H_{p3} - H_{R2} = 0$. Accordingly, we have (Item B 11 for $\Delta \bar{h}_f^\circ$)

$C_8H_{18}(l)]_r$: $171.3 + (-107,532)$ $\quad = -107,361$

$O_2]_r$: $25(11,098.3 - 3725.1)$ $\quad = +184,330$

$N_2]_r$: $\underline{94(10,724.6 - 3729.5)} \quad = +657,539$
H_{R2} $\qquad\qquad\qquad\qquad\qquad\quad = +734,508$ Btu/pmole C_8H_{18}

$H_2O(g)]_p$: $9(\bar{h}_3 - 4258 - 104,036)$ $\quad = 9\bar{h}_3(H_2O) - 974,646$

$CO_2]_p$: $8(\bar{h}_3 - 4030.2 - 169,293) = 8\bar{h}_3(CO_2) - 1,386,584$

$O_2]_p$: $12.3(\bar{h}_3 - 3725.1)$ $\quad = 12.5\bar{h}_3(O_2) - 46,564$

$N_2]_p$: $\underline{94(\bar{h}_3 - 3729.5)} \qquad\qquad = 94\bar{h}_3(N_2) - 350,573$
$H_{p3} = \text{SUM: } 9\bar{h}_3(H_2O) + 8\bar{h}_3(CO_2) + 12.5\bar{h}_3(O_2) + 94\bar{h}_3(N_2) - 2,758,367 = 734,508$

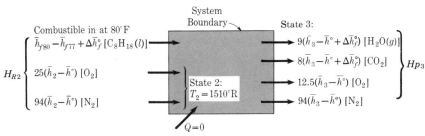

Enthalpy Balance with Enthalpies of Formation. If there are energies involved such as W and K, they should be included.   **Fig. 13/8**

At this point, the procedure is to assume a temperature T_3, look up the corresponding enthalpies, check in the foregoing equation; repeat this process until a balance is found. For a temperature of 3461°R (interpolating to the nearest whole degree), the enthalpy equation nearly balances; therefore, $T_3 \approx 3461$°R. This solution is basically the same as the one in §13.20 because $-H_{rp}^\circ = (\Delta H_f^\circ)_r - (\Delta H_f^\circ)_p$, equation (13-12). It does assume air to be composed entirely of oxygen and nitrogen.

13.31 ENTHALPY OF FORMATION FROM HEATING VALUE

When appropriate heats of reaction are known, these can be used to determine enthalpies of formation. For example, assume that the enthalpy of formation of C_8H_{18} is unknown. Heating values $-\bar{h}_{rp}^\circ$ for the following reactions are taken from Item B 12:

(a) $C_8H_{18}(g) + 12.5O_2 \rightarrow 9H_2O(l) + 8CO_2$ $\bar{h}_{rp}^\circ = -(20{,}747)(114.224)$

(b) $9H_2 + 4.5O_2 \rightarrow 9H_2O(l)$ $H_{rp}^\circ = -(9)(60{,}998)(2.016)$

(c) $8C + 8O_2 \rightarrow 8CO_2$ $H_{rp}^\circ = -(8)(14{,}097)(12.01)$

in which $-H_{rp}^\circ$ is now the value for the number of moles involved. These equations can be manipulated by the rules of algebra. If corresponding sides of equation **(a)** are subtracted from the sum of **(b)** and **(c)**, we get

(d) $9H_2 + 8C \rightleftharpoons C_8H_{18}(g)$

which is the equation for the formation of $C_8H_{18}(g)$ from its elements—$\Delta \bar{h}_f^\circ$ of the *elements* is zero. See Fig. 13/9. The enthalpy change of this process is the enthalpy of formation by definition (and is the enthalpy of the reaction shown) but not of the *combustion* reaction, and is obtained by subtracting the H_{rp}°'s of **(b)** and **(c)** from $\bar{h}_{rp}^\circ$ of **(a)**; thus,

(e) $-\Delta \bar{h}_f^\circ[C_8H_{18}(g)] = \bar{h}_{rp}^\circ - [H_{rp}^\circ(H_2) + H_{rp}^\circ(C)]$

$$= -2{,}369{,}805 + 2{,}461{,}188 = 91{,}383 \text{ Btu/pmole } C_8H_{18}$$

The difference between this value and $\Delta \bar{h}_f^\circ = -89{,}676$ from Item B 11 is attributable to the inconsistency in the basic data behind certain values in Items B 11 and B 12. If $C_8H_{18}(l)$ liquid is formed from its elements, the additional heat rejected is the enthalpy of evaporation h_{fg}. At 77°F, $\bar{h}_{fg}^\circ = 17{,}856$ Btu/pmole from Item B 11; hence $\Delta \bar{h}_f^\circ [C_8H_{18}(l)]$ is $-89{,}676 - 17{,}856 = -107{,}532$ Btu/pmole, where we used the Item B 11 value for $\Delta \bar{h}_f^\circ$. In general,

(f) $\Delta h_f^\circ(\text{liq}) = \Delta h_f^\circ(\text{vap}) - h_{fg\,77^\circ}$

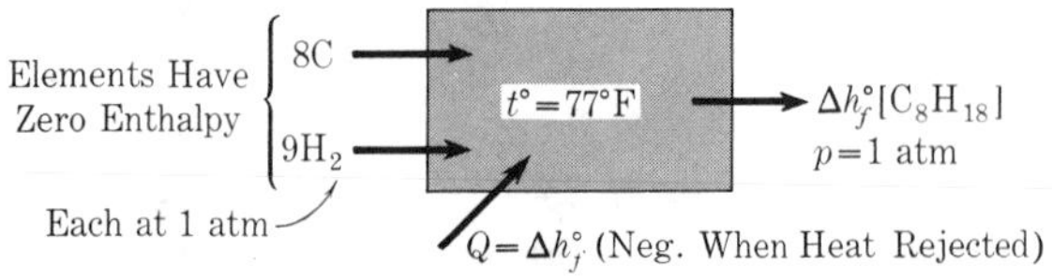

Fig. 13/9 *Enthalpy of Formation. For ideal gases, enthalpy is not a function of p.*

HEATING VALUE FROM ENTHALPY OF FORMATION   13.32

A similar procedure may be used to find heating values when enthalpies of formation are available. Illustrating with CO, we have

(a) $$CO + 0.5O_2 \rightarrow CO_2(g)$$

Applying equation (13-12), we get

$$H_{rp}^{\circ} = (\Delta H_f^{\circ})_p - (\Delta H_f^{\circ})_r = -169,293 - (-47,548) = -121,745 \text{ Btu/pmole CO}$$

(b) $$H_{rp}^{\circ} = \frac{-121,745}{28.01} = 4346 \text{ Btu/lb CO}$$

[COMPARE WITH ITEM B 12]

Some heating values cannot be measured directly, for example, the reaction of C to CO, because some CO_2 is inevitably formed when C is oxidized. By experiment, we can find heating values of C to CO_2 and CO to CO_2.

(c) $$C + O_2 \rightarrow CO_2(g) \qquad \bar{h}_{rp}^{\circ} = -(14,097)(12.01)$$

(d) $$CO + 0.5O_2 \rightarrow CO_2(g) \qquad \bar{h}_{rp}^{\circ} = -(4346)(28.01)$$

Subtracting corresponding sides of **(d)** and **(c)** and combining the O_2 terms, we get

(e) $$C + 0.5O_2 \rightarrow CO \qquad \bar{h}_{rp}^{\circ} = -47,574 \text{ Btu/pmole C}$$

which is the negative of the heating value of C burned to CO and also the enthalpy of formation of CO; $\Delta \bar{h}_f^{\circ} = -47,574 \text{ Btu/pmole CO}$.

GIBBS FUNCTION OF FORMATION   13.33

The values of the Gibbs functions $\Delta \bar{G}_f^{\circ}$ in Item B 11 are the respective changes of the Gibbs function during the formation of the compound from its elements with *each element* and the compound in the standard state of 1 atm and 77°F. This definition is analogous to that for the enthalpy of formation $\Delta \bar{h}_f^{\circ}$; however, the Gibbs function is a function of pressure (entropy is also involved), whereas enthalpy of an ideal gas is not. Since $G = h - Ts$, we find

(a) $$dG_T = dh - T\,ds \qquad \text{at } T = C$$

the integration of which for the formation of the compound at the standard state is (lb, kg, moles, and so on, as convenient)

(b) $$\Delta G = \Delta G_f^{\circ} = \Delta h_f^{\circ} - T^{\circ} S_{rp}^{\circ}$$

where, for the mole basis, the entropy change for the reaction is

$$S_{rp}^{\circ} = n_p \bar{s}_p^{\circ} - \sum_r (n_i \bar{s}_i^{\circ});$$

$\bar{s}_p^\circ$ is the absolute entropy of the product (compound) whose Gibbs function of formation is desired, and the sum covers all i elements of the reactants; $\bar{s}_i^\circ$ for each reactant is the standard value, 1 atm and 77°F.

Considering the Gibbs function further, we note that, for an ideal gas, the change of G with pressure but at constant temperature is $\Delta G = -T \Delta s$ because $\Delta h = 0$; and by equation (6-13), $\Delta \bar{s} = -\bar{R} \ln p_2/p_1$. If $\bar{G}^\circ$ in the standard state is known, then at any other pressure, we have

$$(13\text{-}14) \qquad\qquad \bar{G}_i \text{ (at } p_i) = \bar{G}_i^\circ + \bar{R} T^\circ \ln p_i$$

for p_i atm. Note that if (13-14) is applied to an *element* with $\bar{G}_i$ measured from the base defined for the Gibbs function of formation, $(\Delta \bar{G}_f^\circ)_i = 0$. From (13-14), it follows that for a mixture of i ideal gas components (§ 12.4) the Gibbs function of the mixture is

$$(c) \qquad\qquad G_m^\circ = \sum_i n_i \bar{G}_i = \sum_i n_i (\bar{G}_i^\circ + \bar{R} T^\circ \ln p_i)$$

at standard temperature, each component at its partial pressure p_i; $p_m = \sum p_i$. By the same reasoning, if $\bar{G}_i'$ at any temperature T' and 1 atm pressure is known, the value at any other pressure is

$$(13\text{-}15) \qquad\qquad \bar{G}_i = \bar{G}_i' + \bar{R} T' \ln p_i$$

$$(13\text{-}16) \qquad\qquad G_m' = \sum_i n_i (\bar{G}_i' + \bar{R} T' \ln p_i)$$

where p_i atm is the pressure of the ith component and the temperature is T'.*

13.34   REVERSIBLE REACTIONS

Imagine 1 mole of H_2 and a half mole of O_2 in a reaction chamber at some typical atmospheric pressure and temperature. Any reaction that occurs will be according to

$$(a) \qquad\qquad H_2 + \tfrac{1}{2} O_2 \to H_2O$$

However, as far as any normal measurement shows, nothing happens. Actually, given time, some H_2 and O_2 will get together so that a few molecules of H_2O are formed; the number can be calculated for a mixture in equilibrium, even if it cannot be measured. Let the temperature be raised a few hundred degrees. More H_2O is formed, and the amount soon becomes detectable as the temperature increases. At any particular temperature, the equilibrium mixture consists of a certain amount of each of the reactants and products. (By the judicious use of a catalyst, the reaction can be promoted, as in a fuel cell.) Notice that even if the reactants are not in stoichiometric proportions, say, the amount of H_2 is doubled, any reaction that

---

* Common practice in the literature is to use the symbol G° (or F°) as the Gibbs function at any temperature and 1 atm pressure; the standard state is for 1 atm and any temperature. Considering our limited application, we shall let the superscript ° imply also the standard temperature, 25°C. In the case of extended work in this area, many tabulated values of absolute entropy, enthalpy, and Gibbs functions of formation against temperature are available.

occurs is always according to equation **(a)**—two molecules of H_2 and one of O_2 disappear for every two molecules of H_2O formed. When the mixture arrives at a chemical equilibrium condition, H_2 and O_2 may still react to give H_2O, but for the amount of H_2O formed, an equal amount dissociates into H_2 and O_2; the reaction **(a)** is then proceeding in both directions and is a reversible reaction, written as

$$(13\text{-}17) \qquad\qquad H_2 + \tfrac{1}{2}O_2 \rightleftharpoons H_2O(g)$$

for all substances in the gaseous phase. If the proportion of reactants is changed and if inert gases are included, the extent of the reaction will be different at a particular temperature and total pressure, but the equilibrium reaction remains as given in (13-17). Other equilibrium equations that we shall find useful are

$$(13\text{-}18) \qquad\qquad CO + \tfrac{1}{2}O_2 \rightleftharpoons CO_2$$

$$(13\text{-}19) \qquad\qquad CO_2 + H_2 \rightleftharpoons CO + H_2O \qquad [\text{WATER GAS REACTION}]$$

See also Item B 32. Equilibrium reactions are often written with the equals sign, as $N_2 = 2N$, $H_2 = 2H$, and so on.

To get an aspect about which we are concerned, consider the octane reaction in air, represented as follows:

$$\textbf{(b)} \qquad C_8H_{18} + aO_2 + bD_r \rightarrow cCO_2 + dCO + eH_2O + fH_2 + gO_2 + hD_p$$

where D_r represents all diluting components of the reactants (principally N_2) and D_p represents all other components in the products (again principally N_2), but there are likely to be many others, say C, N, O, H, CH_4, NO, O_3, NO_2, NH_3, C_2N_2, OH, and so on. Most of these will be in small enough quantities that they are negligible for most engineering calculations (but not necessarily so as related to smog problems). Assume that the temperature and available O_2 are such that no C_8H_{18} remains and think of the products as shown in an isolated system in equilibrium at constant p, T. In this condition, all the equilibrium reactions of equations (13-17) through (13-19) are occurring, plus many more. When a molecule of CO_2 combines with a molecule of H_2 to form CO and H_2O, equation (13-19), the opposite reaction is occurring or will soon do so. Typical of the numerous other reversible reactions that are going on: $O_2 = 2O$, $H_2O = \tfrac{1}{2}H_2 + OH$, $N_2 + O_2 = 2NO$, and so on.

When we assumed a simple stoichiometric reaction, § 13.6, atom balances for the different atomic species supplied sufficient conditions to obtain all unknowns. However, there are more unknowns in equation **(b)** above than can be found by atom balances; therefore, additional conditions must be obtained, as explained next.

CONDITION FOR CHEMICAL EQUILIBRIUM, IDEAL GASES   13.35

Suppose we place any quantity of reactants A and B into a reaction chamber to be maintained at constant p, T (think of A and B being H_2 and O_2, if desired). Before anything happens, Gibbs function G_m of this mixture is computed and plotted at r, Fig. 13/10. Different proportions of the components of reactants and products may now be assumed, and the corresponding value of G_m computed. At p, the value of G_m is its value for the mixture being products only, say X and Y. The significant observation is that the curve shows a minimum, defined by $dG_m = 0$. A spontaneous reaction may take place along *rdc*, but no further, because the second law says that a spontaneous reaction with increasing G cannot occur, § 5.26. This observation is significant when ΔG is used as a criterion concerning the probability of a reaction to specified amounts of products.

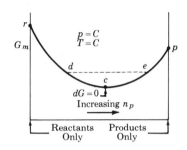

Fig. 13/10 *Variation of Gibbs Function with Change of Composition of Mixture.* Observe that $\Delta G_{d\,to\,e} = 0$ but that this reaction is not an equilibrium reaction; it can only go to *c*. For the specified p, T, point *c* also represents the state of maximum entropy, in accordance with the second law.

Suppose we represent all equilibrium equations (T, p constant) by

(a)
$$\nu_a A + \nu_b B + \cdots = \nu_x X + \nu_y Y + \cdots$$

where the number of reactants A, B and of products X, Y may be more or less than two. Now let a small amount dn_a moles of A react; the corresponding amounts of B, X, Y, and so on, are dn_b, dn_x, dn_y. Note that these reacting quantities are proportional to the stoichiometric coefficients in equation **(a)**, as explained above; that is, $dn_b/dn_a = \nu_b/\nu_a$, $dn_x/dn_a = \nu_x/\nu_a$, $dn_y/dn_a = \nu_y/\nu_a$. For convenience, assume that there are two each of reactants and of products, write the expression for the change of G for the dn reaction, and manipulate as follows;

(b)
$$dG_{rp} = \bar{G}_x\,dn_x + \bar{G}_y\,dn_y - \bar{G}_a\,dn_a - \bar{G}_b\,dn_b$$

$$\frac{dG_{rp}}{dn_a} = \bar{G}_x\frac{dn_x}{dn_a} + \bar{G}_y\frac{dn_y}{dn_a} - \bar{G}_a - \bar{G}_b\frac{dn_b}{dn_a}$$

(c)

$$= \bar{G}_x\frac{\nu_x}{\nu_a} + \bar{G}_y\frac{\nu_y}{\nu_a} - \bar{G}_a - \bar{G}_b\frac{\nu_b}{\nu_a}$$

Multiply through equation **(c)** by $\nu_a\,dn_a$ and equate to zero, since at chemical equilibrium, $dG_{rp} = 0$.

(d)
$$\nu_a\,dG_{rp} = dn_a(\nu_x\bar{G}_x + \nu_y\bar{G}_y - \nu_a\bar{G}_a - \nu_b\bar{G}_b) = 0$$

Inasmuch as $dn_a \neq 0$, the parenthetical part must equal zero. Let the temperature during the reaction be any value T' and substitute for each of the G's in the parentheses its value from equation (13-15);

$$\nu_x\bar{G}'_x + \nu_x\bar{R}T'\ln p_x + \nu_y\bar{G}'_y + \nu_y\bar{R}T'\ln p_y$$

(e)

$$-\nu_a\bar{G}'_a - \nu_a\bar{R}T'\ln p_a - \nu_b\bar{G}'_b - \nu_b\bar{R}T'\ln p_b = 0$$

Let the overall change in G with each constituent at the standard 1 atm and temperature T' be represented by G'_{rp};

(f)
$$G'_{rp} = \nu_x \bar{G}'_x + \nu_y \bar{G}'_y - \nu_a \bar{G}'_a - \nu_b \bar{G}'_b$$

From equation **(e)**, we then have

(13-20)
$$G'_{rp} = -T'\bar{R} \ln \frac{p_x^{\nu_x} p_y^{\nu_y}}{p_a^{\nu_a} p_b^{\nu_b}} \qquad \text{or} \qquad G'_{rp} = -T'\bar{R} \ln K_p$$

where K_p is called the **pressure equilibrium constant**, and where the pressures p_x, p_y, p_a, and p_b are the actual pressures of the particular constituents in the actual mixture of gases, say in the products of equation **(b)**, § 13.34, and ν_i is the number of moles of constituent i in the equilibrium equation, say (13-17) through (13-19). The equilibrium constant K_p is defined by

(13-21)
$$K_p = \frac{p_x^{\nu_x} p_y^{\nu_y}}{p_a^{\nu_a} p_b^{\nu_b}} = \frac{\prod_j p_j^{\nu_j}}{\prod_i p_i^{\nu_i}}$$

where the second form generalizes the definition for any number of constituents, j representing "products," i representing "reactants"; all the i's and j's are ideal gases. The "products" and "reactants" are *defined by the equilibrium equation*; the equilibrium constant is not the same for

(g)
$$CO_2 = CO + 0.5O_2 \qquad \text{as for} \qquad CO + 0.5O_2 = CO_2$$
$$[K_{p1}] \qquad\qquad\qquad\qquad [K_{p2}]$$

because reactants and products have exchanged places. Evidently, $K_{p2} = 1/K_{p1}$; it follows that $\log K_{p2} = \log(1/K_{p1}) = -\log K_{p1}$. The logarithm of the equilibrium constant, $\log K_{p1}$, for $\frac{1}{2}O_2 = 0$ is half $\log K_{p2}$ for the equilibrium equation $O_2 = 2O$; that is, $2 \log K_{p1} = \log K_{p2}$, or $K_{p2} = K_{p1}^2$.

The definition of the equilibrium constant gives additional conditions that may be used to determine unknown coefficients in the chemical equation, say **(b)**, § 13.34. For example, two independent equilibrium equations make possible the determination of two more coefficients than can be found from the atom balances alone, a statement that becomes more evident if the partial pressures are expressed in terms of the total pressure p_m.

By equations (12-2) and (6-10), we have the volumetric fraction of the jth component in a mixture of gases as

(h)
$$X_j = \frac{p_j}{\sum_j p_j} = \frac{p_j}{p_m} = \frac{n_j}{\sum_j n_j} \qquad \text{or} \qquad p_j = \frac{n_j}{\sum n_j} p_m = \frac{n_j}{\sum n} p_m$$

where the final form is written simply for convenience in handling. Let the reaction equation be represented by

(i)
$$C_8H_{18} + aO_2 + bD_r \rightarrow n_a A + n_b B + n_x X + n_y Y + n_0 O + n_d D_p$$

Then $\Sigma\, n_j = n_a + n_b + \cdots$ for all coefficients on the product side. Using the value of p_j from **(h)** in equation (13-21), we get

(j)
$$K_p = \frac{[(n_x/\Sigma\, n)p_m]^{\nu_x}[(n_y/\Sigma\, n)p_m]^{\nu_y}}{[(n_a/\Sigma\, n)p_m]^{\nu_a}[(n_b/\Sigma\, n)p_m]^{\nu_b}} = \frac{n_x^{\nu_x} n_y^{\nu_y}}{n_a^{\nu_a} n_b^{\nu_b}}\left(\frac{p_m}{\Sigma\, n}\right)^{\nu_x + \nu_y - \nu_a - \nu_b}$$

In terms of the volumetric fraction X in equation **(h)**, equation **(j)** reduces to

(k)
$$K_p = \frac{X_x^{\nu_x} X_y^{\nu_y}}{X_a^{\nu_a} X_b^{\nu_b}}(p_m)^{\nu_x + \nu_y - \nu_a - \nu_b}$$

In general, the exponent of p_m for any number of reactants and products is the algebraic sum of the ν's with the ν's for the reactants of the equilibrium equation being negative, symbolized by $\Sigma\, \nu$.

The amount of dissociation of the products of a combustion equation depends on the temperature, which means that, in an engineering analysis, a decision must be made as to the composition of the mixture. This decision is based on experience, and making it is practicing the art of engineering. For example, the temperature must be rather high for the reaction $\frac{1}{2}O_2 = 0$ to be significant. With respect to energy balances, the dissociation of CO_2 begins to be important above about 3500°F; of H_2O, above about 4500°F.

13.36   Example—Dissociation of CO₂

Carbon monoxide is burned in 200% ideal air during a steady-flow process at $p_m = 5$ atm. If the temperature of the products is maintained at 5400°R (3000 K), what is the heat transferred (a) when the reactants enter the process at 77°F and (b) when the reactants enter at 2540°F?

Solution.   (a) Before the heat can be computed, the composition of the products must be determined. In Item B 32, we find $\log K_p = 0.485$ for the following equilibrium reaction at 3000°K;

(a)
$$CO + 0.5O_2 = CO_2$$

From this equation, we see that the stoichiometric amount of O_2 is 0.5 mole; therefore, for 200% air, there will be 1 mole of O_2 per mole of CO. The reaction for all gaseous components may then be assumed to be

(b)
$$CO + O_2 + 3.76N_2 \rightarrow n_1 CO_2 + n_2 CO + n_3 O_2 + 3.76N_2$$

The total moles of products are $\Sigma\, n_j = n_1 + n_2 + n_3 + 3.76$. Using the carbon and oxygen balances,

(c)
$$C:\ 1 = n_1 + n_2 \qquad O_2:\ 1 + 2 = 2n_1 + n_2 + 2n_3$$

we find

(d)
$$n_1 = 1 - n_2 \quad \text{and} \quad n_3 = \frac{1 + n_2}{2} \qquad \text{also } \Sigma\, n_j = 5.26 + 0.5n_2$$

where the coefficients n_1 and n_3 are expressed in terms of n_2, which is the moles of CO in the products and in this context is called the **degree of dissociation** (n_1 would be called the **degree** or *extent of reaction*). The various partial pressures, with n_1 and n_3 from (d), are:

(e)
$$p(CO_2) = \frac{n_1}{\Sigma n_i} p_m = \frac{n_1 p_m}{5.26 + 0.5n_2} = \frac{(1 - n_2)5}{5.26 + 0.5n_2}$$

(f)
$$p(CO) = \frac{n_2 p_m}{5.26 + 0.5n_2} = \frac{n_2 5}{5.26 + 0.5n_2}$$

(g)
$$p(O_2) = \frac{n_3 p_m}{5.26 + 0.5n_2} = \frac{(1 + n_2)5}{2(5.26 + 0.5n_2)}$$

From $\log K_p = 0.485$, we have $K_p = 10^{0.485} = 3.055$. Notice that tabulating $\log K_p$ instead of K_p reduces the range of numbers considerably, yet leaves it easy to find K_p. Substituting the foregoing values into equation (13-21) for K_p, taking the exponents ν from (a), we get

(h)
$$K_p = 3.055 = \frac{p(CO_2)}{[p(CO)][p(O_2)]^{1/2}}$$

$$= \left[\frac{(1 - n_2)5}{5.26 + 0.5n_2}\right]\left[\frac{5.26 + 0.5n_2}{5n_2}\right]\left[\frac{2(5.26 + 0.5n_2)}{(1 + n_2)5}\right]^{1/2}$$

For the occasional problem, simplify this equation and solve for n_2 by trial; say, from

(i)
$$3.055 = \frac{1 - n_2}{n_2}\left[\frac{10.52 + n_2}{5(1 + n_2)}\right]^{1/2}$$

After several trials, we find that $n_2 = 0.297$ nearly satisfies (i). From equations (d), $n_1 = 0.703$ and $n_3 = 0.6485$. The reaction equation (b) now becomes

(j) $CO + O_2 + 3.76N_2 \rightarrow 0.703CO_2 + 0.297CO + 0.6485O_2 + 3.76N_2$

An enthalpy balance for the process is shown, Fig. 13/11, where $\bar{h}_1 = \bar{h}°$ because the reactants enter at 77°F; moles of air $= 1 + 3.76 = 4.76$. The enthalpies of formation, as shown, are probably safer and more convenient than heating values when stoichiometric quantities are not involved; $\bar{h}_2$ values at 5400°R are by extrapolation from the gas tables; $\Delta \bar{h}_f°$ values from Item B 11. Substituting into the expressions on the diagram of Fig. 13/11, we have the enthalpies departing as:

CO_2: $0.703(-169,293 + 69,975.9 - 4030.2) = -72,653$

CO: $0.297(-47,560 + 43,951.8 - 3729.5) = -2,179$

O_2: $0.6485(45,872.2 - 3725.1) = +27,332$

N_2: $3.76(43,612.9 - 3729.5) = +149,962$

Total enthalpy leaving $= +102,462$ Btu/pmole CO

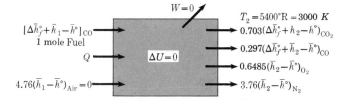

Energy Diagram Using Enthalpies of Formation.   **Fig. 13/11**

From the enthalpy balances of Fig. 13/11 and for $\Delta h_f^\circ = -47{,}560$ for the entering CO,

$$Q = H_p - H_R = 102{,}462 - (-47{,}560) = 150{,}022 \text{ Btu/pmole CO}$$

or $150{,}022/28 = 5358$ Btu/lb CO. The positive sign indicates that the heat must be added to this system in order to maintain a $T_2 = 5400°$R. It also follows that if we were searching for the adiabatic flame temperature, Q should be zero and the next guess of T_2 should be substantially lower than 5400°R.

(b) The fact that the *reactants* are heated to $2540°F \approx 3000°$R $= 1666.7$ K does not affect the partial pressures of the products, the equilibrium constant, and the percentage of each component in the products. Thus, the only effect is that the reactants enter the system with more enthalpy than before. As measured from the 77°F datum of the elements, the entering enthalpy H_R at 3000°R is found by equation (13-13):

$$\text{CO:} \quad -47{,}560 + 22{,}973.4 - 3725.1 = -28{,}312$$

$$\text{Air:} \quad (4.76)(29)(790.68 - 128.34) = +91{,}429$$

for which $H_R = +63{,}117$ Btu/pmole CO. In this case, Q for the system of Fig. 13/11 is

$$Q = 102{,}462 - 63{,}117 = +39{,}345 \text{ Btu/pmole CO}$$

13.37   Example—Effect of Pressure on Dissociation of CO_2

Find the composition of the products of combustion of the reaction in the preceding example when $p_m = 1$ atm (instead of 5 atm).

Solution.   The equilibrium constant, being a function of temperature only, is the same as before; $K_p = 3.055$. The partial pressures are given by the same equations, with $p_m = 1$. Substitution of these pressures in equation (13-21) yields

(a)
$$3.055 = \frac{1 - n_2}{n_2}\left[\frac{10.52 + n_2}{1 + n_2}\right]^{1/2}$$

$n_2 = 0.472$ closely satisfies this equation. The products are therefore $0.528CO_2$, $0.472CO$, $0.736O_2$, $3.76N_2$. Observe that the extent of the reaction for CO to CO_2 is greater at the higher pressure in the preceding example, $0.703CO_2$ versus $0.528CO_2$. If the reaction occurs in 100% ideal air, the extent of the reaction will not be so large as with excess air. Try it. Also, for more engineering experience, check § 13.36 with no N_2.

13.38   Example—Dissociation of H_2O

If pure $H_2O(g)$ is heated to 5000°R (2778 K) and maintained there at 1 atm pressure, what is the equilibrium composition if there are only H_2O, H_2, and O_2?

Solution.   If dissociation occurs (and it does), the composition of the products is

(a)
$$H_2O(g) \rightarrow n_1 H_2O(g) + n_2 H_2 + n_3 O_2$$

if the dissociation of O_2 is negligible. Taking atom balances and solving for mole fractions in terms, of, say, n_2, we find

(b)
$$H_2: \quad 2 = 2n_1 + 2n_2 \qquad n_1 = 1 - n_2$$

(c)
$$O_2: \quad 1 = n_1 + 2n_3 \qquad n_3 = \frac{n_2}{2}$$

From the total number of moles and $p_m = 1$, we get

(d)
$$\sum n_j = n_1 + n_2 + n_3 = 1 + \frac{n_2}{2} = \frac{2 + n_2}{2}$$

(e)
$$p(H_2O) = \frac{n_1 p_m}{\sum n_j} = \frac{2(1 - n_2)}{2 + n_2}$$

(f)
$$p(H_2) = \frac{2n_2}{2 + n_2} \qquad p(O_2) = \frac{2n_3}{2 + n_2} = \frac{n_2}{2 + n_2}$$

For the equilibrium reaction of equation (13-17), we find in Item B 32 that $\log K_p = 1.697$ (use 1.7) at 2778 K; $K_p = 10^{1.7} = 50$. Then

(g)
$$K_p = 50 = \frac{p(H_2O)}{p(H_2)[p(O_2)]^{1/2}} = \left[\frac{2(1 - n_2)}{2 + n_2}\right]\left[\frac{2 + n_2}{2n_2}\right]\left[\frac{2 + n_2}{n_2}\right]^{1/2}$$

where the exponents of the partial pressures are given by the equilibrium reaction, equation (13-17). Simplified for a trial solution, we could use

(h)
$$50 = \frac{1 - n_2}{n_2}\left(\frac{2 + n_2}{n_2}\right)^{1/2}$$

for $n_2 = 0.0885$, this equation is closely satisfied; and the composition is $0.9115 H_2O$, $0.0885 H_2$, and $0.0443 O_2$. Notice that even at this temperature, the dissociation of H_2O is not very large.

VAN'T HOFF EQUILIBRIUM BOX   13.39

It will be helpful to set up a physical picture of ideal processes back of the conclusions of § 13.35. Figure 13/12 represents an equilibrium reaction box. It is maintained at constant pressure p_m, and the reaction proceeds from, at, and to a temperature T'. The components A and B are delivered to the box through

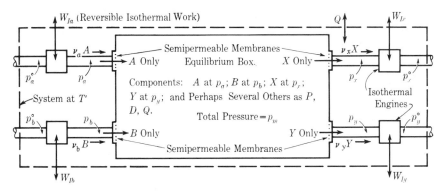

Van't Hoff Equilibrium Box. Equilibrium reaction: $\nu_a A + \nu_b B \rightleftarrows \nu_x X + \nu_y Y$. **Fig. 13/12**
A and B enter the box in strictly stoichiometric proportions ν_b molecules of B for ν_a molecules of A; also X and Y depart in stoichiometric proportions; the mass of each element in the box remains constant; steady flow, $\Delta K = 0$. x passes only gas X; and so on.

semipermeable membranes, membranes that permit the passage of only one component. The semipermeable membranes on the right-hand side permit passage of X, Y only, as seen.*

Suppose the properties of the components are known at some standard pressure, say $p° = 1$ atm. The isothermal engines, Fig. 13/12, which may be compressors or expanders, provide the work needed to change the original pressure of the components $p°$ to the partial pressure of the same component in the box, an increase or decrease as required. The work of an isothermal process for an ideal gas, from § 7.12, is $W = n\bar{R}T \ln (p_1/p_2)$ for n moles. Thus, the net work of the system in Fig. 13/12 ($\Delta K = 0$, $\Delta P = 0$, but they can be included if significant) is the sum of the works; with $\bar{R}T'$ factored,

(a)
$$W = \bar{R}T'\left[\ln\left(\frac{p_a°}{p_a}\right)^{\nu_a} + \ln\left(\frac{p_b°}{p_b}\right)^{\nu_b} + \ln\left(\frac{p_x}{p_x°}\right)^{\nu_x} + \ln\left(\frac{p_y}{p_y°}\right)^{\nu_y}\right]$$

$$= \bar{R}T'\left[\ln\frac{p_x^{\nu_x}p_y^{\nu_y}}{p_a^{\nu_a}p_b^{\nu_b}} - \ln\frac{(p_x°)^{\nu_x}(p_y°)^{\nu_y}}{(p_a°)^{\nu_a}(p_b°)^{\nu_b}}\right]$$

If all the pressures $p°$ are unity (1 atm) as previously suggested, the second "ln" is zero and equation **(a)** is the same as (13-20) because the maximum (or minimum negative) work on the overall reversible process is $-G'_{rp}$ (§ 13.33). Either derivation indicates that a particular equilibrium reaction is unconcerned about the presence of other substances, including catalysts (which affect rate).

13.40   EQUILIBRIUM CONSTANT

As previously mentioned, the pressure equilibrium constant for ideal gases is a function of temperature only. This is seen from equation (13-20), which is, with the primes dropped for convenience,

(13-20)
$$\ln K_p = -\frac{G_{rp}}{T\bar{R}} \quad\quad \text{or} \quad\quad K_p = \exp\left(-\frac{G_{rp}}{T\bar{R}}\right)$$

where G_{rp} is the change in Gibbs function for a stoichiometric reaction with each pure component involved being at 1 atm and temperature T; in the literature this is called the standard Gibbs (free energy) function at temperature T. Equation (13-20) may be used to determine K_p from a known G_{rp}, or G_{rp} from a known K_p.

Let h_{rp} represent the enthalpy of the reaction per mole of a reactant (at $T = C$), and similarly, $\bar{s}_{rp}$ for entropy. Then since the change of Gibbs function at temperature T is $\Delta G_T = \Delta H - T\Delta S$, we have $\bar{G}_{rp} = \bar{h}_{rp} - T\bar{s}_{rp}$. This value of $\bar{G}_{rp}$ in (13-20) gives

(a)
$$\ln K_p = -\frac{\bar{h}_{rp}}{\bar{R}T} + \frac{\bar{s}_{rp}}{\bar{R}}$$

---

* These are, of course, ideal semipermeable membranes, but many actual membranes, with varying degrees of effectiveness, exist. There are many in living matter, including humans. Membranes permeable to positive ions and others permeable to negative ions are necessary parts of one method of desalting water. The fact that hot palladium foil is permeable (virtually) only to hydrogen provides a means of producing high purity H_2, for example, from refinery off-gases. The details of the mass exchanges are not known with certainty.

Differentiate this equation with respect to T and find

(b)
$$\frac{d \ln K_p}{dT} = \frac{\bar{h}_{rp}}{\bar{R}T^2} - \frac{\bar{h}_{rp}}{\bar{R}T\,dT} + \frac{d\bar{s}_{rp}}{\bar{R}\,dT}$$

From equation (4-15), $T\,dS = dH - V\,dp$, we have $T\,dS = dH$ for $p = C$; or for the reaction $T\bar{s}_{rp} = \bar{h}_{rp}$. If this value of $\bar{h}_{rp}$ is substituted into equation **(b)**, the last two terms cancel, being numerically equal but of opposite sign; thus,

(13-22)
$$d \ln K_p = -\frac{\bar{h}_{rp}}{\bar{R}T^2}\,dT \quad \text{or} \quad \frac{d \ln K_p}{d(1/T)} = -\frac{\bar{h}_{rp}}{\bar{R}}$$

Since $\bar{h}_{rp}$ does not vary rapidly with temperature, (13-22) can be integrated for small temperature differences with $\bar{h}_{rp}$ held constant. The second form of (13-22) shows that if $\bar{h}_{rp}$ is constant, a plot of $\ln K_p$ against $1/T$ is a straight line with a negative slope for exothermic reactions; K_p values are often given graphically on this basis in the literature. Integrating (13-22) from state 1 to state 2, we get

(c)
$$\ln K_{p2} - \ln K_{p1} = \ln\frac{K_{p2}}{K_{p1}} = -\frac{\bar{h}_{rp}}{\bar{R}}\left(\frac{1}{T_2} - \frac{1}{T_1}\right)$$

where, for example, K_{p2} at T_2 can be determined if $\bar{h}_{rp}$ and K_{p1} are known.

As observed, the foregoing applications require an equilibrium constant for each reaction at the temperature of the reaction. Since the number of possible reactions with relatively common elements and compounds is quite large and since the number of reactions in general is indefinitely large, we find in the literature[0.22] tabulations of the equilibrium constant of compounds and of other substances. The equilibrium constant of a compound, for instance, is at chemical equilibrium between the compound and the *elements* from which it is formed. Take, for example, the equilibrium equation (13-17); the equilibrium constant for this reaction at temperature T is the equilibrium constant for the compound H_2O at temperature T. If the reaction involves compounds in reactants i and products j at a particular temperature, then

(13-23)
$$\log (K_p)_{\text{reaction}} = \sum_p \log (K_p)_j - \sum \log (K_p)_i$$

where the stable elements have zero equilibrium constants.

Example—Equilibrium Constant from ΔG_f 13.41

(a) Determine the $\log K_p^\circ$ at $77°F$ (≈ 298 K) for the reaction

(13-17)
$$H_2 + \tfrac{1}{2}O_2 \rightleftharpoons H_2O(g)$$

(b) Using the value found in (a), compute $\log K_p$ at $1080°R = 600$ K, assuming constant $\bar{h}_{rp}$.

Solution.   (a) For the given reaction, the Gibbs function of the reaction G_{rp}° is the same as the Gibbs function of formation ΔG_f°, since the reactants are elements, each in the standard state. Therefore, from Item B 11, we find $\Delta \bar{G}_f^\circ = \bar{G}_{rp}^\circ = -98{,}344$ Btu/mole; then at $537°R$, approximately 77°F, we get from (13-20)

(a)
$$\log K_p^\circ = \frac{\ln K_p^\circ}{2.3026} = -\frac{\bar{G}_{rp}^\circ}{2.3026\,T^\circ\bar{R}} = \frac{98,344}{(2.3026)(537)(1.986)} = 40$$

(b) Divide each term of equation **(c)**, § 13.40, by 2.3026 to convert natural logarithms to a base of 10 and get

(b)
$$\log K_{p2} = \log K_p^\circ + \frac{-\bar{h}_{rp}}{2.3026\bar{R}}\left(\frac{1}{T_2} - \frac{1}{T_1}\right)$$

$$= 40 + \frac{104,036}{(2.3026)(1.986)}\left(\frac{1}{1080} - \frac{1}{537}\right) = 18.7$$

The temperatures must be T°R for $\bar{R} = 1.986$ Btu/pmole-°R; the enthalpy of the reaction $\bar{h}_{rp}$ is for 1 pmole of $H_2[= -(2.016)(51,605)$, Item B 12], the same as $\Delta\bar{h}_f^\circ$ for H_2O, Item B 11. The order of this approximation is revealed by comparing the answers above with values given in Item B 32; to wit, 40.048 at 298 K ($\approx$77°F), 18.633 at 600 K (1080°R).

13.42   **Example—Adiabatic Flame Temperature with Dissociation**

If $C_8H_{18}(g)$ is burned in steady flow from a state 2 at 1510°R and 25-atm pressure in 200% ideal air, what is the adiabatic flame temperature T_3?

Solution.   This problem will not be solved completely, but additional light can be thrown on the subject by setting it up and discussing a few more points. First, one decides what products are significant. If there is an appreciable amount of dissociation for hydrocarbon fuels, it should be assumed that there are CO, H_2, and probably OH, as well as the usual products of complete combustion. At higher temperatures than can be obtained adiabatically in 200% air, the reactions $\frac{1}{2}O_2 = O$, $\frac{1}{2}H_2 = H$, and $\frac{1}{2}N_2 + \frac{1}{2}O_2 = NO$ may be significant, as well as several others. To keep the detail at a minimum, we shall assume that the reaction is (all gases)

(a) $C_8H_{18} + 25O_2 + 94N_2 \rightarrow aCO_2 + bCO + cH_2O + dH_2 + eO_2 + 94N_2$

Since there are three appropriate atom balances, the first step is to obtain three of the unknown coefficients in terms of the other two. For best accuracy in the inevitable trial-and-error solutions, one should keep the symbols for what will be the smallest numbers; and these are likely to be the amounts dissociated b and d. Thus

$$\text{C:} \quad 8 = a + b \quad a = 8 - b \qquad H_2: \quad 18 = 2c + 2d \quad c = 9 - d$$

$$O_2: \quad 50 = 2a + b + c + 2e \quad \text{or} \quad e = 12.5 + \frac{b+d}{2}$$

(b) $$\sum_p n_i = a + b + c + d + e + 94 = 123.5 + \frac{b+d}{2}$$

Since the temperature is unknown, there are now three unknowns, b, d, and T_3. Conditions for two of these can be found from the equilibrium constants of two independent equilibrium reactions; say,

Reaction A: $CO + 0.5O_2 \rightleftharpoons CO_2$ [EQUATION (13-18)]

Reaction B: $CO_2 + H_2 \rightleftharpoons CO + H_2O$ [EQUATION (13-19)]

(If it were decided, for example, that the NO content were significant, an additional equilibrium equation would be needed.) One of the three conditions needed must be the energy balance for adiabatic combustion because this balance for $Q = 0$ tells whether or not one has found the right temperature. The procedure is to assume a temperature, look up the corresponding values of $\log K_p$ for the equilibrium reactions involved, and solve for b and d (in this case). Then, knowing the composition of the products for the assumed temperature, the terms of the energy equation can be evaluated as in § 13.36; if $H_p - H_R \neq 0$, assume another temperature and repeat. See the enthalpy balance of Fig. 13/13. Following this plan, assume $T_3 = 3400°R$; from Item B 32 (K_{pA} refers to reaction A above),

$$\log K_{pA} = 3.314 \qquad K_{pA} = 2061 \qquad \log K_{pB} = 0.582 \qquad K_{pB} = 3.819$$

$$K_{pA} = \frac{p(CO_2)}{p(CO)[p(O_2)]^{1/2}}$$

(c)

$$= \left[\frac{(8 - b)p_m}{123.5 + \frac{b + d}{2}}\right] \left[\frac{123.5 + \frac{b + d}{2}}{bp_m}\right] \left\{\frac{123.5 + \frac{b + d}{2}}{\left[12.5 + \frac{b + d}{2}\right]p_m}\right\}^{1/2}$$

(d)

$$2061 = \frac{8 - b}{b} \left\{\frac{123.5 + (b + d)/2}{[12.5 + (b + d)/2]p_m}\right\}^{1/2}$$

(e)

$$K_{pB} = \frac{p(CO)p(H_2O)}{p(CO_2)p(H_2)} = \frac{b(9 - d)}{(8 - b)d} = 3.819$$

Since the exponents are all unity for equation (e), Σn and p_m cancel. Solve these equations simultaneously for b and d; then determine a, c, and e. Now we know the moles of each of the products in Fig. 13/13; compute H_R at $T_2 = 1510°R$ and H_p at the assumed $T_3 = 3400°R$; if $H_p - H_R$ is not zero, assume another T_3, and repeat. For the C_8H_{18} in Fig. 13/13, $\bar{C}_p$ Btu/°R-pmole is the mean value between $T°$ and T_2; for estimation purposes, the value in Item B 1 may be used. (See also Table I, p. 53.) Try to straddle the expected answer. A helpful technique may be to solve for b, say, from the energy equation after setting $H_R = H_p$. Plot the values of b against the temperature, obtaining one curve from the equilibrium conditions and another from the energy equation; a satisfactory value of T_3 would probably be where these intersect.

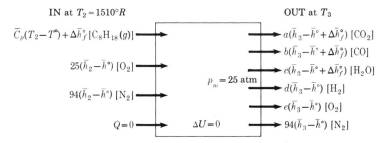

IN at $T_2 = 1510°R$

$\bar{C}_p(T_2 - T°) + \Delta\bar{h}_f° [C_8H_{18}(g)]$

$25(\bar{h}_2 - \bar{h}°) [O_2]$

$94(\bar{h}_2 - \bar{h}°) [N_2]$

$Q = 0$ $\Delta U = 0$ $p_m = 25$ atm

OUT at T_3

$a(\bar{h}_3 - \bar{h}° + \Delta\bar{h}_f°) [CO_2]$

$b(\bar{h}_3 - \bar{h}° + \Delta\bar{h}_f°) [CO]$

$c(\bar{h}_3 - \bar{h}° + \Delta\bar{h}_f°) [H_2O]$

$d(\bar{h}_3 - \bar{h}°) [H_2]$

$e(\bar{h}_3 - \bar{h}°) [O_2]$

$94(\bar{h}_3 - \bar{h}°) [N_2]$

Adiabatic Combustion Chamber. **Fig. 13/13**

The data of this example were purposely chosen to highlight the statement previously made that the dissociation for a considerable excess air is insignificant. Make an order of magnitude solution for b from (d) with $p_m = 25$ atm, and find that b is about 0.0026 (and d is smaller).

In short, the temperature is practically the same as previously computed without considering dissociation.

13.43   FUGACITY AND ACTIVITY

In dealing with mixtures of other than ideal gases that might undergo a chemical process, there are several concepts that are particularly helpful. While it is beyond the intended scope of this text to investigate these situations in detail, it is fitting to make the reader aware of ideas that he may find elaborated elsewhere, principally in chemical engineering thermodynamics.

Recall in § 10.15 we introduced the property fugacity f, as defined by G. N. Lewis,[1.18] and the generalized fugacity coefficient f/p-chart, Item B 37. For imperfect gases we note that

$$d\bar{G} = \bar{R}T\,d(\ln f)$$

and

$$\bar{G} = \bar{R}T \ln f + C(T)$$

during a $T = C$ change of state.

For known fugacities, the Gibbs function at temperature T' then becomes

(a) $$\bar{G} = \bar{G}' + \bar{R}T' \ln f/f^{\circ}$$

where $f^{\circ} = p^{\circ}$, a reference state at sufficiently low pressure such that the fugacity and the pressure are equal as for an ideal gas. Now introducing the activity $a = f/f^{\circ}$, equation **(a)** becomes

(b) $$\bar{G} = \bar{G}' + \bar{R}T' \ln a$$

where the units of $(\bar{R}T')$ are the same as those for G. By analogy with the results obtained in equations (13-20) and (13-21), we write expressions for equilibrium constants as

(c) $$K_a = \frac{a_x^{\nu_x} a_y^{\nu_y}}{a_a^{\nu_a} a_b^{\nu_b}} \quad \text{and} \quad K_f = \frac{f_x^{\nu_x} f_y^{\nu_y}}{f_a^{\nu_a} f_b^{\nu_b}}$$

for the case of products X and Y, reactants A and B, § 13.35. If the reference state for the gases is at 1 atm, then $f^{\circ} = p^{\circ} = 1$ atm, and $K_a = K_f$, and the activity is equal to the fugacity.

13.44   CHEMICAL POTENTIAL

Consider a mixture of phases, a solution of liquids and vapors, of solids, or a mixture of gases, and so on, consisting of n_a, n_b, ... moles of A, B, ... If the amount of one or more of the constituents changes while independent properties are unchanged, the change of any extensive property, say the Gibbs function, may be written for total quantities as [after equation (11-1)]

(a)
$$dG = \left(\frac{\partial G}{\partial n_a}\right)_{p,T,n} dn_a + \left(\frac{\partial G}{\partial n_b}\right)_{p,T,n} dn_b + \cdots = \sum_i \left(\frac{\partial G}{\partial n_i}\right)_{p,T,n} dn_i$$

where $G(n_a, n_b, \ldots)$ and where the subscript n is to say that all mole values are held constant except n_i. In this context, the partials of equation **(a)** are called *partial molal Gibbs functions* $\bar{G}$; $\bar{G}_i = (\partial G/\partial n_i)_{p,T,n}$. This partial is called the **chemical potential**, a name first used by Gibbs. This molal property is evaluated at the p, T of the mixture, but at its actual condition in the mixture. We shall explain further by an analogy. Any extensive property may be expressed as shown for G in equation **(a)**. If the property is volume, a familiar one, then the *partial* molal volume, by analogy, is $\bar{V}_i = (\partial V/\partial n_i)_{p,T,n}$. From prior experience, the reader knows that a gallon each of two liquids (say, alcohol and water) mixed together do not always add to 2 gal of mixture. That is, the *molal partial volume* of i is now always the same volume as a mole of the pure i. This observation applies to all other partial molal properties.

It happens that from the differentials of the Holmholz function A and of the enthalpy, written analogously to equation **(a)**, we can say that the chemical potential of the ith constituent is

(b)
$$\mu_i = \left(\frac{\partial G}{\partial n_i}\right)_{T,p,n} = \left(\frac{\partial A}{\partial n_i}\right)_{T,V,n} = \left(\frac{\partial H}{\partial n_i}\right)_{S,p,n}$$

Perhaps μ_i from the Gibbs function is most used. Let the partial molal Gibbs function be represented by $\bar{G}_i$ for the ith constituent, $\mu_i = \bar{G}_i$, and write the function $G(T, p, n_i)$, total quantities;

(c)
$$dG = \left(\frac{\partial G}{\partial T}\right)_{p,n_i} dT + \left(\frac{\partial G}{\partial p}\right)_{T,n_i} dp + \sum_i \left(\frac{\partial G}{\partial n_i}\right)_{T,p,n} dn_i$$

$$= -S\,dT + V\,dp + \sum_i \mu_i\,dn_i$$

where equivalent values of the partials from Table V, § 11.3, are used. Equation **(c)** is evidently a more complete definition of dG than we have previously used. Observe that if T and p are constant, $dG = \sum \mu_i\,dn_i$; but $dG = 0$ at equilibrium conditions. From the foregoing, we can conceive of $\mu_i = \bar{G}_i$ (an intensive property by being limited to a mole) as a driving potential for mass transfer, in the sense of T being a potential for the transfer of heat, $\mathscr{E}$ for electricity, and so on. For equilibrium, all potentials, pressure (force), temperature, Gibbs μ, and others, must be in balance.

Consider a two-phase mixture, as liquid-vapor or solid-vapor, for which p, T are constant. Then equation **(a)** or **(c)** holds, or

(d)
$$dG = \sum_i \left(\frac{\partial G}{\partial n_i}\right)_{p,T,n} dn_i = \sum_i \bar{G}_i\,dn_i = \sum_i \mu_i\,dn_i$$

Let primed symbols represent the different phases, as dn' and dn'' being infinitesimal quantities of each of the two phases. Suppose dn' moles of one phase pass to the other phase, increasing its quantity by dn'' moles; that is, $dn'' = -dn'$. From equation **(d)** we now get

(e) $$dG = \bar{G}'' \, dn'' + \bar{G}' \, dn' = \bar{G}'' \, dn'' - \bar{G}' \, dn'' = dn''(\bar{G}'' - \bar{G}')$$

If the two phases are in equilibrium at this p, T, then $d\bar{G} = 0$, and we conclude from **(e)** that $\bar{G}'' = \bar{G}'$, which is to say that the molal (or specific) Gibbs function of each phase (at a particular p, T) is the same, a conclusion reached in a different context in § 3.7. From **(d)**, we see also that $\mu' = \mu''$ by the same logic (or $\mu = \bar{G}$). If $\mu' > \mu''$, there is a potential for mass to move from phase $'$ to phase $''$. Even in an equilibrium state, some of the liquid molecules at the interface of a liquid-vapor mixture will have enough energy to escape to the vapor phase; but simultaneously, on a macroscopic scale, a balancing number of vapor molecules will be "captured" by the liquid phase, maintaining $dG = 0$.

It is beyond the intended scope of this text to proceed to the detailed development of these concepts, but we should point out a certain characteristic. For illustration, consider liquefied air to be a mixture of O_2 and N_2. If this liquid is in equilibrium with its vapor, there are two components, each in two phases. The proportions of O_2 and N_2 in the liquid are different from the proportions in the vapor; hence the mixture is not only not a pure substance, but it cannot be treated as a pure substance, as ordinary air is handled. (See § 18.9.)

13.45   CLOSURE

Perhaps the most important idea for the reader to carry away from all chapters having to do with applications of theory is that the aim here is to highlight the use of the most basic thermodynamic principles, and that in the case of all applications mentioned in this book there is much more to be learned. Since chemical processes occur in so many contexts, several books would be needed to cover all.

Notice that thermodynamics gives no help concerning the rate of a reaction, yet this is one of the most important factors in any commercial reaction. The thrust of a rocket is directly dependent on the rate at which the reaction can be made to occur, perhaps helped along by use of a catalytic agent.

PROBLEMS

SI UNITS

13.1 **(a)** The volumetric analysis of air is 21% O_2, 79% N_2. Find the moles of N_2 per mole O_2. What is the ratio kg N_2/kg O_2? **(b)** Air is mixed with propane in the ratio 15 kg air/kg C_3H_8. In 1 kg of this mixture, find the grams of C, H_2, O_2, N_2.

13.2 Gaseous fuel flowing in a pipeline is composed volumetrically of 75% CH_4, 15% C_2H_6, 6% O_2, and 4% CO. **(a)** Convert this to a percentage mass basis. **(b)** In 1 kg of this fuel mixture, find the grams of C, H_2, O_2.

13.3 A gravimetric analysis of a typical automotive gasoline gives 86% C and 14% H_2. What average chemical formula in the form $C_x H_y$ approximates this fuel?

13.4 An ultimate analysis of a bituminous coal as received is 77.5% C, 3.7% H_2, 1.5% N_2, 4.3% O_2, 0.5% S, 6.5% ash, 6.0% H_2O. **(a)** Convert this analysis to a dry basis. **(b)** Find the analysis on a dry-and-ashless basis. **(c)** If this fuel as received is burned at the rate of 100 metric tons/hr and the refuse is analyzed as being 10% combustible (carbon in the ash), how many kilograms of refuse must be handled each hour?

Ans. **(a)** 82.45% C, **(b)** 88.57% C, **(c)** 7222 kg/hr.

13.5 Set up the simple combustion equation for each of the basic combustibles C, H_2, S in stoichiometric air and note the amount of air required to burn each element.

Ans. Air required: $C/11.5$, $H_2/34.3$, $S/4.3$ kg air/kg element.

13.6 Assume 2 mol of CO are burned in 5.76 mol air. Write and balance the theoretical combustion equation and find **(a)** the mass of CO_2 formed, **(b)** the percentage excess air, **(c)** the air/fuel ratio.

13.7 Pure carbon reacts with all the O_2 in 80% of ideal air. Balance the resulting theoretical combustion equation, and find the masses of CO and CO_2 formed per 100 kg of air supplied.

Ans. 10.21 kg CO, 24.05 kg CO_2.

13.8 Set up the necessary combustion equations and determine the amount of air theoretically required to burn 1 kg of pure carbon **(a)** to equal masses of CO and CO_2, **(b)** to where the mass of CO_2 is double that of the CO. *Ans.* **(a)** 7.98 kg air.

13.9 If $1000\ \ell$ of a gaseous mixture has the following gravimetric analysis: 30% O_2, 70% CO, is there sufficient oxygen present to support complete combustion of the CO?

13.10 A hydrocarbon fuel $(CH_x)_n$ requires equal masses of oxygen for a complete reaction of each of its components carbon and hydrogen, respectively. Determine x and n on the basis of 1 mol of $(CH_x)_n$.

Ans. 4, 1.

13.11 Consider the process wherein octane is burned with 80% stoichiometric air, and assume that CO is the only combustible appearing in the products. Find the products analysis by volume and by mass.

Ans. CO_2: 5.49% vol., 8.86% grav.

13.12 Assume 5 mol/hr of propane C_3H_8 are completely burned in the stoichiometric amount of air. Determine **(a)** the volume (m^3/min) of air required measured at 1 atm, 25°C, **(b)** the partial pressure of the CO_2 in the products, measured at 1 atm, 149°C, **(c)** the volume (m^3/min) of the products measured at 1 atm and 149°C, and **(d)** the dew point of the H_2O in the products.

Ans. **(a)** 48.50 m^3/min, **(b)** 11.8 kPaa, **(c)** 74.43 m^3/min, **(d)** 55°C.

13.13 The same as **13.18** except that the fuel is nonane C_9H_{20}, a rocket fuel.

13.14 Hydrogen peroxide $H_2O_2(g)$, sometimes used as an oxidizer for rocket fuels, reacts with nonane $C_9H_{20}(g)$ to produce only CO_2 and $H_2O(g)$ products at 1 atm presssure. Balance the equation for a stoichiometric reaction and determine **(a)** the mass of H_2O_2 used per kg fuel, **(b)** the mass of H_2O formed per kg fuel, **(c)** the partial pressure of the H_2O in the products, **(d)** the dew point of the water in the products, **(e)** the volume of the products at 226.7°C, **(f)** the gravimetric percentage of carbon in the nonane.

Ans. **(a)** 7.43, **(b)** 5.34 kg, **(c)** 81.91 kPaa, **(d)** 94.2°C, **(e)** 1930 m^3, **(f)** 84.4%.

13.15 There are burned 141.6 m^3/min of coke oven gas (measured at 1 atm, 65.6°C) in the stoichiometric amount of air. The gas has the following volumetric composition: 36.9% CH_4, 52% H_2, 5% CO, 0.5% O_2, 4.2% N_2, 1.4% CO_2. What volume of air measured at 1 atm, 65.6°C is required?

Ans. 625.6 m^3/min.

13.16 Two hundred metric tons per hour of coal are burned in 125% (25% excess) stoichiometric air; the as-fired ultimate analysis is 75% C, 4% H_2, 0.5% S, 6% O_2, 1.5% N_2, 8% H_2O, 5% ash. Find **(a)** the mass of air required, kg/hr, **(b)** the mass of refuse collected if all ash shows in the refuse which tests 25% combustible. **(c)** What is the carbon content of the coal on a dry basis?

Ans. **(a)** 2.442×10^6 kg/hr, **(b)** 13,340 kg/hr, **(c)** 81.5%.

13.17 An average Texas lignite coal, as-received, has the following ultimate analysis: 40% C, 3.0% H_2, 1.0% S, 11% O_2, 1.5% N_2, 32% H_2O, 11.5% ash. Balance the combustion equation for this coal burned as received in 85% stoichiometric air. Find **(a)** the air/fuel ratios as received and for dry coal, **(b)** the mass of dry gaseous products per kilogram of the as received fuel burned.

13.18 An anthracite coal has the following dry-basis ultimate analysis: 81.63% C, 2.23% H_2, 0.48% S, 2.92% O_2, 0.80% N_2, 11.94% ash. **(a)** Balance the combustion equation for stoichiometric air, and find the air/fuel ratio on the dry basis; also on the as-received basis if the moisture content is 3.43%. **(b)** The same as **(a)** except that the air supplied is 90% of the stoichiometric amount. **(c)** Assume that CO is the only combustible element in the products and find its mass per kilogram of dry fuel and per kilogram of fuel as received.

Ans. **(a)** 10.06 (dry), **(b)** 9.05 kg a/kg f (dry), **(c)** 0.408 kg CO/kg f.

13.19 The following is a gravimetric (ultimate) analysis of a coal: 70.85% C, 4.48% H_2, 2.11% S, 6.36% O_2, 1.38% N_2, 12.3% ash, 2.52% H_2O. During actual combustion, the following volumetric analysis of the stack gases was obtained: 12.1% CO_2, 0% CO, 7.2% O_2, 80.7% N_2. Determine **(a)** the percentage excess or deficiency of air and **(b)** the mass of dry products per kilogram of coal fired. *Ans.* **(a)** 49.8% excess, **(b)** 14.78 kg.

13.20 The burning of a hydrocarbon fuel $(CH_x)_n$ in an automotive engine results in a dry exhaust gas analysis, percentage by volume, of: 11% CO_2, 0.5% CO, 2% CH_4, 1.5% H_2, 6% O_2, and 79% N_2. Find **(a)** the actual air/fuel ratio, **(b)** the percentage excess air, and **(c)** the mass of water vapor formed per kilogram of fuel.

Ans. **(a)** 15.35, **(b)** 4.71%, **(c)** 0.719 kg.

13.21 The dry products of combustion from a hydrocarbon fuel burned in air, percentages by volume, are: 13.6% CO_2, 0.8% CO, 0.4% CH_4, 0.4% O_2, 84.8% N_2. Write the theoretical chemical equation, and find **(a)** the values of x and n in $(CH_x)_n$, **(b)** the mass of air supplied per kilogram of fuel, **(c)** the percentage of excess or deficiency of air, and **(d)** the mass of dry products per kilogram of fuel burned.

Ans. **(a)** $[CH_{2.31}]_{14.8}$, **(b)** 14.75 kg, **(c)** 2.7% def., **(d)** 14.25 kg.

13.22 Start with the two equalities $-h_{rp}^\circ = q_p^\circ$ and $-u_{rp}^\circ = q_v^\circ$, and show that the difference between the two heating values is given by the expression $q_p^\circ - q_v^\circ = (\bar{R}T^\circ/M_f)(n_r - n_p)$ where n_r and n_p are the moles of reactants and products, respectively, and M_f is the molecular weight of the fuel.

13.23 Develop the relation between fugacity f and the compressibility factor Z as follows: $\ln f/p = \int_0^p (Z - 1)\, dp/p$. Start with the change of Gibbs function for an isothermal process considering ideal and real gas conditions.

13.24 Show that for a reversible isothermal steady-flow, steady-state process, the work is given by $W = G_1 - G_2 = -RT \ln(f_2/f_1)$ where $\Delta K = 0$, $\Delta P = 0$, G is the Gibbs function, and f is the fugacity.

13.25 There are compressed 4.54 kg/min of propane from 1282 kPaa, 60°C to 12,790 kPaa in a reversible isothermal, steady-state, steady-flow manner with $\Delta P = 0$, $\Delta K = 0$. Find the work W by means of fugacities; see **13.58** and Item B 37.

Ans. −206.6 kJ/min.

13.26 Prove that the Orsat analyzer ignores the water vapor and reports the exact analysis of the gaseous mixture as if for a dry gas. *Hint*: Assume a mixture of CO_2, N_2 and water vapor—then follow one of the components (say the CO_2) through the Orsat absorption process.

ENGLISH UNITS

13.27 One pound of carbon is burned so that 1/2 lb of C goes into CO_2 and 1/2 lb into CO. Set up the combustion equation and find **(a)** the pounds of air used per pound of carbon, **(b)** the volume of this air at 65°F and 14.7 psia, **(c)** the volumetric and gravimetric composition of the products, **(d)** the volume of the products at 65°F and 14.7 psia, and **(e)** the partial pressure of each of the products.

Ans. **(a)** 8.63, **(b)** 114 ft³, **(c)** 13%, 19% for CO_2, **(e)** 1.91 psia for CO_2

13.28 In a rigid vessel at 1 atm pressure and 100°F, there are 1 lb of H_2 and 28 lb of O_2. The H_2 reacts completely to H_2O. Determine **(a)** the volume of the vessel, **(b)** the temperature at which the H_2O is on the point of condensing (use steam tables), **(c)** the amount of condensation when the contents are cooled to 80°F, **(d)** the partial pressure of the O_2 at the final state.

Ans. **(a)** 562 ft³, **(b)** 170°F, **(c)** 8.11 lb, **(d)** 0.511 atm.

13.29 A gaseous mixture of 4 moles CH_4 and 9 moles O_2 is ignited. Write the theoretical combustion equation and find **(a)** the volumetric analysis of the products, **(b)** the equivalent amount of air represented by the O_2, **(c)** the percentage excess of O_2 in the mixture and, if excess, how much more CH_4 could have been burned to completion, **(d)** the dew point of the products if the mixture pressure is 20 psia.

Ans. **(a)** $X_{CO} = 30.76\%$, **(b)** 42.84 moles equivalent air, **(c)** 12.5% excess, 0.5 mole CH_4, **(d)** 203.2°F.

13.30 In the combustion of a hydrocarbon fuel at 1 atm, 19 lb da/lb f are supplied.

The humidity ratio of the air supply is $\omega = 0.015$ lb v/lb da. The combustion process produces 1.4 lb H_2O/lb f. For the products from dry air, $M = 28.9$. What is the dew point of the products?

Ans. 123°F.

13.31 **(a)** Set up the chemical equation for the combustion of propane C_3H_8 in stoichiometric air. Show the relative masses and volumes, and compute the air/fuel ratio. **(b)** The same as **(a)** except that combustion occurs in 15% excess air. **(c)** The same as **(a)** except that combustion occurs in 90% of the stoichiometric air and the H_2 reacts completely to H_2O. **(d)** Determine the gravimetric percentages of carbon and hydrogen in the fuel. **(e)** For the reaction in **(a)**, determine the gravimetric and volumetric analyses of the products with $H_2O(g)$.

Ans. **(a)** 15.61, **(b)** 17.95, **(c)** 14.05 lb a/lb f, **(d)** 81.8% C, **(e)** CO_2: 11.62% vol, 18.07% grav.

13.32 The same as **13.31** except that the fuel is nonane, C_9H_{20}, a rocket fuel.

13.33 A Signal Hill, California, gas sample has the following volumetric analysis: 62.5% CH_4, 32.9% C_2H_6, 3.6% H_2, 1.0% CO_2. Balance the chemical equation for this gas burned in the stoichiometric amount of air, and find **(a)** the air/fuel ratio, by mass and volume, and **(b)** the volumetric and gravimetric analyses of the products.

Ans. **(a)** 16.4 lb a/lb f, 11.52 ft³/ft³ f, **(b)** CO_2: 10.21% vol, 16.12% grav.

13.34 The ultimate gravimetric analysis of a coal as received is: 74% C, 1.5% H_2, 1% S, 6% O_2, 2.5% N_2, 5.5% H_2O, 9.5% ash. **(a)** What is the percentage carbon on a dry, ashless basis? **(b)** Find the air/fuel ratio (100% ideal air) required to burn the fuel as-received. **(c)** If 100 ton/hr of the as received coal are burned and the refuse shows 20% combustible, find the pounds per hour refuse collected.

13.35 Assume 100 ton/hr of pulverized coal are burned. Coal analysis by mass showed: 76% C, 6% H_2, 7% O_2, 2% N_2, 4% H_2O, 5% ash. Stack gas analysis by volume showed: 13% CO_2, 1.5% CO, 5.5% O_2, 80% N_2. Refuse pit analysis by mass showed: 23% C, 77% ash. Find **(a)** the volume of actual air measured at 14.7 psia and 90°F, **(b)** the refuse collected each hour, and **(c)** the

volume of dry products at $p_m = 14.7$ psia and 310°F.

Ans. **(a)** 605,000 cfm, **(b)** 6.49 tons/hr, **(c)** 880,000 cfm.

13.36 **(a)** If propane $C_3H_8(g)$ is mixed with the stoichiometric air, what are the higher and lower heating values $(p = C)$ per pound of the mixture and per cu ft of mixture at $p_m = 14.7$ psia and 77°F? **(b)** Find q_h and q_l $(p = C)$ per lb fuel, at 200°F and at 0°R. **(c)** Find q_h and q_l per lb fuel, at 77°F when $V = C$.

Ans. **(a)** q_h: 1294 Btu/lb, 97.8 Btu/ft³, **(b)** $q_h = 21,624$ Btu/lb at 200°F, **(c)** $q_{hv} = 21,417$, $q_{lv} = 19,798$ Btu/lb.

13.37 The same as **13.36** except that the fuel is nonane $C_9H_{20}(g)$.

13.38 The volumetric analysis of a natural gas fuel is 22.6% C_2H_6 and 77.4% CH_4. Find **(a)** the mass of stoichiometric air per pound of fuel, **(b)** the mass of CO_2 and H_2O formed per pound of fuel, **(c)** the gravimetric percentage of C and H_2 in the fuel and in the stoichiometric air-fuel mixture, and **(d)** the higher and lower heating values at 77°F, per pound of fuel and per pound of stoichiometric air-fuel mixture.

Ans. **(a)** 16.8 lb, **(b)** 2.81 lb CO_2, **(c)** 4.31% C, 1.30% H_2, **(d)** $q_h = 25,050$ Btu/lb f, $q_h = 1406$ Btu/lb mix.

13.39 Compare the maximum adiabatic flame temperature of H_2 versus CO when each fuel is burned to completion (no dissociation) in a steady flow system with 200% air; the reactants enter at 1 atm, 77°F.

Ans. 2967°R (H_2), 3185°R (CO).

13.40 Gaseous *n*-butane (C_4H_{10}) is burned at constant pressure with 400% ideal air in a steady flow system. The reactants enter at 77°F. Determine the adiabatic flame temperature using Item B 38.

13.41 Gaseous octane and 200% air react in a steady flow manner; the reactants cross the boundary at 100°F and the products leave at 2000°F. For the process, $\Delta K = 0$, $W = 0$. Determine the heat for a fuel flow of 1 mol/sec. Do not neglect the sensible enthalpy of the fuel; use Item B 9.

13.42 A Diesel engine burns dodecane $C_{12}H_{26}$ in 200% stoichiometric air (100% excess). At the end of the compression process (initial point of fuel injection—fuel at 77°F) the air temperature is 1080°F. Using

Item B 9, compute the temperature of the products after constant-pressure combustion if the combustion efficiency is 94%. Sketch energy diagrams.

13.43 Consider a stoichiometric reaction of nonane $C_9H_{20}(l)$ and hydrogen peroxide $H_2O_2(g)$ to $CO_2(g)$ and $H_2O(l)$. If the process begins and ends at 77°F and 1 atm, determine **(a)** the enthalpy of reaction, **(b)** the change of entropy. **(c)** If the nonane reacts instead with pure O_2, is more or less energy released? Compare this $-h_{rp}$ with that given in Item B 12.
Ans. **(a)** −34,800 Btu/lb f, **(b)** −504 Btu/°R for 1 mole nonane.

13.44 A torpedo propulsion system involves a reaction of methyl alcohol $CH_4O(l)$ and hydrogen peroxide $H_2O_2(g)$. Using Δh_f° and/or ΔG_f°, determine the maximum work that can be done for constant p, T, at $p_m = 1$ atm and 77°F. What are the entropy change and heat? Compare the heat of this reaction with the heating value from Item B 12.

13.45 Let octane $C_8H_{18}(l)$ be burned adiabatically in 400% air with the reactants initially at 77°F; steady flow obtains. Determine the theoretical flame temperature **(a)** using Item B 8, **(b)** using the enthalpies of formation of each compound.

13.46 **(a)** Using the enthalpies of formation, determine the standard enthalpy of combustion, $-h_{rp}^\circ$, of ethane $C_2H_6(g)$ with liquid H_2O in the products. Compare with Item B 12. **(b)** Using enthalpies of reaction, compute the enthalpy of formation. Compare with Item B 11.
Ans. **(a)** −22,320 Btu/lb, **(b)** −36,400 Btu/mole.

13.47 Into a 0.5-ft³ bomb calorimeter, filled with air at 1 atm and 77°F, is placed 0.0001 pmoles of methanol (methyl alcohol, CH_4O). Complete adiabatic combustion occurs. For no dissociation, compute the final pressure.

13.48 Using Item B 11, calculate the enthalpy of combustion and the internal energy of combustion for CO at 1 atm, 77°F.
Ans. −121,745, −121,203 Btu/lb-mol.

13.49 Compute the Gibbs function of formation of CO_2 at 1 atm and 77°F, using $G = H - TS$ and the enthalpies of formation and compare with the values given in Item B 11.

13.50 Determine the Gibbs function of methane $CH_4(g)$ at 5 atm, 140°F measured from the same datum as the Gibbs function of formation given in Item B 11. Assume the specific heat to be constant as given in Item B 1.   *Ans.* −22,786 Btu/mol.

13.51 Benzene $C_6H_6(g)$ is burned with 300% air in a steady flow manner. The reactants are at 1 atm, 77°F. Determine the maximum work possible if the reaction occurs isothermally.

13.52 **(a)** Using Δh_f°, compute ΔG_f° for hydroxyl OH(g) and check with value in Item B 11. **(b)** Using the value from **(a)**, compute the Gibbs function of formation at 400 K. **(c)** Using enthalpy of formation, Gibbs function of formation, and the absolute entropies of the elements, compute the absolute entropy of OH(g) in the standard state; compare with table value.

13.53 If a way were found to make gaseous ethane C_2H_6 react with O_2 in an ideal fuel cell at 140°F and 1 atm, compute ΔS for the reaction and determine the ideal work for $H_2O(l)$.
Ans. −69.5 Btu/°R-mole f, 20, 860 Btu/lb.

13.54 Determine the equilibrium constant K_p for the reaction at 77°F: **(a)** CO + H_2O = CC_2 + H_2; **(b)** CO_2 + H_2 = CO + H_2O. See Item B 11 for $\Delta \tilde{G}_f^\circ$.
         Ans. **(a)** 98,100.

13.55 Calculate the equilibrium constants for the following reactions occurring at 77°F: **(a)** $H_2 + 0.5\,O_2 = H_2O$; **(b)** $CO_2 = CO + 0.5\,O_2$.

13.56 When an object enters the earth's atmosphere at high speed, the temperature behind the shock wave in the vicinity of the object may be quite high. Suppose that the temperature is 7200°R, the total pressure is 0.034 psia, and it is desired to estimate the extent of dissociation of the oxygen. For the reversible reaction $0_2 \rightleftharpoons 2\,O$ at 7200°R, the equilibrium constant is $K_p = 2.4094$. *Hint*: Let the reaction equation be

$$3.76\,N_2 + O_2 \rightarrow n_1 O + n_2 O_2 + 3.76\,N_2$$

which assumes that the dissociation of the N_2 is negligible. Define K_p in terms of n_2.
         Ans. mols O_2 = 0.0007.

13.57 Methane $CH_4(g)$ is burned in "90% air" in a steady flow adiabatic process;

the reactants enter at 77°F. Consider dissociation of CO_2 and H_2O, except assume that there is no O_2 in the products (a simplification involving small error in this case). Estimate the adiabatic flame temperature. *Hint:* Use only the water gas reaction for the solution.

13.58 An object falls to earth creating a shock wave and resulting high temperatures in its path. At one point of this path in the earth's atmosphere, the temperature is 5400°R; the pressure is 1 atm. Estimate the dissociation of oxygen at this point. For the reversible reaction $O_2 \rightleftharpoons 2\,O$ at 5400°R, $K_p = 0.01441$. See **13.56**.

Ans. 0.248 mols O.

13.59 Compute the adiabatic flame temperature for the combustion of CO in stoichiometric oxygen at a constant pressure of 5 atm, starting from 77°F and allowing dissociation. *Suggestion.* Try 5400°R first.

13.60 During the recent energy crisis it was proposed to utilize cattle manure from large feed-lots by hydrogasifying it and burning the resulting gaseous mixture in power plants. The following compositions are known: raw cattle manure before gasification, % mass: 35.4 C, 4.6 H_2, 0.2 S, 30.1 O_2, 3.4 H_2O, 0.7 N_2, 25.6 ash; product gas resulting from gasification of raw manure, %

volume: 78.76 CH_4, 19.02 C_2H_6, 1.07 H_2, 1.15 CO_2. Find the heating value of each composition and determine the stoichiometric air required for each.

13.61 In industry, the CO_2 content of the products of combustion is conveniently used as an index of the combustion efficiency and/or the percentage excess air for a given fuel. Plot the curve "percentage CO_2 (by volume) versus percentage air (theoretical)" for methane CH_4, allowing the amount of air to vary from 50% ideal to 150% ideal. For deficient air, assume that by volume the CO content is twice that of the H_2. Note that the resulting curve may be used to indicate the percentage air supplied when the CO_2 content only is known.

13.62 It is desired to determine the effect of excess air on the dewpoint temperature of the products resulting from burning a selected hydrocarbon in air at constant pressure ($p = 14.7$ psia). Select a fuel from Item B 12 and burn it theoretically in air varying from that of stoichiometric to 400% excess. Plot a curve depicting dewpoint temperature vs percentage excess air. Write the computer program for this problem.

13.63 Find the fugacity f (see Item B 37) for each of the following fluids at 2940 psia, 44°F; **(a)** water, **(b)** propane, **(c)** hydrogen.

14

GAS COMPRESSORS

14.1 INTRODUCTION

The compression of gases is so important in industry that a separate brief chapter devoted to this topic is deemed appropriate. Compressed air, at some pressure above atmospheric, has many practical uses, a few of which are operation of small air engines and pneumatic tools, operation of air hoists, cleaning by air blast, tire inflation (every filling/service station has an air compressor!), paint spraying, air lifting of liquids, and many other specialized industrial applications.

Gas transmission pipelines that extend across our nation and move huge volumes of gaseous fuels require booster compressor stations to maintain this daily flow. The compression of certain gaseous chemical compounds to extreme pressures, often of the order of 35,000 psi, for the making of plastics is common to the chemical industry. Although our discussion will center around the compression of gases (primarily air), the results obtained apply in part to the compression of vapors.

14.2 TYPES OF COMPRESSORS

Compressors are of two general types—the reciprocating piston-in-cylinder and the rotary; the numbers of rotary type in operation predominate. Each of these will be discussed later in some detail; however, it is not intended to cover completely all the design aspects of either since this is a special course within itself. In order to gain some familiarity with the mechanism of each, several compressors are shown in the figures that follow, Fig. 14/1–14/6.

14.3 COMPRESSOR WORK

If the compressor is of the rotary type Fig. 14/2, it would be treated as a steady-state, steady-flow system with flow through the control volume whose boundaries are at the intake and discharge; apply the steady-flow energy equation with negligible energy terms deleted. Let the process be adiabatic with $\Delta P = 0$, $\Delta K \approx 0$; which gives $W = -\Delta H$. Let the mass that enters and leaves be represented by m', or

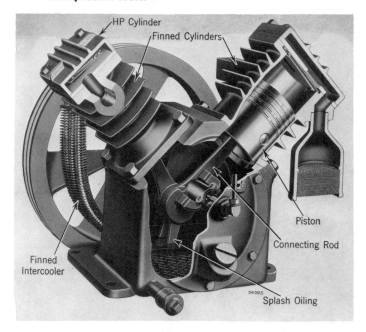

Two-Stage, Air-Cooled Compressor. Observe the finned intercooler, partly visible on the left. Two-stage compression is sometimes recommended for a discharge pressure as low as 80 psia. Several more stages of compression are needed for high discharge pressures, as 100–200 atm. (*Courtesy Ingersoll-Rand Co., New York, N.Y.*)

Fig. 14/1

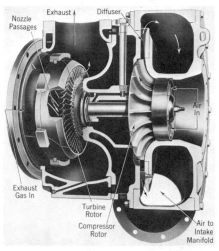

Centrifugal Compressor for Supercharger. Exhaust gases from a Diesel engine (at some 1200°F) expand through the nozzle passages, pass through the turbine wheel, and produce the work to drive the centrifugal compressor. With a greater mass of compressed air in the cylinder, the power output may be increased some 50% for a given size cylinder. (*Courtesy Elliott Co., Jeannette, Pa.*)

Fig. 14/2

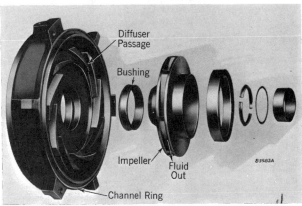

Diffuser and Impeller. This is actually one stage of a centrifugal pump, but a stage in a multistage blower or compressor would look much the same (*Courtesy Ingersoll-Rand Co., New York, N.Y.*)

Fig. 14/3

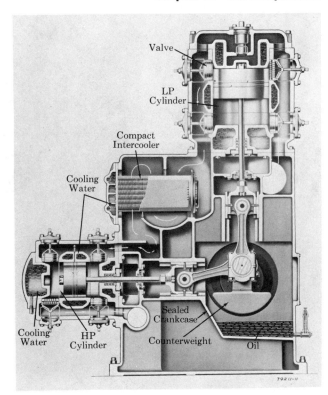

Fig. 14/4   *Two-Stage,   Water-Cooled Compressor*. See Fig. 14/7 for names of principal parts. (*Courtesy Ingersoll-Rand Co., New York, N.Y.*)

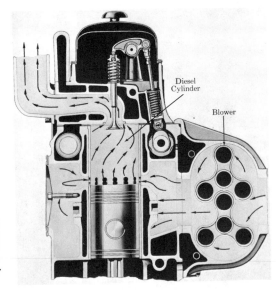

Fig. 14/5   *Positive Displacement Blower*. Gas trapped between lobes and enclosing surface on the intake side (right) is compressed and delivered to the discharge side. This idea is also used as a liquid pump. (*Courtesy General Motors Corp., Detroit, Mich.*)

Sliding-Blade Compressor. Since the rotor is mounted eccentric to the casing, gas entering on the side of greatest clearance is kept between the blades, compressed, and delivered from the side with minimum clearance. (*Courtesy Fuller Co., Catasauqua, Pa.*)

Fig. 14/6

$\dot{m}'$ if a particular rate is considered. Then

(a)
$$W = -\Delta H = -m'c_p(T_2 - T_1) = -m'c_p T_1\left(\frac{T_2}{T_1} - 1\right)$$

[ANY ADIABATIC, IDEAL GAS, $\Delta K = 0$]

Introduce into **(a)** the values in **(b)** and get equation **(c)**;

(b)
$$c_p = \frac{kR}{k-1} \qquad \frac{T_2}{T_1} = \left(\frac{p_2}{p_1}\right)^{(k-1)/k} \qquad p_1 V_1' = m'RT$$

(c)
$$W = \frac{km'RT_1}{1-k}\left[\left(\frac{p_2}{p_1}\right)^{(k-1)/k} - 1\right] = \frac{kp_1 V_1'}{1-k}\left[\left(\frac{p_2}{p_1}\right)^{(k-1)/k} - 1\right]$$

[ISENTROPIC ONLY, IDEAL GAS, $\Delta K = 0$]

where V_1' is the volume measured at p_1 and T_1 corresponding to the mass m'. A similar approach to a steady-flow polytropic yields an equation of the same form, only n is in place of k.

If the compressor is reciprocating, Fig. 14/7, the events within the cylinder are described in Fig. 14/8. If first we imagine the compressor with zero clearance, work is done on the piston from a to 1 by the constant pressure gas at p_1; work is done by the piston from 2 to b on the gas at p_2. The total work done on the gas is now seen to be $-\int V\,dp$, the area behind the curve 1-2; that is, this integral accounts for the work of drawing in, compressing (nonflow), and discharging the gas. By the same logic, the area b-3-4-a is also $-\int V\,dp$, but, of course, for a different mass. Since areas on a pV plane represent energy, the area enclosed 1-2-3-4 represents the work of this conventional compressor. Starting with equation (7-5), we find

(d)
$$-\int_{1}^{2} V\,dp = \frac{k(p_2 V_2 - p_1 V_1)}{1-k} = \frac{kp_1 V_1}{1-k}\left[\left(\frac{p_2}{p_1}\right)^{(k-1)/k} - 1\right]$$
$$[s = C,\ k = C]$$

where $V_2/V_1 = (p_1/p_2)^{1/k}$ was used in the brackets. Observe that this equation is of the same form as **(c)** above. The same final form naturally applies to process 3-4, Fig. 14/8; therefore an algebraic sum gives the enclosed area (work of the diagram). However, to avoid some extra algebra, we shall subtract [4 to 3] instead of adding [3 to 4];

$$W = \frac{kp_1 V_1}{1-k}\left[\left(\frac{p_2}{p_1}\right)^{(k-1)/k} - 1\right] - \frac{kp_4 V_4}{1-k}\left[\left(\frac{p_3}{p_4}\right)^{(k-1-/k)} - 1\right]$$

(14-1)　　$$W = \frac{kp_1(V_1 - V_4)}{1-k}\left[\left(\frac{p_2}{p_1}\right)^{(k-1)/k} - 1\right] = \frac{kp_1 V_1'}{1-k}\left[\left(\frac{p_2}{p_1}\right)^{(k-1)/k} - 1\right]$$

[ISENTROPIC PROCESSES]

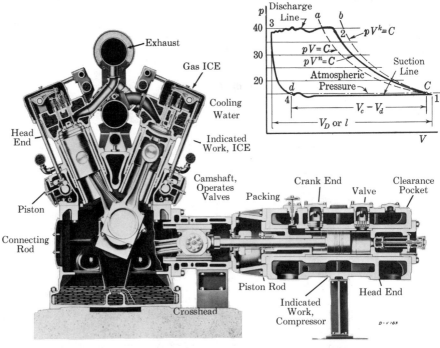

Fig. 14/7　*Gas Engine Drive Compressor.* This machine, designed particularly for natural-gas pumping stations, is built in sizes up to 5500 bhp (ICE), in which there are 16 power cylinders and 6 compressor cylinders. The nominal heat rate of the ICE is 6500 Btu/bhp-hr. A few of the principal parts are named for informational value. The inset at the upper right is a typical indicator card from a compressor cylinder.

since $p_4 = p_1$ and $p_3 = p_2$. For $V_1 - V_4$, the volume of gas drawn in, we use V_1' (perhaps $\dot{V}'$ cfm); and it will often be convenient to let $p_1 V_1' = m'RT_1$ and $p_1 \dot{V}_1' = \dot{m}'RT$, where as previously defined m' is the mass of gas passing through the compressor. If the processes 1-2 and 3-4 are polytropics with the same value of n, substitute n for k in equation (14-1). An actual indicator card is shown in the inset of Fig. 14/7.

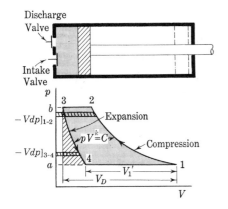

Conventional Diagram for Compressor. An idealized picture of the variation of p and V in the cylinder. Start at 4; 4-1 represents the intake and is called the suction line; 1-2 represents all valves closed, and there is some kind of thermodynamic compression process (here $s = C$); 2-3 is the discharge, the piston pushing the gas from the cylinder through the open discharge valve; 3-4 is an expansion of the gas left in the cylinder's clearance space. A clearance between the piston and cylinder head at the end positions is a necessity. Notice that the greater the clearance volume V_3, the less gas drawn in V_1.   **Fig. 14/8**

If the compression is in two or more stages (Figs. 14/1 and 14/4), the work of the conventional diagram for each stage is given by equation (14-1), where p_2/p_1 is the pressure ratio for that stage and p_1 and V_1' are measured at the particular stage's intake.

PREFERRED COMPRESSION CURVES   14.4

Inasmuch as the isentropic curve 1-2', Fig. 14/9, is steeper than the isothermal on the pV-plane, it takes more work to compress and deliver a gas when the compression is adiabatic than when it is isothermal, the difference being represented by the shaded area. Compression curves with values of n between unity and k will fall within the shaded area. Thus, it is obvious that the work necessary to drive a compressor decreases as the value of the exponent n decreases.

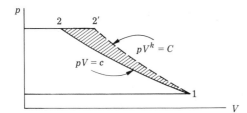

Comparison of Work for Isothermal and for Isentropic Compression.   **Fig. 14/9**

Polytropic compression and values of n less than k are brought about by circulating cooling water or air around the compressor cylinder to conduct away some of the heat generated by the compression work. In small, inexpensive compressors of the type found in automotive service centers, the cooling is always inadequate; values of n will be 1.35 or higher. Under favorable circumstances, a value of $n = 1.3$ or less may be expected. Values of n from 1.25 to 1.30 represent the best results for water jacketed compressors.

These observations are noteworthy: in isothermal compression, all the heat equivalent of the compression work is carried away, the gas then leaving the compressor with the same internal energy as it had upon entering; in isentropic compression, no heat is carried away and the gas leaves with an increase in internal energy equivalent to the compression work; and in polytropic compression, there is some heat carried away and some increase in internal energy and temperature. The heat rejected during the polytropic compression process is $Q = mc_p(T_2 - T_1)$ where m is the total mass of gas in the cylinder.

We remark here that there is no advantage to compressed air leaving the compressor with an internal energy greater than that from isothermal compression. The compressed air is normally delivered to a storage tank (receiver) where it may remain (and cool) for some time before it can be conveniently used. For this reason the isothermal compression process is considered ideal.

Example—Compressor work

A rotary compressor receives $6 \, \text{m}^3/\text{min}$ of a gas ($R = 410 \, \text{J/kg K}$, $c_p = 1.03 \, \text{kJ/kg K}$, $k = 1.67$) at 105 kPaa, 27°C and delivers it at 630 kPaa; $\Delta P = 0$, $\Delta K = 0$. Find the work if the process is (a) isentropic, (b) polytropic with $pV^{1.4} = c$, (c) isothermal.

Solution. For all parts we will use the steady flow energy equation

$$W_{SF} = Q - \Delta H$$

(a) Isentropic work ($pV^k = c$, $Q = 0$)
Solve first for the mass of gas being compressed.

$$\dot{m} = p_1 \dot{V}_1 / RT_1 = (105)(6)/(0.410)(300) = 5.12 \, \text{kg/min}$$

Next find T_2

$$T_2 = T_1(p_2/p_1)^{(k-1)/k} = (300)(630/105)^{(1.67-1)/1.67} = 615.6 \, \text{K}$$

Then $W = -\Delta H = -(5.12)(1.03)(615.6 - 300) = -1664 \, \text{kJ/min}$.

(b) Polytropic work ($pV^n = c$)
Find $T_2 = T_1(p_2/p_1)^{(n-1)/n} = (300)(630/105)^{(1.4-1)/1.4} = 500.5 \, \text{K}$.

$$c_v = c_p/k = 1.03/1.67 = 0.617 \, \text{kJ/kg K}$$

$$c_n = c_v\left(\frac{k-n}{1-n}\right) = (0.617)\left(\frac{1.67-1.4}{1-1.4}\right) = -0.416 \, \text{kJ/kg K}$$

$$\dot{W} = -\dot{m}c_p(T_2 - T_1) + \dot{m}c_n(T_2 - T_1) = \dot{m}(c_n - c_p)(T_2 - T_1)$$

$$= (5.12)(-0.416 - 1.03)(500.5 - 300) = -1484 \, \text{kJ/min}$$

(c) Isothermal work ($pV = c$). Here

$$T_2 = T_1 = 300 \, \text{K} \qquad \text{and} \qquad \Delta H = 0$$

Therefore $W = Q$ and also $W = -\int V \, dp = \int p \, dV$ in this instance.

$$\int_1^2 p \, dV = p_1 V_1 \ln V_2/V_1 = p_1 V_1 \ln p_1/p_2$$

Thus

$$\dot{W} = (105)(6)\ln 1/6 = -1129\,\text{kJ/min}$$

Note that the cooling of the gas during compression lowers the work requirement.

FREE AIR   14.5

Free air is air at normal atmospheric conditions in a particular geographical location. Since both pressures and temperatures vary with altitude, a compressor designed and adjusted to deliver a particular mass of air at a certain discharge pressure from a sea level location will not deliver the same mass when intake is at a 7000-ft location; also, without valve changes, the delivery pressure will be lower. Thus, the compressor on a jet engine inducts and delivers less mass of air at high altitudes than at low altitudes.

The variation of the NASA (USA) standard atmospheric pressure is given in Fig. 14/10. The NASA standard temperature varies linearly from 59°F at sea level (40° latitude) to −67°F at 35,332-ft altitude; this amounts to 0.003566°F/ft, or

(a) $dT/dz = 0.003566$

The Army Summer Standard temperature averages a little over 40° higher at a particular altitude. The NASA standard temperature in the stratosphere is assumed constant at −67°F (there is no fixed line of division, but the atmosphere up to about 32,332 ft is called the *troposphere*; next is the *stratosphere*). Observe in Fig. 14/10

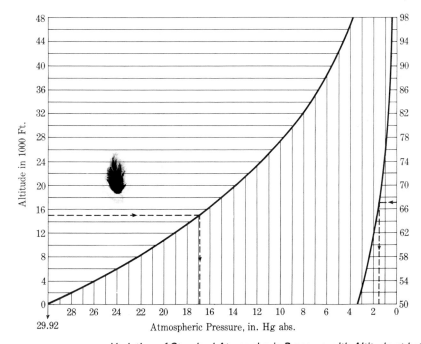

Variation of Standard Atmospheric Pressure with Altitude at Latitude 40°. Enter chart at the ordinate, which represents altitude. Move to curve, as along the dotted line, then downward to the abscissa and read the standard atmospheric pressure at that altitude.   **Fig. 14/10**

that the pressure approaches zero asymptotically. At $400\,\mathrm{mi}$ from the surface of the earth, a molecule travels an average distance (mean free path) of $40\,\mathrm{mi}$ before colliding with another; at $30\,\mathrm{mi}$, about $1\,\mathrm{in}$.

14.6   VOLUMETRIC EFFICIENCY

Since a reciprocating compressor has a clearance volume V_3, the volume of gas intake is always less than the displacement volume of the piston; the larger the clearance volume, the less gas discharged. The actual volumetric efficiency η_v' is the actual volume drawn in at p_0, T_0 (say, as measured by an orifice at intake) *divided by* the displacement volume (§ 14.7), a dimensionless ratio. Since for "displacement volume" we may say "the volume of gas at p_0, T_0 that would occupy the displacement volume V_D," the ratio as defined may be stated in terms of mass;

(a) $$\eta_v' = \frac{m_{\text{in}}}{m_D} = \frac{\text{mass drawn in (say, per stroke or per min)}}{\text{mass at } p_0,\ T_0 \text{ that would occupy } V_D}$$

which is sometimes the more convenient definition.

The **conventional volumetric efficiency**, that is, from the conventional diagram, Fig. 14/11, is $V_1'/V_D = (V_1 - V_4)/V_D$. The **clearance ratio** $c \equiv V_3/V_D$; or $V_3 = cV_D$. Then

(b) $$V_4 = V_3\left(\frac{p_3}{p_4}\right)^{1/n} = cV_D\left(\frac{p_2}{p_1}\right)^{1/n} = cV_D\left(\frac{V_1}{V_2}\right)$$

Since the total volume $V_1 = V_D + cV_D$, we have

$$\eta_v = \frac{V_1 - V_4}{V_D} = \frac{V_D + cV_D - cV_D(p_2/p_1)^{1/n}}{V_D}$$

(14-2) $$\eta_v = 1 + c - c\left(\frac{p_2}{p_1}\right)^{1/n} = 1 + c - c\frac{V_1}{V_2}$$

If the compression process is isentropic, let $n = k$. The actual volumetric efficiency may be much lower than the conventional because of fluid friction of flow (the pressure in the cylinder is less than the pressure of free air) and because the cylinder walls, being relatively hot, heat the incoming air (less mass of hot air can occupy a given space). Since, in equation (14-2), p_2 is greater than p_1 the volumetric efficiency decreases as the clearance increases; and as the volumetric efficiency of a compressor decreases, the capacity decreases. The clearance may become so large that no air is discharged by the compressor. This characteristic is used to control the output of a compressor by increasing the clearance when a reduced output is desired.

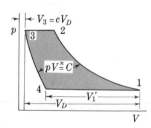

Fig. 14/11   *Conventional Diagram.* Observe that if state 3 were coincident with 2 and if 1-2 were an internally reversible process, the reexpansion 3-4 would coincide with 2-1 if it were the same kind of process and internally reversible (same n in this case). The volume swept by the piston in one stroke is $V_1 - V_3$, called the displacement volume V_D.

The work of compressing and delivering a certain quantity of gas is theoretically independent of the clearance volume (and volumetric efficiency), as is seen from equation (14-1). But as the pressure ratio goes up for a particular clearance ratio, the amount of gas that can be delivered decreases. This effect can be counteracted by *multi-stage compression*, breaking down the overall compression ratio into two or more parts, compressing in a series of cylinders. It happens too that for high-pressure ratios in a single cylinder the gas being discharged may be hot enough to ruin the lubrication, even to get an explosion if air and hydrocarbon lubricants are involved. Moreover, in large installations, *considerable power can be saved* in multistage compression just by cooling the gas between stages (for air, approximately back to its original intake temperature). The foregoing remarks directed to ideal-gas compressors are easily adapted to the compression of imperfect gases (Chapter 10).

DISPLACEMENT VOLUME 14.7

The displacement volume V_D corresponding to one diagram, as in Fig. 14/8 or 14/11, is the volume swept by the piston in one stroke. If D is the diameter of the cylinder and L is the length of the stroke of the piston, $V_D = (\pi D^2/4)(L)$. If the engine is double acting, a diagram is made on each side of the piston. While a compressor diagram is completed in 2 strokes, or one revolution, some internal combustion engines take 4 strokes to complete a diagram and in addition are multicylinder. Since it is often convenient to express the displacement volume in cubic feet per minute, multiply the displacement for one stoke by N diagrams per minute (cfm = ft³/min);

$$(14\text{-}3) \qquad V_D = \frac{\pi D^2}{4} L \text{ ft}^3 \quad \text{or} \quad \dot{V}_D = \frac{\pi D^2}{4} LN \text{ cfm}$$

$$[\text{1 DIAGRAM}] \qquad [N \text{ DIAGRAMS/MIN}]$$

Determine the value of N to accord with the actual engine being studied.

COMPRESSOR EFFICIENCY 14.8

There are a number of defined efficiencies of compressors, basically of the form

$$(\text{a}) \qquad \eta_c = \frac{\text{ideal work, } W}{\text{actual work, } W'}$$

the difference being in where the actual work is measured and what ideal work is the standard of comparison. Unless we define it otherwise locally, we shall use the *adiabatic compressor efficiency*, as

$$(14\text{-}4) \qquad \eta_c \equiv \frac{\text{isentropic work} = W}{\text{actual fluid work} = W'} = \frac{(h_1 - h_2)_s}{h_1 - h_{2'}}$$

where, for steady-flow machines, it is presumed in (14-4) that the kinetic energy change is negligible (or that stagnation enthalpies, § 7.23, are used), where the numerator is $-\Delta h_s$ for an isentropic process and the denominator is the actual enthalpy change, say, along the adiabatic 1-2′, Fig. 14/12. Equation (14-4) is applied to reciprocating compressors when W' is the actual work of an indicator diagram, or the shaft work input.

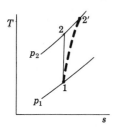

Fig. 14/12

14.9   Example—Air Compressor

A 14 × 15-in., double-acting air compressor, whose clearance is 4%, runs at $n = 150$ rpm. At state 1, Fig. 14/13, the air is at 14 psia and 80°F; discharge is at 56 psia; compression and reexpansion are isentropic. The state of the surrounding atmosphere is $p_a = 14.7$ psia and $t_a = 70°F$. (a) Estimate the amount of free air (cfm), using the conventional volumetric efficiency. (b) Compute the horse-power for a compressor efficiency of 75% based on shaft work input.

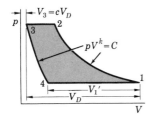

Fig. 14/13

Solution.   (a) Since one diagram, Fig. 14/13, is completed by each side of the piston in one revolution for a double-acting engine, there are 2 diagrams per revolution; and $N = 2n = (2)(150) = 300$ diagrams per minute. By equations (14-3) and (14-2), with $c = 0.04$,

(a)
$$\dot{V}_D = \left(\frac{\pi D^2}{4}\right)(L)(N) = \frac{\pi(14^2)(15)(300)}{(4)(1728)} = 401 \text{ cfm}$$

(b)
$$\eta_v = 1 + c - c\left(\frac{p_2}{p_1}\right)^{1/n} = 1.04 - 0.04\left(\frac{56}{14}\right)^{1/1.4} = 0.963 \text{ or } 96.3\%$$

The volume drawn in as measured at state 1 is

(c)
$$\dot{V}'_1 = \eta_v \dot{V}_D = (0.963)(401) = 386 \text{ cfm}$$

Measured at p_a, t_a, the volume of free air is

(d)
$$\dot{V}'_a = \frac{p_1 \dot{V}'_1 T_a}{T_1 p_a} = \frac{(14)(386)(530)}{(540)(14.7)} = 361 \text{ cfm}$$

CONSTANT k.   (b) We observed that equation (14-1), obtained from the analysis of the conventional diagram, is the same as equation (c), § 14.3, obtained from $W = -\Delta H = -m'c_p \Delta T$. Therefore, we may if we wish use $-\Delta H$.

(e)
$$\dot{m}' = \frac{p_1 \dot{V}_1'}{RT_1} = \frac{(14)(144)(386)}{(53.3)(540)} = 27 \text{ lb/min}$$

(f)
$$T_2 = T_1\left(\frac{p_2}{p_1}\right)^{(k-1)/k} = 540\left(\frac{56}{14}\right)^{(1.4-1)/1.4} = 802°\text{R}, 342°\text{F}$$

(g) $\dot{W} = -\Delta H = -\dot{m}'c_p(T_2 - T_1) = -(27)(0.24)(802 - 540) = -1698 \text{ Btu/min}$

GAS TABLE.   At $T_1 = 540°\text{R}$, from Item B 2, find $h_1 = 129.06$, $p_{r1} = 1.386$. For $r_p = 56/14 = 4$, we get $p_{r2} = 4p_{r1} = (4)(1.386) = 5.544$. Entering Item B 2 with this value of p_{r2}, read $T_2 = 801°\text{R}$ and $h_2 = 192.05$, to the nearest whole degree.

(h) $\dot{W} = -\Delta H = -(27)(192.05 - 129.06) = 1700 \text{ Btu/min}$

Since the gas table answer is more accurate, we find ($\eta_c = W/W'$)

$$\dot{W}' = \frac{\dot{W}}{\eta_c} = \frac{1700}{(0.75)(42.4)} = 53.3 \text{ hp}$$

the power to be delivered to the compressor's shaft to meet the requirements.

Example—Final State and Entropy Change from Efficiency   14.10

A rotary type of compressor compresses air from 14 psia and 525°R through a compression ratio of 5 and with a compressor efficiency of 75%. What is the discharge temperature if $\Delta K = 0$ (same as stagnation temperature) and the change of entropy of the actual process?
Solution.   From the air table at 525°R, we get $h_1 = 125.47$ and $v_{r1} = 154.84$. For $r_k = v_1/v_2 = 5$, we have

(a)
$$r_k = \left(\frac{v_1}{v_2}\right)_s = 5 = \frac{v_{r1}}{v_{r2}} \quad \text{or} \quad v_{r2} = \frac{154.84}{5} = 30.968$$

Entering the air table with this value of v_{r2}, we find properties at 2, Fig. 14/8, of $h_2 = 238.37 \text{ Btu/lb}$ (by interpolation) and $T_2 = 990°\text{R}$ (to the nearest whole degree). From the definition of compressor efficiency,

(b)
$$h_1 - h_{2'} = \frac{h_1 - h_2}{\eta_c} = \frac{125.47 - 238.37}{0.75} = 125.47 - h_{2'}$$

from which $h_{2'} = 275.97 \text{ Btu/lb}$. Interpolate for this value of h and find $T_{2'} = 1139.6°\text{R}$, say 1140°R.

The change of entropy is found either from equation (6-12) or by noting that $s_2' - s_1 = s_2' - s_2 = \phi_2' - \phi_2$, the last expression being valid because $p_2 = p_{2'}$. Look up ϕ at 1140 and 990°R and get

$$s_2' - s_1 = 0.78326 - 0.74792 = 0.03534 \text{ Btu/lb°R}$$

OTHER EFFICIENCIES   14.11

In addition to volumetric efficiency, § 14.6, and compressor efficiency, § 14.8, there are several other efficiencies that are pertinent to the compressor.

Mechanical efficiency is defined by the Compressed Air and Gas Institute in two ways: (1) when the drive is by internal combustion engine or by reciprocating steam engine,

(a) $\eta_m = \dfrac{\text{ihp of the compressor}}{\text{ihp of the driving engine}}$

an expression that applies to the unit; or (2) when the drive is by electric motor,

(b) $\eta_m = \dfrac{\text{ihp of the compressor}}{\text{bhp of the compressor}}$

The frictional horsepower of the compressor proper may be estimated by the empirical equation,*

$$\text{fhp} = 0.105(V_D)^{0.75}$$

where V_D is the piston displacement in cfm.

Another efficiency similar to compressor efficiency (adiabatic) as defined in § 14.8, is isothermal compression efficiency defined as

(c) $\eta_{ic} = \dfrac{\text{isothermal ideal work}}{\text{actual fluid work}}$

a value that often falls within the range of 70–75%.

The overall efficiency η_0 is the product of the mechanical efficiency and the particular compression efficiency, adiabatic or isothermal,

(d) $\eta_0 = \eta_m \eta_c$

where efficiencies based upon adiabatic compression are usually favored.

14.12   MULTISTAGE COMPRESSION

Multistaging is simply the compression of the gas in two or more cylinders in place of a single-cylinder compressor. It is used in reciprocating compressors in order to (1) save power, (2) limit the gas discharge temperature, and (3) limit the pressure differential per cylinder. The latter reason is very important because volumetric efficiency not only is affected by clearance but also by the pressure ratio across the cylinder; see equation (14-2). Moreover, excessively high temperatures caused by compression of the gas to high pressures may cause difficulty in the internal lubrication of the cylinder and piston. Finally, much power can be saved through multi-staging if the discharge pressure is above 75 psia and if the displacement is above 300 cfm. Study Fig. 14/14 and read the caption carefully.

It is common practice to cool the gas between stages of compression in an intercooler—see Figs. 14/1, 14/14, and 14/15; and it is this cooling that effects considerable saving in power.

---

* Suggested by Arthur Korn in a private communication.

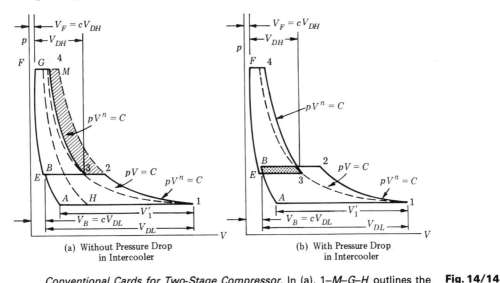

(a) Without Pressure Drop
in Intercooler

(b) With Pressure Drop
in Intercooler

Conventional Cards for Two-Stage Compressor. In (a), 1–*M*–*G*–*H* outlines the **Fig. 14/14**
conventional card for single-stage compression to p_4 for a given per cent
clearance. If the *per cent* clearance is the same in both cylinders, re-expansion in
the high-pressure cylinder starts at some point *F* instead of *G*. Reexpansion in
the low-pressure cylinder starts at *B*, where $V_B = V_G$. For a two-stage
machine, suction starts at *A*. For a single-stage machine, suction starts at *H*. The
capacity of the two-stage compressor is greater than that of the single-stage by
the amount $V_H - V_A$. Observe that it is possible to make the clearance so large
that no air would be delivered. As a matter of fact, varying the clearance is an
efficient manner of governing the output of the compressor. The work saved by
the two-stage compression is represented by the shaded area 2–*M*–4–3. Since
there is some pressure drop during the flow of air through the intercooler, the
conventional cards may show this drop as in (b). The shaded area between *B*
and 3 in (b) represents lost (or repeated) work due to this pressure drop.

Typical horizontal duplex arrangement for engine-type synchronous motor **Fig. 14/15**
drive. 400 to 1000 HP.

Figure 14/14 shows the events of the conventional indicator cards of a two-stage machine (Fig. 14/15), with the high pressure (HP) superposed on the low pressure (LP). Suction in the LP cylinder begins at A and the volume V_1' is drawn in. Compression 1-2 occurs and the gas is discharged along 2-B. The discharged gas passes through the intercooler and is cooled by circulating water through the intercooler tubes. Conventionally, it is assumed that the gas leaving the intercooler and entering the HP cylinder has the same temperature as it had upon entering the LP cylinder; see state point 3. The gas is then drawn into the HP cylinder along E-3, is compressed 3-4, and finally discharged from the compressor unit 4-F. Note that residual gas always remains in each cylinder because of clearance and must reexpand F-E (HP cylinder) and B-A (LP cylinder).

Equation (14-1) gives the work of an indicator card like 1-2-B-A or 3-4-F-E and may be used to obtain the total work for this two-stage compressor; we simply add the LP work to the HP work and have

$$(14\text{-}5) \quad W = \frac{nm'RT_1}{1-n}[(p_2/p_1)^{(n-1)/n} - 1] + \frac{nm'RT_3}{1-n}[(p_4/p_3)^{(n-1)/n} - 1]$$

for the two polytropic compression processes. It is common practice to adjust the operation of multistage compressors so that approximately equal works are done in the cylinders, a practice that results in minimum work for compressing a given amount of air.* Thus for the particular case of $T_1 = T_3$ and of $p_2 = p_3 = p_i$, say, we have the work of the LP stage equal to that of the HP stage, or,

(a) $$\frac{nm'RT_1}{1-n}[(p_i/p_1)^{(n-1)/n} - 1] = \frac{nm'RT_1}{1-n}[(p_4/p_i)^{(n-1)/n} - 1]$$

from which we find

(b) $$p_i/p_1 = p_4/p_i \quad \text{or} \quad p_i = [p_1 p_4]^{0.5}$$

which is the proper value for the intermediate pressure for the conditions specified. Also, since the work of each cylinder is the same, the total work for the two-stage machine is twice the work in either cylinder, or,

$$W = \frac{2nm'RT_1}{1-n}[(p_2/p_1')^{(n-1)/n} - 1]$$

(c) $$= \frac{2nm'RT_1}{1-n}[(p_4/p_1)^{(n-1)/2n} - 1]$$

an equation that applies only when $T_1 = T_3$ (Fig. 14/14) and $p_2 = p_3$ as found from (b).

For three or more stages of compression (more than three stages are uncommon), the method of analysis is similar to that given for the two-stage machine.

---

* This may be proved for the case of no pressure drop between cylinders and for $T_1 = T_2$ by letting, in equation (14-5) $p_2 = p_3 = p_i$, a variable intermediate pressure to be determined. Then differentiate W from equation (14-5) with respect to this pressure p_i and equate the result to zero. The value of p_i found after this differentiation is $p_i = (p_1 p_4)^{1/2}$, the same as that found by equating the work of the first stage to that of the second stage.

HEAT TRANSFERRED IN INTERCOOLER 14.13

If the intercooler (see Fig. 14/15) is analyzed as a steady-flow steady-state device, an equation is obtained that gives the transferred heat whether the process is reversible ($p = c$) or irreversible (with pressure drop). Consider the steady-state, steady-flow energy equation applied to the intercooler process B-E shown in Fig. 14/14.

(a)
$$H_B + K_B + Q = H_E + K_E + W$$

For the process, $W = 0$ and $\Delta K = 0$. It follows that

(b)
$$Q = H_E - H_B = \Delta H$$

or since we are analyzing the compression of a gas,

(c)
$$Q = m'c_p(T_E - T_B)$$

where m' is the mass of gas passing through the intercooler (also the mass drawn in by the LP cylinder and delivered by the HP cylinder).

Example—Two-Stage Compressor 14.14

A two-stage, double-acting compressor operating at 150 rpm takes in air at 14 psia, 80°F. The LP cylinder is 14×15-in., the stroke of HP cylinder is 15 in., and the clearance of both cylinders is 4%. Air is discharged at 56 psia from the LP cylinder, passes through the intercooler, and enters the HP cylinder at 53.75 psia, 80°F; it leaves the HP cylinder at 215 psia. The polytropic exponent $n = 1.3$ for both cylinders. Neglect effect of piston rods on the crank end. Environmental atmospheric conditions are 14.7 psia, 70°F. See Fig. 14/16.

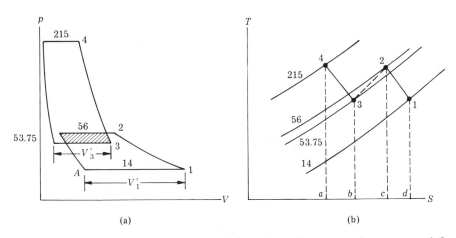

(a)   (b)

Two-Stage Compression. In figure (b) are shown the compression processes 1–2 and 3–4 and the cooling in the intercoolar 2–3. Since a pressure drop due to *friction* has occurred in the inter-cooler, this process is irreversible, and for this reason is shown dotted. **Fig. 14/16**

Find (a) volume of free air compressed, (b) heat transferred during compression to cooling water around the LP cylinder, (c) heat rejected during intercooling, (d) diameter of HP cylinder, and (e) work required for the compressor.

Solution.

(a) For the LP cylinder, the displacement is,

$$V_{DL} = \frac{(\pi D_L^2)}{4}(L)(2n) = \frac{\pi(14)^2(15)(2)(150)}{(4)(1728)} = 401 \text{ cfm}$$

The volume drawn in is,

$$V_1' = \eta_v V_{DL} = [1 + c - c(p_2/p_1)^{1/n}]V_{DL}$$

$$= [1.04 - 0.04(56/14)^{1/1.3}](401) = (0.924)(401) = 370 \text{ cfm}$$

measured at 14 psia, 80°F.

From $pV/T = C$, the volume of free air is,

$$V_a = \frac{p_1 V_1' T_a}{p_a T_1} = \frac{(14)(370)(530)}{(14.7)(540)} = 346 \text{ cfm}$$

(b) The heat transferred in the LP cylinder during compression is shown in Fig. 14/16(b) as process 1-2; further, the air rejecting the heat is that drawn in plus that in the clearance. Now

$$T_2 = T_1(p_2/p_1)^{(n-1)/n} = 540(56/14)^{(1.3-1)/1.3} = 744°R$$

and

$$V_1 = (1 + c)(V_{DL}) = (1.04)(401) = 417 \text{ cfm}$$

from which

$$\dot{m}_1 = \frac{p_1 \dot{V}_1}{RT_1} = \frac{(14)(144)(417)}{(53.3)(540)} = 29.2 \text{ lb/min}$$

and the heat transferred during the compression process, LP cylinder is

$$Q_{12} = m_1 c_v \left(\frac{k-n}{1-n}\right)(T_2 - T_1)$$

$$= (29.2)(0.1715)\left(\frac{1.4 - 1.3}{1 - 1.3}\right)(744 - 540) = -341 \text{ Btu/min}$$

(c) The mass of air rejecting heat in the intercooler corresponds to volume $V_1' = 370$ cfm and equals

$$\dot{m}_1' = \frac{p_1 \dot{V}_1'}{RT_1} = \frac{(14)(144)(370)}{(53.3)(540)} = 25.9 \text{ lb/min}$$

and the heat rejected is

$$Q_{23} = \Delta H_{23} = \dot{m}_1' c_p (T_3 - T_2)$$

$$= (25.9)(0.24)(540 - 744) = -1268 \text{ Btu/min}$$

(d) The mass of air passing through the HP cylinder is $m_1' = 25.9$ lb/min; so the corresponding volume is

$$\dot{V}_3' = \frac{\dot{m}_1' R T_3}{p_3} = \frac{(25.9)(53.3)(540)}{(144)(53.75)} = 96.3 \text{ cfm}$$

and the displacement is

$$\dot{V}_{DH} = \dot{V}_3'/\eta_v = 96.3/0.924 = 104.2 \text{ cfm}$$

$$= \left(\frac{\pi D_H^2}{4}\right)\left(\frac{15}{12}\right)(300)$$

from which $D_H = 0.595$ ft $= 7.14$ in. or say $7\frac{1}{4}$ in.

(e) From the LP cylinder, the work is,

$$\dot{W}_{LP} = \frac{n \dot{m}_1' R T_1}{(1 - n)}[(p_2/p_1)^{(n-1)/n} - 1]$$

$$\dot{W}_{LP} = \frac{(1.3)(25.9)(53.3)(540)}{(33,000)(1 - 1.3)}[(56/14)^{(1.3-1)/1.3} - 1] = -37 \text{ hp}$$

The negative sign indicates that work is done on the air. Since the pressure ratio for the HP cylinder is $p_4/p_3 = 215/53.75 = 4$ (same as for LP cylinder) and $T_3 = T_1 = 540°R$, then the $\dot{W}_{HP} = \dot{W}_{LP} = -37$ hp. So, the total work for the compressor is

$$\dot{W} = 2(-37) = -74 \text{ hp}$$

according to the conventional card.

PISTON SPEEDS 14.15

The piston speeds may be as much as 350 fpm for small compressors, say, with a stroke of about 6 in., to more than 700 fpm in large compressors, say, with a stroke of about 36 in.

CLOSURE 14.16

It would be remiss if this chapter closed without focusing attention on two matters. First, the conventional card for a reciprocating compressor is not a thermodynamic cycle. The two constant pressure lines for suction and discharge do not represent $p = c$ heating or cooling thermodynamic processes. They depict no more than a change of location of the mass of gas with no change of thermodynamic properties. Secondly, if the compression is an irreversible adiabatic (and all rotary compressors are) then we must forego the use of either $-\int V\,dp$ or $\int p\,dV$ to solve for work. However, don't forget that the steady-flow, steady-state energy equation is always good, reversible or irreversible, and is useful even in analyzing reciprocating compressors; that is, we can always move the system boundaries away from the piston/cylinder arrangement and thus avoid using these two integrals.

PROBLEMS

SI UNITS

14.1 A steady flow compressor handles 113.3 m³/min of nitrogen measured at intake where $p_1 = 97.22$ kPaa and $t_1 = 26.7°C$. Discharge is at 310.27 kPaa. The change of kinetic energy is negligible. For each of the following cases, determine the temperature t_2 and the work if the process is **(a)** an isentropic, **(b)** an internally reversible polytropic with $n = 1.34$, **(c)** an irreversible adiabatic with a compressor efficiency of $\eta_c = 80\%$, **(d)** an isothermal. Solve this problem from the view point of energy diagrams. Does $-\int V\,dp$ represent anything asked for?

Ans. **(a)** $-15,454$, **(b)** $-15,168$, **(c)** $-19,318$, **(d)** $-12,783$ kJ/min.

14.2 There are required 1902.3 kW of compressor power to handle air adiabatically from 1 atm, 26.7°C to 304.06 kPaa. The initial air velocity is 21 m/s; the final is 85 m/s. **(a)** If the process is isentropic, find the volume of air handled, m³/min measured at inlet conditions. **(b)** If the compression is an irreversible adiabatic to a temperature of 157.2°C, with the capacity found in **(a)**, find the power input.
Ans. **(a)** 899 m³/min, **(b)** 2377.9 kW.

14.3 A small blower handles 43.33 m³/min of air whose density is $\rho = 1.169$ kg/m³. The static and velocity heads are 16.38 and 1.22 cm wg (at 15.6°C), respectively. Local gravity acceleration is $g = 9.741$ m/s². **(a)** Find the power input to the air from the blower. **(b)** If the initial velocity is negligible, find the final velocity.
Ans. **(a)** 1.24 kW, **(b)** 854 m/min.

14.4 A large forced draft fan is handling air at 1 atm, 43.3°C under a total head of 26.6 cm wg (at 43.3°C). The power input to the fan is 224 kW and the fan is 75% efficient. Compute the volume of air handled each minute. Local gravity acceleration is $g = 9.71$ m/s². *Ans.* 3908 m³/min.

14.5 Find the cylinder dimensions of a single-cylinder, double-acting compressor handling 28.32 ℓ/revolution of air from 99.975 kPaa to 723.954 kPaa. Compression and reexpansion are in accordance with $pV^{1.35} = C$. Use the conventional volumetric efficiency; $c = 5\%$ and $L/D = 1$.
Ans. 26.77 × 26.77 cm.

14.6 There are compressed 6.542 m³/min of oxygen from 1 atm, 26.7°C to 310.27 kPaa by a 35.56 × 35.56-cm single-stage, double-acting compressor operating at 100 rpm. Compression and reexpansion processes are isentropic and $\Delta K = 0$. Find the volumetric efficiency, the work done on the oxygen, and the heat removed. Solve first by using the conventional indicator card; check by using an energy diagram and considering steady flow. *Ans.* 92.6%, 14.53 kW, 14.57 kW.

14.7 Find the volumetric efficiency and estimate the approximate clearance of a 45.75 × 45.75-cm, double-acting, single-cylinder compressor that is turning at 150 rpm and pumping 19.82 m³ of a gas from 1 atm, 26.7°C to 675.7 kPaa. Compression and reexpansion are polytropic with $pV^{1.32} = C$.

14.8 A reciprocating air compressor with a clearance of 6% draws in 4.25 m³/min of air measured at suction conditions of 100 kPaa, 57.2°C. For a discharge pressure of 300 kPaa and an overall adiabatic efficiency of 68%, determine the power of the driving motor. *Ans.* 13.43 kW.

14.9 For a two-stage reciprocating compressor with intercooler, prove that the work will be minimum when the pressure between cylinders (intercooler pressure) is $p_i = (p_1 p_2)^{0.5}$. Here p_1 is compressor inlet pressure and p_2 is compressor discharge pressure. Further, let the intercooler return the air to the compressor inlet temperature.

14.10 Air is compressed in a two-stage, double-acting compressor which is electrically driven at 165 rpm. The low pressure cylinder (30.5 × 35.5-cm) receives 6.85 m³/min of air at 96.53 kPaa, 43.3°C, and the high pressure cylinder (20.3 × 35.5-cm) discharges the air at 717.06 kPaa. Piston rods are 5.1 cm in diameter and the isothermal overall efficiency is 74%. Find **(a)** the volumetric efficiency, **(b)** the power of the driving motor.

14.11 There are compressed 11.33 m³/min of air from 103.42 kPaa, 26.7°C to 827.36 kPaa. All clearances are 8%. **(a)** Find the isentropic power and piston displacement required for a single stage compression. **(b)** Using the same data, find the minimum ideal power for two-stage compression when the intercooler cools the air to the initial temperature. **(c)** Find the displacement of each cylinder for the conditions of part **(b)**. **(d)** How much heat is exchanged in the intercooler? **(e)** For a compressor efficiency of 78%, what driving motor output is required?

Ans. **(a)** 55.4 kW, 15.88 m³/min, **(b)** 47.46 kW, **(c)** 12.43 m³/min, 4.39 m³/min, **(d)** 1423 kJ/min, **(e)** 61.5 kW (2-stage).

14.12 Write this computer program. The effect of the polytropic exponent n on the conventional volumetric efficiency of a given compressor is under study. Select a specific percentage clearance c and an anticipated pressure ratio p_2/p_1 and calculate the volumetric efficiencies letting n vary in the range 1–1.4 (say for air).

ENGLISH UNITS

14.13 The steady flow compressor of a Brayton-cycle gas turbine breathes in 45,000 cfm of air at 15 psia, 60°F and compresses it through a pressure ratio of 9.5; the compressor efficiency is 82%. Use Item B 2 and find **(a)** discharge temperature, **(b)** compression power, hp, **(c)** irreversibility and change in availability for $p_0 = 15$ psia, $t_0 = 60°F$.

14.14 A low-pressure, water-jacketed, steady flow compressor compresses 15 lb/min of air from 14.7 psia and 70°F to 5 psig and 110°F. **(a)** Let the process be polytropic, neglect the change in kinetic energy, and find the power. What mass of water is circulated if the temperature rise of the cooling water is 6°F? How much is $-\int V\,dp$ in this system and what does it represent? **(b)** Consider the process as an irreversible adiabatic (no water jacket) with a final temperature of 130°F (instead of 110°F) and find the value of m in $pV^m = C$, and the work. What does $-\int V\,dp$ represent in this system? Compare works.

Ans. **(a)** − 129,000 ft-lb/min, 3.64 lb/min of H_2O, $-\int V\,dp = -165.8$ Btu/min. **(b)** 1.577, − 216 Btu/min.

14.15 Air is removed from a large space and given a velocity of 63 fps by a fan. The air density is $\rho = 0.075$ lb/ft³ and the work done on the air is 0.0155 hp-min/lb air. Find the static head on the fan, in. wg (at 100°F).

14.16 The adiabatic work input required to compress 50 lb/min of air from 14 psia,

80°F to a higher pressure is 60.4 hp where $\Delta K = 0$. **(a)** If the process is reversible, find the discharge pressure. **(b)** If the compression to this pressure is irreversible, where $n_c = 84\%$, find the adiabatic work input. **(c)** Which of these two processes will cause the greatest increase in availability?

14.17 Water, circulating at the rate of 52 lb/min around the cylinder of an air compressor, enters at 70°F and leaves at 80°F, all heat received coming from the air in the cylinder. The compression is internally reversible from 14.7 psia, 80°F to 330°F; $\Delta K = 0$. For an air flow of 50 lb/min, find **(a)** the power, **(b)** ΔS for the air, **(c)** the available part of the heat with respect to the air and again as it was received by the water if $t_0 = 60°F$.

Ans. **(a)** 83 hp, **(b)** −0.791 Btu/°R–min, **(c)** 109, 14.3 Btu/min.

14.18 A 14 × 14-in., horizontal double-acting air compressor with 5% clearance operates at 120 rpm, drawing in air at 14.4 psia and 88°F, and discharging it at 57.6 psia. The compression and reexpansion processes are polytropic with $n = 1.33$. Sketch the conventional card and determine **(a)** the conventional volumetric efficiency, **(b)** the mass of air discharged, **(c)** the horsepower input to the air. **(d)** How much is $-\int V\,dp$ for the compression process?

Ans. (B 1), **(a)** 90.8%, **(b)** 19.3 lb/min, **(c)** 29 hp.

14.19 A 14 × 12-in., single-cylinder double-acting air compressor with 5.5% clearance operates at 125 rpm. The suction pressure and temperature are 14 psia and 100°F, respectively. The discharge pressure is 42 psia. Compression and reexpansion processes are isentropic. Considering the conventional compressor and neglecting the piston-rod effect, determine **(a)** the volumetric efficiency, **(b)** the mass and volume at suction conditions handled each minute, **(c)** the work in horsepower, **(d)** the heat rejected, and **(e)** the indicated air horsepower developed if the compression efficiency is 75%.

14.20 Carbon dioxide is handled by a rotary type compressor from 15 psia, 90°F through a compression ratio of 7 with a compressor efficiency of 75%; $\Delta K = 0$. Find the discharge temperature. Solve using Item B 1 and check answer using gas tables.

Ans. (B 1) 1100°R, (B 3) 1009°R.

14.21 A compressor handles 3500 cfm of carbon dixoide measured at intake where $p_1 = 14.2$ psia and $t_1 = 75°F$. At discharge, $p_2 = 28.4$ psia and $t_2 = 178°F$. The initial velocity is 40 fps and the final velocity is 150 fps. The process is an irreversible adiabatic. Find **(a)** $\Delta H'$, $\Delta U'$, $\Delta S'$, **(b)** $\dot{W}'$, **(c)** η_c.

Ans. **(a)** 8150, 6400 Btu/min, 2.06 Btu/°R-min, **(b)** -8309 Btu/min, **(c)** 82%.

14.22 A steady flow compressor compresses 2300 cfm of methane from 15 psia and 75°F to 15 psig. The process is an irreversible adiabatic and the change of kinetic energy is negligible. If the compression efficiency is $\eta_c = 82.7\%$, find **(a)** t_2, **(b)** ΔS, **(c)** W. **(d)** How much is $-\int V\,dp$? What is the change of availability per pound of methane flowing if the lowest available temperature is $t_0 = 40°F$?

14.23 A compressor is to be designed with 6% clearance to handle 500 cfm of air at 14.7 psia and 70°F, the state at the beginning of the compression stroke. The compression is isentropic to 90.3 psig. **(a)** What displacement in cfm is necessary? **(b)** If the compressor is used at an altitude of 6000 ft, and if the initial temperature and the discharge pressure remain the same as given above, by what percentage is the capacity of the compressor reduced? **(c)** What should be the displacement of a compressor at the altitude of 6000 ft to handle the same mass of air as in **(a)**?

14.24 A two-stage air compressor without clearance delivers 90 lb/min at 140 psia. At the intake, $p_1 = 14.3$ psia and $t_1 = 60°F$. Compression follows $pV^{1.31} = C$, and the intercooler cools the air back to 60°F. Find **(a)** the optimum intermediate pressure, **(b)** the conventional power, **(c)** the heat of the various processes (sketch these processes on the TS plane). **(d)** What horsepower would be required for isentropic compression in a single-stage machine? **(e)** What is the saving due to the cooling process? Does this appear to be worth while? **(f)** If the temperature of the cooling water in the intercooler rises 15°F, what mass of water is required?

Ans. **(a)** 44.75 psia, **(b)** -198 hp, **(c)** -717, -3483, -717 Btu/min, **(d)** -243 hp, **(e)** 45 hp, **(f)** 232 lb/min.

14.25 A single-cylinder, double-acting air compressor running at 200 rpm has a piston speed of 600 fpm. It compresses 60 lb/min from 14 psia and 60°F to 95 psia. Clearance is 5.5%. For an isentropic compression, find **(a)** η_v, $\dot{V}_D$, and $\dot{W}$; **(b)** p_m of the conventional card; and **(c)** the bore and stroke of the compressor. Solve for $\dot{W}$ twice using air properties from both Item B 1 and Item B 2. Compare.

Ans. (B 1), **(a)** $\dot{V}_D = 983$ cfm, **(b)** 29.9 psi, **(c)** 17.35×18 in.

15

GAS TURBINES AND
JET ENGINES

INTRODUCTION 15.1

Hero, a scientist of Alexandria, is credited with inventing the first gas turbine (circa 120 B.C.). The device consisted of a small metal globe called an aeolipile mounted on a pipe leading from a steam kettle. Steam from the kettle escaped from two L-shaped pipes fastened to opposite sides of the globe and whirled it around; it produced no shaft work.

The first gas turbine to produce work was probably the windmill. It came into use in the Middle East in the 900's and in Europe in the 1100's. In the 1600's fans were mounted over a cooking fire to turn roasting meat on a spit; the hot gases rising from the fire spun the fan that was geared to the spit.

The characteristic features of a gas turbine as we think of the name today include a compression process and a heat addition (or combustion) process. These features are not new, although a practical machine is a relatively recent development. Joule and Brayton* independently proposed the cycle that is the ideal prototype of the actual unit. An unsuccessful turbine unit was built as far back as 1872, and by 1906 a unit that produced net power had been built. There were two principal obstacles to be overcome, as revealed by thermodynamic analysis. In order for practical amounts of power to be delivered: (1) the temperature at the beginning of expansion must be high (until a few years ago, the highest permissible temperatures were about 700 to 800°F), and (2) the compressor and turbine must operate at a high efficiency. Metallurgical developments are raising the highest permissible temperatures (say, to 1500–2000°F, and more if high cost and/or if short life is acceptable). A better knowledge of aerodynamics has been partly responsible for improving the efficiency of both the centrifugal and axial flow compressors. See Fig. f/6, Foreword, p. xx, for an illustration of a gas turbine designed to deliver shaft work.

STEADY-FLOW BRAYTON CYCLE 15.2

Layouts with an explanation of the operation of this cycle are shown in Fig. 15/1. It is designed as both an open and a closed cycle, depending upon its use, with by far

* See footnote, p. 29. George Brayton, a contemporary of Otto (§ 16.1) was a Boston engineer. However, his engine was a reciprocating type, rather than rotary.

the majority of operating gas turbines being of the open variety with intake from the atmosphere and the burning of a fuel in this air. That part of the engine up to the turbine is called the _gasifier._ Closed gas turbine cycles may use a working substance other than air; for example, argon or helium for use with a nuclear-reactor source of energy, because, for one reason, they are not so prone to become radioactive (and helium has a much higher heat transfer coefficient than air, allowing a smaller heat exchanger). The working substance in the closed cycle may be at pressures higher than atmospheric at all points; the denser gas serves to reduce the volume occupied by the equipment, and the work per unit mass can be arranged to be about the same. The net work W is the total turbine work W_t minus the total compressor work W_c (including all stages), and it may drive a propeller, a generator, or other machine. There are work consuming auxiliaries in the actual engine that we choose to ignore.

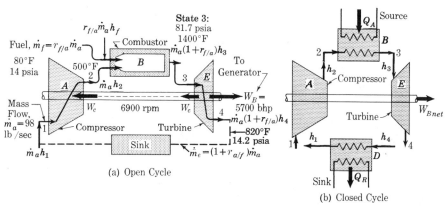

Fig. 15/1   *Layouts of Gas Turbine Units.* In the open cycle (a), air enters the compressor *A*, which delivers it to the combustors (Fig. 15/17), into which fuel is injected and burned; $r_{f/a}$ = fuel–air ratio. The hot products of combustion enter the turbine, expand to approximately the pressure of the surroundings, and are discharged. The properties shown are sample values for stationary power plants. In (b), the working substance circulates, picking up heat and discharging heat in heat exchangers *B* and *D*.

It is common practice, especially for first approximations, to analyze the open cycle as an _air standard cycle_ in which: (1) the mass and properties of the fuel are neglected, (2) the heat added is the energy released by the reaction of the fuel, (3) the substance has the properties of air throughout, and (4) sometimes even constant specific heats are assumed. These assumptions are equivalent to the air passing through a closed cycle. We shall write equations in the basic form (usually stagnation enthalpies) and also on the assumption of constant specific heats. If the working substance is a monatomic gas, the specific heats are essentially constant (for the range of this application), § 2.22, and conclusions corresponding to this assumption are valid. For air and other gases, the specific heats will vary markedly. For the case of the air standard, there is little reason to assume constant specific heats with the availability of tables of properties that allow for their variation, Item B 2. If one is computing Δh between *any* states *a* and *b*, better results will be obtained if the average specific heat between *a* and *b* is used; $\Delta h = c_{pab}(T_b - T_a)$.

For the ideal cycle, which consists of two isobaric and two isentropic processes (nonflow or steady flow as far as the *cycle* work is concerned) and which is shown on the pv and Ts planes in Fig. 15/2, the work is obtained from either $\oint dQ$ or $\oint dW$.

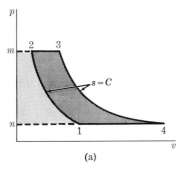

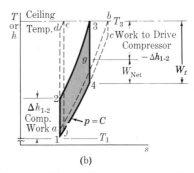

(a)   (b)

Brayton Cycle. Also known as the Joule cycle. The ceiling temperature is   **Fig. 15/2**
determined by the capacity of materials used for resisting the effects of heat.
$\oint dQ = \oint dW = W_{\text{net}}$ = area 1-2-3-4.

From $\oint dW$, we have the fluid work by equation (4-10) for each adiabatic steady-flow machine as $W = -\Delta h - \Delta K = -\Delta h_0$ in terms of stagnation enthalpies. Or, if ΔK across the control volume for each piece of equipment is negligible,

(a) $$W = W_t - |W_c| = h_3 - h_4 - (h_2 - h_1) = c_p(T_3 - T_2 + T_1 - T_4)$$

[ANY PURE SUBSTANCE]   [CONSTANT k]

(b) $$e = \frac{W}{Q_A} = \frac{h_3 - h_2 + h_1 - h_4}{h_3 - h_2} = 1 - \frac{h_4 - h_1}{h_3 - h_2} = 1 - \frac{T_4 - T_1}{T_3 - T_2}$$

[ANY PURE SUBSTANCE]   [CONSTANT k]

This equation for the thermal efficiency may be put into various significant forms. First, using the definition of the pressure ratio, $r_p = p_2/p_1 = p_3/p_4$, § 7.31, and the Tp relation for an isentropic, we have

(c) $$\frac{T_2}{T_1} = \left(\frac{p_2}{p_1}\right)^{(k-1)/k} = r_p^{(k-1)/k}$$

(d) $$\frac{T_3}{T_4} = \left(\frac{p_3}{p_4}\right)^{(k-1)/k} = \left(\frac{p_2}{p_1}\right)^{(k-1)/k} = r_p^{(k-1)/k}$$

from which $T_2/T_1 = T_3/T_4$ for constant k. Rearranging, we get

(e) $$\frac{T_4}{T_1} = \frac{T_3}{T_2} \quad \text{or} \quad \frac{T_4}{T_1} - 1 = \frac{T_3}{T_2} - 1 = \frac{T_4 - T_1}{T_1} = \frac{T_3 - T_2}{T_2}$$

(f) $$\frac{T_4 - T_1}{T_3 - T_2} = \frac{T_1}{T_2} = \frac{T_4}{T_3}$$

Using this relation in equation **(b)**, we get the thermal efficiency as

(g) $$e = 1 - \frac{T_1}{T_2} = \frac{T_2 - T_1}{T_2} = \frac{T_3 - T_4}{T_3} = 1 - \frac{T_4}{T_3}$$

Equation **(g)** reminds one of the Carnot efficiency, but it is not so. A Carnot cycle operating through the same temperature limits, T_3 and T_1, Fig. 15/2, has a greater efficiency, $(T_3 - T_1)/T_3$. The TV relation for the isentropic is

(h)
$$\frac{T_2}{T_1} = \left(\frac{V_1}{V_2}\right)^{k-1} = r_k^{k-1}$$

where $r_k = V_1/V_2$, the compression ratio. Using r_p from equation **(c)** and r_k from equation **(h)**, in equation **(g)** we find

(15-1)
$$e = 1 - \frac{1}{r_k^{k-1}} = 1 - \frac{1}{r_p^{(k-1)/k}} \qquad [\text{CONSTANT } k]$$

Since the pressure ratio r_p for the ideal cycle is the same during compression and expansion, it is customarily used, rather than r_k. The expansion ratio r_e is different from r_k even in the ideal cycle unless $k_{3\text{-}4} = k_{1\text{-}2}$.

An examination of equations **(g)** and (15-1) suggests that the thermal efficiency of the ideal cycle is improved by: (1) increasing T_2, (2) decreasing T_4, and (3) increasing the compression ratio or pressure ratio; but beyond a certain point, this improvement is accompanied by a decreasing work per unit mass. One of the facts of life that the gas turbine engineer must contend with is the ceiling temperature, Fig. 15/2. Consider the cycle 1-*def*, wherein the pressure ratio (1 to *d*) is large; its efficiency is approaching the Carnot efficiency (if *de* is very small heat is supplied at $T \approx C$), but its net work, that which is desired, is approaching zero. Also, the work of the cycle decreases if the pressure ratio decreases below a certain value, as suggested by the cycle 1-*abc*.

15.3   INTERMEDIATE TEMPERATURE FOR MAXIMUM WORK

From the foregoing discussion, we see that for fixed initial and final temperatures, T_1 and T_3, there is some intermediate temperature T_2 that would result in the maximum work. This we are interested in because we may care to keep the size of engine as small as possible. Substitute the value of T_4 from equation **(f)** into equation **(a)**, § 15.2, differentiate W with respect to T_2, and equate the differential to zero (T_1 and T_3 are constants).

(a)
$$\frac{dW}{dT_2} = \frac{mc_p d[T_3 - T_2 + T_1 - (T_1 T_3)/T_2]}{dT_2} = 0$$

(15-2)
$$-1 + \frac{T_1 T_3}{T_2^2} = 0 \qquad \text{or} \qquad T_2 = (T_1 T_3)^{1/2} \qquad [\text{CONSTANT } k]$$

the value of T_2 that will result in the maximum work of the ideal cycle that is limited by the temperatures T_1 and T_3.

15.4   GAS TURBINE WITH FLUID FRICTION

Considering the actual cycle in simple form, Fig. 15/3, let the compressor and turbine processes be adiabatic and let the net ΔK for each be negligible (or use stagnation enthalpies, § 7.23). From a simple energy diagram, or equation (4-10), we have fluid works as follows:

(a) $$W'_c = -\Delta h = h_1 - h_{2'} = c_p(T_1 - T_{2'})$$

(b) $$W'_t = -\Delta H = (1 + r_{f/a})(h_3 - h_{4'}) = (1 + r_{f/a})c_p(T_3 - T_{4'})$$

where $r_{f/a}$ is the air-fuel ratio ($r_{f/a} = 0$ for closed cycle); its magnitude is small, of the order of 0.02, probably less. Moreover, a small quantity of air bled before combustion for cooling parts of the turbine closely equals the mass of fuel. For "400% air" (§ 13.5), $r_{f/a} \approx 0.016$, varying a little for different hydrocarbon fuels. The large quantity of air is necessary to keep the ceiling temperature in bounds (§ 13.20). In terms of stagnation enthalpies, the compressor efficiency η_c and the turbine efficiency η_t are given in §§ 7.31 and 14.8.

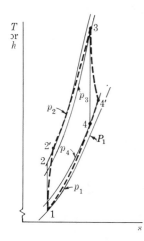

With Fluid Friction. The area enclosed within *irreversible* processes does *not* represent work. Because of friction, the pressure ratio in the actual turbine is less than that in the compressor. In design, one must estimate the pressure drop through the combustors 2-3 (or heat changer for the closed cycle). See § 15.6.   **Fig. 15/3**

The actual heat added to the fluid with $\Delta K = 0$ is $Q'_A = h_3 + h_{2'} = c_{p2'\text{-}3}(T_3 - T_{2'})$; the thermal efficiency is

(15-3A) $$e' = \frac{h_3 - h_{4'} - (h_{2'} - h_1)}{h_3 - h_{2'}} = \frac{T_3 - T_{4'} - (T_{2'} - T_1)}{T_3 - T_{2'}}$$

$$\qquad\qquad [\text{ANY FLUID}] \qquad\qquad [\text{CONSTANT } k]$$

(15-3B) $$e' = \frac{(h_3 - h_4)\eta_t - \dfrac{h_2 - h_1}{\eta_c}}{h_3 - h_{2'}} = \frac{(T_3 - T_4)\eta_t - \dfrac{T_2 - T_1}{\eta_c}}{T_3 - T_{2'}}$$

$$\qquad\qquad [\text{ANY FLUID}] \qquad\qquad [\text{CONSTANT } k]$$

In the further manipulation of the constant-k form without pressure drops, let $T_2/T_1 = T_3/T_4 = r_p^{(k-1)/k}$ in this equation to eliminate T_2 and T_4, use the value of $T_{2'}$ obtained from the definition of compressor efficiency, equation (14-4), namely,

(c) $$T_{2'} = \frac{T_2 - T_1(1 - \eta_c)}{\eta_c}$$

and rearrange to get

(d) $$e' = \left(\frac{\eta_t T_3 - T_1 r_p^{(k-1)/k}/\eta_c}{T_3 - T_1 - T_1(r_p^{(k-1)/k} - 1)/\eta_c} \right)\left(1 - \frac{1}{r_p^{(k-1)/k}}\right)$$

Notice that the last parentheses enclose a term that is the efficiency of an ideal Brayton cycle with a pressure ratio of r_p. The purpose of arriving at this form is to show that the efficiency of the actual cycle depends upon the high and low temperatures as well as on the pressure ratio. If it is desired to know the maximum efficiency for a particular temperature range, differentiate the foregoing equation with respect to r_p and equate to zero. There is little or nothing that can be done about T_1; so let it be, say, 540°R. Now for a particular turbine inlet temperature (say, 1400°F = 1860°R), assume various pressure ratios and plot a curve. You will find a curve such as the one labeled 1400°F in Fig. 15/4, which points up an important fact: *For each turbine inlet temperature, there is a certain pressure ratio that results in maximum thermal efficiency*; that is, the actual thermal efficiency does not go up indefinitely with pressure ratio as in the case of the dotted curve of the ideal cycle.

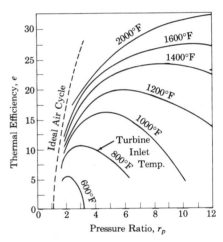

Fig. 15/4 *Efficiency Versus Pressure Ratio. After Zucrow.*[15,1] Each curve is for a particular value of T_3 as labeled, for the simple cycle with friction, Fig. 15/3, but with no pressure drops 2'-3 and 4'-1; combustor efficiency, 100%; $\eta_t = 85\%$; $\eta_c = 84\%$; constant $k = 1.4$ for air. Curves for the variation in work output against r_p vary similarly, but the optimum r_p for maximum work is lower than the optimum r_p for maximum efficiency. For example, at $T_3 = 1400°F$; the optimum $r_p \approx 8$ for $e_{max} \approx 24.5\%$; the optimum $r_p \approx 5$ for $W_{max} \approx 54$ Btu/lb; at $r_p = 8$, $W \approx 50$ Btu/lb; at $r_p = 5$, $e \approx 20\%$.

15.5   Example—Gas Turbine With and Without Friction, Air Standard

The intake of the compressor of an *air-standard* Brayton cycle (variable specific heat) is 40,000 cfm at 15 psia and 40°F. The compression ratio $r_k = 5$ and the temperature at the turbine inlet is 1440°F. Neglect the pressure drop between compressor and turbine, and let the exit pressure of the turbine be 15 psia; kinetic energy changes for each process, Fig. 15/5, are negligible. (a) For the ideal cycle, determine the net horsepower output and the thermal

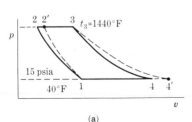

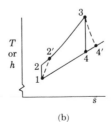

Fig. 15/5   (a)   (b)

efficiency. (b) For a turbine efficiency $\eta_t = 85\%$ and a compressor efficiency $\eta_c = 83\%$, compute the net output and the thermal efficiency. What is the percentage reduction in power? (c) Determine the mep of the ideal cycle. (d) What is the availability of the exhaust air from the actual cycle with respect to a sink at $p_0 = 15$ psia and $T_0 = 500°R$?

Solution. The cycle is represented on the pv and Ts planes in Fig. 15/5. To allow for variable specific heats, use Item B 2; from it

For 500°R: $\quad h_1 = 119.48 \quad p_{r1} = 1.059 \quad v_{r1} = 174.90 \quad \phi_1 = 0.58233$

For 1900°R: $\quad h_3 = 477.09 \quad p_{r3} = 141.51 \quad v_{r3} = 4.974 \quad \phi_3 = 0.91788$

For a compression ratio $r_k = 5$, we get $v_{r2} = v_{r1}/5 = 174.9/5 = 34.98$. Corresponding to this value, we read (T to nearest whole degree is usually sufficiently accurate)

$$T_2 = 944.4°R \quad h_2 = 227.2 \quad p_{r2} = 10 \quad \phi_2 = 0.73624$$

Since the expansion ratio in the Brayton cycle with variable specific heats is different from the compression ratio, find the properties at 4 from p_{r4} as follows:

(a) $\qquad r_{p34} = r_{p12} = \dfrac{p_{r2}}{p_{r1}} = \dfrac{10}{1.059} = 9.44 \qquad p_{r4} = \dfrac{p_{r3}}{r_{p34}} = \dfrac{141.51}{9.44} = 14.99$

$$T_4 = 1056 \quad h_4 = 254.96 \quad v_{r4} = 26.09 \quad \phi_4 = 0.76401$$

(a) For $\Delta K = 0$ in each case, we have

(b) $\qquad W_c = -(h_2 - h_1) = -227.2 + 119.48 = -107.72 \text{ Btu/lb}$

(c) $\qquad W_t = h_3 - h_4 = 477.09 - 254.96 = 222.13 \text{ Btu/lb}$

(d) $\qquad Q_A = h_3 - h_2 = 477.09 - 227.2 = 249.89 \text{ Btu/lb}$

(e) $\qquad e = \dfrac{\oint dW}{Q_A} = \dfrac{222.13 - 107.72}{249.89} = \dfrac{114.41}{249.89} = 45.8\%$

For a mass of air of

(f) $\qquad \dot{m}_a = \dfrac{p_1 \dot{V}_1}{RT_1} = \dfrac{(15)(144)(40,000)}{(53.3)(500)} = 3240 \text{ lb/min}$

we find the net horsepower output as

(g) $\qquad \dot{W} = \dfrac{(114.41)(3240)}{42.4} = 8743 \text{ hp}$

(b) The actual work quantities are

(h) $\qquad W_c' = -\dfrac{107.72}{0.83} = -130 \text{ Btu/lb} \qquad W_t' = (0.85)(222.13) = 188.8 \text{ Btu/lb}$

(i) $\qquad W' = 188.8 - 130 = 58.8 \text{ Btu/lb}$

The actual work delivered by the shaft would be of the order of 3% less than this value—allowing for mechanical efficiency. To get the thermal efficiency, find the enthalpy at state 2′, Fig. 15/5. From $W'_c = -130$, we have

(j) $W'_c = h_1 - h_{2'} = -130$ or $h_{2'} = 119.48 + 130 = 249.48 \text{ Btu/lb}$ $T_{2'} = 1038°R$

(k) $$e' = \frac{W'}{Q_{A'}} = \frac{58.8}{477.09 - 249.48} = 25.9\%$$

(l) $$\text{percentage loss} = \frac{114.41 - 58.8}{114.41} = 48.5\%$$

This large loss of work ignores losses from incomplete combustion, the pressure drops through the combustor and other passages, and so on. An important conclusion is that a small increase in η_c and η_t results in a significant increase in thermal efficiency—and net work.

(c) The "displacement" of the ideal cycle is the largest volume minus the smallest, $V_4 - V_2$. To keep the numbers small, use 1 lb; $v_4 - v_2$. Hence,

(m) $$v_1 = \frac{V_1}{m} = \frac{40,000}{3240} = 12.34 \text{ ft}^3/\text{lb}$$ $$v_2 = \frac{12.34}{5} = 2.468 \text{ ft}^3/\text{lb}$$

(n) $$v_4 = \frac{RT_4}{p_4} = \frac{(53.3)(1055.7)}{(15)(144)} = 26.05 \text{ ft}^3/\text{lb}$$

(o) $$p_m = \frac{W}{V_D} = \frac{(114.41)(778)}{(26.05 - 2.468)(144)} = 26.2 \text{ psi}$$

the mep of this Brayton cycle. Compare it with the mep of the Carnot cycle (11 psi, § 8.9) of the same compression ratio. Low mep can be tolerated in rotary machines that turn at high speed.

(d) The enthalpy at 4′, Fig. 15/5, from an energy balance for the turbine is

(p) $$h_{4'} = h_3 - W'_t = 477.09 - 188.8 = 288.29 \text{ Btu/lb}$$

Accordingly, from Item B 2, we get $T_{4'} = 1188°R$, $\phi_{4'} = 0.79373$, which for the data at hand is an estimate of the actual exhaust state. Assume that K_4 is negligible and note that $p_0 = p_1 = p_{4'}$; then equation (6-13) yields

(q) $$s_{4'} - s_0 = \phi_{4'} - \phi_0 = 0.79373 - 0.58233 = 0.2114 \text{ Btu/lb-°R}$$

The availability at 4′ by equation (5-8) is

(r) $-(\Delta \mathscr{A}_{f/0})_{4'} = h_{4'} - h_0 - T_0 \Delta s = 288.29 - 119.48 - 500(0.2114) = 63.11 \text{ Btu/lb}$

which is 107% of the actual work W'. The high temperature of the exhaust and this availability suggest that something further might be gained from the energy of the exhaust; for one thing, it may be used for regenerative heating, § 15.8.

15.6   ENERGY BALANCE FOR COMBUSTOR

In general, in a combustion process not all the chemical energy is converted into sensible molecular internal energy or sensible enthalpy. The combustion (burning) efficiency may be simply defined as

(a)
$$\eta_b = \frac{\text{actual increase in sensible energy } (\Delta H)}{\text{ideal increase for complete ideal combustion}}$$

where the enthalpies should be *stagnation* enthalpies if ΔK is significant. For an efficiency of η_b, then effectively a mass of unburned fuel of $(1 - \eta_b)$ passes out of the combustor per pound of fuel. An energy diagram, on the basis of unit mass of air and using the enthalpy of combustion $-h_{rp}^\circ = q_i^\circ$, is shown in Fig. 15/6 (see § 13.20). Enthalpies of formation, § 13.30, may be used if desired. The time for the passage of the gases through the combustor is so short that the heat may be taken as zero per unit mass; $Q = 0$. We then have the energy balance,

(15-4) $\qquad r_{f/a}(h_x - h^\circ)_f + \eta_b r_{f/a} q_i^\circ + (h_{2'} - h^\circ)_a = (1 + r_{f/a})(h_3 - h^\circ)_p$

where $(h_x - h^\circ)_f$ is the sensible enthalpy of the fuel; $q_i^\circ = -h_{rp}^\circ$. Imagine equation (15-4) solved for η_b and compare with the definition in **(a)**. If the datum is $0°R$, equation (15-4) reduces to (see § 13.28)

(15-5) $\qquad r_{f/a}(h_{fx} + \eta_b q_{l0}) + h_{a2'} = (1 + r_{f/a})h_{p3}$

It is not unusual for this form of the equation to be used with the standard heating value instead of q_{l0}, with small error. If $r_{f/a}$ is a small number, then $1 + r_{f/a} \approx 1$ and $r_{f/a}h_{fx}$ is negligible, and equation (15-5) becomes

(b) $\qquad \eta_b r_{f/a} q_{l0} + h_{a2'} = h_{p3} \qquad$ or $\qquad h_{p3} - h_{a2'} = \eta_b r_{f/a} q_{l0}$

If the combustion efficiency is 100% and if we consider the products to be air (air standard), we have

(c) $\qquad\qquad h_3 - h_{2'} = r_{f/a} q_{l0} = Q_A'$

which was used as the heat added for the fluid cycle with friction, § 15.5. For convenience, the energy charged against the actual engine is taken as the standard heating value, say $E_c = r_{f/a} q_i^\circ$, which gives

(d) $\qquad\qquad e' = \dfrac{W'}{m_f q_i^\circ} \qquad$ or $\qquad e' = \dfrac{W'}{r_{f/a} q_i^\circ}$

a dimensionless number. The work W' may be fluid work, the brake work (giving e_b), or the combined work (giving e_k). If m_f is (lb fuel/hp-hr), the corresponding

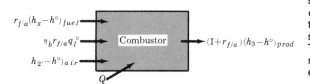

Energy Diagram for Combustor. The **Fig. 15/6**
datum for measuring enthalpies is the
standard state of $77°F$ and 1 atm;
$q_1^\circ = -h_{rp}^\circ$, the standard enthalpy of
the reaction. The fuel enters in some
state x different from that of the air.
The enthalpies used should be stagnation values if ΔK is significant; for
example, $h_{02'} = h_{2'} + K_{2'}$.

$W' = 2544$ Btu/hp-hr and the denominator $m_f q_i^o$ is the *heat rate*. For your convenience, the sensible enthalpy of a liquid hydrocarbon fuel is given closely by equation (13-7), § 13.19,

$$(13\text{-}7) \qquad\qquad h_f = 0.5T - 287 \text{ Btu/lb fuel}$$

and the molecular mass of the products of combustion is closely $M_p = 28.92$.[0.6] As needed, $h_p = \bar{h}_p/M_p$, and so on.

The combustor efficiency at rated loads or thereabouts should be greater than 95%, except at some high altitude.[15.1] The pressure drop in the combustor will be some 2 to 10% of the pressure as it enters.

Let the ideal cycle, corresponding to an actual gas turbine, have the same T_1, p_1, p_2, and T_3 as the actual engine; no pressure drops, $p_4 = p_1$; the subscripts are as defined in Fig. 15/5. Note that the energy chargeable is not the same in ideal and actual engines.

15.7   Example—Combustor

Let the unit be as described in § 15.5(b), and compute the required air-fuel ratio for $\eta_b = 95\%$ and $\Delta K \approx 0$. The fuel is C_8H_{18}, injected at 84°F.

Solution.   For octane, $q_{lo} = 19,330$ Btu/lb from § 13.28. From § 15.5, we have $h_{2'} = 249.48$ Btu/lb of air leaving the compressor. For the first trial calculation, assume that the products have the properties corresponding to 400% air, Item B 8, and find $\bar{h}_3 = 14,087.2$ at 1900°R. By equation (13-7), the sensible enthalpy of the fuel is

(a) $h_f = 0.5T - 287 = (0.5)(544) - 287 = -15$ Btu/lb fuel

By equation (15-5) with enthalpies on the 0°R datum, we have ($M_p = 28.92$)

$$r_{f/a}(h_{fx} + \eta_b q_{lo}) + h_{a2'} = (1 + r_{f/a})h_{p3}$$

(b)
$$r_{f/a}[-15 + (0.95)(19,330)] + 249.48 = (1 + r_{f/a})\frac{14,087.2}{28.92}$$

from which $r_{f/a} = 0.0133$ lb f/lb a, which is closer to 500% air than to 400%. Iteration may be conducted in accordance with the instructions for interpolation in the *Gas Tables*,[0.6] but there will be little change in the result. For further computations, the enthalpy and other properties of the actual products mixture may be determined by methods described in Chapter 13.

15.8   IDEAL CYCLE WITH IDEAL REGENERATIVE HEATING

Having observed that considerable available energy is dumped into the sink and that the temperature of the turbine exhaust 4 (and 4′), Fig. 15/5, is higher than the temperature at the end of compression 2, we might happen to think of applying the Ericsson and Stirling notion of regeneration (§ 8.10). Let the exhaust gas at 4 and the discharged air at 2 be led to a heat exchanger (regenerator), Fig. 15/7, so that the hot exhaust 4 gives up heat to the air 2. Theoretically, if the heat exchanger were large enough and the flow were slow enough, the air from the compressor could be heated reversibly to temperature 4 at state b, Fig. 15/8, while the exhaust cools to temperature 2 at state a (§ 8.11). Some of the formerly discharged heat, to wit $h_4 - h_a$, is *exchanged within* the system and the heat to the sink is not only $h_a - h_1$.

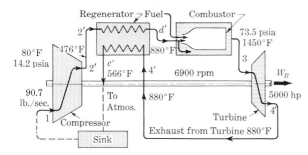

Diagrammatic Layout with Regener-ation. The station names agree with states shown on Fig. 15/9. The numerical properties are adapted from test values. Only part of the air enters the combustion space in order for that mixture to be readily combustible.

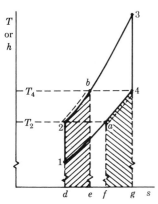

Regeneration—Ideal Cycle. With the s axis at absolute zero, area 4-*afg* = 2-bed, each representing heat interaction.

Moreover, it is necessary to add only the heat equal to $h_3 - h_b$, instead of $h_3 - h_2$ as formerly. Since less fuel is needed, this addition should increase the efficiency of the ideal cycle, which it does [by the ratio $(h_3 - h_2)/(h_3 - h_4)$]. From Fig. 15/8, we find the thermal efficiency for a constant mass of 1 lb ($r_{f/a} = 0$) as,

(a) $$e = \frac{\oint dW}{Q_A} = \frac{\oint dQ}{Q_A} = \frac{h_3 - h_b + (h_a - h_1)}{h_3 - h_b} = \frac{h_3 - h_4 - h_2 + h_1}{h_3 - h_4}$$

where $\Delta K \approx 0$ for each process and, for ideal gases, h is a function of temperature only, so that in the *closed cycle*, $h_b = h_4$ and $h_a = h_2$. Substitute $\Delta h = c_p \Delta T$ and, for present purposes, assume c_p to be constant throughout; equation **(a)** then becomes

(b) $$e = 1 - \frac{T_2 - T_1}{T_3 - T_4} = 1 - \frac{T_1}{T_3}\left(\frac{T_2/T_1 - 1}{1 - T_4/T_3}\right) = 1 - \frac{T_1}{T_3}r_p^{(k-1)/k}$$

where we have used $T_2/T_1 = T_3/T_4 = r_p^{(k-1)/k}$ [equation **(c)**, § 15.2]. With a fixed initial temperature T_1, equation **(b)** shows that with a regenerator the thermal efficiency increases as T_3 increases and *decreases as the pressure ratio increases.* Note in contrast that without the regenerator [equation (15-1)], the cycle efficiency *increases* as the pressure ratio increases. *With* regeneration, the cycle 1-*abc*, Fig. 15/2, approaches the Carnot efficiency. Regeneration is impossible with cycle 1-*def*, Fig. 15-2.

The regenerator is relatively large (with cost of a conventional type roughly proportional to size) principally because of the low heat transfer coefficient that gases have across an interface. They must also be large enough for the gases not to have excessive pressure drops during passage.

15.9   EFFECTIVENESS OF REGENERATOR

The *effectiveness of the regenerator* is defined as

$$\text{(15-6)} \qquad \varepsilon_r = \frac{\text{actual amount of heat transferred}}{\text{amount that could be transferred reversibly}}$$

For a particular mean temperature difference between the gases in the regenerator, the effectiveness increases with size (area of heat transfer surface) for a given type of heat exchanger—but compact types of exchangers are being continually improved (Fig. 15/14). For certain comparative data, the size needed for $\varepsilon_r = 90\%$ is over 6 times that needed for $\varepsilon_r = 50\%$. Considering the actual states 2′ and 4′, Fig. 15/9, a reversible transfer of heat (§ 8.11) would result in the gas being heated from 2′ to d, and the exhaust gases being cooled from 4′ to c. Actually, however, the gas is heated only to some state d' and the products are cooled to some state c'. Thus the effectiveness of the generator in terms of the states shown in Fig. 15/9 is

$$\text{(a)} \qquad \varepsilon_r = \frac{h_{d'} - h_{2'}}{(1 + r_{f/a})(h_{4'} - h_c)} \qquad [t_c = t_{2'}]$$

which applies with or without pressure drop in the heat exchanger (regenerator) if d' is the actual state of the departing air (because $Q = \Delta H$ for steady flow with $\Delta K = 0$). From a practical viewpoint $r_{f/a} = 0$ in open cycles. While $h_{d'} - h_{2'} = (h_{4'} - h_{e'})(1 + r_{f/a})$, it is generally true that the specific heats of these processes are different.

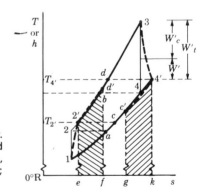

Fig. 15/9   *Imperfect Regeneration with Fluid Friction in Machines.* Area e-2′-d′f = area k-4-c′g; $\Delta p = 0$ in regenerator and pipelines. If the inevitable pressure drops are involved, imagine the lines 2-3 and 4′-1 dotted and not isobaric; then the areas marked do not represent heat.

At a particular ceiling temperature, the *efficiency* curve plotted against pressure ratio for a Brayton cycle with regenerator and with fluid friction rises to a peak at some pressure ratio and then decreases, as it does for the simple cycle, Fig. 15/4; but it peaks at a much *lower* pressure ratio. For the same efficiencies, same T_3 but no pressure drops, comparable pressure ratios for maximum efficiency are 11.5 for simple cycle and 6 for a regenerative cycle with $\varepsilon_r = 60\%$. Comparative specific fuel consumptions for a 300-hp (relatively small) gas turbine are: without regeneration, 0.7 lb/bhp-hr ($e_b \approx 19.7\%$); with regeneration, 0.46 lb/bhp-hr ($e_b \approx 30\%$).[15.24]

OTHER VARIATIONS OF THE BRAYTON CYCLE   15.10

Intercooling in the compression process, described in § 14.12, is used to save work. This feature, together with regeneration, is pictured in Fig. 15/10. The various energy quantities can be written by inspection of the individual systems in Fig. 15/10 or from a Ts diagram, Fig. 15/11. In terms of stagnation or total enthalpies, we have

(a)
$$|W'_c| = h_{a'} - h_1 + h_{2'} - h_b = \frac{h_a - h_1}{\eta_{c1}} + \frac{h_2 - h_b}{\eta_{c2}}$$

where η_{c1}, η_{c2} are the compressor efficiencies of the LP and HP stages, respectively.

(b)
$$W'_t = (1 + r_{f/a})(h_3 - h_{4'}) = (1 + r_{f/a})(h_3 - h_4)(\eta_t)$$

for unit mass entering the engine—$r_{f/a} = 0$ *for a closed cycle.*

(c)
$$|Q'_R| = h_{a'} - h_b + (1 + r_{f/a})(h_{c'} - h_1)$$

(d)
$$Q'_A = (1 + r_{f/a})(h_3 - h_{d'})$$

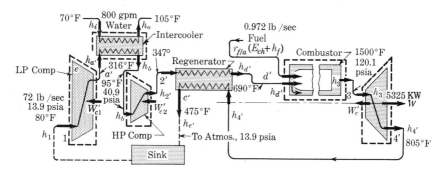

Intercooling and Regeneration. The properties shown are rounded off from test data. See Fig. 15/11. **Fig. 15/10**

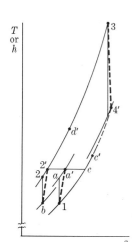

Compression Intercooling and Regeneration. The cycle 1-a'b-2'-d'-3-4'-c'-1 shows the Ts representation of the cycle in Fig. 15/10. Interstage cooling may result in a material increase in the thermal efficiency.[15.7] Actually, t_b is rarely equal to t_1; see Fig. 15/10. The regenerator effectiveness for this cycle is   **Fig. 15/11**

$$\varepsilon_r = \frac{h_{d'} - h_{2'}}{(h_{4'} - h_c)(1 + r_{f/a})}.$$

Often, $r_{f/a}$ is taken to be zero. The effectiveness of the regenerator in Fig. 15/10 could be increased by making it larger; the engineering answer is a compromise of size and effectiveness against cost.

Another variation that has advantages, especially a better fuel economy at less than full load, is the use of two turbines instead of one, in which case one turbine is used to drive the compressor and the other produces the net work. In between these two turbines, we may or may not arrange to burn more fuel by installing another combustor, as in Figs. 15/12 and 15/13, for **reheating**. In Fig. 15/13, the reheat is 4'-5; the expansion in the load turbine is 5-6'. The work of the high-pressure (HP) turbine is numerically equal to the work of compression plus mechanical friction E_f, or

(e) $W'_{HP} = W'_c = (1 + r_{f/a})(h_3 - h_{4'}) = h_{2'} - h_1 + E_f$

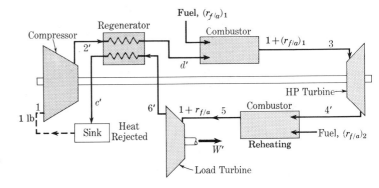

Fig. 15/12   *Diagrammatic Arrangement of Compound Cycle with Regeneration and Reheat.*

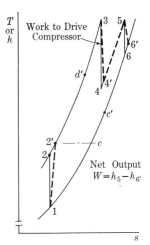

Fig. 15/13   *Gas Turbine Cycle with Regeneration and Reheat.* Two turbines are also often used without reheat; see Fig. 15/14.

The Brayton cycle, with numerous variations, is the ideal prototype for a wide variety of applications: for industrial and marine use; as a power source for space projects, with the basic source of energy being nuclear or solar; there is a design that puts part of the output from a compressor through a combustor chamber and a turbine and part through an after-cooler and a turbine with the attainment of refrigeration at cryogenic temperatures. See a particular application in Fig. 15/14.

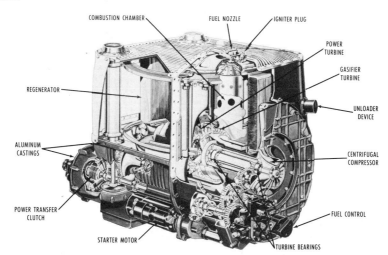

COMBUSTION CHAMBER FUEL NOZZLE IGNITER PLUG

POWER TURBINE

GASIFIER TURBINE

REGENERATOR

UNLOADER DEVICE

ALUMINUM CASTINGS

CENTRIFUGAL COMPRESSOR

POWER TRANSFER CLUTCH

FUEL CONTROL

STARTER MOTOR

TURBINE BEARINGS

Regenerative Gas Turbine for Vehicle Drives. Rated power at sea level and 80°F, 280 hp at 33,860 rpm of power turbine; centrifugal compressor r_p = 3.8; t_3 (Fig. 15/11) = 1725°F; exhaust temperature is approximately 525°F. Notice the two turbine wheels. The turbine that drives the compressor is called the *gasifier turbine*. The regenerator is a rotating (Ljungstrom—Fig. f/3) type, said to have 90% effectiveness. The particular unit incorporates an interconnection between power and gasifier turbines that operates when the consequences are advantageous. (*Courtesy General Motors Research Laboratories, Warren, Mich.*) **Fig. 15/14**

JET PROPULSION 15.11

The ideal cycle for a turbojet engine is the Brayton cycle, Fig. 15/15, but, since this engine does not deliver shaft work, the details of the actual events are different from those of the gas turbine. Let the engine be moving through the atmosphere. First, there is the ram effect $i0$ that occurs in a ***diffuser*** section (§ 15.15), Fig. 15/17. At 0, compressor work raises the state to b, combustion brings it to c ("heat added" in the equivalent air standard), isentropic expansion cd in the turbine that drives the compressor; then the gases enter the nozzle and accelerate to a high speed and kinetic energy at e; and then to the surroundings. For an adiabatic nozzle,

(a)
$$h_d - h_e = K_e - K_d \qquad\qquad [Q = 0]$$

While the nozzle stream moves with a velocity v, macroscopically, the molecules in it have their usual random motions such that, for an ideal gas, their mean-square speed is proportional to the absolute temperature of the gas (§ 1.17)—thermometer moving with the gas. The actual compression and expansion are irreversible; to $0'$ in the diffuser; $0'b'$ in the compressor; cd' in the turbine; $d'e'$ in the nozzle. See the caption to Fig. 15/15 for certain helpful relations.

We shall first make a thermodynamic analysis, followed by the conventional analysis from Newton's second law. In the thermodynamic analysis, one needs certain principles of kinematics, which we shall use in the simplest form for rectilinear motion with all velocity vectors collinear and with the gas streams moving across the boundaries perpendicular to the plane of the openings. Also, there are no energy quantities for body forces (which is to say, $\Delta P = 0$ for horizontal motion),

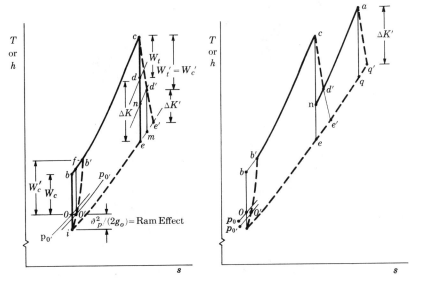

Fig. 15/15   *Turbojet Engine Processes on Ts Plane.* Process $i0$ (or $i0'$ actual), virtually to stagnation temperature (ϑ_p is the air speed), occurs in the diffuser section, Fig. 15/17; the compressor brings the ideal state to b (b' actual). Observe that to achieve pressure $p_b = p^b$, from $0'$, the pressure ratio r_p of the actual compressor must be greater than that of the isentropic compressor. In (a), the compressor efficiency $\eta_c = \Delta h_{0'd}/\Delta h_{0'b}$: turbine $\eta_t = \Delta h_{cd'}/\Delta h_{cn}$; nozzle $\eta_n = \Delta h_{d'e'}/\Delta_{d'm}$. The location of d' satisfies the relation $W'_t = W'_c$ or $(1 + r_{f/a})(h_c - h_{d'}) = h_{b'} - h_{0'}$. If pressure drops are counted, r_p for ce is less than r_p for ib ($p_c < p_b$ and $p_e > p_i$). An after-burner, additional combustion after expansion in the turbine, as in (b), is for the purpose of increasing the thrust force; processes $d'aq'$.

and the states where enthalpies are used are locally equilibrium states. If a jet engine drives a plane at an air speed of v_p, this speed is the velocity v_i of the air relative to axes attached to the engine's frame; $v_i = v_p$. The relative effect is the same as if the engine were on a stationary test block with air blown into it with a velocity v_i (one-dimensional flow). Since velocity is always relative, so is kinetic energy; the *specific* kinetic energy of the entering stream relative to the engine is $K_i = mv_i^2/2 = v_i^2/(2\mathbf{k})$. Similarly, the specific kinetic energy of the exit stream relative to the plane is $K_e = v_e^2/(2\mathbf{k})$, where v_e is the relative exit velocity of the gases. Let the basis of the equations be 1 lb of air entering the engine per unit time; $r_{f/a}$ lb f/lb a. Then the energy diagram that an observer on the engine "sees" is Fig. 15/16(a). He does not see any work done (no shafts leaving the engine), only the entering and exit gases flowing and the chemical process of combustion, which represents a decrease of stored energy of $r_{f/a}q_l$ for each pound of air. On the other hand, the energy

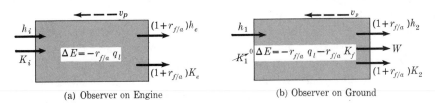

(a) Observer on Engine   (b) Observer on Ground

Fig. 15/16   *Energy Diagrams for Turbojet Engine*

diagram that an observer on the ground sees, Fig. 15/16(b), does have a work term, the work done to overcome the resistance to motion (drag). We remember that the gases are in steady flow and the engine is in static equilibrium (constant velocity). Also, the ground observer sees absolute velocities (meaning relative to the ground) v_1 ($= 0$ in still air) and v_2 and the corresponding specific kinetic energies, and he sees a decrease of stored energy, not only the chemical energy $r_{f/a}q_l$ passing to the surroundings, but also the loss of stored kinetic energy corresponding to the mass of fuel that departs; to wit, $r_{f/a}K_f = r_{f/a}v_p^2/(2\mathbf{k})$, since the fuel is moving with the speed of the engine v_p before it leaves. The energy balances for Fig. 15/16 are:

(b)
$$h_i - (1 + r_{f/a})h_e = (1 + r_{f/a})K_e - K_i - r_{f/a}q_l \qquad [\text{FIG. } 15/16(a)]$$

(c)
$$h_1 - (1 + r_{f/a})h_2 = (1 + r_{f/a})K_2 + W - r_{f/a}q_l - r_{f/a}K_f \qquad [\text{FIG. } 15/16(b)]$$

where $q_l = -h_{rp}$, the sensible enthalpies are for the same datum as q_l, and all kinetic energies K are specific values. Since enthalpy $h = h(T)$ for ideal gases and is a thermodynamic property, $h_1 = h_i$ and $h_2 = h_e$; therefore, the right-hand sides of **(b)** and **(c)** are equal. Setting them so and solving for W gives

(d)
$$W = (1 + r_{f/a})K_e - K_i - (1 + r_{f/a})K_2 + r_{f/a}K_f$$

where the units are consistent (which means that the force unit is involved in the unit of the work W). From the relative velocity principle, $v_A = v_B + v_{A/B}$, where v_A is the absolute velocity of particle A, v_B is the absolute velocity of B, $v_{A/B}$ is the velocity of A relative to B. Then, for the exit jet, $v_2 = v_p + v_e$, applicable, for one-dimensional flow, to any chosen particles of the jet and engine frame. For the rightward direction as positive in Fig. 15/16 and for collinear vectors, vector v_p is negative; we therefore write $v_2 = v_e - |v_p|$. Square this equation, use the result for

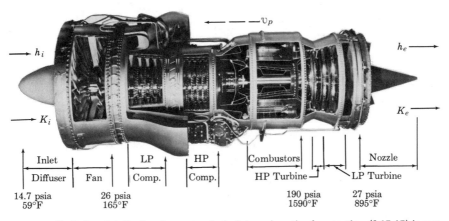

Turbofan Jet Engine. In a pure turbojet engine, the fan section (§ 15.15) is not used. The data shown are characteristic of takeoff conditions; thrust is 18,000 lb for a total air flow of 460 lb/sec, with 265 lb/sec through fan, 195 lb/sec primary (LP and HP compressors). The kinetic energies shown entering and departing are relative to the engine. Nearly all of the outer cover has been removed. (*Courtesy Pratt & Whitney Aircraft, East Hartford, Conn.*)   **Fig. 15/17**

v_2^2 in $K_2 = v_2^2/(2\mathbf{k})$ in equation **(d)**, collect terms, and so on and find

(15-7A) $$W = \frac{v_e v_p}{\mathbf{k}}\left(1 + r_{f/a} - \frac{v_p}{v_e}\right)$$

where, if mass and force are measured in the same units, **k** converts the unit mass to consistent units; v_p is the speed of the air relative to the engine. If the flow is 1 lb/sec of air, the units of W are ft-lb/lb-sec for length in feet and force in pounds; multiply the right-hand side of equation (15-7A) by $\dot{m}_a$ lb/sec of air to get the work W ft-lb/sec. See the Turbofan engine in Fig. 15/17.

15.12   WORK FROM THE IMPULSE-MOMENTUM PRINCIPLE

The mechanics analysis reveals something of the phenomenon that results in a driving force and is the usual approach to this problem. Let Fig. 15/18 represent a free body of the *gases* in the control volume (engine). For a pressure p_1 at the entrance opening of area A_1, there is a force $p_1 A_1$; similarly at the exhaust boundary, there is a force $p_2 A_2$. As the fluid passes through the engine, it imposes on all wetted surfaces a normal force, $dN = p\,dA_w$ on a differential surface area dA_w, and a frictional force dF, owing to relative motion and the viscosity of the fluid. If these forces dN and dF are summed in the x direction, the resultant, represented by the vector F_w in Fig. 15/18, is called the **wall force**, the net force of the wall on the fluid in the control volume.

Since we now have all forces acting on the fluid stream (or symbols therefor), we may apply the principle, proved in mechanics texts, that *the resultant force on the fluid in a control volume is equal to the time rate of momentum leaving minus the rate of momentum entering plus the rate of change of stored momentum in the control volume*. In our simplified analysis, with steady-state operation, the stored momentum remains constant, the force and momentum vectors are collinear, and the properties across the entrance and exit boundaries are uniform; hence we may write the algebraic sum

(a) $$\Sigma F = \frac{\dot{m}_a}{\mathbf{k}}(1 + r_{f/a})v_e - \frac{\dot{m}_a}{\mathbf{k}}v_i = \dot{\mathcal{M}}_{\text{out}} - \dot{\mathcal{M}}_{\text{in}}$$

where ΣF is the resultant in the x direction, Fig. 15/18, and which is the magnitude of the steady force needed to accelerate the stream from relative velocity v_i to relative velocity v_e. Making the sum ΣF, we have

(b) $$F = F_w + p_1 A_1 - p_2 A_2 = \frac{\dot{m}_a}{\mathbf{k}}[(1 + r_{f/a})v_e - v_i]$$

Fig. 15/18   *Forces on Fluid Stream.* The wall force of fluid on wall is equal and opposite to the vector F_w shown; dF is an element of fluid friction on all internal surfaces; body force (gravity) is inapplicable for level flight; drag force and ambient pressure forces are external; the area openings A_1 and A_2 are normal to the direction of flow.

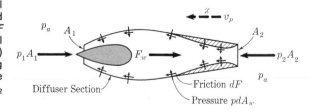

By Newton's third law, the force F_w of the internal surfaces on the fluid is equal and opposite to the force of the fluid on the internal surfaces; and it is by far the major driving force on the engine.

In defining the thrust available to propel the plane, engine manufacturers consider the *engine* (not the fluid) as the *free body* and correct for the pressure of the surroundings p_a. Since p_a acts in all directions, its effect on the engine cancels except at the entrance and exit areas. Thus $F_w - p_a A_1 + p_a A_2 = F_t$, the effective thrust of the engine; using the value of F_w from this expression in equation **(b)**, we have

$$(15\text{-}8) \qquad F_t = \frac{\dot{m}_a}{k}[(1 + r_{f/a})v_e - v_i] + A_2(p_2 - p_a) - A_1(p_1 - p_a)$$

Since the effect of the last two terms of this equation is frequently small, especially in subsonic motion, we shall omit them in the following. In usual units, $\dot{m} = \rho A v$ lb/sec; for $\dot{m}_a/k$ slugs/sec and v fps, the force F is in pounds. Thus for v_p air speed,

$$(15\text{-}9) \qquad F_t = \dot{m}_a\left[\frac{(1 + r_{f/a})v_e}{k} - \frac{v_p}{k}\right] = \frac{\dot{m}_a v_e}{k}\left(1 + r_{f/a} - \frac{v_p}{v_e}\right) \text{ lb}$$

When $\dot{m}_a = 1$ lb/sec, the corresponding F_t is called the **specific thrust** F_{sp}. The distance "moved" by force F in 1 sec is v_p, so that the corresponding work, sometimes called the **thrust power**, is

$$(15\text{-}7\text{B}) \qquad \dot{W} = F_t v_p = \frac{\dot{m}_a v_e v_p}{k}\left(1 + r_{f/a} - \frac{v_p}{v_e}\right) \text{ ft-lb/sec}$$

which is equation (15-7A) above. If the mass of the fuel is negligible ($r_{f/a} \approx 0$), the fluid force on the engine and the work are approximately

$$(c) \qquad F_t = \frac{\dot{m}_a}{k}(v_e - v_p) = \frac{\dot{m}_a v_e}{k}\left(1 - \frac{v_p}{v_e}\right) \text{ lb} \qquad [\text{STILL AIR}]$$

$$(15\text{-}10) \qquad \dot{W} = \frac{\dot{m}_a v_e v_p}{k}\left(1 - \frac{v_p}{v_e}\right) = \dot{m}_a v_e v_p\left(1 - \frac{v_p}{v_e}\right) \text{ ft-lb/sec}$$

where F_t and $\dot{m}_a$ are both in pounds (or, in general, in the same unit) and k is the matching Newton's constant to make the units consistent (§ 1.6). Obtaining force and work from a change of momentum is the principle involved in a propeller drive and in the blades of a turbine. For all situations of steady, one-dimensional flow, the continuity equation (1-17) applies; $\dot{m} = \rho A v$ across any section of area A at a velocity v when the fluid has a density ρ. Even though the jet engine does no shaft work, we observe that the power being developed at a particular speed by an engine with a constant net thrust is directly proportional to speed. For $F_t = 10,000$ lb, the power at 550 fps (375 mph) is 10,000 hp; at 1100 fps (750 mph), $\dot{W} = 20,000$ hp.

If a force greater or less than that required to maintain a constant speed is obtained, the body will have acceleration, will no longer be in equilibrium, and transient conditions will exist. If flight is other than horizontal, gravitational forces will have an effect, and also the force vectors of the engine may not be collinear with the motion.

15.13 MAXIMUM POWER

The net thrust is not constant and equation (15-10) shows that the power developed in overcoming drag at a particular speed is proportional to the difference of inlet and exit gas speeds. To get the speed v_p corresponding to maximum power for a particular nozzle expansion (v_e), find dW/dv_p, say, from equation (15-7), and equate to zero; thus,

(a)
$$\frac{d\dot{W}}{dv_p} = \frac{\dot{m}_a v_e}{k}\left(1 + r_{f/a} - \frac{1}{v_e} \times 2v_p\right) = 0$$

Since the other terms are finite, the parenthetic term is zero, or

(15-11)
$$v_p = \frac{v_e(1 + r_{f/a})}{2}$$

If $r_{f/a} = 0$, we see that $v_p = v_e/2$ for maximum power, not necessarily a point of good thermal efficiency.

15.14 RAM EFFECTS

As the air enters the moving engine, its ideal process in the diffuser is isentropic to the stagnation state. Review § 7.23 on stagnation properties, where the diffuser efficiency is given (more completely in § 18.15). However, the **pressure coefficient** κ_p, also called the *ram recovery ratio*, is more convenient and is repeated for convenience;

(15-12)
$$\kappa_p = \frac{\text{actual pressure rise } \Delta p'}{\text{ideal pressure rise } \Delta p_s} = \frac{p_{0'} - p_i}{p_0 - p_i}$$

from which the actual stagnation pressure $p_{0'}$ may be computed, and where the subscripts are defined in Fig. 15/19. If the kinetic energy at exit from the diffuser is negligible, the actual exit pressure is virtually equal to a stagnation pressure $p_{0'}$. In terms of Mach number **M**, the stagnation temperature T_0 is determined from equation (7-14);

(7-14)
$$\frac{T_0}{T_1} = 1 + \frac{k - 1}{2}\mathbf{M}^2 \qquad \text{[IDEAL GAS, ADIABATIC]}$$

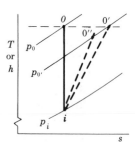

Fig. 15/19   *Ram Processes, Ideal Gas.* For process io', $\Delta s = s_{o'} - s_i = s_{o'} - s_o = R \ln(p_o/p_{o'})$.

for an ideal gas with k constant or for a good average value of k. Repeated for convenience, the sonic speed in a gas is $a = (kkRT)^{1/2}$ where the symbols have the usual units; $M \equiv v/a$. The ideal stagnation pressure p_0, by equation (7-3), is obtained from

(7-3)
$$\frac{T_0}{T_1} = \left(\frac{p_0}{p_1}\right)^{(k-1)/k}$$

There may be enough of a heat loss that the actual temperature $T_{0''}$, Fig. 15/19, is somewhat less than $T_{0'} = T_0$ for adiabatic conditions. This correction is by a *temperature recovery factor* f_t, defined by

(a)
$$f_t = \frac{\text{actual temperature rise}}{\text{adiabatic temperature rise}} = \frac{T_{0''} - T_i}{T_0 - T_i}$$

If the speed of the engine is supersonic, there are problems in the design of the diffuser because of shock waves that occur in the actual diffuser process, a topic left to books specializing in jet engines; but see § 18.15.

Example—Ram Effects   15.15

The jet-propelled La France Concorde SST is flying at 42,000-ft altitude where conditions are 18.8 kPaa, 222 K; its air speed is 2900 km/hr. Air is scooped into the jet engines and brought to a stagnation state within the diffuser that has a pressure coefficient $K_p = 85\%$ and a temperature recovery factor $f_t = 92\%$. Find the (a) stagnation properties p_0, T_0, (b) pressure rise due to ram effects, (c) temperature at final state in the diffuser.

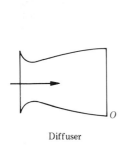 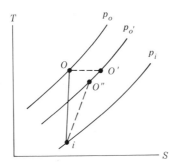

Diffuser   *Diffuser and TS-Diagram.*   **Fig. 15/20**

Solution.   (a) For the diffusing process,

$$h_0 = h_i + K_i$$

$$T_0 = T_i + \frac{v_i^2}{2Jc_p}$$

$$= 222 + \frac{(2643)^2}{(2)(0.24)(1.802)(32.174)(778)} = 222 + 323 = 545 \text{ K}$$

where

$$v_i = \left(\frac{2900}{1.609}\right)\left(\frac{88}{60}\right) = 2643 \text{ fps}$$

$$p_0 = p_i(T_0/T_i)^{k/(k-1)} = 18.8(545/222)^{1.4/(1.4-1)} = 436 \text{ kPaa}$$

(b)
$$K_p = \frac{\Delta p'}{\Delta p}$$

$$\Delta p' = K_p \Delta p = (0.85)(436 - 18.8) = 355 \text{ kPa}$$

(c)
$$f_t = \frac{T_{0''} - T_i}{T_0 - T_i}$$

$$T_{0''} = T_i + f_t(T_0 - T_i)$$

$$= 222 + (0.92)(545 - 222) = 222 + 297 = 519 \text{ K}$$

15.16 PERFORMANCE PARAMETERS FOR JET ENGINES

The **propulsive efficiency** is the work divided by the available energy E_a developed in the engine, which is the work done in flight, say W from equation (15-10), plus the kinetic energy lost in the exhaust jet, $K_2 = v_2^2/(2k)$, where $v_2 = v_e - |v_p|$. The prevalent view ignores the fuel-air ratio $r_{f/a}$. For $\dot{m}_a = 1$ lb/sec,

(a)
$$E_a = \frac{v_p(v_e - v_p)}{k} + \frac{(v_e - v_p)^2}{2k} = \frac{(v_e - v_p)(v_e + v_p)}{2k} \text{ ft-lb/sec-lb}$$

(15-13)
$$\eta_p = \frac{W}{E_a} = \frac{v_p(v_e - v_p)2k}{k(v_e + v_p)(v_e - v_p)} = \frac{2v_p/v_e}{1 + v_p/v_e} = \frac{2}{1 + v_e/v_p}$$

the propulsive efficiency. This equation shows that this efficiency is zero when $v_p = 0$ and is 100% when $v_p = v_e$. Since the condition $v_p = v_e$ is the condition for zero available energy generated [equation **(a)**], there would be no work or thrust at 100% efficiency.

FANJET. An aside may be of interest. The observation from equation (15-13) that high propulsive efficiencies are obtained when the jet velocity v_e is only a little larger than v_p is certainly a factor that led to the development of the fanjet. In the fanjet engine, Fig. 15/17, the fan, which may be forward or aft, pumps a portion of the intake air through a pressure ratio of the order of $r_p = 2$; the remainder of the intake undergoes the usual events of a turbojet. A particular turbine power can pump a larger flow through a small pressure ratio than it can through a large r_p. With a larger flow, the average exit velocity v_e will be less, resulting in an increase of propulsive efficiency. Equation (15-10) shows that the power decreases (as $v_e - v_p$ decreases), but equation (15-11) shows that for a particular air speed v_p, the maximum power is obtained when $v_e \approx 2v_p$. Therefore, the fanjet produces a low-speed takeoff thrust materially larger than a simple jet. Finding the optimum operating parameters is a complex affair. The amount of bypass air (fan air) is some one to two times the primary air which may pass through a pressure ratio of the order of 13. The propulsion from the fan is due to the time rate of change of momentum imparted to the fan air.

For the **engine thermal efficiency**, also called the *overall efficiency* in the industry, the numerator is the work output and the denominator or the energy chargeable is the equivalent of the heat added from an external source, which, in combustion engines, is taken as the heating value of the fuel supplied q_l; or

(15-14)
$$e_{\text{eng}} = \frac{\dot{W}}{\dot{m}_a r_{f/a} q_l} \qquad \text{[DIMENSIONLESS]}$$

where q_l may be taken at the standard state for convenience, and $\dot{W}$ is the power being used when the engine is in motion, say, equation (15-10) in matching units.

For any cycle, or equivalent closed cycle, we have written the thermal efficiency as $e = \oint dQ/Q_A$, applicable to a pure substance, where the numerator was said to be the net shaft work. Since the jet engine does not do shaft work, we broaden the definition of $\oint dQ$ by recognizing that in general it is the available energy generated, in this case the change of kinetic energy between inlet and exit section, $K_e - K_i$, Fig. 15/15(a), *or as given in equation* **(a)** *above*. The heat added in the *equivalent* cycle is the appropriate chemical energy of combustion, $m_a r_{f/a} q_l$ for complete combustion. In terms of properties defined on Fig. 15/15(a), $Q_A = h_c - h_b$ or $Q'_A = h_c - h_{b'}$, when kinetic energy changes are negligible or stagnation enthalpies are used. Also, $\oint dQ = h_c - h_b - (h_e - h_i)$ for unit mass of a pure substance in the ideal cycle. The cycle thermal efficiency (for the equivalent cycle) may then be written

$$(15-15) \qquad e_{cyc} = \frac{K_e - K_i \text{ for } m_a}{m_a r_{f/a} q_l} \quad \text{or} \quad e_{cyc} = \frac{\oint dQ}{Q_A} = 1 - \frac{h_e - h_i}{h_c - h_b}$$

[PURE SUBSTANCE]

which, of course, is dimensionless. These equations are readily adapted to the actual cycle *ib'ce'*, Fig. 15/15(a). Observe that in addition to the heat rejected $h_e - h_i$, Fig. 15/15(a), from the ideal cycle, for example, the jet engine discharges kinetic energy (all available) to the surroundings—all of which is dissipated into the sink.

The *thrust specific fuel consumption* (*TSFC*) is the rate of consumption of fuel for unit thrust force; in English units $\dot{m}_f$ lb/hr and F lb; F is given by, say, equation (15-9); $TSFC = \dot{m}_f/F$ $\text{lb}_m/\text{lb}_f\text{-hr}$.

The *specific impulse* of air-breathing jet engines may be expressed on either of two bases: (1) the impulse ($\int F\, d\tau$) per unit mass of air, I_a; or (2) the impulse per unit mass of fuel, I_f. If the force is constant, $I = F\,\Delta\tau$; if, during the time interval $\Delta\tau$, m_a lb of air pass through, then $I_a = F\,\Delta\tau/m_a = F/\dot{m}_a$. Similarly, $I_f = F\,\Delta\tau/m_f = F/\dot{m}_f$, which is seen to be the reciprocal of the *TSFC*. See I_s for rockets, § 15.18.

The *nozzle efficiency*, by equation **(b)**, § 5.11, is

$$(15-16) \qquad \eta_n = \frac{\text{actual } \Delta K'}{\text{ideal } \Delta K_s} = \frac{h_1 - h_{2'}}{(h_1 - h_2)_s}$$

where states 1 and 2, 2' are at entrance and exit boundaries of the nozzle, respectively. Equation (15-16) with an energy balance for the actual nozzle [see equation **(a)**, § 15.11] may be used to estimate the actual exit velocity $v_{e'}$; state e' is defined in Fig. 15/15.

Solving the Problem. In cycle analysis, Fig. 15/15(a), first find p_0 (or $p_{0'}$) with known air speed and atmospheric pressure; then p_b (or $p_{b'}$) with known r_p in the compressor. With pressure ratios and T_i known, the corresponding temperatures are determined. The ceiling temperature T_c is known. Point d (or d') is located to make $W_t = W_c$ (or $W'_t = W'_c$). With r_p for ce (or ce') known, find T_e (or $T_{e'}$). Other properties may be obtained from gas tables or computed for constant specific heats.

15.17   RAMJETS

Different types of engines have particular advantages under certain operating conditions. For example, the reciprocating-engine-propeller drive is favored for low-speed operation of airplanes. Above about 400 mph, the efficiency of a propeller drive drops rapidly; the fanjet is best for speeds of about Mach 0.7 to 0.9. Next would be the straight turbojet engine.

As the translational speed of the airplane increases, the stagnation pressure at the entrance to the compressor increases; if the speed is high enough, this pressure can be equal to p_b, Fig. 15/15, at which point no compressor is needed (and consequently no turbine) for the ceiling temperature for the turbine blades and the same "heat added". This speed in combination with a compressor may result in the exit temperature from the compressor being at or close to the permissible ceiling temperature. The closer it is, the less fuel can be burned, and with no "heat added," no net work can be obtained. It is evident then that with the air-breathing engine moving fast enough, fuel may be injected, the hot gases expanded, and thrust obtained without the complications and limitations of a compressor and turbine. The result is a ramjet engine, Fig. 15/21. Where the ramjet engine is used, some means must be employed to bring the engine to an operating speed, say a rocket motor for missiles (§ 15.18). Operation of air-breathing engines is limited to an altitude of some 150,000 ft.

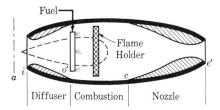

Fig. 15/21   *Representation of a Ramjet Engine.* For more discussion of diffusers, see § 18-15.

For steady-flow operation, the Brayton cycle is the ideal prototype; the *Ts* diagram is shown in Fig. 15/22. Elimination of the turbine permits much higher ceiling temperatures, say to 3500°F, because the walls of the engine can be cooled to

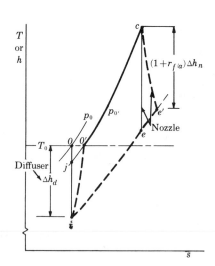

Fig. 15/22   *Ramjet Engine Processes on Ts Plane.* For an ideal gas without Δp between o' and c; p_o is the stagnation pressure that would be attained with an isentropic process in the diffuser for a particular ϑ_i. The entropy increase of the actual process is $s_{o'} - s_o$, where $s_o = s_i$.

some extent. As a result, more fuel is burned and greater thrust obtained. For a stagnation temperature of 2000°R at the exit of the diffuser, the corresponding Mach number, from equation (5-18), is about 3.8. The equations obtained for the turbojet are seen to apply to the ramjet operating in steady state. Evidently, the ideas of turbojet and ramjet may be combined in a single engine.

The next step for an air-breathing engine is the *scramjet*, in which the pressure rise (and therefore the temperature rise) in the diffuser section is small, the air speed remains supersonic throughout the engine, including the combustion process; the fuel is liquid hydrogen doing double duty as a coolant. This engine is visualized as attaining a speed of some 15,000 mph ($\mathbf{M} \approx 20$) at 180,000 ft. With the addition of a rocket motor, it may be boosted into orbit with a relatively large payload.

ROCKETS 15.18

A rocket is a jet propulsion device with no inlet. The propellant may be ions and electrons, hydrogen, hydrogen peroxide, etc., or the products of any one of several reactions. The chemical rocket, the kind that we shall discuss, carries both its fuel and oxidizer, the basic energy phenomenon being the conversion of chemical energy into kinetic energy at the exit section of the nozzle. The amount of the energy of the reaction may be determined from the enthalpies of formation, § 13.29, and we shall presume that the products can be treated as a pure substance. Because of storage capacity, rocket engines cannot operate for very long before the reactants are consumed, but where the atmosphere is quite thin or nonexistent and for many missiles, the rocket is an effective engine.

To determine the work equation for a moving rocket by a thermodynamic analysis, we assume the same conditions as assumed for the jet engine, § 15.12. Consider energy diagrams from the viewpoints of an observer on the engine and of one on the ground (reference axes attached to earth), Fig. 15/23, with the direction of motion collinear with the exit velocity of the gases. In each case the stored energy decreases by the amount of the departing energy, since none enters. For Fig. 15/23(a), we write, for a flow of 1 lb/sec,

(a)
$$-\Delta E = h_e + K_e = h_e + \frac{v_e^2}{2\mathbf{k}} \text{ ft-lb/lb-sec}$$

of gases crossing the exit section. The ground observer sees work W being done against drag, enthalpy h_2 departing, kinetic energy K_2 departing, where v_2 is the velocity relative to the ground. But since departing gases had absolute kinetic energy K_p when they were in the system, he sees a net loss of kinetic energy of $K_2 - K_p$, as

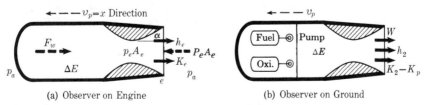

(a) Observer on Engine   (b) Observer on Ground

Energy Diagrams for Rocket. The dotted vectors represent the forces acting on the fluid. The change in stored energy is the enthalpy of the reaction (with sensible enthalpies measured above the same datum); oxi. = oxidizer.   **Fig. 15/23**

shown. From the energy balance for Fig. 15/23(b), for unit mass, we find

(b) $- \Delta E = W + h_2 + K_2 - K_p = W + h_2 + \dfrac{v_2^2 - v_p^2}{2\mathbf{k}}$

where, as before, $v_2 = v_e - |v_p|$. Equating the right-hand sides of **(a)** and **(b)**, noting that the thermodynamic properties h_e and h_2 are identical, and solving for W for $\dot{m} = 1$ lb/sec, we get

$$\dot{W} = \frac{v_e^2}{2\mathbf{k}} - \frac{v_e^2 - 2v_e v_p + v_p^2 - v_p^2}{2\mathbf{k}} = \frac{v_e v_p}{2\mathbf{k}} \text{ ft-lb/lb-sec}$$

(15-17) $\dot{W} = \dfrac{\dot{m}_p v_e v_p}{\mathbf{k}}$ ft-lb/sec

for a flow of $\dot{m}_p$ lb/sec of propellant across the exit section.

Using the principles discussed in § 15.12, we have the net force producing the change of momentum of the fluid in the x direction as

(15-18) $F = \dfrac{\Delta(\dot{m}_p v)}{\mathbf{k}} = \dfrac{\dot{m}_p v_e}{\mathbf{k}}$ or $F = \dfrac{\dot{m}_p(v_2 + |v_p|)}{\mathbf{k}} = \dfrac{\dot{m}_p v_e}{\mathbf{k}}$

[AXES ON ENGINE]   [AXES ON GROUND]

The power of the fluid is then $Fv_p = \dot{m}_p v_e v_p/\mathbf{k}$, as found in equation (15-17).

Since the foregoing analysis assumes that the jet is moving in parallel streams, and since the nozzle is flared with a relatively large angle α, Fig. 15/23(a), it is customary to correct for this effect by multiplying F in (15-18) by a factor. The factor is greater than 0.95 if $\alpha < 30°$, Fig. 15/23; hence, this effect may be small.

The engine manufacturer considers the net thrust of the engine, allowing for the atmospheric pressure surrounding the rocket motor. The forces acting *on the fluid* in the rocket are shown dotted in Fig. 15/23(a); thus the momentum principle applied to the fluid at a constant flow rate gives

$$F_w - p_e A_e = \Delta \dot{m} v = \dot{m} v_e$$

where F_w is the force of the wall on the fluid. Now think of the *engine* (and attached pay-load) *as the free body*. Since the force of the working fluid on the wall (engine) is equal and opposite to F_w on the fluid, we can say that $F_w = \Delta \dot{m} v + p_e A_e$ (acting leftward now) is that force. Since the atmospheric pressure p_a acts on all surfaces of the rocket *except* the exit nozzle area A_e (it thus cancels elsewhere), the resultant force on the *rocket* in horizontal flight, called the **thrust**, is $F_t = F_w - p_a A_e$, or

(15-19) $F_t = \dot{m} v_e + (p_e - p_a)A_e = \dfrac{\dot{m} v_e}{\mathbf{k}} + (p_e - p_a)A_e$

one of the forces acting on the rocket for a mass rate of flow of $\dot{m}/\mathbf{k}$ slugs/sec from the rocket. (If one were studying the *motion* of the rocket, this force, the components of gravitational attraction, and the frictional drag are involved in acceleration. Moreover, the situation is often one of varying atmospheric pressure and gravity—movement through large altitude.) The exit pressure p_e depends on the design of the nozzle (Chapter 18) and the state of the gases entering the nozzle, and it is different from p_a except at particular combinations of altitude and motor operating conditions. With no atmosphere the engine thrust force F_t is greater (for given v_e) than with an atmosphere, whereas the thrust of an air-breathing engine decreases with altitude. Another worthy observation is that air-breathing engines always propel the body at a speed less than the exhaust-gas velocity (horizontal motion), whereas the flight speed of a rocket engine can be greater than the velocity of the exhaust gas. When the resistance to motion is low, as in outer space, flight speed could be immensely greater.

The propulsive efficiency, defined in § 15.16, is

(c) $$\eta_p = \frac{F v_p}{F v_p + \dot{m}_p v_2^2/(2\mathbf{k})} = \frac{\dot{m} v_e v_p}{\dot{m}_p v_e v_p + \dot{m}_p v_2^2/(2\mathbf{k})}.$$

Using (15-18) when $p_e = p_a$ for its simplicity, and letting $v_2 = v_e - |v_p|$, we find

(15-20) $$\eta_p = \frac{2 v_p/v_e}{1 + (v_p/v_e)^2}$$

Equation (15-20) shows that the maximum efficiency occurs for $v_p = v_e$.

An important performance parameter for rockets is the **specific impulse** I_s, which is the impulse per unit mass of propellant departing in the same duration $\Delta\tau$;

(15-21) $$I_s = \frac{F \Delta\tau}{m_p} = \frac{F}{\dot{m}_p} \quad [\text{IN ENGLISH UNITS, } \text{lb}_f\text{-sec/lb}_m]$$

where the units are inconsistent and it is common practice to cancel these dissimilar pounds and call the units of I_s seconds. Comparing equation (15-21) with (15-18), we see that $I_s = v_e/\mathbf{k}$. Some typical values of specific impulse in the units of (15-21) are: liquid H_2 and liquid O_2 (LOX), 340; gasoline and LOX, 242; fluorine and hydrogen, 375; JPL-121, a solid propellant, 182.[15.1] Since the requirement for a high specific impulse is a high exit velocity v_e, a dominant characteristic of rocket propellants is their reaction rate.

For some applications the density of the propellants for a particular thrust is significant, inasmuch as a denser propellant occupies less volume, resulting in smaller drag and other advantages. For this reason, the product of the density and specific impulse ρI_s is a parameter of note. For a typical hydrocarbon with 2.2 lb LOX/lb f, the engine may have $I_s = 264$ and $\rho I_s = 259$; $k_p = 1.24$, $M_p = 22$. For a ratio of 1 lb H_2 with 3.5 lb O_2, the values may be $I_s = 364$, $\rho I_s = 95$, $k_p = 1.22$, and $M_p = 16$. While the greater thrust is obtained from H_2, the necessary volume to contain the propellant for a particular objective is larger.

The exhaust gases are generally handled as ideal gases through the nozzle with suitable average values of k. Let the energy balance for the ideal nozzle, with reference axes attached to the nozzle, be

(d) $$K_e = \frac{v_e^2}{2\mathbf{k}} = -\Delta h = -c_p \Delta T$$

matching units. Substitute $c_p = k_p R_p/(k_p - 1) = k_p \bar{R}/[M_p(k_p - 1)]$ in **(d)** and then substitute the value of v_e from **(d)** into $I_s = v_e/\mathbf{k}$ and get

(e) $$I_s = \frac{v_e}{\mathbf{k}} = \left[\frac{2 k_p \bar{R}(-\Delta T)}{\mathbf{k} M_p(k_p - 1)}\right]^{1/2} = C\left[\frac{\text{energy released}}{\text{molecular weight}}\right]^{1/2}   [s = C]$$

where $-\Delta T$ is positive and $\bar{R} = 1545$ ft-lb/pmole-°R for the units previously indicated. We see that low values of $k_p = c_p/c_v$ and of M_p for the products are desirable, as is a large temperature range (which also means a large pressure range).

The flame temperatures in the combustion zone of a rocket may be upwards of 5500°R to even 8000°R,[15.1] which means that there is significant dissociation (§ 13.35 *et seq.*). The combustion pressures are, say, 300 to 1000 psia (subject to details of design), which means that the nozzle is converging-diverging with the expansion always through the so-called critical pressure at the throat; see § 18.6.

Pressurized monopropellants are used for small thrusts, as nitrogen and hydrogen peroxide for attitude control of space satellites.[15.21] Liquid hydrogen, vaporized by a nuclear reactor, is a prospective propellant for space vehicles. Because of the low molecular mass of 2 for H_2, the specific impulse may be about 800 lb-sec/lb [equation **(e)**]. The ion rocket, with a specific impulse of some 10,000 lb-sec/lb, is envisaged as the propulsion for long space trips; but the actual value of the thrust force for this type of rocket is small; perhaps it will be of the order of 1 lb.[15.22]

15.19   CLOSURE

There are many applications of the ideas of this chapter for which space is not available. One plan uses a free-piston ICE as a gasifier to provide gas to the turbine.[15.20,15.21] Another plan is to burn the fuel in a pressurized combustion space of a steam generator to produce steam for power and then let the partially cooled furnace gases expand through a turbine.

PROBLEMS

SI UNITS

15.1 Sketch the Brayton cycle on the pV and Ts planes; number the corners clockwise and consecutively starting with point 1 at the beginning of the isentropic compression. **(a)** Show that the efficiency of the cycle is given by $e = 1 - 1/r_k^{(k-1)} = 1 - T_1/T_2 = 1 - T_4/T_3$. **(b)** Prove that the optimum intermediate temperature for maximum work is $T_2 = (T_1 T_3)^{1/2}$ where T_1 and T_3 are fixed.

15.2 An ideal gas-turbine unit (Brayton cycle) operating on a closed cycle is to deliver 5968 kW and receive heat from a nuclear reactor. Different gases are to be studied. Let the gas be argon; $p_1 = 482.64$ kPaa, $t_1 = 43.3$°C. The argon enters the turbine at

1093.3°C and the cycle should be such as to deliver maximum work for these constraints. Use Item B 1. **(a)** At what rate should the argon circulate? **(b)** Let $\eta_t = 86\%$, $\eta_c = 84\%$ and compute the argon flow rate for 5968 kW fluid work.

Ans. **(a)** $T_2 = 657.3$ K, 1867 kg/min, **(b)** 3383 kg/min.

15.3 There are required 2238 net kW from a gas turbine unit for the pumping of crude oil from the North Alaskan Slope. Air enters the compressor section at 99.975 kPaa, 278 K; the pressure ratio is $r_p = 10$. The turbine section receives the hot gases at 1111 K. Assume the closed Brayton cycle, and find **(a)** the required air flow and **(b)** the thermal efficiency. **(c)** For maximum work, what should be the temperature of the air leaving the compressor section?

15.4 Neon is used as a working fluid in a nuclear reactor-gas turbine system operating on a closed cycle. At the beginning of compression, state 1, $p_1 = 620.53$ kPaa, $t_1 = 37.8°C$; $r_p = 4.5$. The neon leaves the reactor and enters the turbine portion at $t_3 = 1260°C$; there is no regenerator. These efficiencies apply: turbine $\eta_t = 87\%$, compressor $\eta_c = 84\%$. For a net output of 5968 kW, compute the rate of flow of neon and the mean effective pressure. Use Item B 1.

15.5 It is desired to obtain maximum work in a gas turbine operating on the air-standard Brayton cycle between the temperature limits of 37.8°C and 704.4°C; $p_1 = 103.42$ kPaa. Using air properties found in Item B 1, calculate (a) the temperature at the end of compression, (b) the compression ratio r_k, (c) the pressure ratio r_p, and (d) the thermal efficiency.

Ans. (Item B 1) (a) 278.8°C, (b) 4.18, (c) 7.4, (d) 43.6%.

15.6 The "heat" received by an air standard gas turbine is 25,953 kJ/s for fluid work of 5000 kW; the turbine efficiency is $\eta_t = 84\%$. In the corresponding ideal cycle, $p_1 = 1$ atm, $t_1 = 26.7°C$, $W_c = 201.13$ kJ/kg, $W_t = 414.01$ kJ/kg; the maximum cycle temperature is $t_3 = 732.2°C$. Compute (a) the compressor efficiency, (b) the rate of air flow, (c) the cycle efficiency, (d) the part of Q_A which is unavailable, (e) that part of the heat rejected which is available, (f) the change of availability for each process.

Ans. (a) 80.2%, (b) 51.7 kg/s, (c) 19.3%, (d) 198.87, (e) 146.90, (f) $\Delta \mathscr{A}_{1\text{-}2'} = +222.30$, $\Delta \mathscr{A}_{2'\text{-}3} = +302.63$, $\Delta \mathscr{A}_{3\text{-}4'} = -378.02$, $\Delta \mathscr{A}_{4'\text{-}1} = -146.90$ kJ/kg.

15.7 Air leaves the compressor in a gas turbine and enters the combustor at 482.64 kPaa, 204.4°C, and 45 m/s; gases leave the combustor at 468.85 kPaa, 893.3°C, and 152 m/s. Liquid fuel enters at 15.6°C with a heating value of 42,920 kJ/kg; combustor efficiency $\eta_f = 94\%$. Find the fuel flow per kilogram of entering air. For the products, $M_p = 28.9$, $k = 1.36$.

Ans. 0.0206 kg fuel/kg air.

15.8 A gas-turbine compressor supplies a regenerator with 82,388 kg/hr of air at

405.75 kPaa, 182.2°C, which leaves the regenerator at 396.86 kPaa, 393.3°C. Hot products from the burning of 832.2 kg/hr of fuel enter the regenerator at 104.53 kPaa, 548.9°C. Determine (a) the effectiveness of the regenerator, and (b) the availability of the products from the regenerator with respect to a sink of $p_0 = 101.70$ kPaa and $t_0 = 21.1°C$. Assume the products to have properties similar to those of air. See Item B 1.

Ans. (a) 57%, (b) 103.6 kJ/kg.

15.9 There are absorbed 11,678,850 kJ/hr by 22.68 kg/s of air passing through a gas-turbine regenerator; the fuel-air ratio is $r_{f/a} = 0.017$; $M_p = 28.8$ kg/mol, $k_p = 1.37$. If the air and products enter the regenerator at 182.2°C and 426.7°C, respectively, compute (a) the regenerator effectiveness and (b) the exit air temperature and the exit products temperature.

15.10 The study of nuclear reactor–gas turbine combination in a closed cycle continues (**15.2**). Helium leaves the compressor ($\eta_c = 100\%$) at 4137 kPaa, 416.7 K; it leaves the reactor at 1088.9 K (ignore pressure drops) entering the turbine; it leaves the turbine ($\eta_t = 100\%$) at 2241 kPaa and passes via a regenerator, thence through another heat exchanger (sink) to the compressor which compresses it to 4137 kPaa. From the compressor, it passes through the regenerator (100% effectiveness), whence it enters the reactor and the cycle repeats. For a net work of 16,412 kW, at what rate should the helium flow? Also compute the thermal efficiency.

Ans. 1297 kg/min, 61.8%.

15.11 A plane is powered by a jet propulsive system. The fluid jet leaves the nozzle (tail) end of the system with an absolute speed (with respect to the ground) of 853.7 m/s; the plane moves at 1569 km/hr. If the system is developing 7460 kW, find (a) the thrust, (b) the mass flow (neglecting $r_{f/a}$), (c) the propulsive efficiency.

Ans. (a) 17,117 N, (b) 20.05 kg/s, (c) 50.5%.

15.12 A jet engine driven plane is expected to travel in dynamic equilibrium at 965.4 km/hr when the net thrust force is 17,793 N and the ratio air to fuel is $r_{a/f} = 50$. The exhaust gases depart the nozzle at 610 m/s (relative to engine). Compute (a) the

flow rate of air $\dot{m}_a$ and of fuel $\dot{m}_f$, **(b)** the power being developed. **(c)** If the exhaust state temperature is 944.4 K, the sink 277.8 K, determine the availability and the entropy production for the universe and the irreversibility from the exit section of the nozzle to the dead state, per unit mass of air. (Use specific heat of exhaust as $c_p = 1.089$ kJ/kg-K.) **(d)** Calculate the propulsive efficiency, the *TSFC*, and the air specific impulse.
   Ans. **(a)** 50.3, 1.006 kg/s, **(b)** 4772 kW, **(c)** 515 kJ/kg air, $\Delta S_p = 1.181$ kJ/K-kg air, **(d)** 61.1%, 0.204 kg$_{fuel}$/N-hr, 353 N-s/kg.
   15.13 A plane has an air speed of 274.4 m/s where the atmospheric air is at 27.58 kPaa, 222 K. **(a)** The air is brought virtually to rest relative to the plane in a diffuser. Compute the corresponding stagnation enthalpy, temperature, and pressure. **(b)** Let the pressure coefficient for the diffuser be $K_p = 96\%$. Specify the actual state at the diffuser exit when $K_p \approx 0$. **(c)** The air from the diffuser is compressed adiabatically through a pressure ratio of 4.5 with $\Delta K = 0$. If the compressor efficiency is 80%, determine the work for unit mass. What power is required for an air flow of 81.63 kg/s? **(d)** For $T_0 = 222$ K, compute the overall irreversibility for the event (diffuser and compressor).
   15.14 Show, for a rocket nozzle (or any nozzle), for an ideal gas with properties k, M, that the specific impulse for isentropic flow is given by

$$I = \left\{ \frac{2k\bar{R}T_1}{k(k-1)M} \left[1 - \left(\frac{p_2}{p_1} \right)^{(k-1)/k} \right] \right\}^{1/2}$$

where T_1, p_1 give the stagnation initial state, and p_2 is the static pressure at the exit section.
   15.15 A research rocket delivers a thrust of 17,793 N for a period of 30 sec when steadily and uniformly consuming its 425.9 kg of propellants (red fuming nitric acid and aniline furfuryl). **(a)** With what constant speed do the products leave the nozzle? **(b)** Estimate the temperature in the combustion chamber if the pressure there is 2206.3 kPaa nd the expansion (ideal gas, $k = 1.2$) through the nozzle is to 103.4 kPaa, 1560°C.   *Ans.* **(a)** 1253 m/s, **(b)** 3056 K.
   15.16 A gas turbine operating on the air

standard Brayton cycle is being analyzed. The effect of the intermediate temperature T_2 on the maximum work is being studied. The cycle operates between the minimum temperature $T_1 = 300$ K and the maximum temperature $T_3 = 1400$ K. Write a computer program for this study.

ENGLISH UNITS

   15.17 A blast furnace needs a supply of hot gas pressurized to 30 psia for its operation. The gas is furnished by the exhaust from a gas turbine which delivers no power except that required to furnish the gas. For the turbine: at state 1, $p_1 = 15$ psia, $t_1 = 60°F$; the turbine inlet temperature is 1500°F; both efficiencies, turbine and compressor, are 100%. Find the pressure ratio r_p for the compressor. Use Item B 2.
   15.18 A gas-turbine unit consists of two turbines A and B (see figure), two combustors, and a single compressor C; turbine A drives compressor C to furnish all air required while turbine B delivers 2000 hp, the net output of the unit. Turbine and compressor efficiencies are each 83%. At state 1, $p_1 = 15$ psia, $t_1 = 100°F$; $r_p = 5$. The inlet condition for each turbine is 75 psia, 1300°F; both turbines exhaust at 15 psia. Determine **(a)** the heats Q_1 and Q_2, **(b)** the thermal efficiency for the unit, and **(c)** p_m.
   Ans. **(a)** 341,700, 135,600 Btu/min, **(b)** 17.75%.

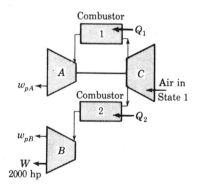

Problem 15.18

   15.19 Hydrogen is being studied as a fluid for the closed Brayton cycle of a nuclear reactor–gas turbine combination. See **15.2**. Hydrogen enters the compressor section at 20 psia, 540°R and leaves at 150 psia,

1040°R to enter the reactor heat source. At the turbine section, the hydrogen is at 2000°R; after expansion it leaves at 1280°R. For a net work output of 10,000 hp and using Item B 5, find (a) compressor efficiency η_c, (b) turbine efficiency η_t, (c) hydrogen flow, lb/sec.

Ans. (a) 84%, (b) 84.1%, (c) 8.54 lb/sec.

15.20 In an air-standard gas-turbine cycle at state 1, $p_1 = 14.5$ psia, $t_1 = 60°F$; $r_p = 6$; The maximum cycle temperature is $t_3 = 1440°F$. Efficiencies are: turbine $\eta_t = 82\%$, compressor $\eta_c = 82\%$. What decrease in turbine efficiency will have the same effect on the cycle thermal efficiency as does a decrease in compressor efficiency to $\eta_c = 75\%$, other values remaining unchanged?

15.21 The pumping of natural gas across the country is accomplished by an air-standard Brayton cycle turbine producing 5000 hp. Air at 14.7 psia, 100°F (point 1) is drawn into the compressor section; the pressure ratio is $r_p = 10$. At the turbine section inlet (point 3), the temperature is 1540°F; $\eta_c = 85\%$, $\eta_t = 85\%$. Use Item B 2 and compute (a) the air flow required, (b) the compressor power, hp, (c) the thermal efficiency.

Ans. (a) 62.33 lb/sec, (b) 12,920 hp, (e) 25.30%.

15.22 A gas turbine unit receives 70 lb/sec of air at 80°F (also sink temperature t_0) and 1 atm. The pressure ratio in the compressor is 6; lower heating value of the fuel is 18,500 Btu/lb. There are two turbines; one (HP) just large enough to drive the compressor, the other (LP) delivers the net power. Between the turbines additional fuel is burned. The inlet temperature for each turbine is 1440°F. The LP turbine exhausts through a regenerator. Ignore pressure drops in the computations, neglect the mass of fuel. For the equivalent air-standard cycle and all ideal processes, compute (a) the thermal efficiency, (b) the net power, (c) specific fuel consumption, lb/hp-hr, (d) heat rate, (e) and the irreversibility of the heat rejection process. If there were no irreversibility *here*, what would be the percentage increase in the work done?

Ans. [Item B 2] (a) 58.4%, (b) 12,050 Btu/sec, (c) 0.236 lb/hp-hr, (d) 4360, (e) 1410 Btu/sec, 16.5%.

15.23 Answer the questions for **15.22** if the departures from the ideal are defined by the following parameters: compressor $\eta_c = 83\%$, each turbine $\eta_t = 85\%$, combustion $\eta_b = 96\%$, regeneration $\varepsilon_r = 50\%$.

Ans. (a) 26.7%, (b) 7240, (c) 0.515, (d) 9520, (e) 99%.

15.24 A gas turbine receives 40,000 cfm of air at 14 psia, 60°F and compresses it to 98 psia. Liquid octane fuel at 70°F is injected into the combustors at a rate of $r_{f/a} = 0.01635$ lb fuel/lb air; see equation (13.7, *Text*). Combustion is complete with a release of $E_c = 19,300$ Btu/lb of fuel (corrected to 0°R) and the resulting products are similar to "400% air" (Item B 8). For ideal processes, determine (a) the air temperature at compressor exit, (b) the gas temperature at turbine discharge, (c) the net horsepower delivered, (d) the cycle efficiency, (e) p_m, (f) the unavailable part of Q_A (1 lb of air and $t_0 = 60°F$) and the available part of Q_R.

Ans. (a) 903°R, (b) 1247°R, (c) 8630 hp, (d) 39.8%.

15.25 Compressed air is to be delivered by a simple gas turbine with an over-size compressor that supplies both the compressed air and the air required for the turbine; all turbine work is used to drive the compressor. At state 1, $p_1 = 14.7$ psia, $t_1 = 60°F$; the pressure ratio is $r_p = 6$; the maximum cycle temperature is 1500°F; the fuel energy release is 18,500 Btu/lb fuel (corrected to 0°R) and the products are similar to those of "400% air," Item B 8; $t_{fuel} = 60°F$. These efficiencies apply: compressor $\eta_c = 82\%$, turbine $\eta_t = 84\%$. Neglect the mass of fuel and determine (a) the pounds of compressed air delivered per pound of products through the turbine, (b) the air-fuel ratio $r_{a/f}$ for the turbine part of the cycle, (c) the compressed air delivered per pound of fuel burned each second.

Ans. (a) 0.613, (b) 64.5, (c) 39.6 lb/sec.

15.26 A combustor receives air at 480°F and liquid fuel at 80°F [see equation (13-7)]; $r_{a/f} = 60$; the heating value is 18,450 Btu/lb fuel (0°R basis); combustor efficiency is $\eta_b = 94\%$. The gases enter the turbine at 90 psia, leave at 15 psia; $\eta_t = 85\%$. Find (a) the temperature of the gases leaving the combustor, and (b) their temperature leaving the turbine.

15.27 Air enters a stationary gas engine at 500°R and 14.7 psia; r_p = 7.95. The actual temperature at discharge from the compressor is 1000°R. The products of combustion enter the turbine at 2000°R and expand to atmospheric pressure. Regeneration with an effectiveness of 60% is used. (No pressure drops). The actual exhaust temperature is 1320°R. Show all states involved in the solution on a *Ts* diagram. **(a)** For the ideal Brayton cycle, compute the net work per lb and thermal efficiency. Compute **(b)** the compression and turbine efficiencies, **(c)** the actual net work, **(d)** the actual heat added and thermal efficiency. What would be the actual thermal efficiency without regeneration? **(e)** If the output of the actual unit is 5000 hp, what volume of air enters the compressor (cfm at state 1)?

Ans. **(a)** 124.2 Btu/lb, 43.1%, **(b)** 79.7%, 82.6%, **(c)** 61.1, **(d)** 215 Btu/lb, 28.4%, 23.2%, **(e)** 43,600 cfm.

15.28 A regenerative gas turbine is constructed in two parts, similar to **15.18** with the addition of a regenerator. One turbine, whose combustor receives 113,600 lb/hr of air and 1150 lb/hr of fuel, drives the compressor. The other turbine, whose combustor receives 68,065 lb/hr of air and 685 lb/hr of fuel, delivers the net power output, 3500-hp capacity. The entire amount of air, 181, 665 lb/hr, enters the compressor at 14.49 psia and 60°F and is discharged at 58.85 psia and 363°F to the regenerator. Leaving the regenerator at 57.56 psia and 750°F, the air goes to the respective combustors and leaves each combustor at 56.54 psia and 1500°F, also taken as the throttle state to each turbine. Each turbine exhaust is at 15.16 psia and 1024°F, whence all the products enter the regenerator. Leaving the regenerator, the products are at 14.75 psia and 637°F. For the fuel, q_l = 18,200 Btu/lb. Assume no heat losses between the pieces of equipment. Sketch the arrangement of the equipment and find **(a)** the compressor efficiency, **(b)** each turbine efficiency, **(c)** the regenerator effectiveness, and **(d)** the actual thermal efficiency.

15.29 Atmospheric air at 4 psia, −60°F enters the diffuser section of a jet plane (v_p = 960 mph) and is brought almost to rest at the compressor entrance of the jet engine.

The pressure coefficient is κ_p = 92% and the temperature recovery factor is f_T = 94% for the diffuser. Find **(a)** the air pressure and temperature at the diffuser exit, **(b)** the propulsive horsepower developed if the absolute exit velocity of the products from the tail section is v_2 = 2500 fps and the air flow is 100 lb/sec; neglect the added mass of fuel. Sketch the *TS* diagram for the diffuser portion of the cycle.

15.30 A jet-engine driven plane has an air speed of 800 fps; for the engine, the air specific impulse is 60 sec, the air-fuel ratio is 60, the heating value of the fuel is 18,600 Btu/lb f. Compute the propulsive efficiency, the engine thermal efficiency, and the cycle thermal efficiency.

Ans. 45.2%, 19.9%, 43.8%.

15.31 A jet engine is operating in an environment at − 48°F and 8.88 in. Hg abs; its speed is 630 mph; $r_{f/a}$ = 0.017 lb f/lb a; ram pressure coefficient = 90%; combustion efficiency = 94%; let q_1 = 18,800 Btu/lb f; *TSFC* = 0.8 lb f/hr-lb thrust; thrust F_t = 3000 lb. Compute **(a)** mass rate of air flow, **(b)** propulsive power being developed, **(c)** jet velocity at the nozzle exit section (relative to engine), **(d)** propulsive efficiency, **(e)** engine thermal efficiency, **(f)** cyclic thermal efficiency, **(g)** the pressure at the exit of the diffuser if the kinetic energy there is negligible, **(h)** Δh in the nozzle if the entering gases have negligible kinetic energy.

Ans. [Item B 2] **(a)** 39.2 lb/sec, **(b)** 5050 hp, **(c)** 3330 fps, **(d)** 49.8%, **(e)** 28.5%, **(f)** 63.5%, **(g)** 7.28 psia, **(h)** − 222 Btu/lb a.

15.32 The air speed of a turbojet engine is 900 fps in air at 4 psia and − 60°F; for the compressor, r_p = 4.5; let $r_{f/a}$ = 0.01635 (virtually "400% air," fuel mass to be accounted for); at 0°R, $-h_{rp}$ = q_1 = 18,500 Btu/lb f; for the products, M_p = 28.9. Consider as negligible the sensible enthalpy of the fuel as it enters the combustor. All processes are ideal. Kinetic energies are negligible as follows: at exit from diffuser, the changes through combustor and turbine. **(a)** Determine the state (p, T, and other needed properties) at exit from diffuser, from compressor, from combustors, from the turbine, and from the nozzle. Compute **(b)** the mass flow rate of air to provide a 10,000-lb thrust, **(c)** power developed, **(d)** the propulsive efficiency, **(e)** the engine thermal efficiency, **(f)** the cycle thermal efficiency. [*Note:* $\oint dQ/Q_A$ gives exactly the same number as the definition of e_{cyc} only for a pure substance over the whole cycle. Try it.] **(g)** the *TSFC*, **(h)** the irreversibility of the exhaust passing to the dead state; $T_0 = 400°R$.

Ans. [Items B 2, B 8] **(a)** $T_3 = 1838°R$, $T_e = 1088°R$, **(b)** 181.2 lb/sec, **(c)** 16,360 hp, **(d)** 50.3%, **(e)** 21.1%, **(f)** 42.5%, **(g)** 1.07.

15.33 It is possible that hydrogen, heated via a nuclear reactor as shown, may be used as a rocket propellant in the upper stage. Assume that during the reactor heating of the H_2 from $T_2 = 150°R$ to $T_3 = 4500°R$ (see figure), the pressure drops from $p_2 = 1000$ psia to $p_3 = 400$ psia. The H_2 then enters the nozzle with negligible kinetic energy and expands isentropically to an exit pressure of 2 psia; the surroundings are also at virtually the same pressure and 380°R. The thrust is to be 10^5 lb. Compute **(a)** the exit speed of the H_2, **(b)** specific impulse, and **(c)** the mass flow rate for the desired thrust. **(d)** When the propulsive efficiency of the rocket is 88%, what is its Mach number?

Ans. [Item B 5.] **(a)** 25,500 fps, **(b)** 792, **(c)** 126 lb/sec, **(d)** 15.9 or 46.4.

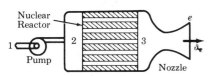

Problem 15.33

15.34 It is desired that a nozzle used for attitude control in a space ship exert a force of 1 lb. Hydrogen is to expand from a stagnation state of 300 psia and 1180°R; the exit section of the nozzle is designed for a pressure of 3 psia; nozzle efficiency $\eta_n = 93\%$. Neglect the effect of the difference between the exit pressure and the surroundings pressure. **(a)** What mass of H_2 should be in

storage for the equivalent of 10 hr of use? Compute **(b)** the specific impulse and **(c)** the throat diameter for an isentropic flow if the pressure there is $p^* = 158$ psia.

Ans. **(a)** 98.5 lb, **(b)** 365 sec, **(c)** 0.0534 in.

15.35 A rocket motor generates combustion gases at 320 psia, 5200°R at entrance to the nozzle where the gas velocity is low (relative to nozzle). The nozzle is designed to expand the gases to the ambient pressure of 5 psia; ambient temperature is 400°R. The expanding gases (considered ideal) have the characteristics $k = 1.2$, $M_g = 15.45$ lb/pmole. For a mass flow of 325 lb/sec and a vehicular speed of 4000 mph, determine **(a)** the relative exit gas speed from the nozzle for steady flow through it, **(b)** the thrust force, the specific impulse, **(c)** the momentary power developed, and **(d)** the propulsive efficiency. **(e)** What is the Mach number of the vehicle speed?

Ans. **(a)** 10,050 fps, **(b)** 101,300 lb, 312 lb-sec/lb, **(c)** 1,080,000 hp, **(d)** 87%, **(e)** 5.98.

15.36 **(a)** The vertical travel of a rocket in a vacuum may be expressed mathematically as $dv = g\,d\tau - v_r\,dm/m$, where v is the rocket speed, g is the local gravity acceleration, $d\tau$ is the interval of burning time for the propellant, v_r is the exit gas velocity relative to the rocket, and m is the combined mass of the rocket and propellant at any given instant. If a rocket is required to escape the earth's gravitational field, determine how much of the total rocket is propellant if the time required to burn the propellant is negligible and if the relative gas velocity $v_r = 8500$ fps is constant during burning. The escape velocity is 25,000 mph; also, let g be constant. **(b)** Four tons of propellant are placed in a 2-ton rocket engine to form a complete 6-ton rocket. It is estimated that the relative jet velocity will be 7500 fps, and that the propellant will burn for 40 sec during flight. Estimate the maximum speed attained by the rocket. Let $g = 32.174$ fps^2 = constant.

16

INTERNAL COMBUSTION ENGINES

16.1 INTRODUCTION

While the gas turbine is an internal combustion engine, the name usually applies specifically to reciprocating internal combustion engines, a type of engine with which we all have some acquaintance. The fuels used are natural or manufactured gas, gasoline, kerosene, oil, alcohol, and others. The common fuels are natural gas, gasoline, and fuel oil.

The earliest attempts to build such an internal combustion engine (ICE) were based on the use of gunpowder. Barsanti and Matteucci built a free-piston engine in 1857, which operated as follows: An explosion drove a piston vertically upward. As it started down under the action of gravity, it engaged a ratchet so connected as to turn a shaft. Such a clumsy machine was doomed to failure, although Otto and Langen successfully marketed a number of free-piston engines about 1867. In 1860, Lenoir proposed and built an engine without compression, but as we shall see, some compensation before combustion is necessary for the production of significant power with acceptable efficiency.

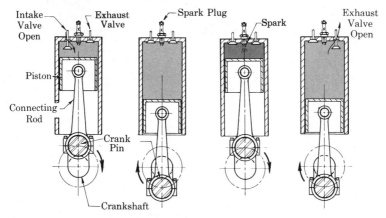

Fig. 16/1 *Four-Stroke Cycle.* Ignition may be by spark as shown, or by self-ignition as in the Diesel engine. (a) Top dead center (TDC); also called head-end dead center; suction stroke begins. (b) Bottom dead center (BDC); also called crank-end dead center, compression stroke begins. (c) Power stroke begins. (d) Exhaust stroke begins.

Although Beau de Rochas, a Frenchman, worked out the theory and gave the conditions for high efficiency in 1862, it remained for Nicholas A. Otto (1832–1891) to build a successful engine in 1876 after he had independently invented the same cycle. This engine was called the silent Otto engine, but the word "silent" should not be taken literally. Otto was born in Holzhausen, Germany, and was a partner in a gas engine manufacturing plant at the time of his famous invention.

Sir Dugald Clerk, born in Glasgow in 1854, invented the two-stroke-cycle engine, which was first exhibited in 1881. In these early stages of the internal combustion engine, rotative speeds of the order of 200 rpm were typical. Gottlieb Daimler (1834–1899), a German, was the first to conceive of small, relatively high speed engines for greater power from a particular size, say 1000 rpm (versus 4000 rpm and more for today's automotive engines), and he made them work by improved hot-bulb ignition. The "high speed" engine made the automobile a practicable idea.

The sequence of events of a four-stroke cycle is explained in Fig. 16/1.

THE OTTO CYCLE 16.2

The Otto cycle is the ideal prototype of spark-ignition (SI) engines, Fig. 16/2. As in the gas turbine and jet engines, it is convenient to analyze an *equivalent air standard*, especially for a first approximation. In a four-stroke ideal Otto cycle, there are no pressure drops because of friction, so that the work quantities associated with intake (suction) 0-1 and discharge 1-0 cancel. During the nonflow isometric processes, $Q = \Delta u$ for unit mass, regardless of working substance. For the purpose of arriving at an important conclusion, we shall assume constant specific heats. For the net work $W = \oint dQ$, the thermal efficiency $e = \oint dQ/Q_A$ is

$$(16\text{-}1) \qquad e = \frac{U_3 - U_2 - (U_4 - U_1)}{U_3 - U_2} = \frac{c_{v23}(T_3 - T_2) - c_{v41}(T_4 - T_1)}{c_{v23}(T_3 - T_2)}$$

$$[\text{ANY SUBSTANCE}] \qquad [\text{CONSTANT } c_v]$$

where $Q_A = U_3 - U_2$, $Q_R = U_1 - U_4$. For the ideal open Otto cycle, the net work is $(U_3 - U_4)_p - (U_2 - U_1)_r$; $U_1 = m_r u_{r1}$, $U_2 = m_r u_{r2}$, $U_3 = m_p u_{p3}$, and $U_4 = m_p u_{p4}$, the subscript r designating reactants and the subscript p the products. If table values of internal energies are available, the first forms of the preceding equations are appropriate. If we assume that $c_{v23} = c_{v41}$ (they are quite different), the c_v's cancel and we get

(a)
$$e = 1 - \frac{T_4 - T_1}{T_3 - T_2}$$

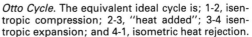

Otto Cycle. The equivalent ideal cycle is; 1-2, isentropic compression; 2-3, "heat added"; 3-4 isentropic expansion; and 4-1, isometric heat rejection. **Fig. 16/2**

Use the TV relation for an isentropic process, equation (7-2); to wit, $T_4/T_3 = (V_3/V_4)^{k-1}$ and $T_1/T_2 = (V_2/V_1)^{k-1}$; or, since $V_3 = V_2$ and $V_4 = V_1$,

(b) $\qquad T_4 = T_3\left(\dfrac{V_3}{V_4}\right)^{k-1} = \dfrac{T_3}{r_k^{k-1}} \quad$ and $\quad T_1 = T_2\left(\dfrac{V_2}{V_1}\right)^{k-1} = \dfrac{T_2}{r_k^{k-1}}$

where we have used the compression ratio $r_k = V_1/V_2 = V_4/V_3$. Substituting these values into **(a)** and simplifying, we find

(16-2) $\qquad\qquad\qquad\qquad e = 1 - \dfrac{1}{r_k^{k-1}} \qquad\qquad\qquad$ [CONSTANT k]

which shows an important characteristic of the Otto cycle; to wit, its efficiency for a particular value of k depends only on the compression ratio r_k. Because of this, the spark-ignition engine has evolved to higher and higher compression ratios. The compression ratio is varied by varying the **clearance volume**, which is the volume of the combustion space when the piston is on TDC position, Fig. 16/2. It is usually expressed as the clearance fraction or **per cent clearance**, c. Thus, the clearance volume is cV_D, where V_D is the displacement volume. For the ideal open cycle,

(c) $\qquad\qquad\qquad\qquad r_k = \dfrac{V_1}{V_2} = \dfrac{V_D + cV_D}{cV_D} = \dfrac{1 + c}{c}$

We should not forget that work is also given by $\oint V\,dp$ or $\oint p\,dV$;

(d) $\qquad\qquad\qquad W = \oint p\,dV = \dfrac{p_2 V_2 - p_1 V_1}{1 - k} + \dfrac{p_4 V_4 - p_3 V_3}{1 - k}$

which, with $pV = mRT$ and $c_v = R/(k - 1) = $ constant, can be shown to be the same as $\oint dQ$. Since the real engine has an intake of air and fuel, it also has a volumetric efficiency, perhaps handily used as a mass ratio; see §§ 14.7 and 14.6 for detail.

16.3   IDEAL STANDARDS OF COMPARISON

Inasmuch as an internal combustion process is involved, there is the school of thought that the ideal standard should correspond to a reversible reaction; that is, $W_{max} = -\Delta A$ at constant temperature, where A is the Helmholtz function, § 5.24. From this point of view, the engine is not a heat engine. The same idea applies to the gas turbine, except that $W_{Tmax} = -\Delta G$; G is the Gibbs function. However, a constant temperature event is so different from the actual happenings that it does nothing to lead one to ideas for improvement of the real engine. Therefore, equivalent cycles, as in the above analysis and in other more realistic analyses, are used.

The ideal Otto engine corresponding to an actual engine has *the same compression ratio, the same initial temperature T_1, and the heat added is the same as that released by the combustion ($m_f q_l$).* Since the specific heats vary widely during the operation, the answers obtained for the air standard depend upon the assumptions made. For

$k = 1.4$, we have the **cold-air standard**, which has little practical value. By using a lower average k, say $k = 1.3$, we have the **hot-air standard**. See Fig. 16/3. With the availability of tabular properties of air (Item B 2 and elsewhere), there is little reason to use constant k's for the air standard, except when the average k fits the particular process.

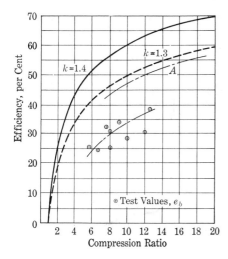

Efficiency Versus Compression Ratio, Otto Cycle. **Fig. 16/3** The solid curve is the ideal efficiency for cold air, $k = 1.4$; dotted curve for hot air, average $k = 1.3$; curve *A*, with allowances for combustion characteristics and for actual mixtures, from charts.[0.24] The test values, computed on the lower heating value and taken at random from the literature, suggest that the actual efficiency tends to improve with ideal efficiency. The practical limit of compression ratio for an SI engine is when detonation starts, which is a rapid, violent, oscillating pressure wave that becomes audible.

The more realistic ideal is the **real-mixture analysis**, which accounts for the products of combustion with dissociation (§ 13.34), the unpurged fraction of burned gases left in the clearance space, and the air-fuel ratio. There are tables and charts available to reduce the computational work.[0.24,0.25]

Example—Air-Standard Otto Cycle   16.4

An ideal Otto engine with 15% clearance operates on 20 lb/min of air, initially at 14 psia and 120°F. It uses $r_{f/a} = 0.0556$ lb fuel/air with a lower constant volume heating value at 0°R of $q_{lo} = 19,000$ Btu/lb fuel (§ 13.27). Determine **(a)** its displacement volume and the mass drawn in during suction for a volumetric efficiency of 75%; **(b)** the temperatures and pressures at the corners of the cycle, Fig. 16/4; **(c)** the horsepower and thermal efficiency; **(d)** the mep and the brake engine efficiency for a brake thermal efficiency of 24%; and **(e)** the availability at state 4 with respect to a sink at 14 psia and 120°F. Use air table values for the air standard.

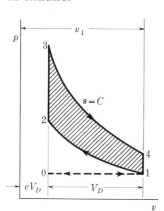

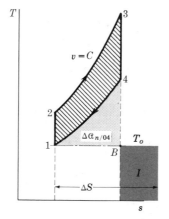

The irreversibility I of process 4-1 is T_o times the change of entropy of the universe.   **Fig. 16/4**

Solution.   (a) First, find r_k; $V_1 = V_4$.

(a)
$$r_k = \frac{V_1}{V_2} = \frac{V_D + cV_D}{cV_D} = \frac{1+c}{c} = \frac{1.15}{0.15} = 7.67$$

(b)
$$V_1 = \frac{mRT_1}{p_1} = \frac{(20)(53.3)(580)}{(14)(144)} = 307 \text{ cfm}$$

(c)
$$V_D = V_1 - V_2 = V_1 - \frac{V_1}{r_k} = 307 - \frac{307}{7.67} = 267 \text{ cfm}$$

For a total mass of $m_1 = 20$ lb, the mass in V_D is $m_D = 20/(1 + c)$; for $\eta_v = V_1'/V_D = m_{in}'/m_D = 0.75$—§ 14.6—the mass drawn in is $m_{in}' = (0.75)(20)/1.15 = 13.03$ lb/min.

(b) For $T_1 = 120 + 460 = 580°$R, we get the following from Item B 2: $p_{r1} = 1.7800$, $u_1 = 98.90$, $v_{r1} = 120.70$, $\phi_1 = 0.61793$. Then, for $s = C$,

(d)
$$v_{r2} = v_{r1}\left(\frac{V_2}{V_1}\right) = \frac{120.7}{7.67} = 15.71$$

Enter the air table with this value of v_{r2} and find, to the nearest whole degree, $T_2 = 1274°$R $(834°$F$)$; $p_{r2} = 30.02$, $u_2 = 222.91$, $\phi_2 = 0.81160$. (By way of comparison: For $k = 1.4$, $T_2 = T_1 r_k^{k-1} = (580)(7.67)^{0.4} = 1310°$R.)

(e)
$$p_2 = p_1\left(\frac{p_{r2}}{p_{r1}}\right) = 14\left(\frac{30.02}{1.78}\right) = 236 \text{ psia}$$

To determine state 3, Fig. 16/4, we note that the energy released on a $0°$R datum is

(f)
$$Q_A = r_{f/a}q_{l0} = (0.0556)(19,000) = 1056 \text{ Btu/lb air}$$

This energy goes solely toward increasing the molecular internal energy during the constant volume combustion. For the air standard,

(g) $Q_A = u_3 - u_2$ or $u_3 = u_2 + Q_A = 222.91 + 1056 = 1278.9$ Btu/lb

To the nearest whole degree corresponding to u_3, we find $T_3 = 5950°$R, $p_{r3} = 19,371$, $v_{r3} = 0.11379$, $\phi_3 = 1.25508$. For a slightly richer mixture, extrapolation will be necessary. The pressure at 3 by Charles' law is

(h)
$$p_3 = p_2\left(\frac{T_3}{T_2}\right) = 236\left(\frac{5950}{1274}\right) = 1101 \text{ psia}$$

Since the expansion ratio V_4/V_3 is the same as the compression ratio V_1/V_2, state 4 is located in the air table (for $s = C$),

(i)
$$v_{r4} = (v_{r3})(r_k) = (0.11379)(7.67) = 0.8728$$

To the nearest whole degree, we find $T_4 = 3288°$R, $p_{r4} = 1395.8$, $u_4 = 650.1$, $\phi_4 = 1.07477$. The pressure at 4 is (for $s = C$)

(j)
$$p_4 = p_3\left(\frac{p_{r4}}{p_{r3}}\right) = 1101\left(\frac{1395.8}{19,371}\right) = 79.3 \text{ psia}$$

(c) The heat rejected as a positive number ($v = C$) is

(k)
$$-Q_R = u_4 - u_1 = 650.1 - 98.9 = 551.2 \text{ Btu/lb}$$

(l)
$$W = \oint dQ = 1056 - 551.2 = 504.8 \text{ Btu/lb}$$

(m)
$$\dot{W} = \frac{(504.8)(20)}{42.4} = 238 \text{ hp}$$

(n)
$$e = \frac{W}{E_c} = \frac{W}{Q_A} = \frac{504.8}{1056} = 47.8\%$$

(d) Use the basic definition for mep, equation (8-6); the work is $(504.8)(20) = 10,096$ Btu/min for a displacement of 267 cfm. Converting units to get psi, we have

(o)
$$p_m = \frac{(10,096)(778)}{(267)(144)} = 204 \text{ psi}$$

In this cycle, the brake engine efficiency is the ratio of thermal efficiencies, since the actual "heat added" Q'_A is equal to the ideal Q_A; thus

(p)
$$\eta_b = \frac{W_B}{W} = \frac{W_B/Q_A}{W/Q_A} = \frac{e_b}{e} = \frac{0.24}{0.478} = 50.2\%$$

(e) In applying the nonflow availability equation (5-6) for $\Delta\mathscr{A}_{n/o}$ given below, we note that the defined dead state is the same as state 1, Fig. 16/4; therefore $v_1 = v_0$ and $v_4 - v_1 = 0$ for the isometric process. The change of entropy from state 4 is

(q)
$$\Delta s = s_1 - s_4 = \phi_1 - \phi_4 - \frac{R}{J}\ln\frac{p_1}{p_4}$$

$$= 0.61793 - 1.07477 - 0.6855\ln\frac{14}{79.3} = -0.3381 \text{ Btu/lb-}°\text{R}$$

(5-6)
$$-\Delta\mathscr{A}_{n/o} = u_4 - u_1 + p_0(v_4 - v_1) - T_0(s_4 - s_1)$$

$$= 650.1 - 98.9 + 0 - 580(0.3381) = 355.2 \text{ Btu/lb}$$

as a positive number or $(355.2)(20)/42.4 = 167.5$ hp, which is additional ideal work that could be obtained, some 64% of the ideal work delivered. The reader should satisfy himself that for the specified relationships this $\Delta\mathscr{A}_{n/o}$ is the same as the irreversibility of the process 4-1, equation (5-11).

ENERGY CONSIDERATIONS OF THE OPEN OTTO CYCLE 16.5

To get the full theory of the ICE, one must go to specialized books on the subject, but the beginner may gain a further appreciation of the problems of thermodynamic analysis by considering an elementary energy balance for events as they may be in an ideal open Otto engine. The energy quantities involved in intake 0-1 and discharge 1-0 cancel for zero pressure drop; so we start at 1, Fig. 16/5. The reactants at state 1 consist of air, fuel, and the unpurged products that were left in the clearance volume during the preceding cycle. Therefore the stored energy at 1 for 1 lb of reactants is

(a) $$E_{s1} = bu_{b1} + r_{f/r}u_{f1} + (1 - b - r_{f/r})u_{a1} + r_{f/r}q_{l0}$$

where all energy quantities are measured from a common datum—0°R as implied by the symbol q_{l0}—and where b is the unpurged fraction of products left in the clearance volume, u_b is the sensible internal energy of these products, $r_{f/r}$ is the fuel-reactants ratio (not fuel-air ratio), u_f is the sensible internal energy of the fuel, u_a is the internal energy of the air, and $q_{l0} = -u_{rp}$ = the constant volume internal energy of the reaction at 0°R. (Another datum may be used if desired.) The sensible internal energy at 1 is then

(b) $$u_{r1} = bu_{b1} + r_{f/r}u_{f1} + (1 - b - r_{f/r})u_{a1}$$

The change of internal energy from 1 to 2, Fig. 16/5, is $u_{r2} - u_{r1}$, and this is the work (except for sign, $W = -\Delta u$) of the nonflow isentropic 1-2. Between 2 and 3, we have

(c) $$u_{r2} + r_{f/r}q_{l0} = u_{p3}$$ [ADIABATIC COMBUSTION]

without dissociation. In equation **(c)**, the detail of u_{r2} is analogous to equation **(b)**; the properties of the products may be obtained via available charts or tables.[0.24] The work of the nonflow isentropic 3-4 is $u_{p3} - u_{p4}$. The net work per unit mass is

(d) $$W = u_{p3} - u_{p4} - (u_{r2} - u_{r1})$$

and we are now back to equation **(d)**, § 16.2, except for symbols. The constant entropy conditions along 1-2 and 3-4 are necessary for complete definitions of the states 2 and 4.

Fig. 16/5   *Open Cycle. The pv diagram is a picture of the variation of pressure with volume in the cylinder—except 4-f-6. State e is the condition of the exhaust if the discharge is via a large plenum where the products can be assumed to be in equilibrium.*

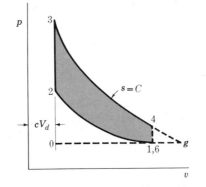
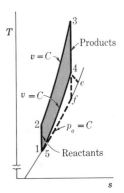

The fraction b of unpurged gases may be calculated by assuming that the products *left* in the cylinder, after the constant volume discharge 4-6, have undergone an isentropic expansion equivalent to state g for 1 lb, § 7.25; that is, imagine the expansion continuing at $s = C$ to g, Fig. 16/5. Then, with an appropriate value of k, we have $T_g/T_4 = (p_g/p_4)^{(k-1)/k} = (p_1/p_4)^{(k-1)/k}$, from which T_g may be calculated.

The process from g to 0, Fig. 16/5, is ideally a transport only, with the result that the mass is proportional to the volume. Therefore, $m_0/m_g = v_0/v_g = b$, where $m_g = m_1 = 1$ lb in the foregoing analysis. Finally, $v_g = RT_g/p_1$.

The actual events differ materially from the above assumptions. (1) There are pressure drops with gas flow in passages and through valves; intake is at a pressure below and discharge is at a pressure above that of the surroundings; consequently, less charge is drawn in. (2) Since the internal surfaces and passages of the engine are relatively hot, the incoming charge is heated and the gases expand in accordance with Charles' law (if ideal), which also reduces the charge that otherwise would be drawn in. (3) And, of course, none of the actual states are equilibrium states because temperature and pressure gradients are inevitable.

Example—Computing the Size of Engine   16.6

(a) Determine the approximate cylinder dimensions of a four-cylinder, four-stroke gasoline engine to deliver 100 bhp at 2000 rpm; diameter/stroke ratio $= D/L = 1$. From experience, it is expected that the bmep ≈ 120 psi, the mechanical efficiency $\eta_m \approx 80\%$, specific fuel consumption $m_{fb} \approx 0.55$ lb/bph-hr, lower heating value of fuel $-u_{rp} = q_t \approx 19{,}000$ Btu/lb. Let $r_k = 6.5$ and use an average $k = 1.32$ for e. (b) Calculate the expected e_b and the indicated engine efficiency.

Solution. (a) The number of engine diagrams (cycles) per minute for the four-stroke cycle is

(a) $$N = (4\ \text{cyl})\left(\frac{1\ \text{diagram}}{2\ \text{rev-cyl}}\right)(2000\ \text{rpm}) = 4000\ \text{diag./min}$$

The volumetric displacement, by equation (14-3), is

(b) $$\dot{V}_D = \frac{\pi D^2 L N}{4} = \frac{\pi D^3\,4000}{4} = \pi 10^3 D^3\ \text{cfm}$$

with the dimension D in feet. For 33,000 ft-lb/min-hp, the work is $\dot{W}_B = (100\ \text{hp})(33{,}000) = 33 \times 10^5$ ft-lb/min. By equation (8-6), the mep in lb/ft^2 is

(c) $$p_{mb} = (120)(144) = \frac{\dot{W}_B}{\dot{V}_D} = \frac{33 \times 10^5}{10^3\,\pi D^3}$$

(d) $$D^3 = \frac{33 \times 10^5 \times 1728}{10^3 \pi (120)(144)} = 105\ \text{in.}^3$$

where 1728 converts D^3 to cubic inches; $D = 4.72$ in.; use $D \times L = 4\frac{3}{4} \times 4\frac{3}{4}$ in., the engine size.

(b) For the specified data, the heat rate is $(0.55)(19{,}000)$ Btu/bhp-hr, and

(e) $$e_b = \frac{2544}{m_{fb}q_l} = \frac{2544}{(0.55)(19{,}000)} = 24.4\%$$

Since the energy charged against the actual and ideal engines is the same, represented by Q_A, the engine efficiency is a ratio of thermal efficiencies.

(f) $$\eta_i = \frac{W_I}{W} = \frac{W_I/Q_A}{W/Q_A} = \frac{e_i}{e}$$

(g) $$e = 1 - \frac{1}{r_k^{(k-1)}} = 1 - \frac{1}{6.5^{0.32}} = 45.1\%$$

The mechanical efficiency is $\eta_m = W_B/W_I$; by the same reasoning as for η_i, $\eta_m = e_b/e_i$ also; therefore $e_i = 24.4/0.80 = 30.5\%$. Then $\eta_i = e_i/e = 30.5/45.1 = 67.6\%$

16.7   DIESEL CYCLE

The final outcome of Rudolf Diesel's* work was a four-stroke-cycle engine in which air only is taken into the cylinder on the suction stroke, and a liquid fuel is later injected, the injection starting theoretically at the end of the compression stroke and continuing at such a rate that burning proceeds at constant pressure 2-3, Fig. 16/6. The remainder of the cycle is as in the Otto, and the air-standard (closed) cycle is 1-2-3-4, Fig. 16/6. The ideal open air cycle would look the same except that 0-1 is the suction and 1-0 is the discharge. The temperature of compression T_2 is high enough that the fuel is self-ignited; this type of engine is called a compression-ignition (CI) engine. The thermal efficiency of the equivalent closed Diesel cycle, $e = W/Q_A = \oint dQ/Q_A$, where $Q_A = h_3 - h_2$ and $Q_R = u_4 - u_1$, is

(a) $$e = \frac{W}{Q_A} = 1 - \frac{u_4 - u_1}{h_3 - h_2} = 1 - \frac{c_v(T_4 - T_1)}{c_p(T_3 - T_2)} = 1 - \frac{T_4 - T_1}{k(T_3 - T_2)}$$

[CONSTANT k]

To put the temperature form of equation **(a)** into a more convenient and revealing form, eliminate the temperatures by expressing three of them in terms of the fourth, say in terms of T_1. Along the isentropic 1-2, Fig. 16/6,

(b) $$T_2 = T_1\left(\frac{V_1}{V_2}\right)^{k-1} = T_1 r_k^{k-1}$$

Along the constant pressure line 2-3 for a constant mass of pure substances, Charles' law gives $T_3/T_2 = V_3/V_2$. Let $r_c \equiv V_3/V_2$, a ratio termed the fuel *cutoff ratio*; then,

(c) $$T_3 = T_2\left(\frac{V_3}{V_2}\right) = T_1 r_k^{k-1} r_c$$

by using equation **(b)**. Write the TV relation for the isentropic 3-4 and manipulate it as follows:

(d) $$T_4 = T_3\left(\frac{V_3}{V_4}\right)^{k-1} = T_3\left(\frac{r_c V_2}{V_1}\right)^{k-1} = T_1 r_k^{k-1} r_c \left(\frac{r_c V_2}{V_1}\right)^{k-1} = T_1 r_c^k$$

---

* Rudolf Diesel (1858–1913), born in Paris of German parents who later moved to London because of the Franco–German War (1870), educated in Germany, obtained in 1893 a patent on the type of engine that now bears his name. After some difficulty in financing the project, he built an engine which blew up at the first injection of fuel. Diesel narrowly escaped being killed. Four years of tedious and costly experiment elapsed before he produced a successful engine. He inexplicably disappeared in 1913 while crossing the English Channel during a storm.

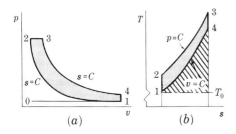

Fig. 16/6

Diesel Cycle. Between the same temperature limits, the constant volume curve on the *Ts* plane is steeper than the constant pressure curve (§ 7.5). However, both curves get steeper as the temperature increases.

where we used the definition of r_c, T_3 from equation **(c)**, and $V_4 = V_1$. Substitute the above values of T_2, T_3, and T_4 into equation **(a)** with $k = C$ and get

$$(16\text{-}3) \qquad e = 1 - \frac{T_1 r_c^k - T_1}{k(T_1 r_k^{k-1} r_c - T_1 r_k^{k-1})} = 1 - \frac{1}{r_k^{k-1}}\left[\frac{r_c^k - 1}{k(r_c - 1)}\right]$$

$$[\text{CONSTANT } k]$$

This expression differs from the efficiency of the Otto cycle, equation (16-2), in the bracketed factor, a factor always greater than 1, because r_c is always greater than 1. It follows that the Otto cycle is more efficient for a particular compression ratio r_k; but since the Diesel cycle compresses only air, with no problem with detonation, its compression ratio may be materially higher, with the end result being that actual thermal efficiencies of the CI (Diesel) engine may be greater than that of the actual SI engine. The compression, cutoff, and expansion ratios are related as follows (Fig. 16/6):

$$(16\text{-}4) \qquad r_k = \frac{V_1}{V_2} = \left(\frac{V_3}{V_2}\right)\left(\frac{V_1}{V_3}\right) = r_c r_e$$

Equation (16-3) shows that as r_c increases the bracketed factor increases, and the efficiency decreases, Fig. 16/7. Therefore, the lower fuel cutoff ratios are conducive to higher theoretical efficiencies, but larger ratios result in greater power. For a given engine running at a particular speed, the friction loss remains relatively constant, so that the *brake* thermal efficiency ceases to improve at some point as the cutoff ratio decreases. In the opposite direction, there is a limit to the amount of fuel that can be injected without excessive "smoking" (unburned combustibles). The maximum practical fuel cutoff is of the order of 10% of the stroke.

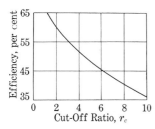

Fig. 16/7

Diesel Thermal Efficiency Versus Cutoff Ratio—Constant Compression Ratio.

Let the standard of comparison be the ideal cycle that has the same compression ratio r_k as the actual engine, the initial state as at intake of the engine, and with "heat added" equal to $r_{f/a}q_l$ Btu/lb air, with enthalpy, internal energy, and q_l measured from an appropriate datum.

16.8   THE IDEAL OPEN-CYCLE DIESEL

The ideal open cycle has a discharge process 1-0 and an intake 0-1 with the only energy quantities being the works of transport that cancel each other. The unpurged products remaining at 0 mix with the air drawn in 0-1. For 1 lb of reactants at 1, there are b lb of unpurged products and $1 - b$ lb of air. Let the energy at state 1, where there is mixture of unpurged products and air, be represented by u_{m1}; then the work of compression is $|u_{m2} - u_{m1}|$. Apply the first law to process 2-3 in the form $E_{\text{stored 2}} + E_{\text{in}} = E_{\text{stored 3}} + E_{\text{out}}$. For the 0°R datum,

(a) $u_{m2} + (r_{f/m}h_{fx} + r_{f/m}q_{l0}) = (1 + r_{f/m})u_{p3} + p_3(1 + r_{f/m})v_{p3} - p_2v_{m2}$

where the fuel is in some state x. Using $h = u + pv$ and simplifying, we get

(b) $h_{m2} + r_{f/m}h_{fx} + r_{f/m}q_{l0} = (1 + r_{f/m})h_{p3}$

where $h_p = \bar{h}_p/M_p$, and so on, also the energy-balance equations may be expressed in mole values. This process is discussed in some detail for a Diesel engine in § 13.20. Equation (13-7) gives a value of the sensible enthalpy h_f of the fuel. Equation **(b)** may be symbolized as $H_{r2} + r_{f/m}q_{l0} = H_{p3}$, where H represents the total sensible enthalpy; $H_{r2} = h_{m2} + r_{f/m}h_{fx}$. The net work is then $\oint dW$, or

$$W = \oint dW = U_{p3} - U_{p4} + p_2(V_3 - V_2) - (U_{r2} - U_{r1})$$

(c) $= H_{p3} - H_{r2} + U_{r1} - U_{p4}$

$$= r_{f/m}q_{l0} + u_{r1} - (1 + r_{f/m})u_{p4}$$

where $H_{p3} - H_{r2} = r_{f/m}q_{l0}$, and the datum is 0°R.

If equation **(b)** is applied to the air standard ($r_{f/a}$ ignored as applied to sensible enthalpies), we get

(d) $r_{f/a}q_l = h_3 - h_2$ [AIR STANDARD]

Equation **(b)** of § 13.27 shows that differences of heating values at different temperatures become meaningless when the reactants and products are taken as the same substance; thus, it is acceptable to use the standard q_l° for the air standard as convenient.

16.9   Example—Diesel Cycle

State 1 for a Diesel engine is $p_1 = 14.4$ psia and $t_1 = 100°F$; $r_k = 13.5$; the air-fuel ratio in the cylinder is $r_{a/f} = 30$, which is closely "200% air," Item B 9; the fuel has a heating value of $q_{l0} = 19,328$ at 0°R; liquid fuel enters at 80°F. The molecular mass of the products is $M_p = 28.84$. Ignore the effect of the unpurged products and compute the work and thermal efficiency. At what fraction of the stroke does cutoff occur?

Solution. For air only during process 1-2, Fig. 16/6, we have from Item B 2 at 560°R, $h_{a1} = 133.86$, $u_{a1} = 95.47$, $v_{r1} = 131.78$.

(a) $$v_{r2} = \frac{v_{r1}}{r_k} = \frac{v_{r1}}{13.5} = \frac{131.78}{13.5} = 9.76$$

Enter Item B 2 with this value of v_{r2} and find $T_2 = 1510°R$, $h_{a2} = 371.82$, $u_{a2} = 268.30$. From $r_{a/f} = 30$, we get $r_{f/a} = 1/30$. The sensible enthalpy of the fuel at $T_f = 540°R$ from the 0°R base, equation (13-7), is

(13-7) $$h_f = 0.5T - 287 = 270 - 287 = -17 \text{ Btu/lb}$$

Using equation (b), § 16.8, we have

(b) $$h_{a2} + r_{f/a}h_f + r_{f/a}q_{lo} = (1 + r_{f/a})h_{p3}$$

$$371.82 + \frac{-17}{30} + \frac{19{,}328}{30} = \left(1 + \frac{1}{30}\right)\frac{\bar{h}_{p3}}{28.84}$$

from which $\bar{h}_{p3} = 28{,}200$ Btu/pmole of products. Entering Item B 9 with this value of $\bar{h}_p$, we get $T_3 = 3461°R$, $\bar{u}_{p3} = 21{,}324.6$, $v_{r3} = 15.141$. The amount of dissociation at this value of T_3 would be small. The cutoff ratio is $r_c = V_3/V_2 \approx T_3/T_2 = 3461/1510 = 2.29$.* Using equation (16-4), we find v_{r4};

(c) $$r_e = \frac{r_k}{r_c} = \frac{13.5}{2.29} = 5.9 \qquad v_{r4} = r_e v_{r3} = (5.9)(15.141) = 89.33$$

For this value of v_{r4}, Item B 9 yields, to the nearest whole degree, $T_4 = 2079°R$, $\bar{u}_{p4} = 11{,}734.1$ Btu/pmole, and $u_{p4} = \bar{u}_{p4}/M_p = 11{,}734.1/28.84 = 406.9$ Btu/lb.
From the energy balance, or the equation (c), § 16.8, we get

(d) $$W = r_{f/a}q_{lo} + u_{r1} - (1 + r_{f/a})u_{p4}$$

$$= \frac{19{,}328}{30} + 95.47 - \left(1 + \frac{1}{30}\right)(406.9) = 319.3 \text{ Btu/lb air at 1}$$

where $u_{r1} = u_{a1}$. The thermal efficiency is

(e) $$e = \frac{W}{r_{f/a}q_{lo}} = \frac{319.3}{19{,}328/30} = 49.6\%$$

The best brake thermal efficiencies for actual CI engines are well over 35%.
As seen in Fig. 16/6, the fraction of the stroke at which cutoff occurs is given by $(V_3 - V_2)/(V_1 - V_2)$, where $V_1 - V_2 = V_D$. Use $V_3 = r_c V_2$ and $V_1 = r_k V_2$, from which

(f) $$\text{fraction of cutoff} = \frac{r_c V_2 - V_2}{r_k V_2 - V_2} = \frac{2.29 - 1}{13.5 - 1} = 0.103$$

---

* For a pure substance process 2-3, this ideal-gas law would be quite close. Considering the chemical reaction, different substances at 2 and 3, and we have $r_c = V_3/V_2 = (m_3 R_3 T_3/p_3)/(m_2 R_2 T_2/p_2)$, which comes out about $r_c = 2.38$. A convenient approximation is $r_c = (1 + r_{f/d})V_3/V_2 = (1 + r_{f/a})T_3/T_2 = 2.365$.

16.10   SECOND LAW ANALYSIS

As mentioned during the discussion of the Otto engine, the most ideal manner of obtaining work as a consequence of combustion is via constant temperature reversible reaction. But the more realistic study involves the changes in pressure and temperature. A second law study may investigate the changes in availability of each process and what becomes of the availability produced by the actual combustion process. The availability remaining in cooling water and in the exhaust and that dissipated in "friction work" that goes into the sink are of interest. Other approaches may be used, the objective being to learn where the greatest irreversibilities occur in the hope of finding means of reducing them.

16.11   DUAL COMBUSTION CYCLE

Actual indicator cards from both modern Otto and Diesel engines show a rounded top, with a shape that suggests that some combustion at constant volume and some at constant pressure would give an ideal cycle more closely resembling the actual events. This observation led to the proposal of a so-called dual combustion cycle, (also called a limited-pressure cycle), indicated on the pv and Ts planes in Fig. 16/8. There is no general understanding as to when this cycle should be used as a standard. For the air standard and constant specific heats, Fig. 16/8, we have

(a)

$$e = \frac{W}{Q_A} = \frac{c_v(T_3 - T_2) + c_p(T_4 - T_3) - c_v(T_5 - T_1)}{c_v(T_3 - T_2) + c_p(T_4 - T_3)}$$

$$= 1 - \frac{T_5 - T_1}{(T_3 - T_2) + k(T_4 - T_3)} = 1 - \frac{1}{r_k^{k-1}}\left[\frac{r_p r_c^{k-1} - 1}{r_p - 1 + r_p k(r_c - 1)}\right]$$

where $r_p \equiv p_3/p_2$ is the pressure ratio during the constant volume portion of the combustion, and other symbols are as previously defined ($r_c = V_4/V_3$).

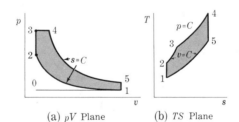

Fig. 16/8   *Dual Combustion Cycle.*   (a) pV Plane   (b) TS Plane

16.12   VARIATIONS IN ACTUAL ENGINES

Many small gasoline engines and a major percentage of Diesels operate on a *two-stroke cycle*. Since the exhaust stroke in the real engine is for the purpose of purging or *scavenging* the cylinder of the products, it is only necessary to provide other means of scavenging in order to be able to complete the cycle in two strokes (one revolution). In the two-stroke cycle, exhaust begins early, at some point e, Fig. 16/9, and scavenging is accomplished by blowing air (or air and fuel) into the cylinder. To allow time for scavenging or for the introduction of fuel mixtures, the valves usually remain open until the piston has moved to some position corresponding to b, where compression begins. While the two-stroke cycle completes twice as many power strokes in a certain number of revolutions as a four-stroke cycle, the two-stroke-cycle engine develops only some 70–90% more power than a four-stroke engine of the same dimensions $D \times L$, rather than 100% more, other operating conditions the same. Basically, this is because the mass of combustible mixture in the cylinder is less for two-stroke engines, which in turn is affected by poorer scavenging. Other factors in varying degree include greater loss of unburned fuel, power consumption in compressing the air for scavenging (commonly, crankcase compression), and early drop of pressure at e, Fig. 16/9.

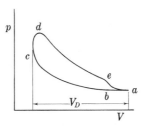

Two-Stroke Cycle. **Fig. 16/9**

The pressure difference between discharge and intake may be varied by a throttling valve at intake, often a means of regulating the power of an SI engine, Fig. 16/10. With only a small opening of the throttle, the intake pressure 7-1 may be only some 5 psia; with discharge 5-6 above atmospheric pressure, this results in a large counterclockwise loop b-6-7-1 (negative work), and a small mass charge that means points 2 and 3 are much lowered. Such operating conditions can be idealized as shown with expansions and compressions isentropic.

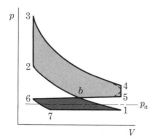

Otto Cycle with Throttling. For a passenger car, $p_5 \approx 0.25$ psig at 20 mph, $p_5 \approx 3$ psig at 70 mph. With wide open throttle at high speed, p_1 is some 2 to 3 in. Hg. vacuum. **Fig. 16/10**

Supercharging, which has the objective of pumping more air (Diesel) or a greater charge (Otto) into the cylinder, is commonly used with Diesel engines, aircraft engines (increasing power at high altitudes where the intake is of low density), and elsewhere. The exhaust pressure at 5, Fig. 16/11, is ideally atmospheric, actually

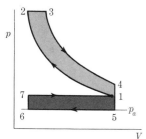

Ideal Supercharged Diesel Cycle. Point 1 is shown slightly displaced to make the cycle easier to follow; $V_1 = V_4$. Example of the effect of supercharging: a diesel engine without supercharging developed 475 bhp with $p_{mb} = 80$ psi; with supercharging, it developed 880 bhp with $p_{mb} = 147$ psi. **Fig. 16/11**

somewhat above. If the supercharger, usually driven by the exhaust, is steady flow with $\Delta K = 0$, the ideal work input is $h_7 - h_a$, where h_a is the enthalpy of the entering air and h_7 is at the discharge state. The charge flowing into the cylinder is represented by 7-1.

16.13   CLOSURE

There are economic and advantageous uses for all kinds of work-producing equipment. The SI type is especially suited to low power (up to several hundred horsepower); one factor in limiting its size is an increasing proneness to detonation, because the flame front has farther to travel in large combustion chambers. Diesel (CI) engines overlap the SI type in size because they burn a cheaper fuel with some improvement in efficiency. The size of CI engines goes up to several thousand horsepower. To keep perspective, new central-station steam power plants are commonly several hundred-thousand horsepower on a single shaft (Fig. 9/17).

While the thermodynamics of ICE's has been well developed for many years, engineering work is still active. There is the Wankel engine, and the spark ignition rotary engine.[16.7] Studies are active toward the use of liquid H_2 and O_2 as ICE fuels in space and under water, with proposals that the same fuel be used for commercial ICE to eliminate smog ingredients in the exhaust.

PROBLEMS

SI UNITS

16.1 An Otto cycle with a compression ratio of 7.5 operates from the suction conditions of 97.91 kPaa, 29.4°C. Find the pressure and temperature at the end of compression **(a)** if cold air ($k = 1.4$) is the working substance, **(b)** if hot air ($k = 1.32$) is the working substance. Compare answers. **(c)** Solve for the ideal thermal efficiency based upon the conditions given in **(a)** and **(b)**.

Ans. **(a)** 1644 kPaa, 404°C, **(b)** 1400 kPaa, 303.3°C, **(c)** 55.35%, 47.5%.

16.2 For an ideal Otto engine with 17% clearance and an initial pressure of 93.08 kPaa, determine the pressure at the end of compression using air-standard properties from Item B 1. If the pressure at the end of the constant volume heating is 3447.4 kPaa, what is the mep in kPa?

Ans. 1385 kPaa, 471 kPa.

16.3 Sketch an Otto cycle on the pV and Ts planes with point 1 being at the beginning of the isentropic compression process. These data apply for the air standard cycle: $p_1 = 101.4$ kPaa, $T_1 = 333.3$ K, $V_1 = 283\ \ell$, $r_k = 5$, $T_3 = 2000$ K. Solve for **(a)** W, **(b)** p_2, V_2, T_2, **(c)** p_3, **(d)** p_4, T_4, **(e)** Q_A, **(f)** Q_R, **(g)** e, **(h)** % clearance c, and **(i)** that portion of Q_A which is unavailable ($T_0 = T_1 = 333.3$ K.)

16.4 A two-stroke, Otto-type, 23.5 × 23.5-cm engine develops 29.85 kW at 200 rpm while using natural gas whose lower heating value is 37,252 kJ/m³ at intake of $p_1 = 1$ atm and $t_1 = 15.6$°C. The heat rate is 14.142 kJ/kW-hr; compression ratio is 6, mechanical efficiency is 81%. Use the air standard with $k = 1.33$ and compute the brake and indicated thermal efficiencies, engine efficiencies, and mep.

Ans. $e_1 = 31.4\%$, $\eta_b = 57.1\%$, $p_{mb} = 878$ kPa.

16.5 During a 1.45-min test on a 7.785 × 8.731-cm, 8-cylinder automotive engine, it used 454 g of fuel ($q_1 = 43,734$ kJ/kg) and developed a torque of 237.3 Nm with a mechanical efficiency of 78%. The engine shaft turned a total of 3520 revolutions. Calculate **(a)** e_b, e_i, **(b)** η_b, η_i for an ideal cycle efficiency of $e = 53.3\%$, and **(c)** p_{mI}.

Ans. **(a)** $e_b = 26.4\%$, **(b)** $\eta_i = 63.6\%$, **(c)** 1150 kPa.

16.6 An engine designer desires to use a fuel tank that will hold a minimum fuel supply for 1 hr operation for a 6-cylinder 9.21 × 8.89-cm, automotive-type gasoline engine which at maximum output develops a brake torque of 267 Nm at 3000 rpm. The corresponding ideal engine showed a thermal efficiency of 56.5%; the brake engine efficiency is expected to be 53%. For the fuel, $q_1 = 43,269$ kJ/kg and its specific gravity is 0.715. Find the smallest tank that will satisfy.

16.7 The 3.44-ℓ British Jaguar four-stroke-cycle automotive engine has 6 cylinders in line, 8.306 × 10.592-cm size, a compression ratio of 8, and develops 156.7 kW at 5500 rpm. For the ideal cycle, the condition at the beginning of compression is 101.35 kPaa, 54.4°C. At 5500 rpm the brake torque of 267 Nm at 3000 rpm. The efficiency are 30% and 78% respectively. The condition at the inlet to the engine is 103.42 kPaa, 26.7°C. Solve for **(a)** percentage clearance, **(b)** p, T at the end of ideal compression, **(c)** η_b using $k = 1.4$, **(d)** mass of air used, **(e)** torque, **(f)** p_{mb}, **(g)** friction power where the mechanical efficiency is $\eta_m = 71\%$, **(h)** $r_{a/f}$ with $q_1 = 44,432$ kJ/kg.

16.8 The Ford Comet engine has 6 cylinders in line, 8.89 × 6.35-cm, a compression ratio of 8.7, a displacement of 2365 cm^3, and develops 67.1 kW at 4200 rpm. While operating at 4200 rpm, it is noted that the brake thermal efficiency is 31%, the volumetric efficiency is 72%, and the mechanical efficiency is 70%. For the ideal cycle, the condition at the beginning of compression (point 1) is 101.35 kPaa, 60°C. At 4200 rpm, find **(a)** the percentage clearance, **(b)** p, T at the end of ideal compression, $k = 1.4$, **(c)** brake engine efficiency, **(d)** mass of air used, kg/min, **(e)** torque, **(f)** p_{mi}, **(g)** fhp, **(h)** $r_{a/f}$ where $q_1 = 44,548$ kJ/kg fuel.

16.9 There are developed 157 kW by an Otto type engine that burns 26.3 kg/hr of fuel ($q_1 = 44,199$ kJ/kg) in 395 kg/hr of air. The air and liquid fuel enter at 43.3°C. For purposes of this problem, use the enthalpy of the products as that of air. The exhaust gases leave at 676.7°C; $\Delta K = 0$. Sketch the energy diagram and find **(a)** the heat rejected (kJ/min) and **(b)** the thermal efficiency.

16.10 A 6-cylinder, 4-stroke cycle, single-acting, spark-ignition engine with a compression ratio of 9.5 is required to develop 67.14 kW with a torque of 194 Nm. Under the conditions, the mechanical efficiency is 78% and the brake mep is 552 kPa. For the ideal cycle, $p_1 = 101.35$ kPaa, $t_1 = 35$°C, and the hot-air standard $k = 1.32$. If $D/L = 1.1$ and the fuel rate is $m_{fi} = 0.353$ kg/kW-hr ($q_1 = 43,967$ kJ/kg), determine **(a)** the bore and stroke, **(b)** the indicated thermal efficiency, **(c)** the brake engine efficiency, and **(d)** the percentage clearance.
Ans. **(a)** 10.11 × 9.19-cm, **(b)** 23.2%, **(c)** 34.4%, **(d)** 11.77%.

16.11 The mep of an ideal Diesel cycle is 758.4 kPa. If $p_1 = 93.1$ kPaa, $r_k = 12.5$, and the overall value of $k = 1.34$, find r_c.
Ans. 2.6.

16.12 For an ideal Diesel cycle with the overall value of $k = 1.33$, $r_k = 15$, $r_c = 2.1$, $p_1 = 97.9$ kPaa, find p_2 and p_m.
Ans. 3596 kPaa, 603 kPa.

16.13 The charge in a Diesel engine consists of 18.34 g of fuel, with a lower heating value of 42,571 kJ/kg, and 409 g of air and products of combustion. At the beginning of compression, $t_1 = 60$°C. Let $r_k = 14$. For constant $c_p = 1.110$ kJ/kg-K, what should be the cutoff ratio in the corresponding ideal cycle? *Ans.* $r_c = 2.96$.

16.14 There are supplied 317 kJ/cycle to an ideal Diesel engine operating on 227 g air; $p_1 = 97.91$ kPaa, $t_1 = 48.9$°C. At the end of compression, $p_2 = 3930$ kPaa. Assume that the air and the products within the cycle have air properties. Determine **(a)** r_k, **(b)** the percentage clearance, **(c)** r_c, **(d)** W, **(e)** e, **(f)** p_m.

16.15 There are developed 1063 kW at 267 rpm by an 8-cylinder, 2-stroke-cycle, 40.64 × 50.80-cm Diesel engine that uses 4.94 kg/min of fuel with a lower heating value of 42,571 kJ/kg. The average indicated mep is 562 kPa. Determine **(a)** e_b, **(b)** e_i, **(c)** η_m. *Ans.* **(a)** 30.3%, **(b)** 47%, **(c)** 80.7%.

16.16 An ideal dual combustion cycle operates on 454 g of air. At the beginning of compression, the air is at 96.53 kPaa, 43.3°C. Let $r_p = 1.5$, $r_c = 1.6$, and $r_k = 11$. Using the air properties in Item B 1, determine **(a)** the percentage clearance, **(b)** the pressure, volume, and temperature at each corner of the cycle, (c) Q_A, Q_R, and W, (d) the thermal efficiency, (e) the mep.

Ans. (a) 10%, (c) W = 277 kJ, (d) 58.5%, (e) 713 kPa.

16.17 The Stirling cycle engine is being studied as a replacement for the conventional open-cycle automotive engine in view of reducing air pollution. An analysis of this cycle follows. Air is made to pass through this cycle, which consists of two isothermal processes and two regenerative isometric processes. At the beginning of the isothermal expansion (point 1), p_1 = 724 kPaa, V_1 = 56.6 ℓ, t_1 = 315.6°C. The isothermal expansion ratio is $r_e = V_2/V_1 = 1.5$; the minimum cycle temperature is t_3 = 26.7°C. Sketch the pV and Ts diagrams and for the cycle find (a) Δs for each isothermal process, (b) Q_A, (c) Q_R, (d) W, (e) e, (f) p_m.

Ans. (a) 0.0282 kJ/K, (b) 16.62 kJ, (c) −8.45 kJ, (d) 8.17 kJ, (e) 49.15%, (f) 288 kPa.

ENGLISH UNITS

16.18 An ideal Otto cycle engine with 15% clearance operates on 0.5 lb/sec of air; intake state is 14.4 psia, 100°F. The energy released during combustion is 525 Btu/sec. Using air-table values, compute (a) r_k, (b) V_D, (c) p and t at each corner, (d) e, (e) the percentage of available energy utilized for a sink temperature of 100°F.

Ans. (a) 7.67, (b) 6.26 cu ft/sec, (c) 243 psia, 773°F, 1161 psia, 5434°F, 83.6 psia, 2795°F, (d) 47.9%, (e) 58.7%.

16.19 The fuel/air ratio for an ideal Otto engine with 12% clearance is $r_{f/a}$ = 0.06 lb fuel/lb air where the lower heating value is q_l = 18,800 Btu/lb fuel. The engine operates on 25 lb/min of air, initially at 14 psia, 110°F. Using the air table (air standard) and neglecting the diluting effect of the mass of fuel, determine (a) V_D, (b) p and T at each corner of the cycle.

16.20 An equivalent air-standard Otto cycle rejects heat in a nonflow isometric process from T_4 = 2100°R and 54 psia to T_1 = 560°R and 14.4 psia. The cycle work is 340 Btu/lb. The surroundings are at 14.4 psia and 60°F. (a) If the heat were rejected reversibly, how much work would be done?

What percentage of the work is this? Determine (b) the irreversibility of process 4-1 and (c) the unavailable portion of the heat (as it leaves the system).

Ans. [Item B 2.] (a) 165.4, (c) 127.7 Btu/lb.

16.21 The Chevrolet Corvair rear mounted air-cooled engine had 6 cylinders, horizontally opposed, 3.375 × 2.60 in., a compression ratio of 8, and developed 80 bhp at 4400 rpm. Operating at 4400 rpm, it was noted that the brake thermal efficiency was 26%, the volumetric efficiency was 74%, and the mechanical efficiency was 71%. For the ideal cycle, the condition at the beginning of compression is 14.7 psia, 140°F. At 4400 rpm find (a) percentage clearance, (b) p, T at end of ideal compression with k = 1.4, (c) brake engine efficiency η_b, k = 1.4, (d) mass of air breathed lb/min, (e) torque, (f) p_{mi}, (g) friction horsepower, (h) $r_{a/f}$ with q_1 = 19,100 Btu/lb fuel.

16.22 The following data apply to a spark-ignition engine that is completing 12,000 cpm: e_b = 31%, $r_{a/f}$ = 15.2, m_f = 0.00015 lb/cycle with q_l = 18,800 Btu/lb; incoming temperature air/fuel is 100°F; exhaust temperature is 1260°F; ΔK = 0. Sketch the energy diagram, use Item B 2 and determine (a) W_B, (b) the air mass flow, (c) Q, assuming that all energy losses leave the system sooner or later as heat.

Ans. (a) 247.5 hp, (b) 1640 lb/hr, (c) 14,780 Btu/min.

16.23 There is required a vertical, 4-cylinder, 4-stroke-cycle, single-acting gas engine to develop 180 bhp at 275 rpm. The probable brake mep is 58 psi. (a) Let L/D = 1.25, and find the bore and stroke. (b) The suction pressure is 14.4 psia, and r_k is 5.75. If the over-all k is 1.32 and the pressure at the end of the combustion process is 460 psia, compute e_b and η_b. (c) If the fuel is natural gas with a lower heating value of 908 Btu/ft³, what will be the fuel consumption in ft³/hr?

Ans. (a) 13.16 × 16.44 in., (b) η_b = 65.3%, (c) 1803 ft³/hr.

16.24 An ideal Diesel engine operates on 1 ft³ (measured at state 1) of air. Let p_1 = 14.4 psia, t_1 = 140°F, r_k = 14, and let the cutoff be at 6.2% of the stroke. Draw the pV and TS diagrams, use the air table, and find (a) t_2, p_2, V_2, t_3, V_3, p_4 and t_4; (b) Q_A and Q_R; (c) W and e; and (d) p_m.

Ans. **(a)** $t_2 = 1167.2°F$, $p_2 = 550$ psia, $t_3 = 2478°F$, $p_4 = 36.4$ psia, $t_4 = 1049°F$; **(b)** $Q_A = 24$ Btu; **(c)** 13.2 Btu, **(d)** 76.8 psi.

16.25 There are supplied 27 Btu/cycle to an ideal 2-stroke, single-cylinder Diesel engine, operating at 280 rpm. At the beginning of compression, $p_1 = 14.7$ psia, $t_1 = 90°F$, and $V_1 = 1.50$ ft^3. At the end of compression $p_2 = 500$ psia. For the air standard of Item B 2 compute **(a)** p, V, and T at each corner of the cycle; **(b)** W and p_m; and **(c)** hp and e.

16.26 A Diesel engine using 29.90 lb air/lb fuel (200% air—Item B 9) yields these operating data: indicated work $W_I = 7300$ Btu/lb fuel; brake work $W_B = 5500$ Btu/lb fuel; temperature of cooling water in is 86°F, out is 140°F; water flow is 80 lb water/lb fuel. For state point 1, $p_1 = 14.3$ psia, $t_1 = 160°F$, $r_k = 15$; sink temperature $t_0 = 100°F$; exhaust gas temperature $t_g = 700°F$; molecular weight of exhaust gas is $M_p = 28.90$; $q_l = 18,550$ Btu/lb fuel. Use Items B 2 and B 9 where relevant and investigate the major changes in available energy.

16.27 At the beginning of compression in an ideal dual combustion cycle, the working fluid is 1 lb of air at 14.1 psia and 80°F. The compression ratio is 9, the pressure at the end of the constant volume addition of heat is 470 psia, and there are added 100 Btu during the constant pressure expansion. Using Item B 2, find **(a)** r_p, **(b)** r_c, **(c)** the percentage clearance, **(d)** the thermal efficiency, and **(e)** the mep.

Ans. **(a)** 1.58, **(b)** 1.187, **(c)** 12.5%, **(d)** 54.4%, **(e)** 57.4 psia.

16.28 The study of the Stirling engine in **16.17** continues. Full-load performance data taken from road tests conducted on a 6-cylinder automotive-type Stirling engine follow: cylinder size 3.47×2.37 in., indicated power developed 240 hp, speed 2500 rpm, mechanical efficiency 75%, p_{mb} 317 psi, brake fuel rate 0.418 lb/bhp-hr, q_l 18,200 Btu/lb fuel. Comparable open-cycle automative engines when tested gave these data: Gasoline Otto type (242 bhp, 4600 rpm, 0.468 lb/bhp-hr, $q_l = 18,900$ Btu/lb fuel, 0.65 bhp/in.^{3} displacement); Diesel type (181 bhp, 2100 rpm, 0.41 lb/bhp-hr, $q_l = 18,200$ Btu/lb fuel, 0.49 bhp/in.^{3} displacement). Analyze the performance of the Stirling engine and compare with those of the other two engines where relevant.

17

REVERSED CYCLES

17.1 INTRODUCTION

The reversed cycle is a system that receives heat from a colder body and delivers heat to a hotter body, not in violation of the second law, but by virtue of a work input. In addition to its well-known uses in preserving foods, in manufacturing ice, and in conditioning air for summer comfort, the refrigeration cycle has many other industrial applications, as in making "cold rubber" (to improve wearing quality), in oil refinery processes, in the treatment of steel, in the manufacture of chemicals, and in liquefying the gases being increasingly used for industrial and power purposes. There are so many applications and so much work going on at very low temperatures that that area of science has been given the name, *cryogenics*. The reversed cycle is also used for heating buildings, and in this application it is popularly called a *heat pump*. In a technical sense, *heat pump* is a general name for all reversed cycles. Of the reversed cycles discussed in this chapter, only the Carnot is reversible (§ 8.13).

17.2 THE REVERSED CARNOT CYCLE

At the outset, since the Carnot cycle is quite revealing and is reversible (the most efficient), we should review § 8.14. If the cycle is used for ordinary refrigeration, Fig. 17/1(a), the refrigerant is isentropically compressed ab from a cold temperature T_1

Fig. 17/1 *Reversed Cycles.* These cycles approach external reversibility as $\Delta T \to 0$. If a reversed-cycle installation is used for cooling in the summer and heating in the winter, as in Fig. 17/2, the sink temperature T_o is lower for heating than for cooling, an unfavorable shift.

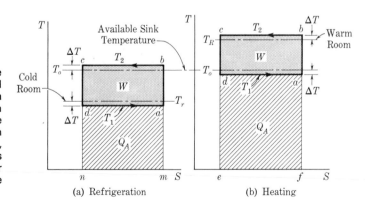

(a) Refrigeration (b) Heating

to a temperature T_2, say slightly above that of some naturally available sink T_0. The refrigerant then discharges heat at some constant temperature T_2 along bc. At some state c, an isentropic expansion cd lowers the temperature to T_1, which is lower than the temperature T_r of the refrigerated room or body to be cooled, so that now heat may flow from the refrigerated room to the refrigerant, thus cooling the room. The refrigerant receives the heat along the path da, from which the cycle repeats.

If the reversed cycle is used for heating, Fig. 17/1(b), the sequence of operations is the same as before, the difference, as far as the system is concerned, being in the temperatures. For heating, the temperature T_2 must be above T_R of the room to be heated, so that heat can flow from the refrigerant to the room. Also, the source of heat is now some naturally available reservoir (the air, a well, or lake, and so on), which is essentially the sink. In both cases, the work is represented by

(a) $$W = (T_2 - T_1)\,\Delta S \text{ Btu}$$

as a positive number, where $\Delta S = S_a - S_d = S_b - S_c$. Although work done *on* a system is conventionally negative, engineers usually find it convenient to ignore this for reversed-cycle work, since the sign tells only the direction of the energy flux. The output of the refrigeration cycle is refrigeration, which is heat added to the system from the cold room, area $ndam$, Fig. 17/1(a);

(b) $$Q_A = T_1(S_a - S_d) = T_1\,\Delta S \text{ Btu}$$

The output of the heating cycle is the heat delivered to the space to be heated, heat rejected by the system, area $ecbf$, Fig. 17/1(b);

(c) $$Q_R = T_2(S_b - S_c) = T_2|\Delta S| \text{ Btu}$$

as a positive number. The coefficients of performance (COP), § 8.14, are

(17-1) $$\gamma_r \equiv \frac{\text{refrigeration}}{\text{work}} = \frac{Q_A}{W} \quad \text{and} \quad \gamma_h \equiv \frac{\text{heating effect}}{\text{work}} = \frac{Q_R}{W}$$

$$[\text{COOLING}] \qquad\qquad\qquad [\text{WARMING}]$$

in general terms. For the Carnot cycle, from **(a)**, **(b)**, and **(c)**,

(d) $$\gamma_r = \frac{Q_A}{W} = \frac{T_1}{T_2 - T_1} \quad \text{and} \quad \gamma_h = \frac{Q_R}{W} = \frac{T_2}{T_2 - T_1} \qquad [\text{CARNOT}]$$

$$[\text{COOLING}] \qquad\qquad\qquad [\text{WARMING}]$$

These values of the COP are the highest possible for all cycles operating between the temperatures T_1 and T_2. Irreversible *ideal* cycles will have a lower COP.

Considering a refrigerating operation, we might go briefly to the job to be done, to wit, removing heat from some substance, and investigate the numerically minimum work required. If the substance is initially at a temperature above that of the environment or sink, it is natural that it be cooled by the sink to temperature T_0 (ideally). Let the cooling process for the substance be $e1$, Fig. 17/2. To cool reversibly, we use an infinite number of reversible engines, one of which has the cycle *abcd*. This engine removes heat $dQ_A = T\,ds$ (heat rejected by substance) and rejects heat $T_0\,ds$ to the sink. The work of this Carnot engine is $dW = dQ_R - dQ_A = T_0\,ds - dQ_A$, where each number and the work are positive. By integration, we get ($Q = Q_A$ for engine)

(e) $$W_{\text{rev}} = T_0(s_e - s_1) - |Q|$$

a positive number, where $|Q|$ is the heat to be removed (internally reversible) from unit mass of the substance being cooled, represented by area *e*-1-*rk*, Fig. 17.2. Since the work above is reversible work, it is the minimum conceivable, represented by area *e*-1-*f*.

If the cooling is a steady-flow process with $\Delta K = 0$ and heat exchanged only with the surroundings, the numerical minimum work is $-\Delta \mathscr{A}_f$; after equation (5-8) for process *e*-1,

(f) $$W_{\text{min}} = -\Delta \mathscr{A}_f = -[h_1 - h_e - T_0(s_1 - s_e)]$$

which, being algebraically consistent, will give a negative number. The heat removed from the substance is $|Q| = h_e - h_1$ for this steady-flow operation; and equation **(f)** agrees with **(e)** except for sign. It often pays to look at the whole process, considering means of reducing irreversibilities.

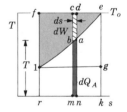

Fig. 17/2   *Reversible Work to Cool a Substance.* Observe that if a single Carnot engine were to do this cooling *e*-1, its cycle must be 1-*gef*, with irreversible heat transfer to the engine along 1-*g* everywhere except at 1! $dQ_A = abmn$; $dQ_R = mcdn$; dQ_A for the engine is heat rejected by substance being cooled.

17.3   CONCLUSIONS FROM THE CARNOT CYCLE

It is fairly obvious that the work W to activate the cycle should be a minimum, since it must be manufactured and paid for. With this in mind, we draw certain important conclusions of general validity from the Carnot cycle. The basic ideas below are easily extended to reversed cycles that operate in cryogenic temperature ranges.

1. The work will be reduced as the temperature T_2 is lowered. (a) In a refrigerating cycle, Fig. 17/1(a), the lowest temperature T_0 that is attainable by a natural coolant, such as the atmosphere or a lake, is the most economical. There is thus a natural lower limit for T_2 set by T_0. In practice, T_2 is some 5 to 20°F greater than T_0. (b) In a warming cycle, Fig. 17/1(b), T_2 must be some 10–20°F or more above the room temperature. Thus, if it is desired to keep a room at 70°F, the refrigerant must at at, say, 90°F or more. Smaller temperature differences than mentioned could be used, but as the temperature difference decreases, the surface area needed in the heat exchanger increases in order to maintain the same rate of heat flow, thereby increasing the cost of the heat exchanger. In practice, it is a matter of getting an economical balance.

2. The work will be reduced as the temperature T_1 is increased. (a) In a refrigerating cycle, Fig. 17/1(a), it is therefore economical of work to carry on the desired *refrigeration at as high a temperature as practicable.* To freeze water, temperatures below 32°F are essential; but to cool air for comfort air conditioning, much higher temperatures may be used. (b) In a warming cycle, Fig. 17/1(b), the temperature T_1 must be below some naturally available temperature. The presence of warm springs is very helpful. Deep wells or the earth itself may provide natural sources of heat at temperatures higher than winter atmospheric temperatures. In a particular geographic location we would tend to use the coldest convenient sink for cooling and the warmest convenient sink for heating.

3. For particular temperature limits, the heat exchanges should take place at constant temperature for the most effective use of the work. In the case of vapor refrigerants, § 17.6, the refrigerant will be at a constant temperature during much of the heat transfer process; but this is not true when a noncondensing gas is the refrigerant.

The reversed cycle makes it possible to utilize work to produce heating without being too lavish with available energy. For an electrical work input of 10 kW-hr and a COP of 3, the heat obtained would be 30 kW-hr; whereas the 10 kW-hr delivered to a resistance heater would give 10 kW-hr of heat. A diagrammatic layout of a heat pump used for heating and cooling is shown in Fig. 17/3.

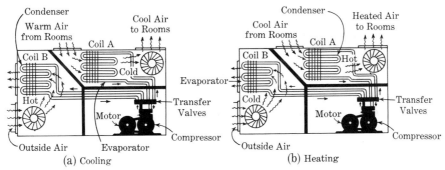

Heat Pump. During the cooling season, the outside coil *B* acts as the condenser, receiving the discharge from the compressor and delivering heat to the outside; coil *A* is the evaporator, cooling the air flowing over it. For heating, coil *A* is the condenser from which heat flows to the air that warms the house; coil *B* is the evaporator, which picks up heat from the outside.

Fig. 17/3

UNIT FOR REFRIGERATING EFFECT   17.4

The historical definition of a capacity unit is the amount of heat that must be extracted to freeze 1 ton of water at 32°F into ice at 32°F (at 1 atm) in 1 day. Since the latent heat of fusion of water is closely 144 Btu/lb, we find (144)(2000) = 288,000 Btu/day. This steady rate of refrigeration is called a *ton* of refrigeration. For time elements more convenient than a day, we have

(a)

$$\frac{288,000}{24} = 12,000 \text{ Btu/hr} \quad \text{or} \quad \frac{12,000}{60} = 200 \text{ Btu/min} \quad \text{or} \quad 211 \text{ kJ/min}$$

that is, a 10-ton plant is capable of doing (10)(200) = 2000 Btu/min of refrigeration at the defined operating states.

A common method of expressing the actual efficiency of a compression system (§ 17.6) is to give the horsepower used per ton of refrigeration. Let the refrigeration be N tons or $200\,N$ Btu/min. Let the horsepower required for this N tons of refrigeration be represented by hp; then, the corresponding work is $42.4(\text{hp})$ Btu/min and the coefficient of performance is

(b)
$$\text{COP} = \gamma = \frac{\text{refrigeration}}{\text{work}} = \frac{200N}{42.4\,\text{hp}}$$

(17-2)
$$\frac{\text{hp}}{N} = \frac{200}{42.4\gamma} = \frac{4.72}{\gamma}$$

the horsepower per ton of refrigeration, good for either actual or ideal cycles.

17.5   Example—Carnot Refrigerator

There are required 1.2 kJ/sec-ton refrigeration for a Carnot refrigerator to maintain a low temperature region at 250 K. If the refrigerator is producing 4 tons of refrigeration, find (a) power requirements, hp, (b) T_2, (c) COP, γ_r, (d) heat rejected, Btu/min, (e) COP, γ_h.

Solution.   See Fig. 17/1(a) for appropriate Ts diagram.

Power requirements for 4 tons,

(a)
$$P = (1.2)(4) = 4.8 \text{ kJ/sec}$$

$$= \left(\frac{4.8 \text{ kJ}}{\text{sec}}\right)\left(\frac{1 \text{ kW-sec}}{\text{kJ}}\right)\left(\frac{\text{hp}}{0.746 \text{ kW}}\right) = 6.43 \text{ hp}$$

For the upper temperature T_2,

(b)
$$\Delta S_{da} = \frac{Q_A}{T_1} - \frac{P}{T_2 - T_1}$$

or

$$\frac{T_2 - T_1}{T_1} = \frac{P}{Q_A} = \left(\frac{6.43 \text{ hp}}{4 \text{ ton}}\right)\left(\frac{\text{min-ton}}{200 \text{ Btu}}\right)\left(\frac{42.4 \text{ Btu}}{\text{hp-min}}\right) = 0.341$$

$$T_2 = 0.341 T_1 + T_1 = (1.341)(250) = 335 \text{ K}$$

(c)
$$\text{COP} = \gamma_r = \frac{Q_A}{P} = \frac{T_1}{T_2 - T_1} = \frac{250}{335 - 250} = 2.94$$

Heat rejected that could be used on reversed heating,

(d)
$$Q_R = Q_A + P$$

$$= (4)(200) + (6.43)(42.4) = 1073 \text{ Btu/min}$$

(e)
$$\text{COP} = \gamma_h = \frac{Q_R}{P} = \frac{Q_A + P}{P} = \gamma_r + 1 = 3.94$$

VAPOR COMPRESSION REFRIGERATION   **17.6**

The most common method of securing refrigeration is by a *vapor-compression system*, represented diagrammatically in Fig. 17/4. See also Figs. 17/14 and 17/15.

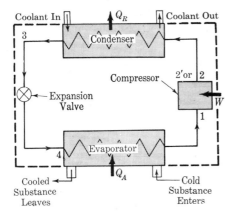

Vapor-Compression Refrigerating System.   **Fig. 17/4**

In the ideal case, all flow is without friction, except flow through the expansion valve, and all processes except those in the condenser and evaporator (cold room) are adiabatic. Figure 17/5 shows the reversed idealized vapor cycle 1-2-3-4 on the *Ts* plane with numbers corresponding to those on Fig. 17/4. Starting at state 1, the vaporous refrigerant enters the compressor, which may be either a rotating or a reciprocating machine; state 1 on the saturated vapor curve is preferred, but it may be elsewhere in actual operation. The pressure p_2 must be such that the corresponding saturation temperature is above the available sink temperature (in general, above the temperature of the body to which heat is to be rejected). The condenser generally subcools the liquid a small amount, say from f to 3, Fig. 17/5. In state 3 as it leaves the condenser, the liquid enters an **expansion valve**, which is a throttling valve separating the region of higher pressure from that of lower pressure.

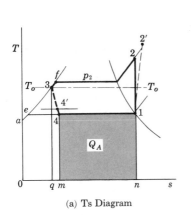

(a) Ts Diagram

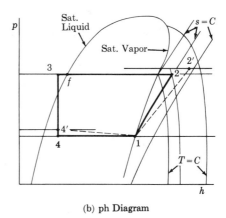

(b) ph Diagram

Refrigeration Cycles, Ideal and Actual. Properties of refrigerants and other   **Fig. 17/5**
substances are often pictured on pressure-enthalpy *ph* diagrams; see item
B 31. Because of friction of flow, there are small pressure drops between actual
states 2′, 4′ and the states 3, 1, respectively, which are both actual and ideal.
State 3 is a state of compressed liquid (§3.11); rarely, if ever, is the subcooling
enough to be worth the computation of accounting for it.

At 4 (or 4'), the very wet mixture enters the evaporator, absorbing heat Q_A from (doing refrigeration *on*) the surroundings, process 4-1 (or 4'-1).

The energy diagram of Fig. 17/4 shows that $Q_A - |Q_R| = W$, as usual. If the system is operating in steady flow, with $\Delta K = 0$ and $W = 0$ in the condenser and in the evaporator, then $Q = \Delta h$ (§ 7.5). For the ideal case, either for constant pressure, as in Fig. 17/5, or for steady flow ($h_3 = h_4$ during the isenthalpic expansion),

(a)
$$Q_A = h_1 - h_4 = h_1 - h_3 \qquad \text{[REFRIGERATION]}$$

(b)
$$|Q_R| = h_2 - h_3$$

(c)
$$W = |Q_R| - Q_A = h_2 - h_3 - (h_1 - h_3) = h_2 - h_1$$

(d)
$$\text{COP} = \gamma = \frac{h_1 - h_4}{h_2 - h_1} = \frac{h_1 - h_3}{h_2 - h_1}$$

where the work is written as a positive number. Observe that $h_2 - h_1$ is the difference of the enthalpies at the ends of the isentropic compression ($s_1 = s_2$). Area m-4-1-n represents the refrigeration; area n-2-f-3-q represents the heat rejected in the condenser.

Since the actual compression process is irreversible to some final state 2', Fig. 17/5 (the final entropy at 2' is always greater than it would be for the corresponding reversible process), we have, for $\Delta K = 0$,

(e)
$$W' = h_{2'} - h_1 - Q \text{ Btu/lb}$$

If the compression is adiabatic, $Q = 0$; otherwise, the usual convention of signs holds for Q in equation **(e)**. The ***compression efficiency*** is $\eta_c = W/W'$. A cutaway view of a refrigerating machine for an air conditioning installation is shown in Fig. f/11, Foreword.

17.7   DISPLACEMENT OF THE COMPRESSOR

For a particular refrigerating capacity, the size of the compressor depends upon (1) the number of pounds of refrigerant that must be circulated per unit of time to obtain the desired refrigerating effect, and (2) the specific volume of the substance at the intake state of the compressor. A plant with a capacity of N tons can refrigerate at the rate of $200N$ Btu/min. Then, with refrigeration of $h_1 - h_4$ Btu/lb, the mass of refrigerant circulated is

(a)
$$\dot{m} = \frac{200N \text{ Btu/min}}{h_1 - h_4 \text{ Btu/lb}} \text{ lb/min}$$

State 1, Fig. 17/5, is known or assumed, so that the specific volume v_1 may be determined. Evidently ($\dot{m}$ lb/min) times (v_1 ft³/lb) is the needed displacement volume V_D cfm for 100% volumetric efficiency (§ 14.6). For a volumetric efficiency of η_v,

(b)
$$V_D = \frac{\dot{m} v_1}{\eta_v} = \frac{v_1}{\eta_v} \left(\frac{200N}{h_1 - h_4} \right) \text{ cfm}$$

Practical values of the volumetric efficiency should usually fall within the range 60–85%. The factors affecting volumetric efficiency in vapor compressors are much the same as those discussed in § 14.6.

Example 17.8

An ammonia compressor receives wet vapor at 10°F and compresses it adiabatically to a saturated state at 190 psia. The temperature at the expansion valve is 85°F. The compressor is single stage, double-acting, 12×14 in., runs 200 rpm, and its volumetric efficiency at normal operating conditions is 78%. Let the compression efficiency be $\eta_c = 80\%$. (a) For the ideal cycle, determine the coefficient of performance γ and the horsepower per ton. (b) For the actual cycle as defined, compute the COP, the hp/ton, and the size of the system in tons and the horsepower input. (c) Determine the temperature at $2'$, Fig. 17/6, if the compression is adiabatic.

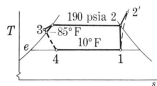

Fig. 17/6

Solution. (a) For the states named in Fig. 17/6, we find

$$h_{f1} = 53.8 \qquad s_{g1} = 1.3157 \qquad h_{g2} = h_2 = 632.4$$

$$h_{fg1} = 561.1 \qquad v_{f1} = 0.02446 \qquad s_{g2} = s_2 = 1.1802$$

$$s_{f1} = 0.1208 \qquad v_{g1} = 7.304 \qquad h_3 = h_4 = 137.8$$

From $s_1 = s_2$, we get

(a) $$1.1802 = 0.1208 + 1.1949 x_1 \qquad \text{or} \qquad x_1 = 88.7\%$$

(b) $$h_1 = 53.8 + (0.887)(561.1) = 551.5 \text{ Btu/lb}$$

(c) $$v_1 = 0.0244 + (0.887)(7.28) = 6.48 \text{ ft}^3/\text{lb}$$

(d) $$\gamma = \frac{Q_A}{W} = \frac{h_1 - h_4}{h_2 - h_1} = \frac{551.5 - 137.8}{632.4 - 551.5} = \frac{413.7}{80.8} = 5.11$$

(e) $$\frac{\text{hp}}{\text{ton}} = \frac{4.72}{\gamma} = \frac{4.72}{5.11} = 0.819$$

(b) For $\eta_c = 0.80$ and $W = 80.9$ Btu/lb,

(f) $$\gamma' = \frac{Q_A}{W'} = \frac{413.7}{80.9/0.80} = \frac{413.7}{101.1} = 4.09$$

$$\frac{\text{hp}}{\text{ton}} = \frac{4.72}{4.09} = 1.154$$

In a double-acting compressor, the number of diagrams completed per minute is $2n$ for n rpm (§ 14.7). Therefore

(g) $$V_D = \frac{\pi D^2}{4} L(2n) = \frac{\pi (144)(14)(2 \times 200)}{(4)(1728)} = 366 \text{ cfm}$$

For $\eta_v = 78\%$, the volume V_1' *drawn in* is $V_1' = (0.78)(366)$ cfm; and for $v_1 = 6.48$ ft^3/lb, the mass of refrigerant circulated per minute is

(h) $$\dot{m} = \frac{V_1'}{v_1} = \frac{(0.78)(366)}{6.48} = 44.1 \text{ lb/min}$$

giving a total refrigeration of $(44.1)(413.7)$ Btu/min.

(i) $$\text{tons} = \frac{(44.1)(413.7)}{200} = 91.2$$

For 44.1 lb/min of refrigerant and $W' = 101.1$ Btu/lb from equation (f), the power is

(j) $$\dot{W}' = \frac{(44.1)(101.1)}{42.4} = 105.1 \text{ hp}$$

(c) Since the fluid work is $h_{2'} - h_1 = (h_2 - h_1)/\eta_c = 101.1$ Btu/lb from equation (f), we have

(k) $$h_{2'} = h_1 + W' = 551.5 + 101.1 = 652.6 \text{ Btu/lb}$$

In the full NH_3 superheat table[0.10] at 190 psia and $h = 652.6$ Btu/lb, we find $t_{2'} = 122.4°$F.

17.9   REFRIGERANTS

Table VI shows a comparison of some characteristics of popular refrigerants. Some of the desirable qualities are:

1. Refrigerants are preferably *nontoxic*, so that in case of leakage no one is in danger of injury. This attribute is of paramount importance in air conditioning systems and home refrigerators, for example, but toxic refrigerants are tolerable in commercial installations where reasonable precautions are taken. The Carrenes, Freons (trade names), and carbon dioxide are not toxic, unless so much is present as to result in an oxygen deficiency, but the others in Table VI are, in more or less degree. Methyl chloride is not only toxic but also is practically odorless; such working substances should contain a warning agent (for example, acrolein, which is irritating to the eyes and nose).
2. Refrigerants should be *economical*, both in initial cost and in maintenance. Maintenance problems include: controlling leakage (there is less trouble with leakage of large molecules than of small ones); providing adequate lubrication (the refrigerant should not react with the oil to destroy its lubricating qualities); and avoiding corrosion (the refrigerant should not corrode the materials it contacts). Also the refrigerant should be readily available for recharging the system when necessary.
3. Refrigerants should be *nonflammable*. A number of hydrocarbons have been and are used as refrigerants, examples of which in Table VI are butane and propane. These and others (ammonia, methyl chloride, and so on) constitute a fire and explosion hazard. The other refrigerants in Table VI are nonflammable.
4. Refrigerants preferably have high latent heat at the evaporator temperature (see Fig. 3/2) and a low specific volume. The type and size of compressor is a function of these physical traits. If the latent heat is high, much refrigeration is done by each pound of substance circulated; if in addition the specific volume is low, the volume of substance to be circulated and therefore the size of compressor and passageways are small. Notice that when the ideal displacement volume V_D is small, reciprocating compressors are feasible; when V_D is large (Carrene 1, F 11, and F 113), centrifugal compressors (which can be run at high speed) become necessary.

5. Refrigerants should have low saturation pressures at normal operating temperatures. The cost of design, manufacture, and operation is involved. The high pressure for CO_2, which has a low COP besides, means heavy parts and thick-walled pipes. Also, it is preferable that the saturation pressure at the evaporator temperature be above atmospheric pressure in order to avoid leakage of air into the system; with the best of modern shaft seals, this is not as serious a problem as formerly.

6. While the foregoing attributes are perhaps the most significant, there are miscellaneous others that are desirable: good thermal conductivity for rapid heat transfer, wetting ability, inertness (the refrigerant should not react in any way with the materials it touches), stability (the refrigerant should not break down into different matter of smaller molecules), low viscosity (for ease of movement), high critical temperature, and a high dielectric strength (in hermetically sealed units where the refrigerant contacts the motors). Also, it stands to reason that the refrigerant should not solidify at any temperature in its cycle.

A commercial advantage of the newer refrigerants is that a manufacturer may use the same compressor for different capacities by changing the refrigerant and installing a different size of motor. For example, for a particular displacement, a greater refrigerating effect is obtained from F 22 than from F 12, accompanied by a greater power need for the F 22.

Characteristics of Refrigerants   **TABLE VI**

Selected from reference [12.1]. The condenser and evaporator temperatures are 86 and 5°F, respectively; Pres. = saturation pressure, psia; the fluid leaves the evaporator as saturated vapor; saturated liquid enters throttling valve at 86°F; displacement is for 100% volumetric efficiency. For comparison, the Carnot COP = 5.74. Under "Type of Compressor": Rec. = reciprocating; Rot. = rotary; Cen. = centrifugal.

Notes: (a) An azeotropic mixture of Freon 12 and $C_2H_4F_2$. (b) Methylene chloride. (c) Trichloromonofluoromethane. (d) Dichlorodifluoromethane. (e) Monochlorodifluoromethane. (f) Trichlorotrifluoroethane. (g) Dichlorotetrafluoroethane. Freon is a well-entrenched trade name (Du Pont). The same refrigerants are available under other trade names.

Name (ASHRAE standard number)	Formula	Pres. at 5°F	Pres. at 86°F	Q_A, Btu/lb	w, lb/ min-ton	V_D, cfm/ ton	COP	Type of Compressor
Ammonia (717)	NH_3	34.3	169.2	474.4	0.42	3.44	4.76	Rec.
Butane (600)	C_4H_{10}	8.2	41.6	128.6	1.56	15.52	4.95	Rec., Rot.
Carbon dioxide (774)	CO_2	332.2	1045.7	55.5	3.62	0.96	2.56	Rec.
Carrene (30) (b)	CH_2Cl_2	1.16	10.6	134.6	1.49	74.3	4.90	Rot., Cen.
Carrene (500) (a)	(a)	31.1	128.1	61.1	3.27	4.97	4.61	Rec., Rot.
Freon (11) (c)	CCl_3F	2.93	18.3	67.5	2.96	36.3	5.09	Rot., Cen.
Freon (12) (d)	CCl_2F_2	26.5	107.9	51.1	3.92	5.8	4.7	Rec., Rot.
Freon (22) (e)	$CHClF_2$	43.0	174.5	69.3	2.89	3.60	4.66	Rec.
Freon (113) (f)	$C_2Cl_3F_3$	1.01	7.5	53.7	3.73	100.7	4.92	Cen.
Freon (114) (g)	$C_2Cl_2F_4$	6.8	36.7	43.1	4.64	20.1	4.49	Rot.
Methyl chloride (40)	CH_3Cl	21.2	94.7	150.2	1.33	5.95	4.90	Rec., Rot.
Sulfur dioxide (764)	SO_2	11.8	66.4	141.4	1.41	9.09	4.87	Rec., Rot.
Propane (290)	C_3H_8	41.9	155.2	121.0	1.65	4.09	4.58	Rec.

VACUUM REFRIGERATION   **17.10**

When the desired cold room temperature is about 40°F, water and steam may be used as the refrigerant in a system that is called *vacuum refrigeration*, Fig. 17/7.

Warm water from the cooling system (say, at 50°F) is sprayed into the evaporator (where the pressure is maintained at, say, 0.248 in. Hg abs., the saturation pressure for 40°F). The entering 50°F water undergoes a throttling process, $h = C$. Since water is continuously introduced in steady operation, the resulting vapor must be continuously removed. Because of the large volumes to be handled, a reciprocating compressor is out of the question. Steam-jet pumps (ejectors, Fig. f/13) or centrifugal compressors are used. In Fig. 17/7, steam enters a group of nozzles, expands to the low pressure of 0.248 in. Hg, entrains the vapor formed by the evaporation of entering water, and carries the entrained vapor via the diffuser to the primary condenser. The pressure in the condenser, which in this case is a four-pass surface condenser, is as low as available cooling water permits, in this instance 1.932 in. Hg abs. With the ejector discharging into a condenser, the work required to maintain the vacuum in the evaporator is very much less than if the discharge were to the atmosphere. The vapor is pumped from a pressure of 0.248 in. Hg (in this case) to only 1.932 in. Hg (which is a pressure ratio of 1.932/0.248 = 7.8), rather than to atmospheric pressure of 29.92 in. Hg. The steam used in the ejector and that from the evaporator is condensed in the primary condenser. The work required to pump the resulting *water* to atmospheric pressure is relatively insignificant. The noncondensable gases in the primary condenser are pumped to atmospheric pressure by two stages of secondary ejectors. The steam used in these secondary ejectors is condensed in the inter- and after-condenser.

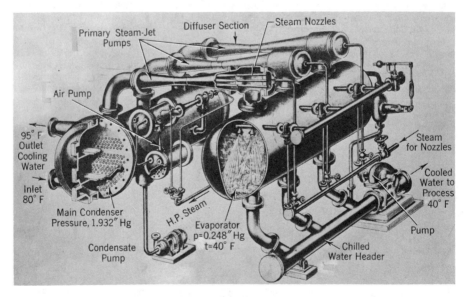

Fig. 17/7   *Elements of a Vacuum Refrigeration System.* The temperatures and pressures noted are typical, not fixed values. (*Courtesy Foster Wheeler Corp., New York.*)

The foregoing temperatures suggest that this system is suitable particularly for moderate refrigerating temperatures, such as those encountered in air conditioning, and in some industrial processes. Moreover, vacuum refrigeration is likely to be economical only for relatively large installations where steam is available and is needed for other purposes.

Example—Vacuum Refrigeration   17.11

A steam-jet pump maintains a temperature of 40°F in the evaporator, Fig. 17/8. The cooling water leaves at the same temperature via 1, and warms to 50°F as it does its refrigeration. The amount of makeup water m_5 at 70°F exactly equals the mass of vapor leaving m_4. There are required 2.8 lb nozzle steam/lb vapor pumped. (It is a local convenience to call the nozzle fluid *steam* and the steam leaving the evaporator *vapor*.) For a 50-ton plant, compute the volume of vapor handled and the amount of steam needed.

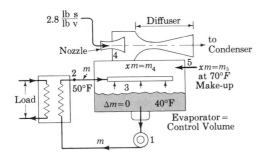

Mass Balance for Evaporator.   **Fig. 17/8**

Solution. Let the evaporator bound the system, Fig. 17/8. Assume that the mass of H_2O in the evaporator is constant and that, effectively, there is equilibrium for both the water and steam. Let the mass of water circulated for cooling purposes be represented by m; let the fraction of the throttled water "flashing" to vapor be x lb vapor/lb chilled water. Then the amount of vapor leaving the system is xm, which is balanced by a makeup rate at 5 of the same amount. First, write an energy balance for $m = 1$; subscripts designate the stations in Fig. 17/8.

(a)
$$mh_{f2} + xmh_{f5} = mh_{f1} + xmh_{g4}$$

(b)
$$18.05 + 38.05x = 8.03 + 1079.0x$$

from which $x = 0.00963$. For a capacity of 50 tons, the load is $\dot{Q}_A = (50)(200) = 10,000$ Btu/min. Also, it is

(c)
$$\dot{m}(h_{f2} - h_{f1}) = \dot{m}(18.05 - 8.03) = 10,000$$

from which $\dot{m} = 1000$ lb/min (closely). The total vapor removed is $x\dot{m} = 9.63$ lb/min. If the vapor at 4 is saturated, we have $v_{g4} = 2446$ ft^3/lb and

(d)   Volume of vapor removed: $\dot{V} = (9.63)(2446) = 23,555$ cfm

(e)   Amount of steam needed: $\dot{m}_s = (2.8)(9.63) = 26.96$ lb/min

which is equivalent to $(26.96)(60)/50 = 32.4$ lb/hr-ton.

ABSORPTION SYSTEMS OF REFRIGERATION   17.12

Absorption systems are characterized by the fact that the refrigerant is absorbed by an absorbent on the low-pressure side of the system and is given up on the high-pressure side. The advantage derived from the absorption routine is that a liquid, rather than a gaseous substance, is pumped from the low-pressure to the high-pressure region, with the consequence of considerably less work.

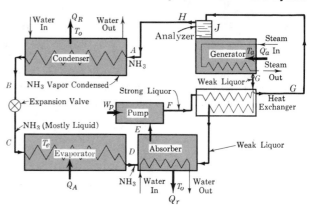

Fig. 17/9   *Ammonia–Water   Absorption System of Refrigeration, Schematic.* As usual, each piece of apparatus may be analyzed separately as a steady-flow device during steady-state operation. A reflux from the condenser is not shown.

The essential elements of an ammonia-water (refrigerant-absorbent) absorption system are suggested by Fig. 17/9. In place of the compressor of Fig. 17/4, there is a relatively complicated array of devices. Starting at the ammonia condenser at A, we find the processes at first the same as those for any vapor refrigerating system: the condensation of the vapor, A to B; the isenthalpic expansion to a lower pressure, B to C; the process of refrigeration in the cold room, C to D. After leaving the evaporator, the vapor enters an absorber, where the water in the absorber absorbs the ammonia. A strong solution of ammonia is called a ***strong liquid***; a weak solution, ***weak liquid***. The absorption process releases heat Q_r (that includes the latent heat of the NH_3 absorbed), which is carried away by cooling water. The strong liquid is pumped E to F from the absorber, through a heat exchanger F to G to the ***generator*** via a small rectifying column (called an *analyzer*) that separates the NH_3 from the NH_3-H_2O mixture leaving the generator. Steam coils, or other heat source, in the generator, evaporate the strong liquid, whose vapor is mostly NH_3 entering the analyzer. (See § 17.20 for some information on rectifying columns.) The ammonia vapor passes to the condenser at A, from which this cycle is repeated. Considering the generator again, we note that the process of boiling off the ammonia leaves a weak solution. The weak solution then passes from the generator, through the heat exhanger back to the absorber, where it again absorbs ammonia. This *regenerative heat exhanger* serves the purpose of cooling the weak solution on its return to the absorber and of heating the strong solution on the way to the generator, thus saving heat in the generator and reducing the rejected heat in the absorber.

The coefficient of performance, as defined in equation (17-1), does not apply to these cycles because the significant energy supplied to the cycle to keep it going is Q_a in the generator. Thus, a suitable performance factor for such cycles is the Btu of refrigeration (or other output) divided by the heat transferred in the generator (input); or if the energy to drive the pumps is accounted for, the denominator is increased by the amount of these works; that is, the performance factor η_p is (Fig. 17/9)

(a)
$$\eta_p = \frac{Q_A}{Q_a} \quad \text{or} \quad \eta_p = \frac{Q_A}{Q_a + \sum W}$$

where $\sum W$ may include the circulating-pump works, according to the desired accuracy. Sometimes in practice, performance factors are expressed in pounds of steam supplied to the system per ton of refrigeration, a value that may fall between 30 and 40 lb/hr-ton.

The highest possible value of the performance factor is obtained by the use of reversible cycles. Let the heat Q_a transferred at T_a without a temperature drop in the generator be used in a Carnot engine. If this reversible engine rejects its heat to the sink at temperature T_0, Fig. 17/10, its thermal efficiency is

(b)
$$e = \frac{T_a - T_0}{T_a} = \frac{W}{Q_a} \quad \text{or} \quad Q_a = \frac{T_a W}{T_a - T_0}$$

If the work $W = abcT_0$, Fig. 17/10, from this Carnot engine is used without loss to drive a reversed Carnot engine $efcT_0$, then the work of the reversed engine is the same as in equation **(b)** and the ratio of the areas $efcT_0$ and $efgh$ (refrigeration) is

(c)
$$\frac{W}{Q_A} = \frac{T_0 - T_e}{T_e} \quad \text{or} \quad Q_A = \frac{T_e W}{T_0 - T_e}$$

Thus, the maximum conceivable performance factor Q_A/Q_a for the specified temperature limits becomes (W cancels)

(d)
$$\eta_p = \frac{Q_A}{Q_a} = \frac{T_e(T_a - T_0)}{(T_0 - T_e)T_a}$$

where T_a is the constant temperature at which heat is supplied in the generator and T_e is the constant temperature in the evaporator. A comparison of the actual performance factor with the value from equation **(d)** indicates how near to (or how far from) perfection the actual cycle is.

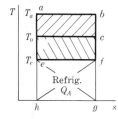

Fig. 17/10

The Planten-Munters (Servel) system of ammonia-water absorption refrigeration cleverly does away with the pumps (no moving parts). There are liquid seals between the condenser (high pressure) and the evaporator (low pressure), but the total pressures on each side of such seals are kept virtually equal by including hydrogen on the low-pressure side. There is enough hydrogen that the partial pressure of the NH_3 is comparable to its pressure in the evaporator of Fig. 17/9; thus it will evaporate and do refrigeration. A gas flame is used in the generator.*

Other substances are used in absorption systems; examples are methylene chloride as the refrigerant and dimethyl ether or tetraethylene glycol as the absorbent; water as the refrigerant and a lithium bromide salt solution as the absorbent; water as the refrigerant and lithium chloride brine as the absorbent.

* See the second or third edition of this text for more detail.

It is noteworthy that manufacturers of air conditioning equipment are marketing and installing lithium bromide absorption generators that operate entirely upon solar energy. The generator sizes vary from 20 to 1800 tons and are of single and multiple stage. The water is solar heated to temperatures within the range of 165–200°F, by means of solar collectors, before entering the concentrator of the generator.

Because of its important role in the use of solar energy, a schematic depicting the principle of operation of a lithium bromide generator is shown in Fig. 17/11 with a brief discussion following.

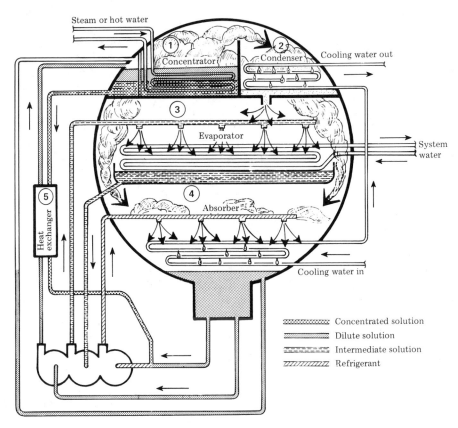

Fig. 17/11   *Principle of the Absorption Cycle.* The production of chilled water from heat (solar energy) using lithium bromide and water as the refrigerant. (*Courtesy Trane Air Conditioning Company.*)

Principle of operation

Evaporator: System water flows through the evaporator tubes. The refrigerant is sprayed over the surface of the tubes and vaporizes, removing heat from the system water.

Absorber: The refrigerant that was vaporized in the evaporator is "absorbed" by a spray of lithium bromide solution. The heat given up, as the vapor is condensed and absorbed, is removed by the cooling water that flows through the tubes in the absorber.

Concentrator: Steam or solar heated water, flowing through the tubes, provides the heat required to separate (boil out) the refrigerant from the dilute solution received from the absorber. The refrigerant vapor travels to the condenser and the concentrated solution returns to the absorber.

Condenser: Cooling water flowing through the tubes condenses the refrigerant coming from the concentrator. The condensed refrigerant flashes through an orifice to the evaporator where the cycle is repeated.

For such air conditioning systems, a salt absorbent and water refrigerant has certain advantages, not the least of which is the complete absence of danger from leakage of the refrigerant. The low efficiency of the NH_3-H_2O system is due in part to concomitant and *unavoidable evaporation in the generator of the absorbent H_2O.* If, say, lithium bromide or other solid salt is the absorber, no energy is lost in evaporating it and a higher efficiency is obtained. Properties of binary mixtures (NH_3-H_2O) and salt solutions are found in the literature.[12.1] A treatment of such mixtures is beyond the intended scope of this book.

GAS REFRIGERATING CYCLE 17.13

Since the temperature of a gas during an isentropic expansion may undergo a large drop, this phenomenon may be used to obtain low temperatures for refrigerating purposes. The working substance in such gas cycles is usually air, especially appropriate where leakage may occur into spaces occupied by living beings, say in air conditioning an airplane. Other gases may be advantageous under other circumstances. The disadvantage of gas refrigeration is its low COP; hence it is used only where there is a substantiating reason.

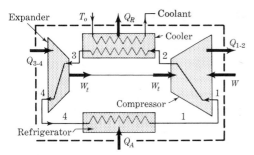

Layout for Air Refrigeration. All the output of the expander or turbine W_t is used to help drive the compressor; $W_{net} = W$.   **Fig. 17/12**

Diagrammatically, the operation is pictured in Fig. 17/12 and the processes are shown in Fig. 17/13. The ideal cycle, consisting of two isentropic and two constant pressure processes, is a reversed Brayton cycle (§ 15.2). The compression process

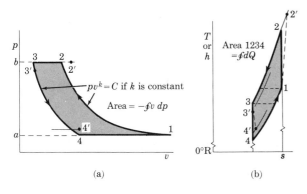

Gas Refrigeration (Reversed Brayton) Cycle. Heat from 2 to 3 (or 2′ to 3′) goes to the sink; therefore, T_0 must be less than T_3 (or $T_{3'}$).   **Fig. 17/13**

1-2, ideally isentropic, may actually closely approach an irreversible adiabatic 1-2′. For a steady-flow compressor, the work is given by an energy balance from an energy diagram or by equation (4-10), which integrates to

$$(4\text{-}10) \qquad W_c = -\Delta h - \Delta K + Q_{1\text{-}2} = h_1 - h_2 + K_1 - K_2 + Q_{1\text{-}2} \qquad [\Delta P = 0]$$

a negative number; ΔK may be approximately zero. Also, for an internally reversible process, the compressor work, represented by the area a-1-2-b, Fig. 17/13(a), is $W_c = -\int v \, dp$. See equations (7-5) and (7-9).

From the compressor the gas flows through the cooler, which reduces its temperature to t_3, perhaps a bit above the available sink temperature. Actually, of course, a pressure drop to some pressure $p_{3'}$ occurs. Recalling that an ideal gas undergoes no change of temperature during a throttling process, we see why an expander (reciprocating or rotary) is necessary. In the expander the ideal is 3-4, with an actual expansion represented by 3′-4′, which is necessarily with increasing entropy *if it is adiabatic*. State 4′ allows for a pressure drop as the gas does its refrigeration 4′-1. Observe that the illustration suggests clearly the loss of refrigeration owing to the irreversibility here. The work of the expander W_t helps drive the compressor, so that the net work W that must be supplied to the cycle is $W = |W_c| - W_t$ as a positive number. If the expander is a steady-flow turbine, equation (4-10), given above for states 3 and 4 or 3′ and 4′, gives the work W_t. If the gas is air, it may be expanded to the pressure of the surroundings at 4, pass into the space to be cooled, and allowed to pass on to the atmosphere—in which case, the compressor draws its supply from the atmosphere (in jet-propulsion engines, the compressed air may be bled from the main compressor). Or the system may be closed, with the working substance being at all points at pressures greater than that of the surroundings; in this case, with the working substance being denser, more refrigeration is obtained from a particular volume of intake at the compressor (for an ideal gas, the refrigeration per pound between specified temperatures is independent of its pressure).

In any case, the net work W is $-\oint dQ$ as a positive number. For isentropic processes 1-2 and 3-4 and $\Delta K \approx 0$,

(a) $$W = -\oint dQ = h_2 - h_3 - (h_1 - h_4) = c_p(T_2 - T_3 - T_1 + T_4)$$

$$[\text{CONSTANT } c_p]$$

for unit mass, as a positive number. For the refrigeration of $h_1 - h_4$, the COP = Q_A/W is

(b) $$\gamma = \frac{h_1 - h_4}{h_2 - h_1 - h_3 + h_4} = \frac{T_1 - T_4}{T_2 - T_1 - T_3 + T_4}$$

$$[\text{CONSTANT } c_p]$$

As previously mentioned, the reversed Stirling cycle is also used for refrigeration (§ 8.11).

VARIATIONS OF BASIC REFRIGERATION CYCLES   17.14

The basic reversed cycles may be modified for several reasons, principally to gain a better COP and/or a lower refrigerating temperature. As pointed out in § 14.6 (see also §§ 15.10 and 17.19), work input may be saved and volumetric efficiency may be improved by using two or more stages of compression with intercooling between stages. Since the refrigerant is at a temperature lower than that of the environment much of the time, all or part of the intercooling may be regenerative, as illustrated in Fig. 17/14 for a situation requiring a large temperature difference between the condenser and the evaporator. If it should happen that the temperature at *b* (or *b′*) is significantly above T_0, some cooling may be done with, say, sink water [not so pictured in Fig. 17/14(a)]. As shown, the first expansion valve *A* throttles the refrigerant to the intermediate pressure between stages; the *m* lb of vapor, the amount being indicated by the quality at *r*, Fig. 17/14(b), mixes with the superheated vapor at *b* (or *b′*) in the mixing chamber *M*, and reduces the temperature at entry to the second stage to some value t_c. If a temperature lower than this t_c is desired, a measured amount of liquid refrigerant could be made to pass to *M*.

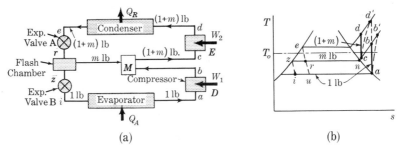

(a)   (b)

Two-Stage Compression with Regenerative Intercooling. The discharge from   **Fig. 17/14**
the low-pressure stage at *b* may also be discharged into the flash chamber,
with a resulting lower temperature at *c*. Moreover, refrigeration at two temperature levels can be obtained with this plan, at t_z and t_i.

Since the mass is different in different parts of the system, one must decide where to let it be the unit mass. In Fig. 17/14, the choice is unit mass in the evaporator. The work in compressor *E* is $(1 + m)(h_d - h_c)$ in, say, Btu/lb in the evaporator $(\Delta K \approx 0)$. Also $1 + m$ lb is throttled through *A* and 1 lb through *B*, if only the vapor at *F* is used for regenerative cooling. The states for the following equations are defined in Fig. 17/14(b).

(a) $$h_b + mh_n = (1 + m)h_c \qquad \text{[ENERGY BALANCE, } M\,]$$

(b) $$(1 + m)h_e = h_z + mh_n \qquad \text{[ENERGY BALANCE } A, F\,]$$

(c) $$Q_A = h_a - h_i = h_a - h_z \qquad \text{[REFRIGERATION]}$$

(d) $$W = h_b - h_a + (1 + m)(h_d - h_c) \qquad \text{[WORK, } W_1 + W_2]$$

Theoretically the lowest usable refrigerating temperature for a particular refrigerant is its triple point (§ 3.6)—throttling below which would produce a solid. If the coldest temperature is low enough, a series arrangement of refrigerating cycles, each with a different refrigerant, called a ***cascade system***, may be used. A binary-vapor system is shown schematically in Fig. 17/15, but additional subsystems may be used. The successive refrigerants should be such that the triple-point temperature of the "higher" one is lower than the critical temperature of the "lower" one. Let m_1 be the mass rate of circulation of refrigerant in the evaporator B, Fig. 17/15, a quantity determined by the amount of refrigeration required; then the mass rate of circulation in the upper cycle $\dot{m}_2$ must be such that

(e) $\dot{m}_1(h_b - h_c) = \dot{m}_2(h_e - h_1)$

the energy balance for the condenser-evaporator A. The actual state b' (and others) may be used in **(e)** if appropriate. The irreversibility accompanying the necessary temperature drop in the condenser-evaporator may be the most significant one in the cascade system.

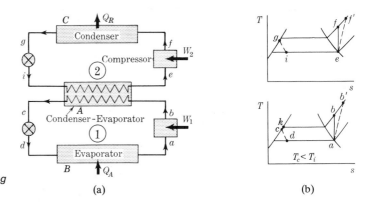

Fig. 17/15   *Cascade   Refrigerating*
System.
                                              (a)   (b)

17.15   CRYOGENICS

Cryogenics has to do with phenomena and operations that occur at temperatures below about $-100°F$, an arbitrary dividing line defining "very cold". Only about a generation ago, liquefied gases (those that act nearly ideally at environmental temperatures) were produced principally in laboratories. But now, the many industrial uses that have evolved, not to mention their utility in rocketry and frozen foods, have made the production of liquefied gases a large and rapidly growing industry. To obtain the low temperatures needed for this and other purposes, we may use the cascading idea discussed above. In one such arrangement for liquefying N_2, ammonia is used to liquefy ethylene (at 19 atm), which in turn liquefies methane (at 25 atm), which liquefies the N_2 (at 18.6 atm).[17.14] Each one arrives at its coldest temperature via an isenthalpic expansion to 1 atm; that portion of the N_2 that remains a liquid after throttling to 1 atm is the production. The work required for obtaining a unit mass of gas by this system may be significantly less than in the systems described next, but the mechanical complications make the required machinery much more expensive. Incidentally, the lowest refrigerating temperature attainable by compression and two-phase throttling is with the use of nitrogen, whose triple-point temperature is $-387°F$. No other substance has a lower triple-point temperature and also a critical temperature greater than $-387°F$.

A phenomenon widely used in the commercial attainment of low temperatures is the Joule-Thomson effect (§ 11.19), from which we recall that for a temperature drop to occur on throttling, the initial temperature should be less than the inversion temperature; and the Joule-Thomson coefficient $\mu = (\partial T/\partial p)_h$ is positive. The maximum temperature drop, for a particular initial temperature, occurs when the throttling starts at a state on the inversion curve, Fig. 11/4. For nitrogen, for example, such a state is defined by 80°F and 375 atm.[1.16] Since a fluid's temperature must be reduced below its critical temperature in order for it to exist as a liquid, the throttling process must start from a sufficiently high pressure that the final temperature meets this requirement. The lower the temperature before throttling, the less the pressure ratio must be to obtain liquid. Also, the fluid's initial temperature must be below its maximum inversion temperature, which is no problem for nearly all gases, whose inversion temperatures are well above normal atmospheric temperatures; the exceptions are H_2 (363°R) and He (~ 72°R).

THE LINDE LIQUEFIER   17.16

The simplest equipment for liquefying gases is shown schematically in Fig. 17/16. The gas being processed is presumed to be clean; for example, atmospheric air must be cleansed of its CO_2, H_2O, and other contaminants that may solidify en route to F (perhaps by solidification at chosen points in its passage). Follow both the layout and the Ts diagram. The compressor A (in several stages) receives the gas at 1 and delivers it at high pressure, perhaps as much as about 200 atm. If the compressor's discharge temperature at 2 is greater than that of the available sink, an after-cooler reduces it, ideally to the sink $T_0 = T_C$. The regenerative heat exchanger D lowers the temperature of the compressed gas to T_3, at which point it is throttled, say, to atmospheric pressure at 4. For 1 lb at state 3, there will be y lb of liquid and x lb of vapor at state 4.

At start-up, no liquid is likely to be produced; but since the temperature of the gas at 4 is less than at 3, it is circulated back through the system 4-FG-1. This action cools the gas passing through C to 3, which further reduces T_4. Eventually, liquid (or solid, if the temperature T_4 is below the triple point) is produced in F, and if it is removed at a steady mass rate, the whole operation becomes virtually steady state, the condition for which we shall make the thermodynamic calculations.

In actual problem work, one may choose to make energy balances for several systems in Fig. 17/16(a): the compressors, the after-cooler, the throttling valve E, and so on (but not all would be independent equations). The following energy balance is for D, E, F, bounded by heavy dashes, assumed to be an adiabatic control volume ($\Delta K = 0$, or stagnation enthalpies):

(a)
$$h_C = y h_H + (1 - y) h_B$$

from which, the mass of liquid per pound compressed is

(b)
$$y = \frac{h_B - h_C}{h_B - h_H}$$

If an estimate of the heat Q for the control volume were made, it would appear on the left side of **(a)**, since the heat would flow into the system. The state in F is 1 atm and saturation temperature of the two-phase mixture; h_H is thus fixed. The temperature at C is at best the sink temperature, as is also T_B (which actually is some 5 or more degrees below T_C); the pressure at B is 1 atm and h_B is thus fixed. The only variable is therefore h_C, which is a function of pressure (as well as T); the production of liquid gas per unit work increases with increase of pressure $p_2 \approx p_C$. For air, Ruhemann[17.14] reports: at $p_2 = 50$ atm, $W = 2.52$ kW-hr/kg; at 100 atm, $W = 1.56$ kW-hr/kg; at 200 atm, $W = 1.16$ kW-hr/kg of liquid air.

An increased production per unit of work W may also be achieved by using vapor refrigeration to lower T_3 and therefore $h_3 = h_4$. Moreover, if refrigeration is introduced, the pressure p_2 of the gas does not need to be so high to obtain liquid, which saves work, but not necessarily the work per unit mass of liquid; the work of the refrigeration cycle is chargeable against the production of liquid. There should be a net improvement because the vapor refrigeration cycle is basically a more efficient way of obtaining "cold" than the isenthalpic process from a gaseous state 3. The refrigeration may be done after a partial regenerative cooling and followed by additional regenerative cooling (regenerator D, Fig. 17/16, in two parts).

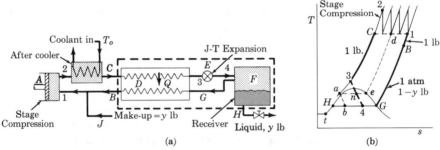

(a)   (b)

Fig. 17/16   *Gas Liquefaction—Linde.* The cycle is sometimes modified by doing refrigeration (by another system) on this working substance between D and E, in order to lower its temperature at its entry to E; say to state a in (b) (or according to the light dotted line for a lower compression pressure). The ideal compression is sometimes taken as isothermal, the dotted line 1-C.

Another means of reducing the work per unit mass of liquid is to use a modified Linde plan with two throttling stages, Fig. 17/17. Most of the cooling effect is achieved during 3-4, with some 80–90% vapor returning 5-6, depending on the intermediate pressure in F, and needing to be compressed only from this intermediate pressure to p_1; only a small portion $(1 - m - y)$ needs to be compressed from 1 atm.

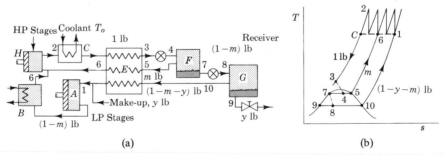

(a)   (b)

Fig. 17/17   *Two Joule–Thomson Expansions.* The pressure 2-3 in E may be some 200 atm; in F, 20–50 atm; in G, 1 atm.

In the manufacture of solid CO_2, whose critical state is defined by 88°F and 72.9 atm, much the same layout as in Fig. 17/16(a) is used (perhaps a cooling curve as the dotted one *den*); after throttling, it snows $CO_2(s)$ in the receiver *F*. The *Ts* diagram will look different because the final temperature T_4 must be below the triple-point temperature T_t in Fig. 17/16(b) of −69.9°F (see Fig. 3/1). At 1 atm, $t_{sat} = 109.3°F$. On the high-pressure side, if some cooling is done by refrigeration, a greater percentage of solid CO_2 is obtained. If a sink temperature below 88°F is available, regenerative heat exchange is not necessary for the production of $CO_2(s)$, but without it the efficiency is reduced.

CLAUDE LIQUEFIER   17.17

If an expander, ideally isentropic, is inserted in place of the expansion valve *E*, Fig. 17/16(a), a larger quantity of liquid is obtained and the work generated may be used to help drive the compressor. Both the turbine and reciprocating expanders have been used, but there have generally been operating troubles with the two phases in the engine, especially in reciprocating ones. For this reason, the expander *M* receives a proportion of the fluid at an intermediate state, such as *a*, Fig. 17/18(a), after a partial regenerative cooling in *D*; the gas from *M* mixes with the gas *d* leaving *J*; the greater quantity of cooling gas cools the compressed gas to lower temperatures at *e* and 3 than would otherwise be attained. Also, the higher pressure may be lower than it otherwise would be, but this change does not necessarily result in improved efficiency. Assuming steady-state operation, one may write energy balances for several chosen subsystems; for example: the various heat exchangers, the receiver *F*, the mixing at *K*, the expander, and overall *DKMJF*. The overall balance in the form of (energy in) = (energy out) for adiabatic boundaries is

(a) $$h_C = (1 - y)h_B + yh_H + m(h_a - h_b)$$

where $W_t = m(h_a - h_b)$. By an empirical approach, it has been found that for particular values of p_2 and m, there is a certain T_a at which the compressed gas should enter the expander for maximum production.[17.14] For air at $p_2 = 40$ atm, then $T_a \approx -80°C$, with 80% of air through the expander; for $p_2 = 200$ atm, then $T_a \approx$ room temperature, with 55% of air through expander. The latter plan is the more productive and, interestingly, the exchanger *D* may be eliminated (but more stages of compression are required).

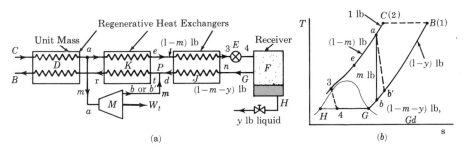

(a)   (b)

Gas Liquefaction—Claude. Since George Claude, a Frenchman, first successfully used an expander, processes that include one are often given his name. Several arrangements are possible. There may be more than one expander. Various cleaning operations and controls are not indicated.   **Fig. 17/18**

With the widespread use of liquefied gases, the problem of storage and transportation had to be satisfactorily solved. A dewar, named after its inventor Dewar, is a vessel within a vessel with a vacuum between. Filling the evacuated space with a powder increases its insulating property. So-called "Superinsulation" (a trade name) consists of alternate layers of aluminum foil and glass fiber mats (40 to 80 layers per inch) filling the evacuated space. Aluminized Mylar film is used similarly.

17.18   MINIMUM WORK FOR LIQUEFYING GASES

Given a gas at environmental conditions, the basic objective in liquefying it is to cool it until it has condensed, say along path 1-*bGH*, Fig. 17/18. The logic of deciding upon the numerically minimum work is no different from that in § 17.2, equations (e) and (f). It is $-\Delta \mathscr{A}_f$ for steady flow, stagnation enthalpies, or

(a)
$$W_{rev} = -\Delta \mathscr{A}_f = -[h_H - h_1 - T_0(s_H - s_1)]$$

where T_0 is about the same as T_1. The ratio W_{rev}/W', where W' is the actual work expenditure, is analogous to an engine efficiency for a power engine, and it gives an idea of how good the liquefaction process is. Since this ratio for the traditional systems is quite low (~8% for a Linde system)[17.14] there is much irreversibility and much room for improvement.

17.19   STAGE COMPRESSION

The discussion of compressors in Chapter 14 is applicable to each individual stage in a multistage compressor, except as the equations apply only to an ideal gas; *pv* and *Ts* diagrams for a three-stage compression are shown in Fig. 17/19. For the imperfect gas, equations (4-10) or (4-11) apply; for stagnation enthalpies or $\Delta K \approx 0$,

(a) $$W = -\Delta H + Q \qquad \text{or} \qquad W = -\Delta H$$
$$[\text{ADIABATIC}]$$

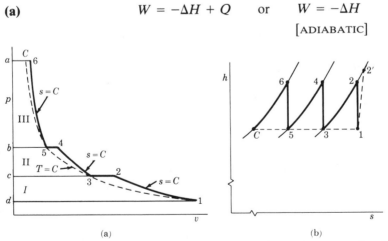

(a)   (b)

Fig. 17/19 *Three-Stage Compression.* Isentropic compression, ideal interstage cooling. If an *hs* diagram to scale is available, the equal stage works may be approximated by trial. On the *pv* diagram, *d*-1 is the intake, 2-*c* discharge from stage 1; stage II intake is *c*-3, discharge is 4-6; and so on. No pressure drop between stages is assumed. Only compression and interstage cooling (2-3, 4-5) can be shown on *Ts* plane, cooling 6-*C* in the after-cooler.

The minimum total compressor work for an ideal gas, with the same compressor efficiency for each stage, and with intercooling between stages to the original intake temperature T_1, occurs when the work of each stage is the same. With these boundary conditions and with no pressure drops between stages, the intermediate pressure between two stages for equal works is $p_i = (p_1 p_4)^{1/2}$, where p_1 is the intake pressure for the lower-pressure stage and p_4 is the discharge pressure of the higher-pressure stage—applicable to any two consecutive stages; $p_i = p_2 = p_3$. Fig. 17/19, for stages I and II; $p_i = p_4$ for stages II and III. The equation for p_i may also be written $p_i/p_1 = (p_4/p_1)^{1/2}$. This relationship may be used as a first approximation for suitable intermediate pressures when the gas is not ideal.

If the isothermal compression is taken as ideal, then the work is

(b) $$W_T = H_1 - H_C - T_1(s_1 - s_C) \qquad \text{[ISOTHERMAL]}$$

where the subscripts are also defined by Fig. 17/18; $T_1(s_1 - s_C)$ is the heat. Change signs in **(b)** if a positive number is desired. The ideal work of equation **(b)** may be converted to actual work by an isothermal (instead of adiabatic) compression efficiency η_{ct}, values of which are available from engineering studies; $\eta_{ct} = W_T/W'$.

SEPARATION OF BINARY MIXTURES 17.20

As previously mentioned, it is beyond the scope of this text to go into the thermodynamic details for mixtures, except where the mixture is acting as a pure substance. However, a brief picture of the rectification process will serve to satisfy a momentary curiosity, to explain a principle used to separate gases, and to point again to a large area of thermodynamics that is covered elsewhere.[1.16,17.14]

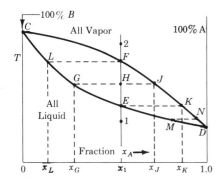

Temperature-Composition Diagram for a Binary Mixture. This may represent air, considered as O_2 and N_2, or the NH_3–H_2O of §17.12, and so on. The shapes of the vapor and liquid curves for all binary mixtures are not pictured here.   **Fig. 17/20**

Consider a binary mixture of substances A (say, N_2) and B (say, O_2), with x_1 fraction of A and $1 - x_1$ fraction of B, as shown in Fig. 17/20 on a temperature-composition chart. If the temperature is below T_E, the mixture is all liquid; if it is above T_F, it is all vapor. If the temperature is raised from T_1 to T_H, with internal equilibrium at all times, vaporization starts at E and the initial macroscopic quantity of *vapor* has the composition indicated by K, or x_K of the more volatile component. Boiling points at 1 atm: N_2, 139.6°R; O_2, 162.8°R. Observe that this vapor has a much larger fraction of A than does the liquid. At temperature T_F or T_2, the mixture is all gaseous; T_F is the dew point (on cooling at constant pressure from 2). The initial macroscopic quantity of liquid on condensation at F has the composition of x_L of component A at temperature $T_L \approx T_F$; it is much richer in component B. At temperature $T_H = T_G = T_J$, the vapor J, with the composition x_J, is in equilibrium with the liquid G whose composition is x_G of A; H does not represent a state, but for a mixture specified by x_1, the temperature must be some value between F and E in order for liquid and vapor to coexist. The coexisting states G and J are ones for which the chemical potentials μ balance; $\mu_i = (\partial G/\partial n_i)_{T,p}$, for each component i must be the same in both phases (§ 5.26). The characteristic of differing compositions of liquid and vapor mixtures in thermal equilibrium is utilized in the rectification process.

In Fig. 17/21, the compressed gas mixture, already quite cold (§§ 17.16 and 17.17), liquefies C to D, passes through the throttling valve E, and is then directed into the column as mostly liquid L_1. Assume for the moment that the vapor coming from the sump without a state change mixes with this liquid. If the temperature of the liquid is, say, T_E in Fig. 17/20, and the temperature of the vapor is T_J, there is a difference in chemical potentials that causes the composition of the *vapor* to move toward K and the composition of the *liquid* to move toward G. By letting this process proceed gradually as the liquid and vapor move in counterflow, the vapor gradually becomes rich in A (N_2 for air); but for liquid *air* at level P, Fig. 17/21, the maximum percentage of N_2 is 93%, which is the equilibrium composition of the vapor with liquid air (21% O_2, 79% N_2) at 1 atm.[17.14] This situation is suggested by equilibrium compositions M and N, Fig. 17/20. To get purer N_2, extend the column, Fig. 17/21, say from P to Q. The vapor V_2 may then be condensed, and returned to the column as L_2, called *reflux*, which by virtual differential distillation between levels P and Q, may approach pure N_2. There are numerous complications and many plans in use which are not even suggested by Fig. 17/21. The column, perhaps under a pressure greater than 1 atm, by its design may produce impure N_2 and pure (almost) O_2; or impure O_2 and pure N_2; or both pure O_2 and N_2; and some designs are such that the rate atmospheric gases, as argon, neon, xenon, and so on, are removed at certain stages of the operation. Reference [17.14] is concerned primarily with liquefying and rectifying air.

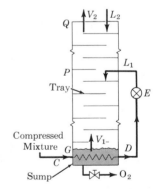

Fig. 17/21   *Rectifying Column.* The trays increase the interface between liquid and vapor, thereby increasing their rates of interactions.

17.21   CLOSURE

While this chapter contains no new thermodynamics, we observe that there are numerous phenomena that one must understand in order to apply the principles successfully. In the application of the ideas of this chapter, we should keep in mind that the irreversibility, measured by the increase of entropy, of an isenthalpic process of a liquid is *less* than that of a vapor. Therefore, the throttling of liquids is to be preferred, when throttling is used, even though it often seems expedient now to throttle from the gaseous state in some liquefying processes. See reference [0.9] for the properties of a number of refrigerants.

PROBLEMS

SI UNITS

17.1 A refrigeration system operating on the reversed Carnot cycle rejects 5800 kJ/min. The minimum and maximum temperatures are 250 K and 345 K, respectively. Sketch the *Ts* diagram and find **(a)** the power input required, **(b)** the refrigeration tonnage developed (211 kJ/min-ton), **(c)** the available part of the heat rejected for a sink temperature of 5°C.

17.2 A reversed Carnot cycle has a refrigerating COP of $\gamma = 4$. **(a)** What is the ratio T_{max}/T_{min}? **(b)** If the work input is 6 kW, what will be the maximum refrigeration effect, kJ/min and tons (1 ton = 211 kJ/min)? **(c)** What are the COP and the delivered heat if this cycle is used for heating?
Ans. **(a)** 1.25, **(b)** 1440 kJ/min, **(c)** COP = 5.

17.3 There are removed 528 kJ/min of heat from a body by a refrigerator operating between the limits of 244.5 K and 305.5 K. If its coefficient of performance is three fourths of that of a Carnot refrigerator working between the same temperature limits, find **(a)** the heat rejected and **(b)** the power input, kW. **(c)** What are the COP and the heat if this device is used to deliver heat?

17.4 The power requirement of a Carnot refrigerator in maintaining a low temperature region at 238.9 K is 1.1 kW per ton. Find **(a)** COP, **(b)** T_2, and **(c)** the heat rejected. **(d)** When this device is used to deliver heat, what is its COP?
Ans. **(a)** 3.2, **(b)** 313.3 K, **(d)** 4.2.

17.5 A Carnot refrigerating system picks up heat at 272 K and requires a power input of 1.75 kW/ton. Find **(a)** the COP and **(b)** the temperature at which heat is discharged. **(c)** If the upper temperature in **(b)** is changed to 311 K, what will be the power input, kW/ton?

17.6 A manufacturer of Carnot refrigerators has the choice of varying either the evaporator temperature $T_1 = 250$ K or the condenser temperature $T_2 = 500$ K by ± 25 K; no restrictions are placed on the operating temperatures. In the best interest of economy, what would you recommend to this manufacturer?

17.7 Air is used in the basic system of a cryogenic refrigeration system. Saturated air vapor enters the ideal compressor at 3 atm and leaves at 50 atm to enter the constant pressure condenser. Pressurized boiling nitrogen condenses the air to 124 K at which point it is throttled to the evaporator pressure of 3 atm. Sketch the *Ts* diagram, use Item B 26, and for a 5-ton (211 kJ/min-ton) capacity find **(a)** the mass of air circulated, kg/min, **(b)** COP, **(c)** heat absorbed by the boiling nitrogen, kJ/min.
Ans. **(a)** 9.51 kg/min, **(b)** 1.13, **(c)** 1991 kJ/min.

17.8 Prove (demonstrate) that the COP(γ_h) of a reversed cycle used for heating, is equivalent to its COP(γ_c), when used for refrigeration, plus one; that is, $\gamma_h = \gamma_c + 1$.

17.9 An oxygen cryogenic refrigeration system circulates 10 kg/min of O_2. The compressor receives saturated oxygen vapor at 4 atm and discharges it at 60 atm, 256 K. Pressurized boiling nitrogen reduces the temperature of the oxygen to 155 K in the condenser; the oxygen is then throttled to the evaporator pressure of 4 atm. Sketch the *Ts* diagram, use Item B 27, and find **(a)** the ton (211 kJ/min-ton) rating of the oxygen system, **(b)** the COP, **(c)** the compressor efficiency, **(d)** the compressor displacement, ℓ/min.

17.10 Assume that there is a planet where the average environmental temperature is 82°C and where the evaporator temperature 30°C may be utilized for cooling. A vapor-compression system is to be designed using water as the refrigerant, with saturated vapor at 100°C leaving the compressor and with liquid at 95°C ($h_f = 395.5$ kJ/kg) entering the expansion valve. For a cooling capacity of 10 tons (211 kJ/min-ton) in an ideal cycle, find **(a)** the mass of refrigerant circulated, **(b)** the required power, and **(c)** the COP. On the basis of the actual cycle with a compression efficiency of 78% and a volumetric efficiency of 75%, find **(d)** kW/ton, **(e)** V_D, and **(f)** the temperature of the vapor from the compressor.

17.11 Water for drinking and air conditioning purposes is chilled from 25°C(h_f = 105 kJ/kg) to 10°C(h_f = 42 kJ/kg) by a steam refrigeration unit; the 25°F water is admitted to a chamber in which low pressure is maintained by means of steam jet ejectors. Part of the incoming water flashes to vapor and the rest is chilled to 10°C. Find **(a)** the pressure maintained in the low pressure chamber, **(b)** the mass of water vapor flashed per kilogram of entering water.

17.12 There are removed 183.52 m³/min of vapor from the water evaporator of a vacuum refrigeration system; the warm water enters the evaporator at 17.8°C(h_f = 74.6 kJ/kg) and chilled water leaves at 10°C(h_f = 42 kJ/kg, v_g = 106 m³/kg); makeup water enters at 17.8°C. Determine the refrigerating capacity.   *Ans.* 20 tons.

17.13 An air-refrigerating system with a 10-ton capacity consists of a centrifugal compressor, an aftercooler, and an air turbine. The turbine is directly connected to the compressor. Both processes, compression and expansion, are irreversible adiabatics. At the compressor inlet, p_1 = 82.74 kPaa, t_1 = 21.1°C; at compressor exit, $t_{2'}$ = 90°C; at the turbine inlet, p_3 = 144.79 kPaa, t_3 = 37.8°C; at turbine exit, $t_{4'}$ = 0°C. Sketch the flow diagram and the Ts plane; steady flow processes; ΔK = 0. Find **(a)** the compressor efficiency, **(b)** the turbine efficiency, **(c)** the mass flow of air, **(d)** the net power, and **(e)** the COP.

Ans. **(a)** 74%, **(b)** 82.4%, **(c)** 99.5 kg/min, **(d)** 51.83 kW, **(e)** 0.678.

17.14 In a "dense-air" refrigerating system, operating on a reversed Brayton cycle, the air enters the compressor from the refrigerator at 344.74 kPaa, 4.44°C; it is compressed to 1,516.86 kPaa, where the air passes through a heat exchanger and is cooled to 29.44°C. The air then passes through the expander, goes through the refrigerator, and repeats the cycle. Amount of refrigeration desired is 10 tons. **(a)** Determine the COP and the power required for operation. **(b)** If the compressor efficiency

and expander efficiency is each 78%, compute the COP and power; $\Delta p \approx 0$ in heat exchange processes.

Ans. **(a)** 1.9, 18.5 kW, **(b)** 0.534, 66 kW.

17.15 Air is to be liquefied by the Claude process; see Fig. 17/18. It is compressed to 50 atm, from 1 atm, 25°C, arriving at C from the aftercooler at t_c = 25°C; p_F = 1 atm. For the first trial, assume that the temperature at the expansion valve is t_3 = −154°C, and with help from Item B 26, we assume t_a = −90°C, from which it is expected that the expander discharge contains little or no liquid; t_B = 20°C; negligible pressure drops in exchangers; steady state, steady flow. **(a)** For an expansion of unit mass to 1 atm and η_t = 75% for the expander M, determine the adiabatic work and the enthalpy and temperature of the exhaust b'. Compute **(b)** the mass fractions y and m (y = fraction of liquid), **(c)** the enthalpies and temperatures of the stream at t and r. Make energy balances for systems not used in the foregoing as a check. *Suggestion:* Try an overall energy balance and one for the receiver F first. Solve the problem using the units of Item B 26; then convert final answers to desired units.

Ans. **(a)** 18.7, 48.7 cal/gm, −192°C, **(b)** 0.152, 0.74, **(c)** −47.8 cal/gm, −192°C, −139 cal/gm, −139°C.

ENGLISH UNITS

17.16 A reversed Carnot cycle, acting as a heat pump, is to supply 16,500 Btu/min to heat rooms of a building to 75°F when the outside temperature is 60°F. **(a)** Since the processes are externally reversible, compute the heat obtained from the atmospheric air and the horsepower required to operate the cycle. **(b)** If the heating is done directly by an electric coil, what amount of electricity expressed in horsepower is consumed?

Ans. **(a)** 16,037 Btu/min, 10.92 hp, **(b)** 389.

17.17 In a vapor-compression refrigerating system using SO_2, saturated vapor at 20°F (state 1) is received by the compressor, and the actual state at discharge is 100 psia and 260°F (state 2'). The liquid SO_2 enters the expansion valve at 100°F (state 3). For both the ideal and actual cycles, sketch the Ts and *ph* diagrams and find **(a)** the required hp/ton, **(b)** the COP, **(c)** the compression efficiency η_c and **(d)** the mass flow of SO_2 required for a 20-ton capacity. See Item B 36.

Ans. **(a)** 1.012, 1.214, **(b)** 4.66, 3.89, **(c)** 83.4%, **(d)** 28.6 lb/min.

17.18 Assume 50 lb/min of Freon F12 refrigerant vapor at 15 psia −10°F are drawn into a compressor (point 1) and compressed adiabatically to 300 psia, 230°F. The liquid refrigerant leaves the condenser and enters the expansion valve saturated at 300 psia. Sketch the *TS* and *ph* diagrams for the cycle, use Item B 35, and find **(a)** ton rating of the system, **(b)** COP, **(c)** compressor efficiency, **(d)** work, *hp*.

17.19 It is desired to replace an existing oil heating system rated at 90,000 Btu/hr (input) with an NH_3 reversed cycle in order to benefit from the refrigeration; the existing system is 80% efficient. For the NH_3 cycle, the evaporator pressure is 37.7 psia and the condenser pressure is 300 psia; the refrigerating effect is 431.7 Btu/lb NH_3 and there is no subcooling at the condenser outlet. The NH_3 leaves the compressor at 340°F; the compressor displacement is 18.63 cfm. Sketch the *TS* and *ph* diagrams. Assume no change in the heating requirements and determine **(a)** the NH_3 circulation requirement, based upon the rated output of the existing heating system, **(b)** the compressor efficiency, **(c)** the tons of refrigeration for the same cycle, **(d)** the volumetric efficiency, **(e)** COP heating, **(f)** COP cooling.

Ans. **(a)** 2 lb/min, **(b)** 81.5%, **(c)** 4.137 tons, **(d)** 80%, **(e)** 3.57, **(f)** 2.57.

17.20 An automobile engine idling at 1700 rpm drives the Freon 12 compressor of its air conditioner. Saturated vapor enters the compressor at 20 psia and is compressed to 300 psia; no subcooling occurs in the condenser. For the 4.5-ton *ideal* unit, sketch the *TS* and *ph* diagrams and find **(a)** Freon circulated, lb/min, **(b)** power, *hp*, **(c)** COP, **(d)** heat removed by the condenser, Btu/min. **(e)** In the actual case the compressor efficiency is 80%; find the temperature of the Freon at compressor discharge.

17.21 Freon 12 at 0°F and 100% quality enters the suction side of a compressor and is discharged at 120 psia. Liquid refrigerant at 80°F enters the expansion valve. For a flow of 20 lb/min in an ideal cycle, sketch the *ph* and *TS* diagrams and find **(a)** the tons of refrigeration, **(b)** the horsepower input, **(c)** the COP, **(d)** the compressor cylinder size if there are twin cylinders with $L/D = 1$, $\eta_v = 80\%$, and a shaft speed of 450 rpm. **(e)** From the viewpoint of a heating cycle, find capacity (same compressor) and COP.

Ans. **(a)** 5.09 tons, **(b)** 5.78 hp, **(c)** 4.15, **(d)** 4.6 × 4.6-in., **(e)** 1264 Btu/min, 5.15.

17.22 There are circulated 65 lb/min Freon 12 in a vapor-compression refrigeration system. At the compressor suction (point 1) the vapor is saturated at 0°F; at exit the Freon is at 400 psia, 240°F. Liquid refrigerant enters the expansion valve at 100°F. Sketch the *Ts* diagram and find **(a)** the capacity tons, **(b)** the compression efficiency, **(c)** COP, **(d)** the heating available, Btu/min.

17.23 **(a)** What is the heating capacity of a 9 × 9-in., twin-cylinder, single-acting compressor that turns at 380 rpm and has a volumetric efficiency of 85% when saturated liquid ammonia from the condenser enters the expansion valve at 85°F and saturated vapor at 10°F enters the compressor to leave at 180 psia? **(b)** Find the isentropic compressor power. **(c)** Compute each COP based on both heating and cooling.

Ans. **(a)** 16,800 Btu/min, **(b)** 67 hp, **(c)** 5.92, 4.92.

17.24 In a vapor-compression refrigerating system using Freon 12, saturated vapor at 12°F enters the compressor and its actual state at discharge is defined by 160 psia and 150°F. Liquid F12 at 100°F enters the expansion valve. The double-acting compressor turns at 300 rpm, and has a volumetric efficiency of 80%. **(a)** For both the ideal and actual cycles, find the coefficients of performance and the hp/ton. Determine **(b)** the compressor efficiency, **(c)** the mass rate of flow of Freon 12 for a 50-ton capacity, **(d)** the required bore D and stroke L of the compressor if $L = D$. **(e)** Compute the change of steady flow availability through the compressor for a sink temperature of 500°R.

Ans. **(a)** 1.62 hp/ton (actual), **(b)** 79.8%, **(c)** 211 lb/min, **(d)** 11 × 11 in., **(e)** 13.3 Btu/lb.

17.25 In a 10-ton vacuum refrigerating system, warm water at 56°F enters the evaporator wherein the temperature is 44°F. **(a)** What volume of vapor must be removed from the evaporator (neglect the effect of make-up water)? What volume if make-up water enters the evaporator at 90°F? **(b)** If the steam consumption of the ejector is 35 lb/hr-ton of saturated steam at 100 psia, find the Btu input per Btu of refrigeration.

Ans. **(a)** 4005 cfm, 4120 cfm, **(b)** 3.465.

17.26 The ejector of a vacuum refrigerating system maintains a pressure of 0.178 psia in the water evaporator. After doing 100 tons of refrigeration, the water returns to the evaporator at 64°F. The mass in the evaporator remains constant during steady operation with the make-up water entering at 84°F. What volume of vapor must be removed from the evaporator each minute? If the steam consumption of the ejector is 2.75 lb steam/lb vapor, determine the energy consumption and the Btu of refrigeration per Btu into the ejector; the steam enters the nozzle at 100 psia, saturated vapor.

17.27 In an aircraft refrigerating unit using the air cycle, 11 lb/min of air at 103 in. Hg abs and 527°F are bled from the air compressor serving the jet engine of the plane. The air then passes through a heat exchanger, leaving at 100 in. Hg abs and 167°F, at which point it is expanded through a small turbine to 22 in. Hg abs and 15.8°F. Ultimately, the air leaves the plane at 90°F. See figure. Find **(a)** the tons of cooling (refrigeration) with respect to the 90°F. **(b)** If the compressor receives the air in the stagnant state of 31 in. Hg abs and 209°F, and if the small air turbine drives a centrifugal fan for passing coolant air through the heat exchanger, find the horsepower input. **(c)** What is the COP?

Ans. **(a)** 0.978 tons, **(b)** 19.8 hp, **(c)** 0.233.

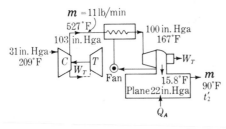

Problem 17.27

17.28 In an ideal NH_3 refrigerating system, two stages of compression are used with regenerative cooling between stages (see figure). Saturated vapor at −49°F enters the *LP* compressor C_1 and is discharged at 40 psia into a cooler wherein circulating water cools it to 80°F. There is next a regenerative cooling to state 3, $p_3 = 40$ psia, $t_3 = 30$°F, at which state the NH_3 enters the HP compressor C_2 and is compressed to 180 psia; for regeneration, m lb, including all vapor, is drawn off after passage through the first expansion valve *A*. Condensation occurs without pressure loss, with saturated liquids entering the expansion valves. For ideal processes, find **(a)** the mass of NH_3 entering the *LP* compressor per lb NH_3 through the condenser, **(b)** the COP, **(c)** the mass of NH_3 through the condenser for 80 tons refrigeration, **(d)** the displacement for each compressor for $\eta_v = 65\%$. **(e)** Let the compressor efficiency be 77% for each and find the hp/ton of refrigeration and the actual COP.

Ans. **(a)** 0.8042 lb, **(b)** 2.51, **(c)** 37 lb/min, **(d)** 1470, 420 cfm, **(e)** 2.44 hp/ton, 1.932.

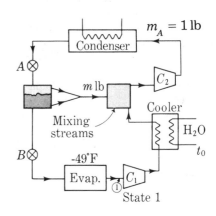

Problem 17.28

17.29 There are produced 100 lb/hr of dry ice by a system that throttles saturated liquid CO_2 at 600 psia to 15 psia in the collection chamber. See figure. Make-up CO_2 gas is supplied at 15 psia, 75°F. Use Item B 31, Appendix, and determine **(a)** the amount of CO_2 liquid throttled each hour, **(b)** the temperature t_1 of the gas at the compressor inlet, **(c)** the work input if the compressor is single stage and has an adiabatic efficiency of 85%, **(d)** the work input if the compression is accomplished in two stages, each stage 90% efficient, and with constant pressure intercooling between stages occurring at 150 psia to 50°F.

Ans. **(a)** 274 lb/hr, **(b)** −25°F, **(c)** 13.05 hp.

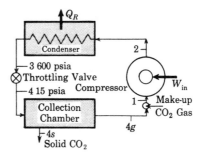

Problem 17.29

18

NOZZLES, DIFFUSERS, AND FLOWMETERS

18.1 INTRODUCTION

Considering the number of fluid flow problems encountered up to this point, you may have wondered what in detail was occurring in the fluid stream. These phenomena are studied extensively in a separate course under the general heading of fluid mechanics. It is fortunate that thermodynamics and fluid mechanics are interrelated since both are usually required for a full investigation of many problems. In this chapter our discussion will be limited to the case of simple one-dimensional flow applied to steady-flow, steady-state devices that either control or meter the flowing fluid.

18.2 SOME BASIC PRINCIPLES

At this point we will present some alternative forms of principles with which you are already familiar. From § 1.22 we have the continuity equation for steady-flow,

$$\dot{m} = \rho A v = \text{constant}$$

Take the logarithm of this equation, differentiate the result, note that $d(\ln A) = dA/A$, and so on, and obtain

(18-1)
$$\frac{d\rho}{\rho} + \frac{dA}{A} + \frac{dv}{v} = 0 \qquad \begin{bmatrix} \text{STEADY-FLOW} \\ \text{STEADY-STATE} \end{bmatrix}$$

Using specific volume $1/v$ in place of density ρ, we have $\dot{m} = Av/v$, which leads to,

(18-2A)
$$\frac{dA}{A} + \frac{dv}{v} - \frac{dv}{v} = 0$$

For an incompressible fluid like liquid water, $dv = 0$ and we have

(18-2B)
$$\frac{dA}{A} + \frac{dv}{v} = 0 \qquad \text{[INCOMPRESSIBE FLUID]}$$

490

Handling the ideal-gas equation similarly, $pv/T = R$ or $p/(\rho T) = R$, we get

(18-3) $$\frac{dp}{p} + \frac{dv}{v} = \frac{dT}{T} \quad \text{or} \quad \frac{dp}{p} - \frac{d\rho}{\rho} = \frac{dT}{T} \qquad \text{[IDEAL GAS]}$$

where $\rho = 1/v$ and $d\rho = -dv/v^2$.

Equations for stagnation properties previously derived are:

(7-14) $$\frac{T_0}{T} = 1 + \frac{k-1}{2}\mathbf{M}^2 \qquad \frac{T}{T_0} = \frac{2}{2 + (k-1)\mathbf{M}^2}$$

(7-2, 7-3) $$\frac{T_0}{T} = \left(\frac{p_0}{p}\right)^{(k-1)/k} = \left(\frac{v}{v_0}\right)^{k-1} = \left(\frac{\rho_0}{\rho}\right)^{k-1} \qquad [s = C]$$

(7-11) $$h_0 = h + K$$

The Mach number is $\mathbf{M} = v/a$, equation (7-13), where a is the sonic speed (§ 18.3).

As usual, the law of conservation of energy is needed. For the case of adiabatic steady flow, $Q = 0$, no shaft work, $W = 0$, *and* $\Delta P \approx 0$, we write

(18-4) $$h_0 = h_1 + K_1 = h_2 + K_2 \quad \text{or} \quad dh + \frac{v\,dv}{kJ} = 0$$

where $K = v^2/2$ for consistent units. Equation (18-4) comes directly from equation (4-9). Consider equation (4-9D),

(4-9D) $$dQ = du + p\,dv + v\,dp + dK + dP + dW$$

If the process is *reversible*, $du + p\,dv = dQ$ and these terms cancel. Then if we let $W = 0$, $v = 1/\rho$, and $dK = v\,dv$, equation (4-9) reduces to

(18-5) $$\frac{dp}{\rho} + v\,dv + g\,dz = 0 \qquad \text{[CONSISTENT UNITS]}$$

an equation for unit mass known as Euler's equation for steady flow. Observe that Bernoulli's equation (4-24) reduces to this form if expressed in terms of differentials. Another form of Euler's equation where ΔP is zero or negligible is

(18-6) $$\frac{dp}{\rho} + v\,dv = 0$$

$$\text{[CONSISTENT UNITS]}$$

Historically, the Euler equations were derived from principles of mechanics, but the results by the simpler thought processes of thermodynamics are just as useful. If each term in equation (18-6) is multiplied by ρA, the equation is recognized as a momentum equation for mass flow ρAv and force $A\,dp$; hence it is commonly called the *momentum equation*. In passing, note that by integrating the terms of equation (18-5) for an *incompressible* fluid, we get Bernoulli's equation, § 4.18.

From $\dot{m} = \rho A v$, we have $\rho v = \dot{m}/A$, which, used in equation (18-6) gives $dp + (\dot{m}/A)\,dv = 0$. Integration for constant area A gives

(18-7) $p_1 + \dot{m}v_1/A = p_2 + \dot{m}v_2/A = \text{constant}$

[FRICTIONLESS STEADY-FLOW, AREA CONSTANT]

The impulse-momentum principle was convenient in obtaining the thrust for a jet. [By rearranging (18-7), you will find it there.] From § 15.12, we recall that for one-dimensional flow in a horizontal straight line and negligible body forces the vectors of force and momentum are collinear; also that for no change in the momentum content of the control volume, the rate of change of momentum is the rate at exit minus that at entry. With motion in the x direction, we sum the components in this direction and have

(18-8) $\sum F_x = \dot{m}v_{x2} - \dot{m}v_{x1} = \rho_2 A_2 v_{x2}^2 - \rho_1 A_1 v_{x1}^2$

18.3   ACOUSTIC SPEED

Although the speed of sound (***acoustic*** or ***sonic speed***) results from a three-dimensional effect, accurate results are obtained from a one-dimensional concept for small disturbances. (Large disturbances, as a shock wave, propagate at a much greater speed.) Some medium is necessary for the propagation of these small disturbances, which are pressure waves, and the medium with which we are primarily concerned is gaseous.

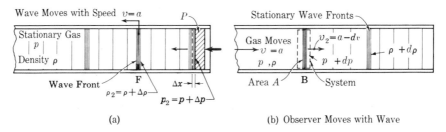

(a)   (b) Observer Moves with Wave

Fig. 18/1   *Wave Propagation.*The piston P could be a vibrating diaphragm. The continuity of mass requires that $\rho v A = (\rho + d\rho)(v + dv)A$.

In Fig. 18/1(a), let the piston P be given a sudden small movement Δx toward the left that initiates a pressure rise Δp in the adjacent gas, causing a pressure wave to begin a leftward movement with sonic speed a; one such wave front is at F. A simple analysis is now obtained by imagining the wave to be stationary with the gas moving rightward with the speed a, Fig. 18/1(b)—the same situation as for the observer (reference axes) moving with the wave in (a). Let the control surface at B bound the wave and consider the disturbance to be small enough that, on the right-hand side of the wave, the pressure is $p + dp$, the density $\rho + d\rho$, and the speed $a - dv$. The channel has a constant area and fluid shear forces are negligible. Applying $\sum F_x = \dot{m}\,\Delta v_x$ with $\dot{m} = \rho A v = \rho A a$, we have

(a) $pA - (p + dp)A = \rho A a[(a - dv) - a]$

(b) $$dp = \rho a\, dv \qquad\qquad \text{[NO FRICTION]}$$

The continuity equation for this case reduces to $\rho v = $ constant, or

(c) $$\rho a = (\rho + d\rho)(a - dv) \qquad \text{or} \qquad dv = \frac{a\, d\rho}{\rho}$$

Compare with equation (18-1) and substitute this value of dv into **(b)** and get

(d) $$a = \left(\frac{dp}{d\rho}\right)^{1/2} \qquad\qquad \text{[CONSISTENT UNITS]}$$

Newton first came to a conclusion equivalent to that in equation **(d)** and assumed that the process was isothermal; but Laplace[1.1] later showed that an ordinary sound wave involves more nearly an adiabatic compression (largely because the pressure and temperature gradients across the wave are small) and the compression is virtually isentropic. The acoustic speed in a gas then is

(18-9) $$a = \left(\frac{\partial p}{\partial \rho}\right)_s^{1/2} = \left[-\frac{v}{\rho}\left(\frac{\partial p}{\partial v}\right)_s\right]^{1/2} = \left(\frac{1}{\rho \kappa_s}\right)^{1/2} \qquad \text{[CONSISTENT UNITS]}$$

where the compressibility coefficient $\kappa_s = (\partial v/\partial p)_s/v$, equation (11-13), § 11.7. For known values of κ_s, the last form of equation (18-9) is applicable to liquids.†
For the speed of sound in an ideal gas with constant k, use $pv^k = C$ or $p = C\rho^k$. Differentiate the latter;

(e) $$dp = kC\rho^{k-1}\, d\rho = \frac{kp}{\rho}\, d\rho = kpv\, d\rho$$

(18-10) $$a = \left(\frac{\partial p}{\partial \rho}\right)_s^{1/2} = \left(\frac{kp}{\rho}\right)^{1/2} = (kpv)^{1/2} = (kRT)^{1/2} = \left(\frac{k\bar{R}T}{M}\right)^{1/2}$$

For atmospheric air at typical pressure and temperature, $a \approx 49\, T^{1/2}$ fps for $T°$R; check this by equation (18-10).
The Mach number $\mathbf{M} = v/a$ is now a convenient parameter; when $\mathbf{M} < 1$, the corresponding speed is said to be *subsonic*; when $\mathbf{M} > 1$, *supersonic* (or *hypersonic* when the speed is several times $\mathbf{M}$); when $\mathbf{M} = 1$ and the speed is sonic and in the vicinity of 1, it is spoken of as *transonic*.

Example—Acoustic Speed   18.4

The state of a given fluid is 15 psia, 250°F. Find the velocity of sound in the fluid if that fluid is (a) air, (b) water.
Solution.   (a) In the case of air,

$$a = (kRT)^{0.5} = [(1.4)(53.34)(710)(32.174)]^{0.5}$$

$$= 1306 \text{ fps}$$

† In a solid, $a = (E/\rho)^{1/2}$, where E is Young's modulus of elasticity and the units *must* be consistent.

(b) For this state, the water is vapor and,

$$a = \left(\frac{1}{\rho K_s}\right)^{0.5}$$

Using Item B 15,

$$\rho = 1/v = 1/27.837 = 0.0359 \,\text{lb/ft}^3$$

To obtain K_s, note that entropy s is essentially the same between the given state and that of 20 psia, 300°F. We can now calculate K_s as,

$$K_s = -\frac{1}{v}(\partial v/\partial p)_s = -\frac{1}{(27.837)}\left(\frac{22.356 - 27.837}{20 - 15}\right)_{1.7809}$$

$$= 0.0394$$

So,

$$a = \left[\frac{(144)(32.174)}{(0.0359)(0.0394)}\right]^{0.5} = 1810 \,\text{fps}$$

18.5 ISENTROPIC FLOW WITH VARYING AREA

Consider the reversible flow of an expanding substance with no work being done (nozzle). First, solve for $\rho = -\mathbf{k}\,dp/(v\,dv)$ from the momentum equation (18-6), which is limited to frictionless flow; substitute this value in the continuity equation (18-1) after solving for dA/A, and get

(18-11) $$\frac{dA}{A} = -\frac{v\,dv}{-\mathbf{k}(\partial p/d\rho)_s} - \frac{dv}{v} = \frac{dv}{v}(\mathbf{M}^2 - 1)$$

Let the speed v be increasing, as in a nozzle; then the change dv is positive. It follows that if $\mathbf{M} < 1$ (subsonic), the change of area A is negative, equation (18-11)—area must decrease. On the other hand, still with speed increasing but with $\mathbf{M} > 1$, the right-hand side of (18-11) is positive and the area must be increasing in size. In other words, the channel confining an isentropic flow with positive acceleration must first converge, then diverge, Fig. 18/2, *when the expansion goes far enough that the speed of flow is greater than Mach* 1. If the expansion proceeds to some speed equal to or *less than* $\mathbf{M} = 1$, then the channel is converging only, Fig. 18/3. Since, by equation (18-11), the channel ceases to converge at $\mathbf{M} = 1$, this occurs at the section of smallest area, called the **throat**.

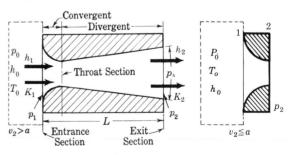

Fig. 18/2 *Convergent-Divergent Nozzle,* $\mathbf{M}_2 >$ 1. The pressure drop per unit of length is not constant. (See §18.10.) An angle of flare θ that is too large results in excessive turbulence.

(a) Convergent—Divergent Nozzle   (b) Convergent Nozzle

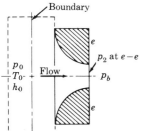

Convergent Nozzles, $M_2 \gtrless 1$. Convergent nozzles in turbines are typically foil nozzles—see Fig. 18/4.

If the speed is decreasing instead of increasing as assumed above, the channel is called a *diffuser*; more detail in § 18.15. The increasing-speed channels are called **convergent-divergent nozzles** and **convergent nozzles**. They may have any cross-section shape to suit the application. The elements of the surface of the divergent section, Fig. 18/2, are generally straight, for convenience in manufacture, but many rocket nozzles are being made with "bell-shape" sides in order to control the expansion better. The customary manner of proportioning a convergent-divergent nozzle is to determine the throat area, as explained later, to provide a well-rounded entrance, and to choose a nozzle length L, Fig. 18/2, such that the flare of the sides of the divergent section is within good limits in accordance with experience; θ is some 12° to 15° in turbine nozzles, upwards of 30° to 35° in rockets.

MASS RATE OF FLOW THROUGH A NOZZLE 18.6

For any substance in steady flow, the rate of flow is computed from the continuity equation $\dot{m} = \rho A v$ applied to any section where the average state across the section is known (state assumed uniform across the section). The density $\rho = 1/v$, where the properties are given by p, T, h, and so on, without subscripts, can be computed for an ideal gas from $\rho = p/(RT)$. From an initial stagnation state, Fig. 18/2, and an isentropic expansion, ideal-gas properties at any section are calculated from equations (7-1), (7-2), and (7-3); say,

(a) $v = \left(\dfrac{p_0}{p}\right)^{1/k} v_0 = \left(\dfrac{p_0}{p}\right)^{1/k} \dfrac{RT_0}{p_0}$ and $\dfrac{T}{T_0} = \left(\dfrac{p}{p_0}\right)^{(k-1)/k} = \left(\dfrac{v_0}{v}\right)^{k-1}$

From the energy equation (18-4), we have

(18-12) $v = [2\mathbf{k}J(h_0 - h)]^{1/2} = 223.8(h_0 - h)^{1/2}$ fps $[2\mathbf{k}J \approx 50{,}000]$

[GAS OR VAPOR, $Q = 0$, $W = 0$, $\Delta P = 0$; $s = C$ OR $s \neq C$]

For an ideal gas with constant specific heats, use $\Delta h = c_p \, \Delta T$ and get, with help from the relations in **(a)**,

(b) $v = \left[2\mathbf{k}Jc_pT_0\left(1 - \dfrac{T}{T_0}\right)\right]^{1/2} = \left\{2\mathbf{k}Jc_pT_0\left[1 - \left(\dfrac{p}{p_0}\right)^{(k-1)/k}\right]\right\}^{1/2}$ fps

[IDEAL GAS, $Q = 0$]   [IDEAL GAS, $s = C$]

where the symbols have our usual units, applicable at any section where the state is defined by p, T. The maximum speed, used as a reference speed, attainable during an isentropic expansion from a given state T_0, p_0 occurs when $p = p_2 = 0$; for constant k,

$$\text{(c)} \qquad v_{max} = (2kJc_pT_0)^{1/2} = \left(\frac{2kkRT_0}{k-1}\right)^{1/2} = a_0\left(\frac{2}{k-1}\right)^{1/2}$$

where $a_0 = (kkRT_0)^{1/2}$ is the sonic speed at the stagnation state p_0, t_0. We observe too that the volume goes to infinity at zero pressure, from which it is inferred that the cross-sectional area to pass the flow as the pressure approaches zero is necessarily quite large. Notice how rapidly the area curve moves upward in Fig. 18/6 at the lower pressures.

The properties at the throat of a convergent-divergent nozzle where $\mathbf{M} = 1$ are particularly significant, since this state is used as a reference state even when those properties do not occur during a process. To distinguish these particular properties, use an asterisk, as p^*, T^*, $v^* = a^*$. Let $\mathbf{M} = 1$ in equation (7-14), repeated in § 18.2 above, and obtain T^*;

$$\text{(18-13)} \qquad T^* = T_0\left(\frac{2}{k+1}\right) \qquad\qquad [s = C, k = C]$$

Then use the p, T relation of equation (a) in (18-3) and obtain

$$\text{(18-14)} \qquad p^* = p_0\left(\frac{2}{k+1}\right)^{k/(k-1)}$$

where p^* is called the **critical pressure**.†

Since, as we learned from equation (18-11), the cross-sectional area A decreases to a minimum (at throat) and then increases, the maximum flow per unit area $\dot{m}/A$ occurs at the throat and is $\dot{m}/A^* = \rho^*v^*$. To get the velocity at the throat in terms of the stagnation state, substitute $T^*/T_0 = 2/(k+1)$, equation (18-13) where $\mathbf{M} = 1$, and $c_p = k(R/J)(k-1)$ into equation (b), and find

$$\text{(d)} \qquad v^* = a^* = \left(\frac{2kkRT_0}{k+1}\right)^{1/2} = a_0\left(\frac{2}{k+1}\right)^{1/2}$$

where a^* is the acoustic speed corresponding to the state at the throat and a_0 is that corresponding to the stagnation state (R ft-lb/lb-°R). By equation (7-1), $\rho^* = \rho_0(T^*/T_0)^{1/(k-1)}$; with $\rho = p/(RT)$,

$$\text{(e)} \qquad \rho^* = \rho_0\left(\frac{T^*}{T_0}\right)^{1/(k-1)} = \frac{p_0}{RT_0}\left(\frac{2}{k+1}\right)^{1/(k-1)}$$

The product of these values of ρ^* and v^*, simplified, yields

$$\text{(18-15)} \qquad \frac{\dot{m}}{A^*} = \rho^*v^* = p_0\left[\frac{kk}{RT_0}\left(\frac{2}{k+1}\right)^{(k+1)/(k-1)}\right]^{1/2}$$

† But see § 3.6. The word *critical* is overworked, but this usage is fairly well entrenched.

which is lb/ft²-sec for p_0 lb/ft², R ft-lb/lb °R, **k** = 32.2; or delete **k** and use any consistent system of units. In equation (18-15), $\dot{m}$ is called the mass flow rate for **choked flow**. If the pressure at the exit of a nozzle or other channel is less than the critical pressure from equation (18-14), then equation (18-15) gives the mass flow governed by the smallest cross section A^*. *The pressure just inside the exit section of a converging nozzle, Fig. 18/3, will not fall below the critical pressure p^*, no matter how low the pressure of the surroundings.*

For a particular substance, equation (18-15) may be simplified. For air, the resulting equation, called the Fliegner equation, is

(f) $$\dot{m} = \frac{0.532A^*p_0}{T_0^{1/2}} \text{ lb/sec} \qquad [\text{AIR, } S = C]$$

CRITICAL PRESSURE RATIO 18.7

The ratio of (critical pressure)/(stagnation pressure) is called the **critical pressure ratio**; p^*/p_0 from equation (18-14) for an ideal gas. Since this ratio does not vary greatly from one expansible fluid to another, equations (18-13) and (18-14) are applied to actual gases with suitable values of k. The range of the ratio is shown in the following:

$k = 1.12$: $p^* = 0.58p_0$ wet steam.
$k = 1.2$: $p^* = 0.564p_0$ typical value for hot gas in rocket nozzle.
$k = 1.3$: $p^* = 0.545p_0$ for superheated steam (especially in the 200–300 psia range) and for supersaturated steam.
$k = 1.4$: $p^* = 0.528p_0$ air and diatomic gases at ordinary temperatures.
$k = 1.67$: $p^* = 0.487p_0$ monatomic gases.

CORRECTION FOR INITIAL VELOCITY 18.8

When the initial state is taken as the stagnation state, as done above, correction for initial velocity is inherently included. However, an equation for v_2 is found by using the steady-flow continuity equation (1-17) and equation (18-4) to get

(18-16A) $$v_2 = \left[\frac{2\text{k}J(h_1 - h_2)}{1 - [A_2v_1/(A_1v_2)]^2} \right]^{1/2} \text{ fps}$$

[GAS OR VAPOR, IDEAL OR ACTUAL FLOW]

where the states 1 and 2 are at any two sections 1 and 2, wherever taken. Comparing equation (18-16A) with (18-12) wherein the flow started from a stagnation condition, initial velocity is negligible, we can then state that the correction for initial velocity is

(18-16B) $$C_f = \left[\frac{1}{1 - \{A_2v_1/(A_1v_2)\}^2} \right]^{0.5}$$

Where the drop in pressure is very small as in most flowmeters, see §§ 18.17–18.20,

$v_1 \approx v_2$ and the correction factor becomes,

$$(18\text{-}16\text{C}) \qquad C_f = \left[\frac{1}{1 - (A_2/A_1)^2}\right]^{0.5} = \left[\frac{1}{1 - (D_2/D_1)^4}\right]^{0.5}$$

where D_1 and D_2 are the respective diameters.

18.9   EFFICIENCY AND COEFFICIENTS FOR NOZZLE

The nozzle efficiency, previously defined in §§ 7.31 and 15.16, is

$$(18\text{-}17) \qquad \eta_n \equiv \frac{\text{actual } \Delta K'}{\text{ideal } \Delta K_s} = \frac{h_1 - h_{2'}}{(h_1 - h_2)_s}$$

In convergent-divergent nozzles and in convergent nozzles that provide a substantial expansion, the initial kinetic energy K_1 is often negligible compared to other relevant energy quantities. If so, the nozzle efficiency reduces to $\eta_n = K_{2'}/K_2 = v_{2'}^2/v_2^2$, with states 2 and 2' defined in Figs. 18/4 and 18/5. Nozzles typically have good efficiencies, say 94% or better for nozzles whose axes are straight. Quite small nozzles have the lowest efficiency because the boundary layer, where the major fluid friction occurs, may occupy a sizable proportion of the channel; very large nozzles have the highest efficiency, as over 99% for some wind tunnels, because the boundary layer friction becomes relatively almost insignificant.[18.1] The *velocity coefficient* or *nozzle coefficient*

$$(18\text{-}18) \qquad \kappa_v \equiv \frac{v_{2'}}{v_2} = \sqrt{\eta_n}$$

is another performance factor sometimes used.

The ideal rate of flow is related to the actual flow through a nozzle (or other flow device, as an orifice) by a *coefficient of discharge* κ_d, defined as

$$(18\text{-}19) \qquad \kappa_d \equiv \frac{\text{actual mass flow per unit time}}{\text{isentropic mass flow per unit time}} = \frac{\dot{m}'}{\dot{m}}$$

the denominator being computed for an ideal nozzle with the same minimum cross-sectional area and with the initial state the same as the actual initial state. This coefficient can often be estimated with considerable accuracy from past data for similar devices.

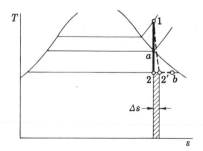

Fig. 18/4   *Equilibrium Expansion.*

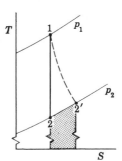

Equilibrium Expansion. **Fig. 18/5**

Example—Elementary Design Calculations and Variation of Velocity, 18.10 Specific Volume, and Area of Section, Ideal Gas

At the rate of 5 lb/sec, air in a stagnant state of 200 psia and 60°F expands through a nozzle to 15 psia. (a) For an isentropic expansion, determine the velocity, specific volume, and section area at the exit and at the throat. (b) If the nozzle efficiency is 95%, compute the velocity, temperature, specific volume, section area, and Mach number at the exit. (c) Repeat the calculations for several downstream pressures during isentropic expansion and plot curves of A, v, $\dot{m}/A$, and v against an abscissa of pressure.

Solution. (a) The problem can be worked either by ideal-gas laws with k constant or from the air table. From Item B 2 for $T_1 = 520$, $p_{r1} = 1.2147$, $h_1 = 124.27$, $v_{r1} = 158.58$; then $p_{r2} = p_{r1}(p_2/p_1) = 1.2147(15/200) = 0.091$. For this p_{r2}, we find $T_2 = 248°R$ (nearest whole degree), $h_2 = 59.16$, $v_{r2} = 1003.4$. Also

(a)
$$T_2 = T_1\left(\frac{p_2}{p_1}\right)^{(k-1)/k} = (520)\left(\frac{15}{200}\right)^{0.4/1.4} = 248°R \qquad [k = C]$$

a close check because for the temperature range, $k \approx C$ closely. Observe that although the temperature is low enough to do refrigeration, some of the kinetic energy must first be converted into work because the stagnation temperature has not changed, equation (18-4).

From the energy equation (18-12) with $\Delta h = c_p \Delta T$, we find

(b)
$$v_2 = 223.8\sqrt{(0.24)(520 - 248)} = 1808 \text{ fps} \qquad [k = C]$$

(c)
$$v_2 = 223.8\sqrt{124.27 - 59.16} = 1805 \text{ fps} \qquad [\text{GAS TABLE}]$$

The specific volume at state 2, Fig. 18/5, is

(d)
$$v_2 = \frac{RT_2}{p_2} = \frac{(53.3)(248)}{(15)(144)} = 6.11 \text{ ft}^3/\text{lb} \qquad [k = C]$$

As before, v_2 from gas table data is virtually the same. From the continuity equation for $\dot{m} = 5$ lb/sec, we get the exit area as

(e)
$$A_2 = \frac{\dot{m}v_2}{v_2} = \frac{(5)(6.11)}{1805} = 0.0169 \text{ ft}^2 \quad \text{or} \quad 2.44 \text{ in.}^2$$

At the throat, $p^* = (0.53)(200) = 106$ psia. Proceeding as before, we find

$$T^* = 434°R, \quad v^* = a^* = 1020 \text{ fps} \quad v^* = 1.515 \text{ ft}^3/\text{lb} \quad A^* = 1.07 \text{ in.}^2$$

These values are italicized in the tabulation under Fig. 18/6.

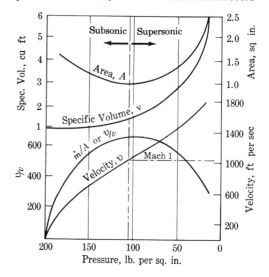

Fig. 18/6   *Specific   Volume,   Area,   and v/v Plotted against Pressure.* Ideal air, $s = C$. Vapors and other gases flowing through nozzles have similar attributes. Observe that the velocity may be large, the maximum in this case being 1805 fps = 1230 mph. The pressure drop in a nozzle with straight sides is not constant per unit length of nozzle.

p_2, psia	v_2, ft³/lb	v_2, fps	A_2, in.²	$v_2/v_2 = \dot{m}/A_2$, A_2 ft²
200	0.962			
175	1.059	483	1.578	456
150	1.18	700	1.212	593
125	1.346	887	1.091	659
106	*1.515*	*1020*	*1.070*	*673*
100	1.58	1060	1.072	671
50	2.59	1431	1.302	553
25	4.24	1675	1.821	395
15	6.11	1805	2.44	296

(b) With $\eta_n = 95\%$, the actual exit speed and temperature are found as follows:

(f)
$$v_{2'} = 223.8\sqrt{(0.95)(0.24)(520 - 248)} = 1760 \text{ fps}. \qquad [k = C]$$

(g)
$$\eta_n = 0.95 = \frac{c_p(T_0 - T_{2'})}{c_p(T_0 - T_2)_s} = \frac{520 - T_{2'}}{520 - 248} \qquad [k = C]$$

from which $T_{2'} = 262°R$. The actual specific volume at exit is

(h)
$$v_{2'} = \frac{RT_{2'}}{p_2} = \frac{(53.3)(262)}{(15)(144)} = 6.46 \text{ ft}^3/\text{lb}$$

(i)
$$A_{2'} = \frac{\dot{m}v_{2'}}{v_{2'}} = \frac{(5)(6.46)}{1760} = 0.01835 \text{ ft}^2 \quad \text{or} \quad 2.64 \text{ in.}^2$$

the area actually needed to pass 5 lb/sec. The local speed of sound at state $2'$ and the Mach number are

(j)
$$a_{2'} = (\mathbf{k}kRT_{2'})^{1/2} = [(32.2)(1.4)(53.3)(262)]^{1/2} = 793 \text{ fps}$$

(k)
$$\mathbf{M}_{2'} = \frac{v_{2'}}{a_{2'}} = \frac{1760}{793} = 2.22$$

(c) The foregoing computations for an isentropic expansion are repeated for various pressures between p_1 and p_2 and the results are plotted, Fig. 18/6. Since the flow is steady, the $\dot{m}/A$ curve comes to a maximum at minimum A, as concluded in § 18.5, and then decreases as A increases. The specific volume is increasing at a faster rate than is the velocity. By way of illustration, suppose that an expansion to $p_2 = 50$ psia were desired instead of to 15 psia. Then the exit area, as obtained from the tabulation with Fig. 18/6, should be 1.302 in.^{2} for isentropic expansion. That part of the original nozzle beyond this area would not be used at all.

Special Solution. The Keenan and Kaye *Gas Tables* contains a number of tables designed to reduce the computational detail for various flow problems, which therefore are valuable for repetitive calculations. To obtain the most advantage from these tables, the equations used should be arranged in appropriate forms, something left largely to the reader (or to a course on fluid flow). For example, one could use Table 30 (Keenan and Kaye), which gives certain ratios for isentropic compressible flow, in the solution of this example. For those who have the complete tables, the numbers below in brackets are ones from this table. Enter table with (stagnation state designated p_0, T_0)

(l) $\qquad \dfrac{p_2}{p_0} = \dfrac{15}{200} = 0.075 \qquad$ also $\qquad \dfrac{p^*}{p_0} = 0.528$

(m) $\qquad \dfrac{T_2}{T_0} = [0.477] \qquad$ or $\qquad T_2 = (0.477)(520) = 248°R$

(n) $\qquad v_0 = \dfrac{RT_0}{p_0} = \dfrac{(53.3)(520)}{(200)(144)} = 0.961 \text{ ft}^3/\text{lb}$

(o) $\qquad \dfrac{v_0}{v_2} = \dfrac{p_2}{p_0} = [0.1573] \qquad$ or $\qquad v_2 = 6.11 \text{ ft}^3/\text{lb}$

(p) $\qquad v_2 = 1808 \text{ fps} \qquad$ and $\qquad A_2 = 2.44 \text{ in.}^2$, computed as before

(q) $\qquad \dfrac{T^*}{T_0} = [0.833] \qquad$ or $\qquad T^* = 433°R$

(r) $\qquad \dfrac{A_2}{A^*} = [2.274] \qquad$ or $\qquad A^* = 1.07 \text{ in.}^2$

Example—Equilibrium Flow in a Steam Nozzle   **18.11**

Steam enters a nozzle with negligible speed at 10 bar, 200°C, flowing at the rate of 4.5 kg/s. It expands isentropically to 0.5 bar. Determine the throat and exit areas.

Solution. Process 1-2, Fig. 18/4, represents the reversible expansion.

$$p^* = (0.545)(10) = 5.45 \text{ bar}$$

Locating the pertinent state points on Item B 16 (SI) we read these properties:

$$h_1 = 2874 \text{ kJ/kg} \quad s_1 = 6.79 \text{ kJ/kg K} \qquad v_1 = 0.217 \text{ m}^3/\text{kg}$$

$$h^* = 2752 \text{ kJ/kg} \quad s_1^* = s_1 = 6.79 \text{ kJ/kg K} \quad v^* = 0.34 \text{ m}^3/\text{kg}$$

$$h_2 = 2360 \text{ kJ/kg} \quad s_2 = s_1 = 6.79 \text{ kJ/kg K} \quad v_2 = 2.70 \text{ m}^3/\text{kg}$$

Using these data, we get:

$$v^* = [2(h_1 - h^*)]^{0.5} = [2(2874 - 2752)(1000)]^{0.5} = 493 \text{ m/s}$$

$$A^* = \dot{m}v^*/v^* = (4.5)(0.34)/493 = 0.0031 \text{ m}^2 = 31 \text{ cm}^2$$

$$v_2 = [2(h_1 - h_2)]^{0.5} = [2(2874 - 2360)(1000)]^{0.5} = 1014 \text{ m/s}$$

$$A_2 = \dot{m}v_2/v_2 = (4.5)(2.70)/1014 = 0.012 \text{ m}^2 = 120 \text{ cm}^2$$

18.12   SUPERSATURATED FLOW

In an actual expansion of a steam or other vapor, condensation does not start as soon as the pressure falls to a value that puts the equilibrium state in the wet region. That is, condensation does not start immediately after *a* or *a'*, Fig. 18/7, is passed. Instead, for isentropic flow, no drops of liquid are formed until some state *c* is reached, where condensation suddenly occurs, a phenomenon called *condensation shock*. The steam in states between those represented by *a* and *c* is **supersaturated**, a *metastable state* (§ 5.25). There is required more than a very small action to achieve the internal equilibrium states along *ac*—say, an action as big as injecting fine, charged dust particles into the stream ahead of state *a* (as in "seeding" a cloud to bring about rain). This is not to imply that such an action is desired in the nozzle. Because the time of flow of a particular group of molecules through a nozzle is small, say less than 0.001 sec, some lag in condensation might be expected; but quiescent supersaturated states may exist indefinitely. The formation of a drop of liquid is a process opposite to the evaporation of a drop of liquid. In the presence of a big body of liquid (and a big drop), there is a larger molecular attraction by neighboring molecules to restrain a particular molecule from passing into vapor (where the mean free paths are vastly larger and the molecular forces much smaller). The number of molecules in a *droplet* (which is a very small drop) is few enough that the force of attraction on a particular molecule is so small that it will evaporate with less provocation (lower temperature) than a large drop. The droplets can and do exist in equilibrium with vapor at temperatures less than the saturation temperature corresponding to the pressure. Thus, in the reverse phenomenon, a strong provocation is needed to start the formation of a drop, something in the way of nuclei. Once the drop begins to form, it grows rapidly.†

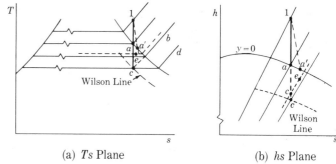

Fig. 18/7   *Representation of Super-saturated Flow (no scale).*   (a) *Ts* Plane   (b) *hs* Plane

† Very pure water has been maintained in a metastable state at −42°F without freezing (B. J. Mason, *Scientific American*, Vol. 204, No. 1, p. 120). Also, review "Superheated Liquids" (R. C. Reid, *American Scientist*, Vol. 64, No. 2 March–April, 1976 pp. 146–156).

Returning to Fig. 18/7, note that the actual pressure at the supersaturated state c is some value p_b; but the saturation pressure corresponding to the temperature t_c is p_d. The ratio of these pressures p_b/p_d, is called the *degree of supersaturation* or *supersaturation ratio*. The point c, where conditions are right for the sudden condensation, is located by the *Wilson "line"* (a narrow region). Yellott and Holland[18.5] locate the Wilson line at approximately 60 Btu below the saturated vapor line on the Mollier diagram; that is, ac in Fig. 18/7(b) is equal to about 60 Btu (for Δh).

The process of passing to the equilibrium state is irreversible and therefore with increasing entropy, say to some state e after c. If the expansion continues after state e, phase equilibrium thereafter is a reasonable assumption. Properties of supersaturated steam are in the ASME S.T. In the absence of these tables, the following procedure may be used. Assume that the expansion up to point c follows the relation $pv^k = C$, where $k \approx 1.3$ in the region of moderate pressures (but $pv \neq RT$). Letting $W = 0$ and $\Delta P = 0$ in equation (4-16), § 4.12, and integrating $-\int v \, dp$, we get

$$(18\text{-}20) \qquad \Delta K = \frac{v_2^2 - v_1^2}{2k} = -\int_1^2 v \, dp = \frac{kp_1v_1}{k-1}\left[1 - \left(\frac{p_2}{p_1}\right)^{(k-1)/k}\right]$$

[SUPERSATURATED STEAM, SUPERHEATED STEAM, IDEAL GAS]

from which the *ideal* velocity (v fps for customary units) of supersaturated steam (or superheated steam through moderate pressure drops with suitable average values of k) can be computed. If the initial velocity v_1 is significant, one may use stagnation properties p_0, v_0 ($v_0 = 0$) for p_1, v_1.

Since the collapse of the metastable state has not been observed in a converging nozzle but always in the diverging part, one is probably safe in assuming that supersaturation, if it occurs at all, will persist to some point beyond the throat, even though the initial superheat is small or nonexistent. The flow is computed for the throat section, since this is where it is limited (choked).

Example—Supersaturated Flow 18.13

(a) On the basis of supersaturation, find the throat area of a nozzle for the conditions given in § 18.11; $p_1 = 160$ psia, $t_1 = 400°F$, $p_2 = 10$ psia, 10 lb/sec, isentropic process, $v_1 \approx 0$. (b) If the nozzle efficiency is 98%, compute the actual flow for the area found in (a) and the coefficient of discharge.

Solution. (a) From § 18.11, we have: $v_1 = 3.006 \text{ ft}^3/\text{lb}$, $p^* = 87$ psia. Then

(a) $$v^* = v_1\left(\frac{p_1}{p^*}\right)^{1/k} = 3.006\left(\frac{160}{87}\right)^{1/1.3} = 4.81 \text{ ft}^3/\text{lb}$$

which is observed to be smaller than for the equilibrium flow (4.95 in § 18.11). From equation (18-20), we get (result with supersaturated steam table nearly the same)

(b) $$v^* = \left\{\frac{(2)(32.2)(160 \times 144)(3.006)(1.3)}{1.3 - 1}\left[1 - \left(\frac{87}{160}\right)^{0.3/1.3}\right]\right\}^{1/2} = 1590 \text{ fps}$$

(c) $$A^* = \frac{\dot{m}v^*}{v^*} = \frac{(10)(4.81)(144)}{1590} = 4.36 \text{ in}^2$$

for $\dot{m} = 10$ lb/sec. Compare with 4.47 in.^{2} from § 18.11. The difference between the answers would have been greater if the expansion had been farther below the saturated vapor line.

(b) The actual velocity at the throat is

(d) $$v'^* = (\eta_n)^{1/2} v^* = (0.98)^{1/2}(1590) = 1570 \text{ fps}$$

Because of friction, which results in a different state at the throat, the actual specific volume at that point is not quite the same as the ideal. However, for the high efficiencies that are likely to apply up to the throat, the difference between the actual and ideal volumes will be small; thus, let $v'^* \approx v^* = 4.81 \text{ ft}^3/\text{lb}$. Then

(e) $$\dot{m}' = \frac{A^* v'^*}{v'^*} = \frac{(4.36)(1570)}{(144)(4.81)} = 9.88 \text{ lb/sec}$$

(f) $$\kappa_d = \frac{\dot{m}'}{\dot{m}} = \frac{9.88}{10} = 98.8\%$$

the coefficient of discharge, by equation (18-19). (For situations of low nozzle efficiency, use the actual specific volume at the section for which the efficiency applies.)

18.14 FLOW FROM NOZZLES WITH VARYING PRESSURE RATIOS

Consider a *convergent nozzle* with flow from p_0, T_0 to some region where the back pressure is p_b, Fig. 18/3. If $p_b < p^*$, there is choked flow and the pressure p_2 at the exit section is $p_2 = p^*$; it continues to be p^* as the back pressure increases until p_b becomes greater than p^*. For this regime of $p_b \gtreqless p^*$, other properties at the exit section for isentropic flow are also critical properties; $v_2 = v^* = a^*$, $v_2 = v^*$, $T_2 = T^*$, and the flow is given by equation (18-15). If the back pressure $p_b > p^*$, the exit pressure is $p_2 = p_b$, effectively, and the equations relating any two states 1 and 2 in nozzle flow apply.

A *convergent-divergent* nozzle is designed to handle an expansion between certain specified states p_0, T_0 and $p_2 < p^*$. If the back pressure p_b of the discharge region is less than the design pressure at the discharge boundary, so-called **underexpansion** has occurred; there is a free expansion after the substance leaves the nozzle, and since it is moving supersonically, complex shock or expansion waves are initiated. No matter how low the external pressure is, the ideal nozzle expansion goes to the design pressure p_2. (If the difference between pressures p_2 and p_b is large, this effect will, of course, be felt significantly inside the exit section.)

If the back pressure p_b is greater than p_2, **overexpansion** occurs; that is, the pressure before the substance leaves the nozzle will become lower than the back pressure. Depending on the pressure ratio p_0/p_b, the details of the events are different. Consider Fig. 18/8, where the curves are for steam, for a 10-in. long nozzle designed for $p_2/p_0 = 0.2$; $p^*/p_0 = 0.545$; and the labels on the curves are the ratios p_b/p_0. If the back pressure is only slightly below p_0, the channel is acting as a venturi (§ 18.18)—similar to *tqr*, Fig. 18/8. If, approximately, $p_b/p_0 > 0.82$, the stream does not reach sonic speed at the throat. If $p_b/p_0 < 0.82$, the speed at the throat is sonic, yet the speed at exit is subsonic; the flow is the maximum. If $p_b/p_0 = 0.77$, the variation of p_2/p_0, where p_2 is at any section, follows the curve *nadew*. At point d in the expansion, there is a standing shock front across which there is a sharp irreversible increase in pressure, decrease in speed, and, naturally, an increase in entropy. This is the event in each curve where a portion is nearly vertical, as *de*. For the design condition, the curve is *nadgh*, where the pressure jump

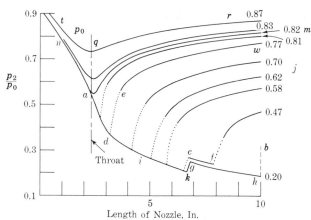

0.87
0.83   0.82 *m*
0.77 0.81
0.70
   j
0.62
0.58

0.47

b

0.20

Throat   *i*

Length of Nozzle, In.

Effect of Varying Discharge Pressure, Convergent-Divergent Nozzle. After Binnie and Woods.[18.6]

kg is the condensation shock when the steam passes from supersaturation to a two-phase equilibrium (almost) state. Notice that even when the back pressure is almost as large as the throat pressure p^* (0.47 shown) the expansion follows the area change of the nozzle to f; the pressure just before the shock has almost reached the design exit pressure.

In substantitation of the events described, consider the stream in the diverging part $(\mathbf{M} > 1)$ at some pressure p_i, Fig. 18/8, with a pressure at the exit section of p_j. If the nozzle flow is isentropic and the pressure increases, some of the kinetic energy must be used to do the compression; hence there is necessarily a decrease in speed for the fluid. With respect to equation (18-11), $dA/A = (\mathbf{M}^2 - 1) dv/v$ with $s = C$, we note that with Δv negative and $\mathbf{M} > 1$, ΔA *must* also be negative (decreasing A). But the nozzle has an increasing area between i and j; one may therefore conclude that an isentropic process between i and j is impossible. A shock occurs with property changes as mentioned. At some back pressure p_b, Fig. 18/8, there is a standing shock at the exit section that brings the pressure to the back pressure. After the shock has occurred, as de, Fig. 18/8, the succeeding part of the nozzle *acts as a subsonic diffuser.* A gas would exhibit the features as described except for the condensation shock kg.

DIFFUSER   **18.15**

As previously noted, §§ 7.31 and 15.14, the ideal diffuser is the reverse of the ideal nozzle. In Fig. 18/9(b), along the isentropic, $h_1 + K_1 = h_2 + K_2 = h_{01} = h_{02}$,

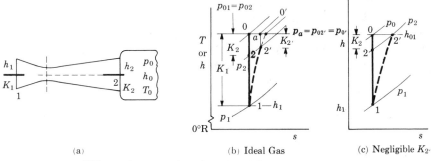

(a)   (b) Ideal Gas   (c) Negligible $K_{2'}$

Diffuser. A converging-diverging diffuser implies that v_1, is supersonic.   **Fig. 18/9** and the fluid is depicted as leaving the diffuser with kinetic energy K_2. The actual change of state 1-2′ is with increasing entropy; the fluid leaves with kinetic energy $K_{2'}$. Typically, the exit kinetic energy is negligible, so that $K_{2'}$ is exaggerated in Fig. 18/9(b). If $K_{2'}$ is nearly zero, points a and $0'$ approach coincidence, and $h_{2'} \approx h_0$ as in Fig. 18/9(c), where K_2 is the kinetic energy the fluid would have if it left the diffuser after an isentropic to p_2. For adiabatic flow and $W = 0$, equation (18-12) adapted to the diffuser becomes

(a) $$v_1 = [2\mathbf{k}J(h_0 - h_1)]^{1/2}$$

From $\int v \, dp = J(h_0 - h_1)$ for $pv^k = C$ (ideal gas), we have

(b) $$-\Delta K = \frac{v_1^2 - v_2^2}{2\mathbf{k}} = \int_1^2 v \, dp = \frac{kp_2v_2}{k-1}\left[1 - \left(\frac{p_1}{p_2}\right)^{(k-1)/k}\right] \qquad [s = C]$$

analogous to equation (18-20) for a nozzle. In general, actual diffuser efficiencies are lower than those for nozzles operating between the same end states; and for supersonic diffusers, there are troubles with shocks, especially for variable operating conditions, as on a supersonic jet engine.[18.1] The diffuser efficiency is, Fig. 18/9,

(18-21) $$\eta_d \equiv \frac{(h_2 - h_1)_s}{h_{2'} - h_1} = \frac{(h_2 - h_1)_s}{K_1 - K_{2'}} = \frac{T_2 - T_1}{T_{2'} - T_1}, \qquad \text{[DIMENSIONLESS]}$$

$$[\text{CONSTANT } k]$$

where $K_{2'}$ may be negligible. The numerator of the efficiency is noted to be the isentropic change of kinetic energy.

Diffusers may be diverging only when the entering fluid is moving subsonically, or, theoretically, converging-diverging, when the entering fluid is moving supersonically. Equation (18-11), repeated for convenience,

(18-11) $$\frac{dA}{A} = (\mathbf{M}^2 - 1)\frac{dv}{v}$$

shows how the area A must vary for isentropic flow. If v is decreasing, as in a diffuser, the change of v is negative; then with $\mathbf{M}_1 < 1$, the right-hand side of (18-11) is positive, which means that the change of area is positive (diverging passage). With $\mathbf{M}_1 > 1$, the area must be decreasing to $\mathbf{M} = 1$, the throat condition. The equations for properties of an ideal gas at the throat of a nozzle in § 18.6 are applicable to a diffuser when the process is isentropic; note the definition of p_0 in Fig. 18/9 and by equations (7-2) and (18-14). Let $p^* = Bp_0$, where $B = [2/(k+1)]^{k/(k-1)}$ for an ideal gas with $k = C$, from equation (18-14), and compare a nozzle and a diffuser as follows:

NOZZLE ($p_2 < p_1$)   DIFFUSER ($p_2 < p_1$)

If $\dfrac{p_2}{p_1} = \dfrac{p_2}{p_0} < \dfrac{p^*}{p_0}$,   If $\dfrac{p_1}{p_2} = \dfrac{p_1}{p_0} < \dfrac{p^*}{p_0}$

then $v_2 >$ sonic velocity, converging-diverging nozzle.

then $v_1 >$ sonic velocity, converging-diverging diffuser.

If $\dfrac{p_2}{p_1} = \dfrac{p_2}{p_0} > \dfrac{p^*}{p_0}$

If $\dfrac{p_1}{p_2} = \dfrac{p_1}{p_0} > \dfrac{p^*}{p_0}$

then $v_2 <$ sonic velocity, converging nozzle.

then $v_1 <$ sonic velocity, diverging diffuser.

The *pressure coefficient* κ_p for diffusers in general is given by equation (15-12),

(15-12) $\kappa_p \equiv \dfrac{\text{actual pressure rise}}{\text{ideal pressure rise}} = \dfrac{p_2 - p_1}{p_0 - p_1}$

with subscripts defined in Fig. 18/9. Manipulate the efficiency equation (18-21) for an ideal gas;

(c) $(h_2 - h_1)_s = c_p T_1\left(\dfrac{T_2}{T_1} - 1\right) = c_p T_1\left[\left(\dfrac{p_2}{p_1}\right)^{(k-1)/k} - 1\right]$

where $p_2 = p_{2'} =$ actual final pressure. Note that $T_{2'} = T_0$ and get

(d) $h_{2'} - h_1 = c_p(T_0 - T_1) = c_p T_1\left(\dfrac{k-1}{2}\right)M_1^2$

where T_0 from equation (7-14) is used. Substitute the values in equations (c) and (d) into (18-21) and get

(e) $\eta_d = \dfrac{(p_2/p_1)^{(k-1)/k} - 1}{\dfrac{k-1}{2}M_1^2}$ [IDEAL GAS]

An example of the use of a nozzle and a diffuser in one device is the ejector, Fig. f/13 Foreword. Most diffusers in practice are subsonic.

Example—Diffuser   18.16

Air enters a diffuser at 600°R and 15 psia. If the exit pressure is to be 41 psia with a diffuser efficiency of 80%, determine the initial velocity, Mach number, and the pressure coefficient. The final state is virtually stagnation. Let $k = 1.4$, a constant.

Solution.   In Fig. 18/9, $p_2 = p_{2'} = 41$ psia. Thus

(a) $T_2 = T_1\left(\dfrac{p_2}{p_1}\right)^{(k-1)/k} = 600\left(\dfrac{41}{15}\right)^{0.286} = 800°R$

(b) $\eta_d = \dfrac{T_2 - T_1}{T_{2'} - T_1} = 0.80 = \dfrac{800 - 600}{T_{2'} - 600}$ or $T_{2'} = T_0 = 850°R$

(c) $K_1 = \dfrac{v_1^2}{2kJ} = c_p(T_{2'} - T_1) = 0.24(850 - 600)$

from which $v_1 = 1728$ fps.

(d) $a_1 = (\mathbf{k}kRT_1)^{1/2} = [(32.2)(1.4)(53.3)(600)]^{1/2} = 1200$ fps

which gives $\mathbf{M}_1 = 1728/1200 = 1.44$. The pressure p_0 along 1-0 is

(e) $$p_0 = p_1\left(\frac{T_{2'}}{T_1}\right)^{k/(k-1)} = 15\left(\frac{850}{600}\right)^{3.5} = 51 \text{ psia}$$

(f) $$\kappa_p = \frac{p_2 - p_1}{p_0 - p_1} = \frac{41 - 15}{51 - 15} = 0.722$$

18.17 APPROXIMATE EQUATIONS FOR A SMALL PRESSURE CHANGE

When the pressure drop is small, as happens in flow-metering devices (and elsewhere) and p_2 is nearly the same as p_0 or p_1, the evaluation of $(p_2/p_0)^{(k-1)/k}$ is difficult to make accurately.† The bionomial theorem,

$$(a + b)^n = a^n + na^{n-1}b + \frac{n(n-1)}{2!}a^{n-2}b^2 + \frac{n(n-1)(n-2)}{3!}a^{n-3}b^3 + \cdots$$

can be used, after adding and subtracting 1 as follows:

(a) $$\left(\frac{p_2}{p_0}\right)^{(k-1)/k} = \left(1 + \frac{p_2}{p_0} - 1\right)^{(k-1)/k} = \left(1 + \frac{p_2 - p_0}{p_0}\right)^{(k-1)/k}$$

Use the first three terms of the binomial theorem, in which we note that $a = 1$, $b = (p_2 - p_0)/p_0$, and $n = (k - 1)/k$; substitute the resulting value of $(p_2/p_0)^{(k-1)/k}$ into the second part of equation **(b)**, § 18.5, and obtain

(b) $$v_2 = \left\{2\mathbf{k}RT_0\left[\frac{p_0 - p_2}{p_0} + \frac{1}{2k}\left(\frac{p_0 - p_2}{p_0}\right)^2\right]\right\}^{1/2} \text{ fps}$$

which is quite accurate when $p_2 > 0.85p_0$. If the pressure drop is of the order of 1% of the initial pressure or less, excellent accuracy is obtained using only the first term in the bracket; to wit,

(c) $$v_2 = \left[2\mathbf{k}RT_0\left(\frac{p_0 - p_2}{p_0}\right)\right]^{1/2} = [2\mathbf{k}v_0(p_0 - p_2)]^{1/2} \text{ fps}$$

The steady-flow energy equation gives the same result as equation **(c)** for frictionless flow when $v = C$ (exactly), $\Delta u = 0$, $Q = 0$, $W = 0$, $\Delta P = 0$. Usually, in cases where these approximate equations are appropriate, the effect of the initial velocity must be included by using the stagnation properties as shown, or by using an initial velocity correction factor, equation (18-16).

---

† But Table 25, *Gas Tables*, helps in this regard.

As is true for the use of any measuring instrument, one needs to know in detail the characteristics of the device.[18.22,18.23] The following brief discussions serve the purpose of indicating the principles involved. Basically, the plan is to determine the velocity of flow by measuring a pressure change and then computing the mass flow from the continuity equation, $\dot{m} = \rho A v$, plus, of course, modification by experience factors such as a coefficient of discharge.

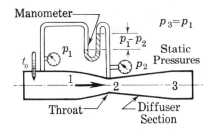

Venturi. Schematic pressure gages are not ordinarily attached as shown. Instead of single tapped holes in the venturi as indicated, the better practice is to drill a number of holes about the circumference at 1 and 2, which lead into an annular ring called a piezometer ring. The manometer leads come from the piezometer ring.

Fig. 18/10

A venturi is a combination of nozzle 1-2 and diffuser 2-3, Fig. 18/10. It is installed as part of the pipeline. In any case, the flow is taken as one of constant stagnation enthalpy. When the pressure drop from 1 to 2 is small, the expansion of the gas may be approximated as for an incompressible fluid; if not, $p_1 v_1^k = p_2 v_2^k$ for an ideal gas. Typically, the change in velocity is small enough that the initial kinetic energy is included (or correction is made for it); thus,

(a) $K_2 = K_1 + h_1 - h_2 = h_{01} - h_2$ or $v_2 = [v_1^2 + 2kJc_p(T_1 - T_2)]^{1/2}$

[IDEAL GAS]

For incompressible flow or for a small gas pressure drop during which $v \approx C$, we have from equation (4-16), $\int -v\,dp = \Delta K$,

(b) $v_2 = [v_1^2 + 2kv(p_1 - p_2)]^{1/2}$ or $v_2 = [2kv(p_0 - p_2)]^{1/2}$

in which v_1 may be eliminated by use of the continuity equation $\rho_1 A_1 v_1 = \rho_2 A_2 v_2$; the subscripts are as defined in Fig. 18/10; detail in the next section. With some area A_2 and v_2 known, the flow is now obtained from $\dot{m} = Av/v$.

Example—Venturi 18.19

A $3 \times 1\frac{1}{2}$-in. venturi, whose coefficient of discharge is $\eta_d = 0.98$, is used to measure the flow of air in a 3-in. ID pipe. The barometer is 30.05 in. Hg; room temperature is 96°F (556°R). The temperature of the air flowing in the pipe is 114°F (574°R) and its static pressure is $p_1 = 3.9$ in. H_2O gage (0.1408 psig). The pressure drop in the venturi is $-\Delta p = p_1 - p_2 = 11.4$ in. H_2O, Fig. 18/10. (a) What is the air flow in cfm? (b) Estimate the velocity pressure and the stagnation pressure ahead of the venturi. Assume that manometer readings are at standard temperature for conversion purposes.

Solution. Use conversion factors from Item B 38. The atmospheric pressure is (30.05) (13.6 in. H_2O/in. Hg) = 408.7 in. $H_2O = p_a$; $p_1 = 408.7 + 3.9 = 412.6$ in. H_2O or (412.6) (0.0361) = 14.89 psia; $T_1 = 114 + 460 = 574°R$.

(a) $v_2 \approx v_1 = \dfrac{RT_1}{p_1} = \dfrac{(53.3)(114 + 460)}{(14.89)(144)} = 14.27 \text{ ft}^3/\text{lb}$

For $v_2 = v_1$ in $A_1v_1/v_1 = A_2v_2/v_2$, we have $v_1 = A_2v_2/A_1 = v_2(D_2/D_1)^2$. Adapting equation **(b)** § 18.17, to states 1 and 2 with v_1^2 included in the radical, and using the foregoing value of v_1 to eliminate it, we have

$$v_2 = \left\{ \frac{2kRT_1}{1 - (A_2/A_1)^2}\left[\frac{p_1 - p_2}{p_1} + \frac{1}{2k}\left(\frac{p_1 - p_2}{p_1}\right)^2 \right]\right\}^{1/2}$$

(b)

$$= \left\{ \frac{(2)(32.2)(53.3)(574)}{1 - (1.5/3)^4}\left[\frac{11.4}{412.6} + \frac{1}{2.8}\left(\frac{11.4}{412.6}\right)^2 \right]\right\}^{1/2} = 242 \text{ fps}$$

(c)

$$\dot{m}' = \eta_d \dot{m} = \frac{\eta_d A_2 v_2}{v_2} = \frac{(0.98)(\pi\,1.5^2)(242)}{(4)(144)(14.27)} = 0.204 \text{ lb/sec}$$

(d)

$$\dot{V}_1 = \frac{\dot{m}' R T_1}{p_1} = \frac{(0.204)(60)(53.3)(574)}{(14.89)(144)} = 175 \text{ cfm}$$

(b) An average velocity of the air in the pipeline is

(e)

$$v_1 = \frac{\dot{m}' v_1}{A_1} = \frac{(0.204)(14.27)(144)}{\pi 9/4} = 59.3 \text{ fps}$$

Apply equation **(c)**, § 18.17, to states 0 and 1 and get

(f)

$$v_1 = [2kv_0(p_0 - p_1)]^{1/2}$$

To get this particular form, we also used $c_p = kR/(k-1)$ and $p_0v_0 = RT_0$.

(g)

$$p_0 - p_1 = \frac{v_1^2}{2kv_0} = \frac{59.3^2}{(64.4)(14.27)(144)(0.0361)} = 0.736 \text{ in. } H_2O$$

which is the velocity pressure. The stagnation pressure is

$$p_0 = p_1 + (p_0 - p_1) = 14.89 + (0.736)(0.0361) \approx 14.916 \text{ psia}.$$

18.20 FLOW NOZZLE, ORIFICE AND PITOT TUBE

A flow nozzle is a converging nozzle installed in a line with taps for obtaining pressure and temperature readings, Fig. 18/11. If the diameter of the nozzle at 2 is $D_2 > 0.22D_1$, and it ordinarily is, the effect of the initial velocity should be included. The coefficients of discharge η_d are fairly high, 96–99%.

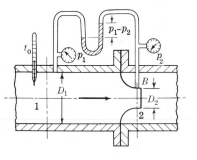

Fig. 18/11 *Flow Nozzle.* There is an advantage in having the nozzles conform to the ASME references [18.22] and [18.23], because suitable experience coefficients are available, including an initial velocity correction factor. This is also true for the other flow-measuring devices.

Since an orifice is usually a hole in a thin plate, Fig. 18/12, it is sometimes called a **thin-plate orifice**. After the stream leaves the orifice, it continues to contract in cross-sectional area until a minimum area is reached at a section called the **vena contracta**, Fig. 18/12(a). At some downstream section, the stream again fills the pipe at a pressure slightly lower than that ahead of the orifice, although there is a larger pressure drop across the orifice, with the minimum pressure at the vena contracta. The downstream tap is not necessarily located as drawn in Fig. 18/12(a) at the vena contracta, whose location is variable but is roughly a function of the pipe diameter and diameter ratio. The stated coefficients of discharge are low because they include the effect of the contraction of the jet to the vena contracta, an amount of the order of 40%.

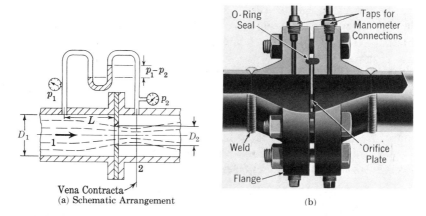

Orifices. In general, the distance L, in (a), should preferably be at least one pipe diameter. However, there are several approved locations of taps for manometer connections, one arrangement being shown in (b).   **Fig. 18/12**

The pitot tube, Fig. 18/13, points into the gas or liquid stream; the stream comes to rest against 0. The ideal change of state for a compressible fluid is isentropic, with the result that the properties at 0 are stagnation properties (closely true for moderate gas speed). In this context the pressure p_0 is also called **total pressure** and **impact pressure**, terms that apply with irreversibility present. An opening parallel to

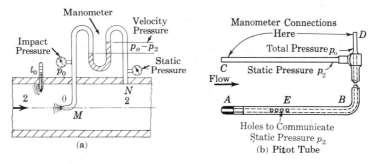

Pitot Tube. The arrangement in (a) is schematic. The pitot tube in (b) consists of two concentric tubes; the inner one opens at A where there is the impact pressure p_0; the outer one has openings as at E parallel to the stream and is subjected to the static pressure p_2. The length AB is inside the pipe and parts C and D are outside for manometer connections.   **Fig. 18/13**

the stream, as at N and E, is subjected to the so-called ***static pressure*** p_2. Similarly, a stationary thermometer in the stream registers the stagnation temperature t_0, ideally. Several readings of the pressure difference $p_0 - p_2$ are made across a diameter, the traverse being made in an approved manner to approximate the average speed. If the speed is low, as usual, the pressure change is small with the result that the volume can be taken constant, that is, as though the fluid is incompressible. See § 4.18. The applicable equation may be found now in any of several approaches, say equation **(b)**, § 18.18.

18.21   CLOSURE

Even something as simple as fluid flow in constant area passages, subsonic or supersonic, raises many questions not treated here. Also, some details of flow through turbine blades, which is inevitably a part of the thermodynamic study of internal events in turbines, are absent because of space limitations.

A reminder on the use of the critical pressure p^* is now in order. The first step in solving a nozzle problem is to calculate p^*; see §§ 18.6, 18.7. Next, compare this value of p^* with the given exit nozzle pressure. If the two are equal, the nozzle is convergent and the throat section a^* is the exit section. If the exit pressure is greater than p^* (as calculated), the fluid expansion never reached the critical pressure; hence, the nozzle is convergent but a^* is never reached. If the exit pressure is less than p^*, the fluid expansion passed through p^* and the nozzle is convergent-divergent with minimum section being a^*.

PROBLEMS

SI UNITS

18.1 Show that for a flowing gas stream, the ratio of its stagnation temperature T_0 to its flow temperature T is given by $T_0/T = 1 + M^2(k - 1)/2$ where k is the ratio c_p/c_v for the gas and M is its speed expressed as a Mach number.

18.2 Derive the Bernoulli equation for frictionless, incompressible flow starting with the steady flow energy equation: $Q = \Delta P + \Delta K + \Delta u + \Delta(pv) + W$. State all assumptions.

18.3 The Euler equation for steady flow is expressed as $dp/\rho + v\,dv + g\,dz = 0$. **(a)** Show that this is a form of the Bernoulli equation. **(b)** Demonstrate how the Euler equation may be easily converted into the momentum equation. State all assumptions.

18.4 Demonstrate that the reaction force F of an incompressible fluid undergoing a reversible flow through a converging nozzle is $F = 2A_t(p_0 - p_t)$, where A_t = throat area, p_0 = stagnation pressure, and p_t = throat pressure.

18.5 Oxygen at 172.37 kPaa is flowing in a pipe. The pressure and temperature of the O_2 on the nose of a small stationary object positioned in the stream are 193.05 kPaa and 65.6°C, respectively. Find the velocity.
Ans. 141 m/s.

18.6 Two water reservoirs are connected by an underground pipeline; the water surface in one is 10 m above that of the other. For a flow of 0.6 m³/s through the pipe, find the losses in N·m/kg.

18.7 A useful parameter for work in fluid mechanics flow is T/T^*, where T is the temperature at any section of the stream and T^* is the temperature at Mach 1. Show that it is equal to

$$\frac{T}{T^*} = \frac{(k + 1)/2}{1 + M^2(k - 1)/2}$$

18.8 The same as **18.7** except that the parameter is p/p^*; to show that

$$\frac{p}{p^*} = \frac{1}{M}\left[\frac{(k+1)/2}{1 + M(k-1)/2}\right]^{1/2}$$

18.9 A chemical reaction produces 113.4 kg/s of propellant within a rocket. What minimum effective nozzle exit velocity is required to develop a thrust of 177,930 N?

18.10 The speed of a jet-propelled plane is 1125 km/hr. The plane breathes in 75 kg/s air and burns 1.4 kg/s fuel while developing a thrust of 35,585 N. Find the exit speed of the exhaust gas.

18.11 Air at 861.85 kPaa and 32.2°C enters a nozzle and expands isentropically to the exit pressure of 172.37 kPaa. Calculate the temperature and the velocity at the exit section **(a)** when the entrance state is stagnant and **(b)** when $v_1 = 183$ m/s. **(c)** What are the stagnation enthalpy and temperature in **(b)**?
Ans. **(a)** 193 K, 476 m/s, **(b)** 510 m/s, **(c)** $t_0 = 48.8°C$.

18.12 A frictionless, adiabatic nozzle receives 2.25 kg/s of steam at 90 bar 480°C and discharges it at 50 bar. For the exit section find t_2, v_2, and A_2 **(a)** where $K_1 = 0$ and **(b)** where $v_1 = 150$ m/s. **(c)** Is this nozzle convergent? **(d)** Find the stagnation properties h_0, t_0, s_0, p_0 in **(b)**. Use Item B 16 (SI).
Ans. **(a)** 595 m/s, 2.117 cm².

18.13 A nozzle receives 907 kg/hr of steam at 25 bar, 380°C and discharges it at 7 bar. Let the initial kinetic energy be negligible and the nozzle efficiency be 93%. Use $p_t = 0.56p_1$ and determine **(a)** the throat area and **(b)** the exit area. Use Item B 16 (SI).

18.14 A nozzle expands 2.27 kg/s of steam in ideal supersaturated flow from 11 bar, 193.3°C to a discharge pressure of 4.14 bar. For the throat section, determine the pressure, velocity, specific volume, area, and Mach number **(a)** when the initial state is stagnant and **(b)** when $v_1 = 198$ m/s. **(c)** Find the velocity, volume, and area at the exit section in **(a)**.
Ans. **(a)** 6.01 bar, 476 m/s, 0.29 m³/kg, 13.80 cm², 1.0.

18.15 Air at 103.7 kPaa, 60°C enters a frictionless diffuser and is decelerated from Mach 3 at entrance to Mach 1 at exit. For a flow of 13.61 kg/s, determine **(a)** the temperature and pressure at exit, **(b)** the inlet area, **(c)** the exit area.

18.16 Steam enters a diffuser at 0.07 bar, dry-saturated, and is discharged at 0.40 bar with negligible velocity. **(a)** For an isentropic process, determine the initial velocity in mps. **(b)** The same as **(a)** except that the diffuser efficiency is 85%. What will be the temperature of the steam at discharge?

18.17 A pitot tube is used to measure the flow of air in a 10.16-cm ID pipe. The impact (stagnation) temperature is $t_1 = 37.8°C$, the static manometer reads 5.59 cm H_2O gage, and the mean velocity differential is 2.34 cm. H_2O. The barometric pressure is 1 atm and the room temperature is 36.7°C. Find **(a)** the mean velocity of the air in the pipe and **(b)** the equivalent volume of free air flowing in m³/s. **(c)** What is the difference between the stagnation enthalpy of the stream and the flowing enthalpy, J/kg?
Ans. **(a)** 20 m/s, **(b)** 9.74 m³/min.

18.18 A 15.25 × 10.15-cm venturi receives air at 344.75 kPaa, 93.3°C. The manometer differential across the venturi is 11.56 cm of water and the discharge coefficient is 0.935 (not including the initial-velocity effect). Environmental conditions are: a barometer of 1 atm and a temperature of 40°C. Find **(a)** the mass of air flowing and **(b)** the equivalent m³/min of free air.
Ans. **(a)** 43.6 kg/min, **(b)** 40.6 m³/min.

18.19 An 20.32 × 15.25-cm venturi is metering oxygen in a steady flow manner. Oxygen enters the venturi at 827.4 kPaa, 4.45°C and the manometer differential, entrance to throat, is 20.3 cm H_2O and located in an atmosphere of 1 atm, 43.3°C. The coefficient of discharge is 0.965. Find the flow of oxygen kg/min accounting for the initial velocity correction.

18.20 A 5.1-cm flow nozzle is used to measure the flow from a drum of air wherein the pressure is constant at 791 kPaa and the temperature is 32.2°C. The discharge is to the atmosphere, and the coefficient of discharge is 97%. Find the exit velocity and the mass of air discharged. *Ans.* 3.54 kg/s.

18.21 The fuel consumption of a gas engine is 180.5 kg/min of methane gas as metered by a 15.25-cm orifice placed in a 20.3-cm ID flow line. Just ahead of the orifice, the static pressure 552 kPaa and the stagnation temperature is 48.9°C; the manometer differential across the orifice is 55.9 cm H_2O; environmental conditions are 1 atm, 36.7°C. Find the discharge coefficient for the orifice.

18.22 The effect of discharge pressure on the mass rate of a gas through an ideal nozzle is desired. Select an initial state for a given gas, let the minimum cross-sectional area in the nozzle remain constant (discharge and/or throat area), and vary the discharge pressure from that of the initial state to some value well below critical. Write the program for this study.

ENGLISH UNITS

18.23 Water at 40 psig, 60°F is contained by a smooth 90°-bend in a 6-in. pipe. Find the magnitude and direction of the force on the bend **(a)** if there is no flow, **(b)** if the flow is 10 cfs.   *Ans.* **(a)** 1600 lb, **(b)** 2995 lb.

18.24 A stagnation probe made of pure tin is inserted in a high speed air jet whose temperature measured by a thermometer moving with the stream is 100°F. Melting and ablation of the probe occurs. Estimate the least stream velocity. Tin melts at 450°F.

18.25 The adiabatic bulk modulus for saturated liquid water at 100°F is $\zeta = 331{,}000$ psi. **(a)** Determine the acoustic velocity in this fluid. **(b)** Estimate the pressure necessary to reduce its volume by 2%.

18.26 Prove that the change in kinetic energy of a compressible fluid flowing through a frictionless nozzle is given by $\Delta K = -\int v\, dp$, whether or not the nozzle is adiabatic.

18.27 Show that the isentropic flow of an ideal gas with k constant through a nozzle across any section of area A_2 is given by

$$\dot{m} = A_2 p_0 \left\{ \frac{2kk}{(k-1)RT_0} \left[\left(\frac{p_2}{p_0}\right)^{2/k} - \left(\frac{p_2}{p_0}\right)^{(k+1)/k} \right] \right\}^{1/2} \text{lb/sec}$$

18.28 A convergent-divergent nozzle with a throat area of 0.6 in.^{2} expands helium isentropically from 40 psia and 85°F to atmospheric pressure. What are the pressure,

temperature, specific volume, and velocity at the throat **(a)** when the entrance state is stagnant and **(b)** when $v_1 = 500$ fps? **(c)** What mass will be discharged in lb/sec in **(a)**? In **(b)**? **(d)** What are the stagnation properties h_0, t_0, p_0 in **(b)**? **(e)** Find the Mach number at the throat and at the exit.

Ans. **(a)** 19.56 psia, 2910 fps; **(b)** 20.5 psia, 2915 fps; **(c)** 2.16 lb/sec ($v_1 = 0$); **(d)** 549°R, 42 psia ($v_1 = 500$); **(e)** 1.1 (exit).

18.29 Steam at 400 psia and 580°F enters a nozzle with a velocity of 250 fps and is discharged at 215 psia. For a flow of 4600 lb/hr, $\eta_n = 0.92$ and $p^* = 0.5375p_1$, determine **(a)** the exit velocity, **(b)** the throat area in in.^{2}, **(c)** the actual temperature at discharge, and **(d)** the increase in unavailability for $t_0 = 80$°F.

Ans. **(a)** 1701 fps, **(b)** 0.254 in.^{2}, **(c)** 448.6°F, **(d)** 60 Btu/lb.

18.30 Steam at 600 psia and 500°F enters a nozzle and leaves at 400 psia. Neglecting initial kinetic energy and considering supersaturation, determine the discharge area for a flow of 23,000 lb/hr and a nozzle coefficient of 96%. What is the nozzle efficiency?

Ans. 0.803 in.^{2}, 92.2%.

18.31 Stagnant steam at 330 psia and 440°F is expanded through a nozzle to an exit pressure of 160 psia. The flow is 2400 lb/hr and the initial velocity is negligible. Consider each of the two separate cases and determine the throat diameter in inches. **(a)** The nozzle is ideal and supersaturation is considered. **(b)** Consider supersaturation and a nozzle coefficient of 0.96.

Ans. **(a)** 0.422 in., **(b)** 0.433 in.

18.32 An airplane powered by a jet engine is moving with an airspeed of 1200 fps (about 820 mph) where the atmospheric pressure is $p_a = 7$ psia and $t_a = 40$°F. **(a)** If the air is brought to rest (relative to the plane) at the exit from the diffuser by an isentropic compression, what are its (stagnation) temperature, enthalpy, and pressure? **(b)** Solve **(a)** with a diffuser efficiency of 82%.   *Ans.* **(a)** 160°F, 14.85 psia.

18.33 A traverse of a 15-in. ID duct with a pitot tube shows a mean velocity differential of 0.75 in. H_2O. The stagnation manometer reading and temperature are 12.5 in. H_2O gage and 105°F, respectively. The barometer reads 29.78 in. Hg and the room temperature is 90°F. Calculate the mean velocity of the air in the duct and the equivalent quantity of free air flowing in cfm.

Ans. 59 fps, 4344 cfm.

18.34 Air at 50 psia, 200°F is flowing in a 20-in. ID duct. A pitot tube traverse indicates an average manometer differential reading of 7.5 in. H_2O. The manometer and duct are located in an environment of 14.7 psia, 80°F. Calculate the mass of air flowing in the duct, lb/hr.

18.35 There are flowing 389 cfm of equivalent free air through a 3×2-in. venturi, entering at 20 psig and 170°F. The differential manometer reads 7.55 in. H_2O. The surroundings are at 29.10 in. Hg and 102°F. Determine the discharge coefficient.

Ans. 0.95.

18.36 A $3 \times 1\frac{1}{2}$-in. venturi tube, with $\eta_d = 0.98$, is used to measure the flow of air in a 3-in. ID pipe. The barometer is 30.05 in. Hg and the room temperature is 96°F. The impact temperature of the air is 114°F, and the static manometer reading ahead of the venturi is 3.9 in. H_2O. The differential in the venturi is 11.4 in. H_2O. **(a)** How much free (room) air is flowing in cfm? **(b)** Estimate the impact pressure ahead of the venturi.

Ans. **(a)** 165 cfm, **(b)** 4.74 in. H_2O gage.

18.37 The velocity through the throat of a carburetor venturi is 150 fps. For the venturi, $\eta_d = 0.80$, $t_1 = 70$°F, and $p_1 = 30$ in. Hg. **(a)** Find the pressure at the throat if the flow is considered incompressible. **(b)** If the diameter of the fuel jet is 0.045 in., find the air/fuel ratio where $\eta_d = 0.75$ for the fuel jet and the fuel weighs 45 lb/ft^3; venturi ID, 0.8 in.

18.38 A 5×4-in. flow nozzle in a 5-in. ID pipe receives 180.2 lb/min of air at 40 psia and 175°F and causes a manometer differential of 15 in. water. The surroundings are at 88°F and 14.50 psia. Determine the coefficient of discharge.

Ans. 0.92.

18.39 Gaseous methane ($R = 96.33$, $k = 1.303$) at 100 psia, 200°F flows in a 10-in. ID pipe and is metered through an 8-in. diameter orifice ($\eta_d = 0.8$) positioned in the pipe. A differential manometer across the orifice reads 30 in. H_2O (at 110°F); the barometer is 14.7 psia. Find the mass flow of gas, lb/min. [*Note*: the value of $\eta_d = 0.8$ does not correct for the approach velocity.]

18.40 It is desired to relate the velocity manometer differentials for these three flowmeters that are positioned in a 6-in. ID line in which air is flowing;—a pitot tube, a 6×4-in. venturi ($\eta_d = 0.95$), and a 6×4-in. orifice ($\eta_d = 0.62$). Let the air ahead of each meter be at 50 psia, 100°F and let the flow vary from 10 lb/min to 100 lb/min. Compare the three manometer differentials as they vary with air flow. Program this study.

19

HEAT TRANSFER

19.1 INTRODUCTION

Time and again we have spoken of and solved for the heat Q in a given process without a thought as to the details of the phenomena by which it was transferred. The purpose of this chapter is to give a general idea of the nature of the problems involved and to provide an introductory background for further study. In no way is this brief chapter intended to be a full coverage of this somewhat sophisticated topic; however, there is sufficient information here to give good familiarity with heat transfer.

The three modes of heat transfer will be studied—conduction, radiation, and convection. The first two modes are true heat transfer processes since they depend for their operation on the simple existence of a temperature difference. The latter mode, convection, does not strictly adhere to the definition of heat transfer since it depends upon mass transport for its operation. However, it does accomplish the transmission of energy from a high level energy region to a lower level one and is therefore generally accepted as a separate mode.

The term heat exchanger will be used often in this chapter as the name of a device in which a transfer of energy takes place from one substance to another; see Figs. f/3, f/4 (Foreword), and Figs. 9/5 and 9/6 for excellent examples. Each of these keep the two fluids separated by means of a surface and is called a *closed heat exchanger*. There are numerous devices wherein the hot and cold fluids mix, thereby effecting an energy interchange; these are called *open heat exchangers*. Our study will involve the closed type primarily.

19.2 CONDUCTION

Modes of heat transfer are defined in § 2.17 which should be reviewed now. We recall that in conduction heat flow, the energy is transmitted by direct molecular communication without appreciable displacement of the molecules. The faster moving atoms, or electrons, or molecules in the hotter part induce by impacts an increased activity of adjacent atoms, or electrons, or molecules; thus heat flows from the hotter to the colder parts. In opaque solids, conduction is the only mode by which heat can flow.

Heat may be conducted at uneven rates, the rate increasing or decreasing; that is, the temperature difference in the path of heat flow may be increasing or decreasing. This condition is an unsteady state, as during the warm-up period of an engine, and it must be analyzed via infinitesimal heat flows $dQ/d\tau$, where $d\tau$ is the duration of time during which heat dQ passes a certain section. However, this study will be confined to systems in a steady state, wherein each point of the system remains at a constant temperature and heat transfer is constant at a rate of Q energy units per unit of time. Also, it is assumed that the flow of heat is unidirectional.

FOURIER'S LAW 19.3

For steady state, unidirectional flow, Fourier's equation (1822) gives the heat conduction as

(19-1)
$$Q = -kA\left(\frac{dt}{dL}\right) \text{Btu/hr}$$

where Q Btu/hr is the heat conducted across a surface of A ft^2, through a wall thickness of dL in., and with a temperature drop of $dt°$F through the distance dL. The dt/dL is called the **temperature gradient** along the path. The negative sign in equation (19-1) is used because the temperature decreases in the direction of heat flow (that is, $\int dt = t_2 - t_1$ is a negative number and the negative sign makes Q positive for convenience).

The symbol k represents the **thermal conductivity,**[*] which is the amount of heat (Btu/hr) transmitted in unit time across unit area (ft^2) through unit thickness (both in. and ft are used) for unit temperature change (1°F). Any set of units for k, as defined by equation (19-1), may be used, but the foregoing units are the usual ones. Values of k given and used in this chapter are for a 1-in. thickness; or

(a)
$$k \rightarrow \frac{\text{Btu-in.}}{\text{ft}^2\text{-°F-hr}}$$

If the thickness L is in feet, then

(b)
$$k\left(\frac{\text{Btu-in.}}{\text{ft}^2\text{-°F-hr}}\right)\left(\frac{\text{ft}}{12 \text{ in.}}\right) = k\frac{\text{Btu-ft}}{\text{ft}^2\text{-°F-hr}}$$

It will soon be evident that extraordinary care must be exercised in the matter of units in heat transfer, and because units are diverse in the literature, the reader is cautioned to check thoroughly the units of all factors entering into any equation. We note here that in the SI system of units, the units of k become

$$k \rightarrow \frac{\text{W/m}^2}{\text{K/m}} \rightarrow \frac{\text{W}}{\text{m-K}}$$

and 1 W/m-K = 0.578 Btu/hr-ft-°F.

---

[*] Do not confuse the symbol k with $k = c_p/c_v$ or with **k**, the Newton proportionality constant. Familiarity with the subject will make it easy to attach the correct meaning to the symbol by its context.

19.4   **Example—Heat Flux**

Compare the heat fluxes that result from a 50°F temperature difference existing across the respective surfaces of these separate 1-in. thick slabs of magnesia (85%), brickwork, and steel. Report answers in Btu/hr-ft^2 and W/m^2 (SI).

Solution. From Table VII, we find the following values of k, Btu-in./hr-ft^2-°F: magnesia (85%), 0.43; brickwork, 5; steel, 312. Now from equation (19-1), $Q = -kA(dt/dL)$, we get heat flux, Q/A as

$$Q/A \int_0^1 dL = -k \int_{t_1}^{t_2} dt$$

or

$$(Q/A)(L) = -k \, \Delta t$$

from which

$$Q/A = -k(\Delta t/L)$$

For magnesia:

$$Q/A = -(0.43)(-50/1) = 21.5 \text{ Btu/hr-ft}^2$$

$$= \left(\frac{21.5 \text{ Btu}}{\text{hr-ft}^2}\right)\left(\frac{1055 \text{ J}}{\text{Btu}}\right)\left(\frac{\text{W-s}}{1 \text{ J}}\right)\left(\frac{\text{hr}}{3600 \text{ s}}\right)\left(\frac{10.76 \text{ ft}^2}{\text{m}^2}\right)$$

$$= 67.8 \text{ W/m}^2$$

For brick work (using the above information):

$$Q/A = -(5)(-50/1) = 250 \text{ Btu/hr-ft}^2$$

$$= (250)(1055)(1/3600)(10.76) = 788 \text{ W/m}^2$$

For steel:

$$Q/A = -(312)(-50.1) = 15,600 \text{ Btu/hr-ft}^2$$

$$= (15,600)(1055)(1/3600)(10.76) = 49,191 \text{ W/m}^2$$

19.5   **VARIATION OF THERMAL CONDUCTIVITY**

Thermal conductivity varies widely, just as electrical conductivity does. Study the values in Table VII. Not only is there great variation between materials, but particular materials may also have widely different conductivities in different states. For example, k for aluminum at 212°F = 672°R is about 1440 Btu-in. per ft^2-hr-°F, but at about 18°R it goes up to about 35,000 Btu-in. per ft^2-hr-°F. However, the conductivity of solids varies so little with the more usual temperatures that a single average or typical value may be used with small error. In most applications, the *thermal resistivity*, which is the reciprocal of the conductivity, is so small for metals as compared with other resistances to heat flow that a small variation of k from its true value has little effect on the over-all conductance or resistance. If the relevant temperatures are extreme, high or low, it is advisable to look into more detailed sources for test values. As we shall see, the largest resistance is the fluid film adjacent to the solid. King[19.2] makes the following generalizations concerning conductivities of solids:

1. The conductivities of all homogeneous solid materials are relatively high; practically all good (heat) insulators are porous, cellular, fibrous, or laminated materials.
2. In general, the conductivity increases with the density and elasticity.
3. With rare exceptions, the conductivity of insulating materials increases very materially with the temperature.
4. The absorption of moisture greatly impairs the insulating value of porous materials.

The conductivities of liquids and gases are more sensitive to temperature changes. Moreover, the difficulty of eliminating convection currents in tests on liquids and gases has somewhat complicated the determination of conductivities. For relatively small temperature variations, the conductivity of solids, liquids, and gases may be assumed to vary linearly with the temperature. For this assumption the k in equation (19-1) may be taken as the arithmetic average for the temperature range involved (or the value of k for the mean temperature), and the integration made with k constant. If the variation is not linear, and a plot of k values is available, the mean value between two temperatures may be estimated from the curve (as suggested for specific heats in connection with Fig. 2/9). Lacking other data, interpolate between the values given in Table VII for k at the average temperature of the conducting body.

CONDUCTION THROUGH A PLANE WALL 19.6

After a steady state of unidirectional heat flow has been reached in a single homogeneous material whose thermal conductivity k is constant, the temperature gradient dt/dL for a plane wall is constant (a straight line—not true when k varies with temperature). Accordingly, integration of equation (19-1) gives

$$Q \int_0^L dL = -kA \int_{t_a}^{t_b} dt$$

(a)
$$Q = \frac{kA(t_a - t_b)}{L} = \text{Btu/hr}$$

where t_a and t_b are the surface temperatures of a partition such as A (Fig. 19/1), and k is the average value for the given conditions. We see that the rate of flow of heat depends (1) directly upon the temperature difference (potential) between the two surfaces of the wall, (2) directly upon the area of surface A through which transmission occurs, (3) indirectly upon the thickness L of the wall, and (4) directly upon the value of the thermal conductivity k. Applying equation (a) to a composite wall made up of three homogeneous materials A, B, C (Fig. 19/1), we have

(b)
$$Q_A = \frac{k_1 A(t_a - t_b)}{L_1} \qquad Q_B = \frac{k_2 A(t_b - t_c)}{L_2} \qquad Q_C = \frac{k_3 A(t_c - t_d)}{L_3} \text{Btu/hr}$$

Solving for the temperature difference from each of these equations, and noting that

$Q_A = Q_B = Q_C = Q$ Btu/hr for steady-state flow, we find

(c) $\qquad t_a - t_b = \dfrac{QL_1}{k_1 A} \qquad t_b - t_c = \dfrac{QL_2}{k_2 A} \qquad$ and $\qquad t_c - t_d = \dfrac{QL_3}{k_3 A}$

Next, equating the sum of the left-hand terms to the sum of the right-hand terms of these equations, we get

(d) $$t_a - t_d = \frac{Q}{A}\left[\frac{L_1}{k_1} + \frac{L_2}{k_2} + \frac{L_3}{k_3}\right]$$

or

(e) $$Q = \frac{A(t_a - t_d)}{L_1/k_1 + L_2/k_2 + L_3/k_3} = \frac{A\,\Delta t}{\Sigma(L/k)}\text{ Btu/hr}$$

for the composite wall in which the temperature drop from surface to surface is $\Delta t (\Delta t = t_a - t_d$ for the wall of Fig. 19/1; observe that in this usage, the Δt is the *first* temperature minus the second, counting in the direction in which heat flows). It is readily seen that if another section were added to the wall, the only change necessary in equation **(e)** would be the addition of another term L_4/k_4 to the sum of the other L/k's and the interpretation of t_d as being the temperature of the final surface. This sum of L/k values is simply represented by $\Sigma\,L/k$, as shown.

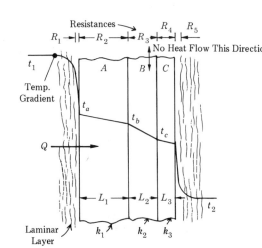

Fig. 19/1   *Temperature Gradients in a Composite Wall.*

For any one section of the wall (Fig. 19/1), $C = kA/L$ in equation **(b)** is called the *conductance*, which is seen to include the effect of size and shape of the conducting body. The conductance is the rate of heat transfer per unit temperature difference, $Q/\Delta t$. The reciprocal of the conductance is the *thermal resistance*, $R = L/(kA) = \Delta t/Q$. The *conductance per unit area*, or the *unit conductance* k/L, is often given for particular bodies. See Table VIII.) The *unit resistance* (per unit area) is L/k. For the composite plane wall (Fig. 19/1), the conductance from surface to surface is $A/\Sigma(L/k)$ [equation **(e)**], the unit conductance is $1/\Sigma(L/k)$, and the resistance is $\Sigma(L/k)/A$.

The units of k^* are Btu-in./ft^2-hr-°F, at atmospheric pressure; k for a solid changes little with pressures below 200 psi, but significant changes of k for liquids and vapors may accompany pressure changes. The values are from various sources, here selected largely from McAdams[19.1]. Straight-line interpolations are permissible between the temperatures given. Nearby extrapolations may give satisfactory results.

Material	Temp. °F	k^*	Material	Temp. °F	k^*
Solids			*Liquids (continued)*		
Aluminum	32	1400	Kerosene	68	1.03
	392	1490		167	0.97
Aluminum piston alloy	0–400	1290*	Petroleum oil,	68	1.0
			average		
Asbestos, 29 lb./ft.^{3}	−200	0.865	Sodium	212	590
				410	550
Asbestos, corrugated, 4	300	0.828	Sulfur dioxide	5	1.53
plies/in.				68	1.33
Bearing metal, white	68	164			
Brickwork, low density	68	5	Water	32	4.1
Cast iron, grey	0–400	360*		200	4.7
Copper, pure	32	2690*		300	4.75
	212	2616		620	3.3
Concrete, 1-4 dry		5.4			
Cork Board	86	0.3	*Gases*		
Glass window		3.6–7.4	Air	−148	0.109
Gold	64	2028		32	0.168
Graphite	32	1165		572	0.312
	392	910	Ammonia	−58	0.107
Gypsum	68	3		32	0.151
Gypsum plaster		3.3		212	0.23
Magnesia (85%)	100–300	0.43*	Carbon dioxide	−58	0.077
Mineral wool (glass and	86	0.27		32	0.101
rock wool)				212	0.154
Monel	68	242	Freon F12	32	0.057
				212	0.096
Plaster on wood lath,		2.5	Hydrogen	32	1.06
$\frac{3}{4}$-in. total thickness				572	2.04
Steel	0–400	312*			
Wallboard, insulating	70	0.34	Nitrogen	32	0.167
Wood, balsa	86	0.32		572	0.306
Oak, maple	59	1.44	Oxygen	32	0.17
White pine	59	1.05		212	0.226
			Steam	212	0.163
Liquids				932	0.394
Ammonia	68	3.13	Sulfur dioxide	32	0.06
	140	3.48		212	0.0827

* Average for temperature range given. Multiply given value by 0.14419 to obtain W/m-K units.

FILM COEFFICIENT 19.7

On each side of the composite wall of Fig. 19/1 is a fluid, the nature of which is of no concern at the moment. On the hot side of the wall, the fluid is hotter than the surface at some temperature $t_1 > t_a$. On the cold side, the fluid is colder than the surface, its temperature being $t_2 < t_d$. Thus, through thin films of the fluids adjacent to the surfaces, there are temperature drops $t_1 - t_a$ and $t_d - t_2$. The unit rate of heat flow through these films is called the ***film coefficient*** (and other names such as *film*

TABLE VIII   Conductances and Transmittances

The units of k/L and U are Btu per hr-ft^2-°F. These values are intended as representative, suggesting the order of magnitude in the various situations. They are not to be used in actual design unless it is known that they apply. Since the values of U are not particularized, the reference area is not meaningful. However, if these values are used for problem work, let the reference area be the internal tube or pipe area in such cases. (a) From McAdams[19.1]. (b) From ASHRAE[19.3].

Construction and Materials		$\frac{k}{L}$	U
Air space, $\frac{3}{4}$ in. or more in width	(b)	1.10	
Air space $\frac{3}{4}$ in. or more in width, bounded by aluminum foil	(b)	0.46	
Ammonia condenser, 2 × 3-in. double pipe, water inside at v = 6 fps, NH$_3$ in annular space, Δt_m = 3.5°F, clean	(a)		320
Asphalt shingles	(b)	6.50	
Brick wall, 8 in. thick, plaster inside	(b)		0.46
Brick veneer, frame wall with wood sheathing, $\frac{1}{2}$-in. plaster on gypsum lath	(b)		0.27
Brick veneer as above, plus 2-in. mineral wool insulation	(b)		0.097
Concrete blocks, 8 in., hollow gravel aggregate	(b)	1.00	
Feedwater heaters, closed; steam condensers, free convection	(a)		50–200
Feedwater heaters, closed; steam condensers, forced convection	(a)		150–800
Insulating board, $\frac{1}{2}$ in. thick	(b)	0.66	
Heat exchanger, air in tube, condensing steam outside tubes (on outside surface area)	(a)		8
Heat exchanger, cooling oil with water in tubes			50
Steam condensing, to air, free convection	(a)		1–2
Steam condensing, to air, forced convection	(a)		2–10
Steam condensing, to boiling water, free convection	(a)		300–800
Steam condensing, to liquid oil, free convection	(a)		10–30
Steam condensing, to liquid oil, forced convection	(a)		20–60
Superheaters, steam, free convection	(a)		1.6–2
Superheaters, steam, forced convection	(a)		2–6
Water to gas and liquid to gas (hot water radiators, air coolers, economizers, steam boilers), free convection	(a)		1–3
Water to gas and liquid to gas (hot water radiators, air coolers, economizers, steam boilers), forced convection	(a)		2–10
Water to water, free convection	(a)		25–60
Water to water, forced convection	(a)		150–300

conductance, surface conductance) and is represented by h, whose units in this book are Btu per hr-ft^2-°F; that is, h is the rate of heat flow (Btu in 1 hr.) through an area of 1 ft^2 when the temperature potential across the film is 1°F. Its magnitude depends on many variables, to be covered later in § 19.28. At this moment we are interested in the film coefficient purely in order to obtain a picture of the over-all, fluid-to-fluid flow of heat. From the definition of h, we see that the heat passing across the films of Fig. 19/1 is

(a) $Q = h_1 A(t_1 - t_a)$ and $Q = h_2 A(t_d - t_2)$ Btu/hr

As examples of the surface coefficients, we have from ASHRAE[19.3]

Inside building walls, still air, design value h = 1.65 Btu/hr-ft^2-°F

Outside walls, 15-mph wind, design value h = 6.00 Btu/hr-ft²-°F
Evaporating refrigerants in tube, typical value h = 200 Btu/hr-ft²-°F
Condensing steam in tube, typical value h = 2000 Btu/hr-ft²-°F

Actual values of h may be quite different from the foregoing.*

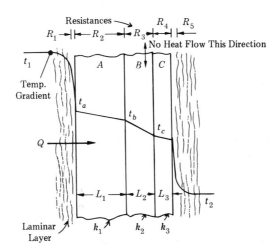

Repeated.   **Fig. 19/1**

HEAT TRANSFER FROM FLUID TO FLUID   19.8

If it were simple and convenient to measure surface temperatures, equation (**e**, § 19.6) would be the one to use. However, in practice, the fluid temperatures are easy to find, unless the speed of the fluid is high, and it is therefore desirable to express the heat flow in terms of these temperatures. To find the corresponding equation, solve for the temperature differences in equation (**a**, § 19.7), adding them to those in equation (**c**, § 19.6) as previously explained, and obtain

(19-2A)
$$Q = \frac{A(t_1 - t_2)}{1/h_1 + L_1/k_1 + L_2/k_2 + L_3/k_3 + 1/h_2} \text{ Btu/hr}$$

Generalizing from this equation, we write for the steady-state, unidirectional flow of heat,

(19-2B)
$$Q = \frac{A \, \Delta t}{\Sigma(1/h) + \Sigma(L/k)} = \frac{\Delta t}{\Sigma R} \text{ Btu/hr}$$

where Q is the rate of heat transfer from fluid to fluid through a wall of area A ft². When the temperature difference between the fluids is $\Delta t = t_1 - t_2$, ΣR is the total thermal resistance, and the total unit resistance to heat flow is

(**a**)
$$\frac{1}{U} = \Sigma \frac{1}{h} + \Sigma \frac{L}{k}$$

* Do not confuse the symbol h with that for enthalpy h.

in which $\Sigma(1/h)$ is the sum of all the surface-layer fluid resistances, and $\Sigma(L/k)$ is the sum of all the unit resistances of the materials of the wall. The over-all unit resistance is represented by $1/U$ and the reciprocal of this over-all resistance is called the **transmittance** U; *also the* **over-all coefficient of heat transfer** *and the* **over-all** (unit) **conductance**.* Since from equation **(a)**,

$$U = \frac{1}{\Sigma(1/h) + \Sigma(L/k)}$$

equation (19-2B) may be written

(19-2C) $$Q = UA \, \Delta t \; \text{Btu/hr}$$

(See § 19.12.) Table VIII shows some representative values of U.

It will be helpful to draw an analogy between the flow of heat and the flow of electricity. You recall Ohm's law as $I = E/R$, where I is the current flowing (analogous to the rate of heat flow), E is the electromotive force or electrical potential (analogous to the temperature difference or thermal potential which is the "driving force" for heat), and R is the resistance to the flow of electricity (analogous to the thermal resistances of $1/hA$ and L/kA). Note that the total thermal resistance is the equivalent of a series connection of electrical resistances, in which case the total resistance is the sum of the series connected resistances; $R = R_1 + R_2 + R_3 + \cdots$, where R_1, R_2, R_3, and so on are the individual resistances. Compare with Fig. 19/1. For R constant, the greater the voltage drop, the greater the current flow (the greater the temperature drop, the greater the heat flow). If the total resistance is increased by adding more resistances in the denominator of equation (19-2B), or by increasing the size of one or more resistance, heat flows at a slower rate. This effect is the desired one in providing insulation for steam lines, cold-storage rooms, and so on. On the other hand, if a greater rate of heat flow is desired, as in apparatus designed for the purpose of transferring heat, we endeavor to reduce the resistance. If one resistance is very much larger than any of the others, there will be little benefit derived from reducing any of the resistances except the largest; thus, if the largest resistance cannot be reduced, practically the best conditions have been attained.

The engineer has to keep in mind that the transmittance may *decrease* significantly during the operation of a heat exchanger because of the accumulation of deposits on the transmitting surfaces. On the other hand, the transmittance of insulated walls may *increase* due to deterioration of the insulating material. These materials often lose a large part of their insulating value because they become wet from seepage or from moisture deposited from air cooled below the dew point.

---

* "Terms ending in -*ivity* designate characteristics of materials, normally independent of size or shape; sometimes called *specific properties*. Examples are *conductivity* and *resistivity*. Terms ending in -*ance* designate properties of a particular object, depending not only on the material, but also upon size and shape: sometimes called *total quantities*. Examples are *conductance* and *transmittance*. Terms ending in -*ion* designate time rate of the process of transfer ... Examples are *conduction* and *transmission*. *Transmission, transmissivity, transmittance* usually refer to transfer by one or more of the processes of conduction, convection, and radiation." From A.S.A. Standard, Z 10.4–1943.

The plane wall of a building is composed of the following materials: 1 in. white pine wood, 3.5 in. mineral wool insulation, and 0.75 in. gypsum plaster. On the outer pine wood surface, the film coefficient is $h_o = 6$ Btu/hr-ft^2-°F; on the inner plaster surface, $h_i = 1.65$. Find (a) the transmittance U for the composite wall (fluid to fluid), (b) Q per ft^2 of wall surface for an overall temperature difference of $\Delta t = 50$°F (fluid to fluid), (c) the temperature drop across each resistance for the total $\Delta t = 50$°F.

Solution. From Table VII, we find these values of k, Btu-in./hr-ft^2-°F; white pine, 1.05; mineral wool, 0.27; gypsum plaster, 3.3.

(a) The transmittance U from equation (**a**, § 19.8),

$$\frac{1}{U} = \Sigma \frac{1}{h} + \Sigma \frac{L}{k}$$

$$= \frac{1}{6} + \frac{1}{1.65} + \frac{1}{1.05} + \frac{3.5}{0.27} + \frac{0.75}{3.3}$$

$$= 0.167 + 0.606 + 0.952 + 12.962 + 0.227 = 14.914$$

$$U = 1/14.914 = 0.067 \text{ Btu/hr-ft}^2\text{-°F}$$

(b) From equation (19-2C),

$$Q = AU \, \Delta t = (1)(0.067)(50) = 3.35 \text{ Btu/hr}$$

(c) Various temperature drops,

Air-film: $\Delta t = Q/Ah_o = (3.35)/(1)(6) = 0.56$°F

Pine wood: $\Delta t = QL/kA = (3.35)(1)/(1.05)(1) = 3.19$°F

Mineral wool: $\Delta t = QL/kA = (3.35)(3.5)/(0.27)(1) = 43.44$°F

Gypsum plaster: $\Delta t = QL/kA = (3.35)(0.75)/(3.3)(1) = 0.77$°F

Air film: $\Delta t = Q/Ah_i = 3.35/(1)(1.65) = 2.04$°F

These temperature drops add up to be 50°F. Note that the major resistance and the largest Δt are presented by the mineral wool.

CONDUCTION THROUGH CURVED WALL 19.10

Since the area through which heat flows in a curved wall is not constant, we must reconsider Fourier's equation (19-1) in this connection. Consider a thick cylinder (Fig. 19/2) for which the temperature on the inside surface is t_a, the temperature on the outside surface is t_b, and the thermal conductivity is k. The heat flows radially, say, from the inside to the outside, and in doing so, a given quantity of heat passes across larger and larger areas, since the cylindrical area increases with the radius of the cylinder. Consider a length of cylinder z and take a very thin element of the cylinder of thickness dr with a radius of r inches. The area of this thin cylindrical surface is $2\pi rz$. The change in temperature across dr is a differential amount dt.

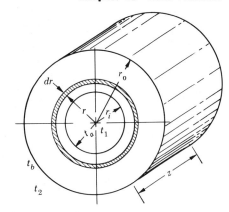

Fig. 19/2   *Curved Wall.* The ends are insulated so that there is no longitudinal flow of heat.

Thus, Fourier's equation gives

(19-3)
$$Q = -k2\pi rz \frac{dt}{dr}$$

Separating the variables and integrating, we get

$$Q \int_{r_i}^{r_o} \frac{dr}{r} = -2\pi zk \int_{t_a}^{t_b} dt,$$

$$Q \ln \frac{r_o}{r_1} = 2\pi zk(t_a - t_b),$$

or

(a)
$$Q = \frac{2\pi zk(t_a - t_b)}{\ln (r_o/r_i)} = \frac{2\pi zk \, \Delta t}{\ln (D_o/D_i)}$$

where r_o is the outside radius of the pipe, r_i is the inside radius, and $r_o/r_i = D_o/D_i$. In equation (a), note that the resistance for the curved wall is

(b)
$$R = \frac{\Delta t}{Q} = \frac{\ln (r_o/r_i)}{2\pi zk}$$

Setting up an equation for the heat flow through a composite curved wall, such as an insulated pipe, is simple when it is done by summing resistances ΣR in accordance with equation (b), $Q = \Delta t/\Sigma R$. Consider Fig. 19/3, which represents a pipe X with insulation Y, wherein the inside film coefficient is h_i and the outside value is h_o; the temperature of the fluid on the inside is t_1, which, in the following discussion, is greater than t_2, the temperature of the fluid on the outside. The film resistances are (assuming no film resistance at the intermediate area A)

(c)
$$R_i = \frac{1}{A_i h_i} \quad \text{and} \quad R_o = \frac{1}{A_o h_o}$$

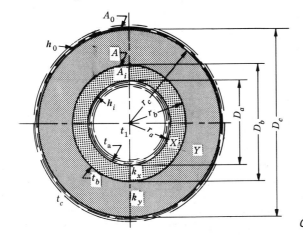

Composite Curved Wall. **Fig. 19/3**

Applying equation **(b)** to X and Y (see the dimensions in Fig. 19/3—$D_b/D_a = r_b/r_a$), we get their resistances as

(d) $$R_z = \frac{\ln (D_b/D_a)}{2\pi z k_x} \quad \text{and} \quad R_y = \frac{\ln (D_c/D_b)}{2\pi z k_y}$$

Thus, the heat flow is $Q = \Delta t/\Sigma R$, or

(19-4A) $$Q = \frac{\Delta t}{\dfrac{1}{A_i h_i} + \dfrac{\ln (D_b/D_a)}{2\pi z k_x} + \dfrac{\ln (D_c/D_b)}{2\pi z k_y} + \dfrac{1}{A_o h_o}}$$

or generalizing for any number of laminated cylinders, each with certain inside and outside diameters of D_i and D_o, with a conductivity of k, and z ft. long, we have for a steady state

(19-4B) $$Q = \frac{\Delta t}{\Sigma \dfrac{1}{Ah} + \Sigma \dfrac{\ln (D_o/D_i)}{2\pi z k}}$$

where $\Delta t = t_1 - t_2$ is the temperature drop from fluid to fluid, $\Sigma(1/Ah)$ is the sum of all film resistances in the path of heat flow, and the other term in the denominator sums all other resistances. The heat is flowing radially only.

IMPORTANT NOTE: The value of k in equation (19-4B) must have the units Btu-ft per hr-ft²-°F in order to get heat in Btu per hr. Compare with the units given in Table VII.

In applying $Q = UA \Delta t$ to curved walls, the area A becomes a convenient reference area; for a single pipe, either the outside or the inside pipe area. Inasmuch as a certain amount of heat Q is passing through the wall under steady-state operation no matter which area is the reference area, we have

(19-4C) $$Q = U_o A_o \Delta t \quad \text{or} \quad Q = U_i A_i \Delta t$$

wherein, by comparison with equation (19-4B),

(19-5) $$U_o A_o = U_i A_i = \cfrac{1}{\sum \cfrac{1}{Ah} + \sum \cfrac{\ln (D_o/D_i)}{2\pi z k}}$$

Thus, if an over-all transmittance U is given for a curved wall, the corresponding reference area should be stated.

19.11   Example

An insulated steam pipe, located where the ambient temperature is 90°F, has an internal diameter of 2 in. and an external diameter of $2\frac{1}{2}$ in. The outside diameter of the corrugated asbestos insulation is 5 in. and the surface coefficient on the outside is h_o = 2 Btu/hr-ft²-°F, a value intended to include the effect of radiated heat where the movement of the air is that due to natural circulation (free convection). On the inside, the steam has a temperature of 300°F and h_i = 1000 Btu/hr-ft²-°F. (This value of h_i is commonly used for *saturated* or *wet* steam flowing in a pipe. Note that this value corresponds to a low resistance to heat flow, so that some inaccuracy here has little effect on the transmittance.) Compute (a) the heat loss per foot of pipe length, and (b) the surface temperature on the outside of the insulation.

   Solution.   (a) Refer to Fig. 19/3. From Table VII, we find

$$k_x = 312 \text{ Btu-in./hr-ft}^2\text{-}°F \text{ for steel,}$$

$$k_y = 0.828 \text{ Btu-in./hr-ft}^2\text{-}°F \text{ for the corrugated asbestos insulation}$$

Converting the inch unit to feet, we have k_x = 312/12 and k_y = 0.828/12. The true value of k for steel will vary with the composition of the steel, but since the resistance of any steel to the flow of heat is relatively small in a case like this, whatever error is involved will be seen to have negligible effect after numerical results have been obtained. First, compute the various resistances per foot of pipe length. The area of a cylinder is $\pi D z$, where z = 1 ft is its length and the area is to be in square feet. (For R_1 and R_4 below, 12 converts the diameters to feet.)

Inside film: $R_1 = \cfrac{1}{A_i h_i} = \cfrac{12}{(\pi 2)(1000)} = 0.00191$ hr-°F/Btu

Pipe: $R_2 = \cfrac{\ln (D_o/D_i)}{2\pi z k} = \cfrac{\ln (2.5/2)}{(2\pi)(312/12)} = 0.001365$ hr-°F/Btu

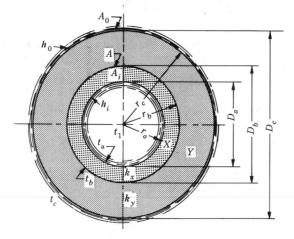

Fig. 19/3   Repeated.

Insulation: $R_3 = \dfrac{\ln(D_o/D_i)}{2\pi z k} = \dfrac{\ln(5/2.5)}{(2\pi)(0.828/12)} = 1.6$ hr-°F/Btu

Outside film: $R_4 = \dfrac{1}{A_o h_o} = \dfrac{12}{(\pi 5)(2)} = 0.382$ hr-°F/Btu

Summing these resistances, we have $R = 1.985275$ hr-°F/Btu, where it is evident that the last several digits are not significant. Observe the small effect of the first two resistances. For $\Delta t = 300 - 90 = 210°F$, we get

$$Q = \frac{\Delta t}{\sum R} = \frac{210}{1.985} = 106 \text{ Btu/hr for each foot of length}$$

(b) The heat $Q = 106$ Btu/hr flows through each cylindrical lamination and will be equal to

$$Q = \frac{\text{temperature difference between two sections}}{\text{sum of the resistances between the same sections}}$$

For a temperature difference to the outside surface of $t_1 - t_c = 300 - t_c$ (Fig. 19/3), and a resistance to the outside surface of $R_1 + R_2 + R_3 = 1.603$, we get

$$106 = \frac{300 - t_c}{1.603}$$

from which $t_c = 130°F$, a safe value for the outside surface.

LOGARITHMIC MEAN TEMPERATURE DIFFERENCE   19.12

Our discussion of the transfer of heat through walls from one fluid to another has been on the tacit assumption that the hot fluid remains at a constant temperature t_1 and the cold fluid remains at a constant temperature t_2. However, in many instances, either the cold fluid or the hot, or both, undergo a change of temperature in passage through the heat exchanger.

For purposes of explanation, consider a double-pipe type of exchanger (Fig. 19/4). Let the hot fluid flow through the annular space and the cool fluid through the inner pipe. As the temperatures of the fluids change, the difference between the temperatures of the fluids changes. Thus we find different differences between the temperatures of the hot and cold fluids at different sections of the exchanger. In such cases, the *logarithmic mean temperature difference* (LMTD) represented by Δt_m, is used in place of $t_1 - t_2$. Since the rate of change of temperatures of the substances is not constant, the LMTD is not the same as the arithmetic mean, as shown in equation (19-6).

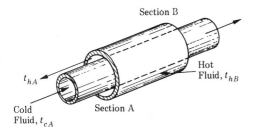

Double-Pipe Arrangement. Observe the notation. One end of the exchanger is designated as section *A* and the other end, section *B*. In this illustration, a counterflow arrangement, t_{cA} is the *initial* temperature of the cold fluid and t_{hA} is the *final* temperature of the hot fluid.

Fig. 19/4

At this time, we shall consider two types of flow, *parallel flow*, in which case the fluids flow in the same direction through the heat exchanger, and *counterflow*, where the fluids flow in opposite directions. In a general way, the variation of temperatures is as indicated in Fig. 19/5. In case of parallel flow, both fluids enter at section A, and as they pass to section B their temperatures approach one another. At and near section A [Fig. 19/5(a)], the temperature difference is a maximum, and consequently the rate of flow of heat and the rate of decrease of temperature difference are a maximum. As the temperatures of the fluids approach one another, the rates of change of the temperatures decrease and the curves flatten out. In Fig. 19/5(b), a counterflow arrangement, the cold fluid enters at section B, leaves at A. In contrast to parallel flow, heat transfer is taking place between the fluids at the moment when each is in its coldest state, and when each is in its hottest state. The transfer of heat in a counterflow exchanger tends to conserve available energy and makes possible a higher final temperature of the fluid being heated than obtainable with parallel flow.

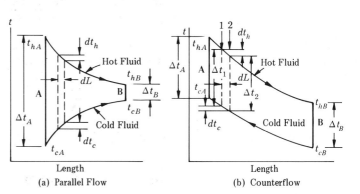

Fig. 19/5   *Temperature Variations, Flow and Counterflow.* In each case, the temperature difference at section A is $\Delta t_A = t_{hA} - t_{cA}$; and at section B, the temperature difference is $\Delta t_B = t_{hB} - t_{cB}$.

(a) Parallel Flow   (b) Counterflow

To find the equation for the mean temperature difference, consider a differential area of transmitting surface dA in a *counterflow* exchanger. As the cold fluid flows past this area, its temperature increases a differential amount dt_c due to the transfer of a differential quantity of heat dQ, the amount of which is $dQ = m_c c_c \, dt_c$, where m_c is the mass of cold fluid flowing in a unit time and c_c is the specific heat of the cold fluid. Assuming no heat loss from the exchanger and no change of kinetic energy, we conclude that the heat loss of the hot fluid $dQ = m_h c_h \, dt_h$ is the same as the gain of the cold fluid. Note that the $\int dt_h$ is a negative number, so that

(a) $$dt_c + dt_h = d(\Delta t)$$

is the change in the difference of temperatures; $\int d(\Delta t) = \Delta t_2 - \Delta t_1$ (Fig. 19/5) across a differential area dA of infinitesimal length dL between sections 1 and 2. From the definition of the transmittance U, we have also

(b) $$dQ = U \, dA \, \Delta t$$

Using the conclusions of the foregoing discussion, we get

(c) $$dt_c + dt_h = d(\Delta t) = dQ\left(\frac{1}{m_c c_c} - \frac{1}{m_h c_h}\right) = U \, dA \, \Delta t\left(\frac{1}{m_c c_c} - \frac{1}{m_h c_h}\right)$$

where dQ is a positive number and the negative sign on the right-hand side is necessary to produce an arithmetic difference. Rearranging the foregoing equation, we have

(d)
$$\int_{\Delta t_A}^{\Delta t_B} \frac{d(\Delta t)}{\Delta t} = U\left(\frac{1}{m_c c_c} - \frac{1}{m_h c_h}\right) \int_0^A dA$$

wherein the transmittance U and specific heats c are taken as constant. Integration of this equation gives

(e)
$$\ln \frac{\Delta t_B}{\Delta t_A} = -\ln \frac{\Delta t_A}{\Delta t_B} = \frac{UA}{m_c c_c} - \frac{UA}{m_h c_h} = -\left(\frac{UA}{m_h c_h} - \frac{UA}{m_c c_c}\right)$$

Now using $Q = UA \Delta t_{mc} = m_c c_c (t_{cA} - t_{cB}) = m_h c_h (t_{hA} - t_{hB})$ in this equation, we find

(f)
$$\ln \frac{\Delta t_A}{\Delta t_B} = \frac{t_{hA} - t_{hB}}{\Delta t_m} - \frac{t_{cA} - t_{cB}}{\Delta t_m} = \frac{(t_{hA} - t_{cA}) - (t_{hB} - t_{cB})}{\Delta t_m}$$

from which

(19-6)
$$\Delta t_m = \frac{\Delta t_A - \Delta t_B}{\ln (\Delta t_A / \Delta t_B)}$$

the value of the log mean temperature difference between sections A and B when the difference in temperature at A is Δt_A, and the difference at B is Δt_B. It does not matter which end of the heat exchanger is taken as section A and B; the same result will be obtained from equation (19-6). A similar analysis of the case of parallel flow will yield the same equation (19-6). Hence this equation applies to either case and it serves also whenever the temperature of either fluid is constant, as during evaporation or condensation. When Δt_A is equal to or nearly equal to Δt_B, use the arithmetic average for Δt_m. Where there is cross flow, as around baffles in heat exchangers, and where there are fins, the mean temperature difference is not given by equation (19-6); nor in case m, c, or U varies. For such other situations, see McAdams[19.1]. Equation (19-2B) may now be written,

(19-7)
$$Q = UA \Delta t_m$$

In taking values of U for curved walls from the literature, notice whether they are based on the inside A_i or outside A_o surfaces.

Example—Logarithmic Mean Temperature Difference 19.13

An economizer (see Figs. f/2 and f/7, Foreword) receives hot gas ($c_p = 0.27$ Btu/lb-°R) and water in the ratio 1.5 lb gas/lb water. The gas enters at 850°F and leaves at 355°F; the water enters at 120°F. Find the exit temperature of the water and the LMTD (a) for parallel flow, (b) for counterflow. There are no external energy losses.

Solution. The energy given up by the hot gas is that absorbed by the water, that is,

$$Q_w = Q_g$$

$$m_w c_{pw}(t_2 - t_1) = m_g c_{pg}(t_b - t_a)$$

$$(1)(t_2 - 120) = (1.5)(0.27)(850 - 355)$$

Exit water temperature:

$$t_2 = (1.5)(0.27)(495) + 120$$

$$= 200 + 120 = 320°F$$

(a) COUNTER FLOW: From equation (19-6),

$$LMTD = \frac{\Delta t_A - \Delta t_B}{\ln \Delta t_A / \Delta t_B}$$

$$= \frac{530 - 235}{\ln (530/235)} = 363°F$$

(b) PARALLEL FLOW:

$$LMTD = \frac{730 - 35}{\ln (730/35)}$$

$$= 229°F$$

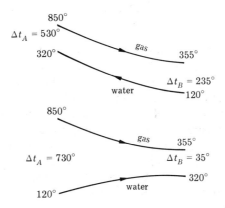

Fig. 19/6   *Temperature gradients.*

19.14   THERMAL RADIATION

All bodies *radiate* heat. If two bodies were completely isolated in a vacuum from all other bodies, but not from each other, the colder body would radiate heat to the hotter body, and the hotter body would radiate heat to the colder body. For purposes of explanation, let all the radiant heat from the hotter body go to the colder body, and all the radiant heat from the colder body to the hotter body. Then the colder body would be heated because, due to its lower temperature, it radiates less heat than it receives from the body with the higher temperature. Radiant heat or thermal radiation is an emanation of the same nature as light and radio waves. Like other waves in striking a body, some will be reflected (the fraction being indicated by the reflectance ρ), some will be absorbed (the fraction being indicated by the absorptance α), and if the body is transparent to the waves, some will be transmitted through the body (the fraction being indicated by the transmittance T). The sum of these three fractions must be $\rho + \alpha + T = 1$. Bodies which are opaque to light are for the most part opaque to radiant heat (true of most solids). For opaque substances, the energy of the radiant heat is either absorbed or reflected, in which case *the reflectance plus absorptance equals unity*. At the other extreme, some gases and some glass *transmit* nearly all the radiant heat. Brightly polished metals are such good reflectors that most of the radiant heat is reflected.

A body which absorbs all the impinging radiant heat is called a **black body**, a hypothetical conception in which the absorptivity is unity. A black body is also the best radiator. At a particular body temperature, actual bodies radiate less heat than a black body and are called gray bodies. The ratio of the radiation from an actual body to the radiation from a black body is called the **emittance** ε. The emittance is not a constant property (see Table X) but usually it increases with the temperature of the radiating body.

Kirchoff's law may be stated as follows: the absorptance α and the emittance ε of a body are the same when the body is in thermal equilibrium with its surroundings. From this statement we can see that the emittance of highly polished metals may be quite low.

The nearest approach to a black body is obtained by a hollow vessel penetrated only by a small pin hole through which radiant heat may pass to the inside. Once inside, the radiant heat has little chance of being reflected back out the pin hole. Thus, for practical purposes this vessel absorbs all the energy entering the hole.

STEFAN–BOLTZMANN LAW   19.15

The Stefan–Boltzmann law states that the amount of radiation from a black body is proportional to the fourth power of the absolute temperature, $Q_R = \sigma A T^4$, where $\sigma = (0.1713)(10^{-8})$ is the Stefan–Boltzmann constant when Q_R is in Btu per hour, and A is the radiating area in square feet of the black body whose surface temperature is $T°$R. As usual, this equation for an ideal situation must be modified to account for actual situations. As previously stated, net heat transferred by radiation is the result of an interchange of radiation, the radiation from the hot body to the cold body minus the radiation from the cold body. Also, the effective emittance and the fact that all the radiation from a certain source will not strike the surface must be considered. We shall proceed to develop the means by which these effective emittances and the surface configurations may be handled in a given situation.

SOME BASIC DEFINITIONS   19.16

The following definitions are required in order to express a given amount of radiant heat q in terms of its several variables.

Radiosity, denoted by J, is the total radiant energy leaving a unit of surface area per unit time; it includes the emitted and reflected energy.

Irradiation, denoted by G, is the total radiant energy incident upon a unit of surface area per unit of time.

Total emissive power, denoted by w, is the total emitted thermal radiation leaving a unit of black body surface area per unit of time; reflected energy is not included.

These three definitions combined with those in § 19.14 lead to this useful relation:

(a) $$J = w + \rho G$$

or

$$J = w + (1 - \alpha)G = w + (1 - \varepsilon)G$$

Intensity of radiation, denoted by I, is the thermal radiation flux from a surface included in a unit solid angle per unit area of the surface projected normal to the line connecting the area and the observer. The relation between radiosity J and intensity I can be shown to be $J = \pi I$.

19.17   CONFIGURATION FACTOR

Most radiation problems involve two or more surfaces and require something to account for the fraction of diffuse radiant energy that leaves one surface and falls directly on the other(s). This fractional part is caused by the spatial configurations of two (or more) surfaces and is termed *the configuration factor*. Taking the general case, consider two areas A_1 and A_2 arbitrarily located in space; see Fig. 19/7.

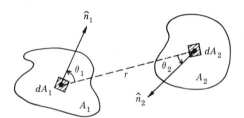

Fig. 19/7   *Two areas in space and radiating—$\hat{n}$'s are the surface normals.*

Radiant energy leaving dA_1 and striking dA_2 must travel along a straight line connecting these two areas. The quantity of radiant energy leaving dA_1 directly toward dA_2 per unit of solid angle is

(a) $$dq_1 = I_1 \, dA_1 \cos \theta_1$$

Now defining the solid angle subtended by dA_2 as

(b) $$d\phi = (dA_2 \cos \theta_2)/r^2$$

we have for the total radiation striking dA_2 from dA_1 as

(c) $$d^2q = \frac{I_1 \cos \theta_1 \cos \theta_2 \, dA_1 \, dA_2}{r^2}$$

which upon integration gives

(d) $$q = \int_{A_1} \int_{A_2} I_1 \frac{\cos \theta_1 \cos \theta_2}{r^2} \, dA_1 \, dA_2$$

Now the total radiant energy leaving A_1 is $A_1 J_1 = \pi I_1 A_1$; therefore, the fraction which strikes dA_1 is

(e)
$$F_{12} = q/\pi I_1 A_1 \quad \text{or} \quad I_1 = \frac{q}{F_{12}\pi A_1}$$

and, when substituted in **(d)**, gives

(19-8)
$$F_{12} = \frac{1}{A_1} \int_{A_1} \int_{A_2} \frac{\cos \theta_1 \cos \theta_2}{\pi r^2} dA_1 \, dA_2$$

Here F_{12} is called the configuration factor—also, angle factor, geometric factor, shape factor, and so on. Thus, the configuration factor is the fraction of diffuse radiant energy that leaves one surface in space and directly strikes another surface. Its value is based solely upon the manner in which the two surfaces are positioned.

We may now say that the radiant energy striking one black body from another is

$$q_{12} = F_{12}\sigma T_1^4 A_1$$

Also, the amount of radiant energy striking A_1 from A_2 is

$$q_{21} = F_{21}\sigma T_2^4 A_2$$

and the net heat is

(19-9)
$$q_{12\text{net}} = q_{12} - q_{21} = F_{12}\sigma A_1 (T_1^4 - T_2^4)$$

where $A_1 F_{12} = A_2 F_{21}$ since $q_{12\text{net}} = 0$ if $T_1 = T_2$.

It is also true that

(f)
$$\sum_i F_{1_i} = 1$$

for i surfaces seen by area A_1 since all energy emitted by A_1 must go somewhere.

Table IX and the several charts that follow it list numerous situations and the means for determining the value of F_{12} in each case.

RADIATION BETWEEN GRAY BODIES   19.18

Since all bodies (surfaces) are actual, we must account for their respective emittances in addition to their configurations.

Let the amount of radiant heat q_1 leaving surface A_1 be,

(a)
$$q_1 = A_1(J_1 - G_1)$$

or

$$q_1/A_1 = J_1 - G_1$$

But

$$J_1 = \varepsilon_1 w_1 + \rho G_1 = \varepsilon_1 w_1 + (1 - \alpha)G_1 = \varepsilon_1 w_1 + (1 - \varepsilon_1)G_1$$

from which

$$G_1 = [1/(1 - \varepsilon_1)](J_1 - \varepsilon_1 w_1)$$

Introducing this last expression into equation **(a)** yields

$$q_1/A = J_1 - \left(\frac{1}{1 - \varepsilon_1}\right)(J_1 - \varepsilon_1 w_1)$$

$$= \frac{w_1 - J_1}{(1 - \varepsilon_1)/\varepsilon_1}$$

or

(b)
$$q_1 = \frac{w_1 - J_1}{(1 - \varepsilon_1)/\varepsilon_1 A_1}$$

Equation **(b)** suggests an electrical analogy with the current q_1 being analogous to the potential difference $(w_1 - J_1)$ divided by the resistance $(1 - \varepsilon_1)/\varepsilon_1 A_1$. In an analogous circuit we have the form shown in Fig. 19/8.

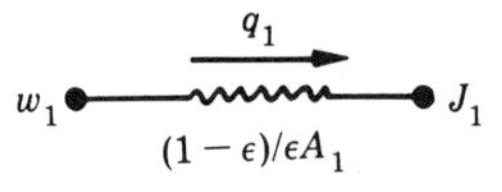

Fig. 19/8

If we now extend this idea to two surfaces A_1 and A_2 located in space relative one with another and exchanging radiant heat, it follows that

$$q_{12} = J_1 A_1 F_{12} - J_2 A_2 F_{21} = A_1 F_{12}(J_1 - J_2)$$

(c)

$$= \frac{J_1 - J_2}{1/(A_1 F_{12})} = \frac{J_1 - J_2}{1/(A_2 F_{21})}$$

Here the resistance due to the space configuration is $1/(A_1 F_{12})$, or $1/(A_2 F_{21})$, where that resistance due to the gray effect is $(1 - \varepsilon)/\varepsilon A$. When combined into an electrical analogy, even the most complex situation can be solved with some ease.

19.19   Example—Two Gray Surfaces in Space

Two parallel gray planes $(\varepsilon_1, \varepsilon_2)$ in space are exchanging radiant heat. Use the electrical analogy method and solve for the net radiant heat leaving surface A_1.

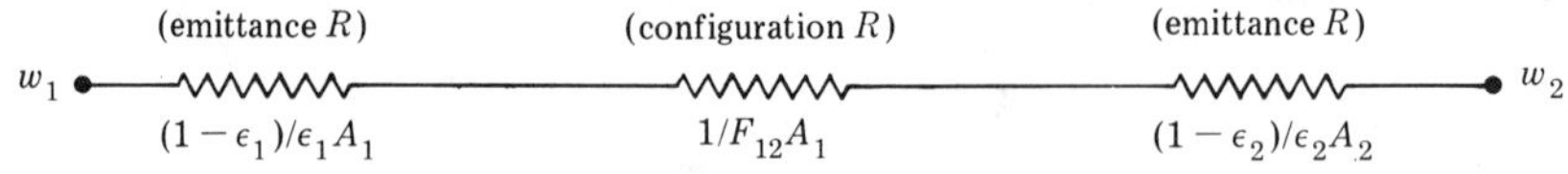

Fig. 19/9

Solution.   The electrical analogy would appear as three resistances in series as in Fig. 19/9. The total resistance is

$$R_T = (1 - \varepsilon_1)/\varepsilon_1 A_1 + 1/F_{12}A_1 + (1 - \varepsilon_2)/\varepsilon_2 A_2$$

From $I = E/R$ (OHM's LAW)

$$q_{12} = \frac{w_1 - w_2}{(1 - \varepsilon_1)/\varepsilon_1 A_1 + 1/F_{12}A_1 + (1 - \varepsilon_2)/\varepsilon_2 A_2}$$

$$= \frac{\sigma T_1^4 - \sigma T_2^4}{(1 - \varepsilon_1)/\varepsilon_1 A_1 + 1/F_{12}A_1 + (1 - \varepsilon_2)/\varepsilon_2 A_1(F_{12}/F_{12})}$$

$$q_{12} = \frac{A_1\sigma(T_1^4 - T_2^4)}{(1 - \varepsilon_1)/\varepsilon_1 + 1/F_{12} + (1 - \varepsilon_2)F_{21}/\varepsilon_2 F_{12}}$$

If the planes are infinite, $F_{12} = 1 = F_{21}$ and we have

$$q_{12\text{net}} = \frac{A_1\sigma(T_1^4 - T_2^4)}{(1 - \varepsilon_1)/\varepsilon_1 + 1 + (1 - \varepsilon_2)/\varepsilon_2}$$

$$= \frac{A_1\sigma(T_1^4 - T_2^4)}{1/\varepsilon_1 + 1/\varepsilon_2 - 1}$$

Configuration (Space) Factors for Radiation Equation*   TABLE IX

Case	Configuration of Surface	Area, A	F_{12}
1	Infinite parallel planes.	Either	1
2	Completely enclosed body 1, small compared with enclosing body 2.	A_1	1
3	Completely enclosed body 1, large compared with enclosing body 2.	A_1	1
4	Concentric spheres or infinite cylinders	A_1	1
5	Element dA and rectangular surface above and parallel to it, with one corner of rectangle contained in normal to dA.	dA	Given in Fig. 19/10.
6	Two parallel circular disks of same diameter with centers on same normal to their planes, or	Either	Given in Fig. 19/11.
7	Two equal rectangles in parallel planes and directly opposite one another.	Either	Given in Fig. 19/12.
8	Two perpendicular rectangles with a common edge.	A_1	Given in Fig. 19/13.

* Hottel[19.9].

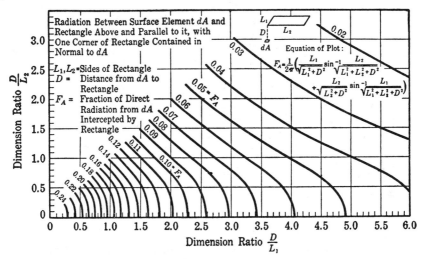

Radiation Between Element and Rectangular Parallel Surface Above Element.   **Fig. 19/10**
(Hottel[19.9].)

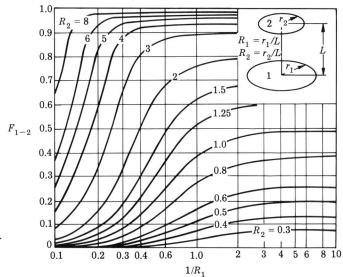

Fig. 19/11   *Radiation shape factor for parallel, concentric, discs. (Chapman[19.14])*

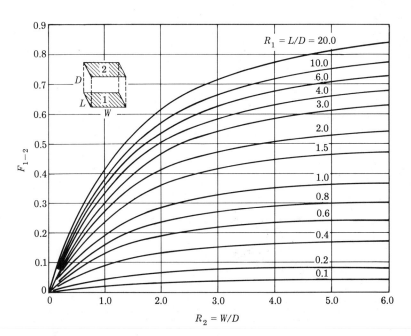

Fig. 19/12   *Radiation shape factor for parallel, directly opposed, rectangles. (Chapman[19.14])*

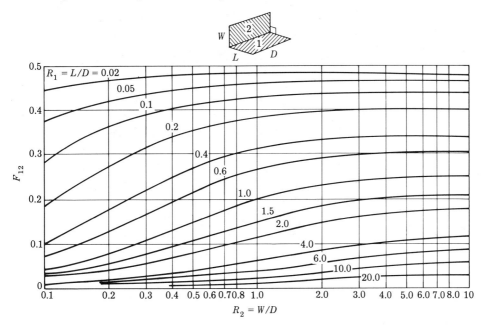

Radiation shape factor for perpendicular rectangles with a common edge. **Fig. 19/13**
(Chapman[19.14])

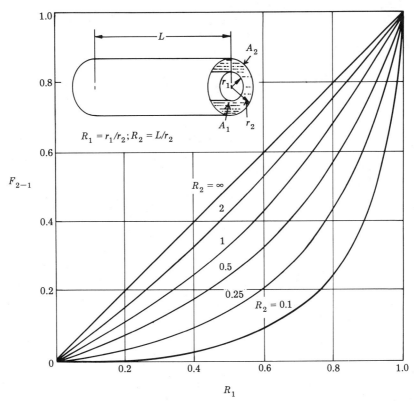

Radiation shape factor for concentric cylinders of finite length. (Chapman[19.14]) **Fig. 19/14**

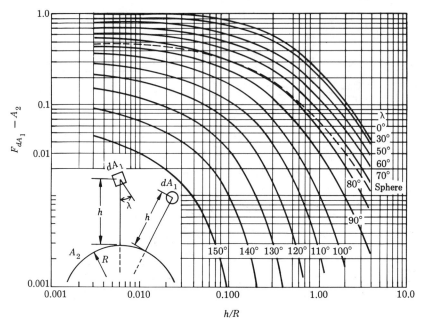

Fig. 19/15   *Radiation shape factor for infinitesimal planes and spheres in the presence of a large sphere.* (Chapman[19.14])

19.20   Example—Radiation from Steam Line

The steam line in § 19.11 has an outer insulation surface temperature of 130°F and is located in a large room where the ambient temperature is 90°F. Find the heat radiated per foot of pipe length.

Fig. 19/16

emittance R space R emittance R

w_1 •——〰〰〰——————〰〰〰——————〰〰〰——• w_2

$(1 - \epsilon_1)/\epsilon_1 A_1$ $1/F_{12}A_1$ $(1 - \epsilon_2)/\epsilon_2 A_2$

Solution. The external area of the insulation is $A_1 = 5\pi/12 = 1.31$ ft^2/ft length. From Table IX, Case 2, $F_{12} = 1$. For the electrical analogy see Fig. 19/16. Now $A_1 F_{12} = A_2 F_{21} = A_1$ since $F_{12} = 1$. So $F_{21} = A_1/A_2$, a very small value since A_2 is large compared to A_1. Summing the resistances, we get,

$$(1 - \varepsilon_1)\varepsilon_1 A_1 + 1/A_1 + (1 - \varepsilon_2)F_{21}/\varepsilon_2 A_1$$

We can drop the term $(1 - \varepsilon_2)F_{21}/\varepsilon_2 A_1$ since it is small compared to the other two terms and get $(1 - \varepsilon_1)/\varepsilon_1 A_1 + 1/A_1$ as the total resistance.
Then

(a)
$$q_{12} = \frac{\sigma[T_1^4 - T_2^4]}{(1 - \varepsilon_1)/\varepsilon_1 A_1 + 1/A_1}$$

$$= \sigma A_1 \varepsilon_1 [T_1^4 - T_2^4]$$

From Table X, $\varepsilon_1 = 0.93$ (approx.) and

$$q_{12} = (0.1713 \times 10^{-8})(1.31)(0.93)[590^4 - 550^4]$$

$$= (0.1713)(1.31)(0.93)[5.90^4 - 5.50^4]$$

$$= 62 \text{ Btu/hr (per ft of length)}$$

Total Normal Emittances* TABLE X

Material		Temperature of Radiating Body, °F (S = Solar rad.)	Emittance
Aluminum, oxidized	(a)	100–1000	0.11–0.18
Aluminum foil	(b)	212	0.087
with 1 coat linseed oil paint	(b)	212	0.561
Aluminum paint			
26% Al, 27% lacquer	(b)	212	0.3
Asbestos paper	(b)	100–700	0.93–0.945
Brass, dull plate	(b)	120–660	0.22
Brick, red	(a)	S	0.70
silica	(a)	2500	0.84
refractory, black and chrome	(a)	200–1000–2000	0.92–0.97–0.98
Chromium, polished	(b)	100–2000	0.08–0.36
Concrete	(a)	2500–S	0.63–0.65
Galvanized sheet iron			
bright	(b)	82	0.23
oxidized	(b)	75	0.28
Glass	(b)	72	0.94
Lampblack, rough deposit	(b)	212–932	0.84–0.78
Paint			
black	(a)	200–600–1000–S	0.92–0.95–0.97–0.90
green	(a)	200–600–1000–S	0.93–0.90–0.80–0.50
white	(a)	200–600–1000–S	0.92–0.84–0.68–0.30
Stainless steel 301, cleaned	(b)	450–1740	0.57–0.55
Steel, oxidized	(a)	100–1000	0.79–0.79
polished	(a)	100–1000–S	0.07–0.14–0.45
rolled sheet	(b)	70	0.66
rough plate	(b)	100–700	0.94–0.97
Water	(b)	32–212	0.95–0.963

* Selected values from (a) Croft[19.27] and (b) McAdams[19.1]

RADIATION OF GASES 19.21

In boiler furnaces and other industrial furnaces, the heat transferred by radiation is a major part of the total, a likely circumstance in general when the temperature of the radiating surface is high, because the radiation is a function of the fourth power of the absolute temperature. The processes of transfer are complicated. The walls of the furnace become very hot and therefore radiate large quantities of heat. However, since the walls are usually insulated so that the conduction losses are insignificant, they are usually considered as nonconducting. Thus, after steady operation has been attained, the walls reradiate as much heat as they receive.

The amount of radiant heat which reaches the surfaces to be heated from sources such as the fuel bed and the hot walls is affected by the properties of the gases in between. Some gases, carbon dioxide and water vapor, for example, readily absorb radiant heat. If there should be a thick layer of such gases between the radiating source and the heated surface, a large part of the radiation from the source may be absorbed by these gases. However, gases which absorb radiant energy readily are also good radiators (absorptance equal to emittance). Other gases with absorptances and emittances of significant values include the hydrocarbons, ammonia, carbon monoxide, and sulfur dioxide. On the other hand, the diatomic gases absorb very little radiation and are therefore poor radiators. It is a common observation that the air is warmed very little by the radiant heat from the sun or a fireplace. The ever-present water vapor in atmospheric air may result in the absorption of up to some 20% of the incident radiation, the amount absorbed depending on the relative humidity; see references [19.1], [19.2], [19.27].

19.22   REVIEW OF UNITS

Now that we have covered conduction and radiation, we are ready for convection. Before proceeding with this study, you should review §§ 1.6–1.12 on the matter of units. Convection of heat (the film coefficient h) is concerned with dimensionless numbers; that is, groups of parameters, which together have no units when the units for the individual parameters are in accord with a consistent system of units. Remember—a consistent system is one based upon Newton's $F = ma/k$ law which mutually defines force and mass when the proportionality constant is equal to unity.

While it is not the intent to belabor the topic of units, one must have a clear concept of the English fps-system (both pound-mass and pound-force), the metric cgs-system, and the SI mks-system. Unfortunately, tables, charts, and curves may be based on any of these systems and the values taken therefrom must be resolved when used in solving heat transfer problems, especially when convection is included. Because convection depends much on viscosity, dimensional analysis, and the flow of fluids, these topics will be touched upon in the paragraphs which follow.

19.23   VISCOSITY

The surface coefficient of heat transfer h depends, among other factors, on the **absolute viscosity** of the fluid. Since a casual perusal of engineering literature will reveal five or more different units used for absolute viscosity, since writers are sometimes not explicit concerning units, and since consistency in units must be observed in heat transfer problems, it is desirable to know the physical concept of viscosity.

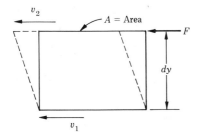

Fig. 19/17

Take an element of a flowing fluid (Fig. 19/17). If there is any *shearing* of the fluid, there will be a relative movement of one layer with respect to another layer of the fluid. Thus, the lower surface of this element is moving with a velocity v_1, and the upper surface is moving with a velocity v_2. The change in velocity is $v_2 - v_1 = dv$ for a distance of dy between the planes. The velocity gradient is dv/dy, and one of Newton's laws is that the *shearing stress* on the fluid, *F/A, is proportional to the velocity gradient,* or

$$\frac{F}{A} = \mu \frac{dv}{dy} \qquad \text{[NEWTONIAN FLUID]}$$

where μ (mu), the proportionality constant, is called the **absolute viscosity** or just the **viscosity**. Solving for μ and indicating its dimensions, we have

$$\mu = \frac{F \, dy}{A \, dv} \rightarrow \frac{(\text{force})(L)}{L^2 (L/\tau)}$$

where the unit of force has not been indicated. Now if we use the system of units based on the **pound force**, the units of μ are

(b) $$\mu' \rightarrow \frac{PL}{L^2(L/\tau)} = \frac{P\tau}{L^2} \rightarrow \frac{\text{lb-sec}}{\text{ft}^2} \qquad \text{[POUND FORCE]}$$

If the unit of *mass* is taken as the *pound P*, the unit of force ($F = ma$) is obtained from PL/τ^2; thus, for μ, we have

(c) $$\mu \rightarrow \frac{(PL)(L)}{(\tau^2)(L^2)(L/\tau)} = \frac{P}{\tau L} \rightarrow \frac{\text{lb}}{\text{sec-ft}} \qquad \text{[POUND MASS]}$$

The relation between these values of viscosity is $\mu = k\mu'$ or $\mu' = \mu/k$, where the time and length units are the same in each case. These units of viscosity have no generally recognized names, but the *poisel* is a name that has been suggested for the unit lb/ft-sec.

The cgs units of viscosity, the poise (dynes-sec/cm^2) and the centipoise ($= 0.01$ poise), are also frequently used. Conversions may be made as follows:

$$(\text{viscosity, lb/sec-ft})(1490) = (\text{viscosity, centipoises})$$

$$(\text{viscosity, lb-sec/ft}^2)(47{,}800) = (\text{viscosity, centipoises})$$

Viscosities of a few fluids are shown in Figs. 19/18, 19/19, and 19/20. The variation of viscosity with pressure may be neglected as long as the pressure does not vary much from usual practices. Time and length units other than seconds and feet are used for viscosity; hence the reader must be alert to units of viscosity.

KINEMATIC VISCOSITY   19.24

Test values of viscosity are often presented in the literature as the *kinematic viscosity* v (nu), which is defined as the absolute viscosity divided by the density; $v = \mu/\rho$. The kinematic viscosity is the same in either of the foregoing systems of units; thus,

$$v = \frac{\mu'}{\rho'} \rightarrow \frac{P\tau/L^2}{P\tau^2/L^4} = \frac{L^2}{\tau} \qquad \text{and} \qquad v = \frac{\mu}{\rho} \rightarrow \frac{P/\tau L}{P/L^3} = \frac{L^2}{\tau}$$

$$\text{[POUND FORCE]} \qquad\qquad\qquad \text{[POUND MASS]}$$

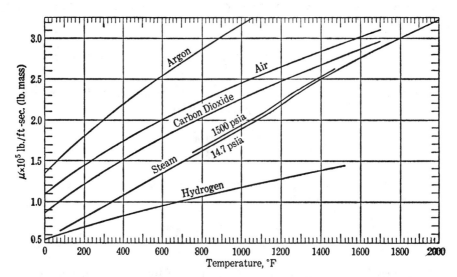

Fig. 19/18   *Absolute Viscosities of Gases.* Divide by **k** to get lb-sec/ft.² (pound-force system). Plotted from Hilsenrath and Touloukian [19.26]. Except for the 1500-psia steam curve, the pressure is 1 atm. At higher pressures, the viscosity of steam increases materially. The effect of pressure on viscosity generally becomes more and more pronounced as the pressure increases, and to different degrees at different temperatures. It would seem that moderate extrapolation would be safe.

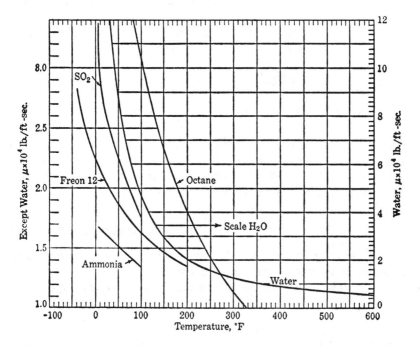

Fig. 19/19   *Absolute Viscosity of Some Liquids.* Divide by **k** to convert to lb-sec./ft.² (pound-force system). Plotted from data in Keenan and Keyes, McAdams [19.1], and ASHRAE[19.3].

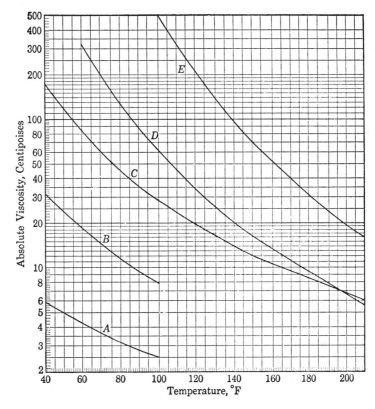

Absolute Viscosities of Selected Crude Oils. **Fig. 19/20**

In the English system, these units may be square feet per second, but kinematic viscosity is usually given in the cgs system as *stokes* or *centistokes*. Hereafter, we shall drop the use of the prime marks: the units of the symbol will be indicated or defined by the context.

DIMENSIONAL ANALYSIS AND DIMENSIONLESS NUMBERS 19.25

The following discussion suggests the nature of many of the deductions regarding the flow of fluids and the convection of heat. The method of attack is known as *dimensional analysis.* Suppose we wish to study the factors which define the drop in pressure of a fluid flowing in a pipe. From a consideration of the pressure drop per unit length of pipe $\Delta p/x$ would be a function of (1) the size (or diameter) D of the pipe, because we expect the pressure drop to increase as the pipe becomes smaller, other factors remaining the same, (2) the velocity of flow v, since we are sure that the faster a fluid is pumped through a pipe, the greater must be the pressure drop, (3) the viscosity of the fluid μ, because we have observed that, for example, viscous oils do not flow as readily as less viscous gasoline, and (4) the density of the fluid ρ, because if there are any accelerations of particles of the fluid, as in turbulent flow, we expect the energy to overcome these effects to follow from the work done as represented by the pressure drop. The foregoing statement may be written in the form

(a)
$$\frac{\Delta p}{x} = \phi(D^a v^b \mu^c \rho^d)$$

where a, b, c, and d are exponents of unknown value, ϕ means *function of*, and x is the length of the pipe. Now equation **(a)** must have the same dimensions on each side if it is to be physically correct. The next step is to substitute into equation **(a)** the dimensions of the various quantities in place of the quantities themselves. The units or dimensions must be consistent; either system yields the same result. If the *force* unit is taken as the *pound*, density ρ has the units $P\tau^2/L^4$ [equation **(b)**, § 19.24], and viscosity has the units $P\tau/L^2$ [equation **(b)**, § 19.23]. Since the units of pressure are P/L^2, we may write the dimensional equation corresponding to equation **(a)** as

$$\frac{P}{L^3} = L^a \left(\frac{L}{\tau}\right)^b \left(\frac{P\tau}{L^2}\right)^c \left(\frac{P\tau^2}{L^4}\right)^d$$

If this equation balances dimensionally, the exponents of the P's on the right side must reduce to 1, because the exponent of P on the left side is 1. Moreover, the exponent of L must be -3 on the right side, since it is -3 on the left. The exponent of τ must be zero. Thus, using the exponents of the P's, we write

(b)
$$1 = c + d$$

From the exponents of the L's, we have

(c)
$$-3 = a + b - 2c - 4d$$

and for the τ's,

(d)
$$0 = -b + c + 2d$$

In these three equations, there are four unknowns. Hence, in this instance, numerical values of a, b, c, and d cannot be found. We may, however, reduce the number of unknown exponents to one by solving for a, b, and d in terms of c as follows. From equation **(b)**,

(e)
$$d = 1 - c$$

from equations **(d)** and **(e)**,

(f)
$$b = 2 - c$$

and from equations **(c)**, **(e)**, and **(f)**,

$$a = -1 - c$$

Substituting these values of a, b, and d, into $\Delta p / x = \phi(D^a v^b \mu^c \rho^d)$, we get

$$\frac{\Delta p}{x} = \phi(D^{-1-c} v^{2-c} \mu^c \rho^{1-c})$$

$$= \phi\left(\frac{1}{DD^c} \times \frac{v^2}{v^c} \times \mu^c \times \frac{\rho}{\rho^c}\right) = \phi\left[\frac{v^2 \rho}{D}\left(\frac{\mu}{Dv\rho}\right)^c\right]$$

This is as far as the dimensional analysis will carry us. The form of the function ϕ must be found from experimental data. Assume that the form of the function is

(g)
$$\frac{\Delta p}{x} = C\frac{v^2 \rho}{D}\left(\frac{\mu}{Dv\rho}\right)^c = C\frac{v^2 \rho}{D}\left(\frac{Dv\rho}{\mu}\right)^{c'}$$

where C is a dimensionless constant. With two unknowns left, the constant C and the exponent c' ($= -c$), it may appear at first that little has been accomplished. On the contrary, we have discovered a very significant grouping of the variables. The group $Dv\rho/\mu$, called the **Reynolds number** Re, should evidently be used as a group variable in analyzing experimental data. If this is done and the proper experimental data are available, we may thereby determine the constant C and c', thus establishing a general equation for the flow of fluids in accordance with the original assumptions. We see then that the success of dimensional analysis depends not only upon the experimental determination of certain constants, but also upon the correctness of the assumptions in the very beginning [equation (a)]. One factor omitted in the foregoing analysis is the degree of roughness of the inside of the pipe, since we would expect smooth pipes to offer less resistance to flow (that is, have a smaller pressure drop) than rough pipes. Hence, equation (g) above should be applied to data concerning pipes that have a similar degree of roughness, that are geometrically similar.

Checking the dimension of the Reynolds number by substituting the dimensions of the various quantities which make up this number, we have

$$\text{Re} = \frac{Dv\rho}{\mu} \rightarrow \frac{(L)(L/\tau)(P\tau^2/L^4)}{P\tau/L^2}$$

where the units cancel and Re is therefore a dimensionless number. It follows that the group $(\Delta p)(D)/xv^2\rho$ from equation (g) above, is also a dimensionless group, a fact that the reader may check for himself. Moreover, since the units of D and x are the same, $\Delta p/(v^2\rho)$ must also be a dimensionless group. For practice the reader should check the Reynolds number using a consistent system based on the *pound mass*, and see that it is dimensionless in this system. The Reynolds number is important in convected heat problems as well as in the flow of fluids. It may be expressed in terms of kinematic viscosity ($\nu = \mu/\rho$) or in terms of the so-called *mass velocity* $G(G = \rho v$-lb/ft^2-sec. for the pound-mass system):

$$\text{Re} = \frac{Dv}{\nu} = \frac{DG}{\mu}$$

We see from the foregoing discussion that the dimensional analysis results in a logical grouping of the involved variables into dimensionless groups, a result that is invaluable in analyzing phenomena that we have not been able to analyze by a direct mathematical approach, phenomena such as the flow of fluids and heat.

19.26   FLOW OF FLUIDS

Consider a fluid flowing in a pipe. If the Reynolds number is below about 2100, the flow is said to be *laminar*, also called *streamline* or *viscous* flow. In laminar flow, all particles of the fluid move in parallel paths in the direction of flow. The velocity of the stream at the center of the pipe is a maximum and it is zero at the pipe surface, the variation being parabolic during *isothermal* flow, as shown in Fig. 19/21. There are no component motions in a radial direction. If the fluid is being heated or cooled, the velocity distribution is no longer parabolic and some radial movement of the fluid occurs. Nevertheless, for laminar flow, the transfer of heat from fluid to pipe, or vice versa, is substantially by *conduction* through the thin surface film, which has little or no motion.

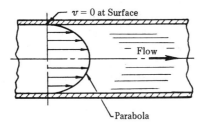

Fig. 19/21   *Laminar Flow.*

Suppose there are small rough projections on the inside surface of the pipe which extend into the moving stream. When the fluid strikes these projections, eddies are formed, so that a portion of the fluid will move through the stagnant surface film and strike the wall. The particles of the fluid which strike the wall give up energy (or receive energy), and thus some heat is carried to (or from) the wall by convection. Eddies are caused also by any sort of obstruction in the line, such as valves and pipe bends, but it should be noted that the effect of these eddies in transferring heat by convection to or from the wall is limited and local, because the flow remains streamline outside of the region of the disturbance.

If the Reynolds number is greater than about 10,000 (greater than 3000 for the less viscous fluids and gases), the flow is said to be *turbulent*. In turbulent flow, there is a thin layer (several thousandths of an inch thick) of fluid adjacent to the surface in laminar motion, indicated by a (Fig. 19/22). Next, there is a buffer layer, of the order of a few hundredths of an inch thick, indicated by b (Fig. 19/22). The main

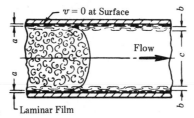

Fig. 19/22   *Turbulent Flow.*

body of fluid is turbulent, with particles moving in eddies, many with large component velocities in a radial direction. Thus, many particles, with component velocities toward the wall or surface, break through the laminar surface film and carry heat to or from the wall by *convection*. The heat convected to or from the wall in turbulent flow is very much larger than that *conducted* to or from the wall during laminar flow, so that, if rapid transfer of heat is desired, turbulent flow would in general be desired (but sometimes it is too costly in pump work).

Between the conditions of laminar flow and turbulent flow, there is a transition stage, which apparently is part laminar and part turbulent. At any rate, at the beginning of this stage, particles are beginning to move radially to the wall and convect heat, an action which reaches its full effect with turbulent flow. In some instances, laminar flow may exist at Reynolds numbers ordinarily indicative of turbulent flow. In the laboratory, with carefully controlled conditions, laminar flow has been obtained at quite high Reynolds numbers, but this condition was unstable, turbulent flow being brought about by jarring the apparatus, for example. If the length of straight section is short, with elbows or valves nearby, turbulence may exist with Reynolds numbers below 2100.

There is an economic limit to which turbulence may be utilized to increase the rate of heat transfer. As we have learned, the frictional loss increases with the turbulence and the work of friction increases with the Reynolds number. At some point, the gain in heat-transmitting capacity is offset by the increased power for operation.

CONVECTED HEAT   19.27

We recall that convected heat is a combination of radiation, conduction, and the transport of a fluid. The convecting fluid is heated by radiation and conduction from a hot body; then the fluid, or part of it, with the stored energy moves to a cooler body or space and heats the body or space by radiation and conduction. *Free convection* occurs when the fluid circulates, because differences in temperatures result in differences in densities, the heavier parts of the fluid moving downward under the force of gravity. When work is done to blow or pump the fluid, the fluid is said to be in *forced convection.*

FILM COEFFICIENT   19.28

No single equation expresses the film coefficient for all situations. This chapter will deal with three conditions of flow: forced convection with turbulent flow, forced convection with streamline flow, and free convection.

A dimensional analysis, after the pattern of § 19.25, of the phenomenon of heat transfer across a bounding layer of fluid, in which it is assumed that the film coefficient h is a function of a characteristic dimension of the containing surfaces D, the density of the fluid ρ, the conductivity of the fluid k, the velocity of the fluid v, the absolute viscosity μ, the specific heat of the fluid (at constant pressure in case of a gas) c_p, and the length L of the pipe or duct in which flow and transfer of heat occurs, yields

(a)
$$\frac{hD}{k} = C\left(\frac{D\rho v}{\mu}\right)^a\left(\frac{c_p\mu}{k}\right)^b\left(\frac{D}{L}\right)^c$$

in which the constant C and the exponents a, b, and c are set in accordance with experimental data. The dimensionless group hD/k is called the **Nusselt number** Nu, the dimensionless group Dpv/μ is recognized as the Reynolds number Re, and the dimensionless group $c_p\mu/k$ is called the **Prandtl number** Pr. Since the ratio of D/L = diameter/length often exerts a negligible effect on the film coefficient, many authorities prefer the expression in the form

(b) $$\text{Nu} = C(\text{Re})^a(\text{Pr})^b$$ [NUSSELT'S EQUATION]

The reciprocal of the Prandtl number is called the **Stanton number** St; St $= k/(c_p\mu) = 1/\text{Pr}$). Hence, the foregoing equation is often written with the Stanton number instead of the Prandtl number. Observe that, since each of the symbol groups named is dimensionless, a consistent system of units must be employed unless a counteracting adjustment is made in the constant C. One system of units may be employed for Nu, another for Re, and so on, but within the group there must be no inconsistency. Thus, if the unit of *force* is a pound, the units of c_p, which are fundamentally Btu per unit of mass per degree, must be Btu-ft/(lb-sec^2-°F).

In case there is any doubt concerning the consistency of the units, make a dimensional check. Let B represent the heat unit and θ the temperature unit. Then, for example,

$$\frac{c_p\mu}{k} \rightarrow \left(\frac{BL}{P\theta\tau^2}\right)\left(\frac{P\tau}{L^2}\right)\bigg/\left(\frac{BL}{\theta\tau L^2}\right) \quad \text{dimensionless}$$

Moreover, note that the time unit for the values of k given in Tables VII and VIII is the *hour*, and that this time unit must be the same throughout the terms of any dimensionless group.

19.29   FILM COEFFICIENT, TURBULENT FLOW INSIDE PIPE

The literature on heat transfer abounds in special forms of Nusselt's equation and in derived or empirical equations. The equations below are taken from McAdam's[19.1]. For fluids being heated or cooled during turbulent flow inside round pipes of internal diameter D,

(19-10) $$\frac{h_iD}{k_b} = 0.023\left(\frac{Dv\rho}{\mu}\right)^{0.8}_b\left(\frac{c_p\mu}{k}\right)^{0.4}_b$$

[LIMITS: $0.7 < \text{Pr} < 120$; $10{,}000 < \text{Re} < 120{,}000$ FOR HIGH VISCOSITY LIQUIDS; $\text{Re} > 2100$ FOR LOW VISCOSITY LIQUIDS AND GASES; $L/D > 60$; MODERATE Δt]

where the subscript b indicates that the properties k, μ, and ρ should be evaluated at the bulk temperature (for convenience, the average of inlet and outlet temperatures of the fluid), and h_i is the inside film coefficient. The transition range of viscous fluids, such as oils, in changing from streamline to turbulent flow is $2100 < \text{Re} < 10{,}000$. Equation (19-10) should give good results for less viscous liquids, such as gasoline, down to $\text{Re} \approx 3000$. However, for a liquid whose viscosity does not exceed twice that of water, and for all gases and vapors, use equation (19-10) down to Re ≈ 2100.

The value of the Prandtl number, $\text{Pr} = c_p\mu/k$, for gaseous substances at about 1 atm is given by an empirical relation [19.1],

(a)
$$\text{Pr} = \frac{4}{9 - 5/k} = \frac{c_p\mu}{k}$$

in which $k = c_p/c_v$. For $k = 1.4$ (diatomic gases at usual temperatures), $\text{Pr} = 0.74$; for $k = 1.3$, $\text{Pr} \approx 0.78$; for $k = 1.2$, $\text{Pr} \approx 0.83$. After Pr is raised to the 0.4 power, the range of the resulting values is even smaller. Thus, to a close approximation

(19-11)
$$\frac{h_i D}{k_b} = 0.021 \left(\frac{D v\rho}{\mu}\right)_b^{0.8} = 0.021\, \text{Re}^{0.8}$$

[GASES AND VAPORS, Re > 2100]

Since data on the viscosities of gases are not always available, it may be important to note that a reasonable value of μ can be computed from equation **(a)**.

Example—Air Preheater   19.30

A counterflow air preheater receives 330,000 lb/hr of flue gas at 509°F, for which $c_p = 0.243$ Btu/lb and $R = 56$. The exit temperature of the gas is 289°F. This gas flows through 3-in. OD tubes (2.782 in. ID) with an average velocity of 15 fps at atmospheric pressure of 14.7 psia. What is the film coefficient on the inside of the tubes?

Solution. The properties of the gas in equation (19-10) are based on the average bulk temperature, which is $(509 + 289)/2 = 399°F$. Using the pound-mass system of units, we get the average density as

$$\rho = \frac{p}{RT} = \frac{(14.7)(144)}{(56)(399 + 460)} = 0.044\ \text{lb/ft}^3$$

The conductivity of the flue gas cannot be found with the data at hand. However, we observe from Table VII that the principal constituents of a typical flue gas, nitrogen, oxygen, carbon monoxide, have conductivities much the same as that of air. While the carbon dioxide and water vapor have somewhat lower values, we will probably get a reasonable estimate of k for the flue gas by considering it as air. Thus, interpolating between the values in Table VII, we have, for 399°F,

$$k_b = \frac{(399 - 32)(0.312 - 0.168)}{572 - 32} + 0.168 = 0.266\ \text{Btu-in./ft}^2\text{-hr-°F}$$

$$k_b = \frac{0.266}{(12)(3600)}\ \text{Btu-ft/ft}^2\text{-sec-°F}$$

Assuming that $k = c_p/c_v = 1.4$ for the flue gases, we estimate μ from equation **(a, § 19.29)** as

$$\mu = \frac{4k_b}{(9 - 5/k)c_p} = \frac{(4)(0.266)}{(12)(3600)(9 - 5/1.4)(0.243)} = 18.7 \times 10^{-6}\ \text{lb/ft-sec}$$

Check the units for yourself. The Reynolds number is

$$\text{Re} = \frac{Dv\rho}{\mu} = \frac{(2.782)(15)(0.044)}{(12)(18.7 \times 10^{-6})} = 8180$$

where 12 converts the diameter to feet. Since Re > 2100, equation (19-11) applies and we get

$$h_i = \left(\frac{k_b}{D}\right)(0.021)(\text{Re})^{0.8} = \frac{0.266)(0.021)(8180)^{0.8}}{(12)(3600)(2.782/12)} = \frac{2.67}{3600} \text{ Btu/ft}^2\text{-sec-}°\text{F}$$

$$h_i = 2.67 \text{ Btu/ft}^2\text{-hr-}°\text{F}$$

The beginner would be wise to fix in mind the system of units to be employed and to include all conversion factors for units in order to be sure of consistency. Since some factors often cancel, one could carry them along as in the foregoing computation for h_i.

The design of the air preheater may proceed as follows: By a procedure similar to that above, but one which allows for the effects of adjacent tubes, determine h_o from data in books which give a full treatment of the subject. The total thermal resistance can then be computed for one foot of pipe or tube length ($z = 1$), after which equation (19-4B) will yield the heat transferred per foot of tube in one hour. Knowing the total heat to be transferred, say, from the temperature drop of the gases, divide it by the heat per foot to obtain the length of tubes needed in feet. In this application $\Delta t = \Delta t_m$, the LMTD. In practice, such a design is one of trial and error. Perhaps the velocity has to be adjusted because the total flow passes through more or fewer tubes than originally estimated. Also, the length of the tube bundle should have a reasonable relation to the cross-sectional area of the bundle.

19.31   Example—Film Coefficient for Steam

Steam at 200 psia and 500°F flows at 100 fps in a 4-in. pipe (approximate ID). What is the surface coefficient on the inside of the pipe?

Solution. First, determine the Reynolds number, so that the nature of the flow may be predicted. Using the pound mass, we get the density

$$\rho = \frac{1}{v} = \frac{1}{2.725} \text{ lb/ft}^3$$

where the value of v is taken from the steam tables, Item B 15. Figure 19/18 indicates that there will be little error in using the viscosity at atmospheric pressure and 500°F, $(1.2)(10)^{-5}$ lb/ft-sec. The Reynolds number is

$$\text{Re} = \frac{Dv\rho}{\mu} = \frac{(4)(100)(10)^5}{(12)(1.2)(2.725)} = 1,020,000$$

Thus, turbulent flow exists and equation (19-10) applies. The conductivity k of steam at 500°F is found from the values in Table VII by interpolation:

$$k = 0.163 + \tfrac{288}{720}(0.231) = 0.255 \text{ Btu-in./ft}^2\text{-hr-}°\text{F}$$

The specific heat of this steam may be estimated by using enthalpies from the steam tables, at 200 psia and 400°F and at 600°F, a range of 200°F (a smaller range is desirable if the full tables are at hand:

$$c_p = \frac{1322.1 - 1210.3}{200} = 0.559 \text{ Btu/lb-}°\text{F}$$

The Prandtl number is (using the pound-mass system)

$$\text{Pr} = \frac{c_p \mu}{k_b} = \frac{(0.559)(1.2)(12)(3600)}{(0.255)(10)^5} = 1.135$$

where the $(12)(3600)$ converts k to Btu-ft/ft^2-sec-°F. Using equation (19-10), we get

$$h_i = \left(\frac{k_b}{D}\right)(0.023)(\text{Re})^{0.8}(\text{Pr})^{0.4}$$

$$= \frac{(0.255)(12)}{(12)(3600)(4)}(0.023)(1{,}020{,}000)^{0.8}(1.135)^{0.4} = \frac{113}{3600} \text{ Btu/ft}^2\text{-sec-°F}$$

$$= 113 \text{ Btu/ft}^2\text{-hr-°F}$$

FILM COEFFICIENT, LAMINAR FLOW OF LIQUIDS IN PIPES 19.32

In the isothermal laminar flow of a fluid, the ideal shape of the velocity gradient in the pipe is parabolic (Fig. 19/21). However, when the fluid is being heated or cooled, this is no longer true. If a *liquid* is being heated, its viscosity next to the wall is less than that at a point nearer the center of the stream (temperature higher); it follows from the lowered viscosity that the velocity of those particles close to the wall is greater than that for isothermal flow. An opposite effect occurs when the liquid is being cooled. The effects of the variation of the viscosity of the liquid and the free convection currents in laminar flow are thought to be responsible in part for the difficulty in obtaining a simple relation which correlates available data satisfactorily. McAdams[19.1] suggests the following equation for small D and small Δt, fluid being heated,

(19-12)
$$\frac{h_i D}{k_b} = 2.02\left(\frac{\mu_b}{\mu_w}\right)^{0.14}\left(\frac{\dot{m}c}{kL}\right)^{1/3}_b$$

[VISCOUS LIQUIDS, Re < 2100; $\dot{m}c/(kL) > 10$]

where $\dot{m}$ is the mass flow per unit of time, usually given in pounds per hour, which corresponds to the pound-mass system; μ_w is the viscosity of the fluid measured at the temperature of the inside wall of the pipe; L is the heated length of a straight pipe; c is the specific heat of the liquid; and the other symbols have the meanings previously defined. *All symbol groups are dimensionless.* The group $\dot{m}c/(kL)$ is called the **Graetz number** (Gz). Equation (19-12) makes no allowance for natural convection currents, which may result in significantly larger film coefficients at very low velocities.

An empirical equation by Sieder and Tate[19.25] may also be used for determining the film coefficient h for laminar flow of liquid in a pipe and follows:

(19-13)
$$\frac{hD}{k} = 1.86\left(\frac{D\upsilon\rho}{\mu}\right)^{0.333}\left(\frac{c_p\mu}{k}\right)^{0.333}\left(\frac{D}{L}\right)^{0.333}\left(\frac{\mu}{\mu_s}\right)^{0.14}$$

or expressed in named dimensionless groups we have

$$\text{Nu} = 1.86 \, \text{Re}^{0.333} \text{Pr}^{0.333} \left(\frac{D}{L}\right)^{0.333} \left(\frac{\mu}{\mu_s}\right)^{0.14}$$

where the dimensionless groups are Nusselt number (Nu), Reynolds number (Re) and Prandtl number (Pr). The fluid properties are evaluated at the bulk fluid temperature except for the viscosity μ_s which is evaluated at the inside pipe surface temperature. Equations (19.12) and (19.13) should give comparable results.

19.33   Example—Oil Heater

An oil heater has $\frac{3}{4}$-in. (OD), No. 16 B.W. gage steel tubes, through which flows an oil with the properties of oil C (Fig. 19/20). The oil enters at a temperature of $t_1 = 60°F$ and flows with an average velocity of 4 fps. The inside diameter of the tubes is 0.620 in., the inside surface area is 0.1613 ft^2/ft of length, and the inside sectional area is 0.3019 in.^{2} The length of one pass of the tubes is 10 ft. Condensing steam at 215°F surrounds the tubes. If the specific heat of the oil is 0.5 Btu/lb-°F and its conductivity is $k = 0.08$ Btu-ft/hr-ft^2-°F (compare with Table VII), what is its temperature at the end of the 10-ft pass?

Solution.   Many problems of this nature require a trial-and-error solution. Assume the final bulk temperature as $t_2 = 76°F$. Then check the Reynolds number at this highest bulk temperature. The specific gravity of oil C at 76°F is

$$SG_{76} = SG_{60} - 0.00035(76 - 60) = 0.899 - 0.0056 = 0.893$$

from the data of Fig. 19/20. The density is then

$$\rho_2 = (0.893)(62.3) = 55.6 \, \text{lb/ft}^3 \text{ at } 76°F$$

Also from Fig. 19/20 for oil C at 76°F, $\mu_2 = 50$ centipoises, or $\mu_2 = \dfrac{50}{1490}$ lb/ft-sec. Substituting known values, we get

$$\text{Re} = \frac{D_i v \rho}{\mu} = \frac{(0.62/12)(4)(55.6)}{50/1490} = 343$$

Since Re = 343 < 2100, the flow is viscous. The mass of flow is

$$\dot{m} = \frac{Av}{v} = \rho A v = (55.6)\left(\frac{0.3019}{144}\right)(4) = 0.466 \, \text{lb/sec}$$

Since the velocity for a particular quantity varies somewhat as the temperature and the density vary, this answer would be different for another section of the tube. The change in this example is not significant. We now find the dimensionless number

$$\frac{\dot{m}c}{k_b L} = \frac{(0.466)(0.5)}{(0.08/3600)(10)} = 1049$$

which, being greater than 10, indicates that equation (19-12) may be used.

For the assumed $t_2 = 76°F$, the average bulk temperature is $(60 + 76)/2 = 68°F$. Thus, the average bulk viscosity of oil C is $\mu_b = 65$ centipoises (Fig. 19/18). Next, we may estimate the inside surface temperature of the tubes at about 5° below the steam temperature, or $t_s = 210°F$. Corresponding to this temperature, $\mu_w = 6$ centipoises. Solving for h_i from equation (19-12), we have

$$h_i = \left(\frac{2.02 k_b}{D}\right)\left(\frac{\mu_b}{\mu_w}\right)^{0.14}\left(\frac{mc}{k_b L}\right)^{1/3}$$

$$= \frac{(2.02)(0.08)}{(0.62/12)}\left(\frac{65}{6}\right)^{0.14}(1049)^{1/3}$$

$$= 43.7 \text{ Btu/hr-ft}^2\text{-}°\text{F}$$

the film coefficient on the inside. The heat transmitted from the *inside* surface of the tube through the surface film to the bulk of the oil is $Q = h_i A(t_s - t_o)$ where t_0 is the average bulk temperature of the oil, 68°F. For a surface temperature $t_s = 210°F$, we get

$$Q = h_i A_i(t_s - t_o) = (43.7)(10 \times 0.1613)(210 - 68)$$

$$= 10,000 \text{ Btu/hr}$$

This same quantity of heat is received by the oil in accordance with $Q = mc(t_2 - t_1)$, where t_1 and t_2 are, respectively, the initial and final temperatures of the oil. Using a mass flow $\dot{m} = (0.466)(3600)$ lb/hr, we get

$$10,000 = (0.466)(3600)(0.5)(t_2 - 60)$$

from which, $t_2 = 72°F$. This computation shows that our assumption of $t_2 = 76°F$ was high. For practice, the student should rework this example, assuming a t_2 of, say 73°F.

Another assumption that should be checked is the inside surface temperature, taken as 210°F. This computation may be made by setting up the expression for the flow of heat from the steam *to* the inside surface from equation (19-4B), and proceeding as previously explained (§ 19.12). The outside film coefficient may be taken as 1000, the tube thickness is 0.065 in., and the quantity of heat is as computed in this example.

As a point of interest, let's find h using equation (19-13) and compare with the answer found earlier. We need the Pr number,

$$\text{Pr} = \frac{c_p \mu}{k} = \left(\frac{0.5 \text{ Btu}}{\text{lb-}°\text{F}}\right)\left(\frac{\text{hr-ft}^2\text{-}°\text{F}}{0.08 \text{ Btu-ft}}\right)\left(\frac{65 \text{ cp}}{1490 \text{ cp}}\right)\left(\frac{\text{lb}}{\text{sec-ft}}\right)\left(\frac{3600 \text{ sec}}{\text{hr}}\right) = 981$$

Then

$$\frac{hD}{k} = (1.86)(343)^{0.333}(981)^{0.333}\left(\frac{0.62}{120}\right)^{0.333}\left(\frac{65}{6}\right)^{0.14} = 31.2$$

so that

$$h = (31.2)\left(\frac{0.08}{0.62}\right)(12) = 48.3 \text{ Btu/hr-ft}^2\text{-}°\text{F}$$

an answer that is within 10% of the above value $h_i = 43.7$ and that is good.

FILM COEFFICIENTS FOR ANNULAR SPACE   19.34

For turbulent flow through a concentric annular area (Fig. 19/4), McAdams recommends that the film coefficient h for both the inner and outer surfaces of the annulus be determined from

(19-14) $\text{Nu} = 0.023 \, \text{Re}^{0.8} \text{Pr}^n$

or

$$\frac{hD}{k} = 0.023 \left(\frac{D_e v \rho}{\mu}\right)_b^{0.8} \left(\frac{c_p \mu}{k}\right)_b^n$$

where $n = 0.4$ for heating and $n = 0.3$ for cooling.

The various groups are dimensionless and the subscript b means that the properties are evaluated at the bulk temperature; D_e is the equivalent diameter which is four times the hydraulic radius.

Since the hydraulic radius is the sectional area of the stream divided by the wetted perimeter, we have the value of D_e to be used in computing the Reynolds number as

(a) $D_e = 4\left(\dfrac{\pi D_2^2/4 - \pi D_1^2/4}{\pi D_2 + \pi D_1}\right) = D_2 - D_1$

The other symbols in equation (19-14) have their usual meanings. If a hot gas transparent to radiant heat is in the annular space, with heat being transferred to a cold fluid in the inside pipe, the *radiation* from the inside surface of the larger pipe to the smaller pipe may be a substantial part of the total transferred heat, and it should be added to that obtained by convection. The coefficient h is heat transferred across the film by convection and conduction only.

In case of laminar flow in the annulus, use equation (19-12) with the equivalent diameter D_e used for D.

19.35   FLOW OVER OUTSIDE OF A SINGLE PIPE, FORCED CONVECTION

The multiplicity of cases of flow over the outside of pipes, such as various tube banks and finned tubes, precludes a significant discussion in this book. For diatomic gases, and approximately for others, flowing across a single pipe, we have

(19-15) $\dfrac{hD_o}{k_f} = 0.24 \left(\dfrac{D_o v \rho}{\mu_f}\right)^{0.6}$

one of the equations given by McAdams. The value of $(D_o v \rho)/\mu_f$ should be between 1000 and 50,000; D_o is the outside diameter; and k_f and μ_f are evaluated at the film temperature, which is taken as

(a) $t_f = \dfrac{t_b + t_w}{2}$

where t_b is the bulk temperature and t_w is the temperature of the pipe wall on the side whose film temperature is desired.

FILM COEFFICIENTS WITH FREE CONVECTION   19.36

The form of equation applicable to many problems in free convection is

(a)
$$\frac{h_c D}{k_f} = C\left[\left(\frac{D^3 \rho^2 g \beta \Delta t}{\mu^2}\right)_f \left(\frac{c_p \mu}{k}\right)_f\right]^m$$

where the subscript f indicates that the properties of the fluid are evaluated at the film temperature, C and m are dimensionless constants, β is the coefficient of thermal expansion ($= 1/T$ for ideal gases), and the other symbols have the usual meanings. Refer to McAdams[19.1] for values of C and m under various circumstances. All symbols are dimensionless; the first group on the right-hand side is called the **Grashof number** (Gr); the other two are recognized as the Nusselt and Prandtl numbers. As illustrations of how the foregoing equation **(a)** may be simplified for a particular fluid in free convection, we have for a hot surface in atmospheric air,

(b)   Vertical plates over 1 ft high:

$$h_c = 0.27\,\Delta t^{0.25} \text{ Btu/ft}^2\text{-hr-}°\text{F}$$

(c)   Horizontal pipes and vertical pipes over 1 ft high:

$$h_c = 0.27\left(\frac{\Delta t}{D_o}\right)^{0.25} \text{ Btu/ft}^2\text{-hr-}°\text{F}$$

where the subscript c is used as a reminder that only convected heat is involved and that the radiant heat should also be considered (§ 19.20); D_o is the outside diameter in feet; Δt is the temperature difference between the air and the surface. If repeated calculations are to be made with a particular convecting fluid, one should aim to simplify the various equations previously given when the properties of the fluid do not vary widely, as was done in obtaining equations **(b)** and **(c)**.

Example—Heat Loss from Pipe   19.37

A single 2-in. pipe (OD = 2.375 in.) has a surface temperature of 240°F in a room where the ambient temperature is 80°F. The air is at standard atmospheric pressure. Determine the transferred heat per square foot of external surface.

Solution.   From equation (**c**, § 19.36), we have

$$h_c = 0.27\left(\frac{\Delta t}{D_o}\right)^{0.25} = 0.27\left(\frac{160}{2.375/12}\right)^{0.25} = 1.44 \text{ Btu/ft}^2\text{-hr-}°\text{F}$$

$$Q_{cv} = h_c A\,\Delta t = (1.44)(160) = 230 \text{ Btu/ft}^2\text{-hr}$$

which is that portion of the transferred heat lost by convection only. The amount of heat radiated must be added to this to obtain the total. The pipe is a completely enclosed surface (case 2, Table IX), where we find $A = A_1$, and $F_{12} = 1$. Assume that the surface of the pipe is equivalent to rough plate steel and let $\varepsilon_1 = 0.94$, from Table X. Then by equation (**a**, § 19.20), the radiated heat per square foot of pipe area is

$$Q_r = (0.1713)(0.94)\left[\left(\frac{700}{100}\right)^4 - \left(\frac{540}{100}\right)^4\right] = 252 \text{ Btu/ft}^2\text{-hr}$$

The total transferred heat is

$$Q_T = Q_{cv} + Q_r = 230 + 252 = 482 \text{ Btu/ft}^2\text{-hr}$$

19.38   CONDENSING VAPORS

When vapors condense on a surface they may do so either by *dropwise condensation* or by *film-type condensation*. If the surface is oiled, condensation occurs in droplets because the water does not *wet* the oily surface. Since oil is eventually washed off, there are other "promoters" which are added to steam, such as oleic acid, for the purpose of reducing or eliminating the wettability of the surface. The reason that dropwise condensation is desired is that the corresponding film coefficients are some four to eight times those for film-type condensation. Whenever the surface is clean and the steam is clean (filtered and free of contaminants), film-type condensation is obtained. When the surface is contaminated with a "promoter" which prevents wetting, dropwise condensation occurs. Promoters which are more than temporarily effective are ones which are adsorbed or held by the surface [19.1]. The reason that the "film" coefficient is higher for drop condensation is the absence of the insulating effects of the liquid film through which heat flows by conduction.

Unless steps are taken to obtain drop condensation, we should assume that the film type occurs, for which the average value of the film coefficient may be computed from the following theoretical equation derived by Nusselt, applicable to horizontal tubes and reasonably agreeable with experimental observations:

(19-16) $$\boldsymbol{h} = 0.725\left(\frac{k^3\rho^2 gh_{fg}}{ND_o\mu\,\Delta t}\right)^{0.25}_f \text{ Btu/hr-ft}^2\text{-}°\text{F}$$

where the symbol group is not dimensionless but where the same system of units is used for every term, conveniently the pound-mass system except with the time unit as the hour; g ft/hr^2; k Btu-ft/hr-ft-°F; ρ lb/ft^3; h_{fg} Btu/lb; μ lb per hr-ft; Δt°F; the constant 0.725 is dimensionless; D_o is the outside diameter of the tube; and N is the number of tubes in a vertical row of horizontal tubes. The subscript f indicates that the fluid properties at the film temperature are to be used [equation (**a**)].

In a condenser wherein film condensation occurs, it is reasonably expected that the major resistance to heat flow is in the water film on the water side. Even if the surface factor on the steam side varies, the resistance of the condensing film and of the metal wall and of the dirt deposit will ordinarily be relatively small, so that all these resistances might be lumped together as a constant. For turbulent flow of the water in an old (dirty) tube, experimental results suggest the following over-all resistance [19.1]:

(19-17) $$\frac{1}{U_0} = 0.00092 + \frac{1}{268v^{0.8}} \frac{\text{hr-ft}^2\text{-}°\text{F}}{\text{Btu}}$$

where U_o is based on the outside tube area, 0.00092 is the sum of the small resistances previously mentioned, v fps is the water velocity on the inside of the tube, 268 is an experimental constant, and $1/(268v^{0.8})$ is the film resistance on the water side (inside). The experimental data were for a 1-in. OD tube. Economic water velocities are usually between 6 and 10 fps. At higher speeds, pumping costs will increase rapidly; at lower speeds, the rate of heat transfer will become uneconomically small (too large heat exchanger needed).

CLOSURE 19.39

At this stage of development of the science of heat transfer, the computation of the heat flux is dependent largely on experimental results, as indicated in the previous discussions. On this account, it is often desirable to search the literature for experiments related to the particular situation in which one is interested; or to verify one's design by experiment. If the effects of contaminants in the fluids and on the surfaces are included, the over-all coefficients vary widely for seemingly similar circumstances, which suggests a thoughtful approach to actual designs.

PROBLEMS

SI UNITS

19.1 A 20.32×20.32-cm test panel, 2.54-cm thick, is placed between two plates, and the whole is properly insulated. The interface surface of one plate is maintained at 79.4°C by an electric energy supply of 50 watts; the other plate has an interface surface temperature of 21.1°C. Find k for the test panel. *Ans.* 0.528 W/m-K.

19.2 It is desired that no more than 1892 W/m^2 be conducted through a 30-cm thick wall whose average thermal conductivity is $k = 0.865$ W/m-K; the conducted heat will be controlled by insulating one side. Find the least thickness of insulating material ($k = 0.346$ W/m-K) that will assure this heat constraint if the surface temperatures of the composite wall are 1150°C and 40°C. *Ans.* 8.3 cm.

19.3 (a) Find an equation for the heat conducted through a plate of area A, thickness L, having surface temperatures of t_1 and t_2, when the conductivity varies in accordance with $k = k_0(1 + at)$. (b) Data for a plane plate glass door are: $A = 1.86$ m^2, $L = 1.905$ cm, $t_1 = 4.4$°C, $t_2 = 26.7$°C, $k_0 = 0.721$ W/m-K, $a = 0.031$/°C. Determine the hourly heat flow.

19.4 Hot water is flowing through an 11.43-cm OD steel pipe ($k = 45$ W/m-K) which is insulated with 5.08 cm of 85% magnesia ($k = 0.062$ W/m-K). Thermocouples embedded in the inner and outer surfaces of the insulation indicate temperatures of 121.1°C and 46.1°C, respectively. Find the hourly heat loss per 61 m of pipe length.

19.5 A cylindrical pipe of length L and radii r_1 and r_o is made of material whose conductivity is k. The inner surface temperature is t_1; the outer is t_o. (a) Determine the rate of heat flow if k remains constant. (b) Determine the rate of heat flow if k varies linearly with temperature and is given as $k = k_o[1 + at]$ where k_o and a are constant.

19.6 A hollow sphere with inner radius r_i, outer radius r_o, inner and outer surface temperatures t_i and t_o, is made of a material whose thermal conductivity is k. Derive the expression for the conducted heat loss, W/m^2, (a) based upon the outer area and (b) based upon the inner area. (c) Let $r_i = 7.62$ cm; $r_o = 12.70$ cm; $k = 46.15$ W/m-K; $t_i = 426.7$°C. If the heat from the sphere is 439.6 W, what is t_o? *Ans.* 422.7°C.

19.7 Air enters a preheater at 25°C and leaves at 110°C. The hot gas leaves at 130°C. Find the temperature of the hot gas entering when the logarithmic mean temperature difference is 67.4°C and **(a)** the flow is parallel, **(b)** the flow is countercurrent.

Ans. **(b)** 160°C.

19.8 The oil (c_p = 1759 W-s/kg K) from an oil-cooled electric transformer is cooled from 79.4°C to 29.4°C at the rate of 1360.5 kg/hr. This is done in an oil-water heat exchanger that receives 2948 kg/hr of water at 15.6°C. For the exchanger, U = 295 W/m²K. Find the exit water temperature and heating area required **(a)** for counterflow and **(d)** for parallel flow.

19.9 What surface area must be provided by the filament of a 100-W evacuated light globe where t = 248°C and ε = 0.38 for the filament? Assume the ambient temperature to be 25.6°C. *Ans.* 0.806 cm².

19.10 A manufacturer of electrical roaster ovens decided to change the oven cover from aluminum to a colorful porcelain enamel finish for marketing reasons. These data are known: surface area of the cover is 1807 cm²; when 592 W of power are consumed, the aluminum cover is at 204.4°C; environmental temperature is 22.2°C; respective emissivities, ε (aluminum) = 0.08, ε (porcelain cover) = 0.89; assume the convection losses to be the same in each case. If the cover temperature of the porcelain cover is 204.4°C, what power is consumed in steady-state operation?

19.11 The water resistance (drag) R on the submarine Nautilus is a function of its length L, velocity v of the ship, the viscosity μ, and density ρ of the water in which it moves. Determine dimensionless groups that could be used to organize test data.

19.12 Steam at 30 bar, 240°C flows through a 10.2-cm steel pipe (9.73-cm ID) at 2200 m/min. Calculate the film coefficient for the inside surface of the pipe; for the steel, k = 45 W/m-K.

19.13 A single 10.16-cm steel pipe, whose OD is 11.43 cm, has an outer surface temperature of 149°C. The horizontal pipe is located in a large room where the ambient temperature is 25.6°C and the barometer is standard. Determine the total heat (free convection and radiation) for 10 m of pipe length. *Ans.* 7925 W.

ENGLISH UNITS

19.14 Compare the heat fluxes that result from a 25°F temperature difference existing across the respective surfaces of 1-in. layers of aluminum, steel, concrete, and cork.

19.15 A composite plane wall consisting of two layers of materials (1.5-in. steel and 2-in. aluminum) separates a hot gas at t_i = 200°F, h_i = 2, from a cold gas at t_o = 80°F, h_o = 5 Btu/hr-ft-°F. If the hot fluid is on the aluminum side, find **(a)** the transmittance U, **(b)** the resistance R, **(c)** the interface temperature at the junction of the two metals, and **(d)** the heat through 100 ft² of the surface under steady state conditions.

Ans. **(a)** 1.416 Btu/hr-ft²-°F, **(b)** 0.70615, **(c)** 115°F, **(d)** 17,000 Btu/hr.

19.16 Dry saturated steam at 30 psia enters a 50-ft section of steel pipe (OD = 2.375 in., ID = 1.939 in.) and flows at a rate of 10 lb/min; the pipe is covered with 1 in. of 85% magnesia; the film coefficients are: h_i = 1000, h_o = 4. Determine the quality of the steam as it leaves the section. Neglect pressure loss.

19.17 Steam is flowing from a boiler to a small turbine through 200 ft of 3.5-in. steel pipe (4-in. OD, 3.548-in. ID). The steam leaves the boiler saturated at 175 psia and enters the turbine at 173.33 psia and with a moisture content of 1%. The turbine develops 50 bhp with a brake steam rate of 41 lb/bhp-hr. The ambient temperature is 90°F; h_i = 1000 and h_o = 1.9 Btu/hr-ft²-°F. If the pipe is lagged with a 2.5-in. layer of insulation, compute the value of the thermal conductivity for the insulation.

Ans. k = 0.544 Btu-in/hr-ft²-°F.

19.18 A hollow steel sphere contains a 100-watt electrical filament, and these data are known: r_i = 9 in., r_0 = 12 in. The film coefficients for the inner and outer surfaces are h_i = 6, h_o = 2 Btu per hr-ft²-°F; the environmental temperature is 80°F. Assuming steady state, compute the temperature of the inside air. (See **19.6**.) *Ans.* 101.9°F.

19.19 A water cooler uses 50 lb/hr of melting ice to cool running water from 80°F to 42°F. Based on the inside coil area, U_i = 110 Btu/hr-ft²-°F. Find **(a)** the LMTD, **(b)** the inside area of the coil, and **(c)** the gpm of water cooled.

Ans. **(a)** 24.23°F, **(b)** 2.7 ft², **(c)** 0.380 gpm.

19.20 In a 10-ton Freon 12 refrigerating system, liquid refrigerant from the condenser is cooled from 80°F to 70°F in a concentric double-pipe heat exchanger. This is done by passing the liquid through the inner pipe and saturated vapor (after the refrigeration is done) from the 20°F evaporator through the annulus. For the heat exchanger, $U = 110$ Btu per hr-ft^2-°F; for the vapor, $c_p = 0.15$ Btu/lb-°F. Find the required heating surface for **(a)** counterflow, and **(b)** parallel flow.

19.21 Calculate the rate of energy emission from each square foot of a "black body" at a temperature of **(a)** -100°F, **(b)** 0°F, **(c)** 100°F, and **(d)** 2000°F. Does the presence of other bodies affect this rate? Explain.

19.22 A 30×40-ft room is heated by means of hot water coils laid in the concrete floor. The ceiling is 9 ft high and painted white. The respective surface temperatures are 82°F floor, 60°F ceiling. If the connecting walls are nonconducting but reradiating, find the net exchange of radiant energy between floor and ceiling.

19.23 A 20×30-ft furnace floor is lined with refractory brick. The vertical distance from the floor to the water tubes is 22 ft, and the walls are nonconducting but reradiating. For the tube surfaces, $t = 525$°F and the emissivity is $\varepsilon = 0.93$; the floor surface temperature is 2250°F. Find the amount of radiant energy received by the tubes.

Ans. 26,000,000 Btu/hr.

19.24 **(a)** Find the heat loss by radiation only from an 8-in. diameter, polished steel sphere whose outer surface temperature is maintained at 750°F by means of internal electrical heating coils. The sphere is suspended in a large room wherein the environmental temperature is 50°F. **(b)** What will be the heat loss if the surface is oxidized?

Ans. **(b)** 3944 Btu/hr.

19.25 The flow resistance R per unit area encountered by a fluid moving through a closed duct depends principally upon the fluid properties of density ρ, absolute viscosity μ, and velocity v; also, the duct diameter D. In mathematical form, $R = K\rho^a\mu^b v^c D^d$, where K is a dimensionless proportionality constant and a, b, c, and d are constant exponents. Through dimensional analysis, show that $R = f\rho v^2$ where $f = C\,\text{Re}^{-b}$, a dimensionless quantity called the coefficient of friction; C is a constant and Re

the Reynolds number.

19.26 The same as **19.25** except that the velocity of sound a is included for which the assumption is $R = K\rho^a\mu^b v^c D^d a^e$. Now demonstrate that the coefficient of friction f will depend upon the Mach number **M** as well as the Reynolds number Re.

19.27 The main trunk duct of an air conditioning system is rectangular in cross section (16×30-in.) and has air at 15 psia and 40°F flowing through it with a velocity of 1400 fpm. Find h_i. *Ans.* 3.69 Btu/hr-ft^2-°F.

19.28 A manufacturer of gas kitchen stoves desires to substitute fiber glass for 85% magnesia, used as the insulation for the oven. With a maximum oven temperature of 600°F, the top outer surface of the oven is not to exceed 120°F. Neglect the metallic resistance to heat flow and determine **(a)** the thickness of 85% magnesia currently used, **(b)** the thickness of the fiber glass to be used.

19.29 A double-pipe, counterflow heat exchanger contains a 1.5-in. steel pipe (1.90-in. OD, 1.61-in. ID) inside of a 2.5-in. steel pipe (2.88-in. OD, 2.47-in. ID). Hot oil with properties similar to those of oil C, Fig. 19/20, is flowing through the inner pipe with a velocity of 3.5 fps and a bulk temperature of 200°F; also, $c_p = 0.52$ Btu/lb-°R, $k = 0.96$ Btu-in/hr-ft^2-°F. Cold oil with properties similar to those of oil B, Fig. 19/20, is flowing through the annular space with a velocity of 12.5 fps and a bulk temperature of 80°F; also, $c_p = 0.52$ Btu/lb-°R, $k = 0.94$ Btu-in./hr-ft^2-°F. **(a)** Find the film coefficients for the inner and outer surfaces of the inner pipe. **(b)** What is U for the inner pipe?

19.30 A nuclear reactor shell is spherical in shape, has an internal volume of 65.4 cu ft, contains water boiling at 400°F, and is made from stainless steel ($k = 160$) 3 in. thick. The reactor has a 2-in. layer of lead ($k = 230$ Btu-in./hr-ft^2-°F) around it with a layer of 85% magnesia in between the two materials; the maximum power level for the reactor is 5 kW. If the maximum heat loss through the shell is not to exceed 5% of the power, compute the needed thickness of 85% magnesia. What will be the outer surface temperature of the lead under maximum heat loss conditions?

LIST OF
REFERENCES

This is not a bibliography. The selected references with zero prefixes, as 0.1 and 0.2, contain numerical values of properties and references to other sources of properties. The references with numbered prefixes, as 1.1 and 1.2 (Chapter 1), comprise a number of representative books and papers on thermodynamics, which may be useful to the reader in reinforcing notions and in many cases delving further into the subject. This list of references also includes books and papers from which acknowledgment of specific *material could be made, and, of course, works from which the benefits were less tangible and therefore more difficult to identify.*

0.1   Rossini, F. D., *et al.*, *Selected Values of Physical and Thermodynamic Properties of Hydrocarbons and Related Compounds*, Am. Petrol. Inst. Res. Proj. 44, Carnegie Press.

0.2   Rossini, F. D., *et al.*, *Selected Values of Chemical Thermodynamic Properties*, NBS *Circ.* 500.

0.3   *Physical Tables*, 1954, Smithsonian Institution.

0.4   Touloukian, Y. S., *et al.*, *Thermodynamic and Transport Properties of Gases, Liquids, and Solids*, ASME and McGraw-Hill.

0.5   Touloukian, Y. S., *et al.*, *Retrieval Guide to Thermophysical Properties* (3 vols.), McGraw-Hill.

0.6   Keenan and Kaye, *Gas Tables*, Wiley.

0.7   Keenan and Keyes, *Thermodynamic Properties of Steam*, Wiley.

0.8   Reid and Sherwood, *The Properties of Gases and Liquids*, McGraw-Hill.

0.9   *Properties of Commonly-Used Refrigerants*, Air-Conditioning and Refrigeration Institute.

0.10 *Thermodynamic Properties of Ammonia*, NBS Circ. 142.

0.11 Hilsenrath, J., *et al.*, *Tables of Thermal Properties of Gases*, NBS Circ. 564.

0.12 Ellenwood and Mackey, *Thermodynamic Charts*; *Steam, Water, Ammonia, Freon-12, and Mixtures of Air and Water Vapor*, Wiley.

0.13 Tsu and Beecher, *Thermodynamic Properties of Compressed Water*, ASME.

0.14 Hilsenrath and Touloukian, "The viscosity, thermal conductivity, and Prandtl number for air, O_2, N_2, NO, CO, CO_2, H_2O, He and A," *ASME Trans.*, Vol. 76, p. 967.

0.15 *Selected Values of Hydrocarbons*, NBS Circ. C461.

0.16 Wicks and Block, eds., *Thermodynamic Properties of 65 Elements*; *Their Oxides, Halides, Carbides, and Nitrides*, Bur. Mines Bull. 605.

0.17 Stull and Sinke, *Thermodynamic Properties of the Elements*, Am. Chem. Soc.

0.18 Din, F., ed., *Thermodynamic Functions of Gases*, Vols. 1, 2, 3, Butterworth. (Includes NH_3, CO_2, CO, air, C_2H_2, C_2H_4, C_3H_8, A, C_2H_6, CH_4, N_2.)

0.19 Hougen and Watson, *Chemical Process Principles Chart*, Wiley.

0.20 Bloomer and Rao, *Properties of Nitrogen*, Inst. Gas Technol., Res. Bull. No. 18, 1952.

0.21 Lick and Emmons, *Properties of Helium to* 50,000°K, Harvard Univ. Press.

0.22 JANAF, *Thermochemical Tables*, Vols. 1–4, Dow Chemical Co., Midland, Mich., D. R. Stull, director. JANAF = Joint Army Navy Air Force.

0.23 McBride, Heimel, Ehlers, and Gordon, *Thermodynamic Properties to 6000°K for 210 Substances*, NASA SP-3001.

0.24 Hottel, Williams, and Satterfield, *Thermodynamic Charts for Combustion Processes*, Vols. I and II, Wiley.

0.25 General Electric, Powell and Suciu, eds., *Properties of Combustion Gases*, McGraw-Hill.

0.26 Farrell, W. F., chart of "Combustion Process in Diesel Cycles," Thesis, U.S. Naval Postgrad. School.

0.27 Hilsenrath, Joseph, *et al.*, *Tables of Thermodynamic and Transport Properties* (air, A, CO_2, CO, H_2, N_2, O_2, H_2O), Pergamon Press, 1960.

0.28 Beckett and Haar, *Thermodynamic Properties at High Temperature*; *Ideal Gas...to* 25,000°K, Inst. M.E., 1958.

0.29 Geyer and Bruges, *Tables of Properties of Gases, with Dissociation Theory*, Longmans.

0.30 Brinckley, S. R., *et al.*, *Thermodynamics of Combustion Products of Kerosene with Air at Low Pressure*, Bur. Mines, pp. x, 3–107/9.

0.31 Quinn and Jones, *Carbon Monoxide*, Reinhold.

0.32 Meyer, C. A., *et al.*, *ASME Steam Tables*, Am. Soc. Mech. Eng.

1.1 Zemansky, M. W., *Heat and Thermodynamics*, McGraw-Hill.

1.2 Keenan, J. H., *Thermodynamics*, Wiley.

1.3 Lee and Sears, *Thermodynamics*, Addison-Wesley.

1.4 Hatsopoulos and Keenan, *Principles of General Thermodynamics*, Wiley.

1.5 Obert, E. F., *Concepts of Thermodynamics*, McGraw-Hill.

1.6 Obert and Gaggioli, *Thermodynamics*, McGraw-Hill.

1.7 Kiefer, Kinney, and Stuart, *Principles of Engineering Thermodynamics*, Wiley.

1.8 Van Wylen and Sonntag, *Fundamentals of Classical Thermodynamics*, Wiley.

1.9 Saad, M. A., *Thermodynamics for Engineers*, Prentice-Hall.

1.10 Lay, J. E., *Thermodynamics*, Merrill.

1.11 Zemansky and Van Ness, *Basic Engineering Thermodynamics*, McGraw-Hill.

1.12 Tribus, M., *Thermostatics and Thermodynamics*, Van Nostrand.

1.13 Hougen, Watson, and Ragatz, *Chemical Process Principles*, Part II, 2nd ed., Wiley.

1.14 Smith and Van Ness, *Introduction to Chemical Engineering Thermodynamics*, McGraw-Hill.

1.15 Kestin, J., *A Course in Thermodynamics*, Blaisdell.

1.16 Dodge, B. F., *Chemical Engineering Thermodynamics*, McGraw-Hill.

1.17 Sears, F. W., *Thermodynamics, The Kinetic Theory of Gases, and Statistical Mechanics*, Addison-Wesley.

1.18 Lewis and Randall, Thermodynamics, Revised by Pitzer and Brewer, McGraw-Hill.

1.19 Weber and Meissner, *Thermodynamics for Chemical Engineers*, Wiley.

1.20 Sommerfeld, A. W., *Thermodynamics and Statistical Mechanics*, Academic Press.

1.21 Maxwell, James C., *Theory of Heat*, Van Nostrand.

1.22 Baker, Ryder, and Baker, *Temperature Measurements in Engineering*, Vols. I and II, Wiley.

1.23 ASME *Codes on Instruments and Apparatus.*

1.24 Shoop and Tuve, *Mechanical Engineering Practice*, McGraw-Hill.

1.25 Weber, R. L., *Heat and Temperature Measurement*, Prentice-Hall.

1.26 Richtmyer and Kennard, *Introduction to Modern Physics*, McGraw-Hill.

1.27 Smith and Cooper, *Elements of Physics*, McGraw-Hill.

1.28 Kingery, W. D., *Property Measurements at High Temperatures*, Wiley.

1.29   Rubin, L. G., "Measuring temperature," Intern. Sci. Technol., Jan. 1964.

1.30   *Temperature, Its Measurement and Control in Science and Industry* (4 vols.), Reinhold.

1.31   Harrison, R. T., *Radiation Pyrometry and Its Underlying Principles*, Wiley.

1.32   Zimmerman and Lavine, *Conversion Factors and Tables*, Industrial Research Service.

1.33   Rogers and Mayhew, *Engineering Thermodynamics*, Wiley.

2.1   Sweigert and Beardsley, "Empirical specific heat equations based upon spectroscopic data," Ga. Inst. Technol., Eng. Expt. Sta. Bull. No. 2.

2.2   Spencer and Justice, "Empirical heat capacity equations for simple gases," *J. Am. Chem. Soc.*, Vol. 56, p. 2311.

2.3   Spencer and Flannagan, "Empirical heat capacity equations of gases," *J. Am. Chem. Soc.*, Vol. 64, p. 2511.

2.4   Partington and Schilling, *The Specific Heat of Gases*, Van Nostrand.

2.5   Ellenwood, Kulik, and Gay, *The Specific Heat of Certain Gases over Wide Ranges of Pressures and Temperatures*, Cornell Eng. Sta., Bull. No. 30.

2.6   Chipman and Fontana, "A new approximate equation for heat capacities at high temperatures," *J. Am. Chem. Soc.*, Vol. 57, p. 48.

2.7   Spencer, *J. Am. Chem. Soc.*, Vol. 67, p. 1859.

2.8   Corruccini and Gniewek, *Specific Heats and Enthalpies of Technical Solids at Low Temperatures*, NBS Mono. 21.

2.9   Eaton, J. R., *Electrons, Neutrons, and Protons in Engineering*, Pergamon.

2.10   Andronov, Vitt, and Khaikin, *Theory of Oscillators*, Addison-Wesley.

2.11   Baumeister, T., Jr., *Fans*, McGraw-Hill.

2.12   Church, A. H., *Centrifugal Pumps and Blowers*, Wiley.

3.1   *Steam*, 37th ed., The Babcock & Wilcox Co.

3.2   *Combustion Engineering*, Combustion Engineering-Superheater, Inc.

3.3   Peebles, J. H., Jr., Thesis, Tulane University.

3.4   Templerley, H. N. V., "Changes of state," Intern. Sci. Technol., Oct. 1965.

3.5   Drost-Hansen, W., "The puzzle of water," Intern. Sci. Technol., Oct. 1966.

4.0   General texts under Chapter 1 are the best references for this chapter.

4.1   Fay, J. A., *Molecular Thermodynamics*, Addison-Wesley.

4.2   Knuth, E. L., *Statistical Thermodynamics*, McGraw-Hill.

4.3   Vanderslice, Schamp, and Mason, *Thermodynamics*, Prentice-Hall.

4.4   Hirschfelder, Curtiss, and Bird, *Molecular Theory of Gases and Liquids*, Wiley.

4.5   Beer and Johnson, *Mechanics for Engineers*, McGraw-Hill.

4.6   Symon, K. R., *Mechanics*, Addison-Wesley.

4.7   King, A. L., *Thermophysics*, Freeman.

4.8   Furth, H. P., "Magnetic pressure," Intern. Sci. Technol., Sept. 1966.

4.9   Ridenour, L. N., et al., *Modern Physics for the Engineer*, McGraw-Hill.

5.1   Montgomery, S. R., *Second Law of Thermodynamics*, Pergamon.

5.2   Lee, Sears, and Turcotte, *Statistical Thermodynamics*, Addison-Wesley.

5.3   Keenan, J. H., "Availability and irreversibility in thermodynamics," *Brit. J. Appl. Phys.*, Vol. 2, p. 183.

5.4   Buchdahl, H. A., "On the principle of Caratheodory," *Am. J. Phys.*, Vol. 17, No. 1, p. 41.

5.5   Potter, J. H., "On the inequality of Clausius," *Combustion*, June 1964.

5.6   Darrieus, G., "The rational definition of steam turbine efficiencies," *Engineering*, Vol. 130, p. 195.

5.7   Del Duca and Fuscoe, "Electrons, enzymes, and energy," Intern. Sci. Technol., March 1965.

5.8   Bent, H. A., *The Second Law*, Oxford.

5.9   Gibbs, J. W., *The Collected Works of J. Willard Gibbs*, Longmans.

5.10 Planck, M., *Treatise on Thermodynamics*, Longmans.

5.11 Aston and Fritz, *Thermodynamics and Statistical Thermodynamics*, Wiley.

5.12 Fast, J. D., *Entropy*, McGraw-Hill.

5.13 Young, G. J., *Fuel Cells*, Reinhold.

5.14 Mitchell, W., *Fuel Cells*, Academic Press.

5.15 Williams, K. R., ed., *An Introduction to Fuel Cells*, Elsevier.

5.16 Baker, B. S., ed. *Hydrocarbon Fuel Cell Technology*, Academic Press.

5.17 Keenan, J. H., "A steam chart for second law analysis," *Mech. Eng.*, Vol. 54, p. 195.

6.0   Best references for this chapter are other books on thermodynamics; see the references for Chapter 1.

6.1   Gottlieb, M. B., "Plasma—the fourth state," Intern. Sci. Technol., Aug. 1965.

6.2   Frances, G., *Ionization Phenomena in Gases*, Academic Press.

7.0   Best references for this chapter are other books on thermodynamics; see references for Chapter 1.

8.0   See references for Chapter 1.

8.1   Hefner, F. E., "Highlights from 6500 hours of Stirling engine operation," GMR 456, General Motors Research Laboratories.

8.2   Flynn, Percival, and Hefner, "GMR Stirling thermal engine," *SAE Trans.*, Vol. 68, pp. 665–683.

8.3   Carnot, S., *Reflections on the Motive Power of Heat*, ASME.

9.1   Gaffert, G. A., *Steam Power Stations*, McGraw-Hill.

9.2   Morse, F. T., *Power Plant Engineering and Design*, Van Nostrand.

9.3   Zerban and Nye, *Steam Power Plants*, International Textbook.

9.4   Berry, C. H., "Steam turbine testing," *Mech. Eng.*, Nov. 1935.

9.5   Mooney, D. A., *Mechanical Engineering Thermodynamics*, Prentice-Hall.

9.6   Potter, P. J., *Steam Power Plants*, Ronald.

9.7   Kreisinger and Purcell, "Some operating data for large steam generating units," *ASME Trans.*, Vol. 50.

9.8   Meyer, Silvestri, and Martin, "Availability balance of steam power plants," *ASME Trans.*, Vol. 81, p. 35.

9.9   Shepherd, D. G., *Principles of Turbomachinery*, Macmillan.

9.10 Emmet, W. L. R., "Status of the Emmet mercury-vapor process," *Mech. Eng.*, Nov. 1937.

9.11 Sheldon, L. A., "Properties of mercury vapor," *ASME Trans.*, Vol. 46, p. 272.

9.12 Hackett, H. N., "Mercury-steam power plants," *Mech. Eng.*, Vol. 73, p. 559.

9.13 Salisbury, J. K., *Steam Turbines and Their Cycles*, McGraw-Hill.

9.14 Krieg, E. H., "Superposition," *Mech. Eng.*, Sept. 1936.

9.15 Wooten, W. R., *Steam Cycles for Nuclear Power Plants*, Simmons-Boardman.

9.16 Kramer, A. W., *Boiling Water Reactors*, Addison-Wesley.

9.17 *Steam*, 38th ed., The Babcock & Wilcox Co.

10.1 Pitzer, Lippman, *et al.*, "The volumetric and thermodynamic properties of fluids, " *J. Am. Chem. Soc.*, Vol. 77, p. 3433.

10.2   Su and Chang, "A generalized van der Waals' equation of state for real gases," *Ind. Eng. Chem.*, Vol. 38, p. 800.

10.3   Su, G. J., "Modified law of corresponding states for real gases," *Ind. Eng. Chem.*, Vol. 38, p. 803.

10.4   Nelson and Obert, "Laws of corresponding states," *A.I.Ch.E. J.*, Vol. 1, p. 74.

10.5   Nelson and Obert, "Generalized compressibility charts," *Chem. Eng.*, Vol. 61, p. 203.

10.6   Nelson and Obert, "Generalized *pvT* properties of gases," *ASME Trans.*, Vol. 76, p. 1057.

10.7   Hall and Ibele, "Compressibility deviations for polar gases," *ASME Trans.*, Vol. 77, p. 1003.

10.8   Lydersen, Greenkorn, and Hougen, *Generalized Thermodynamic Properties of Pure Fluids*, Report No. 4, Eng. Expt. Sta., Univ. of Wisconsin.

10.9   Watson and Smith, "Generalized high-pressure properties of gases," *Nat. Petrol. News*, Vol. 28, No. 27, p. 29.

10.10 Lewis and Elbe, "Heat capacities and dissociation equilibria of gases," *J. Am. Chem. Soc.*, Vol. 57, p. 612.

10.11 Lype, E. F., "Determination of equation of state from wave front observations," *ASME Trans.*, Vol. 80, p. 1.

10.12 Hall and Ibele, "The tabulation of imperfect-gas properties for air, nitrogen, and oxygen," *ASME Trans.*, Vol. 76, p. 1039.

10.13 Kaw, W. B., "Density of Hydrocarbon Gases and Vapors," *Ind. Eng. Chem.*, Vol. 28, p. 1014.

10.14 Taylor and Glasstone, *States of Matter*, Van Nostrand.

10.15 Tisza, L., *et al.*, *Generalized Thermodynamics*, M.I.T. Press.

10.16 Hsieh, J. S., "Four-parameter generalized compressibility charts for nonpolar fluids," ASME 65-WA/PID-1.

10.17 Shah and Thodos, "A comparison of equations of state," *Ind. Eng. Chem.*, Vol. 57, No. 3.

10.18 Redlich and Kwong, "On thermodynamic solutions V. An equation of state," *Chem. Rev.*, Vol. 44, p. 233.

11.0   Mathematical relations are treated to various extents in chemical thermodynamic texts and modern texts in general; see refs. 1.3, 1.4, 1.10, 1.15, and 1.16.

11.1   Bridgman, P. W., *A Condensed Collection of Thermodynamic Formulas*, Harvard Univ. Press.

11.2   Bridgman, P. W., *The Nature of Thermodynamics*, Harvard Univ. Press.

11.3   Eller, E., "Squeeze electricity," Intern. Sci. Technol., July 1963.

11.4   Potter and Chow, "On Joule's internal energy law," *Combustion*, April 1965.

11.5   Sando, R. M., "Compressibility of fluids," *Machine Design*, Sept. 29, 1960.

11.6   Tilson, S., "Solids under pressure," Intern. Sci. Technol., July 1962.

11.7   International Atomic Energy Agency, *Thermodynamics*, Vols. I and II (proceedings of symposium).

11.8   Corruccini, R. J., *Specific Heats and Enthalpies of Technical Solids at Low Temperature*, NBS Mono. 21.

11.9   Haar, L., *et al.*, *Ideal Gas Functions and Isotope Exchange Functions for Diatomic Hydrides, Deuterides, and Tritides*, NBS Mono. 20.

11.10 Lange, N. A., *Handbook of Chemistry*, Handbook Publishers.

11.11 Giles, Robin, *Mathematical Foundations of Thermodynamics*, Pergamon's Pure Appl. Math. Series.

12.1   ASHRAE, *Guide and Data Book*, American Society of Heating, Refrigerating, and Air-Conditioning Engineers.

12.2 Carrier, Cherne, Grant, *Modern Air Conditioning, Heating and Ventilating*, Pitman.

12.3 Greene, A. M., Jr., *Principles of Heating, Ventilating, and Air Conditioning*, Wiley.

12.4 Goodman, William, *Air Conditioning Analysis*, Macmillan.

12.5 Arnold, J. H., "The theory of the psychrometer," *Physics*, Vol. 4.

12.6 London, Mason, Boelter, "Performance characteristics of a mechanically induced draft, counterflow, packed cooling tower," *ASME Trans.*, Vol. 62, p. 41.

12.7 Lichtenstein, J., "Performance and selection of mechanical-draft cooling towers," *ASME Trans.*, Vol. 65, p. 779.

12.8 Moss, J. F., *Cooling tower performance for air conditioning systems*, Bull. No. 110, Texas Eng. Expt. Sta., College Station.

12.9 Bosnjakovic (trans. by Blackshear), *Technical Thermodynamics*, Holt.

13.1 Lewis and von Elbe, *Combustion Flames and Explosions*, Academic Press.

13.2 Jost, W. (trans. by H. O. Craft), *Explosion and Combustion Processes in Gases*, McGraw-Hill.

13.3 *Combustion Engineering*, Combustion Engineering-Superheater, Inc.

13.4 Pearson and Fellinger, "Thermodynamic properties of combustion products," ASME paper 65-WA/Ener-2.

13.5 Barrow, G. M., *Physical Chemistry*, McGraw-Hill.

13.6 Ellern, H., *Modern Pyrotechnics*, Chemical Publishing.

13.7 Baldwin, L. V., *et al.*, *Rocket and Missile Technology*, A.I.Ch.E.

13.8 Minkoff and Tipper, *Chemistry of Combustion Reactions*, Butterworth.

13.9 Williams, F. A., *Combustion Theory*, Addison-Wesley.

13.10 Nikolaev, B. A., *Thermodynamic Assessment of Rocket Engines*, Pergamon.

13.11 Strehlow, R. A., *Fundamentals of Combustion*, International Textbook.

14.1 *Compressed Air Handbook*, Compressed Air and Gas Institute.

14.2 Wilson and Crocker, "Fundamentals of the Elliott-Lysholm compressor," *Mech. Eng.*, Vol. 68, p. 514.

14.3 ASME *Test Code for Compressors and Exhausters*, PTC 10-1949.

14.4 Gill, T. T., *Air and Gas Compression*, Wiley.

14.5 Stuart and Jackson, "The analysis and evaluation of compressor performance," *Mech. Eng.*, Vol. 76, p. 287.

14.6 *Compressed Air Power*, Compressed Air and Gas Institute.

14.7 Baumeister, T., *Marks' Mechanical Engineers Handbook*, McGraw-Hill.

15.1 Zucrow, M. J., *Aircraft and Missile Propulsion* (2 vols.), Wiley.

15.2 Dusinberre and Lester, *Gas Turbine Power*, International Textbook.

15.3 Godsey and Young, *Gas Turbines for Aircraft*, McGraw-Hill.

15.4 Jennings and Rogers, *Gas Turbine Analysis and Practice*, McGraw-Hill.

15.5 Cohen and Rogers, *Gas Turbine Theory*, Longmans.

15.6 Lewis, A. D., *Gas Power Dynamics*, Van Nostrand.

15.7 von Karmon, T., *et al.*, *High-Speed Aerodynamics and Jet Propulsion*, Princeton Univ. Press.

15.8 Cox, H. R., ed., *Gas Turbine Principles and Practice*, Van Nostrand.

15.9 Hawthorne and Olson, eds., *Design Performance of Gas Turbine Power Plants*, Princeton Univ. Press.

15.10 Hodge, James, *Cycles and Performance Estimation*, Academic Press.

15.11 Foa, J. V., *Elements of Flight Propulsion*, Wiley.

15.12 Hill and Peterson, *Mechanics and Thermodynamics of Propulsion*, Addison-Wesley.

15.13 Hesse and Mumford, *Jet Propulsion for Aerospace Applications*, Pitman.

15.14 Vincent, E. T., *The Theory and Design of Gas Turbines and Jet Engines*, McGraw-Hill.

15.15 Barrère, Marcel, *et al.*, *Rocket Propulsion*, Elsevier.

15.16 Sutton, G. P., *Rocket Propulsion Elements*, Wiley.

15.17 Melik-Pashaev, N. I. (trans. by W. E. Jones), *Liquid-Propellant Engines*, Pergamon.

15.18 Crocco and Cheng, *Theory of Combustion Instability in Liquid Propellant Rocket Motors*, Butterworth.

15.19 London and Oppenheim, "The free-piston engine development," *ASME Trans.*, Vol. 74, p. 1349.

15.20 London, A. L., "Supercharged-and-intercooled free-piston-and-turbine compound engine," *ASME Trans.*, Vol. 78, p. 1757.

15.21 Maes and Sutherland, "Tiny rockets," Intern. Sci. Technol., Aug. 1966.

15.22 Forrester and Eilenberg, "Ion rockets," Intern. Sci. Technol., Jan. 1964.

15.23 Peters, R. L., *Design of Liquid, Solid, Hybrid Rockets*, Heyden.

15.24 *Annual Reports, 1967*, Gas Turbine Division, ASME.

15.25 Mills, R. G., "The combined steam turbine-gas turbine plant for marine use," *ASME Trans.*, Vol. 81, paper 59-A-237.

16.1 Taylor and Taylor, *The Internal Combustion Engine*, rev. ed., International Textbook.

16.2 Hersey, Eberhardt, and Hottel, "Thermodynamic properties of the working fluid in internal combustion engines," *SAE J.*, Vol. 39, p. 409.

16.3 Lichty, L. C., *Internal Combustion Engines*, McGraw-Hill.

16.4 Taylor, C. F., *The Internal Combustion Engine in Theory and Practice*, Technology Press.

16.5 Rogowske, A. R., *Elements of Internal Combustion Engines*, McGraw-Hill.

16.6 Polson, J. A., *Internal Combustion Engines*, 2nd ed., Wiley.

16.7 Dean, W. C., "A new rotary piston engine," *Mech. Eng.*, Aug. 1964.

16.8 Obert, E. F., *Internal Combustion Engines*, 3rd ed., International Textbook.

17.0 See reference 12.1.

17.1 Jordan and Priester, *Refrigeration and Air Conditioning*, Prentice-Hall.

17.2 Jennings and Lewis, *Air Conditioning and Refrigeration*, International Textbook.

17.3 Raber and Hutchinson, *Refrigeration and Air Conditioning Engineering*, Wiley.

17.4 Sparks, N. R., *Theory of Mechanical Refrigeration*, McGraw-Hill.

17.5 Sharpe, N., *Refrigerating Principles and Practices*, McGraw-Hill.

17.6 Penrod, Baker, Levy, and Chung, *Univ. of Kentucky Heat Pump No. 1*, Eng. Expt. Sta. Bull. No. 30, Univ. of Kentucky.

17.7 Barron, R., *Cryogenic Systems*, McGraw-Hill.

17.8 Vance and Duke, eds., *Applied Cryogenic Engineering*, Wiley.

17.9 Bell, J. H., Jr., *Cryogenic Engineering*, Prentice-Hall.

17.10 Sittig, M., *Cryogenics*, Van Nostrand.

17.11 Collins and Cannady, *Expansion Machines for Low Temperature Processes*, Oxford.

17.12 Severns and Fellows, *Air Conditioning and Refrigeration*, Wiley.

17.13 Stoecker, W. F., *Refrigeration and Air Conditioning*, McGraw-Hill.

17.14 Ruhemann, M., *The Separation of Gases*, Clarendon Press.

17.15 Burton, E. F., *et al.*, *Phenomena at the Temperature of Liquid Helium*, Reinhold.

17.16 Davies, M. M., *The Physical Principles of Gas Liquefaction and Low Temperature Rectification*, Longmans.

17.17 Scott, Denton, and Nicholls, *Technology and Uses of Liquid Hydrogen*, Pergamon.

18.1 Shapiro, A. H., *The Dynamics and Thermodynamics of Compressible Fluid Flow* (2 vols.), Ronald Press.

18.2 Dodge and Thompson, *Fluid Mechanics*, McGraw-Hill.

18.3 Binder, R. C., *Fluid Mechanics* and *Advanced Fluid Dynamics and Fluid Machinery*, Prentice-Hall.

18.4	Liepmann and Roshko, *Elements of Gasdynamics*, Wiley.
18.5	Yellott and Holland, "Condensation of steam in diverging nozzles," *ASME Trans.*, Vol. 59, p. 171.
18.6	Binnie and Woods, "The pressure distribution in a convergent-divergent nozzle," *Proc. Inst. Mech. Eng.*, Vol. 138, p. 260.
18.7	Daily and Harleman, *Fluid Dynamics*, Addison-Wesley.
18.8	Sabersky and Acosta, *Fluid Flow*, Macmillan.
18.9	Kenyon, R. A., *Principles of Fluid Mechanics*, Ronald.
18.10	Pao, R. H., *Fluid Mechanics*, Wiley.
18.11	Bird, R. B., *et al.*, *Transport Phenomena*, Wiley.
18.12	Hunt, J. N., *Incompressible Fluid Dynamics*, Wiley.
18.13	Yuan, S. W., *Foundations of Fluid Mechanics*, Prentice-Hall.
18.14	McLeod, E. B., *Introduction to Fluid Dynamics*, Pergamon.
18.15	Lee, John F., *Theory and Design of Steam and Gas Turbines*, McGraw-Hill.
18.16	Stodola-Lowenstein, *Steam and Gas Turbines*, McGraw-Hill.
18.17	Church, E. F., Jr., *Steam Turbines*, McGraw-Hill.
18.18	Bartlet, R. L., *et al.*, *Steam-Turbine Performance and Economics*, McGraw-Hill.
18.19	Skrotzki and Vopat, *Steam and Gas Turbines*, McGraw-Hill.
18.20	Goudie, W. J., *Steam Turbines*, Longmans.
18.21	Newman, Keller, Lyons, Wales, *Modern Turbines*, Wiley.
18.22	ASME, *Fluid Meters, Theory and Application*.
18.23	ASME, *Flow measurements, Part 5, Power Test Codes*.
18.24	Rohsenow and Choi, *Heat, Mass, and Momentum Transfer*, Prentice-Hall.
18.25	Benedict and Steltz, *Handbook of Generalized Gas Dynamics*, Plenum Press.
18.26	Moody, L. F., "Friction factors for pipe flow," *ASME Trans.*, Vol. 66, p. 671.
18.27	Cambel and Jennings, *Gas Dynamics*, McGraw-Hill.

19.1	McAdams, W. H., *Heat Transmission*, McGraw-Hill.
19.2	King, W. J., "The basic laws and data of heat transmission," *Mech. Eng.*, Vol. 54, April 1932.
19.3	ASHRAE, *Guide and Data Book*, American Society of Heating, Refrigerating, and Air Conditioning Engineers.
19.4	Kreith, Frank, *Principles of Heat Transfer*, IEP—A Dun-Donnelley Publisher.
19.5	Jakob, M., *Heat Transfer*, Vol. 2, Wiley.
19.6	Gebhart, B., *Heat Transfer*, 2nd ed., McGraw-Hill.
19.7	Schneider, P. J., *Conduction Heat Transfer*, Addison-Wesley.
19.8	Dusinberre, G. M., *Heat Transfer Calculations by Finite Differences*, International Book Company.
19.9	Hottel, H. C., "Radiant heat transmission," *Mech. Eng.*, Vol. 52, 1930.
19.10	Oppenheim, A. K., "The network method of radiation analysis," ASME Paper 54-A75, 1954.
19.11	Sparrow and Cess, *Radiant Heat Transfer*, Wadsworth Publishing Company.
19.12	Planck, M., *The Theory of Heat Radiation*, Dover Publications.
19.13	Eckert and Drake, *Heat and Mass Transfer*, 2nd ed., McGraw-Hill.
19.14	Chapman, A. J., *Heat Transfer*, Macmillan Company.
19.15	Brown and Marco, *Introduction to Heat Transfer*, 3rd ed., McGraw-Hill.
19.16	Schlichting, H., *Boundary Layer Theory*, 4th ed., McGraw-Hill.
19.17	Rohsenow and Choi, *Heat, Mass, and Momentum Transfer*, Prentice-Hall.
19.18	Kreith, F., *Radiation Heat Transfer*, International Textbook Company.
19.19	Wiebelt, J. A., *Engineering Radiation Heat Transfer*, Holt Publishing Company.
19.20	Boelter, Cherry, Johnson, Martinelli, *Heat Transfer*, Univ. of Calif. Press.
19.21	Jakob and Hawkins, *Elements of Heat Transfer and Insulation*, Wiley.
19.22	Lee and Sears, *Thermodynamics*, Addison-Wesley.

19.23 Hilsenrath and Touloukian, "The viscosity, thermal conductivity, and Prandtl number for air, O_2, N_2, NO, CO, CO_2, H_2O, He and A," *ASME Trans.*, Vol. 76, p. 967.

19.24 Keyes, F. G., "Summary of viscosity and heat conduction data for He, A, H_2, O_2, N_2, CO, CO_2, H_2O, and air," *ASME Trans.*, Vol. 73, p. 589.

19.25 Sieder and Tate, "Heat transfer and pressure drop of liquids in tubes," *Ind. Eng. Chem.*, Vol. 28, 1936, p. 1429.

19.26 Croft, H. O., *Thermodynamics, Fluid Flow, and Heat Transmission*, McGraw-Hill.

19.27 Hottel, H. C., "Radiation from non-luminous flames," *ASME Trans.*, Vol. 57, p. 463.

APPENDIX A

Efficiencies and Performance Information—Reciprocating Power Systems

INTRODUCTION   A.1

The beginning of any actual operating engine was in the initial analysis that the designing engineer performed on the particular ideal cycle. Having computed the ideal work W and thermal efficiency e, he then estimated closely the actual efficiency, size, and other specifications required to build the real engine. He could do this because the work output of many types and sizes of real engines had been measured again and again. At the expense of appearing repetitious, this appendix brings together for convenient reference the various definitions plus some additional power measurement relations because these engineering aspects of thermodynamics pervade all practical considerations.

MEASURING WORK   A.2

The amount of actual work obtained depends upon where it is measured (work is always being degraded). In Fig. A/1, for example, the work that is done in the cylinder can be measured with the use of an indicator, Fig. A/2. This work, called the *indicated work*, is closely equal to the work of the fluid, which is the work obtained by the use of thermodynamic properties. Indicated work W_I is relevant only for a reciprocating machine.

We also have means of measuring the work on the shaft, Fig. A/1. This work is called *brake work* W_B, or *shaft work*. Since the shaft drives *something*, the work can similarly be measured further along, as desired. In electric power generation, the shaft drives a generator, the output of the generator, being electricity, is called *overall* or *combined work* W_K. This work is easily measured with electrical instruments.

THE INDICATOR CARD   A.3

A widely useful means of studying the performance of reciprocating engines is the *indicator card*, or *indicator diagram*. This "card" is a pictured record of the variation of pressure and volume of the working substance in a cylinder as the piston reciprocates. The record is obtained with an *indicator*, a sample of which is shown in

Fig. A/1   *Meaning of Indicated Work, Brake Work and Combined Work; W_I, W_B, W_K.*

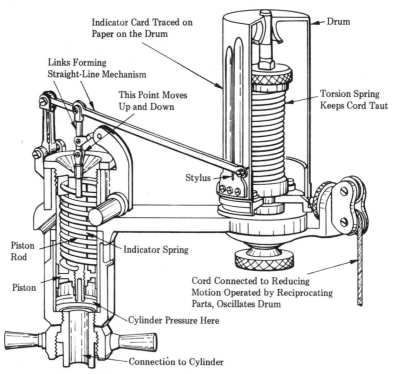

Fig. A/2   *Indicator.* Steam or other fluid enters through the connection to the cylinder and acts on a small piston, whose motion is opposed by the indicator spring. The greater the pressure from the cylinder, the more the piston (and stylus) moves. The stylus, which traces the indicator card, is actuated through an approximate straight-line linkage, and the indicator card is traced. The cord is attached to the drum, which moves in phase (the motion is reduced in magnitude) with the engine piston (and crosshead). No indicator paper is shown on the drum.

Fig. A/2. Other types of indicators that eliminate the piston in Fig. A/2 and its inertia are used for high-speed engines. A card for a well-adjusted reciprocating steam or air engine is shown in Fig. A/3. The area of the card is found by using the planimeter, Fig. A/4.

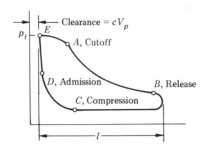

Indicator Card. Fig. A/3

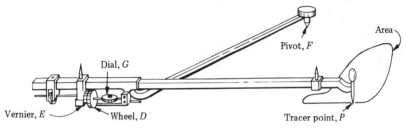

Fig. A/4

Planimeter. Point *F* is fixed. Point *P* is moved clockwise around the boundary of the area to be measured. Wheel *D* turns with the movement of *P*, and after the complete areal outline has been traced, the magnitude of the area may be read from the dial *G* and the vernier *E*.

Although the indicator card is a closed pressure-volume diagram, it is not a thermodynamic cycle. The indicator diagram shows the details of what is necessary for the work process of a cycle. The events of an engine in a power cycle are: Intake valve opens at the ***point of admission*** *D* and fluid flows in until the ***point of cutoff*** *A*, where the intake valve closes; the fluid expands (doing positive work) from *A* to *B*, which is the ***point of release*** where the exhaust valve opens and the fluid begins to flow out of the cylinder; at the ***point of compression***, *C*, the exhaust valve closes, trapping a small amount in the cylinder (for its cushioning effect and other reasons); the intake valve opens at *D* and the diagram repeats. The point of cutoff is usually designated by the ***percentage cutoff***, which is the percentage of the *stroke* traversed by the piston when cutoff occurs. Indicator cards for compressors and internal combustion engines will be shown at appropriate places. Since the indicator card gives (closely) a record of the pressure on the piston, *areas enclosed within indicator cards represent work*, indicated work W_I. This work includes the effects of getting the working substance into and out of the cylinder; the net work on the surroundings (by moving boundary or piston) for the two strokes necessary to complete the diagram is zero.

MEP AND IHP   A.4

The indicator card is taken by an indicator, Fig. A/2, that is fitted with a spring of a certain scale. If the scale is 100 lb *in the case of an indicator*, we mean that a vertical movement of 1 in. of the stylus indicates a pressure of 100 psi.* Therefore the actual heights on the indicator card multiplied by the scale of the indicator spring will be the actual pressures in pounds per square inch (neglecting errors inherent in the indicator). Thus, to find the indicated mep, find the area of the indicator card in square inches, divide the area by the length of the card l in inches, Fig. A/3, and obtain the *average height* of card in inches. This average height multiplied by the scale of the indicator spring is the average pressure, called the *indicated mep* (p_{mI}).

(a) $$p_{mI} = \frac{\text{(area of indicator card)(scale of indicator spring)}}{\text{length of indicator card}}$$

If this pressure acts on the face of the piston during one stroke, the work is the same as produced by the actual pressures throughout all events of the indicator card. See § 8.8 again.

To get horsepower from the mep, divide the work in foot-pounds per minute by 33,000. The force corresponding to a pressure of p_{mI} psi is $p_{mI}A$ lb, where A in.^{2} is the piston area. This force acts through a distance equal to the stroke of the piston, L ft, so that the work per stroke is $p_{mI}AL$ ft-lb, which is also the work for one complete card. If the *number of cards per minute* (or the *number of cycles per minute*) is N, the work per minute is $p_{mI}ALN = p_{mI}LAN$ ft-lb/min. The indicated horsepower is therefore

(A-1) $$\text{ihp} = \frac{p_{mI}LAN}{33,000}$$

where p_{mI} is obtained from the indicator card as explained above. This equation is often called the **mep equation**. The beginner frequently makes an error in applying equation (A-1) because, as derived, A is in square inches (to match pressure in psi), but the stroke L is in feet. However, it is quite correct to use square feet and pounds per square foot together. Observe that, in general, N is *not* the number of revolutions per minute n but the number of cards (or cycles) completed per minute; N and n happen to be the same for a single-cylinder, single-acting machine. If the single cylinder is double-acting, $N = 2n$, where n is rpm. In a **single-acting engine**, the working substance acts on one side of the piston; in a **double-acting engine**, the working substance acts on both sides of the piston.

We have already defined the symbol p_m as the mep of an ideal cycle (or indicator diagram) and hp as ideal horsepower; hence, the mep equation as applied to ideal engines is

(A-2) $$\text{hp} = \frac{p_m LAN}{33,000}$$

When the mep equation is applied to the crank end B of a double-acting cylinder, Fig. 4/13, the A in equation (A-1) should be the area of the piston *minus* the area of the piston rod, which is the net area over which the mep acts.

---

* This statement is true when a standard piston is used in the indicator. If an undersized piston is used, a common practice, the scale stamped on the spring must, of course, be modified.

The engine size is given by two numbers, say 6×8 in., where the first number is always the bore and the second one the stroke; 6-in. diameter, 8-in. stroke.

BRAKE POWER A.6

The name *brake work* came about because in the first determinations the power output was dissipated in the friction of a brake. Brakes are still used for this purpose over suitable ranges of power and speed. A type known as a ***prony brake*** is shown in Figs. A/5 and A/6. When the brake is clamped to the flywheel, Fig. A/5, the frictional force F ($= F/2 + F/2$ at the brake shoes) tends to turn the brake with the flywheel. However, the knife edge on the beam rests on scale and prevents motion of the brake. The force P, which is weighed by the scales, consists of the reaction produced by friction and a portion of the weight of the brake, called the ***tare***—unless there is a counterweight to balance the brake about the center line of the flywheel. To find the tare, support the brake on an *edge* at B and weigh the tare on

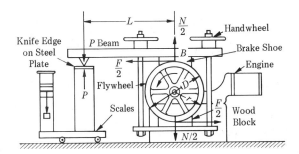

Prony Brake. **Fig. A/5**

Setup for Prony Brake. Compare this photograph with the diagrammatic **Fig. A/6** arrangement of Fig. A/5.

the scales. Thus, the net force on the scales produced by the frictional moment is $(P - \text{tare})$.

The information obtained from a prony brake test is used to compute the brake horsepower. Thus, the frictional force F acting through one revolution of the flywheel does work $W = (F\pi D)/12$ ft-lb, where the flywheel diameter D is expressed in inches (usual practice). The work multiplied by n rpm gives the work in ft-lb/min. Dividing this result by 33,000 converts to horsepower; thus

(A-3)
$$\text{hp} = \frac{F\pi Dn}{(12)(33,000)} = \frac{F2\pi rn}{(12)(33,000)} = \frac{Tn}{63,000}$$

where $T = Fr = FD/2$ in.-lb is the frictional torque on the flywheel (watch units). The sum of the moments of the forces on the brake about the center of the flywheel gives

(b)
$$Fr = \frac{FD}{2} = (P - \text{tare})L = T$$

where L in. is the moment arm of the force P. Thus, knowing the dimension L on the brake, Fig. A/5, we may observe the value of P indicated by the scales and the number of rpm of the engine, and calculate the brake horsepower (bhp) from equation (A-3).

Other instruments for measuring shaft work are a hydraulic brake, not illustrated, a dynamometer, Fig. A/7, and strain gage systems to report the torque. Because the ordinary type of indicator, Fig. A/2, has too much inertia to respond accurately at high speed, it became customary to measure the power output of automotive and

Fig. A/7 *Dynamometer Test of a Stationary Industrial Engine.* The controls for the dynamometer are observed in the background.

other similar engines by dynamometer only. This practice brought about the common use of the **brake mean effective pressure** (bmep) p_{mB}, which is p_m computed from the mep equation (A-2), using bhp instead of ihp;

(c)
$$p_{mB} = \frac{33,000 \text{ bhp}}{LAN}$$

While dynamometer tests are still routine industrial practice, high-speed indicators, utilizing the oscilloscope, and also optical and photographic effects have been developed so that it is now possible to obtain from high-speed engines the valuable information revealed by an indicator card.

THERMAL EFFICIENCIES   A.7

The thermal efficiency $e = W/Q_A$ has already been defined. In more general terms, for a power cycle or engine,

(A-4) $$\text{thermal efficiency} = \frac{\text{work output of the system}}{\text{energy chargeable against the system}}$$

For ideal cycles and engines, the numerator and denominator are ideal values, which are defined for each ideal system taken up. In this expression, the work may be an energy which is 100% available; electricity, for example. Since there are several places where work and power may be measured, there is no one *actual* thermal efficiency. If we let E_c' represent the actual energy to be charged against a system, we have the following three thermal efficiencies, the indicated e_i, the brake e_b, and the combined e_k:

(d)
$$e_i = \frac{W_I}{E_c'} \qquad e_b = \frac{W_B}{E_c'} \qquad e_k = \frac{W_K}{E_c'}$$

The energy E_c' will be defined for each actual engine and cycle analyzed.

HEATING VALUE OF FUELS   A.8

In gas turbines and internal combustion engines, the energy for producing work comes from the reaction of fuel and oxygen. The energy released during the process is commonly called **heating value** by engineers.

The **heating value** of a fuel *is that amount of heat given up by the products of combustion on being cooled to the initial temperature* after complete combustion at constant pressure (or constant volume), §§ 13.15 and 13.23, corrected to a standard state of 1 atm and 77°F. This is not simply a single number because of the different ways in which the test may be run and because of H_2O, which is formed from some fuels. Fuels used are commonly hydrocarbons such as fuel oil, kerosene, and gasoline, the chemical formula of which is in the form C_xH_y. When these fuels react with oxygen, the hydrogen forms H_2O. If the products of combustion are *hot* (above about 125°F), this H_2O is vapor (steam); if the products have been cooled to normal atmospheric temperatures, the H_2O is condensed, or largely condensed, and the H_2O is water. During condensation it gives up the latent heat of evaporation. Thus, considering this factor only, we see that there may be at least two heating values *for fuels containing hydrogen*, the **higher heating value** q_h when the H_2O formed from the fuel is condensed, and the **lower heating value** q_l when the fuel is burned so that the H_2O does not condense. (See Chapter 13 for more detail.) Since tests are run sometimes at *constant volume* and sometimes at *constant pressure*, this gives two more heating values.

In actual engines, the exhaust gases are hot and the steam does not come close to condensing. Since this is so, it is reasoned that it would be unfair to the engine to charge against it the higher heating value: hence the tendency is to use the *lower heating value at constant pressure* when computing the thermal efficiency of actual engines. (Past practice in this country has been to use the higher heating value for the same purpose when computing the thermal efficiency of internal combustion engines. With so many alternatives, courtesy demands that the kind of heating value used be stated.)

A.9 THERMAL EFFICIENCIES IN TERMS OF FUEL CONSUMPTION

The symbol m_f is here used to designate the mass of fuel used. Its units are defined by the context, usually either m_f lb fuel/lb air or m_f lb fuel/unit of work; say, lb f/hp-hr or lb f/kW-hr. In the latter units, it is generally called the **specific fuel consumption** (sfc). In the former units, we shall for convenience on occasion use the symbol $r_{f/a}$ to emphasize that it is a fuel-air ratio.

The **heat rate** of an engine accepting the fuel is defined by

(e) $$\text{heat rate} = \left(m_f \frac{\text{lb f}}{\text{hp-hr}} \right) \left(q_l \frac{\text{Btu}}{\text{lb f}} \right) = m_f q_l \frac{\text{Btu}}{\text{hp-hr}}$$

(f) $$\text{heat rate} = \left(m_f \frac{\text{lb f}}{\text{kW-hr}} \right) \left(q_l \frac{\text{Btu}}{\text{lb f}} \right) = m_f q_l \frac{\text{Btu}}{\text{kW-hr}}$$

The heat rate is the energy charged against such engines when the output of the engine is the appropriate unit work. Since a hp-hr = 2544 Btu and a kW-hr = 3412 Btu, the thermal efficiency is given by

(A-5) $$e = \frac{2544}{m_f q_l} \left(\frac{\text{Btu/hp-hr}}{\text{Btu/hp-hr}} \right) \quad \text{or} \quad e = \frac{3412}{m_f q_l} \left(\frac{\text{Btu/kW-hr}}{\text{Btu/kW-hr}} \right)$$

or

(g) $$e_i = \frac{2544}{m_{fi} q_l} \qquad e_b = \frac{2544}{m_{fb} q_l} \quad \text{and} \quad e_b = \frac{(2544)\,(\text{bhp})}{(\dot{m}_f\ \text{lb/hr}) q_l}$$

where the specific fuel consumption is m_{fi} lb/ihp-hr. and m_{fb} lb/bhp-hr. On a steady-state test, the instrumentation may be such that the rate of fuel flow and the horsepower output bhp at any instant can be observed (say, $\dot{m}_f$ lb/hr), in which case a form like the last one would be useful. For a generator output of kW and for

m_{fk} lb/kW-hr output of the generator, we have

(h) $$e_k = \frac{3412}{m_{fk}q_l} \quad \text{and} \quad e_k = \frac{(3412)\,(kW)}{(\dot{m}_f\,\text{lb/hr})q_l}$$

ENGINE EFFICIENCIES   A.9

The engine efficiency η is defined as

(A-6)   engine efficiency $\eta = \dfrac{W'}{W} = \dfrac{\text{actual work of a system}}{\text{work of the corresponding ideal system}}$

For the three actual works W_I, W_B, and W_K, we get three engine efficiencies. Thus, for the **brake engine efficiency** η_b, the **indicated engine efficiency** η_i, and the **combined engine efficiency** η_k, we have

(i) $$\eta_i = \frac{W_I}{W} \qquad \eta_b = \frac{W_B}{W} \qquad \eta_k = \frac{W_K}{W}$$

where the ideal work W is in the same units as the numerator and is computed for the corresponding ideal system, to be defined for each engine. Since the engine efficiency can often be estimated closely from previous experience with a certain kind of engine, it is a convenient design factor for use in determining the size of an actual engine to produce a specified amount of power.

MECHANICAL EFFICIENCY   A.11

The mechanical efficiency η_m is a number that tells of the mechanical losses in a machine. For the generator, Fig. A/1, it includes internal electrical losses:

(A-7) $$\eta_m = \frac{W_K}{W_B} \qquad\qquad [\text{GENERATOR}]$$

For a reciprocating engine of any type that delivers work,

(A-8) $$\eta_m = \frac{W_B}{W_I} \qquad\qquad [\text{RECIPROCATING ENGINE}]$$

See § 14.11 for η_m for a compressor. The foregoing works may be expressed in any convenient work or power unit or in terms of any numbers proportional to the W's, but remember that engine efficiency is a dimensionless ratio. The difference (ihp) − (bhp), or $W_I - W_R$, represents the loss due to mechanical friction of the moving parts of the engine; expressed in horsepower, it is called the **friction horsepower**; fhp = $(1 - \eta_m)$ (ihp). The mechanical efficiency *is not a fixed number characteristic of the machine*; it depends upon operating conditions, especially output, speed, and lubrication. In reciprocating engines, the compression work is also included as a loss.

A.12   TYPICAL PERFORMANCE CURVES, AUTOMOBILE ENGINE

The performance curves shown in Fig. A/8 are typical of most automotive engines and bring together most of what has been expressed in this appendix. The caption for this figure should be read closely; it has much information relevant to the engine under test.

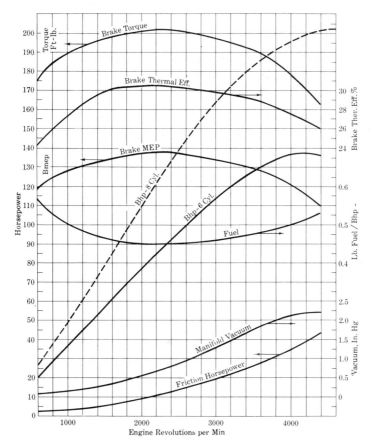

Fig. A/8   *Typical Performance Curves, Automobile Engine.* Plotted from data by courtesy of the Ford Motor Company. The solid curves are for a 6-cylinder, 3.625 × 3.6-in. engine, $r_k = 8$; $V_D = 223\ in^3$. The test is run with *wide-open throttle.* This friction-horsepower curve is obtained by driving the engine with the dynamometer and noting the consumption of power. The thermal efficiency is based on the lower heating value (18,500 Btu/lb of fuel). The horsepower curve of the 8-cylinder engine is included for an interesting comparison.

APPENDIX B

Properties of Substances

NOTE.   Other properties in the text include:

CONTENTS OF APPENDIX B

Gas Constants; Specific Heats at Low Pressure   ITEM B 1

For each substance, the specific gas constant was computed from $R = 1545.32/M$ ft-lb/lb-°R. Instantaneous values of C_p at the standard temperature of 77°F were taken from the literature, from which other values were computed as follows: $c_p = C_p/M$, $C_v = C_p - \bar{R}$, $\bar{R} = 1.986$ Btu/pmole-°R, $k = C_p/C_v$, $c_v = c_p - R$.

Notes. A number of values are brought up to date, principally with data in JANAF,[0.22] but in most cases the change is not of engineering significance. (a) Values at 100°F from Keenan and Kaye Gas Tables. (b) Molal specific heat of all monatomic gases is taken as $C_p = 4.97$ Btu/pmole-°R. (c) From Rossini et al.[0.1,0.2] (d) McBride et al.[0.23] (e) JANAF.[0.22] There is a movement under way to change the basis of molecular masses from $O_2 = 32$ exactly to $C = 12$. Most available data are on the old basis. All molecular masses in this book are for $O_2 = 32$.

Gas		M, lb/pmole	c_p, Btu/lb-°R	c_v, Btu/lb-°R	c_p, kJ/kg-K	c_v, kJ/kg-K	$k = c_p/c_v$	R, ft-lb/lb-°R	R, J/kg-K
Argon (A)	(b)	39.95	0.1244	0.0747	0.5215	0.3132	1.666	38.68	208.17
Helium (He)	(b)	4.003	1.241	0.745	5.2028	3.1233	1.666	386.04	2077.67
Mercury (Hg)	(b)	200.61	0.0248	0.0148	0.1039	0.0624	1.666	7.703	41.45
Neon (Ne)	(b)	20.183	0.246	0.1476	1.0313	0.6188	1.666	76.57	412.10
Xenon (Xe)	(b)	131.30	0.0378	0.0227	0.1585	0.0952	1.666	11.77	63.34
Air	(a)	28.970	0.24	0.1714	1.0062	0.7186	1.4	53.342	287.08
Carbon monoxide (CO)		28.01	0.2487	0.1778	1,0426	0.7454	1.399	55.170	296.92
Chlorine (Cl$_2$)		70.914	0.1144	0.0864	0.4796	0.3622	1.324	21.791	117.28
Fluorine (F$_2$)	(d)	38.00	0.197	0.1447	0.8259	0.6066	1.36	40.67	218.88
Hydrogen (H$_2$)		2.016	3.419	2.434	14.3338	10.2043	1.40	766.54	4125.52
Hydroxyl (OH)	(e)	17.008	0.421	0.3031	1.7650	1.2708	1.383	90.858	488.99
Nitric oxide (NO)		30.008	0.2378	0.1716	0.9969	0.7194	1.386	51.497	277.15
Nitrogen (N$_2$)		28.016	0.2484	0.1775	1.0414	0.7442	1.399	55.158	296.86
Oxygen (O$_2$)	(e)	32	0.2194	0.1573	0.9198	0.6595	1.395	48.291	259.90
Carbon dioxide (CO$_2$)		44.010	0.2016	0.1565	0.8452	0.6561	1.288	35.11	188.96
Hydrogen sulfide (H$_2$S)	(e)	34.086	0.2397	0.1799	1.0049	0.7542	1.321	45.33	243.96
Nitrogen dioxide (NO$_2$)	(e)	46.008	0.1921	0.1489	0.8054	0.6242	1.29	33.59	180.78
Nitrous oxide (N$_2$O)	(e)	44.016	0.2097	0.1646	0.8791	0.6901	1.274	35.11	188.96
Ozone (O$_3$)	(e)	48	0.1954	0.154	0.8192	0.6456	1.269	32.194	173.27
Sulfur dioxide (SO$_2$)	(e)	64.07	0.1487	0.1177	0.6234	0.4934	1.263	24.12	129.81
Water vapor (H$_2$O)	(a)	18.016	0.4454	0.3352	1.8673	1.4053	1.329	85.77	461.61
Acetylene (C$_2$H$_2$)	(e)	26.036	0.4048	0.3285	1.6971	1.3772	1.232	59.35	319.42
Ammonia (NH$_3$)	(e)	17.032	0.499	0.382	2.0920	1.6015	1.304	90.73	488.31
n-Butane (C$_4$H$_{10}$)		58.120	0.4007	0.3665	1.6799	1.5365	1.093	26.59	143.11
Cyanogen (C$_2$N$_2$)	(e)	52.038	0.261	0.2228	1.0942	0.9341	1.172	29.7	159.84
Ethane (C$_2$H$_6$)		30.068	0.4186	0.3526	1.7549	1.4782	1.187	51.39	276.58
Ethylene (C$_2$H$_4$)	(e)	28.052	0.3654	0.2946	1.5319	1.2351	1.240	55.09	296.49
Hydrazine (N$_2$H$_4$)	(e)	32.048	0.393	0.33	1.6476	1.3834	1.195	48.22	259.52
Hydrogen peroxide (C$_2$H$_2$)	(e)	34.016	0.303	0.2446	1.2703	1.0225	1.239	45.43	244.50
Methane (CH$_4$)	(e)	16.043	0.5099	0.3861	2.1377	1.6187	1.321	96.33	518.45
Methanol (CH$_4$O)	(c)	32.042	0.336	0.274	1.4086	1.1487	1.226	48.23	259.57
n-Octane (C$_8$H$_{18}$)		114.224	0.3952	0.3778	1.6568	1.5839	1.046	13.53	72.82
Propane (C$_3$H$_8$)		44.094	0.3985	0.3535	1.6707	1.4820	1.127	35.05	188.64

WATER   1 BTU / lb-°R

ITEM B 2   Properties of Air at Low Pressures (one pound)

Reproduced from Keenan and Kaye, *Gas Tables*, with permission of authors and publisher, John Wiley

T°R	h Btu/lb	p_r	u Btu/lb	v_r	φ Btu/lb-°R	T°R	h Btu/lb	p_r	u Btu/lb	v_r	φ Btu/lb-°R
360	85.97	0.3363	61.29	396.6	0.50369	1460	358.63	50.34	258.54	10.743	0.84704
380	90.75	0.4061	64.70	346.6	0.51663	1480	363.89	53.04	262.44	10.336	0.85062
400	95.53	0.4858	68.11	305.0	0.52890	1500	369.17	55.86	266.34	9.948	0.85416
420	100.32	0.5760	71.52	270.1	0.54058	1520	374.47	58.78	270.26	9.578	0.85767
440	105.11	0.6776	74.93	240.6	0.55172	1540	379.77	61.83	274.20	9.226	0.86113
460	109.90	0.7913	78.36	215.33	0.56235	1560	385.08	65.00	278.13	8.890	0.86456
480	114.69	0.9182	81.77	193.65	0.57255	1580	390.40	68.30	282.09	8.569	0.86794
500	119.48	1.0590	85.20	174.90	0.58233	1600	395.74	71.73	286.06	8.263	0.87130
520	124.27	1.2147	88.62	158.58	0.59173	1620	401.09	75.29	290.04	7.971	0.87462
537	128.34	1.3593	91.53	146.34	0.59945	1640	406.45	78.99	294.03	7.691	0.87791
540	129.06	1.3860	92.04	144.32	0.60078	1660	411.82	82.83	298.02	7.424	0.88116
560	133.86	1.5742	95.47	131.78	0.60950	1680	417.20	86.82	302.04	7.168	0.88439
580	138.66	1.7800	98.90	120.70	0.61793	1700	422.59	90.95	306.06	6.924	0.88758
600	143.47	2.005	102.34	110.88	0.62607	1720	428.00	95.24	310.09	6.690	0.89074
620	148.28	2.249	105.78	102.12	0.63395	1740	433.41	99.69	314.13	6.465	0.89387
640	153.09	2.514	109.21	94.30	0.64159	1760	438.83	104.30	318.18	6.251	0.89697
660	157.92	2.801	112.67	87.27	0.64902	1780	444.26	109.08	322.24	6.045	0.90003
680	162.73	3.111	116.12	80.96	0.65621	1800	449.71	114.03	326.32	5.847	0.90308
700	167.56	3.446	119.58	75.25	0.66321	1820	455.17	119.16	330.40	5.658	0.90609
720	172.39	3.806	123.04	70.07	0.67002	1840	460.63	124.47	334.50	5.476	0.90908
740	177.23	4.193	126.51	65.38	0.67665	1860	466.12	129.95	338.61	5.302	0.91203
760	182.08	4.607	129.99	61.10	0.68312	1880	471.60	135.64	342.73	5.134	0.91497
780	186.94	5.051	133.47	57.20	0.68942	1900	477.09	141.51	346.85	4.974	0.91788
800	191.81	5.526	136.97	53.63	0.69558	1920	482.60	147.59	350.98	4.819	0.92076
820	196.69	6.033	140.47	50.35	0.70160	1940	488.12	153.87	355.12	4.670	0.92362
840	201.56	6.573	143.98	47.34	0.70747	1960	493.64	160.37	359.28	4.527	0.92645
860	206.46	7.149	147.50	44.57	0.71323	1980	499.17	167.07	363.43	4.390	0.92926
880	211.35	7.761	151.02	42.01	0.71886	2000	504.71	174.00	367.61	4.258	0.93205
900	216.26	8.411	154.57	39.64	0.72438	2020	510.26	181.16	371.79	4.130	0.93481
920	221.18	9.102	158.12	37.44	0.72979	2040	515.82	188.54	375.98	4.008	0.93756
940	226.11	9.834	161.68	35.41	0.73509	2060	521.39	196.16	380.18	3.890	0.94026
960	231.06	10.610	165.26	33.52	0.74030	2080	526.97	204.02	384.39	3.777	0.94296
980	236.02	11.430	168.83	31.76	0.74540	2100	532.55	212.1	388.60	3.667	0.94564
1000	240.98	12.298	172.43	30.12	0.75042	2150	546.54	233.5	399.17	3.410	0.95222
1020	245.97	13.215	176.04	28.59	0.75536	2200	560.59	256.6	409.78	3.176	0.95868
1040	250.95	14.182	179.66	27.17	0.76019	2250	574.69	281.4	420.46	2.961	0.96501
1060	255.96	15.203	183.29	25.82	0.76496	2300	588.82	308.1	431.16	2.765	0.97123
1080	260.97	16.278	186.93	24.58	0.76964	2350	603.00	336.8	441.91	2.585	0.97732
1100	265.99	17.413	190.58	23.40	0.77426	2400	617.22	367.6	452.70	2.419	0.98331
1120	271.03	18.604	194.25	22.30	0.77880	2450	631.48	400.5	463.54	2.266	0.98919
1140	276.08	19.858	197.94	21.27	0.78326	2500	645.78	435.7	474.40	2.125	0.99497
1160	281.14	21.18	201.63	20.293	0.78767	2550	660.12	473.3	485.31	1.9956	1.00064
1180	286.21	22.56	205.33	19.377	0.79201	2600	674.49	513.5	496.26	1.8756	1.00623
1200	291.30	24.01	209.05	18.514	0.79628	2650	688.90	556.3	507.25	1.7646	1.01172
1220	296.41	25.53	212.78	17.700	0.80050	2700	703.35	601.9	518.26	1.6617	1.01712
1240	301.52	27.13	216.53	16.932	0.80466	2750	717.83	650.4	529.31	1.5662	1.02244
1260	306.65	28.80	220.28	16.205	0.80876	2800	732.33	702.0	540.40	1.4775	1.02767
1280	311.79	30.55	244.05	15.518	0.81280	2850	746.88	756.7	551.52	1.3951	1.03282
1300	316.94	32.39	227.83	14.868	0.81680	2900	761.45	814.8	562.66	1.3184	1.03788
1320	322.11	34.31	231.63	14.253	0.82075	2950	776.05	876.4	573.84	1.2469	1.04288
1340	327.29	36.31	235.43	13.670	0.82464	3000	790.68	941.4	585.04	1.1803	1.04779
1360	332.48	38.41	239.25	13.118	0.82848	3500	938.40	1829.3	698.48	0.7087	1.09332
1380	337.68	40.59	243.08	12.593	0.83229	4000	1088.26	3280	814.06	0.4518	1.13334
1400	342.90	42.88	246.93	12.095	0.83604	4500	1239.86	5521	931.39	0.3019	1.16905
1420	348.14	45.26	250.79	11.622	0.83975	5000	1392.87	8837	1050.12	0.20959	1.20129
1440	353.37	47.75	254.66	11.172	0.84341	6000	1702.29	20120	1291.00	0.11047	1.25769
						6500	1858.44	28974	1412.87	0.08310	1.28268

Carbon Dioxide Gas at Low Pressures (per p-mole)   ITEM B 3

Reproduced from Keenan and Kaye, *Gas Tables*, with permission of authors and publisher. John Wiley.

$T°R$	$\bar{h}$	p_r	$\bar{u}$	v_r	$\bar{\phi}$	$T°R$	$\bar{h}$	p_r	$\bar{u}$	v_r	$\bar{\phi}$
300	2108.2	0.013705	1512.4	234900	46.353	1380	13101.0	21.33	10360.5	694.3	60.949
320	2256.6	0.017440	1621.1	196920	46.832	1400	13344.7	23.30	10564.5	644.8	61.124
340	2407.3	0.02196	1732.1	166170	47.289	1420	13589.1	25.43	10769.2	599.3	61.298
360	2560.5	0.02738	1845.6	141090	47.728	1440	13834.5	27.72	10974.8	557.5	61.469
380	2716.4	0.03385	1961.8	120480	48.148	1460	14080.8	30.19	11181.4	518.9	61.639
400	2874.7	0.04153	2080.4	103360	48.555	1480	14328.0	32.86	11388.9	483.2	61.808
420	3035.7	0.05062	2201.7	89050	48.947	1500	14576.0	35.74	11597.2	450.4	61.974
440	3199.4	0.06131	2325.6	77000	49.329	1520	14824.9	38.83	11806.4	420.0	62.138
460	3365.7	0.07386	2452.2	66830	49.698	1540	15074.7	42.16	12016.5	392.0	62.302
480	3534.7	0.08852	2581.5	58190	50.058	1560	15325.3	45.74	12227.3	365.9	62.464
500	3706.2	0.10560	2713.3	50820	50.408	1580	15576.7	49.58	12439.0	342.0	62.624
520	3880.3	0.12541	2847.7	44500	50.750	1600	15829.0	53.70	12651.6	319 7	62.783
537	4030.2	0.14471	2963.8	39838	51.032	1620	16081.9	58.13	12864.8	299.0	62.939
540	4056.8	0.14833	2984.4	39070	51.082	1640	16335.7	62.87	13078.9	279.9	63.095
560	4235.8	0.17474	3123.7	34390	51.408	1660	16590.2	67.95	13293.7	262.2	63.250
580	4417.2	0.2051	3265.4	30340	51.726	1680	16845.5	73.39	13509.2	245.6	63.403
600	4600.9	0.2399	3409.4	26840	52.038	1700	17101.4	79.20	13725.4	230.3	63.555
620	4786.8	0.2797	3555.6	23780	52.343	1800	18391.5	114.84	14816.9	168.21	64.292
640	4974.9	0.3252	3704.0	21120	52.641	1900	19697.8	163.89	15924.7	124 41	64.999
660	5165.2	0.3769	3854.6	18796	52.934	2000	21018.7	230.5	17046.9	93.08	65.676
680	5357.6	0.4355	4007.2	16756	53.225	2100	22352.7	320.0	18182.4	70.42	66.327
700	5552.0	0.5019	4161.9	14968	53.503	2200	23699.0	438.7	19330.1	53.82	66.953
720	5748.4	0.5769	4318.6	13393	53.780	2300	25056.3	594.5	20488.8	41.52	67.557
740	5946.8	0.6615	4477.3	12004	54.051	2400	26424.0	797.0	21657.9	32.32	68.139
760	6147.0	0.7568	4637.9	10777	54.319	2500	27801.2	1057.9	22836.5	25.36	68.702
780	6349.1	0.8637	4800.1	9692	54.582	2600	29187.1	1391.2	24023.8	20.06	69.245
800	6552.9	0.9837	4964.2	8729	54.839	2700	30581.2	1813.1	25219.4	15.981	69.771
820	6758.3	1.1177	5129.9	7875	55.093	2800	31982.8	2344	26422.4	12.820	70.282
840	6965.7	1.2677	5297.6	7112	55.343	2900	33391.5	3007	27632.5	10.351	70.776
860	7174.7	1.4347	5466.9	6433	55.589	3000	34806.6	3828	28849.0	8.410	71.255
880	7385.3	1.6206	5637.7	5827	55.831	3100	36227.9	4841	30071.7	6.871	71.722
900	7597.6	1.8276	5810.3	5285	56.070	3200	37654.7	6082	31299.9	5.646	72.175
920	7811.4	2.057	5984.4	4799	56.305	3300	39086.7	7593	32533.3	4.664	72.616
940	8026.8	2.311	6160.1	4363	56.536	3400	40523.6	9425	33771.6	3.872	73.045
960	8243.8	2.594	6337.4	3972	56.765	3500	41965.2	11633	35014.7	3.229	73.462
980	8462.2	2.905	6516.1	3620	56.990	3600	43411.0	14282	36261.9	2.705	73.870
1000	8682.1	3.249	6696.2	3303	57.212	3700	44860.6	17445	37512.9	2.276	74.267
1020	8903.4	3.628	6877.8	3017	57.432	3800	46314.0	21200	38767.7	1.9229	74.655
1040	9126.2	4.046	7060.9	2759	57.647	3900	47771.0	25660	40026.1	1.6311	75.033
1060	9350.3	4.505	7245.3	2525	57.861	4000	49231.4	30910	41287.9	1.3887	75.404
1080	9575.8	5.010	7431.1	2313	58.072	4100	50695.1	37080	42553.0	1.1865	75.765
1100	9802.6	5.563	7618.1	2122	58.281	4200	52162.0	44310	43821.4	1.0172	76.119
1120	10030.6	6.170	7806.4	1948.0	58.485	4300	53632.1	52740	45092.9	0.8749	76.464
1140	10260.1	6.834	7996.2	1790.1	58.689	4400	55105.1	62550	46367.3	0.7548	76.803
1160	10490.6	7.560	8187.0	1646.5	58.889	4500	56581.0	73930	47644.6	0.6533	77.135
1180	10722.3	8.353	8379.0	1515.9	59.088	4600	58059.7	87080	48924.7	0.5669	77.460
1200	10955.3	9.219	8572.3	1396.8	59.283	4700	59541.1	102250	50207.5	0.4933	77.779
1220	11189.4	10.163	8766.6	1288.2	59.477	4800	61024.9	119640	51492.7	0.4305	78.091
1240	11424.6	11.188	8962.1	1189.1	59.668	4900	62511.3	139640	52780.5	0.3766	78.398
1260	11661.0	12.307	9158.8	1098.6	59.858	5000	64000.0	162480	54070.6	0.3303	78.698
1280	11898.4	13.525	9356.5	1015.6	60.044	5100	65490.9	188530	55363.0	0.2904	78.994
1300	12136.9	14.846	9555.3	939.9	60.229	5200	66984.0	218100	56657.5	0.2558	79.284
1320	12376.4	16.273	9755.0	870.4	60.412	5300	68479.1	251800	57954.0	0.2258	79.569
1340	12617.0	17.828	9955.9	806.7	60.593	5380	69676.5	281900	58992.5	0.2048	79.793
1360	12858.5	19.509	10157.7	748.1	60.772						

ITEM B 4 Carbon Monoxide at Low Pressures (per p-mole)

Reproduced from Keenan and Kaye, *Gas Tables*, with permission of authors and publisher, John Wiley.

$T°R$	$\bar{h}$	p_r	$\bar{u}$	v_r	$\bar{\phi}$	$T°R$	$\bar{h}$	p_r	$\bar{u}$	v_r	$\bar{\phi}$
300	2081.9	0.2834	1486.1	11357	43.223	1380	9796.6	65.12	7056.1	227.3	54.021
320	2220.9	0.3553	1585.4	9666	43.672	1400	9948.1	68.80	7167.9	218.4	54.129
340	2359.9	0.4393	1684.7	8305	44.093	1420	10100.0	72.64	7280.1	209.8	54.237
360	2498.8	0.5366	1783.9	7200	44.490	1440	10252.2	76.64	7392.6	201.6	54.344
380	2637.9	0.6484	1883.3	6290	44.866	1460	10404.8	80.81	7505.4	193.89	54.448
400	2776.9	0.7760	1982.6	5532	45.223	1480	10557.8	85.16	7618.7	186.48	54.552
420	2916.0	0.9205	2081.9	4897	45.563	1500	10711.1	89.69	7732.3	179.48	54.655
440	3055.0	1.0833	2181.2	4359	45.886	1520	10864.9	94.41	7846.4	172.78	54.757
460	3194.0	1.2657	2280.5	3900	46.194	1540	11019.0	99.32	7960.8	166.40	54.858
480	3333.0	1.4692	2379.8	3506	46.491	1560	11173.4	104.44	8075.4	160.30	54.958
500	3472.1	1.6950	2479.2	3166	46.775	1580	11328.2	109.76	8190.5	154.47	55.056
520	3611.2	1.9440	2578.6	2869	47.048	1600	11483.4	115.28	8306.0	148.92	55.154
537	3729.5	2.177	2663.1	2648	47.272	1620	11638.9	121.03	8421.8	143.66	55.251
540	3750.3	2.220	2677.9	2611	47.310	1640	11794.7	127.00	8537.9	138.57	55.347
560	3889.5	2.521	2777.4	2384	47.563	1660	11950.9	133.20	8654.4	133.72	55.441
580	4028.7	2.851	2876.9	2183	47.807	1680	12107.5	139 64	8771.2	129.12	55.535
600	4168.0	3.211	2976.5	2005	48.044	1700	12264.3	146 33	8888.3	124.68	55.628
620	4307.4	3.603	3076.2	1846.5	48.272	1800	13053.2	18. 54	9478.6	105.19	56.078
640	4446.9	4.028	3175.9	1704.8	48.494	1900	13849.8	228. _	10076.6	89.38	56.509
660	4586.5	4.488	3275.8	1577.9	48.709	2000	14653.2	280.7	10681.5	76.45	56.922
680	4726.2	4.986	3375.8	1463.6	48.917	2100	15463.3	342.6	11293.0	65.78	57.317
700	4866.0	5.521	3475.9	1360.7	49.120	2200	16279.4	414.7	11910.5	56.92	57.696
720	5006.1	6.098	3576.3	1267.1	49.317	2300	17101.0	498.5	12533.5	49.51	58.062
740	5146.4	6.717	3676.9	1181.9	49.509	2400	17927.4	595.1	13161.3	43.28	58.414
760	5286.8	7.382	3777.5	1104.8	49.697	2500	18758.8	706.1	13794.1	38.00	58.754
780	5427.4	8.094	3878.4	1034.2	49.880	2600	19594.3	832.8	14431.0	33.50	59.081
800	5568.2	8.854	3979.5	969.6	50.058	2700	20434.0	976.9	15072.2	29.66	59.398
820	5709.4	9.666	4081.0	910.4	50.232	2800	21277.2	1140.1	15716.8	26.35	59.705
840	5850.7	10.532	4182.6	855.9	50.402	2900	22123.8	1324.0	16364.8	23.50	60.002
860	5992 3	11.453	4284.5	805.8	50.569	3000	22973.4	1530.8	17015.8	21.03	60.290
880	6134.2	12.434	4386.6	759.5	50.732	3100	23826.0	1762.1	17669.8	18.878	60.559
900	6276.4	13.476	4489.1	716.7	50.892	3200	24681.2	2020	18326.4	16.994	60.841
920	6419.0	14.582	4592.0	677.0	51.048	3300	25539.0	2308	18985.6	15.344	61.105
940	6561.7	15.755	4695.0	640.3	51.202	3400	26399.3	2626	19647.3	13.895	61.362
960	6704.9	16.998	4798.5	606.1	51.353	3500	27261.8	2979	20311.2	12.608	61.612
980	6848.4	18.312	4902.3	574.3	51.501	3600	28126.6	3368	20977.5	11.469	61.855
1000	6992.2	19.700	5006.3	544.7	51.646	3700	28993.5	3796	21645.8	10.460	62.093
1020	7136.4	21.17	5110.8	517.0	51.788	3800	29862.3	4266	22316.0	9.560	62.325
1040	7281.0	22.72	5215.7	491.2	51.929	3900	30732.9	4780	22988.0	8.756	62.551
1060	7425.9	24.36	5320.9	467.0	52.067	4000	31605.2	5343	23661.7	8.035	62.772
1080	7571.1	26.08	5426.4	444.4	52.203	4100	32479.1	5956	24337.0	7.387	62.988
1100	7716.8	27.90	5532.3	423.2	52.337	4200	33354.4	6623	25013.8	6.804	63.198
1120	7862.9	29.81	5638.7	403.2	52.468	4300	34231.2	7349	25692.0	6.279	63.405
1140	8009.2	31.82	5745.4	384.5	52.598	4400	35109.2	8136	26371.4	5.804	63.607
1160	8156.1	33.93	5852.5	366.9	52.726	4500	35988.6	8987	27052.2	5.374	63.805
1180	8303.3	36.15	5960.0	350.2	52.852	4600	36869.3	9907	27734.3	4.983	63.998
1200	8450.8	38.48	6067.8	334.6	52.976	4700	37751.0	10900	28417.4	4.627	64.188
1220	8598.8	40.93	6176.0	319.9	53.098	4800	38633.9	11970	29101.7	4.303	64.374
1240	8747.2	43.49	6284.7	306.0	53.218	4900	39517.8	13121	29787.0	4.008	64.556
1260	8896.0	46.18	6393.8	292.8	53.337	5000	40402.7	14357	30473.4	3.737	64.735
1280	9045.0	48.99	6503.1	280.4	53.455	5100	41288.6	15683	31160.7	3.490	64.910
1300	9194.6	51.93	6613.0	268.6	53.571	5200	42175.5	17104	31849.0	3.263	65.082
1320	9344.6	55.02	6723.2	257.4	53.685	5300	43063.2	18625	32538.1	3.054	65.252
1340	9494.8	58.24	6833.7	246.9	53.799	5380	43774.1	19914	33090.1	2.899	65.385
1360	9645.5	61.61	6944.7	236.9	53.910						

Hydrogen at Low Pressures (per p-mole) ITEM B 5

Reproduced from Keenan and Kaye, *Gas Tables*, with permission of authors and publisher, John Wiley.

$T°R$	$\bar{h}$	p_r	$\bar{u}$	v_r	$\bar{\varphi}$	$T°R$	$\bar{h}$	p_r	$\bar{u}$	v_r	$\bar{\varphi}$
300	2063.5	0.9517	1467.7	3383	27.337	1380	9532.6	183.04	6792.1	80.90	37.781
320	2189.4	1.1678	1553.9	2940	27.742	1400	9673.8	192.66	6893.6	77.98	37.883
340	2317.2	1.4195	1642.0	2570	28.130	1420	9815.1	202.6	6995.2	75.21	37.983
360	2446.8	1.7104	1731.9	2259	28.501	1440	9956.5	213.0	7096.8	72.57	38.083
380	2577.8	2.044	1823.2	1994.6	28.856	1460	10098.0	223.7	7198.7	70.05	38.181
400	2710.2	2.426	1915.8	1769.7	29.195	1480	10239.7	234.8	7300.6	67.64	38.277
420	2843.7	2.858	2009.6	1577.2	29.520	1500	10381.5	246.3	7402.7	65.33	38.372
440	2978.1	3.345	2104.3	1411.5	29.833	1520	10523.4	258.3	7504.9	63.15	38.466
460	3113.5	3.892	2200.0	1268.2	30.133	1540	10665.5	270.6	7607.3	61.06	38.559
480	3249.4	4.503	2296.2	1143.9	30.424	1560	10807.6	283.4	7709.6	59.06	38.651
500	3386.1	5.182	2393.2	1035.4	30.703	1580	10950.0	296.7	7812.3	57.14	38.740
520	3523.3	5.936	2490.6	940.1	30.972	1600	11092.5	310.4	7915.1	55.31	38.830
537	3640.3	6.633	2573.9	869.0	31.194	1620	11235.1	324.5	8018.0	53.57	38.919
540	3660.9	6.762	2588.5	857.0	31.232	1640	11378.0	339.2	8121.2	51.89	39.007
560	3798.8	7.673	2686.7	783.2	31.482	1660	11521.0	354.3	8224.5	50.28	39.093
580	3937.1	8.671	2785.3	717.9	31.724	1680	11664.1	370.0	8327.8	48.73	39.179
600	4075.6	9.758	2884.1	659.8	31.959	1700	11807.4	386.1	8431.4	47.26	39.264
620	4214.3	10.942	2983.1	608.0	32.187	1800	12526.8	474.9	8952.2	40.68	39.675
640	4353.1	12.228	3082.1	561.7	32.407	1900	13250.9	578.5	9477.8	35.25	40.067
660	4492.1	13.617	3181.4	520.0	32.621	2000	13980.1	698.2	10008.4	30.73	40.441
680	4631.1	15.116	3280.7	482.7	32.829	2100	14714.5	836.3	10544.2	26.94	40.799
700	4770.2	16.733	3380.1	448.9	33.031	2200	15454.4	994.7	11085.5	23.74	41.143
720	4909.5	18.469	3479.6	418.4	33.226	2300	16199.8	1175.2	11632.3	21.00	41.475
740	5048.8	20.33	3579.2	390.6	33.417	2400	16950.6	1380.6	12184.5	18.656	41.794
760	5188.1	22.33	3678.8	365.3	33.603	2500	17707.3	1613.0	12742.6	16.633	42.104
780	5327.6	24.46	3778.6	342.2	33.784	2600	18469.7	1875.0	13306.4	14.881	42.403
800	5467.1	26.73	3878.4	321.1	33.961	2700	19237.8	2170	13876.0	13.353	42.692
820	5606.7	29.16	3978.3	301.8	34.134	2800	20011.8	2500	14451.4	12.019	42.973
840	5746.3	31.74	4078.2	284.0	34.302	2900	20791.5	2870	15032.5	10.845	43.247
860	5885.9	34.47	4178.0	267.7	34.466	3000	21576.9	3281	15619.3	9.811	43.514
880	6025.6	37.38	4278.0	252.7	34.627	3100	22367.7	3739	16211.5	8.897	43.773
900	6165.3	40.45	4378.0	238.8	34.784	3200	23164.1	4247	16809.3	8.087	44.026
920	6305.1	43.70	4478.1	225.9	34.938	3300	23965.5	4808	17412.1	7.365	44.273
940	6444.9	47.14	4578.1	214.0	35.087	3400	24771.9	5428	18019.9	6.722	44.513
960	6584.7	50.77	4678.3	202.9	35.235	3500	25582.9	6111	18632.4	6.148	44.748
980	6724.6	54.59	4778.4	192.64	35.379	3600	26398.5	6860	19249.4	5.633	44.978
1000	6864.5	58.62	4878.6	183.06	35.520	3700	27218.5	7682	19870.8	5.170	45.203
1020	7004.0	62.87	4978.8	174.09	35.659	3800	28042.8	8582	20496.5	4.753	45.423
1040	7144.4	67.32	5079.1	165.78	35.795	3900	28871.1	9562	21126.2	4.377	45.638
1060	7284.4	71.99	5179.4	158.02	35.928	4000	29703.5	10632	21760.0	4.037	45.849
1080	7424.5	76.90	5279.8	150.72	36.059	4100	30539.8	11799	22397.7	3.729	46.056
1100	7564.6	82.02	5380.1	143.89	36.188	4200	31379.8	13066	23039.2	3.449	46.257
1120	7704.7	87.42	5480.6	137.51	36.314	4300	32223.5	14438	23684.3	3.196	46.456
1140	7844.9	93.05	5581.0	131.47	36.438	4400	33070.9	15926	24333.1	2.965	46.651
1160	7985.2	98.95	5681.6	125.81	36.560	4500	33921.6	17537	24985.2	2.754	46.842
1180	8125.5	105.11	5782.2	120.46	36.681	4600	34775.7	19275	25640.7	2.561	47.030
1200	8265.8	111.54	5882.8	115.44	36.798	4700	35633.0	21150	26299.4	2.384	47.215
1220	8406.2	118.26	5983.4	110.71	36.913	4800	36493.4	23170	26961.2	2.223	47.396
1240	8546.7	125.25	6084.2	106.24	37.028	4900	37356.9	25340	27626.1	2.075	47.574
1260	8687.3	132.55	6185.1	102.00	37.140	5000	38223.3	27680	28294.0	1.9383	47.749
1280	8828.0	140.16	6286.1	98.00	37.251	5100	39092.8	30190	28964.9	1.8130	47.921
1300	8968.7	148.07	6387.1	94.21	37.360	5200	39965.1	32870	29638.6	1.6973	48.090
1320	9109.5	156.31	6488.2	90.62	37.468	5300	40840.2	35750	30315.1	1.5907	48.257
1340	9250.4	164.89	6589.4	87.20	37.574	5380	41542.1	38200	30858.1	1.5113	48.388
1360	9391.5	173.78	6690.7	83.97	37.678						

ITEM B 6 Nitrogen at Low Pressures (per p-mole)

Reproduced from Keenan and Kaye, *Gas Tables*, with permission of authors and publisher, John Wiley.

$T°R$	$\bar{h}$	p_r	$\bar{u}$	v_r	$\bar{\phi}$	$T°R$	$\bar{h}$	p_r	$\bar{u}$	v_r	$\bar{\phi}$
300	2082.0	0.13130	1486.2	24520	41.695	1380	9747.5	29.45	7007.0	502.8	52.444
320	2221.0	0.16458	1585.5	20870	42.143	1400	9896.9	31.09	7116.7	483.2	52.551
340	2360.0	0.2035	1684.8	17931	42.564	1420	10046.6	32.80	7226.7	464.6	52.658
360	2498.9	0.2485	1784.0	15543	42.962	1440	10196.6	34.57	7337.0	446.9	52.763
380	2638.0	0.3003	1883.4	13579	43.337	1460	10347.0	36.43	7447.6	430.1	52.867
400	2777.0	0.3594	1982.6	11943	43.694	1480	10497.8	38.36	7558.7	414.1	52.969
420	2916.1	0.4263	2082.0	10572	44.034	1500	10648.9	40.38	7670.1	398.8	53.071
440	3055.1	0.5017	2181.3	9410	44.357	1520	10800.4	42.46	7781.9	384.1	53.171
460	3194.1	0.5862	2280.6	8420	44.665	1540	10952.2	44.64	7893.9	370.2	53.271
480	3333.1	0.6805	2379.9	7570	44.962	1560	11104.3	46.90	8006.4	356.9	53.369
500	3472.2	0.7852	2479.3	6835	45.246	1580	11256.9	49.25	8119.2	344.2	53.465
520	3611.3	0.9003	2578.6	6198	45.519	1600	11409.7	51.70	8232.3	332.1	53.561
537	3729.5	1.0083	2663.1	5717	45.743	1620	11562.8	54.23	8345.7	320.5	53.656
540	3750.3	1.0281	2678.0	5637	45.781	1640	11716.4	56.87	8459.6	309.5	53.751
560	3889.5	1.1676	2777.4	5147	46.034	1660	11870.2	59.60	8573.6	298.8	53.844
580	4028.7	1.3200	2876.9	4715	46.278	1680	12024.3	62.44	8688.1	288.8	53.936
600	4167.9	1.4868	2976.4	4331	46.514	1700	12178.9	65.38	8802.9	279.1	54.028
620	4307.1	1.6679	3075.9	3989	46.742	1720	12333.7	68.43	8918.0	269.7	54.118
640	4446.4	1.8645	3175.5	3684	46.964	1740	12488.8	71.59	9033.4	260.8	54.208
660	4585.8	2.077	3275.2	3410	47.178	1760	12644.3	74.87	9149.2	252.3	54.297
680	4725.3	2.307	3374.9	3163	47.386	1780	12800.2	78.26	9265.3	244.1	54.385
700	4864.9	2.554	3474.8	2941	47.588	1800	12956.3	81.77	9381.7	236.2	54.472
720	5004.5	2.820	3574.7	2740	47.785	1820	13112.7	85.41	9498.4	228.7	54.559
740	5144.3	3.106	3674.7	2557	47.977	1840	13269.5	89.17	9615.5	221.4	54.645
760	5284.1	3.411	3774.9	2391	48.164	1860	13426.5	93.07	9732.8	214.4	54.729
780	5424.2	3.738	3875.2	2239	48.345	1880	13583.9	97.11	9850.5	207.8	54.813
800	5564.4	4.088	3975.7	2100	48.522	1900	13741.6	101.29	9968.4	201.3	54.896
820	5704.7	4.461	4076.3	1972.8	48.696	1920	13899.5	105.58	10086.7	195.16	54.979
840	5845.3	4.858	4177.1	1855.6	48.865	1940	14057.8	110.02	10205.2	189.19	55.061
860	5985.9	5.281	4278.1	1747.8	49.031	1960	14216.4	114.61	10324.0	183.48	55.143
880	6126.9	5.729	4379.4	1648.3	49.193	1980	14375.2	119.37	10443.2	178.00	55.223
900	6268.1	6.206	4480.8	1556.4	49.352	2000	14534.4	124.28	10562.6	172.69	55.303
920	6409.6	6.711	4582.6	1471.2	49.507	2100	15334.0	151.27	11163.7	148.97	55.694
940	6551.2	7.246	4684.5	1392.2	49.659	2200	16139.8	182.70	11770.9	129.22	56.068
960	6693.1	7.812	4786.7	1318.7	49.808	2300	16951.2	219.1	12383.7	112.66	56.429
980	6835.4	8.410	4889.3	1250.4	49.955	2400	17767.9	261.0	13001.8	98.68	56.777
1000	6977.9	9.043	4992.0	1186.7	50.099	2500	18589.5	309.0	13624.9	86.82	57.112
1020	7120.7	9.712	5095.1	1127.2	50.241	2600	19415.8	363.8	14252.5	76.69	57.436
1040	7263.8	10.414	5198.5	1071.6	50.380	2700	20246.4	426.0	14884.5	68.02	57.750
1060	7407.2	11.155	5302.2	1019.8	50.516	2800	21081.1	496.4	15520.6	60.54	58.053
1080	7551.0	11.939	5406.2	971.0	50.651	2900	21919.5	575.6	16160.5	54.07	58.348
1100	7695.0	12.759	5510.5	925.2	50.783	3000	22761.5	664.6	16803.9	48.44	58.632
1120	7839.3	13.622	5615.2	882.4	50.912	3100	23606.8	764.0	17450.6	43.53	58.910
1140	7984.0	14.527	5720.1	842.1	51.040	3200	24455.0	875.3	18100.2	39.24	59.179
1160	8129.0	15.479	5825.4	804.1	51.167	3300	25306.0	998.5	18752.7	35.46	59.442
1180	8274.4	16.481	5931.0	768.4	51.291	3400	26159.7	1135.3	19407.7	32.14	59.697
1200	8420.0	17.530	6037.0	734.6	51.413	3500	27015.9	1286.5	20065.3	29.19	59.944
1220	8566.1	18.629	6143.4	702.8	51.534	3600	27874.4	1453.0	20725.3	26.59	60.186
1240	8712.6	19.774	6250.1	672.8	51.653	3700	28735.1	1636.4	21387.4	24.27	60.422
1260	8859.3	20.98	6357.2	644.5	51.771	3800	29597.9	1837.3	22051.6	22.19	60.562
1280	9006.4	22.24	6464.5	617.5	51.887	3900	30462.8	2057	22717.9	20.34	60.877
1300	9153.9	23.56	6572.3	592.0	52.001	4000	31329.4	2298	23386.0	18.681	61.097
1320	9301.8	24.94	6680.4	568.0	52.114	4500	35687.8	3853	26751.4	12.533	62.123
1340	9450.0	26.38	6788.9	545.1	52.225	5000	40079.8	6141	30150.5	8.738	63.049
1360	9598.6	27.88	6897.8	523.4	52.335	5300	42728.3	7956	32203.2	7.148	63.563
						5380	43436.0	8504	32752.1	6.789	63.695

Oxygen at Low Pressures (per p-mole)   ITEM B 7

Reproduced from Keenan and Kaye, *Gas Tables*, with permission of authors and publisher, John Wiley.

$T°R$	$\bar{h}$	p_r	$\bar{u}$	v_r	$\bar{\phi}$	$T°R$	$\bar{h}$	p_r	$\bar{u}$	v_r	$\bar{\phi}$
300	2073.5	0.6684	1477.8	4816	44.927	1480	10855.1	232.5	7916.0	68.32	56.547
320	2212.6	0.8379	1577.1	4099	45.375	1500	11017.1	245.5	8038.3	65.56	56.656
340	2351.7	1.0360	1676.5	3522	45.797	1520	11179.6	259.2	8161.1	62.94	56.763
360	2490.8	1.2658	1775.9	3052	46.195	1540	11342.4	273.4	8284.2	60.43	56.869
380	2630.0	1.5295	1875.3	2666	46.571	1560	11505.4	288.3	8407.4	58.04	56.975
400	2769.1	1.8305	1974.8	2345	46.927	1580	11668.8	303.8	8531.1	55.80	57.079
420	2908.3	2.172	2074.3	2075	47.267	1600	11832.5	320.0	8655.1	53.65	57.182
440	3047.5	2.556	2173.8	1847.1	47.591	1620	11996.6	336.9	8779.5	51.61	57.284
460	3186.9	2.988	2273.4	1652.2	47.900	1640	12160.9	354.4	8904.1	49.65	57.385
480	3326.5	3.470	2373.3	1484.7	48.198	1660	12325.5	372.6	9029.0	47.81	57.484
500	3466.2	4.005	2473.2	1339.7	48.483	1680	12490.4	391.6	9154.1	46.03	57.582
520	3606.1	4.598	2573.4	1213.8	48.757	1700	12655.6	411.4	9279.6	44.34	57.680
537	3725.1	5.151	2658.7	1119.0	48.982	1720	12821.1	431.9	9405.4	42.73	57.777
540	3746.2	5.253	2673.8	1103.2	49.021	1740	12986.9	453.3	9531.5	41.19	57.873
560	3886.6	5.973	2774.5	1006.2	49.276	1760	13153.0	475.5	9657.9	39.72	57.968
580	4027.3	6.764	2875.5	920.2	49.522	1780	13319.2	498.5	9784.4	38.32	58.062
600	4168.3	7.629	2976.8	844.0	49.762	1800	13485.8	522.4	9911.2	36.98	58.155
620	4309.7	8.572	3078.4	776.3	49.993	1820	13652.5	547.2	10038.2	35.69	58.247
640	4451.4	9.600	3180.4	715.3	50.218	1840	13819.6	572.9	10165.6	34.46	58.339
660	4593.5	10.718	3282.9	660.7	50.437	1860	13986.8	599.5	10293.1	33.29	58.428
680	4736.2	11.930	3385.8	611.6	50.650	1880	14154.4	627.2	10421.0	32.16	58.518
700	4879.3	13.244	3489.2	567.1	50.858	1900	14322.1	656.0	10549.0	31.08	58.607
720	5022.9	14.662	3593.1	526.9	51.059	1920	14490.1	685.6	10677.2	30.05	58.695
740	5167.0	16.194	3697.4	490.4	51.257	1940	14658.2	716.4	10805.6	29.06	58.782
760	5311.4	17.843	3802.2	457.1	51.450	1960	14826.6	748.2	10934.3	28.11	58.868
780	5456.4	19.619	3907.5	426.6	51.638	1980	14995.2	781.1	11063.2	27.20	58.954
800	5602.0	21.53	4013.3	398.8	51.821	2000	15164.0	815.1	11192.3	26.33	59.039
820	5748.1	23.57	4119.7	373.3	52.002	2100	16010.9	1003.8	11840.6	22.45	59.451
840	5894.8	25.77	4226.6	349.8	52.179	2200	16862.6	1225.2	12493.7	19.269	59.848
860	6041.9	28.11	4334.1	328.3	52.352	2300	17718.8	1483.9	13151.3	16.632	60.228
880	6189.6	30.62	4442.0	308.4	52.522	2400	18579.2	1784.4	13813.1	14.434	60.594
900	6337.9	33.30	4550.6	290.0	52.688	2500	19443.4	2131	14478.7	12.589	60.946
920	6486.7	36.16	4659.7	273.1	52.852	2600	20311.4	2530	15148.1	11.030	61.287
940	6636.1	39.20	4769.4	257.3	53.012	2700	21182.9	2985	15821.0	9.707	61.616
960	6786.0	42.44	4879.5	242.7	53.170	2800	22057.8	3504	16497.4	8.576	61.934
980	6936.4	45.89	4990.3	229.2	53.326	2900	22936.1	4092	17177.1	7.605	62.242
1000	7087.5	49.55	5101.6	216.6	53.477	3000	23817.7	4756	17860.1	6.770	62.540
1020	7238.9	53.44	5213.3	204.8	53.628	3100	24702.5	5503	18546.3	6.044	62.831
1040	7391.0	57.56	5325.7	193.86	53.775	3200	25590.5	6344	19235.7	5.413	63.113
1060	7543.6	61.94	5438.6	183.67	53.921	3300	26481.6	7282	19928.2	4.863	63.386
1080	7696.8	66.56	5552.1	174.14	54.064	3400	27375.9	8330	20623.9	4.380	63.654
1100	7850.4	71.46	5665.9	165.20	54.204	3500	28273.3	9495	21322.8	3.956	63.914
1120	8004.5	76.63	5780.3	156.85	54.343	3600	29173.9	10788	22024.8	3.581	64.168
1140	8159.1	82.09	5895.2	149.02	54.480	3700	30077.5	12221	22729.8	3.249	64.415
1160	8314.2	87.86	6010.6	141.68	54.614	3800	30984.1	13802	23437.8	2.954	64.657
1180	8469.8	93.94	6126.5	134.79	54.748	3900	31893.6	15545	24148.7	2.692	64.893
1200	8625.8	100.36	6242.8	128.32	54.879	4000	32806.1	17463	24862.6	2.458	65.123
1220	8782.4	107.10	6359.6	122.23	55.008	4100	33721.6	19569	25579.5	2.248	65.350
1240	8939.4	114.19	6476.9	116.52	55.136	4200	34639.9	21870	26299.2	2.060	65.571
1260	9096.7	121.68	6594.5	111.11	55.262	4300	35561.1	24400	27021.9	1.8916	65.788
1280	9254.6	129.55	6712.7	106.04	55.386	4400	36485.0	27150	27747.2	1.7392	66.000
1300	9412.9	137.80	6831.3	101.25	55.508	4500	37411.8	30150	28475.4	1.6016	66.208
1320	9571.6	146.46	6950.2	96.71	55.630	4600	38341.4	33420	29206.4	1.4773	66.413
1340	9730.7	155.54	7069.6	92.44	55.750	4700	39273.6	36960	29940.0	1.3643	66.613
1360	9890.2	165.07	7189.4	88.40	55.867	4800	40208.6	40820	30676.4	1.2621	66.809
1380	10050.1	175.08	7309.6	84.58	55.984	4900	41146.1	44990	31415.3	1.1689	67.003
1400	10210.4	185.54	7430.1	80.98	56.099	5000	42086.3	49500	32157.0	1.0838	67.193
1420	10371.0	196.50	7551.1	77.56	56.213	5100	43029.1	54380	32901.2	1.0064	67.380
1440	10532.0	208.0	7672.4	74.32	56.326	5200	43974.3	59640	33647.9	0.9355	67.562
1460	10693.3	219.9	7793.9	71.25	56.437	5300	44922.2	65320	34397.1	0.8708	67.743
						5380	45682.1	70170	34998.1	0.8227	67.885

ITEM B 8   Products—400% Stoichiometric Air (per p-mole)

Reproduced from Keenan and Kaye, *Gas Tables*, with permission of authors and publisher, John Wiley.

$T°R$	$\bar{h}$	p_r	$\bar{u}$	v_r	$\bar{\phi}$	$T°R$	$\bar{h}$	p_r	$\bar{u}$	v_r	$\bar{\phi}$
380	2644.9	0.3967	1890.3	10280	43.890	1460	10562.4	53.75	7663.0	291.5	53.639
400	2784.7	0.4754	1990.3	9030	44.249	1480	10718.7	56.71	7779.6	280.1	53.745
420	2925.0	0.5646	2090.9	7982	44.591	1500	10875.6	59.80	7896.8	269.2	53.851
440	3065.2	0.6654	2191.4	7096	44.917	1520	11033.1	63.01	8014.6	258.8	53.955
460	3205.7	0.7785	2292.2	6341	45.229	1540	11190.7	66.37	8132.5	249.0	54.057
480	3346.2	0.9050	2393.0	5692	45.528	1560	11348.6	69.86	8250.7	239.6	54.160
500	3486.7	1.0457	2493.8	5131	45.815	1580	11506.8	73.51	8369.1	230.6	54.260
520	3627.4	1.2017	2594.7	4644	46.091	1600	11665.6	77.30	8488.2	222.1	54.360
537	3746.8	1.3469	2680.4	4279	46.318	1620	11824.6	81.24	8607.5	214.0	54.459
540	3768.0	1.3738	2695.6	4218	46.357	1640	11984.1	85.35	8727.3	206.2	54.557
560	3909.2	1.5632	2797.1	3845	46.613	1660	12143.8	89.61	8847.2	198.80	54.654
580	4050.4	1.7709	2898.6	3514	46.861	1680	12303.9	94.05	8967.6	191.71	54.750
600	4191.9	1.998	3000.4	3223	47.101	1700	12464.3	98.64	9088.3	184.93	54.844
620	4333.5	2.246	3102.3	2963	47.333	1720	12625.3	103.43	9209.6	178.46	54.939
640	4475.2	2.515	3204.2	2731	47.558	1740	12786.4	108.40	9331.0	172.25	55.032
660	4617.5	2.807	3306.8	2522	47.777	1760	12947.8	113.55	9452.7	166.33	55.124
680	4759.3	3.124	3408.9	2336	47.989	1780	13109.4	118.90	9574.5	160.66	55.215
700	4901.7	3.467	3511.6	2167.0	48.195	1800	13271.7	124.45	9697.1	155.21	55.306
720	5044.2	3.836	3614.4	2014.4	48.396	1820	13434.3	130.21	9820.0	150.00	55.396
740	5187.0	4.233	3717.5	1876.0	48.591	1840	13597.0	136.18	9943.0	145.01	55.485
760	5330.2	4.660	3820.9	1750.2	48.783	1860	13760.3	142.35	10066.6	140.22	55.573
780	5473.8	5.118	3924.8	1635.4	48.968	1880	13923.6	148.76	10190.2	135.62	55.660
800	5617.5	5.609	4028.8	1530.6	49.150	1900	14087.2	155.39	10314.1	131.22	55.747
820	5761.7	6.134	4133.3	1434.4	49.329	1920	14251.3	162.26	10438.4	126.98	55.833
840	5905.7	6.695	4237.6	1346.3	49.502	1940	14415.7	169.37	10563.1	122.91	55.918
860	6050.5	7.294	4342.6	1265.3	49.672	1960	14580.3	176.73	10688.0	119.01	56.002
880	6195.2	7.932	4447.6	1190.6	49.839	1980	14745.3	184.33	10813.3	115.27	56.086
900	6340.3	8.611	4553.0	1121.6	50.002	2000	14910.3	192.21	10938.6	111.66	56.169
920	6485.9	9.334	4658.9	1057.8	50.162	2020	15075.8	200.35	11064.3	108.20	56.252
940	6631.7	10.101	4765.0	998.9	50.319	2040	15241.6	208.76	11190.4	104.87	56.333
960	6778.3	10.916	4871.9	943.7	50.473	2060	15407.6	217.45	11316.7	101.66	56.414
980	6925.1	11.779	4978.9	892.8	50.624	2080	15574.0	226.42	11443.4	98.58	56.494
1000	7072.1	12.694	5086.2	845.3	50.773	2100	15740.5	235.7	11570.2	95.61	56.574
1020	7219.7	13.662	5194.1	801.1	50.919	2150	16157.9	260.2	11888.3	88.66	56.771
1040	7367.2	14.685	5301.9	760.1	51.062	2200	16577.1	286.7	12208.2	82.33	56.964
1060	7515.6	15.768	5410.6	721.2	51.203	2250	16998.0	315.3	12529.8	76.56	57.152
1080	7664.1	16.909	5519.4	685.5	51.342	2300	17419.8	346.2	12852.3	71.28	57.338
1100	7812.9	18.116	5628.4	651.6	51.479	2350	17843.2	379.5	13176.4	66.46	57.520
1120	7962.3	19.385	5738.1	620.0	51.613	2400	18268.0	415.3	13501.9	62.02	57.699
1140	8112.0	20.723	5848.1	590.5	51.746	2450	18694.1	453.7	13828.7	57.94	57.875
1160	8262.1	22.14	5958.5	562.5	51.877	2500	19121.4	494.9	14156.8	54.20	58.048
1180	8412.5	23.61	6069.2	536.3	52.006	2550	19550.1	539.1	14486.1	50.76	58.217
1200	8563.4	25.17	6180.4	511.6	52.132	2600	19979.7	586.4	14816.5	47.58	58.384
1220	8715.0	26.80	6292.2	488.4	52.257	2650	20410.6	636.9	15148.0	44.65	58.548
1240	8866.6	28.52	6404.1	466.6	52.380	2700	20842.8	690.9	15481.0	41.94	58.710
1260	9018.8	30.32	6516.6	445.9	52.502	2750	21276.0	748.5	15814.8	39.43	58.869
1280	9171.3	32.21	6629.4	426.4	52.622	2800	21709.8	809.9	16149.4	37.10	59.026
1300	9324.1	34.20	6742.5	407.9	52.741	2850	22145.3	875.2	16485.5	34.94	59.180
1320	9477.6	36.28	6856.3	390.5	52.858	2900	22581.4	944.7	16822.4	32.94	59.331
1340	9631.4	38.45	6970.3	374.0	52.973	2950	23018.5	1018.6	17160.2	31.08	59.481
1360	9785.5	40.73	7084.7	358.4	53.088	3000	23456.6	1096.8	17499.0	29.35	59.628
1380	9939.9	43.10	7199.4	343.6	53.201	3100	24335.5	1268.3	18179.3	26.23	59.916
1400	10095.0	45.60	7314.8	329.5	53.312	3200	25217.8	1460.4	18863.1	23.51	60.196
1420	10250.6	48.19	7430.7	316.2	53.422	3300	26102.9	1675.3	19549.5	21.14	60.469
1440	10406.2	50.91	7546.5	303.6	53.531	3400	26991.4	1914.6	20239.6	19.06	60.734

Products—200% Stoichiometric Air (per p-mole)   ITEM B 9

Reproduced from Keenan and Kaye, *Gas Tables*, with permission of authors and publisher, John Wiley.

$T°R$	$\bar{h}$	p_r	$\bar{u}$	v_r	$\bar{\phi}$	$T°R$	$\bar{h}$	p_r	$\bar{u}$	v_r	$\bar{\phi}$
380	2660.3	0.3878	1905.7	10516	43.845	1460	10729.3	57.26	7829.9	273.6	53.764
400	2801.4	0.4655	2007.0	9222	44.208	1480	10889.5	60.49	7950.4	262.5	53.874
420	2943.0	0.5539	2108.9	8138	44.553	1500	11050.2	63.88	8071.4	252.0	53.981
440	3084.8	0.6539	2211.0	7223	44.883	1520	11211.4	67.40	8192.9	242.0	54.088
460	3226.8	0.7663	2313.3	6441	45.198	1540	11372.8	71.08	8314.6	232.5	54.194
480	3368.9	0.8925	2415.7	5772	45.500	1560	11534.7	74.91	8436.7	223.5	54.298
500	3511.2	1.0330	2518.3	5194	45.791	1580	11696.8	78.92	8559.1	214.8	54.401
520	3653.7	1.1893	2621.0	4692	46.070	1600	11859.6	83.10	8682.2	206.63	54.504
537	3774.9	1.3350	2708.5	4317	46.300	1620	12022.7	87.44	8805.6	198.81	54.605
540	3796.3	1.3620	2723.9	4255	46.340	1640	12186.2	91.97	8929.4	191.35	54.705
560	3939.4	1.5526	2827.3	3871	46.600	1660	12350.0	96.69	9053.5	184.25	54.805
580	4082.7	1.7621	2930.9	3532	46.851	1680	12514.2	101.60	9177.9	177.46	54.903
600	4226.3	1.992	3034.8	3233	47.094	1700	12678.6	106.70	9302.6	170.97	55.000
620	4370.1	2.243	3138.9	2967	47.330	1720	12843.8	112.01	9428.1	164.79	55.097
640	4514.0	2.516	3243.0	2730	47.558	1740	13009.0	117.53	9553.6	158.88	55.192
660	4658.6	2.814	3347.9	2517	47.781	1760	13174.6	123.27	9679.5	153.22	55.287
680	4802.8	3.136	3452.4	2326	47.996	1780	13340.3	129.23	9805.5	147.80	55.380
700	4947.7	3.487	3557.6	2154.5	48.207	1800	13507.0	135.43	9932.4	142.63	55.473
720	5092.6	3.864	3662.8	1999.3	48.411	1820	13673.8	141.86	10059.5	137.68	55.566
740	5237.9	4.273	3768.4	1858.7	48.610	1840	13840.5	148.54	10186.5	132.94	55.657
760	5383.7	4.711	3874.4	1731.1	48.804	1860	14008.4	155.46	10314.7	128.40	55.748
780	5529.9	5.183	3980.9	1614.9	48.994	1880	14176.1	162.55	10442.7	124.03	55.837
800	5676.3	5.690	4087.6	1508.7	49.179	1900	14344.1	170.09	10571.0	119.87	55.926
820	5823.1	6.234	4194.7	1411.7	49.360	1920	14512.5	177.82	10699.6	115.87	56.015
840	5969.9	6.815	4301.8	1322.6	49.537	1940	14681.4	185.82	10828.8	112.04	56.102
860	6117.5	7.437	4409.6	1241.0	49.711	1960	14850.4	194.13	10958.1	108.35	56.189
880	6264.9	8.101	4517.3	1165.8	49.881	1980	15019.8	202.71	11087.8	104.82	56.275
900	6413.0	8.808	4625.7	1096.4	50.047	2000	15189.3	211.6	11217.6	101.43	56.360
920	6561.5	9.564	4734.5	1032.4	50.210	2020	15359.3	220.8	11347.9	98.16	56.444
940	6710.4	10.366	4843.7	973.3	50.370	2040	15529.7	230.4	11478.5	95.04	56.529
960	6859.8	11.221	4953.4	918.2	50.528	2060	15700.3	240.2	11609.4	92.03	56.612
980	7009.7	12.126	5063.6	867.2	50.682	2080	15871.3	250.4	11740.7	89.15	56.694
1000	7159.8	13.089	5173.9	819.8	50.833	2100	16042.4	260.9	11872.1	86.36	56.777
1020	7310.5	14.109	5284.9	775.8	50.983	2150	16471.5	288.9	12201.9	79.86	56.978
1040	7461.1	15.189	5395.8	734.9	51.128	2200	16902.5	319.2	12533.6	73.96	57.177
1060	7612.7	16.333	5507.7	696.2	51.273	2250	17335.3	352.0	12867.1	68.57	57.377
1080	7764.3	17.542	5619.6	660.8	51.415	2300	17769.3	387.5	13201.8	63.68	57.562
1100	7916.4	18.822	5731.9	627.1	51.555	2350	18204.9	425.9	13538.1	59.21	57.749
1120	8069.0	20.170	5844.8	595.9	51.692	2400	18642.1	467.4	13876.0	55.12	57.933
1140	8222.1	21.595	5958.2	566.6	51.828	2450	19080.7	511.9	14215.3	51.36	58.114
1160	8375.5	23.10	6071.9	539.0	51.961	2500	19520.7	559.8	14556.0	47.91	58.292
1180	8529.2	24.68	6185.9	513.2	52.093	2550	19962.0	611.3	14898.0	44.76	58.467
1200	8683.6	26.34	6300.6	488.9	52.222	2600	20404.6	666.6	15241.3	41.85	58.639
1220	8838.6	28.09	6415.8	466.0	52.351	2650	20848.4	725.9	15585.8	39.18	58.808
1240	8993.7	29.94	6531.2	444.5	52.477	2700	21293.8	789.4	15932.0	36.71	58.974
1260	9149.3	31.87	6647.1	424.2	52.601	2750	21740.3	857.2	16279.2	34.42	59.138
1280	9305.3	33.90	6763.4	405.1	52.724	2800	22187.5	929.8	16627.1	32.31	59.300
1300	9461.7	36.05	6880.1	387.0	52.845	2850	22636.3	1007.2	16976.6	30.36	59.459
1320	9618.8	38.29	6997.3	370.0	52.965	2900	23086.0	1089.8	17327.0	28.56	59.615
1340	9776.2	40.64	7115.1	353.9	53.084	2950	23536.7	1177.9	17678.4	26.88	59.769
1360	9933.8	43.10	7233.0	338.6	53.200	3000	23988.5	1271.2	18030.9	25.321	59.921
1380	10091.9	45.68	7351.4	324.2	53.316	3100	24895.3	1476.8	18739.1	22.528	60.218
1400	10250.7	48.38	7470.5	310.5	53.430	3200	25805.6	1708.2	19450.8	20.103	60.507
1420	10410.0	51.21	7590.1	297.6	53.543	3400	27636.4	2259.1	20884.4	16.152	61.063
1440	10569.3	54.17	7709.7	285.3	53.654	3600	29479.9	2946	22330.8	13.114	61.590
						3900	32266.2	4284	24521.3	9.770	62.333

ITEM B 10   Water Vapor at Low Pressures (per p-mole)

Reproduced from Keenan and Kaye, *Gas Tables*, with permission of authors and publisher, John Wiley.

$T°R$	$\bar h$	p_r	$\bar u$	v_r	$\bar\phi$	$T°R$	$\bar h$	p_r	$\bar u$	v_r	$\bar\phi$
300	2367.6	0.06975	1771.8	46150	40.439	1380	11441.4	39.68	8700.9	373.2	53.037
320	2526.8	0.09033	1891.3	38010	40.952	1400	11624.8	42.41	8844.6	354.3	53.168
340	2686.0	0.11515	2010.8	31690	41.435	1420	11808.8	45.29	8988.9	336.4	53.299
360	2845.1	0.14479	2130.2	26690	41.889	1440	11993.4	48.34	9133.8	319.7	53.428
380	3004.4	0.17982	2249.8	22680	42.320	1460	12178.8	51.55	9279.4	304.0	53.556
400	3163.8	0.2209	2369.4	19436	42.728	1480	12364.8	54.94	9425.7	289.1	53.682
420	3323.2	0.2686	2489.1	16780	43.117	1500	12551.4	58.53	9572.7	275.1	53.808
440	3482.7	0.3237	2608.9	14585	43.487	1520	12738.8	62.30	9720.3	261.8	53.932
460	3642.3	0.3870	2728.8	12756	43.841	1540	12926.8	66.28	9868.6	249.4	54.055
480	3802.0	0.4592	2848.8	11217	44.182	1560	13115.6	70.47	10017.6	237.6	54.177
500	3962.0	0.5412	2969.1	9914	44.508	1580	13305.0	74.90	10167.3	226.4	54.298
520	4122.0	0.6337	3089.4	8805	44.821	1600	13494.9	79.53	10317.6	215.9	54.418
537	4258.0	0.7217	3191.9	7987	45.079	1620	13685.7	84.43	10468.6	205.9	54.535
540	4282.4	0.7381	3210.0	7851	45.124	1640	13877.0	89.58	10620.2	196.47	54.653
560	4442.8	0.8548	3330.7	7029	45.415	1660	14069.2	94.99	10772.7	187.53	54.770
580	4603.7	0.9853	3451.9	6317	45.696	1680	14261.9	100.69	10925.6	179.04	54.886
600	4764.7	1.1305	3573.2	5696	45.970	1700	14455.4	106.67	11079.4	171.03	54.999
620	4926.1	1.2914	3694.9	5153	46.235	1720	14649.5	112.93	11233.8	163.45	55.113
640	5087.8	1.4696	3816.8	4672	46.492	1740	14844.3	119.54	11388.9	156.20	55.226
660	5250.0	1.6662	3939.3	4250	46.741	1760	15039.8	126.45	11544.7	149.36	55.339
680	5412.5	1.8825	4062.1	3876	46.984	1780	15236.1	133.73	11701.2	142.83	55.449
700	5575.4	2.120	4185.3	3543	47.219	1800	15433.0	141.35	11858.4	136.66	55.559
720	5738.8	2.380	4309.0	3246	47.450	1820	15630.6	149.33	12016.3	130.78	55.668
740	5902.6	2.665	4433.1	2979	47.673	1840	15828.7	157.71	12174.7	125.20	55.777
760	6066.9	2.976	4557.6	2741	47.893	1860	16027.6	166.51	12333.9	119.89	55.884
780	6231.7	3.314	4682.7	2526	48.106	1880	16227.2	175.72	12493.8	114.83	55.991
800	6396.9	3.683	4808.2	2331	48.316	1900	16427.5	185.36	12654.4	110.01	56.097
820	6562.6	4.082	4934.2	2156	48.520	1920	16628.5	195.43	12815.6	105.43	56.203
840	6728.9	4.515	5060.8	1996.3	48.721	1940	16830.0	206.0	12977.4	101.06	56.307
860	6895.6	4.984	5187.8	1851.7	48.916	1960	17032.4	217.1	13140.1	96.89	56.411
880	7062.9	5.491	5315.3	1719.8	49.109	1980	17235.3	228.6	13303.3	92.93	56.514
900	7230.9	6.038	5443.6	1599.4	49.298	2000	17439.0	240.8	13467.3	89.15	56.617
920	7399.4	6.627	5572.4	1489.6	49.483	2100	18466.9	310.0	14296.6	72.69	57.119
940	7568.4	7.263	5701.7	1389.0	49.665	2200	19510.8	396.0	15141.9	59.761	57.605
960	7738.0	7.945	5831.6	1296.6	49.843	2300	20570.6	502.2	16003.1	49.15	58.077
980	7908.2	8.680	5962.0	1211.6	50.019	2400	21645.7	632.4	16879.6	40.73	58.535
1000	8078.9	9.469	6093.0	1133.3	50.191	2500	22735.4	791.5	17770.7	33.89	58.980
1020	8250.4	10.313	6224.8	1061.4	50.360	2600	23839.5	984.8	18676.2	28.33	59.414
1040	8422.4	11.219	6357.1	994.8	50.528	2700	24957.2	1218.4	19595.4	23.78	59.837
1060	8595.0	12.187	6490.0	933.4	50.693	2800	26088.0	1499.4	20527.6	20.04	60.248
1080	8768.2	13.220	6623.5	876.7	50.854	2900	27231.2	1835.6	21472.2	16.953	60.650
1100	8942.0	14.327	6757.5	823.8	51.013	3000	28386.3	2237	22428.7	14.394	61.043
1120	9116.4	15.507	6892.2	775.1	51.171	3100	29552.8	2713	23396.6	12.263	61.426
1140	9291.4	16.766	7027.5	729.6	51.325	3200	30730.2	3276	24375.4	10.481	61.801
1160	9467.1	18.106	7163.5	687.6	51.478	3300	31918.2	3940	25364.8	8.988	62.167
1180	9643.4	19.536	7300.1	648.1	51.630	3400	33116.0	4720	26364.0	7.730	62.526
1200	9820.4	21.05	7437.4	611.7	51.777	3500	34323.5	5633	27373.0	6.668	62.876
1220	9998.0	22.67	7575.2	577.7	51.925	3600	35540.1	6697	28391.0	5.770	63.221
1240	10176.1	24.39	7713.6	545.7	52.070	3700	36765.4	7935	29417.7	5.004	63.557
1260	10354.9	26.21	7852.7	516.0	52.212	3800	37998.9	9368	30452.6	4.353	63.887
1280	10534.4	28.14	7992.5	488.2	52.354	3900	39240.2	11025	31495.3	3.796	64.210
1300	10714.5	30.19	8132.9	462.1	52.494	4000	40489.1	12934	32545.6	3.319	64.528
1320	10895.3	32.37	8274.0	437.7	52.631	4500	46835.9	27530	37899.5	1.7543	66.028
1340	11076.6	34.67	8415.5	414.9	52.768	5000	53327.4	54970	43398.0	0.9760	67.401
1360	11258.7	37.10	8557.9	393.3	52.903	5380	58338.7	89600	47654.7	0.6443	68.371

Enthalpy and Gibbs Function of Formation; Absolute Entropy   ITEM B 11

See also Table III, p. 152, for additional entropy values. The original data of this table were from Rossini,[0.1, 0.2] many of which have been slightly up-dated from JANAF,[0.22] also the source of the new additions; designated by —J. Δh_f° Btu/pmole = enthalpy of formation (§ 13.29), ΔG_f° Btu/pmole = Gibbs function of formation (§13.33), $\bar{s}^\circ$ Btu/pmole-°R = the absolute entropy, each at the standard state of 1 atm and 77°F; $\bar{h}_{fg}^\circ$ Btu/pmole = the enthalpy of vaporization at 77°F.

Substance	Mol. mass, M	$\Delta \bar{h}_f^\circ$	$\Delta \bar{G}_f^\circ$	$\bar{h}_{fg}^\circ$	$\bar{s}^\circ$
Acetylene, $C_2H_2(g)$—J	26.036	+97,542	+89,987		48.004
Ammonia, $NH_3(g)$—J	17.032	−19,746	−7,047	8,545	46.03
Benzene, $C_6H_6(g)$	78.108	+35,676	+55,780		64.34
$C_6H_6(l)$—J		+20,934	+52,920		29.756
n-Butane, $C_4H_{10}(g)$	58.120	−54,270	−7,380	9,063	74.12
Carbon (graphite), $C(s)$—J	12.01	0	0		1.359
Carbon (diamond)—J		+0.453	+0.685		0.583
Carbon dioxide, $CO_2(g)$	44.010	−169,293	−169,668		51.061
Carbon monoxide, $CO(g)$—J	28.01	−47,560	−59,010		47.21
Cyanogen, $C_2N_2(g)$—J	52.038	+132,480	+127,458		57.86
Electron gas—J	0.0005488	0	0		4.988
Ethane, $C_2H_6(g)$	30.068	−36,425	−14,148		54.85
Ethanol, $C_2H_6O(g)$—J	46.068	−101,232	−72,540	18,216	67.4
$C_2H_6O(l)$—J		−119,448	−75,186		38.4
Ethene, $C_2H_4(g)$	28.052	+22,572	+29,408		52.396
n-Heptane, $C_7H_{16}(g)$—J	100.198	−80,802	+3,762	15,728	101.64
$C_7H_{16}(l)$—J		−96,530	+756		77.92
Hydrazine, $N_2H_4(g)$—J	32.048	+41,022	+68,476		57.03
Hydrogen (Mono), $H(g)$—J	1.008	+93,780	+87,453		27.392
Hydrogen peroxide,					
$\quad H_2O_2(g)$—J	34.016	−58,554	−45,374	22,158	55.66
$H_2O_2(l)$		−80,712	−49,032		21.2
Hydrogen sulfide, $H_2S(g)$—J	34.086	−8,784	−14,391		49.15
Hydroxyl, $OH(g)$—J	17.008	+16,978	+14,953		43.88
Methane, $CH_4(g)$—J	16.042	−32,210	−21,862		44.48
Methanol, $CH_4O(g)$	32.042	−86,544	−69,642	16,096	56.8
$CH_4O(l)$—J		−102,640	−71,514		30.3
Methyl chloride, $CH_3Cl(g)$—J	50.49	−37,190	−27,083		55.99
Nitrogen, $N_2(g)$	28.016	0	0		45.767
Nitrogen oxide, $NO(g)$	30.008	+38,880	+37,294		50.339
Nitrogen dioxide, $NO_2(g)$—J	48.008	+14,238	+22,045		57.343
n-Octane, $C_8H_{18}(g)$	114.224	−89,676	+7,452	17,856	111.82
$C_8H_{18}(l)$—J		−107,532	+3,186		85.50
Nonane, $C_9H_{20}(g)$		−98,432	+10,728	19,940	120.86
Oxygen (Mono), $O(g)$—J	16	+107,206	+99,711		38.468
Oxygen, $O_2(g)$	32	0	0		49.004
Ozone, $O_3(g)$—J	48	+61,380	+70,195		57.080
Propane, $C_3H_8(g)$	44.094	−44,676	−10,105	6,489	64.51
Sulfur dioxide, $SO_2(g)$—J	64.07	−127,705	−129,134	9,370	59.298
Water, $H_2O(g)$	18.016	−104,036	−98,344		45.106
Water (l)	18.016	−122,971	−102,042		16.716

ITEM B 12   Enthalpy of Reaction, Typical Fuels

Heating values are for a reference state at 77°F, or essentially so, with combustion in air or oxygen to stable species. The values for the hydrocarbons were taken from reference [0.1], given to more significant figures than are warranted by the accuracy of the individual values in order to retain the significance of small differences between liquid and gaseous states. Miscellaneous values are from references [13.3] and [13.7]. Values marked with an asterisk * are for academic use only, since the compositions and heating values vary markedly. Because of different sources or basic data, some values in this table may not be consistent with some of those in Item B 11. Meanings of symbols: (s) = solid; (l) = liquid; (g) = gas. Unless otherwise indicated, choose values according to the natural state of the fuel; for example, benzene is normally a liquid—therefore, use values opposite (l). In every case, the CO_2 is in a gaseous state. Where applicable, the latent heat of vaporization of the fuel is q_h° in the last column minus q_h° in the fifth column.

Fuel	Formula	M	$-h_{rp}^\circ$, Btu/lb Fuel, Solid or Liquid		$-h_{rp}^\circ$, Btu/lb Fuel, Gaseous Fuel	
			Lower q_l°, $H_2O(g)$	Higher q_h°, $H_2O(l)$	Lower q_l°, $H_2O(g)$	Higher q_h°, $H_2O(l)$
SOLID (s)						
Anthracite coal*			13,330	13,540		
Bituminous coal*			13,100	13,600		
Carbon, graphite (to CO_2)	C	12.01	14,097			
Carbon graphite (to CO)	C	12.01	3,960			
Coke, beehive			12,450	12,530		
Lignite*			6,700	7,350		
Sulfur	S	32.06	3,980			
Wood, air-dried oak*			8,000			
NORMALLY LIQUID (l)						
Benzene	C_6H_6	78.108	17,259	17,986	17,446	18,172
n-Decane	$C_{10}H_{22}$	142.276	19,020	20,483	19,175	20,638
n-Dodecane	$C_{12}H_{26}$	170.328	18,966	20,410	19,120	20,564
Ethyl alcohol	C_2H_6O	46.068	11,929	13,161		
Fuel oil*			18,500	19,700		
Gasoline*			18,800	20,200		
n-Heptane	C_7H_{16}	100.198	19,157	20,668	19,314	20,825
n-Hexadecane	$C_{16}H_{34}$	226.432	18,898	20,318	19,052	20,472
Kerosene*			18,500	19,900		
Methyl alcohol	CH_4O	32.042	9,078	10,259		
Nonane	C_9H_{20}	128.25	19,056	20,531	19,211	20,687
n-Octane	C_8H_{18}	114.224	19,100	20,591	19,256	20,747
Octene	C_8H_{16}	112.208	19,000	20,350	19,157	20,506
n-Pentane	C_5H_{12}	72.146	19,340	20,914	19,499	21,072
NORMALLY GAS (g)						
Acetylene	C_2H_2	26.036			20,734	21,460
Blast-furnace gas*					1,100	1,120
n-Butane	C_4H_{10}	58.120	19,496	21,124	19,655	21,283
Carbon monoxide	CO	28.01			4,346	
Cyanogen [to CO_2 and N_2]	C_2N_2	80.052			5,890	
Ethane	C_2H_6	30.068			20,416	22,304
Ethene	C_2H_4	28.052			20,276	21,625
Hydrogen	H_2	2.016			51,605	60,998
Methane	CH_4	16.042			21,502	23,861
Natural gas*					20,500	23,000
Propane	C_3H_8	44.094	19,774	21,490	19,929	21,646
Refinery gas*					19,600	21,400

Properties of Saturated H$_2$O: Temperatures **ITEM B 13**

Temp Fahr t	Abs Press. Lb per Sq In. p	Specific Volume			Enthalpy			Entropy			Temp Fahr t
		Sat. Liquid v_f	Evap v_{fg}	Sat. Vapor v_g	Sat. Liquid h_f	Evap h_{fg}	Sat. Vapor h_g	Sat. Liquid s_f	Evap s_{fg}	Sat. Vapor s_g	
32.0	0.08859	0.016022	3304.7	3304.7	−0.0179	1075.5	1075.5	0.0000	2.1873	2.1873	32.0
34.0	0.09600	0.016021	3061.9	3061.9	1.996	1074.4	1076.4	0.0041	2.1762	2.1802	34.0
36.0	0.10395	0.016020	2839.0	2839.0	4.008	1073.2	1077.2	0.0081	2.1651	2.1732	36.0
38.0	0.11249	0.016019	2634.1	2634.2	6.018	1072.1	1078.1	0.0122	2.1541	2.1663	38.0
40.0	0.12163	0.016019	2445.8	2445.8	8.027	1071.0	1079.0	0.0162	2.1432	2.1594	40.0
42.0	0.13143	0.016019	2272.4	2272.4	10.035	1069.8	1079.9	0.0202	2.1325	2.1527	42.0
44.0	0.14192	0.016019	2112.8	2112.8	12.041	1068.7	1080.7	0.0242	2.1217	2.1459	44.0
46.0	0.15314	0.016020	1965.7	1965.7	14.047	1067.6	1081.6	0.0282	2.1111	2.1393	46.0
48.0	0.16514	0.016021	1830.0	1830.0	16.051	1066.4	1082.5	0.0321	2.1006	2.1327	48.0
50.0	0.17796	0.016023	1704.8	1704.8	18.054	1065.3	1083.4	0.0361	2.0901	2.1262	50.0
52.0	0.19165	0.016024	1589.2	1589.2	20.057	1064.2	1084.2	0.0400	2.0798	2.1197	52.0
54.0	0.20625	0.016026	1482.4	1482.4	22.058	1063.1	1085.1	0.0439	2.0695	2.1134	54.0
56.0	0.22183	0.016028	1383.6	1383.6	24.059	1061.9	1086.0	0.0478	2.0593	2.1070	56.0
58.0	0.23843	0.016031	1292.2	1292.2	26.060	1060.8	1086.9	0.0516	2.0491	2.1008	58.0
60.0	0.25611	0.016033	1207.6	1207.6	28.060	1059.7	1087.7	0.0555	2.0391	2.0946	60.0
62.0	0.27494	0.016036	1129.2	1129.2	30.059	1058.5	1088.6	0.0593	2.0291	2.0885	62.0
64.0	0.29497	0.016039	1056.5	1056.5	32.058	1057.4	1089.5	0.0632	2.0192	2.0824	64.0
66.0	0.31626	0.016043	989.0	989.1	34.056	1056.3	1090.4	0.0670	2.0094	2.0764	66.0
68.0	0.33889	0.016046	926.5	926.5	36.054	1055.2	1091.2	0.0708	1.9996	2.0704	68.0
70.0	0.36292	0.016050	868.3	868.4	38.052	1054.0	1092.1	0.0745	1.9900	2.0645	70.0
72.0	0.38844	0.016054	814.3	814.3	40.049	1052.9	1093.0	0.0783	1.9804	2.0587	72.0
74.0	0.41550	0.016058	764.1	764.1	42.046	1051.8	1093.8	0.0821	1.9708	2.0529	74.0
76.0	0.44420	0.016063	717.4	717.4	44.043	1050.7	1094.7	0.0858	1.9614	2.0472	76.0
78.0	0.47461	0.016067	673.8	673.9	46.040	1049.5	1095.6	0.0895	1.9520	2.0415	78.0
80.0	0.50683	0.016072	633.3	633.3	48.037	1048.4	1096.4	0.0932	1.9426	2.0359	80.0
82.0	0.54093	0.016077	595.5	595.5	50.033	1047.3	1097.3	0.0969	1.9334	2.0303	82.0
84.0	0.57702	0.016082	560.3	560.3	52.029	1046.1	1098.2	0.1006	1.9242	2.0248	84.0
86.0	0.61518	0.016087	527.5	527.5	54.026	1045.0	1099.0	0.1043	1.9151	2.0193	86.0
88.0	0.65551	0.016093	496.8	496.8	56.022	1043.9	1099.9	0.1079	1.9060	2.0139	88.0
90.0	0.69813	0.016099	468.1	468.1	58.018	1042.7	1100.8	0.1115	1.8970	2.0086	90.0
92.0	0.74313	0.016105	441.3	441.3	60.014	1041.6	1101.6	0.1152	1.8881	2.0033	92.0
94.0	0.79062	0.016111	416.3	416.3	62.010	1040.5	1102.5	0.1188	1.8792	1.9980	94.0
96.0	0.84072	0.016117	392.8	392.9	64.006	1039.3	1103.3	0.1224	1.8704	1.9928	96.0
98.0	0.89356	0.016123	370.9	370.9	66.003	1038.2	1104.2	0.1260	1.8617	1.9876	98.0
100.0	0.94924	0.016130	350.4	350.4	67.999	1037.1	1105.1	0.1295	1.8530	1.9825	100.0
102.0	1.00789	0.016137	331.1	331.1	69.995	1035.9	1105.9	0.1331	1.8444	1.9775	102.0
104.0	1.06965	0.016144	313.1	313.1	71.992	1034.8	1106.8	0.1366	1.8358	1.9725	104.0
106.0	1.1347	0.016151	296.16	296.18	73.99	1033.6	1107.6	0.1402	1.8273	1.9675	106.0
108.0	1.2030	0.016158	280.28	280.30	75.98	1032.5	1108.5	0.1437	1.8188	1.9626	108.0
110.0	1.2750	0.016165	265.37	265.39	77.98	1031.4	1109.3	0.1472	1.8105	1.9577	110.0
112.0	1.3505	0.016173	251.37	251.38	79.98	1030.2	1110.2	0.1507	1.8021	1.9528	112.0
114.0	1.4299	0.016180	238.21	238.22	81.97	1029.1	1111.0	0.1542	1.7938	1.9480	114.0
116.0	1.5133	0.016188	225.84	225.85	83.97	1027.9	1111.9	0.1577	1.7856	1.9433	116.0
118.0	1.6009	0.016196	214.20	214.21	85.97	1026.8	1112.7	0.1611	1.7774	1.9386	118.0
120.0	1.6927	0.016204	203.25	203.26	87.97	1025.6	1113.6	0.1646	1.7693	1.9339	120.0
122.0	1.7891	0.016213	192.94	192.95	89.96	1024.5	1114.4	0.1680	1.7613	1.9293	122.0
124.0	1.8901	0.016221	183.23	183.24	91.96	1023.3	1115.3	0.1715	1.7533	1.9247	124.0
126.0	1.9959	0.016229	174.08	174.09	93.96	1022.2	1116.1	0.1749	1.7453	1.9202	126.0
128.0	2.1068	0.016238	165.45	165.47	95.96	1021.0	1117.0	0.1783	1.7374	1.9157	128.0
130.0	2.2230	0.016247	157.32	157.33	97.96	1019.8	1117.8	0.1817	1.7295	1.9112	130.0
132.0	2.3445	0.016256	149.64	149.66	99.95	1018.7	1118.6	0.1851	1.7217	1.9068	132.0
134.0	2.4717	0.016265	142.40	142.41	101.95	1017.5	1119.5	0.1884	1.7140	1.9024	134.0
136.0	2.6047	0.016274	135.55	135.57	103.95	1016.4	1120.3	0.1918	1.7063	1.8980	136.0
138.0	2.7438	0.016284	129.00	129.11	105.95	1015.2	1121.1	0.1951	1.6986	1.8937	138.0
140.0	2.8892	0.016293	122.98	123.00	107.95	1014.0	1122.0	0.1985	1.6910	1.8895	140.0
142.0	3.0411	0.016303	117.21	117.22	109.95	1012.9	1122.8	0.2018	1.6534	1.8852	142.0
144.0	3.1997	0.016312	111.74	111.76	111.95	1011.7	1123.6	0.2051	1.6759	1.8810	144.0
146.0	3.3653	0.016322	106.58	106.59	113.95	1010.5	1124.5	0.2084	1.6684	1.8769	146.0
148.0	3.5381	0.016332	101.68	101.70	115.95	1009.3	1125.3	0.2117	1.6610	1.8727	148.0
150.0	3.7184	0.016343	97.05	97.07	117.95	1008.2	1126.1	0.2150	1.6536	1.8686	150.0
152.0	3.9065	0.016353	92.66	92.68	119.95	1007.0	1126.9	0.2183	1.6463	1.8646	152.0
154.0	4.1025	0.016363	88.50	88.52	121.95	1005.8	1127.7	0.2216	1.6390	1.8606	154.0
156.0	4.3068	0.016374	84.56	84.57	123.95	1004.6	1128.6	0.2248	1.6318	1.8566	156.0
158.0	4.5197	0.016384	80.82	80.83	125.96	1003.4	1129.4	0.2281	1.6245	1.8526	158.0
160.0	4.7414	0.016395	77.27	77.29	127.96	1002.2	1130.2	0.2313	1.6174	1.8487	160.0
162.0	4.9722	0.016406	73.90	73.92	129.96	1001.0	1131.0	0.2345	1.6103	1.8448	162.0
164.0	5.2124	0.016417	70.70	70.72	131.96	999.8	1131.8	0.2377	1.6032	1.8409	164.0
166.0	5.4623	0.016428	67.67	67.68	133.97	998.6	1132.6	0.2409	1.5961	1.8371	166.0
168.0	5.7223	0.016440	64.78	64.80	135.97	997.4	1133.4	0.2441	1.5892	1.8333	168.0
170.0	5.9926	0.016451	62.04	62.06	137.97	996.2	1134.2	0.2473	1.5822	1.8295	170.0
172.0	6.2736	0.016463	59.43	59.45	139.98	995.0	1135.0	0.2505	1.5753	1.8258	172.0
174.0	6.5656	0.016474	56.95	56.97	141.98	993.8	1135.8	0.2537	1.5684	1.8221	174.0
176.0	6.8690	0.016486	54.59	54.61	143.99	992.6	1136.6	0.2568	1.5616	1.8184	176.0
178.0	7.1840	0.016498	52.35	52.36	145.99	991.4	1137.4	0.2600	1.5548	1.8147	178.0

ITEM B 13   Properties of Saturated H_2O: Temperatures (continued)

Temp Fahr t	Abs Press. Lb per Sq In. p	Specific Volume Sat. Liquid v_f	Evap v_{fg}	Sat. Vapor v_g	Enthalpy Sat. Liquid h_f	Evap h_{fg}	Sat. Vapor h_g	Entropy Sat. Liquid s_f	Evap s_{fg}	Sat. Vapor s_g	Temp Fahr t
180.0	7.5110	0.016510	50.21	50.22	148.00	990.2	1138.2	0.2631	1.5480	1.8111	180.0
182.0	7.850	0.016522	48.172	18.189	150.01	989.0	1139.0	0.2662	1.5413	1.8075	182.0
184.0	8.203	0.016534	46.232	46.249	152.01	987.8	1139.8	0.2694	1.5346	1.8040	184.0
186.0	8.568	0.016547	44.383	44.400	154.02	986.5	1140.5	0.2725	1.5279	1.8004	186.0
188.0	8.947	0.016559	42.621	42.638	156.03	985.3	1141.3	0.2756	1.5213	1.7969	188.0
190.0	9.340	0.016572	40.941	40.957	158.04	984.1	1142.1	0.2787	1.5148	1.7934	190.0
192.0	9.747	0.016585	39.337	39.354	160.05	982.8	1142.9	0.2818	1.5082	1.7900	192.0
194.0	10.168	0.016598	37.808	37.824	162.05	981.6	1143.7	0.2848	1.5017	1.7865	194.0
196.0	10.605	0.016611	36.348	36.364	164.06	980.4	1144.4	0.2879	1.4952	1.7831	196.0
198.0	11.058	0.016624	34.954	34.970	166.08	979.1	1145.2	0.2910	1.4888	1.7798	198.0
200.0	11.526	0.016637	33.622	33.639	168.09	977.9	1146.0	0.2940	1.4824	1.7764	200.0
204.0	12.512	0.016664	31.135	31.151	172.11	975.4	1147.5	0.3001	1.4697	1.7698	204.0
208.0	13.568	0.016691	28.862	28.878	176.14	972.8	1149.0	0.3061	1.4571	1.7632	208.0
212.0	14.696	0.016719	26.782	26.799	180.17	970.3	1150.5	0.3121	1.4447	1.7568	212.0
216.0	15.901	0.016747	24.878	24.894	184.20	967.8	1152.0	0.3181	1.4323	1.7505	216.0
220.0	17.186	0.016775	23.131	23.148	188.23	965.2	1153.4	0.3241	1.4201	1.7442	220.0
224.0	18.556	0.016805	21.529	21.545	192.27	962.6	1154.9	0.3300	1.4081	1.7380	224.0
228.0	20.015	0.016834	20.056	20.073	196.31	960.0	1156.3	0.3359	1.3961	1.7320	228.0
232.0	21.567	0.016864	18.701	18.718	200.35	957.4	1157.8	0.3417	1.3842	1.7260	232.0
236.0	23.216	0.016895	17.454	17.471	204.40	954.8	1159.2	0.3476	1.3725	1.7201	236.0
240.0	24.968	0.016926	16.304	16.321	208.45	952.1	1160.6	0.3533	1.3609	1.7142	240.0
244.0	26.826	0.016958	15.243	15.260	212.50	949.5	1162.0	0.3591	1.3494	1.7085	244.0
248.0	28.796	0.016990	14.264	14.281	216.56	946.8	1163.4	0.3649	1.3379	1.7028	248.0
252.0	30.883	0.017022	13.358	13.375	220.62	944.1	1164.7	0.3706	1.3266	1.6972	252.0
256.0	33.091	0.017055	12.520	12.538	224.69	941.4	1166.1	0.3763	1.3154	1.6917	256.0
260.0	35.427	0.017089	11.745	11.762	228.76	938.6	1167.4	0.3819	1.3043	1.6862	260.0
264.0	37.894	0.017123	11.025	11.042	232.83	935.9	1168.7	0.3876	1.2933	1.6808	264.0
268.0	40.500	0.017157	10.358	10.375	236.91	933.1	1170.0	0.3932	1.2823	1.6755	268.0
272.0	43.249	0.017193	9.738	9.755	240.99	930.3	1171.3	0.3987	1.2715	1.6702	272.0
276.0	46.147	0.017228	9.162	9.180	245.08	927.5	1172.5	0.4043	1.2607	1.6650	276.0
280.0	49.200	0.017264	8.627	8.644	249.17	924.6	1173.8	0.4098	1.2501	1.6599	280.0
284.0	52.414	0.01730	8.1280	8.1453	253.3	921.7	1175.0	0.4154	1.2395	1.6548	284.0
288.0	55.795	0.01734	7.6634	7.6807	257.4	918.8	1176.2	0.4208	1.2290	1.6498	288.0
292.0	59.350	0.01738	7.2301	7.2475	261.5	915.9	1177.4	0.4263	1.2186	1.6449	292.0
296.0	63.084	0.01741	6.8259	6.8433	265.6	913.0	1178.6	0.4317	1.2082	1.6400	296.0
300.0	67.005	0.01745	6.4483	6.4658	269.7	910.0	1179.7	0.4372	1.1979	1.6351	300.0
304.0	71.119	0.01749	6.0955	6.1130	273.8	907.0	1180.9	0.4426	1.1877	1.6303	304.0
308.0	75.433	0.01753	5.7655	5.7830	278.0	904.0	1182.0	0.4479	1.1776	1.6256	308.0
312.0	79.953	0.01757	5.4566	5.4742	282.1	901.0	1183.1	0.4533	1.1676	1.6209	312.0
316.0	84.688	0.01761	5.1673	5.1849	286.3	897.9	1184.1	0.4586	1.1576	1.6162	316.0
320.0	89.643	0.01766	4.8961	4.9138	290.4	894.8	1185.2	0.4640	1.1477	1.6116	320.0
324.0	94.826	0.01770	4.6418	4.6595	294.6	891.6	1186.2	0.4692	1.1378	1.6071	324.0
328.0	100.245	0.01774	4.4030	4.4208	298.7	888.5	1187.2	0.4745	1.1280	1.6025	328.0
332.0	105.907	0.01779	4.1788	4.1966	302.9	885.3	1188.2	0.4798	1.1183	1.5981	332.0
336.0	111.820	0.01783	3.9681	3.9859	307.1	882.1	1189.1	0.4850	1.1086	1.5936	336.0
340.0	117.992	0.01787	3.7699	3.7878	311.3	878.8	1190.1	0.4902	1.0990	1.5892	340.0
344.0	124.430	0.01792	3.5834	3.6013	315.5	875.5	1191.0	0.4954	1.0894	1.5849	344.0
348.0	131.142	0.01797	3.4078	3.4258	319.7	872.2	1191.1	0.5006	1.0799	1.5806	348.0
352.0	138.138	0.01801	3.2423	3.2603	323.9	868.9	1192.7	0.5058	1.0705	1.5763	352.0
356.0	145.424	0.01806	3.0863	3.1044	328.1	865.5	1193.6	0.5110	1.0611	1.5721	356.0
360.0	153.010	0.01811	2.9392	2.9573	332.3	862.1	1194.4	0.5161	1.0517	1.5678	360.0
364.0	160.903	0.01816	2.8002	2.8184	336.5	858.6	1195.2	0.5212	1.0424	1.5637	364.0
368.0	169.113	0.01821	2.6691	2.6873	340.8	855.1	1195.9	0.5263	1.0332	1.5595	368.0
372.0	177.648	0.01826	2.5451	2.5633	345.0	851.6	1196.7	0.5314	1.0240	1.5554	372.0
376.0	186.517	0.01831	2.4279	2.4462	349.3	848.1	1197.4	0.5365	1.0148	1.5513	376.0
380.0	195.729	0.01836	2.3170	2.3353	353.6	844.5	1198.0	0.5416	1.0057	1.5473	380.0
384.0	205.294	0.01842	2.2120	2.2304	357.9	840.8	1198.7	0.5466	0.9966	1.5432	384.0
388.0	215.220	0.01847	2.1126	2.1311	362.2	837.2	1199.3	0.5516	0.9876	1.5392	388.0
392.0	225.516	0.01853	2.0184	2.0369	366.5	833.4	1199.9	0.5567	0.9786	1.5352	392.0
396.0	236.193	0.01858	1.9291	1.9477	370.8	829.7	1200.4	0.5617	0.9696	1.5313	396.0
400.0	247.259	0.01864	1.8444	1.8630	375.1	825.9	1201.0	0.5667	0.9607	1.5274	400.0
404.0	258.725	0.01870	1.7640	1.7827	379.4	822.0	1201.5	0.5717	0.9518	1.5234	404.0
408.0	270.600	0.01875	1.6877	1.7064	383.8	818.2	1201.9	0.5766	0.9429	1.5195	408.0
412.0	282.894	0.01881	1.6152	1.6340	388.1	814.2	1202.4	0.5816	0.9341	1.5157	412.0
416.0	295.617	0.01887	1.5463	1.5651	392.5	810.2	1202.8	0.5866	0.9253	1.5118	416.0
420.0	308.780	0.01894	1.4808	1.4997	396.9	806.2	1203.1	0.5915	0.9165	1.5080	420.0
424.0	322.391	0.01900	1.4184	1.4374	401.3	802.2	1203.5	0.5964	0.9077	1.5042	424.0
428.0	336.463	0.01906	1.3591	1.3782	405.7	798.0	1203.7	0.6014	0.8990	1.5004	428.0
432.0	351.00	0.01913	1.30266	1.32179	410.1	793.9	1204.0	0.6063	0.8903	1.4966	432.0
436.0	366.03	0.01919	1.24887	1.26806	414.6	789.7	1204.2	0.6112	0.8816	1.4928	436.0
440.0	381.54	0.01926	1.19761	1.21687	419.0	785.4	1204.4	0.6161	0.8729	1.4890	440.0
444.0	397.56	0.01933	1.14874	1.16806	423.5	781.1	1204.6	0.6210	0.8643	1.4853	444.0
448.0	414.09	0.01940	1.10212	1.12152	428.0	776.7	1204.7	0.6259	0.8557	1.4815	448.0
452.0	431.14	0.01947	1.05764	1.07711	432.5	772.3	1204.8	0.6308	0.8471	1.4778	452.0
456.0	448.73	0.01954	1.01518	1.03472	437.0	767.8	1204.8	0.6356	0.8385	1.4741	456.0

Properties of Saturated H₂O: Temperatures (concluded)   ITEM B 13

Temp Fahr t	Abs Press. Lb per Sq In. p	Specific Volume			Enthalpy			Entropy			Temp Fahr t
		Sat. Liquid v_f	Evap v_{fg}	Sat. Vapor v_g	Sat. Liquid h_f	Evap h_{fg}	Sat. Vapor h_g	Sat. Liquid s_f	Evap s_{fg}	Sat. Vapor s_g	
460.0	466.87	0.01961	0.97463	0.99424	441.5	763.2	1204.8	0.6405	0.8299	1.4704	460.0
464.0	485.56	0.01969	0.93588	0.95557	446.1	758.6	1204.7	0.6454	0.8213	1.4667	464.0
468.0	504.83	0.01976	0.89885	0.91862	450.7	754.0	1204.6	0.6502	0.8127	1.4629	468.0
472.0	524.67	0.01984	0.86345	0.88329	455.2	749.3	1204.5	0.6551	0.8042	1.4592	472.0
476.0	545.11	0.01992	0.82958	0.84950	459.9	744.5	1204.3	0.6599	0.7956	1.4555	476.0
480.0	566.15	0.02000	0.79716	0.81717	464.5	739.6	1204.1	0.6648	0.7871	1.4518	480.0
484.0	587.81	0.02009	0.76613	0.78622	469.1	734.7	1203.8	0.6696	0.7785	1.4481	484.0
488.0	610.10	0.02017	0.73641	0.75658	473.8	729.7	1203.5	0.6745	0.7700	1.4444	488.0
492.0	633.03	0.02026	0.70794	0.72820	478.5	724.6	1203.1	0.6793	0.7614	1.4407	492.0
496.0	656.61	0.02034	0.68065	0.70100	483.2	719.5	1202.7	0.6842	0.7528	1.4370	496.0
500.0	680.86	0.02043	0.65448	0.67492	487.9	714.3	1202.2	0.6890	0.7443	1.4333	500.0
504.0	705.78	0.02053	0.62938	0.64991	492.7	709.0	1201.7	0.6939	0.7357	1.4296	504.0
508.0	731.40	0.02062	0.60530	0.62592	497.5	703.7	1201.1	0.6987	0.7271	1.4258	508.0
512.0	757.72	0.02072	0.58218	0.60289	502.3	698.2	1200.5	0.7036	0.7185	1.4221	512.0
516.0	784.76	0.02081	0.55997	0.58079	507.1	692.7	1199.8	0.7085	0.7099	1.4183	516.0
520.0	812.53	0.02091	0.53864	0.55956	512.0	687.0	1199.0	0.7133	0.7013	1.4146	520.0
524.0	841.04	0.02102	0.51814	0.53916	516.9	681.3	1198.2	0.7182	0.6926	1.4108	524.0
528.0	870.31	0.02112	0.49843	0.51955	521.8	675.5	1197.3	0.7231	0.6839	1.4070	528.0
532.0	900.34	0.02123	0.47947	0.50070	526.8	669.6	1196.4	0.7280	0.6752	1.4032	532.0
536.0	931.17	0.02134	0.46123	0.48257	531.7	663.6	1195.4	0.7329	0.6665	1.3993	536.0
540.0	962.79	0.02146	0.44367	0.46513	536.8	657.5	1194.3	0.7378	0.6577	1.3954	540.0
544.0	995.22	0.02157	0.42677	0.44834	541.8	651.3	1193.1	0.7427	0.6489	1.3915	544.0
548.0	1028.49	0.02169	0.41048	0.43217	546.9	645.0	1191.9	0.7476	0.6400	1.3876	548.0
552.0	1062.59	0.02182	0.39479	0.41660	552.0	638.5	1190.6	0.7525	0.6311	1.3837	552.0
556.0	1097.55	0.02194	0.37966	0.40160	557.2	632.0	1189.2	0.7575	0.6222	1.3797	556.0
560.0	1133.38	0.02207	0.36507	0.38714	562.4	625.3	1187.7	0.7625	0.6132	1.3757	560.0
564.0	1170.10	0.02221	0.35099	0.37320	567.6	618.5	1186.1	0.7674	0.6041	1.3716	564.0
568.0	1207.72	0.02235	0.33741	0.35975	572.9	611.5	1184.5	0.7725	0.5950	1.3675	568.0
572.0	1246.26	0.02249	0.32429	0.34678	578.3	604.5	1182.7	0.7775	0.5859	1.3634	572.0
576.0	1285.74	0.02264	0.31162	0.33426	583.7	597.2	1180.9	0.7825	0.5766	1.3592	576.0
580.0	1326.17	0.02279	0.29937	0.32216	589.1	589.9	1179.0	0.7876	0.5673	1.3550	580.0
584.0	1367.7	0.02295	0.28753	0.31048	594.6	582.4	1176.9	0.7927	0.5580	1.3507	584.0
588.0	1410.0	0.02311	0.27608	0.29919	600.1	574.7	1174.8	0.7978	0.5485	1.3464	588.0
592.0	1453.3	0.02328	0.26499	0.28827	605.7	566.8	1172.6	0.8030	0.5390	1.3420	592.0
596.0	1497.8	0.02345	0.25425	0.27770	611.4	558.8	1170.2	0.8082	0.5293	1.3375	596.0
600.0	1543.2	0.02364	0.24384	0.26747	617.1	550.6	1167.7	0.8134	0.5196	1.3330	600.0
604.0	1589.7	0.02382	0.23374	0.25757	622.9	542.2	1165.1	0.8187	0.5097	1.3284	604.0
608.0	1637.3	0.02402	0.22394	0.24796	628.8	533.6	1162.4	0.8240	0.4997	1.3238	608.0
612.0	1686.1	0.02422	0.21442	0.23865	634.8	524.7	1159.5	0.8294	0.4896	1.3190	612.0
616.6	1735.9	0.02444	0.20516	0.22960	640.8	515.6	1156.4	0.8348	0.4794	1.3141	616.0
620.0	1786.9	0.02466	0.19615	0.22081	646.9	506.3	1153.2	0.8403	0.4689	1.3092	620.0
624.0	1839.0	0.02489	0.18737	0.21226	653.1	496.6	1149.8	0.8458	0.4583	1.3041	624.0
628.0	1892.4	0.02514	0.17880	0.20394	659.5	486.7	1146.1	0.8514	0.4474	1.2988	628.0
632.0	1947.0	0.02539	0.17044	0.19583	665.9	476.4	1142.2	0.8571	0.4364	1.2934	632.0
636.0	2002.8	0.02566	0.16226	0.18792	672.4	465.7	1138.1	0.8628	0.4251	1.2879	636.0
640.0	2059.9	0.02595	0.15427	0.18021	679.1	454.6	1133.7	0.8686	0.4134	1.2821	640.0
644.0	2118.3	0.02625	0.14644	0.17269	685.9	443.1	1129.0	0.8746	0.4015	1.2761	644.0
648.0	2178.1	0.02657	0.13876	0.16534	692.9	431.1	1124.0	0.8806	0.3893	1.2699	648.0
652.0	2239.2	0.02691	0.13124	0.15816	700.0	418.7	1118.7	0.8868	0.3767	1.2634	652.0
656.0	2301.7	0.02728	0.12387	0.15115	707.4	405.7	1113.1	0.8931	0.3637	1.2567	656.0
660.0	2365.7	0.02768	0.11663	0.14431	714.9	392.1	1107.0	0.8995	0.3502	1.2498	660.0
664.0	2431.1	0.02811	0.10947	0.13757	722.9	377.7	1100.6	0.9064	0.3361	1.2425	664.0
668.0	2498.1	0.02858	0.10229	0.13087	731.5	362.1	1093.5	0.9137	0.3210	1.2347	668.0
672.0	2566.6	0.02911	0.09514	0.12424	740.2	345.7	1085.9	0.9212	0.3054	1.2266	672.0
676.0	2636.8	0.02970	0.08799	0.11769	749.2	328.5	1077.6	0.9287	0.2892	1.2179	676.0
680.0	2708.6	0.03037	0.08080	0.11117	758.5	310.1	1068.5	0.9365	0.2720	1.2086	680.0
684.0	2782.1	0.03114	0.07349	0.10463	768.2	290.2	1058.4	0.9447	0.2537	1.1984	684.0
688.0	2857.4	0.03204	0.06595	0.09799	778.8	268.2	1047.0	0.9535	0.2337	1.1872	688.0
692.0	2934.5	0.03313	0.05797	0.09110	790.5	243.1	1033.6	0.9634	0.2110	1.1744	692.0
696.0	3013.4	0.03455	0.04916	0.08371	804.4	212.8	1017.2	0.9749	0.1841	1.1591	696.0
700.0	3094.3	0.03662	0.03857	0.07519	822.4	172.7	995.2	0.9901	0.1490	1.1390	700.0
702.0	3135.5	0.03824	0.03173	0.06997	835.0	144.7	979.7	1.0006	0.1246	1.1252	702.0
704.0	3177.2	0.04108	0.02192	0.06300	854.2	102.0	956.2	1.0169	0.0876	1.1046	704.0
705.0	3198.3	0.04427	0.01304	0.05730	873.0	61.4	934.4	1.0329	0.0527	1.0856	705.0
705.47*	3208.2	0.05078	0.00000	0.05078	906.0	0.0	906.0	1.0612	0.0000	1.0612	705.47*

*Critical temperature

ITEM B 14 Properties of Saturated H₂O: Pressures

Abs. Press. Lb/Sq In. p	Temp Fahr t	Sat. Liquid v_f	Specific Volume Evap v_{fg}	Sat. Vapor v_g	Sat. Liquid h_f	Enthalpy Evap h_{fg}	Sat. Vapor h_g	Sat. Liquid s_f	Entropy Evap s_{fg}	Sat. Vapor s_g	Abs. Press. Lb/Sq In. p
0.08865	32.018	0.016022	3302.4	3302.4	0.0003	1075.5	1075.5	0.0000	2.1872	2.1872	0.08865
0.25	59.323	0.016032	1235.5	1235.5	27.382	1060.1	1087.4	0.0542	2.0425	2.0967	0.25
0.50	79.586	0.016071	641.5	641.5	47.623	1048.6	1096.3	0.0925	1.9446	2.0370	0.50
1.0	101.74	0.016136	333.59	333.60	69.73	1036.1	1105.8	0.1326	1.8455	1.9781	1.0
5.0	162.24	0.016407	73.515	73.532	130.20	1000.9	1131.1	0.2349	1.6094	1.8443	5.0
10.0	193.21	0.016592	38.404	38.420	161.26	982.1	1143.3	0.2836	1.5043	1.7879	10.0
14.696	212.00	0.016719	26.782	26.799	180.17	970.3	1150.5	0.3121	1.4447	1.7568	14.696
15.0	213.03	0.016726	26.274	26.290	181.21	969.7	1150.9	0.3137	1.4415	1.7552	15.0
20.0	227.96	0.016834	20.070	20.087	196.27	960.1	1156.3	0.3358	1.3962	1.7320	20.0
30.0	250.34	0.017009	13.7266	13.7436	218.9	945.2	1164.1	0.3682	1.3313	1.6995	30.0
40.0	267.25	0.017151	10.4794	10.4965	236.1	933.6	1169.8	0.3921	1.2844	1.6765	40.0
50.0	281.02	0.017274	8.4967	8.5140	250.2	923.9	1174.1	0.4112	1.2474	1.6586	50.0
60.0	292.71	0.017383	7.1562	7.1736	262.2	915.4	1177.6	0.4273	1.2167	1.6440	60.0
70.0	302.93	0.017482	6.1875	6.2050	272.7	907.8	1180.6	0.4411	1.1905	1.6316	70.0
80.0	312.04	0.017573	5.4536	5.4711	282.1	900.9	1183.1	0.4534	1.1675	1.6208	80.0
90.0	320.28	0.017659	4.8779	4.8953	290.7	894.6	1185.3	0.4643	1.1470	1.6113	90.0
100.0	327.82	0.017740	4.4133	4.4310	298.5	888.6	1187.2	0.4743	1.1284	1.6027	100.0
110.0	334.79	0.01782	4.0306	4.0484	305.8	883.1	1188.9	0.4834	1.1115	1.5950	110.0
120.0	341.27	0.01789	3.7097	3.7275	312.6	877.8	1190.4	0.4919	1.0960	1.5879	120.0
130.0	347.33	0.01796	3.4364	3.4544	319.0	872.8	1191.7	0.4998	1.0815	1.5813	130.0
140.0	353.04	0.01803	3.2010	3.2190	325.0	868.0	1193.0	0.5071	1.0681	1.5752	140.0
150.0	358.43	0.01809	2.9958	3.0139	330.6	863.4	1194.1	0.5141	1.0554	1.5695	150.0
160.0	363.55	0.01815	2.8155	2.8336	336.1	859.0	1195.1	0.5206	1.0435	1.5641	160.0
170.0	368.42	0.01821	2.6556	2.6738	341.2	854.8	1196.0	0.5269	1.0322	1.5591	170.0
180.0	373.08	0.01827	2.5129	2.5312	346.2	850.7	1196.9	0.5328	1.0215	1.5543	180.0
190.0	377.53	0.01833	2.3847	2.4030	350.9	846.7	1197.6	0.5384	1.0113	1.5498	190.0
200.0	381.80	0.01839	2.2689	2.2873	355.5	842.8	1198.3	0.5438	1.0016	1.5454	200.0
210.0	385.91	0.01844	2.16373	2.18217	359.9	839.1	1199.0	0.5490	0.9923	1.5413	210.0
220.0	389.88	0.01850	2.06779	2.08629	364.2	835.4	1199.6	0.5540	0.9834	1.5374	220.0
230.0	393.70	0.01855	1.97991	1.99846	368.3	831.8	1200.1	0.5588	0.9748	1.5336	230.0
240.0	397.39	0.01860	1.89909	1.91769	372.3	828.4	1200.6	0.5634	0.9665	1.5299	240.0
250.0	400.97	0.01865	1.82452	1.84317	376.1	825.0	1201.1	0.5679	0.9585	1.5264	250.0
260.0	404.44	0.01870	1.75548	1.77418	379.9	821.6	1201.5	0.5722	0.9508	1.5230	260.0
270.0	407.80	0.01875	1.69137	1.71013	383.6	818.3	1201.9	0.5764	0.9433	1.5197	270.0
280.0	411.07	0.01880	1.63169	1.65049	387.1	815.1	1202.3	0.5805	0.9361	1.5166	280.0
290.0	414.25	0.01885	1.57597	1.59482	390.6	812.0	1202.6	0.5844	0.9291	1.5135	290.0
300.0	417.35	0.01889	1.52384	1.54274	394.0	808.9	1202.9	0.5882	0.9223	1.5105	300.0
350.0	431.73	0.01912	1.30642	1.32554	409.8	794.2	1204.0	0.6059	0.8909	1.4968	350.0
400.0	444.60	0.01934	1.14162	1.16095	424.2	780.4	1204.6	0.6217	0.8630	1.4847	400.0
450.0	456.28	0.01954	1.01224	1.03179	437.3	767.5	1204.8	0.6360	0.8378	1.4738	450.0
500.0	467.01	0.01975	0.90787	0.92762	449.5	755.1	1204.7	0.6490	0.8148	1.4639	500.0
550.0	476.94	0.01994	0.82183	0.84177	460.9	743.3	1204.3	0.6611	0.7936	1.4547	550.0
600.0	486.20	0.02013	0.74962	0.76975	471.7	732.0	1203.7	0.6723	0.7738	1.4461	600.0
650.0	494.89	0.02032	0.68811	0.70843	481.9	720.9	1202.8	0.6828	0.7552	1.4381	650.0
700.0	503.08	0.02050	0.63505	0.65556	491.6	710.2	1201.8	0.6928	0.7377	1.4304	700.0
750.8	510.84	0.02069	0.58880	0.60949	500.9	699.8	1200.7	0.7022	0.7210	1.4232	750.0
800.0	518.21	0.02087	0.54809	0.56896	509.8	689.6	1199.4	0.7111	0.7051	1.4163	800.0
850.0	525.24	0.02105	0.51197	0.53302	518.4	679.5	1198.0	0.7197	0.6899	1.4096	850.0
900.0	531.95	0.02123	0.47968	0.50091	526.7	669.7	1196.4	0.7279	0.6753	1.4032	900.0
950.0	538.39	0.02141	0.45064	0.47205	534.7	660.0	1194.7	0.7358	0.6612	1.3970	950.0
1000.0	544.58	0.02159	0.42436	0.44596	542.6	650.4	1192.9	0.7434	0.6476	1.3910	1000.0
1050.0	550.53	0.02177	0.40047	0.42224	550.1	640.9	1191.0	0.7507	0.6344	1.3851	1050.0
1100.0	556.28	0.02195	0.37863	0.40058	557.5	631.5	1189.1	0.7578	0.6216	1.3794	1100.0
1150.0	561.82	0.02214	0.35859	0.38073	564.8	622.2	1187.0	0.7647	0.6091	1.3738	1150.0
1200.0	567.19	0.02232	0.34013	0.36245	571.9	613.0	1184.8	0.7714	0.5969	1.3683	1200.0
1250.0	572.38	0.02250	0.32306	0.34556	578.8	603.8	1182.6	0.7780	0.5850	1.3630	1250.0
1300.0	577.42	0.02269	0.30722	0.32991	585.6	594.6	1180.2	0.7843	0.5733	1.3577	1300.0
1350.0	582.32	0.02288	0.29250	0.31537	592.3	585.4	1177.8	0.7906	0.5620	1.3525	1350.0
1400.0	587.07	0.02307	0.27871	0.30178	598.8	576.5	1175.3	0.7966	0.5507	1.3474	1400.0
1450.0	591.70	0.02327	0.26584	0.28911	605.3	567.4	1172.8	0.8026	0.5397	1.3423	1450.0
1500.0	596.20	0.02346	0.25372	0.27719	611.7	558.4	1170.1	0.8085	0.5288	1.3373	1500.0
1550.0	600.59	0.02366	0.24235	0.26601	618.0	549.4	1167.4	0.8142	0.5182	1.3324	1550.0
1600.0	604.87	0.02387	0.23159	0.25545	624.2	540.3	1164.5	0.8199	0.5076	1.3274	1600.0
1650.0	609.05	0.02407	0.22143	0.24551	630.4	531.3	1161.6	0.8254	0.4971	1.3225	1650.0
1700.0	613.13	0.02428	0.21178	0.23607	636.5	522.2	1158.6	0.8309	0.4867	1.3176	1700.0
1750.0	617.12	0.02450	0.20263	0.22713	642.5	513.1	1155.6	0.8363	0.4765	1.3128	1750.0
1800.0	621.02	0.02472	0.19390	0.21861	648.5	503.8	1152.3	0.8417	0.4662	1.3079	1800.0
1850.0	624.83	0.02495	0.18558	0.21052	654.5	494.6	1149.0	0.8470	0.4561	1.3030	1850.0
1900.0	628.56	0.02517	0.17761	0.20278	660.4	485.2	1145.6	0.8522	0.4459	1.2981	1900.0
1950.0	632.22	0.02541	0.16999	0.19540	666.3	475.8	1142.0	0.8574	0.4358	1.2931	1950.0
2000.0	635.80	0.02565	0.16266	0.18831	672.1	466.2	1138.3	0.8625	0.4256	1.2881	2000.0
2100.0	642.76	0.02615	0.14885	0.17501	683.8	446.7	1130.5	0.8727	0.4053	1.2780	2100.0
2200.0	649.45	0.02669	0.13603	0.16272	695.5	426.7	1122.2	0.8828	0.3848	1.2676	2200.0
2300.0	655.89	0.02727	0.12406	0.15133	707.2	406.0	1113.2	0.8929	0.3640	1.2569	2300.0
2400.0	662.11	0.02790	0.11287	0.14076	719.0	384.8	1103.7	0.9031	0.3430	1.2460	2400.0
2500.0	668.11	0.02859	0.10209	0.13068	731.7	361.6	1093.3	0.9139	0.3206	1.2345	2500.0
2600.0	673.91	0.02938	0.09172	0.12110	744.5	337.6	1082.0	0.9247	0.2977	1.2225	2600.0
2700.0	679.53	0.03029	0.08165	0.11194	757.3	312.3	1069.7	0.9356	0.2741	1.2097	2700.0
2800.0	684.96	0.03134	0.07171	0.10305	770.7	285.1	1055.8	0.9468	0.2491	1.1958	2800.0
2900.0	690.22	0.03262	0.06158	0.09420	785.1	254.7	1039.8	0.9588	0.2215	1.1803	2900.0
3000.0	695.33	0.03428	0.05073	0.08500	801.8	218.4	1020.3	0.9728	0.1891	1.1619	3000.0
3100.0	700.28	0.03681	0.03771	0.07452	824.0	169.3	993.3	0.9914	0.1460	1.1373	3100.0
3200.0	705.08	0.04472	0.01191	0.05663	875.5	56.1	931.6	1.0351	0.0482	1.0832	3200.0
3208.2*	705.47	0.05078	0.00000	0.05078	906.0	0.0	906.0	1.0612	0.0000	1.0612	3208.2*

*Critical pressure

Properties of Superheated Steam   ITEM B 15

Abs Press. Lb/Sq In. (Sat. Temp)		Sat. Water	Sat. Steam	200	250	300	350	400	450	500	600	700	800	900	1000	1100	1200
1 (101.74)	Sh			98.26	148.26	198.26	248.26	298.26	348.26	398.26	498.26	598.26	698.26	798.26	898.26	998.26	1098.26
	v	0.01614	333.6	392.5	422.4	452.3	482.1	511.9	541.7	571.5	631.1	690.7	750.2	809.8	869.4	929.1	988.7
	h	69.73	1105.8	1150.2	1172.9	1195.7	1218.7	1241.8	1265.1	1288.6	1336.1	1384.5	1431.0	1480.8	1531.4	1583.0	1635.4
	s	0.1326	1.9781	2.0509	2.0841	2.1152	2.1445	2.1722	2.1985	2.2237	2.2708	2.3144	2.3512	2.3892	2.4251	2.4592	2.4918
5 (162.24)	Sh			37.76	87.76	137.76	187.76	237.76	287.76	337.76	437.76	537.76	637.76	737.76	837.76	937.76	1037.76
	v	0.01641	73.53	78.14	84.21	90.24	96.25	102.24	108.23	114.21	126.15	138.08	150.01	161.94	173.86	185.78	197.70
	h	130.20	1131.1	1148.6	1171.7	1194.8	1218.0	1241.3	1264.7	1288.2	1335.9	1384.3	1433.6	1483.7	1534.7	1586.7	1639.6
	s	0.2349	1.8443	1.8716	1.9054	1.9369	1.9664	1.9943	2.0208	2.0460	2.0932	2.1369	2.1776	2.2159	2.2521	2.2866	2.3194
10 (193.21)	Sh			6.79	56.79	106.79	156.79	206.79	256.79	306.79	406.79	506.79	606.79	706.79	806.79	906.79	1006.79
	v	0.01659	38.42	38.84	41.93	44.98	48.02	51.03	54.04	57.04	63.03	69.00	74.98	80.94	86.91	92.87	98.84
	h	161.26	1143.3	1146.6	1170.2	1193.7	1217.1	1240.6	1264.1	1287.8	1335.5	1384.0	1433.4	1483.5	1534.6	1586.6	1639.5
	s	0.2836	1.7879	1.7928	1.8273	1.8593	1.8892	1.9173	1.9439	1.9692	2.0166	2.0603	2.1011	2.1394	2.1757	2.2101	2.2430
14.696 (212.00)	Sh				38.00	88.00	138.00	188.00	238.00	288.00	388.00	488.00	588.00	688.00	788.00	888.00	988.00
	v	0.0167	26.828		28.44	30.52	32.61	34.65	36.73	38.75	42.83	46.91	50.97	55.03	59.09	63.19	67.25
	h	180.07	1150.4		1169.2	1192.0	1215.4	1238.9	1262.1	1285.4	1333.0	1381.4	1430.5	1480.4	1531.1	1582.7	1635.1
	s	0.3120	1.7566		1.7838	1.8148	1.8446	1.8727	1.8989	1.9238	1.9709	2.0145	2.0551	2.0932	2.1292	2.1634	2.1960
15 (213.03)	Sh				36.97	86.97	136.97	186.97	236.97	286.97	386.97	486.97	586.97	686.97	786.97	886.97	986.97
	v	0.01673	26.290		27.837	29.899	31.939	33.963	35.977	37.985	41.986	45.978	49.964	53.946	57.926	61.905	65.882
	h	181.21	1150.9		1168.7	1192.5	1216.2	1239.9	1263.6	1287.3	1335.2	1383.8	1433.2	1483.4	1534.5	1586.5	1639.4
	s	0.3137	1.7552		1.7809	1.8134	1.8437	1.8720	1.8988	1.9242	1.9717	2.0155	2.0563	2.0946	2.1309	2.1653	2.1982
20 (227.96)	Sh				22.04	72.04	122.04	172.04	222.04	272.04	372.04	472.04	572.04	672.04	772.04	872.04	972.04
	v	0.01683	20.087		20.788	22.356	23.900	25.428	26.946	28.457	31.466	34.465	37.458	40.447	43.435	46.420	49.405
	h	196.27	1156.3		1167.1	1191.4	1215.4	1239.2	1263.0	1286.9	1334.9	1383.5	1432.9	1483.2	1534.3	1586.3	1639.3
	s	0.3358	1.7320		1.7475	1.7805	1.8111	1.8397	1.8666	1.8921	1.9397	1.9836	2.0244	2.0628	2.0991	2.1336	2.1665
25 (240.07)	Sh				9.93	59.93	109.93	159.93	209.93	259.93	359.93	459.93	559.93	659.93	759.93	859.93	959.93
	v	0.01693	16.301		16.558	17.829	19.076	20.307	21.527	22.740	25.153	27.557	29.954	32.348	34.740	37.130	39.518
	h	208.52	1160.6		1165.6	1190.2	1214.5	1238.5	1262.5	1286.4	1334.6	1383.3	1432.7	1483.0	1534.2	1586.2	1639.2
	s	0.3535	1.7141		1.7212	1.7547	1.7856	1.8145	1.8415	1.8672	1.9149	1.9588	1.9997	2.0382	2.0744	2.1089	2.1418
30 (250.34)	Sh					49.66	99.66	149.66	199.66	249.66	349.66	449.66	549.66	649.66	749.66	849.66	949.66
	v	0.01701	13.744			14.810	15.859	16.892	17.914	18.929	20.945	22.951	24.952	26.949	28.943	30.936	32.927
	h	218.93	1164.1			1189.0	1213.6	1237.8	1261.9	1286.0	1334.2	1383.0	1432.5	1482.8	1534.0	1586.1	1639.0
	s	0.3682	1.6995			1.7334	1.7647	1.7937	1.8210	1.8467	1.8946	1.9386	1.9795	2.0179	2.0543	2.0888	2.1217
35 (259.29)	Sh					40.71	90.71	140.71	190.71	240.71	340.71	440.71	540.71	640.71	740.71	840.71	940.71
	v	0.01708	11.896			12.654	13.562	14.453	15.334	16.207	17.939	19.662	21.379	23.092	24.803	26.512	28.220
	h	228.03	1167.1			1187.8	1212.7	1237.1	1261.3	1285.5	1333.9	1382.8	1432.3	1482.7	1533.9	1586.0	1638.9
	s	0.3809	1.6872			1.7152	1.7468	1.7761	1.8035	1.8294	1.8774	1.9214	1.9624	2.0009	2.0372	2.0717	2.1046
40 (267.25)	Sh					32.75	82.75	132.75	182.75	232.75	332.75	432.75	532.75	632.75	732.75	832.75	932.75
	v	0.01715	10.497			11.036	11.338	12.624	13.398	14.165	15.685	17.195	18.699	20.199	21.697	23.194	24.689
	h	236.14	1169.8			1186.6	1211.7	1236.4	1260.8	1285.0	1333.6	1382.5	1432.1	1482.5	1533.7	1585.8	1638.8
	s	0.3921	1.6765			1.6992	1.7312	1.7608	1.7883	1.8143	1.8624	1.9065	1.9476	1.9860	2.0224	2.0569	2.0899
45 (274.43)	Sh					25.57	75.57	125.57	175.57	225.57	325.57	425.57	525.57	625.57	725.57	825.57	925.57
	v	0.01722	9.403			9.782	10.503	11.206	11.897	12.584	13.939	15.284	16.623	17.959	19.292	20.623	21.954
	h	243.47	1172.1			1185.4	1210.8	1235.7	1260.2	1284.6	1333.3	1382.3	1432.0	1482.4	1533.6	1585.7	1638.8
	s	0.4021	1.6671			1.6849	1.7174	1.7472	1.7749	1.8010	1.8492	1.8934	1.9345	1.9730	2.0094	2.0439	2.0769
50 (281.02)	Sh					18.98	68.98	118.98	168.98	218.98	318.98	418.98	518.98	618.98	718.98	818.98	918.98
	v	0.1727	8.514			8.769	9.424	10.062	10.688	11.306	12.529	13.741	14.947	16.150	17.350	18.549	19.746
	h	250.21	1174.1			1184.1	1209.9	1234.9	1259.6	1284.1	1332.9	1382.0	1431.7	1482.2	1533.4	1585.6	1638.6
	s	0.4112	1.6586			1.6720	1.7048	1.7349	1.7628	1.7890	1.8374	1.8816	1.9227	1.9613	1.9977	2.0322	2.0652
55 (287.07)	Sh					12.93	62.93	112.93	162.93	212.93	312.93	412.93	512.93	612.93	712.93	812.93	912.93
	v	0.01733	7.787			7.947	8.550	9.134	9.706	10.270	11.385	12.489	13.587	14.682	15.775	16.865	17.954
	h	256.42	1176.0			1182.9	1208.9	1234.3	1259.1	1283.6	1332.6	1381.8	1431.6	1482.0	1533.3	1585.5	1638.5
	s	0.4196	1.6510			1.6602	1.6934	1.7238	1.7518	1.7781	1.8267	1.8710	1.9123	1.9507	1.9871	2.0217	2.0546
60 (292.71)	Sh					7.29	57.29	107.29	157.29	207.29	307.29	407.29	507.29	607.29	707.29	807.29	907.29
	v	0.1738	7.174			7.257	7.815	8.354	8.881	9.400	10.425	11.438	12.446	13.450	14.452	15.452	16.450
	h	262.21	1177.6			1181.6	1208.0	1233.5	1258.5	1283.2	1332.3	1381.5	1431.3	1481.8	1533.2	1585.3	1638.4
	s	0.4273	1.6440			1.6492	1.6830	1.7134	1.7417	1.7681	1.8168	1.8612	1.9024	1.9410	1.9774	2.0120	2.0450
65 (297.98)	Sh					2.02	52.02	102.02	152.02	202.02	302.02	402.02	502.02	602.02	702.02	802.02	902.02
	v	0.01743	6.653			6.675	7.195	7.697	8.186	8.667	9.615	10.552	11.484	12.412	13.337	14.261	15.183
	h	267.63	1179.1			1180.3	1207.0	1232.7	1257.9	1282.7	1331.9	1381.3	1431.1	1481.6	1533.0	1585.2	1638.3
	s	0.4344	1.6375			1.6390	1.6731	1.7040	1.7324	1.7590	1.8077	1.8522	1.8935	1.9321	1.9685	2.0031	2.0361
70 (302.93)	Sh						47.07	97.07	147.07	197.07	297.07	397.07	497.07	597.07	697.07	797.07	897.07
	v	0.01748	6.205				6.664	7.133	7.590	8.039	8.922	9.793	10.659	11.522	12.382	13.240	14.097
	h	272.74	1180.6				1206.0	1232.0	1257.3	1282.2	1331.6	1381.0	1430.9	1481.5	1532.9	1585.1	1638.2
	s	0.4411	1.6316				1.6640	1.6951	1.7237	1.7504	1.7993	1.8439	1.8852	1.9238	1.9603	1.9949	2.0279
75 (307.61)	Sh						42.39	92.39	142.39	192.39	292.39	392.39	492.39	592.39	692.39	792.39	892.39
	v	0.01753	5.814				6.204	6.645	7.074	7.494	8.320	9.135	9.945	10.750	11.553	12.355	13.155
	h	277.56	1181.9				1205.0	1231.2	1256.7	1281.7	1331.3	1380.7	1430.7	1481.3	1532.7	1585.0	1638.1
	s	0.4474	1.6260				1.6554	1.6868	1.7156	1.7424	1.7915	1.8361	1.8774	1.9161	1.9526	1.9872	2.0202

Sh = superheat, F
v = specific volume, cu ft per lb

h = enthalpy, Btu per lb
s = entropy, Btu per F per lb

*Values from STEAM TABLES, Properties of Saturated and Superheated Steam
Published by COMBUSTION ENGINEERING, INC., Copyright 1940
**Values interpolated from ASME STEAM TABLES

ITEM B 15 Properties of Superheated Steam (continued)

Abs Press. Lb/Sq In. (Sat. Temp)		Sat. Water	Sat. Steam	350	400	450	500	550	600	700	800	900	1000	1100	1200	1300	1400
80 (312.04)	Sh			37.96	87.96	137.96	187.96	237.96	287.96	387.96	487.96	587.96	687.96	787.96	887.96	987.96	1087.96
	v	0.01757	5.471	5.801	6.218	6.622	7.018	7.408	7.794	8.560	9.319	10.075	10.829	11.581	12.331	13.081	13.829
	h	282.15	1183.1	1204.0	1230.5	1256.1	1281.3	1306.2	1330.9	1380.5	1430.5	1481.1	1532.6	1584.9	1638.0	1692.0	1746.8
	s	0.4534	1.6208	1.6473	1.6790	1.7080	1.7349	1.7602	1.7842	1.8289	1.8702	1.9089	1.9454	1.9800	2.0131	2.0446	2.0750
85 (316.26)	Sh			33.74	83.74	133.74	183.74	233.74	283.74	383.74	483.74	583.74	683.74	783.74	883.74	983.74	1083.74
	v	0.01762	5.167	5.445	5.840	6.223	6.597	6.966	7.330	8.052	8.768	9.480	10.190	10.898	11.604	12.310	13.014
	h	286.52	1184.2	1203.0	1229.7	1255.5	1280.8	1305.8	1330.6	1380.2	1430.3	1481.0	1532.4	1584.7	1637.9	1691.9	1746.8
	s	0.4590	1.6159	1.6396	1.6716	1.7008	1.7279	1.7532	1.7772	1.8220	1.8634	1.9021	1.9386	1.9733	2.0063	2.0379	2.0682
90 (320.28)	Sh			29.72	79.72	129.72	179.72	229.72	279.72	379.72	479.72	579.72	679.72	779.72	879.72	979.72	1079.72
	v	0.01766	4.895	5.128	5.505	5.869	6.223	6.572	6.917	7.600	8.277	8.950	9.621	10.290	10.958	11.625	12.290
	h	290.69	1185.3	1202.0	1228.9	1254.9	1280.3	1305.4	1330.2	1380.0	1430.1	1480.8	1532.3	1584.6	1637.8	1691.8	1746.7
	s	0.4643	1.6113	1.6323	1.6646	1.6940	1.7212	1.7467	1.7707	1.8156	1.8570	1.8957	1.9323	1.9669	2.0000	2.0316	2.0619
95 (324.13)	Sh			25.87	75.87	125.87	175.87	225.87	275.87	375.87	475.87	575.87	675.87	775.87	875.87	975.87	1075.87
	v	0.01770	4.651	4.845	5.205	5.551	5.889	6.221	6.548	7.196	7.838	8.477	9.113	9.747	10.380	11.012	11.643
	h	294.70	1186.2	1200.9	1228.1	1254.3	1279.8	1305.0	1329.9	1379.7	1429.9	1480.6	1532.1	1584.5	1637.7	1691.7	1746.6
	s	0.4694	1.6069	1.6253	1.6580	1.6876	1.7149	1.7404	1.7645	1.8094	1.8509	1.8897	1.9262	1.9609	1.9940	2.0256	2.0559
100 (327.82)	Sh			22.18	72.18	122.18	172.18	222.18	272.18	372.18	472.18	572.18	672.18	772.18	872.18	972.18	1072.18
	v	0.01774	4.431	4.590	4.935	5.266	5.588	5.904	6.216	6.833	7.443	8.050	8.655	9.258	9.860	10.460	11.060
	h	298.54	1187.2	1199.9	1227.4	1253.7	1279.3	1304.6	1329.6	1379.5	1429.7	1480.4	1532.0	1584.4	1637.6	1691.6	1746.5
	s	0.4743	1.6027	1.6187	1.6516	1.6814	1.7088	1.7344	1.7586	1.8036	1.8451	1.8839	1.9205	1.9552	1.9883	2.0199	2.0502
105 (331.37)	Sh			18.63	68.63	118.63	168.63	218.63	268.63	368.63	468.63	568.63	668.63	768.63	868.63	968.63	1068.63
	v	0.01778	4.231	4.359	4.690	5.007	5.315	5.617	5.915	6.504	7.086	7.665	8.241	8.816	9.389	9.961	10.532
	h	302.24	1188.0	1198.8	1226.6	1253.1	1278.8	1304.2	1329.2	1379.2	1429.4	1480.3	1531.8	1584.2	1637.5	1691.5	1746.4
	s	0.4790	1.5988	1.6122	1.6455	1.6755	1.7031	1.7288	1.7530	1.7981	1.8396	1.8785	1.9151	1.9498	1.9828	2.0145	2.0448
110 (334.79)	Sh			15.21	65.21	115.21	165.21	215.21	265.21	365.21	465.21	565.21	665.21	765.21	865.21	965.21	1065.21
	v	0.01782	4.048	4.149	4.468	4.772	5.068	5.357	5.642	6.205	6.761	7.314	7.865	8.413	8.961	9.507	10.053
	h	305.80	1188.9	1197.7	1225.8	1252.5	1278.3	1303.8	1328.9	1379.0	1429.2	1480.1	1531.7	1584.1	1637.4	1691.4	1746.4
	s	0.4834	1.5950	1.6061	1.6396	1.6698	1.6975	1.7233	1.7476	1.7928	1.8344	1.8732	1.9099	1.9446	1.9777	2.0093	2.0397
115 (338.08)	Sh			11.92	61.92	111.92	161.92	211.92	261.92	361.92	461.92	561.92	661.92	761.92	861.92	961.92	1061.92
	v	0.01785	3.881	3.957	4.265	4.558	4.841	5.119	5.392	5.932	6.465	6.994	7.521	8.046	8.570	9.093	9.615
	h	309.25	1189.6	1196.7	1225.0	1251.8	1277.9	1303.3	1328.6	1378.7	1429.0	1479.9	1531.6	1584.0	1637.2	1691.4	1746.3
	s	0.4877	1.5913	1.6001	1.6340	1.6644	1.6922	1.7181	1.7425	1.7877	1.8294	1.8682	1.9049	1.9396	1.9727	2.0044	2.0347
120 (341.27)	Sh			8.73	58.73	108.73	158.73	208.73	258.73	358.73	458.73	558.73	658.73	758.73	858.73	958.73	1058.73
	v	0.01789	3.7275	3.7815	4.0786	4.3610	4.6341	4.9009	5.1637	5.6813	6.1928	6.7006	7.2060	7.7096	8.2119	8.7130	9.2134
	h	312.58	1190.4	1195.6	1224.1	1251.2	1277.4	1302.9	1328.2	1378.4	1428.8	1479.8	1531.4	1583.9	1637.1	1691.3	1746.2
	s	0.4919	1.5879	1.5943	1.6286	1.6592	1.6872	1.7132	1.7376	1.7829	1.8246	1.8635	1.9001	1.9349	1.9680	1.9996	2.0300
130 (347.33)	Sh			2.67	52.67	102.67	152.67	202.67	252.67	352.67	452.67	552.67	652.67	752.67	852.67	952.67	1052.67
	v	0.01796	3.4544	3.4699	3.7489	4.0129	4.2672	4.5151	4.7589	5.2384	5.7118	6.1814	6.6486	7.1140	7.5781	8.0411	8.5033
	h	318.95	1191.7	1193.4	1222.5	1249.9	1276.4	1302.1	1327.5	1377.9	1428.4	1479.4	1531.1	1583.6	1636.9	1691.1	1746.1
	s	0.4998	1.5813	1.5833	1.6182	1.6493	1.6775	1.7037	1.7283	1.7737	1.8155	1.8545	1.8911	1.9259	1.9591	1.9907	2.0211
140 (353.04)	Sh				46.96	96.96	146.96	196.96	246.96	346.96	446.96	546.96	646.96	746.96	846.96	946.96	1046.96
	v	0.01803	3.2190		3.4661	3.7143	3.9526	4.1844	4.4119	4.8588	5.2995	5.7364	6.1709	6.6036	7.0349	7.4652	7.8946
	h	324.96	1193.0		1220.8	1248.7	1275.3	1301.3	1326.8	1377.4	1428.0	1479.1	1530.8	1583.4	1636.7	1690.9	1745.9
	s	0.5071	1.5752		1.6085	1.6400	1.6686	1.6949	1.7196	1.7652	1.8071	1.8461	1.8828	1.9176	1.9508	1.9825	2.0129
150 (358.43)	Sh				41.57	91.57	141.57	191.57	241.57	341.57	441.57	541.57	641.57	741.57	841.57	941.57	1041.57
	v	0.01809	3.0139		3.2208	3.4555	3.6799	3.8978	4.1112	4.5298	4.9421	5.3507	5.7568	6.1612	6.5642	6.9661	7.3671
	h	330.65	1194.1		1219.1	1247.4	1274.3	1300.5	1326.1	1376.9	1427.6	1478.7	1530.5	1583.1	1636.5	1690.7	1745.7
	s	0.5141	1.5695		1.5993	1.6313	1.6602	1.6867	1.7115	1.7573	1.7992	1.8383	1.8751	1.9099	1.9431	1.9748	2.0052
160 (363.55)	Sh				36.45	86.45	136.45	186.45	236.45	336.45	436.45	536.45	636.45	736.45	836.45	936.45	1036.45
	v	0.01815	2.8336		3.0060	3.2288	3.4413	3.6469	3.8480	4.2420	4.6295	5.0132	5.3945	5.7741	6.1522	6.5293	6.9055
	h	336.07	1195.1		1217.4	1246.0	1273.3	1299.6	1325.4	1376.4	1427.2	1478.4	1530.3	1582.9	1636.3	1690.5	1745.6
	s	0.5206	1.5641		1.5906	1.6231	1.6522	1.6790	1.7039	1.7499	1.7919	1.8310	1.8678	1.9027	1.9359	1.9676	1.9980
170 (368.42)	Sh				31.58	81.58	131.58	181.58	231.58	331.58	431.58	531.58	631.58	731.58	831.58	931.58	1031.58
	v	0.01821	2.6738		2.8162	3.0288	3.2306	3.4255	3.6158	3.9879	4.3536	4.7155	5.0749	5.4325	5.7888	6.1440	6.4983
	h	341.24	1196.0		1215.6	1244.7	1272.2	1298.8	1324.7	1375.8	1426.8	1478.0	1530.0	1582.6	1636.1	1690.4	1745.4
	s	0.5269	1.5591		1.5823	1.6152	1.6447	1.6717	1.6968	1.7428	1.7850	1.8241	1.8610	1.8959	1.9291	1.9608	1.9913
180 (373.08)	Sh				26.92	76.92	126.92	176.92	226.92	326.92	426.92	526.92	626.92	726.92	826.92	926.92	1026.92
	v	0.01827	2.5312		2.6474	2.8508	3.0433	3.2286	3.4093	3.7621	4.1084	4.4508	4.7907	5.1289	5.4657	5.8014	6.1363
	h	346.19	1196.9		1213.8	1243.4	1271.2	1297.9	1324.0	1375.3	1426.3	1477.7	1529.7	1582.4	1635.9	1690.2	1745.3
	s	0.5328	1.5543		1.5743	1.6078	1.6376	1.6647	1.6900	1.7362	1.7784	1.8176	1.8545	1.8894	1.9227	1.9545	1.9849
190 (377.53)	Sh				22.47	72.47	122.47	172.47	222.47	322.47	422.47	522.47	622.47	722.47	822.47	922.47	1022.47
	v	0.01833	2.4030		2.4961	2.6915	2.8756	3.0525	3.2246	3.5601	3.8889	4.2140	4.5365	4.8572	5.1766	5.4949	5.8124
	h	350.94	1197.6		1212.0	1242.0	1270.1	1297.1	1323.3	1374.8	1425.9	1477.4	1529.4	1582.1	1635.7	1690.0	1745.1
	s	0.5384	1.5498		1.5667	1.6006	1.6307	1.6581	1.6835	1.7299	1.7722	1.8115	1.8484	1.8834	1.9166	1.9484	1.9789
200 (381.80)	Sh				18.20	68.20	118.20	168.20	218.20	318.20	418.20	518.20	618.20	718.20	818.20	918.20	1018.20
	v	0.01839	2.2873		2.3598	2.5480	2.7247	2.8939	3.0583	3.3783	3.6915	4.0008	4.3077	4.6128	4.9165	5.2191	5.5209
	h	355.51	1198.3		1210.1	1240.6	1269.0	1296.2	1322.6	1374.3	1425.5	1477.0	1529.1	1581.9	1635.4	1689.8	1745.0
	s	0.5438	1.5454		1.5593	1.5938	1.6242	1.6518	1.6773	1.7239	1.7663	1.8057	1.8426	1.8776	1.9109	1.9427	1.9732

Sh = superheat, F
v = specific volume, cu ft per lb
h = enthalpy, Btu per lb
s = entropy, Btu per F per lb

Abs Press. Lb/Sq In. (Sat. Temp)		Sat. Water	Sat. Steam	400	450	500	550	600	700	800	900	1000	1100	1200	1300	1400	1500
																	Temperature — Degrees Fahrenheit
210 (385.91)	Sh			14.09	64.09	114.09	164.09	214.09	314.09	414.09	514.09	614.09	714.09	814.09	914.09	1014.09	1114.09
	v	0.01844	2.1822	2.2364	2.4181	2.5880	2.7504	2.9078	3.2137	3.5128	3.8080	4.1007	4.3915	4.6811	4.9695	5.2571	5.5440
	h	359.91	1199.0	1208.02	1239.2	1268.0	1295.3	1321.9	1373.7	1425.1	1476.7	1528.8	1581.6	1635.2	1689.6	1744.8	1800.8
	s	0.5490	1.5413	1.5522	1.5872	1.6180	1.6458	1.6715	1.7182	1.7607	1.8001	1.8371	1.8721	1.9054	1.9372	1.9677	1.9970
220 (389.88)	Sh			10.12	60.12	110.12	160.12	210.12	310.12	410.12	510.12	610.12	710.12	810.12	910.12	1010.12	1110.12
	v	0.01850	2.0863	2.1240	2.2999	2.4638	2.6199	2.7710	3.0642	3.3504	3.6327	3.9125	4.1905	4.4671	4.7426	5.0173	5.2913
	h	364.17	1199.6	1206.3	1237.8	1266.9	1294.5	1321.2	1373.2	1424.7	1476.3	1528.5	1581.4	1635.0	1689.4	1744.7	1800.6
	s	0.5540	1.5374	1.5453	1.5808	1.6120	1.6400	1.6658	1.7128	1.7553	1.7948	1.8318	1.8668	1.9002	1.9320	1.9625	1.9919
230 (393.70)	Sh			6.30	56.30	106.30	156.30	206.30	306.30	406.30	506.30	606.30	706.30	806.30	906.30	1006.30	1106.30
	v	0.01855	1.9985	2.0212	2.1919	2.3503	2.5008	2.6461	2.9276	3.2020	3.4726	3.7406	4.0068	4.2717	4.5355	4.7984	5.0606
	h	368.28	1200.1	1204.4	1236.3	1265.7	1293.6	1320.4	1372.7	1424.2	1476.0	1528.2	1581.1	1634.8	1689.3	1744.5	1800.5
	s	0.5588	1.5336	1.5385	1.5747	1.6062	1.6344	1.6604	1.7075	1.7502	1.7897	1.8268	1.8618	1.8952	1.9270	1.9576	1.9869
240 (397.39)	Sh			2.61	52.61	102.61	152.61	202.61	302.61	402.61	502.61	602.61	702.61	802.61	902.61	1002.61	1102.61
	v	0.01860	1.9177	1.9268	2.0928	2.2462	2.3915	2.5316	2.8024	3.0661	3.3259	3.5831	3.8385	4.0926	4.3456	4.5977	4.8492
	h	372.27	1200.6	1202.4	1234.9	1264.6	1292.7	1319.7	1372.1	1423.8	1475.6	1527.9	1580.9	1634.6	1689.1	1744.3	1800.4
	s	0.5634	1.5299	1.5320	1.5687	1.6006	1.6291	1.6552	1.7025	1.7452	1.7848	1.8219	1.8570	1.8904	1.9223	1.9528	1.9822
250 (400.97)	Sh				49.03	99.03	149.03	199.03	299.03	399.03	499.03	599.03	699.03	799.03	899.03	999.03	1099.03
	v	0.01865	1.8432		2.0016	2.1504	2.2909	2.4262	2.6872	2.9410	3.1909	3.4382	3.6837	3.9278	4.1709	4.4131	4.6546
	h	376.14	1201.1		1233.4	1263.5	1291.8	1319.0	1371.6	1423.4	1475.3	1527.7	1580.6	1634.4	1688.9	1744.2	1800.2
	s	0.5679	1.5264		1.5629	1.5951	1.6239	1.6502	1.6976	1.7405	1.7801	1.8173	1.8524	1.8858	1.9177	1.9482	1.9776
260 (404.44)	Sh				45.56	95.56	145.56	195.56	295.56	395.56	495.56	595.56	695.56	795.56	895.56	995.56	1095.56
	v	0.01870	1.7742		1.9173	2.0619	2.1981	2.3289	2.5808	2.8256	3.0663	3.3044	3.5408	3.7758	4.0097	4.2427	4.4750
	h	379.90	1201.5		1231.9	1262.4	1290.9	1318.2	1371.1	1423.0	1474.9	1527.3	1580.4	1634.2	1688.7	1744.0	1800.1
	s	0.5722	1.5230		1.5573	1.5899	1.6189	1.6453	1.6930	1.7359	1.7756	1.8128	1.8480	1.8814	1.9133	1.9439	1.9732
270 (407.80)	Sh				42.20	92.20	142.20	192.20	292.20	392.20	492.20	592.20	692.20	792.20	892.20	992.20	1092.20
	v	0.01875	1.7101		1.8391	1.9799	2.1121	2.2388	2.4824	2.7186	2.9509	3.1806	3.4084	3.6349	3.8603	4.0849	4.3087
	h	383.56	1201.9		1230.4	1261.2	1290.0	1317.5	1370.5	1422.6	1474.6	1527.1	1580.1	1634.0	1688.5	1743.9	1800.0
	s	0.5764	1.5197		1.5518	1.5848	1.6140	1.6406	1.6885	1.7315	1.7713	1.8085	1.8437	1.8771	1.9090	1.9396	1.9690
280 (411.07)	Sh				38.93	88.93	138.93	188.93	288.93	388.93	488.93	588.93	688.93	788.93	888.93	988.93	1088.93
	v	0.01880	1.6505		1.7665	1.9037	2.0322	2.1551	2.3909	2.6194	2.8437	3.0655	3.2855	3.5042	3.7217	3.9384	4.1543
	h	387.12	1202.3		1228.8	1260.0	1289.1	1316.8	1370.0	1422.1	1474.2	1526.8	1579.9	1633.8	1688.4	1743.7	1799.9
	s	0.5805	1.5166		1.5464	1.5798	1.6093	1.6361	1.6841	1.7273	1.7671	1.8043	1.8395	1.8730	1.9050	1.9356	1.9649
290 (414.25)	Sh				35.75	85.75	135.75	185.75	285.75	385.75	485.75	585.75	685.75	785.75	885.75	985.75	1085.75
	v	0.01885	1.5948		1.6988	1.8327	1.9578	2.0772	2.3058	2.5269	2.7440	2.9585	3.1711	3.3824	3.5926	3.8019	4.0106
	h	390.60	1202.6		1227.3	1258.9	1288.1	1316.0	1369.5	1421.7	1473.9	1526.5	1579.6	1633.5	1688.2	1743.6	1799.7
	s	0.5844	1.5135		1.5412	1.5750	1.6048	1.6317	1.6799	1.7232	1.7630	1.8003	1.8356	1.8690	1.9010	1.9316	1.9610
300 (417.35)	Sh				32.65	82.65	132.65	182.65	282.65	382.65	482.65	582.65	682.65	782.65	882.65	982.65	1082.65
	v	0.01889	1.5427		1.6356	1.7665	1.8883	2.0044	2.2263	2.4407	2.6509	2.8585	3.0643	3.2688	3.4721	3.6746	3.8764
	h	393.99	1202.9		1225.7	1257.7	1287.2	1315.2	1368.9	1421.3	1473.6	1526.2	1579.4	1633.3	1688.0	1743.4	1799.6
	s	0.5882	1.5105		1.5361	1.5703	1.6003	1.6274	1.6758	1.7192	1.7591	1.7964	1.8317	1.8652	1.8972	1.9278	1.9572
310 (420.36)	Sh				29.64	79.64	129.64	179.64	279.64	379.64	479.64	579.64	679.64	779.64	879.64	979.64	1079.64
	v	0.01894	1.4939		1.5763	1.7044	1.8233	1.9363	2.1520	2.3600	2.5638	2.7650	2.9644	3.1625	3.3594	3.5555	3.7509
	h	397.30	1203.2		1224.1	1256.5	1286.3	1314.5	1368.4	1420.9	1473.2	1525.9	1579.2	1633.1	1687.8	1743.3	1799.4
	s	0.5920	1.5076		1.5311	1.5657	1.5960	1.6233	1.6719	1.7153	1.7553	1.7927	1.8280	1.8615	1.8935	1.9241	1.9536
320 (423.31)	Sh				26.69	76.69	126.69	176.69	276.69	376.69	476.69	576.69	676.69	776.69	876.69	976.69	1076.69
	v	0.01899	1.4480		1.5207	1.6462	1.7623	1.8725	2.0823	2.2843	2.4821	2.6774	2.8708	3.0628	3.2538	3.4438	3.6332
	h	400.53	1203.4		1222.5	1255.2	1285.3	1313.7	1367.8	1420.5	1472.9	1525.6	1578.9	1632.9	1687.6	1743.1	1799.3
	s	0.5956	1.5048		1.5261	1.5612	1.5918	1.6192	1.6680	1.7116	1.7516	1.7890	1.8243	1.8579	1.8899	1.9206	1.9500
330 (426.18)	Sh				23.82	73.82	123.82	173.82	273.82	373.82	473.82	573.82	673.82	773.82	873.82	973.82	1073.82
	v	0.01903	1.4048		1.4684	1.5915	1.7050	1.8125	2.0168	2.2132	2.4054	2.5950	2.7828	2.9692	3.1545	3.3389	3.5227
	h	403.70	1203.6		1220.9	1254.0	1284.4	1313.0	1367.3	1420.0	1472.5	1525.3	1578.7	1632.7	1687.5	1742.9	1799.2
	s	0.5991	1.5021		1.5213	1.5568	1.5876	1.6153	1.6643	1.7079	1.7480	1.7855	1.8208	1.8544	1.8864	1.9171	1.9466
340 (428.99)	Sh				21.01	71.01	121.01	171.01	271.01	371.01	471.01	571.01	671.01	771.01	871.01	971.01	1071.01
	v	0.01908	1.3640		1.4191	1.5399	1.6511	1.7561	1.9552	2.1463	2.3333	2.5175	2.7000	2.8811	3.0611	3.2402	3.4186
	h	406.80	1203.8		1219.2	1252.8	1283.4	1312.2	1366.7	1419.6	1472.2	1525.0	1578.4	1632.5	1687.3	1742.8	1799.0
	s	0.6026	1.4994		1.5165	1.5525	1.5836	1.6114	1.6606	1.7044	1.7445	1.7820	1.8174	1.8510	1.8831	1.9138	1.9432
350 (431.73)	Sh				18.27	68.27	118.27	168.27	268.27	368.27	468.27	568.27	668.27	768.27	868.27	968.27	1068.27
	v	0.01912	1.3255		1.3725	1.4913	1.6002	1.7028	1.8970	2.0832	2.2652	2.4445	2.6219	2.7980	2.9730	3.1471	3.3205
	h	409.83	1204.0		1217.5	1251.5	1282.4	1311.4	1366.2	1419.2	1471.8	1524.7	1578.2	1632.3	1687.1	1742.6	1798.9
	s	0.6059	1.4968		1.5119	1.5483	1.5797	1.6077	1.6571	1.7009	1.7411	1.7787	1.8141	1.8477	1.8798	1.9105	1.9400
360 (434.41)	Sh				15.59	65.59	115.59	165.59	265.59	365.59	465.59	565.59	665.59	765.59	865.59	965.59	1065.59
	v	0.01917	1.2891		1.3285	1.4454	1.5521	1.6525	1.8421	2.0237	2.2009	2.3755	2.5482	2.7196	2.8898	3.0592	3.2279
	h	412.81	1204.1		1215.8	1250.3	1281.5	1310.6	1365.6	1418.7	1471.5	1524.4	1577.9	1632.1	1686.9	1742.5	1798.8
	s	0.6092	1.4943		1.5073	1.5441	1.5758	1.6040	1.6536	1.6976	1.7379	1.7754	1.8109	1.8445	1.8766	1.9073	1.9368
380 (439.61)	Sh				10.39	60.39	110.39	160.39	260.39	360.39	460.39	560.39	660.39	760.39	860.39	960.39	1060.39
	v	0.01925	1.2218		1.2472	1.3606	1.4635	1.5598	1.7410	1.9139	2.0825	2.2484	2.4124	2.5750	2.7366	2.8973	3.0572
	h	418.59	1204.4		1212.4	1247.7	1279.5	1309.0	1364.5	1417.9	1470.8	1523.8	1577.4	1631.6	1686.5	1742.2	1798.5
	s	0.6156	1.4894		1.4982	1.5360	1.5683	1.5969	1.6470	1.6911	1.7315	1.7692	1.8047	1.8384	1.8705	1.9012	1.9307

Sh = superheat, F
v = specific volume, cu ft per lb
h = enthalpy, Btu per lb
s = entropy, Btu per F per lb

ITEM B 15 Properties of Superheated Steam (continued)

Abs Press. Lb/Sq In. (Sat. Temp)		Sat. Water	Sat. Steam	450	500	550	600	650	700	800	900	1000	1100	1200	1300	1400	1500
400 (444.60)	Sh			5.40	55.40	105.40	155.40	205.40	255.40	355.40	455.40	555.40	655.40	755.40	855.40	955.40	1055.40
	v	0.01934	1.1610	1.1738	1.2841	1.3836	1.4763	1.5646	1.6499	1.8151	1.9759	2.1339	2.2901	2.4450	2.5987	2.7515	2.9037
	h	424.17	1204.6	1208.8	1245.1	1277.5	1307.4	1335.9	1363.4	1417.0	1470.1	1523.3	1576.9	1631.2	1686.2	1741.9	1798.2
	s	0.6217	1.4847	1.4894	1.5282	1.5611	1.5901	1.6163	1.6406	1.6850	1.7255	1.7632	1.7988	1.8325	1.8647	1.8955	1.9250
420 (449.40)	Sh			.60	50.60	100.60	150.60	200.60	250.60	350.60	450.60	550.60	650.60	750.60	850.60	950.60	1050.60
	v	0.01942	1.1057	1.1071	1.2148	1.3113	1.4007	1.4856	1.5676	1.7258	1.8795	2.0304	2.1795	2.3273	2.4739	2.6196	2.7647
	h	429.56	1204.7	1205.2	1242.4	1275.4	1305.8	1334.5	1362.3	1416.2	1469.4	1522.7	1576.4	1630.8	1685.8	1741.6	1798.0
	s	0.6276	1.4802	1.4808	1.5206	1.5542	1.5835	1.6100	1.6345	1.6791	1.7197	1.7575	1.7932	1.8269	1.8591	1.8899	1.9195
440 (454.03)	Sh				45.97	95.97	145.97	195.97	245.97	345.97	445.97	545.97	645.97	745.97	845.97	945.97	1045.97
	v	0.01950	1.0554		1.1517	1.2454	1.3319	1.4138	1.4926	1.6445	1.7918	1.9363	2.0790	2.2203	2.3605	2.4998	2.6384
	h	434.77	1204.8		1239.7	1273.4	1304.2	1333.2	1361.1	1415.3	1468.7	1522.1	1575.9	1630.4	1685.5	1741.2	1797.7
	s	0.6332	1.4759		1.5132	1.5474	1.5772	1.6040	1.6286	1.6734	1.7142	1.7521	1.7878	1.8216	1.8538	1.8847	1.9143
460 (458.50)	Sh				41.50	91.50	141.50	191.50	241.50	341.50	441.50	541.50	641.50	741.50	841.50	941.50	1041.50
	v	0.01959	1.0092		1.0939	1.1852	1.2691	1.3482	1.4242	1.5703	1.7117	1.8504	1.9872	2.1226	2.2569	2.3903	2.5230
	h	439.83	1204.8		1236.9	1271.3	1302.5	1331.8	1360.0	1414.4	1468.0	1521.5	1575.4	1629.9	1685.1	1740.9	1797.4
	s	0.6387	1.4718		1.5060	1.5409	1.5711	1.5982	1.6230	1.6680	1.7089	1.7469	1.7826	1.8165	1.8488	1.8797	1.9093
480 (462.82)	Sh				37.18	87.18	137.18	187.18	237.18	337.18	437.18	537.18	637.18	737.18	837.18	937.18	1037.18
	v	0.01967	0.9668		1.0409	1.1300	1.2115	1.2881	1.3615	1.5023	1.6384	1.7716	1.9030	2.0330	2.1619	2.2900	2.4173
	h	444.75	1204.8		1234.1	1269.1	1300.8	1330.5	1358.8	1413.6	1467.3	1520.9	1574.9	1629.5	1684.7	1740.6	1797.2
	s	0.6439	1.4677		1.4990	1.5346	1.5652	1.5925	1.6176	1.6628	1.7038	1.7419	1.7777	1.8116	1.8439	1.8748	1.9045
500 (467.01)	Sh				32.99	82.99	132.99	182.99	232.99	332.99	432.99	532.99	632.99	732.99	832.99	932.99	1032.99
	v	0.01975	0.9276		0.9919	1.0791	1.1584	1.2327	1.3037	1.4397	1.5708	1.6992	1.8256	1.9507	2.0746	2.1977	2.3200
	h	449.52	1204.7		1231.2	1267.0	1299.1	1329.1	1357.7	1412.7	1466.6	1520.3	1574.4	1629.1	1684.4	1740.3	1796.9
	s	0.6490	1.4639		1.4921	1.5284	1.5595	1.5871	1.6123	1.6578	1.6990	1.7371	1.7730	1.8069	1.8393	1.8702	1.8998
520 (471.07)	Sh				28.93	78.93	128.93	178.93	228.93	328.93	428.93	528.93	628.93	728.93	828.93	928.93	1028.93
	v	0.01982	0.8914		0.9466	1.0321	1.1094	1.1816	1.2504	1.3819	1.5085	1.6323	1.7542	1.8746	1.9940	2.1125	2.2302
	h	454.18	1204.5		1228.3	1264.8	1297.4	1327.7	1356.5	1411.8	1465.9	1519.7	1573.9	1628.7	1684.0	1740.0	1796.7
	s	0.6540	1.4601		1.4853	1.5223	1.5539	1.5818	1.6072	1.6530	1.6943	1.7325	1.7684	1.8024	1.8348	1.8657	1.8954
540 (475.01)	Sh				24.99	74.99	124.99	174.99	224.99	324.99	424.99	524.99	624.99	724.99	824.99	924.99	1024.99
	v	0.01990	0.8577		0.9045	0.9884	1.0640	1.1342	1.2010	1.3284	1.4508	1.5704	1.6880	1.8042	1.9193	2.0336	2.1471
	h	458.71	1204.4		1225.3	1262.5	1295.7	1326.3	1355.3	1410.9	1465.1	1519.1	1573.4	1628.2	1683.6	1739.7	1796.4
	s	0.6587	1.4565		1.4786	1.5164	1.5485	1.5767	1.6023	1.6483	1.6897	1.7280	1.7640	1.7981	1.8305	1.8615	1.8911
560 (478.84)	Sh				21.16	71.16	121.16	171.16	221.16	321.16	421.16	521.16	621.16	721.16	821.16	921.16	1021.16
	v	0.01998	0.8264		0.8653	0.9479	1.0217	1.0902	1.1552	1.2787	1.3972	1.5129	1.6266	1.7388	1.8500	1.9603	2.0699
	h	463.14	1204.2		1222.2	1260.3	1293.9	1324.9	1354.2	1410.0	1464.4	1518.6	1572.9	1627.8	1683.3	1739.4	1796.1
	s	0.6634	1.4529		1.4720	1.5106	1.5431	1.5717	1.5975	1.6438	1.6853	1.7237	1.7597	1.7938	1.8263	1.8573	1.8870
580 (482.57)	Sh				17.43	67.43	117.43	167.43	217.43	317.43	417.43	517.43	617.43	717.43	817.43	917.43	1017.43
	v	0.02006	0.7971		0.8287	0.9100	0.9824	1.0492	1.1125	1.2324	1.3473	1.4593	1.5693	1.6780	1.7855	1.8921	1.9980
	h	467.47	1203.9		1219.1	1258.0	1292.1	1323.4	1353.0	1409.2	1463.7	1518.0	1572.4	1627.4	1682.9	1739.1	1795.9
	s	0.6679	1.4495		1.4654	1.5049	1.5380	1.5668	1.5929	1.6394	1.6811	1.7196	1.7556	1.7898	1.8223	1.8533	1.8831
600 (486.20)	Sh				13.80	63.80	113.80	163.80	213.80	313.80	413.80	513.80	613.80	713.80	813.80	913.80	1013.80
	v	0.02013	0.7697		0.7944	0.8746	0.9456	1.0109	1.0726	1.1892	1.3008	1.4093	1.5160	1.6211	1.7252	1.8284	1.9309
	h	471.70	1203.7		1215.9	1255.6	1290.3	1322.0	1351.8	1408.3	1463.0	1517.4	1571.9	1627.0	1682.6	1738.8	1795.6
	s	0.6723	1.4461		1.4590	1.4993	1.5329	1.5621	1.5884	1.6351	1.6769	1.7155	1.7517	1.7859	1.8184	1.8494	1.8792
650 (494.89)	Sh				5.11	55.11	105.11	155.11	205.11	305.11	405.11	505.11	605.11	705.11	805.11	905.11	1005.11
	v	0.02032	0.7084		0.7173	0.7954	0.8634	0.9254	0.9835	1.0929	1.1969	1.2979	1.3969	1.4944	1.5909	1.6864	1.7813
	h	481.89	1202.8		1207.6	1249.6	1285.7	1318.3	1348.7	1406.0	1461.2	1515.9	1570.7	1625.9	1681.6	1738.0	1794.9
	s	1.6828	1.4381		1.4430	1.4858	1.5207	1.5507	1.5775	1.6249	1.6671	1.7059	1.7422	1.7765	1.8092	1.8403	1.8701
700 (503.08)	Sh					46.92	96.92	146.92	196.92	296.92	396.92	496.92	596.92	696.92	796.92	896.92	996.92
	v	0.02050	0.6556			0.7271	0.7928	0.8520	0.9072	1.0102	1.1078	1.2023	1.2948	1.3858	1.4757	1.5647	1.6530
	h	491.60	1201.8			1243.4	1281.0	1314.6	1345.6	1403.7	1459.4	1514.4	1569.4	1624.8	1680.7	1737.2	1794.3
	s	0.6928	1.4304			1.4726	1.5090	1.5399	1.5673	1.6154	1.6580	1.6970	1.7335	1.7679	1.8006	1.8318	1.8617
750 (510.84)	Sh					39.16	89.16	139.16	189.16	289.16	389.16	489.16	589.16	689.16	789.16	889.16	989.16
	v	0.02069	0.6095			0.6676	0.7313	0.7882	0.8409	0.9386	1.0306	1.1195	1.2063	1.2916	1.3759	1.4592	1.5419
	h	500.89	1200.7			1236.9	1276.1	1310.7	1342.5	1401.5	1457.6	1512.9	1568.2	1623.8	1679.8	1736.4	1793.6
	s	0.7022	1.4232			1.4598	1.4977	1.5296	1.5577	1.6065	1.6494	1.6886	1.7252	1.7598	1.7926	1.8239	1.8538
800 (518.21)	Sh					31.79	81.79	131.79	181.79	281.79	381.79	481.79	581.79	681.79	781.79	881.79	981.79
	v	0.02087	0.5690			0.6151	0.6774	0.7323	0.7828	0.8759	0.9631	1.0470	1.1289	1.2093	1.2885	1.3669	1.4446
	h	509.81	1199.4			1230.1	1271.1	1306.8	1339.3	1399.1	1455.8	1511.4	1566.9	1622.7	1678.9	1735.7	1792.9
	s	0.7111	1.4163			1.4472	1.4869	1.5198	1.5484	1.5980	1.6413	1.6807	1.7175	1.7522	1.7851	1.8164	1.8464
850 (525.24)	Sh					24.76	74.76	124.76	174.76	274.76	374.76	474.76	574.76	674.76	774.76	874.76	974.76
	v	0.02105	0.5330			0.5683	0.6296	0.6829	0.7315	0.8205	0.9034	0.9830	1.0606	1.1366	1.2115	1.2855	1.3588
	h	518.40	1198.0			1223.0	1265.9	1302.8	1336.0	1396.8	1454.0	1510.0	1565.7	1621.6	1678.0	1734.9	1792.3
	s	0.7197	1.4096			1.4347	1.4763	1.5102	1.5396	1.5899	1.6336	1.6733	1.7102	1.7450	1.7780	1.8095	1.8395
900 (531.95)	Sh					18.05	68.05	118.05	168.05	268.05	368.05	468.05	568.05	668.05	768.05	868.05	968.05
	v	0.02123	0.5009			0.5263	0.5869	0.6388	0.6858	0.7713	0.8504	0.9262	0.9998	1.0720	1.1430	1.2131	1.2825
	h	526.70	1196.4			1215.5	1260.6	1298.6	1332.7	1394.4	1452.2	1508.5	1564.4	1620.6	1677.1	1734.1	1791.6
	s	0.7279	1.4032			1.4223	1.4659	1.5010	1.5311	1.5822	1.6263	1.6662	1.7033	1.7382	1.7713	1.8028	1.8329

Sh = superheat, F
v = specific volume, cu ft per lb
h = enthalpy, Btu per lb
s = entropy, Btu per F per lb

Abs Press. Lb/Sq In. (Sat. Temp)		Sat. Water	Sat. Steam	550	600	650	700	750	800	850	900	1000	1100	1200	1300	1400	1500
950 (538.39)	Sh			11.61	61.61	111.61	161.61	211.61	261.61	311.61	361.61	461.61	561.61	661.61	761.61	861.61	961.61
	v	0.02141	0.4721	0.4883	0.5485	0.5993	0.6449	0.6871	0.7272	0.7656	0.8030	0.8753	0.9455	1.0142	1.0817	1.1484	1.2143
	h	534.74	1194.7	1207.6	1255.1	1294.4	1329.3	1361.5	1392.0	1421.5	1450.3	1507.0	1563.2	1619.5	1676.2	1733.3	1791.0
	s	0.7358	1.3970	1.4098	1.4557	1.4921	1.5228	1.5500	1.5748	1.5977	1.6193	1.6595	1.6967	1.7317	1.7649	1.7965	1.8267
1000 (544.58)	Sh			5.42	55.42	105.42	155.42	205.42	255.42	305.42	355.42	455.42	555.42	655.42	755.42	855.42	955.42
	v	0.02159	0.4460	0.4535	0.5137	0.5636	0.6080	0.6489	0.6875	0.7245	0.7603	0.8295	0.8966	0.9622	1.0266	1.0901	1.1529
	h	542.55	1192.9	1199.3	1249.3	1290.1	1325.9	1358.7	1389.6	1419.4	1448.5	1505.4	1561.9	1618.4	1675.3	1732.5	1790.3
	s	0.7434	1.3910	1.3973	1.4457	1.4833	1.5149	1.5426	1.5677	1.5908	1.6126	1.6530	1.6905	1.7256	1.7589	1.7905	1.8207
1050 (550.53)	Sh				49.47	99.47	149.47	199.47	249.47	299.47	349.47	449.47	549.47	649.47	749.47	849.47	949.47
	v	0.02177	0.4222		0.4821	0.5312	0.5745	0.6142	0.6515	0.6872	0.7216	0.7881	0.8524	0.9151	0.9767	1.0373	1.0973
	h	550.15	1191.0		1243.4	1285.7	1322.4	1355.8	1387.2	1417.3	1446.6	1503.9	1560.7	1617.4	1674.4	1731.8	1789.6
	s	0.7507	1.3851		1.4358	1.4748	1.5072	1.5354	1.5608	1.5842	1.6062	1.6469	1.6845	1.7197	1.7531	1.7848	1.8151
1100 (556.28)	Sh				43.72	93.72	143.72	193.72	243.72	293.72	343.72	443.72	543.72	643.72	743.72	843.72	943.72
	v	0.02195	0.4006		0.4531	0.5017	0.5440	0.5826	0.6188	0.6533	0.6865	0.7505	0.8121	0.8723	0.9313	0.9894	1.0468
	h	557.55	1189.1		1237.3	1281.2	1318.8	1352.9	1384.7	1415.2	1444.7	1502.4	1559.4	1616.3	1673.5	1731.0	1789.0
	s	0.7578	1.3794		1.4259	1.4664	1.4996	1.5284	1.5542	1.5779	1.6000	1.6410	1.6787	1.7141	1.7475	1.7793	1.8097
1150 (561.82)	Sh				39.18	89.18	139.18	189.18	239.18	289.18	339.18	439.18	539.18	639.18	739.18	839.18	939.18
	v	0.02214	0.3807		0.4263	0.4746	0.5162	0.5538	0.5889	0.6223	0.6544	0.7161	0.7754	0.8332	0.8899	0.9456	1.0007
	h	564.78	1187.0		1230.9	1276.6	1315.2	1349.9	1382.2	1413.0	1442.8	1500.9	1558.1	1615.2	1672.6	1730.2	1788.3
	s	0.7647	1.3738		1.4160	1.4582	1.4923	1.5216	1.5478	1.5717	1.5941	1.6353	1.6732	1.7087	1.7422	1.7741	1.8045
1200 (567.19)	Sh				32.81	82.81	132.81	182.81	232.81	282.81	332.81	432.81	532.81	632.81	732.81	832.81	932.81
	v	0.02232	0.3624		0.4016	0.4497	0.4905	0.5273	0.5615	0.5939	0.6250	0.6845	0.7418	0.7974	0.8519	0.9055	0.9584
	h	571.85	1184.8		1224.2	1271.8	1311.5	1346.9	1379.7	1410.8	1440.9	1499.4	1556.9	1614.2	1671.6	1729.4	1787.6
	s	0.7714	1.3683		1.4061	1.4501	1.4851	1.5150	1.5415	1.5658	1.5883	1.6298	1.6679	1.7035	1.7371	1.7691	1.7996
1300 (577.42)	Sh				22.58	72.58	122.58	172.58	222.58	272.58	322.58	422.58	522.58	622.58	722.58	822.58	922.58
	v	0.02269	0.3299		0.3570	0.4052	0.4451	0.4804	0.5129	0.5436	0.5729	0.6287	0.6822	0.7341	0.7847	0.8345	0.8836
	h	585.58	1180.2		1209.9	1261.9	1303.9	1340.8	1374.6	1406.4	1437.1	1496.3	1554.3	1612.0	1669.8	1727.9	1786.3
	s	0.7843	1.3577		1.3860	1.4340	1.4711	1.5022	1.5296	1.5544	1.5773	1.6194	1.6578	1.6937	1.7275	1.7596	1.7902
1400 (587.07)	Sh				12.93	62.93	112.93	162.93	212.93	262.93	312.93	412.93	512.93	612.93	712.93	812.93	912.93
	v	0.02307	0.3018		0.3176	0.3667	0.4059	0.4400	0.4712	0.5004	0.5282	0.5809	0.6311	0.6798	0.7272	0.7737	0.8195
	h	598.83	1175.3		1194.1	1251.4	1296.1	1334.5	1369.3	1402.0	1433.2	1493.2	1551.8	1609.9	1668.0	1726.3	1785.0
	s	0.7966	1.3474		1.3652	1.4181	1.4575	1.4900	1.5182	1.5436	1.5670	1.6096	1.6484	1.6845	1.7185	1.7508	1.7815
1500 (596.20)	Sh				3.80	53.80	103.80	153.80	203.80	253.80	303.80	403.80	503.80	603.80	703.80	803.80	903.80
	v	0.02346	0.2772		0.2820	0.3328	0.3717	0.4049	0.4350	0.4629	0.4894	0.5394	0.5869	0.6327	0.6773	0.7210	0.7639
	h	611.68	1170.1		1176.3	1240.2	1287.9	1328.0	1364.0	1397.4	1429.2	1490.1	1549.2	1607.7	1666.2	1724.8	1783.7
	s	0.8085	1.3373		1.3431	1.4022	1.4443	1.4782	1.5073	1.5333	1.5572	1.6004	1.6395	1.6759	1.7101	1.7425	1.7734
1600 (604.87)	Sh					45.13	95.13	145.13	195.13	245.13	295.13	395.13	495.13	595.13	695.13	795.13	895.13
	v	0.02387	0.2555			0.3026	0.3415	0.3741	0.4032	0.4301	0.4555	0.5031	0.5482	0.5915	0.6336	0.6748	0.7153
	h	624.20	1164.5			1228.3	1279.4	1321.4	1358.5	1392.8	1425.2	1486.9	1546.6	1605.6	1664.3	1723.2	1782.3
	s	0.8199	1.3274			1.3861	1.4312	1.4667	1.4968	1.5235	1.5478	1.5916	1.6312	1.6678	1.7022	1.7347	1.7657
1700 (613.13)	Sh					36.87	86.87	136.87	186.87	236.87	286.87	386.87	486.87	586.87	686.87	786.87	886.87
	v	0.02428	0.2361			0.2754	0.3147	0.3468	0.3751	0.4011	0.4255	0.4711	0.5140	0.5552	0.5951	0.6341	0.6724
	h	636.45	1158.6			1215.3	1270.5	1314.5	1352.9	1388.1	1421.2	1483.8	1544.0	1603.4	1662.5	1721.7	1781.0
	s	0.8309	1.3176			1.3697	1.4183	1.4555	1.4867	1.5140	1.5388	1.5833	1.6232	1.6601	1.6947	1.7274	1.7585
1800 (621.02)	Sh					28.98	78.98	128.98	178.98	228.98	278.98	378.98	478.98	578.98	678.98	778.98	878.98
	v	0.02472	0.2186			0.2505	0.2906	0.3223	0.3500	0.3752	0.3988	0.4426	0.4836	0.5229	0.5609	0.5980	0.6343
	h	648.49	1152.3			1201.2	1261.1	1307.4	1347.2	1383.3	1417.1	1480.6	1541.4	1601.2	1660.7	1720.1	1779.7
	s	0.8417	1.3079			1.3526	1.4054	1.4446	1.4768	1.5049	1.5302	1.5753	1.6156	1.6528	1.6876	1.7204	1.7516
1900 (628.56)	Sh					21.44	71.44	121.44	171.44	221.44	271.44	371.44	471.44	571.44	671.44	771.44	871.44
	v	0.02517	0.2028			0.2274	0.2687	0.3004	0.3275	0.3521	0.3749	0.4171	0.4565	0.4940	0.5303	0.5656	0.6002
	h	660.36	1145.6			1185.7	1251.3	1300.2	1341.4	1378.4	1412.9	1477.4	1538.8	1599.1	1658.8	1718.6	1778.4
	s	0.8522	1.2981			1.3346	1.3925	1.4338	1.4672	1.4960	1.5219	1.5677	1.6084	1.6458	1.6808	1.7138	1.7451
2000 (635.80)	Sh					14.20	64.20	114.20	164.20	214.20	264.20	364.20	464.20	564.20	664.20	764.20	864.20
	v	0.02565	0.1883			0.2056	0.2488	0.2805	0.3072	0.3312	0.3534	0.3942	0.4320	0.4680	0.5027	0.5365	0.5695
	h	672.11	1138.3			1168.3	1240.9	1292.6	1335.4	1373.5	1408.7	1474.1	1536.2	1596.9	1657.0	1717.0	1777.1
	s	0.8625	1.2881			1.3154	1.3794	1.4231	1.4578	1.4874	1.5138	1.5603	1.6014	1.6391	1.6743	1.7075	1.7389
2100 (642.76)	Sh					7.24	57.24	107.24	157.24	207.24	257.24	357.24	457.24	557.24	657.24	757.24	857.24
	v	0.02615	0.1750			0.1847	0.2304	0.2624	0.2888	0.3123	0.3339	0.3734	0.4099	0.4445	0.4778	0.5101	0.5418
	h	683.79	1130.5			1148.5	1229.8	1284.9	1329.3	1368.4	1404.4	1470.9	1533.6	1594.7	1655.2	1715.4	1775.7
	s	0.8727	1.2780			1.2942	1.3661	1.4125	1.4486	1.4790	1.5060	1.5532	1.5948	1.6327	1.6681	1.7014	1.7330
2200 (649.45)	Sh					.55	50.55	100.55	150.55	200.55	250.55	350.55	450.55	550.55	650.55	750.55	850.55
	v	0.02669	0.1627			0.1636	0.2134	0.2458	0.2720	0.2950	0.3161	0.3545	0.3897	0.4231	0.4551	0.4862	0.5165
	h	695.46	1122.2			1123.9	1218.0	1276.8	1323.1	1363.3	1400.0	1467.6	1530.9	1592.5	1653.3	1713.9	1774.4
	s	0.8828	1.2676			1.2691	1.3523	1.4020	1.4395	1.4708	1.4984	1.5463	1.5883	1.6266	1.6622	1.6956	1.7273
2300 (655.89)	Sh						44.11	94.11	144.11	194.11	244.11	344.11	444.11	544.11	644.11	744.11	844.11
	v	0.02727	0.1513				0.1975	0.2305	0.2566	0.2793	0.2999	0.3372	0.3714	0.4035	0.4344	0.4643	0.4935
	h	707.18	1113.2				1205.3	1268.4	1316.7	1358.1	1395.7	1464.2	1528.3	1590.3	1651.5	1712.3	1773.1
	s	0.8929	1.2569				1.3381	1.3914	1.4305	1.4628	1.4910	1.5397	1.5821	1.6207	1.6565	1.6901	1.7219

Sh = superheat, F
v = specific volume, cu ft per lb

h = enthalpy, Btu per lb
s = entropy, Btu per F per lb

ITEM B 15　Properties of Superheated Steam (continued)

Abs Press. Lb/Sq In. (Sat. Temp)		Sat Water	Sat Steam	700	750	800	850	900	950	1000	1050	1100	1150	1200	1300	1400	1500
2400 (662.11)	Sh			37.89	87.89	137.89	187.89	237.89	287.89	337.89	387.89	437.89	487.89	537.89	637.89	737.89	837.89
	v	0.02790	0.1408	0.1824	0.2164	0.2424	0.2648	0.2850	0.3037	0.3214	0.3382	0.3545	0.3703	0.3856	0.4155	0.4443	0.4724
	h	718.95	1103.7	1191.6	1259.7	1310.1	1352.8	1391.2	1426.9	1460.9	1493.7	1525.6	1557.0	1588.1	1649.6	1710.8	1771.8
	s	0.9031	1.2460	1.3232	1.3808	1.4217	1.4549	1.4837	1.5095	1.5332	1.5553	1.5761	1.5959	1.6149	1.6509	1.6847	1.7167
2500 (668.11)	Sh			31.89	81.89	131.89	181.89	231.89	281.89	331.89	381.89	431.89	481.89	531.89	631.89	731.89	831.89
	v	0.02859	0.1307	0.1681	0.2032	0.2293	0.2514	0.2712	0.2896	0.3068	0.3232	0.3390	0.3543	0.3692	0.3980	0.4259	0.4529
	h	731.71	1093.3	1176.7	1250.6	1303.4	1347.4	1386.7	1423.1	1457.5	1490.7	1522.9	1554.6	1585.9	1647.8	1709.2	1770.4
	s	0.9139	1.2345	1.3076	1.3701	1.4129	1.4472	1.4766	1.5029	1.5269	1.5492	1.5703	1.5903	1.6094	1.6456	1.6796	1.7116
2600 (673.91)	Sh			26.09	76.09	126.09	176.09	226.09	276.09	326.09	376.09	426.09	476.09	526.09	626.09	726.09	826.09
	v	0.02938	0.1211	0.1544	0.1909	0.2171	0.2390	0.2585	0.2765	0.2933	0.3093	0.3247	0.3395	0.3540	0.3819	0.4088	0.4350
	h	744.47	1082.0	1160.2	1241.1	1296.5	1341.9	1382.1	1419.2	1454.1	1487.7	1520.2	1552.2	1583.7	1646.0	1707.7	1769.1
	s	0.9247	1.2225	1.2908	1.3592	1.4042	1.4395	1.4696	1.4964	1.5208	1.5434	1.5646	1.5848	1.6040	1.6405	1.6746	1.7068
2700 (679.53)	Sh			20.47	70.47	120.47	170.47	220.47	270.47	320.47	370.47	420.47	470.47	520.47	620.47	720.47	820.47
	v	0.03029	0.1119	0.1411	0.1794	0.2058	0.2275	0.2468	0.2644	0.2809	0.2965	0.3114	0.3259	0.3399	0.3670	0.3931	0.4184
	h	757.34	1069.7	1142.0	1231.1	1289.5	1336.3	1377.5	1415.2	1450.7	1484.6	1517.5	1549.8	1581.5	1644.1	1706.1	1767.8
	s	0.9356	1.2097	1.2727	1.3481	1.3954	1.4319	1.4628	1.4900	1.5148	1.5376	1.5591	1.5794	1.5988	1.6355	1.6697	1.7021
2800 (684.96)	Sh			15.04	65.04	115.04	165.04	215.04	265.04	315.04	365.04	415.04	465.04	515.04	615.04	715.04	815.04
	v	0.03134	0.1030	0.1278	0.1685	0.1952	0.2168	0.2358	0.2531	0.2693	0.2845	0.2991	0.3132	0.3268	0.3532	0.3785	0.4030
	h	770.69	1055.8	1121.2	1220.6	1282.2	1330.7	1372.8	1411.2	1447.2	1481.6	1514.8	1547.3	1579.3	1642.2	1704.5	1766.5
	s	0.9468	1.1958	1.2527	1.3368	1.3867	1.4245	1.4561	1.4838	1.5089	1.5321	1.5537	1.5742	1.5937	1.6306	1.6651	1.6975
2900 (690.22)	Sh			9.78	59.78	109.78	159.78	209.78	259.78	309.78	359.78	409.78	459.78	509.78	609.78	709.78	809.78
	v	0.03262	0.0942	0.1138	0.1581	0.1853	0.2068	0.2256	0.2427	0.2585	0.2734	0.2877	0.3014	0.3147	0.3403	0.3649	0.3887
	h	785.13	1039.8	1095.3	1209.6	1274.7	1324.9	1368.0	1407.2	1443.7	1478.5	1512.1	1544.9	1577.0	1640.4	1703.0	1765.2
	s	0.9588	1.1803	1.2283	1.3251	1.3780	1.4171	1.4494	1.4777	1.5032	1.5266	1.5485	1.5692	1.5889	1.6259	1.6605	1.6931
3000 (695.33)	Sh			4.67	54.67	104.67	154.67	204.67	254.67	304.67	354.67	404.67	454.67	504.67	604.67	704.67	804.67
	v	0.03428	0.0850	0.0982	0.1483	0.1759	0.1975	0.2161	0.2329	0.2484	0.2630	0.2770	0.2904	0.3033	0.3282	0.3522	0.3753
	h	801.84	1020.3	1060.5	1197.9	1267.0	1319.0	1363.2	1403.1	1440.2	1475.4	1509.4	1542.4	1574.8	1638.5	1701.4	1763.8
	s	0.9728	1.1619	1.1966	1.3131	1.3692	1.4097	1.4429	1.4717	1.4976	1.5213	1.5434	1.5642	1.5841	1.6214	1.6561	1.6888
3100 (700.28)	Sh				49.72	99.72	149.72	199.72	249.72	299.72	349.72	399.72	449.72	499.72	599.72	699.72	799.72
	v	0.03681	0.0745		0.1389	0.1671	0.1887	0.2071	0.2237	0.2390	0.2533	0.2670	0.2800	0.2927	0.3170	0.3403	0.3628
	h	823.97	993.3		1185.4	1259.1	1313.0	1358.4	1399.0	1436.7	1472.3	1506.6	1539.9	1572.6	1636.7	1699.8	1762.5
	s	0.9914	1.1373		1.3007	1.3604	1.4024	1.4364	1.4658	1.4920	1.5161	1.5384	1.5594	1.5794	1.6169	1.6518	1.6847
3200 (705.08)	Sh				44.92	94.92	144.92	194.92	244.92	294.92	344.92	394.92	444.92	494.92	594.92	694.92	794.92
	v	0.04472	0.0566		0.1300	0.1588	0.1806	0.1987	0.2151	0.2301	0.2442	0.2576	0.2704	0.2827	0.3065	0.3291	0.3510
	h	875.54	931.6		1172.3	1250.9	1306.9	1353.4	1394.9	1433.1	1469.2	1503.8	1537.4	1570.3	1634.8	1698.3	1761.2
	s	1.0351	1.0832		1.2877	1.3515	1.3951	1.4300	1.4600	1.4866	1.5110	1.5335	1.5547	1.5749	1.6126	1.6477	1.6806
3300	Sh																
	v				0.1213	0.1510	0.1727	0.1908	0.2070	0.2218	0.2357	0.2488	0.2613	0.2734	0.2966	0.3187	0.3400
	h				1158.2	1242.5	1300.7	1348.4	1390.7	1429.5	1466.1	1501.0	1534.9	1568.1	1632.9	1696.7	1759.9
	s				1.2742	1.3425	1.3879	1.4237	1.4542	1.4813	1.5059	1.5287	1.5501	1.5704	1.6084	1.6436	1.6767
3400	Sh																
	v				0.1129	0.1435	0.1653	0.1834	0.1994	0.2140	0.2276	0.2405	0.2528	0.2646	0.2872	0.3088	0.3296
	h				1143.2	1233.7	1294.3	1343.4	1386.4	1425.9	1462.9	1498.3	1532.4	1565.8	1631.1	1695.1	1758.5
	s				1.2600	1.3334	1.3807	1.4174	1.4486	1.4761	1.5010	1.5240	1.5456	1.5660	1.6042	1.6396	1.6728
3500	Sh																
	v				0.1048	0.1364	0.1583	0.1764	0.1922	0.2066	0.2200	0.2326	0.2447	0.2563	0.2784	0.2995	0.3198
	h				1127.1	1224.6	1287.8	1338.2	1382.2	1422.2	1459.7	1495.5	1529.9	1563.6	1629.2	1693.6	1757.2
	s				1.2450	1.3242	1.3734	1.4112	1.4430	1.4709	1.4962	1.5194	1.5412	1.5618	1.6002	1.6358	1.6691
3600	Sh																
	v				0.0966	0.1296	0.1517	0.1697	0.1854	0.1996	0.2128	0.2252	0.2371	0.2485	0.2702	0.2908	0.3106
	h				1108.6	1215.3	1281.2	1333.0	1377.9	1418.6	1456.5	1492.6	1527.4	1561.3	1627.3	1692.0	1755.9
	s				1.2281	1.3148	1.3662	1.4050	1.4374	1.4658	1.4914	1.5149	1.5369	1.5576	1.5962	1.6320	1.6654
3800	Sh																
	v				0.0799	0.1169	0.1395	0.1574	0.1729	0.1868	0.1996	0.2116	0.2231	0.2340	0.2549	0.2746	0.2936
	h				1064.2	1195.5	1267.6	1322.4	1369.1	1411.2	1450.1	1487.0	1522.4	1556.8	1623.6	1688.9	1753.2
	s				1.1888	1.2955	1.3517	1.3928	1.4265	1.4558	1.4821	1.5061	1.5284	1.5495	1.5886	1.6247	1.6584
4000	Sh																
	v				0.0631	0.1052	0.1284	0.1463	0.1616	0.1752	0.1877	0.1994	0.2105	0.2210	0.2411	0.2601	0.2783
	h				1007.4	1174.3	1253.4	1311.6	1360.2	1403.6	1443.6	1481.3	1517.3	1552.2	1619.8	1685.7	1750.6
	s				1.1396	1.2754	1.3371	1.3807	1.4158	1.4461	1.4730	1.4976	1.5203	1.5417	1.5812	1.6177	1.6516
4200	Sh																
	v				0.0498	0.0945	0.1183	0.1362	0.1513	0.1647	0.1769	0.1883	0.1991	0.2093	0.2287	0.2470	0.2645
	h				950.1	1151.6	1238.6	1300.4	1351.2	1396.0	1437.1	1475.5	1512.2	1547.6	1616.1	1682.6	1748.0
	s				1.0905	1.2544	1.3223	1.3686	1.4053	1.4366	1.4642	1.4893	1.5124	1.5341	1.5742	1.6109	1.6452
4400	Sh																
	v				0.0421	0.0846	0.1090	0.1270	0.1420	0.1552	0.1671	0.1782	0.1887	0.1986	0.2174	0.2351	0.2519
	h				909.5	1127.3	1223.3	1289.0	1342.0	1388.3	1430.4	1469.7	1507.1	1543.0	1612.3	1679.4	1745.3
	s				1.0556	1.2325	1.3073	1.3566	1.3949	1.4272	1.4556	1.4812	1.5048	1.5268	1.5673	1.6044	1.6389

Sh = superheat, F
v = specific volume, cu ft per lb

h = enthalpy, Btu per lb
s = entropy, Btu per F per lb

Properties of Superheated Steam (continued)   ITEM B 15

Abs Press. Lb/Sq In. (Sat. Temp)		Sat Water	Sat Steam	750	800	850	900	950	1000	1050	1100	1150	1200	1250	1300	1400	1500
4600	Sh																
	v			0.0380	0.0751	0.1005	0.1186	0.1335	0.1465	0.1582	0.1691	0.1792	0.1889	0.1982	0.2071	0.2242	0.2404
	h			883.8	1100.0	1207.3	1277.2	1332.6	1380.5	1423.7	1463.9	1501.9	1538.4	1573.8	1608.5	1676.3	1742.7
	s			1.0331	1.2084	1.2922	1.3446	1.3847	1.4181	1.4472	1.4734	1.4974	1.5197	1.5407	1.5607	1.5982	1.6330
4800	Sh																
	v			0.0355	0.0665	0.0927	0.1109	0.1257	0.1385	0.1500	0.1606	0.1706	0.1800	0.1890	0.1977	0.2142	0.2299
	h			866.9	1071.2	1190.7	1265.2	1323.1	1372.6	1417.0	1458.0	1496.7	1533.8	1569.7	1604.7	1673.1	1740.0
	s			1.0180	1.1835	1.2768	1.3327	1.3745	1.4090	1.4390	1.4657	1.4901	1.5128	1.5341	1.5543	1.5921	1.6272
5000	Sh																
	v			0.0338	0.0591	0.0855	0.1038	0.1185	0.1312	0.1425	0.1529	0.1626	0.1718	0.1806	0.1890	0.2050	0.2203
	h			854.9	1042.9	1173.6	1252.9	1313.5	1364.6	1410.2	1452.1	1491.5	1529.1	1565.5	1600.9	1670.0	1737.4
	s			1.0070	1.1593	1.2612	1.3207	1.3645	1.4001	1.4309	1.4582	1.4831	1.5061	1.5277	1.5481	1.5863	1.6216
5200	Sh																
	v			0.0326	0.0531	0.0789	0.0973	0.1119	0.1244	0.1356	0.1458	0.1553	0.1642	0.1728	0.1810	0.1966	0.2114
	h			845.8	1016.9	1156.0	1240.4	1303.7	1356.6	1403.4	1446.2	1486.3	1524.5	1561.3	1597.2	1666.8	1734.7
	s			0.9985	1.1370	1.2455	1.3088	1.3545	1.3914	1.4229	1.4509	1.4762	1.4995	1.5214	1.5420	1.5806	1.6161
5400	Sh																
	v			0.0317	0.0483	0.0728	0.0912	0.1058	0.1182	0.1292	0.1392	0.1485	0.1572	0.1656	0.1736	0.1888	0.2031
	h			838.5	994.3	1138.1	1227.7	1293.7	1348.4	1396.5	1440.3	1481.1	1519.8	1557.1	1593.4	1663.7	1732.1
	s			0.9915	1.1175	1.2296	1.2969	1.3446	1.3827	1.4151	1.4437	1.4694	1.4931	1.5153	1.5362	1.5750	1.6109
5600	Sh																
	v			0.0309	0.0447	0.0672	0.0856	0.1001	0.1124	0.1232	0.1331	0.1422	0.1508	0.1589	0.1667	0.1815	0.1954
	h			832.4	975.0	1119.9	1214.8	1283.7	1340.2	1389.6	1434.3	1475.9	1515.2	1552.9	1589.6	1660.5	1729.5
	s			0.9855	1.1008	1.2137	1.2850	1.3348	1.3742	1.4075	1.4366	1.4628	1.4869	1.5093	1.5304	1.5697	1.6058
5800	Sh																
	v			0.0303	0.0419	0.0622	0.0805	0.0949	0.1070	0.1177	0.1274	0.1363	0.1447	0.1527	0.1603	0.1747	0.1883
	h			827.3	958.8	1101.8	1201.8	1273.6	1332.0	1382.6	1428.3	1470.6	1510.5	1548.7	1585.8	1657.4	1726.8
	s			0.9803	1.0867	1.1981	1.2732	1.3250	1.3658	1.3999	1.4297	1.4564	1.4808	1.5035	1.5248	1.5644	1.6008
6000	Sh																
	v			0.0298	0.0397	0.0579	0.0757	0.0900	0.1020	0.1126	0.1221	0.1309	0.1391	0.1469	0.1544	0.1684	0.1817
	h			822.9	945.1	1084.6	1188.8	1263.4	1323.6	1375.7	1422.3	1465.4	1505.9	1544.6	1582.0	1654.2	1724.2
	s			0.9758	1.0746	1.1833	1.2615	1.3154	1.3574	1.3925	1.4229	1.4500	1.4748	1.4978	1.5194	1.5593	1.5960
6500	Sh																
	v			0.0287	0.0358	0.0495	0.0655	0.0793	0.0909	0.1012	0.1104	0.1188	0.1266	0.1340	0.1411	0.1544	0.1669
	h			813.9	919.5	1046.7	1156.3	1237.8	1302.7	1358.1	1407.3	1452.2	1494.2	1534.1	1572.5	1646.4	1717.6
	s			0.9661	1.0515	1.1506	1.2328	1.2917	1.3370	1.3743	1.4064	1.4347	1.4604	1.4841	1.5062	1.5471	1.5844
7000	Sh																
	v			0.0279	0.0334	0.0438	0.0573	0.0704	0.0816	0.0915	0.1004	0.1085	0.1160	0.1231	0.1298	0.1424	0.1542
	h			806.9	901.8	1016.5	1124.9	1212.6	1281.7	1340.5	1392.2	1439.1	1482.6	1523.7	1563.1	1638.6	1711.1
	s			0.9582	1.0350	1.1243	1.2055	1.2689	1.3171	1.3567	1.3904	1.4200	1.4466	1.4710	1.4938	1.5355	1.5735
7500	Sh																
	v			0.0272	0.0318	0.0399	0.0512	0.0631	0.0737	0.0833	0.0918	0.0996	0.1068	0.1136	0.1200	0.1321	0.1433
	h			801.3	889.0	992.9	1097.1	1188.3	1261.0	1322.9	1377.2	1426.0	1471.0	1513.3	1553.7	1630.8	1704.6
	s			0.9514	1.0224	1.1033	1.1818	1.2473	1.2980	1.3397	1.3751	1.4059	1.4335	1.4586	1.4819	1.5245	1.5632
8000	Sh																
	v			0.0267	0.0306	0.0371	0.0465	0.0571	0.0671	0.0762	0.0845	0.0920	0.0989	0.1054	0.1115	0.1230	0.1338
	h			796.6	879.1	974.4	1074.3	1165.4	1241.0	1305.5	1362.2	1413.0	1459.6	1503.1	1544.5	1623.1	1698.1
	s			0.9455	1.0122	1.0864	1.1613	1.2271	1.2798	1.3233	1.3603	1.3924	1.4208	1.4467	1.4705	1.5140	1.5533
8500	Sh																
	v			0.0262	0.0296	0.0350	0.0429	0.0522	0.0615	0.0701	0.0780	0.0853	0.0919	0.0982	0.1041	0.1151	0.1254
	h			792.7	871.2	959.8	1054.5	1144.0	1221.9	1288.5	1347.5	1400.2	1448.2	1492.9	1535.3	1615.4	1691.7
	s			0.9402	1.0037	1.0727	1.1437	1.2084	1.2627	1.3076	1.3460	1.3793	1.4087	1.4352	1.4597	1.5040	1.5439
9000	Sh																
	v			0.0258	0.0288	0.0335	0.0402	0.0483	0.0568	0.0649	0.0724	0.0794	0.0858	0.0918	0.0975	0.1081	0.1179
	h			789.3	864.7	948.0	1037.6	1125.4	1204.1	1272.1	1333.0	1387.5	1437.1	1482.9	1526.3	1607.9	1685.3
	s			0.9354	0.9964	1.0613	1.1285	1.1918	1.2468	1.2926	1.3323	1.3667	1.3970	1.4243	1.4492	1.4944	1.5349
9500	Sh																
	v			0.0254	0.0282	0.0322	0.0380	0.0451	0.0528	0.0603	0.0675	0.0742	0.0804	0.0862	0.0917	0.1019	0.1113
	h			786.4	859.2	938.3	1023.4	1108.9	1187.7	1256.6	1318.9	1375.1	1426.1	1473.1	1517.3	1600.4	1679.0
	s			0.9310	0.9900	1.0516	1.1153	1.1771	1.2320	1.2785	1.3191	1.3546	1.3858	1.4137	1.4392	1.4851	1.5263
10000	Sh																
	v			0.0251	0.0276	0.0312	0.0362	0.0425	0.0495	0.0565	0.0633	0.0697	0.0757	0.0812	0.0865	0.0963	0.1054
	h			783.8	854.5	930.2	1011.3	1094.2	1172.6	1242.0	1305.3	1362.9	1415.3	1463.4	1508.6	1593.1	1672.8
	s			0.9270	0.9842	1.0432	1.1039	1.1638	1.2185	1.2652	1.3065	1.3429	1.3749	1.4035	1.4295	1.4763	1.5180
10500	Sh																
	v			0.0248	0.0271	0.0303	0.0347	0.0404	0.0467	0.0532	0.0595	0.0656	0.0714	0.0768	0.0818	0.0913	0.1001
	h			781.5	850.5	923.4	1001.0	1081.3	1158.9	1228.4	1292.4	1351.1	1404.7	1453.9	1500.0	1585.8	1666.7
	s			0.9232	0.9790	1.0358	1.0939	1.1519	1.2060	1.2529	1.2946	1.3371	1.3644	1.3937	1.4202	1.4677	1.5100

Sh = superheat, F   h = enthalpy, Btu per lb
v = specific volume, cu ft per lb   s = entropy, Btu per F per lb

ITEM B 15   Properties of Superheated Steam (concluded)

Abs Press. Lb/Sq In. (Sat. Temp)		Sat. Water	Sat. Steam	Temperature – Degrees Fahrenheit													
				750	800	850	900	950	1000	1050	1100	1150	1200	1250	1300	1400	1500
11000	v			0.0245	0.0267	0.0296	0.0335	0.0386	0.0443	0.0503	0.0562	0.0620	0.0676	0.0727	0.0776	0.0868	0.0952
	h			779.5	846.9	917.5	992.1	1069.9	1146.3	1215.9	1280.2	1339.7	1394.4	1444.6	1491.5	1578.7	1660.6
	s			0.9196	0.9742	1.0292	1.0851	1.1412	1.1945	1.2414	1.2833	1.3209	1.3544	1.3842	1.4112	1.4595	1.5023
11500	v			0.0243	0.0263	0.0290	0.0325	0.0370	0.0423	0.0478	0.0534	0.0588	0.0641	0.0691	0.0739	0.0827	0.0909
	h			777.7	843.8	912.4	984.5	1059.8	1134.9	1204.3	1268.7	1328.8	1384.4	1435.5	1483.2	1571.8	1654.7
	s			0.9163	0.9698	1.0232	1.0772	1.1316	1.1840	1.2308	1.2727	1.3107	1.3446	1.3750	1.4025	1.4515	1.4949
12000	v			0.0241	0.0260	0.0284	0.0317	0.0357	0.0405	0.0456	0.0508	0.0560	0.0610	0.0659	0.0704	0.0790	0.0869
	h			776.1	841.0	907.9	977.8	1050.9	1124.5	1193.7	1258.0	1318.5	1374.7	1426.6	1475.1	1564.9	1648.8
	s			0.9131	0.9657	1.0177	1.0701	1.1229	1.1742	1.2209	1.2627	1.3010	1.3353	1.3662	1.3941	1.4438	1.4877
12500	v			0.0238	0.0256	0.0279	0.0309	0.0346	0.0390	0.0437	0.0486	0.0535	0.0583	0.0629	0.0673	0.0756	0.0832
	h			774.7	838.6	903.9	971.9	1043.1	1115.2	1184.1	1247.9	1308.8	1365.4	1418.0	1467.2	1558.2	1643.1
	s			0.9101	0.9618	1.0127	1.0637	1.1151	1.1653	1.2117	1.2534	1.2918	1.3264	1.3576	1.3860	1.4363	1.4808
13000	v			0.0236	0.0253	0.0275	0.0302	0.0336	0.0376	0.0420	0.0466	0.0512	0.0558	0.0602	0.0645	0.0725	0.0799
	h			773.5	836.3	900.4	966.8	1036.2	1106.7	1174.8	1238.5	1299.6	1356.5	1409.6	1459.4	1551.6	1637.4
	s			0.9073	0.9582	1.0080	1.0578	1.1079	1.1571	1.2030	1.2445	1.2831	1.3179	1.3494	1.3781	1.4291	1.4741
13500	v			0.0235	0.0251	0.0271	0.0297	0.0328	0.0364	0.0405	0.0448	0.0492	0.0535	0.0577	0.0619	0.0696	0.0768
	h			772.3	834.4	897.2	962.2	1030.0	1099.1	1166.3	1229.7	1291.0	1348.1	1401.5	1451.8	1545.2	1631.9
	s			0.9045	0.9548	1.0037	1.0524	1.1014	1.1495	1.1948	1.2361	1.2749	1.3098	1.3415	1.3705	1.4221	1.4675
14000	v			0.0233	0.0248	0.0267	0.0291	0.0320	0.0354	0.0392	0.0432	0.0474	0.0515	0.0555	0.0595	0.0670	0.0740
	h			771.3	832.6	894.3	958.0	1024.5	1092.3	1158.5	1221.4	1283.0	1340.2	1393.8	1444.4	1538.8	1626.5
	s			0.9019	0.9515	0.9996	1.0473	1.0953	1.1426	1.1872	1.2282	1.2671	1.3021	1.3339	1.3631	1.4153	1.4612
14500	v			0.0231	0.0246	0.0264	0.0287	0.0314	0.0345	0.0380	0.0418	0.0458	0.0496	0.0534	0.0573	0.0646	0.0714
	h			770.4	831.0	891.7	954.3	1019.6	1086.2	1151.4	1213.8	1275.4	1332.9	1386.4	1437.3	1532.6	1621.1
	s			0.8994	0.9484	0.9957	1.0426	1.0897	1.1362	1.1801	1.2208	1.2597	1.2949	1.3266	1.3560	1.4087	1.4551
15000	v			0.0230	0.0244	0.0261	0.0282	0.0308	0.0337	0.0369	0.0405	0.0443	0.0479	0.0516	0.0552	0.0624	0.0690
	h			769.6	829.5	889.3	950.9	1015.1	1080.6	1144.9	1206.8	1268.1	1326.0	1379.4	1430.3	1526.4	1615.9
	s			0.8970	0.9455	0.9920	1.0382	1.0846	1.1302	1.1735	1.2139	1.2525	1.2880	1.3197	1.3491	1.4022	1.4491
15500	v			0.0228	0.0242	0.0258	0.0278	0.0302	0.0329	0.0360	0.0393	0.0429	0.0464	0.0499	0.0534	0.0603	0.0668
	h			768.9	828.2	887.2	947.8	1011.1	1075.7	1139.0	1200.3	1261.1	1319.6	1372.8	1423.6	1520.4	1610.8
	s			0.8946	0.9427	0.9886	1.0340	1.0797	1.1247	1.1674	1.2073	1.2457	1.2815	1.3131	1.3424	1.3959	1.4433

Sh = superheat, F
v = specific volume, cu ft per lb

h = enthalpy, Btu per lb
s = entropy, Btu per F per lb

ITEM B 16-B   Mollier Chart for Steam (SI)

Modified and reduced from *Steam and Air Tables in SI Units*, Irvine and Hartnett, published by Hemisphere Publishing Corporation (1976). Reproduced by permission of Springer-Verlag Publishing Company, owners of copyrights on chart.

(See pages 608–609.)

Modified and greatly reduced from Keenan and Keye's *Thermodynamic Properties of Steam*, published (1936) by John Wiley and Sons, Inc. Reproduced by permission of the publishers.

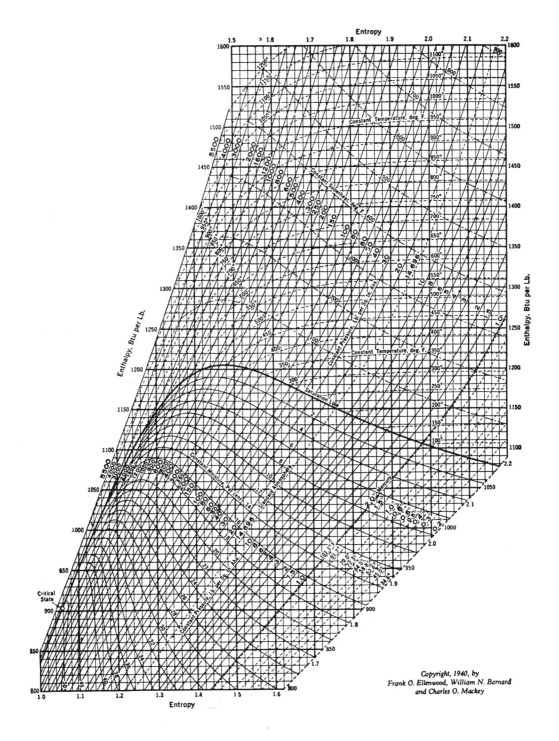

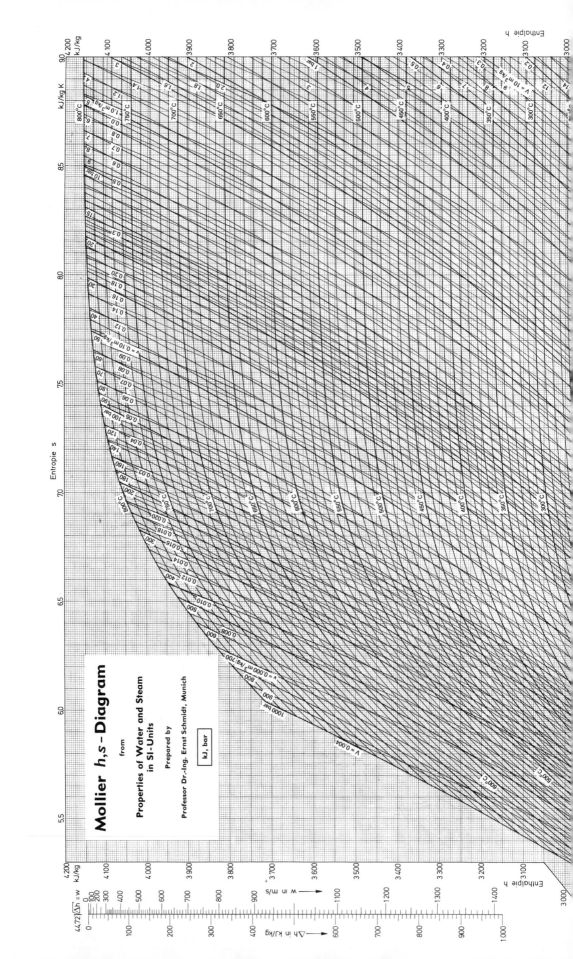

Mollier h,s - Diagram

from

Properties of Water and Steam in SI-Units

Prepared by

Professor Dr.-Ing. Ernst Schmidt, Munich

kJ, bar

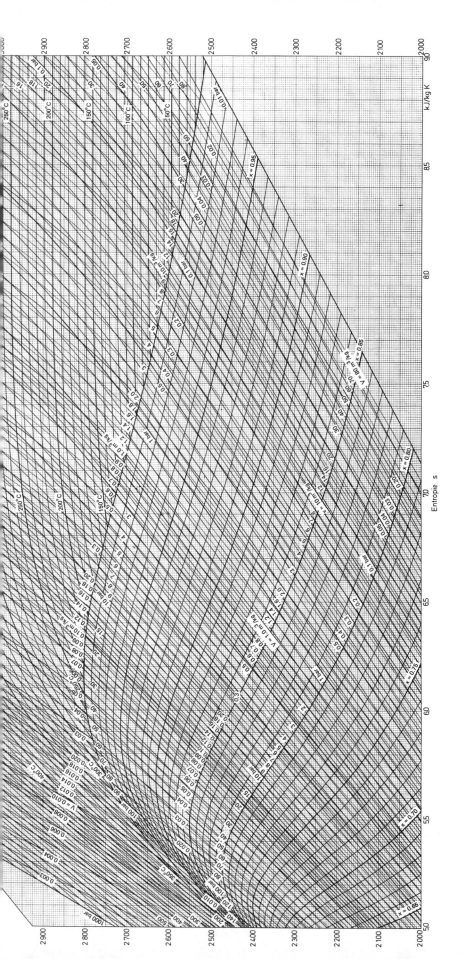

Entropie s

kJ/Kg K

ITEM B 17 Psychrometric Chart for Air-Steam Mixtures

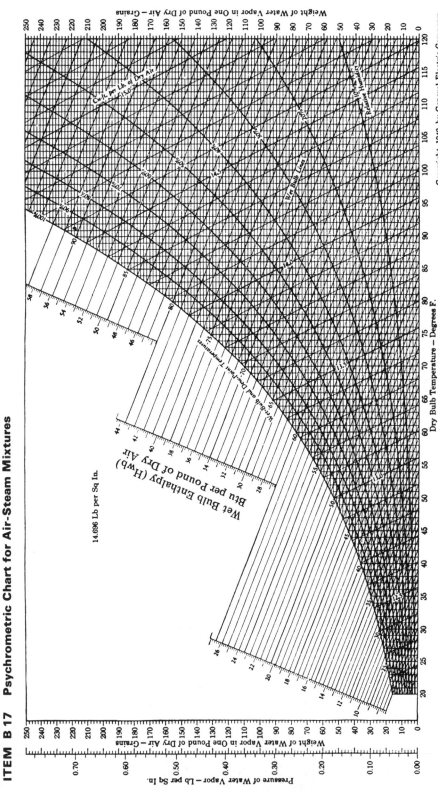

14.696 Lb per Sq In.

Copyright, 1942, by General Electric Company

Critical Properties and van der Waals Constants   ITEM B 18

Most of the values of T_c and p_c are adapted from the collection by Nelson and Obert[10.6]; others from International Critical Tables. Most values of $\bar{v}_c$ from reference [0.3]. Others from International Critical Tables. The values of Z_c, a, and b are computed from these critical data: $a = 27R^2 T_c^2/(64p_c)$ and $b = RT_c/(8p_c)$.

| | | | | | | van der Waals | |
Substance	T_c, °R	p_c, atm	$\bar{v}_c$, ft³/pmole	$\bar{v}_{ci}$, ft³/pmole	$Z_c = p_c v_c/RT_c$	a, atm $\left(\dfrac{ft^3}{pmole}\right)^2$	b ft³/pmole
Acetylene (C_2H_2)	556	62	1.80	6.55	0.274	1121	0.818
Air (equivalent)	239	37.2	1.33	4.69	0.284	345.2	0.585
Ammonia (NH_3)	730	111.3	1.16	4.79	0.242	1076	0.598
Argon (A)	271.6	48.34	1.19	4.10	0.291		
Benzene (C_6H_6)	1011	47.7	4.12	15.47	0.266	4736	1.896
n-Butane (C_4H_{10})	765	37.5	4.13	14.89	0.274	3508	1.919
Carbon dioxide (CO_2)	548	72.9	1.51	5.49	0.276	926	0.686
Carbon monoxide (CO)	239	34.5	1.49	5.06	·0.294	372	0.632
Freon 12 (CCl_2F_2)	692	39.6	3.55	12.76	0.273	2718	1.595
Ethane (C_2H_6)	550	48.2	2.29	8.33	0.284	1410	1.041
Ethylene (C_2H_4)	510	50.5	1.98	7.37	0.270	1158	0.922
Helium (He)	9.33	2.26	0.93	3.01	0.300	8.66	0.376
n-Heptane (C_7H_{16})	972	27	6.86	8.31	0.26	7866	3.298
Hydrogen (H_2)	59.8	12.8	1.04	34.1	0.304	62.8	0.426
Methane (CH_4)	344	45.8	1.59	5.48	0.290	581	0.685
Methyl chloride (CH_3Cl)	749	65.8	2.18	8.31	0.262	1917	1.040
Neon (Ne)	80.3	25.9	0.668	2.26	0.295		
Nitrogen (N_2)	227	33.5	1.44	4.59	0.291	346	0.618
Nonane (C_9H_{20})	1071	22.86	8.86	34.2	0.250		
n-Octane (C_8H_{18})	1025	24.6	7.82	30.42	0.258	9601	3.76
Oxygen (O_2)	278	50.1	1.19	5.05	0.290	348	0.506
Propane (C_3H_8)	666	42.1	3.13	11.55	0.276	2368	1.445
Sulfur dioxide (SO_2)	775	77.7	1.99	7.28	0.268	1738	0.911
Water (H_2O)	1165	218.3	0.896	3.898	0.230	1400	0.488

ITEM B 19 Compressibility Factors Z, Low Pressures

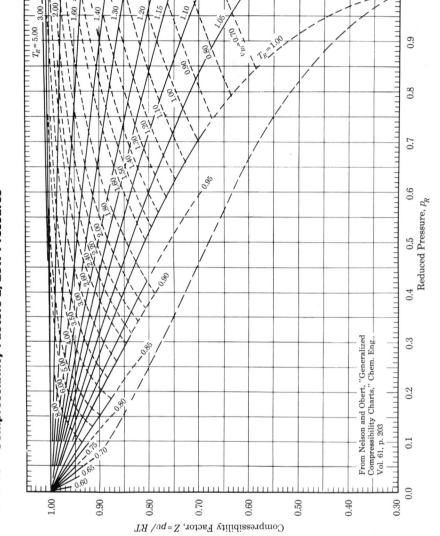

From Nelson and Obert, "Generalized Compressibility Charts," Chem. Eng. Vol. 61, p. 203

ITEM B 20 Compressibility Factors Z, Intermediate Pressures

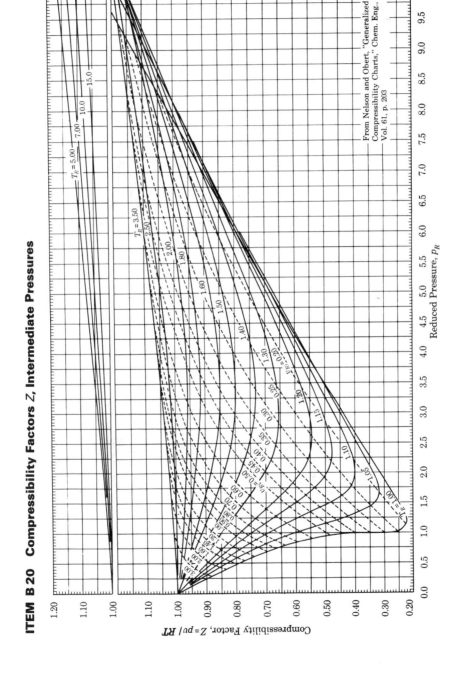

From Nelson and Obert, "Generalized Compressibility Charts," Chem. Eng., Vol. 61, p. 203

ITEM B 21 Compressibility Factors Z, High Pressures

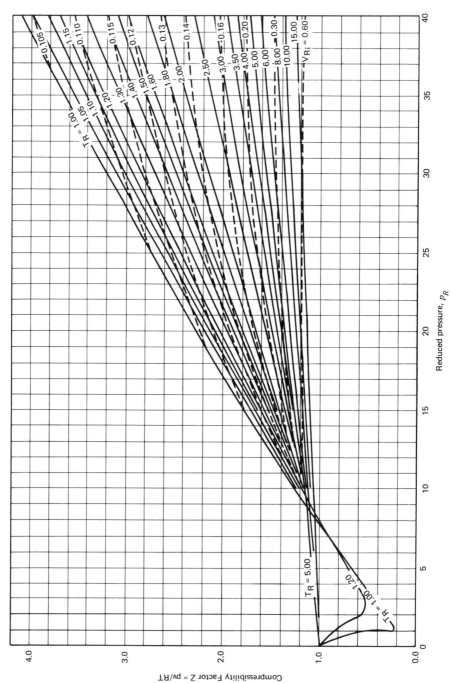

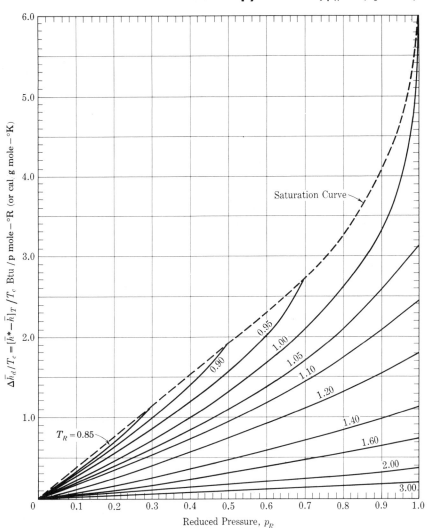

ITEM B 23   **Enthalpy Deviation,** $p_R > 1$ $(Z_c = 0.27)$

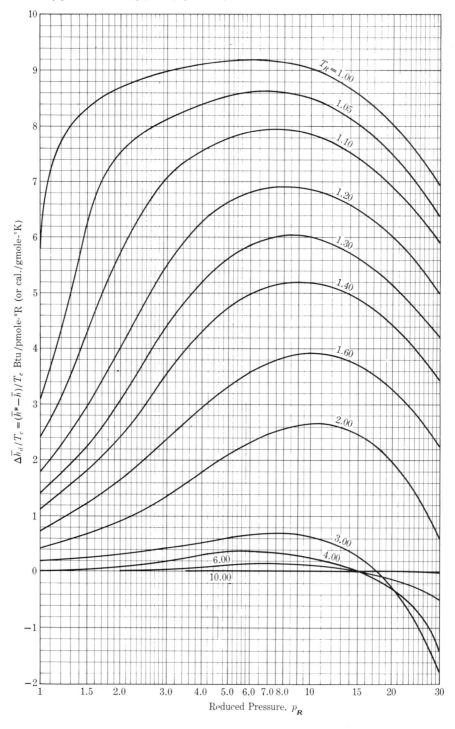

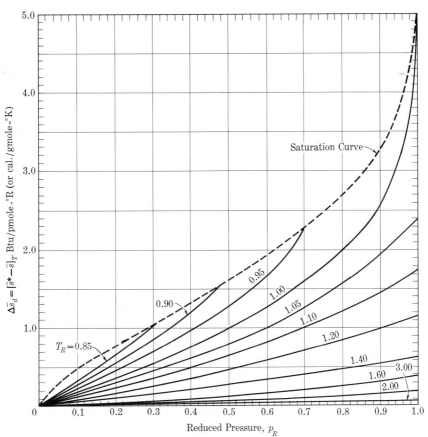

ITEM B 25 **Entropy Deviation,** $p_R > 1$ $(Z_c = 0.27)$

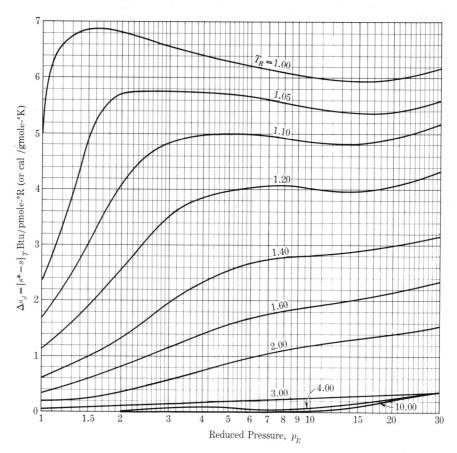

**Temperature-Entropy Chart, Air (courtesy NBS; drawn from data of   ITEM B 26
Michels, Wassenaar, and Wolkers; Claitor and Crawford)**

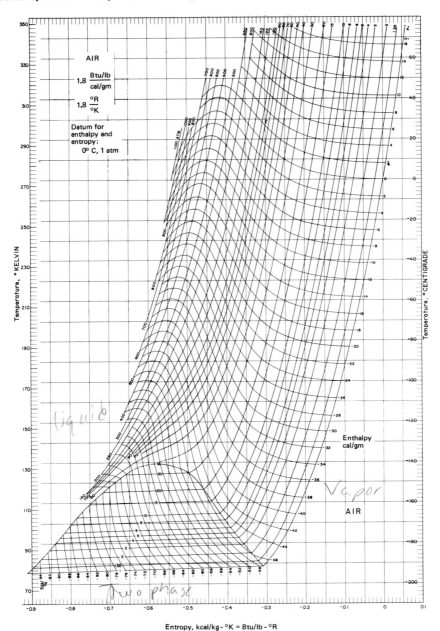

Entropy, kcal/kg - °K = Btu/lb - °R

ITEM B 27 **Temperature-Entropy Chart, Oxygen (courtesy NBS; by Richard B. Stewart, Ph.D. dissertation, Univ. of Iowa)**

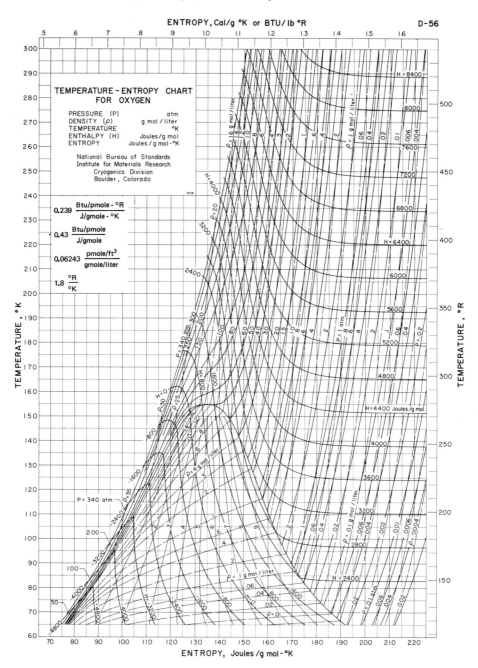

Temperature-Entropy Chart, Hydrogen (courtesy NBS)   ITEM B 28

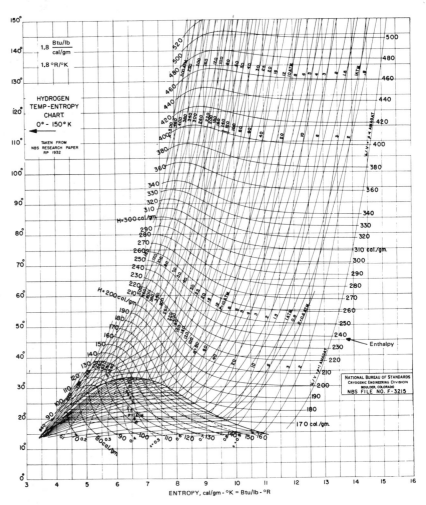

ENTROPY, cal/gm - °K = Btu/lb - °R

ITEM B 29 **Temperature-Entropy Chart, Nitrogen (courtesy NBS; attributed to T. B. Strobridge, NBS TN 129; R. D. McCarty, L. J. Ericks)**

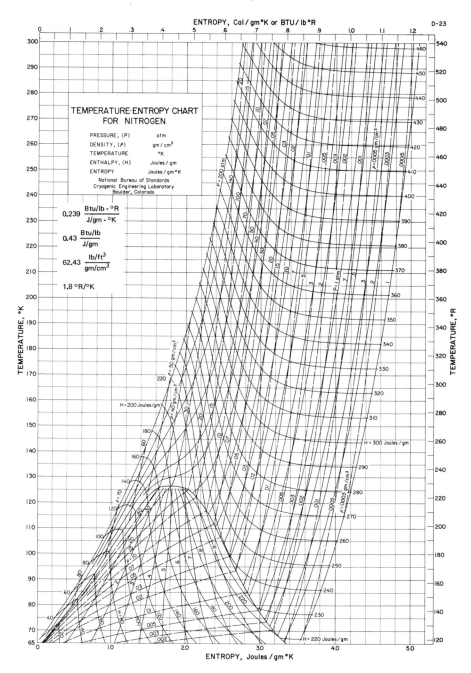

Temperature-Entropy Chart, Helium (courtesy NBS; attributed to   ITEM B 30
D. B. Mann, NBS TN 154; R. D. McCarty, L. J. Ericks)

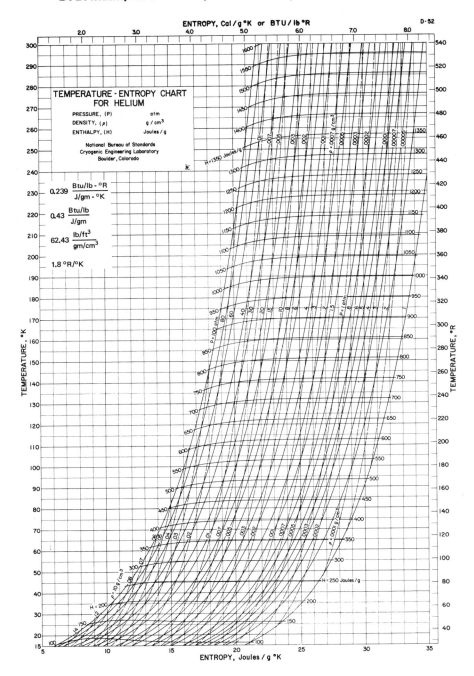

ITEM B 31   Pressure-Enthalpy Chart, Carbon Dioxide

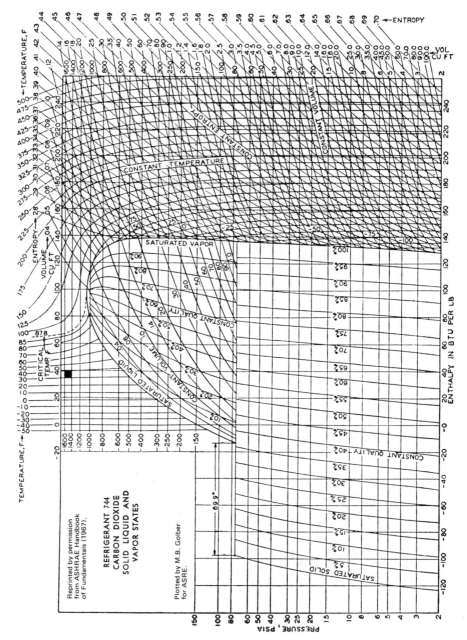

Reprinted by permission
from ASHRAE Handbook
of Fundamentals (1967).

REFRIGERANT 744
CARBON DIOXIDE
SOLID LIQUID AND
VAPOR STATES

Plotted by M.B. Golber
for ASRE.

Values of log$_{10}$ K_p for Certain Reactions of Gases   ITEM B 32

Taken or derived from JANAF.[0.22]

Note: Temperatures are degrees Kelvin.

Reaction → T, °K ↓	$CO + \frac{1}{2}O_2$ $= CO_2$	$H_2 + \frac{1}{2}O_2$ $= H_2O$	$CO_2 + H_2$ $= CO + H_2O$	$H_2O =$ $\frac{1}{2}H_2 + OH$	$\frac{1}{2}N_2 + \frac{1}{2}O_2$ $= NO$	$\frac{1}{2}O_2 = O$	$\frac{1}{2}N_2 = N$
298	45.066	40.048	−5.018	−46.137	−15.171	−40.604	−79.800
300	44.760	39.786	−4.974	−45.832	−15.073	−40.334	−79.289
400	32.431	29.240	−3.191	−33.567	−11.142	−29.473	−58.704
600	20.087	18.633	−1.454	−21.242	−7.210	−18.574	−38.081
800	13.916	13.289	−0.627	−15.044	−5.243	−13.101	−27.744
1000	10.221	10.062	−0.159	−11.309	−4.062	−9.807	−21.528
1200	7.764	7.899	+0.135	−8.811	−3.275	−7.604	−17.377
1300	6.821	7.064	0.243	−7.848	−2.972	−6.755	−15.778
1400	6.014	6.347	0.333	−7.021	−2.712	−6.027	−14.406
1500	5.316	5.725	0.409	−6.305	−2.487	−5.395	−13.217
1600	4.706	5.180	0.474	−5.677	−2.290	−4.842	−12.175
1700	4.169	4.699	0.530	−5.124	−2.116	−4.353	−11.256
1800	3.693	4.270	0.577	−4.631	−1.962	−3.918	−10.437
1900	3.267	3.886	0.619	−4.190	−1.823	−3.529	−9.705
2000	2.884	3.540	0.656	−3.793	−1.699	−3.178	−9.046
2100	2.539	3.227	0.668	−3.434	−1.586	−2.860	−8.449
2200	2.226	2.942	0.716	−3.107	−1.484	−2.571	−7.905
2300	1.940	2.682	0.742	−2.809	−1.391	−2.307	−7.409
2400	1.679	2.443	0.764	−2.535	−1.305	−2.065	−6.954
2500	1.440	2.224	0.784	−2.284	−1.227	−1.842	−6.535
2600	1.219	2.021	0.802	−2.052	−1.154	−1.636	−6.149
2700	1.015	1.833	0.818	−1.837	−1.087	−1.446	−5.790
2800	0.825	1.658	0.833	−1.637	−1.025	−1.268	−5.457
2900	0.649	1.495	0.846	−1.451	−0.967	−1.103	−5.147
3000	0.485	1.343	0.858	−1.278	−0.913	−0.949	−4.858
3100	0.332	1.201	0.869	−1.116	−0.863	−0.805	−4.587
3200	0.189	1.067	0.878	−0.963	−0.815	−0.670	−4.332
3300	0.054	0.942	0.888	−0.821	−0.771	−0.543	−4.093
3400	−0.071	0.824	0.895	−0.687	−0.729	−0.423	−3.868
3500	−0.190	0.712	0.902	−0.559	−0.690	−0.310	−3.656
3600	−0.302	0.607	0.909	−0.440	−0.653	−0.204	−3.455
3700	−0.408	0.507	0.915	−0.327	−0.618	−0.103	−3.265
3800	−0.508	0.413	0.921	−0.220	−0.585	−0.007	−3.086
3900	−0.603	0.323	0.926	−0.120	−0.554	+0.084	−2.915
4000	−0.692	0.238	0.930	−0.022	−0.524	0.170	−2.752
4200	−0.858	+0.079	0.937	+0.158	−0.470	0.330	−2.450
4400	−1.009	−0.065	0.944	0.320	−0.420	0.475	−2.176
4600	−1.147	−0.197	0.950	0.469	−0.375	0.608	−1.924
4800	−1.272	−0.319	0.953	0.606	−0.333	0.730	−1.694
5000	−1.386	−0.430	0.956	0.731	−0.296	0.843	−1.481
5200	−1.492	−0.534	0.958	0.847	−0.261	0.946	−1.284
5400	−1.589	−0.630	0.959	0.954	−0.229	1.042	−1.102
5600	−1.679	−0.719	0.960	1.053	−0.199	1.132	−0.932
5800	−1.763	−0.803	0.960	1.146	−0.172	1.215	−0.774
6000	−1.841	−0.880	0.961	1.232	−0.147	1.292	−0.625

ITEM B 33 Pressure-Enthalpy Chart, Ammonia

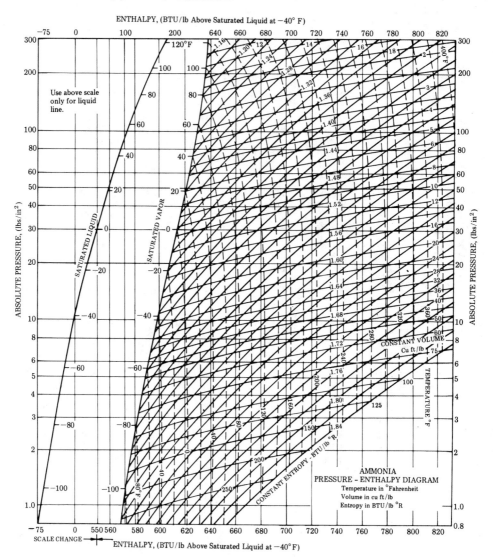

(Reprinted by permission from ASHRAE Handbook of Fundamentals 1967.)

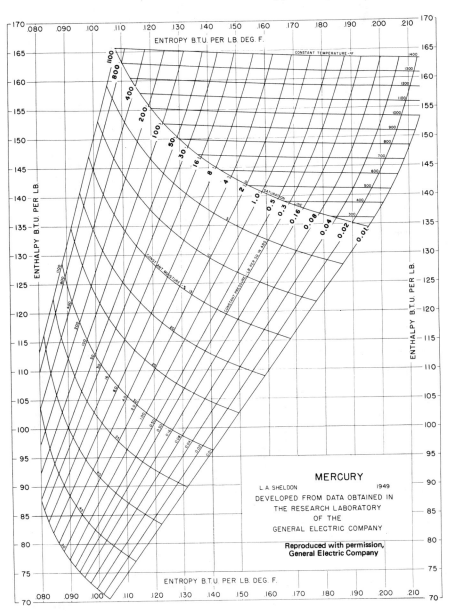

MERCURY

L. A. SHELDON   1949

DEVELOPED FROM DATA OBTAINED IN
THE RESEARCH LABORATORY
OF THE
GENERAL ELECTRIC COMPANY

Reproduced with permission,
General Electric Company

ITEM B 35
Pressure-Enthalpy Diagram of Freon, Refrigerant 12 (Dichloro-difluoromethane)
(courtesy "Freon" Products Division, E. I. du Pont and Co., Inc., copyright 1967)

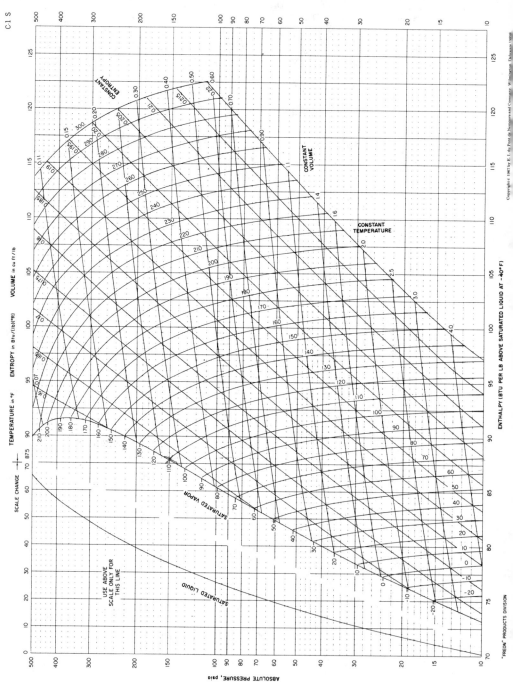

ITEM B 36   Pressure-Enthalpy Chart, Sulfur Dioxide (courtesy Shell Development Co.)

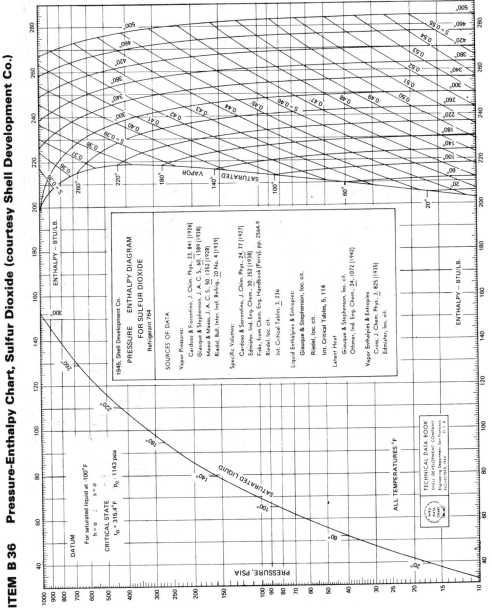

1945, Shell Development Co.

PRESSURE   ENTHALPY DIAGRAM
FOR SULFUR DIOXIDE
Refrigerant 764

SOURCES OF DATA

Vapor Pressures:

Cardoso & Fiorentino, J. Chim. Phys., 23, 841 (1926)
Giauque & Stephenson, J. A. C. S., 60, 1389 (1938)
Maass & Maass, J. A. C. S., 50, 1352 (1928)
Riedel, Bull. Inter. Inst. Refrig., 20 No. 4 (1939)

Specific Volumes:

Cardoso & Sorrentino, J. Chim. Phys., 24, 77 (1927)
Edmister, Ind. Eng. Chem., 30, 352 (1938)
Fiske, from Chem. Eng. Handbook (Perry), pp. 2564-9
Riedel, loc. cit.

Int. Critical Tables, 3, 236

Liquid Enthalpies & Entropies:

Giauque & Stephenson, loc. cit.

Riedel, loc. cit.

Int. Critical Tables, 5, 114

Latent Heat

Giauque & Stephenson, loc. cit.
Ohmer, Ind. Eng. Chem., 34, 1072 (1942)

Vapor Enthalpies & Entropies

Cross, J. Chem. Phys., 3, 825 (1935)
Edmister, loc. cit.

DATUM

For saturated liquid at -100°F
h = 0   ;   s = 0

CRITICAL STATE
t_0 = 315.4°F P_c · 1143 psia

ALL TEMPERATURES °F

TECHNICAL DATA BOOK
SHELL DEVELOPMENT COMPANY
Engineering Department San Francisco
NOVEMBER 1944

ITEM B 37   Generalized Fugacity Coefficient Chart

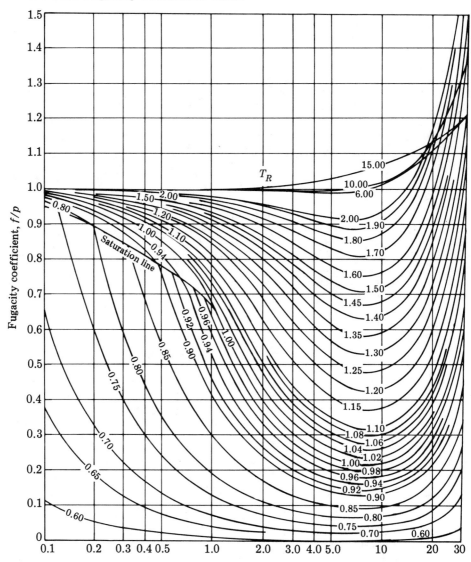

Reduced pressure, P_R

(Reprinted from *Fundamentals of Classical Thermodynamics*, Van Wylen and Sonntag, by permission from John Wiley & Sons, Inc.)

Relations of Units; Basic Constants   **ITEM B 38**

Where relevant, abbreviations given in *The International System of Units*, NASA SP-7012, are used. For the given conversions that are not exact by definition, additional digits may be found in the literature if greater accuracy is desired.

Abbreviations. Å = angstrom, atm = standard atmosphere, 760 mm of Hg at 0°C, cal = calorie (gram), cm = centimeter, deg = degree, gal = gallon, U.S. liquid, gm (and g) = gram, gmole = gram-mole, J = joule, kcal = kilocalorie, kg = kilogram, kJ = kilojoule, km = kilometer, kW = kilowatt, l = liter, lb = avoirdupois pound, m = meter, mi = mile (U.S.), mm = millimeter, N = newton, oz = avoirdupois ounce, pmole = pound mole, pt = pint, rad = radian, rev = revolution, s = second, ton = short U.S. ton, V = volt, W = watt. Others are as usual; see Abbreviations in front matter.

LENGTH

$12 \dfrac{\text{in.}}{\text{ft}}$	$6080.2 \dfrac{\text{ft}}{\text{naut. mi}}$	$5280 \dfrac{\text{ft}}{\text{mi}}$	$0.3937 \dfrac{\text{in.}}{\text{cm}}$	$30.48 \dfrac{\text{cm}}{\text{ft}}$	$10^4 \dfrac{\text{microns}}{\text{cm}}$
$3 \dfrac{\text{ft}}{\text{yd}}$	$1.152 \dfrac{\text{mi}}{\text{naut. mi}}$	$10^{10} \dfrac{\text{Å}}{\text{m}}$	$2.54 \dfrac{\text{cm}}{\text{in.}}$	$3.28 \dfrac{\text{ft}}{\text{m}}$	$1.609 \dfrac{\text{km}}{\text{mi}}$

AREA

$144 \dfrac{\text{in.}^2}{\text{ft}^2}$	$43{,}560 \dfrac{\text{ft}^2}{\text{acre}}$	$640 \dfrac{\text{acres}}{\text{mi}^2}$	$10.76 \dfrac{\text{ft}^2}{\text{m}^2}$	$929 \dfrac{\text{cm}^2}{\text{ft}^2}$	$6.452 \dfrac{\text{cm}^2}{\text{in.}^2}$

VOLUME

$1728 \dfrac{\text{in.}^3}{\text{ft}^3}$	$7.481 \dfrac{\text{gal}}{\text{ft}^3}$	$43{,}560 \dfrac{\text{ft}^3}{\text{acre-ft}}$	$3.7854 \dfrac{\text{l}}{\text{gal}}$	$28.317 \dfrac{\text{l}}{\text{ft}^3}$	$35.31 \dfrac{\text{ft}^3}{\text{m}^3}$
$231 \dfrac{\text{in.}^3}{\text{gal}}$	$8 \dfrac{\text{pt}}{\text{gal}}$	$10^3 \dfrac{\text{l}}{\text{m}^3}$	$61.025 \dfrac{\text{in.}^3}{\text{l}}$	$10^3 \dfrac{\text{cm}^3}{\text{l}}$	$28{,}317 \dfrac{\text{cm}^3}{\text{ft}^3}$

DENSITY

$1728 \dfrac{\text{lb/ft}^3}{\text{lb/in.}^3}$	$32.174 \dfrac{\text{lb/ft}^3}{\text{slug/ft}^3}$	$0.51538 \dfrac{\text{gm/cm}^3}{\text{slug/ft}^3}$	$16.018 \dfrac{\text{kg/m}^3}{\text{lb/ft}^3}$	$1000 \dfrac{\text{kg/m}^3}{\text{gm/cm}^3}$

ANGULAR

$2\pi = 6.2832 \dfrac{\text{rad}}{\text{rev}}$	$57.3 \dfrac{\text{deg}}{\text{rad}}$	$\dfrac{1}{2\pi} \dfrac{\text{rpm}}{\text{rad/min}}$	$9.549 \dfrac{\text{rpm}}{\text{rad/sec}}$

TIME

$60 \dfrac{\text{s}}{\text{min}}$	$3600 \dfrac{\text{s}}{\text{hr}}$	$60 \dfrac{\text{min}}{\text{hr}}$	$24 \dfrac{\text{hr}}{\text{day}}$

SPEED

$88 \dfrac{\text{fpm}}{\text{mph}}$	$0.6818 \dfrac{\text{mph}}{\text{fps}}$	$0.5144 \dfrac{\text{m/s}}{\text{knot}}$	$0.3048 \dfrac{\text{m/s}}{\text{fps}}$	$0.44704 \dfrac{\text{m/s}}{\text{mph}}$
$1.467 \dfrac{\text{fps}}{\text{mph}}$	$1.152 \dfrac{\text{mph}}{\text{knot}}$	$1.689 \dfrac{\text{fps}}{\text{knot}}$	$152.4 \dfrac{\text{cm/min}}{\text{ips}}$	

FORCE, MASS

$16 \dfrac{\text{oz}}{\text{lb}_m}$	$32.174 \dfrac{\text{lb}_m}{\text{slug}}$	$444{,}820 \dfrac{\text{dynes}}{\text{lb}_f}$	$2.205 \dfrac{\text{lb}_m}{\text{kg}}$	$9.80665 \dfrac{\text{N}}{\text{kg}_f}$
$1000 \dfrac{\text{lb}_f}{\text{kip}}$	$32.174 \dfrac{\text{poundals}}{\text{lb}_f}$	$980.665 \dfrac{\text{dynes}}{\text{gm}_f}$	$14.594 \dfrac{\text{kg}}{\text{slug}}$	$4.4482 \dfrac{\text{N}}{\text{lb}_f}$
$2000 \dfrac{\text{lb}_m}{\text{ton}}$	$7000 \dfrac{\text{grains}}{\text{lb}_m}$	$453.6 \dfrac{\text{gm}}{\text{lb}_m}$	$10^5 \dfrac{\text{dynes}}{\text{N}}$	$1 \dfrac{\text{kilopond}}{\text{kg}}$
$14.594 \dfrac{\text{kg}}{\text{slug}}$	$28.35 \dfrac{\text{gm}}{\text{oz}}$	$453.6 \dfrac{\text{gmole}}{\text{pmole}}$	$907.18 \dfrac{\text{kg}}{\text{ton}}$	$1000 \dfrac{\text{kg}}{\text{metric ton}}$

PRESSURE

$14.696 \dfrac{\text{psi}}{\text{atm}}$	$101{,}325 \dfrac{\text{N/m}^2}{\text{atm}}$	$13.6 \dfrac{\text{kg}}{\text{mm Hg(0°C)}}$	$51.715 \dfrac{\text{mm Hg(0°C)}}{\text{psi}}$	$47.88 \dfrac{\text{N/m}^2}{\text{psf}}$
$29.921 \dfrac{\text{in. Hg(0°C)}}{\text{atm}}$	$10^5 \dfrac{\text{N/m}^2}{\text{bar}}$	$13.57 \dfrac{\text{in. H}_2\text{O(60°F)}}{\text{in. Hg(60°F)}}$	$703.07 \dfrac{\text{kg/m}^2}{\text{psi}}$	$6894.8 \dfrac{\text{N/m}^2}{\text{psi}}$
$33.934 \dfrac{\text{ft H}_2\text{O(60°F)}}{\text{atm}}$	$14.504 \dfrac{\text{psi}}{\text{bar}}$	$0.0361 \dfrac{\text{psi}}{\text{in. H}_2\text{O(60°F)}}$	$0.0731 \dfrac{\text{kg/cm}^2}{\text{psi}}$	$760 \dfrac{\text{torr}}{\text{atm}}$
$1.01325 \dfrac{\text{bar}}{\text{atm}}$	$10^6 \dfrac{\text{dynes/cm}^2}{\text{bar}}$	$0.4898 \dfrac{\text{psi}}{\text{in. Hg(60°F)}}$	$9.869 \dfrac{\text{atm}}{10^7 \text{ dyne/cm}^2}$	$133.3 \dfrac{\text{N/m}^2}{\text{torr}}$
$33.934 \dfrac{\text{ft H}_2\text{O(60°F)}}{\text{atm}}$	$760 \dfrac{\text{mm Hg(0°C)}}{\text{atm}}$	$406.79 \dfrac{\text{in. H}_2\text{O(39.2°F)}}{\text{atm}}$	$0.1 \dfrac{\text{dyne/cm}^2}{\text{N/m}^2}$	$1.0332 \dfrac{\text{kg/cm}^2}{\text{atm}}$

ENERGY AND POWER

$778.16 \dfrac{\text{ft-lb}}{\text{Btu}}$	$2544.4 \dfrac{\text{Btu}}{\text{hp-hr}}$	$5050 \dfrac{\text{hp-hr}}{\text{ft-lb}}$	$1 \dfrac{\text{J}}{\text{W-s}}, \dfrac{\text{J}}{\text{N-m}}$	$0.01 \dfrac{\text{bar-dm}^3}{\text{J}}$
$550 \dfrac{\text{ft-lb}}{\text{hp-s}}$	$42.4 \dfrac{\text{Btu}}{\text{hp-min}}$	$1.8 \dfrac{\text{Btu/lb}}{\text{cal/gm}}$	$1 \dfrac{\text{kW-s}}{\text{kJ}}$	$16.021 \dfrac{\text{J}}{10^{12}\text{ MeV}}$
$33{,}000 \dfrac{\text{ft-lb}}{\text{hp-min}}$	$3412.2 \dfrac{\text{Btu}}{\text{kW-hr}}$	$1800 \dfrac{\text{Btu/pmole}}{\text{kcal/gmole}}$	$1 \dfrac{\text{V-amp}}{\text{W-s}}$	$1.6021 \dfrac{\text{erg}}{10^{12}\text{ eV}}$
$737.562 \dfrac{\text{ft-lb}}{\text{kW-s}}$	$56.87 \dfrac{\text{Btu}}{\text{kW-min}}$	$2.7194 \dfrac{\text{Btu}}{\text{atm-ft}^3}$	$10^7 \dfrac{\text{ergs}}{\text{J}}$	$11.817 \dfrac{\text{ft-lb}}{10^{12}\text{ MeV}}$
$1.3558 \dfrac{\text{J}}{\text{ft-lb}}$	$251.98 \dfrac{\text{cal}}{\text{Btu}}$	$4.1868 \dfrac{\text{kJ}}{\text{kcal}}$	$3600 \dfrac{\text{kJ}}{\text{kW-hr}}$	$0.746 \dfrac{\text{kW}}{\text{hp}}$
$1.055 \dfrac{\text{kJ}}{\text{Btu}}$	$101.92 \dfrac{\text{kg-m}}{\text{kJ}}$	$0.4300 \dfrac{\text{Btu/pmole}}{\text{J/gmole}}$	$860 \dfrac{\text{cal}}{\text{W-hr}}$	$1.8 \dfrac{\text{Btu}}{\text{chu}}$

ENTROPY, SPECIFIC HEAT, GAS CONSTANT

$1\dfrac{\text{Btu/pmole-°R}}{\text{cal/gmole-K}}$	$1\dfrac{\text{Btu/lb-°R}}{\text{gal/gm-K}}$	$1\dfrac{\text{Btu/lb-°R}}{\text{kcal/kg-K}}$	$0.2389\dfrac{\text{Btu/pmole-°R}}{\text{J/gmole-K}}$	$4.187\dfrac{\text{kJ/kg-K}}{\text{Btu/lb-°R}}$

UNIVERSAL GAS CONSTANT $\overline{R}$

$1545.32 \dfrac{\text{ft-lb}}{\text{pmole-°R}}$	$8.3143 \dfrac{\text{kJ}}{\text{kgmole-K}}$	$0.7302 \dfrac{\text{atm-ft}^3}{\text{pmole-°R}}$	$82.057 \dfrac{\text{atm-cm}^3}{\text{gmole-K}}$
$1.9859 \dfrac{\text{Btu}}{\text{pmole-°R}}$	$1.9859 \dfrac{\text{cal}}{\text{gmole-K}}$	$10.731 \dfrac{\text{psi-ft}^3}{\text{pmole-°R}}$	$83.143 \dfrac{\text{bar-cm}^3}{\text{gmole-K}}$
$8.3143 \dfrac{\text{J}}{\text{gmole-K}}$	$8.3149 \times 10^7 \dfrac{\text{erg}}{\text{gmole-K}}$	$0.08206 \dfrac{\text{atm-m}^3}{\text{kgmole-K}}$	$0.083143 \dfrac{\text{bar-l}}{\text{gmole-K}}$

Newton's proportionality constant **k** (as a conversion unit)

$32.174 \text{ fps}^2\left(\dfrac{\text{lb}}{\text{slug}}\right)$	$386.1 \text{ ips}^2\left(\dfrac{\text{lb}}{\text{psin}}\right)$	$9.80665 \dfrac{\text{m}}{\text{s}^2}\left(\dfrac{\text{N}}{\text{kg}}\right)$	$980.665 \dfrac{\text{cm}}{\text{s}^2}\left(\dfrac{\text{dynes}}{\text{gm}}\right)$

MISCELLANEOUS CONSTANTS

Speed of Light	Avogadro Constant	Planck Constant
$c = 2.9979 \times 10^8 \dfrac{\text{m}}{\text{s}}$	$N_A = 6.02252 \times 10^{23} \dfrac{\text{molecules}}{\text{gmole}}$	$h = 6.6256 \times 10^{-34}\text{ J-s}$
Boltzmann Constant	Gravitational Constant	Normal mole volume
$\kappa = 1.38054 \times 10^{-23} \dfrac{\text{J}}{\text{K}}$	$G = 6.670 \times 10^{-11} \dfrac{\text{N-m}^2}{\text{kg}^2}$	$2.24136 \times 10^{-2} \dfrac{\text{m}^3}{\text{gmole}}$

INDEX

SYMBOLS (Cont.)

W shaft or fluid work; W_I, indicated work; W_B, brake work; W_K, combined work; W_f, flow work; W_p, pump work.

$\dot{W}$ power.

w total emissive power.

X volumetric or mole fraction.

x quality of a two-phase system; coordinate.

y percentage (or fraction) of liquid in a two-phase system; coordinate; spring deflection.

Z compressibility factor; atomic number.

z altitude; potential energy of a unit weight

α (alpha)   constant in specific heat equation; absorptivity; linear thermal expansion coefficient $(dL/L)/dT$.

β (beta)   constant in specific heat equation; coefficient of volumetric expansion.

γ (gamma)   specific weight; constant in specific heat equation; coefficient of performance; angle.

δ (delta)   angle.

ε (epsilon)   effectiveness; emissivity; energy of photon, of translation of molecule; unit strain; ε_r, effectiveness of regeneration.

ζ (zeta)   bulk modulus

η (eta)   energy ratios; engine efficiency; combustion efficiency; η_b; brake engine efficiency, turbine blade efficiency; η_c, compressor efficiency (adiabatic if not qualified); η_i, indicated engine efficiency; η_k, combined engine efficiency; η_m, mechanical efficiency; η_n, nozzle efficiency; η_p, propulsive efficiency; pump efficiency; η_r, efficiency of reaction blades; η_s, turbine stage efficiency; η_v, volumetric efficiency.

θ (theta)   represents unit of temperature; angle.

κ (kappa)   Boltzmann's constant; κ_d, coefficient of discharge; κ_f, velocity friction coefficient; κ_p, pressure coefficient; κ_s, adiabatic compressibility coefficient; κ_T, isothermal compressibility coefficient; κ_v, velocity coefficient.

λ (lambda)   wave length.

μ (mu)   degree of saturation; absolute viscosity; Joule–Thomson coefficient; micron.

μ_o permeability.

ν (nu)   kinematic viscosity; frequency.

π (pi) $3.1416\ldots$, Peltier coefficient.

ρ (rho)   density; reflectivity.

σ (sigma)   Stefan–Boltzmann constant; unit stress.

τ (tau)   time; represents a unit of time; transmissivity.

ϕ (phi)   relative humidity; angle; $d\phi = c_p \, dT/T$.

ω (omega)   angular velocity; humidity ratio; angle.

Δ (delta)   indicates a difference or a change of value, Δt = change of temperature or difference in temperature, in accordance with the context.

Ω (omega)   thermodynamic probability; electric resistance.